BAYESIAN LINEAR REGRESSION

$$\{A \mid B\} = \frac{\{B \mid A\}\{A\}}{\int \{B \mid \xi\}\{\xi\}\, d\xi}$$

by

DALE MARK NESBITT

APRIL 30, 2025

Nesbitt, Dale Mark
Bayesian Linear Regression/Dale Mark Nesbitt
Self published
Includes illustrations and bibliographical references.
ISBN Number 978-0-578-34132-3

$100.00
ISBN 978-0-578-34132-3
10000>
9 780578 341323

Contents

DEDICATION

This book is dedicated to Prof. Ronald A. Howard of Stanford University, with respect and gratitude.

ACKNOWLEDGEMENTS

Necessity is the mother of taking chances.

Mark Twain

The author wishes to specifically recognize and thank the late Dr. Arnold Zellner of the University of Chicago, one of the most avuncular and professional people around. He tutored us on Bayesian regression, and his publications were truly seminal. Without any obligation to do so, he thoroughly educated us in Bayesian regression methods. I continue to believe that Dr. Zellner might well have deserved the Nobel Prize in Economics for his contributions to econometrics.

The author completed projects on the prediction of the performance of pavement assets in Canada in concert with Dr. Gordon Sparks of the University of Saskatchewan. Dr. Sparks' contribution, particularly on the practical and application sides, was seminal and not to be minimized. Those pavement projects were resourced by the Canadian Strategic Highway Research Program (C-SHRP), administered by the Transportation Association of Canada (TAC), and are hereby recognized. Their seminal project hired us to conceive and perform Bayesian Linear Regression for highway performance prediction. This book is the product of twenty-five years of learning and applying single and multivariate Bayesian models and teaching Decision Analysis at Stanford University. It couldn't have been possible without that.

Dr. Ronald Howard has reviewed several presentations by the author on the topic, and his suggestions and pedantry have been pivotal. Dr. Howard's "Acme Rain Detector" analysis has been conceptually important here, as has his lifelong contribution to Bayesian analysis and probability.[1] The Stanford "Decision Analysis Working Group," or DAWG, has reviewed, critiqued, and helped shape

---

[1]Readings on the Acme Rain Detector are "data" in the pure sense of the word. Prof. Howard's approach to data, imperfect data at that, is seminal. This analysis crystallizes just how important that simple rain detector truly was.

the mathematics and explication. Dr. Howard's contributions to Decision Analysis and probability are peerless, and they have materially contributed to this work. Dr. Ross Shachter's comments and perspective as part of DAWG were material and welcome.

Collaboration with Dr. Thomas Keelin on the use of Bayesian Linear Regression in the formulation and updating of the metalog distribution has contributed materially. His specific application and his persistent questioning and analysis have helped refine and shape this work.

Christian Ostberg (a Stanford graduate student and teaching assistant) has carefully and comprehensively edited, checked, and corrected this textbook and its underlying equations and calculations, and he has contributed materially to the substance and clarity of the work. His technical and presentational contributions have been exemplary, and his substantive contributions have been pivotal.

Madison Coots (a Stanford graduate student and teaching assistant) has carefully and comprehensively edited, checked, and corrected the final textbook and its underlying equations and calculations (second independent review). She has contributed materially to the substance and clarity of the work. Her technical and presentational contributions have been exemplary, and her substantive contributions to the final manuscript in its entirety have been pivotal.

Vince de Masi edited the final manuscript and provided valuable improvements and corrections.

Any errors or opacities are the responsibility of the author.

Dale Nesbitt is President of ArrowHead Economics LLC and an adjunct lecturer of Management Science and Engineering at the Stanford School of Engineering. Material from this book has been used in the master's level course "Bayesian Linear Regression." Supplementary educational materials from the author are available from the author at www.dalemarknesbitt.com by email to the author at dale.nesbitt@arrowheadeconomics.com.

Dale Mark Nesbitt
Stanford, California, April 30, 2025

Chapter 1

OVERVIEW

> ***Probability theory is nothing more than common sense reduced to calculation.***
>
> *Pierre Simon Laplace*

1.1 Objectives

What is Bayesian Linear Regression? For that matter, what is regression? I wanted to answer these questions definitively for myself. I couldn't find a book that integrated the theory, mathematics, interpretation, and practice, using common notation and concepts, to my satisfaction. I started to draft manuscripts and slides focusing on important portions and topics. In so doing, I was episodically walking the path to a focused, step-by-step, fully self-contained, common-notation treatise to elucidate Bayesian Linear Regression for myself. I always hoped to find a book that would publish everything I hadn't written, but I never did. I felt the need to work my way to the end of the path.

Why a new book (and class)? I had been doing regression and probability encoding in my consulting practice and research (e.g., demand elasticity, performance prediction). I was going through the motions of getting linear coefficients, predicting with them, and wowing clients, most often using linear regression software written by others. One day, after a particularly brow-furrowing elasticity estimation project, I looked at the "answer" I was getting and then into the mirror: Dale, you don't really know what the heck you are doing or why:

- You yourself never really learned classical linear regression. Requisite information is dispersed, scattered, fragmented, across a century-long and large literature, but not integrated, step by step, using common notation.

- What do the equations and statistical parameters in that software you are using and reporting even mean?
- Where did they come from?
- You don't actually know what regression jargon means. (I confess, regression jargon was vague and perplexing to me.)
- You don't really know how to use data to make better decisions.
- You are running someone else's regression computer program, and you couldn't even write down the equations they have embedded or the equations you yourself want for things like mean, variance-covariance, t-test, F-test, prediction, etc. As a result, you can't really interpret the results you are getting.
- You don't know what hypothesis testing, confidence intervals, and other concepts from classical regression truly mean. Your Bayesian training may have gotten into your way, and you could have implicitly blown off classical linear regression because you couldn't relate it to Bayesian probability.
- You may be an "ace" at probability, but you have not connected probability with regression. Can regression be interpreted solely from a pure probability perspective? Is probability theory the key to understanding it? (It turns out the answer is a definitive yes.)
- You are not a frequentist.[1] Frequentism simply doesn't seem to jibe with real-world probabilities of real-world events. The real world offers no "continual repeatability" of events in advance. You cannot play the Superbowl 500 times in advance of your decision whether to bet on the Philadelphia Eagles and take the points or bet on the Kansas City Chiefs and give away two points to the casino booking agent. You are forcibly Bayesian by virtue of the problems you are trying to solve and decisions you are trying to make.[2]
- You have been trained, if not indoctrinated, in (Bayesian) Decision Analysis, decision theory, by the great Dr. Ronald A. Howard.

[1] See https://en.wikipedia.org/wiki/Frequentist_probability, which asserts that *Frequentist probability or frequentism is an interpretation of probability; it defines an event's probability as the limit of its relative frequency in infinitely many trials (the long-run probability). Probabilities can be found (in principle) by a repeatable objective process (and are thus ideally devoid of opinion).* This definition of frequentist probability sharply contrasts with subjective or Bayesian probability.

[2] See https://en.wikipedia.org/wiki/Bayesian_probability for a definition of subjective or Bayesian probability.

- You know that computation has been outrunning knowledge for some time, and that has lured people into computation and away from knowledge. Dale, you can no longer just "compute to get an answer" and overlook theory or logic. I needed a book that puts the theory and logic forth sequentially, soup to nuts, in an understandable fashion before I ever touch a computer.

I couldn't find a book that to my satisfaction related it all to probability, save for Zellner[107], which was correct and fairly complete but not at all the didactic book I wanted. I set out on a journey to make a book that relates "regression" to pure probability. Early in my search, I contacted Dr. Arnold Zellner, University of Chicago, and the journey was underway. This monograph is the result of that two decade plus journey.

Classical linear regression (which we term classical statistics or simple linear regression) represents a basic, elementary approach to statistical inference (even though it has not always been framed assiduously as such). Bayesian statistics systematically extends classical statistics, respecting during such extension the firmly established principles of joint and conditional probability and Bayes theorem. Bayesian statistics, in fact, adheres to those principles unabashedly and transparently. Bayes offers a high degree of intellectual power and method relative to classical statistics. Both classical and Bayesian methods are technically sound, yet the former can be regarded as a "special case" of the latter in a sense that will be made clear herein.[3] Bayesian statistics is, in fact, an exercise in pure probability. **Bayesian Linear Regression estimates a probability distribution.**

In the vernacular, Bayesian Statistics "manufactures" a probability distribution out of data, as caricatured in Figure 1.1. Akin to the fabled Rumpelstiltskin, Bayesian Linear Regression spins straw into gold! The straw is the data. (The straw is also judgment, if you want to use it.) The answer is a probability distribution, which is the gold! Bayesian Linear Regression spins straw (data and judgment) into gold (a probability distribution, suitable for decision making, prediction, analysis, etc.)

In contrast to the classical statistical approach, Bayesian Linear Regression makes crystal clear the information brought to the estimation problem by the analyst and the information the solution delivers back to the analyst. From that perspective, classical statistical applications can mislead. With classical statistics, the analyst is allowed to deliver absolutely zero prior knowledge to the problem, not one whit of knowledge aside from the data (which the classical statistician must hold as complete and immutable), and by design. The classical statistician brings only a "diffuse prior," the consummate expression of complete, unequivocal, abject, stubborn ignorance, virtual stupidity, and literal agnosticism, to the problem.[4]

---

[3] How and why classical statistics is a special case of Bayes is profound in its own right, but Bayesian statistics is more general and conceptually rich.

[4] Dr. Ross Shachter of Stanford was the first we know of to suggest that the diffuse prior that we

Figure 1.1: Bayesian Linear Regression Spins Straw into Gold

With classical statistics, everything that is known or knowable resides in the data. Period! The idea that one has absolutely no ex ante idea of any kind, "no clue," no bias, or objective or subjective information above and beyond the data gathered for any given problem strains credulity; it simply does not make intellectual or conceptual sense. Such a view discards valuable, reliable, and valid information, empiricism, judgment, or knowledge from any source other than the data. The classical regression mantra, illustrated in Figure 1.2, is that "everything that is known or knowable is resident within the data."[5] It is all-but chanted, literally over beads! In spite of the "no other knowledge" restriction, classical statisticians have rarely been loath to render judgments ex post facto (e.g., "Hey! That coefficient is likely to lie within a certain range" after they have seen the havoc questionable, incomplete, or flawed data can wreak with an assumed model). Many have not been loath to manipulate, discard, add, or fudge

associate with classical regression is in actuality a "model" of ignorance. His insight is quite correct, and we explore that notion in detail herein. It will not be productive for our purposes to follow the research of people such as Jeffreys[50] or the maximum entropy people regarding diffuse priors. We develop a specific type of diffuse prior and advocate for its use on a conceptual basis, but we do not present a definitive analysis of diffuse priors. We leave that to Jeffreys[50] and his successors. We do, however, assiduously equate our prior with a maximum entropy prior, and that has tremendous conceptual utility.

[5]The author once heard Harvard's great Dr. Dale Jorgenson state from the stage at a conference, "Man made the models, but God made the data!" This is an absolutely wonderful articulation of the sentiment of classical statistics and typical classical statistical thinking. Classicists argue that the data is God-given and the sole and complete embodiment of any and all information and knowledge relevant to the problem at hand. That is completely at odds not only with Bayes but also with reality. People bring knowledge to the problem, and data merely adds to that knowledge. And of course, data gathering in practice is often about as clean and attractive as the proverbial sausage, (an analogy attributed to Otto von Bismarck, perhaps erroneously) fraught with uncertainty, inaccuracy, infractions, implicit subjectivity, implicit assumptions, definitional ambiguity, inconsistency, systematic or accidental omissions, "corner cutting," and other practices.

data or alter models. Bayes offers a more honest approach.

Figure 1.2: Classical Regression Mantra

Bayes allows statisticians and decision analysts to include judgment transparently in a systematic, logical, auditable, and probabilistic fashion. Therein lies the intellectual and practical power of Bayes. The knowledge brought to the problem is explicit and in the form of what Zellner[107] termed an "informative" prior. When we first used the term "informative prior," Zellner's word chosen to stand in contrast with a diffuse or totally ignorant prior, Prof. Ronald Howard interjected: "What other type of prior could there be?" From such a simple comment, great depth ensues. Realistically, there is no such thing in practice as an non-informative prior, i.e., a diffuse prior. A diffuse prior is an abstraction, a veritable unicorn, a **model** of abject ignorance. No person could actually be that ignorant. All priors carry information. The diffuse prior carries a carefully constructed, intellectually appealing definition of zero information! The notion of an informative prior lies at the heart of Bayesian statistics. And yet, the diffuse prior, the quintessential model of abject ignorance, is a linchpin underlying classical statistics and connecting classical statistics and Bayesian statistics. You have to be as ignorant ex ante as the diffuse prior implies if you are to use classical statistics. That simply isn't conceptually attractive; in the vernacular, nobody is that ignorant!

Analysts should take the process of developing a prior seriously. It is not a useless or arbitrary task to be done as a "chore" or a "throwaway" exercise.[6] It should be done well and authentically represent prior knowledge. One is never sure the quality or comprehensiveness of one's data or model, and the prior is what you think the answer is before analyzing the data. Not to overemphasize–

[6]It is not like "priming a diesel truck with ether in cold weather just to get it started," allegedly quite dangerous, which the author was ordered to do fairly frequently when working in a mining operation during his early working days. When the ether burns off, the truck is running and the diesel combusted, or it explodes and people and equipment are incinerated! The author is lucky to be here.

don't give short shrift to the prior. We are going to see that enough good data often overrides the prior, but not always.

Gill[36] has stated on p. 98: "Prior distributions remain the most controversial aspect of Bayesian inference. It is rare to approach a problem with absolutely no prior information about parameters of interest." Absolutely correct! In our view, any such controversy has been largely dispelled since Gill's writing. Every problem of which we are aware is preceded by prior information that resides wholly and completely outside the data and is ported in by analysts or experts. Where is the value in studying various non-informative priors when the rule (not the exception) is that prior information is omnipresent and frequently important?

It is also our objective herein to show intuitively how one gets a "prior" in a straightforward, rational, logical way that is strongly preferable to the classical statistics approach that ignores (or worse, endeavors to sidestep) the entire issue of prior knowledge, which exists above and beyond the data. There has, as is known to just about everyone, been a backlash against "a prior" in statistical methods. Some have wanted to argue that "there is no judgment in statistics." Alas, that is far from true in the real world, the decision-making world, the prediction world. People bring judgment to the problem all the time, often good or experienced judgment, and the analyst most certainly will want to consider it explicitly, systematically, and transparently. Bayes renders such judgment explicit, and that is one of its most important intellectual and practical contributions.

The purpose of this textbook is to present Bayesian Linear Regression (which we sometimes refer to as Bayesian statistics for brevity herein) in technical depth without omitting any steps, algebraic or conceptual, or any interim or final interpretations. Our objective here is to offer a comprehensive articulation of the fundamental assumptions and, thereafter, all the steps along the way from assumptions to conclusions or results. An important purpose of this book is to "democratize Bayes," or more specifically Bayesian Linear Regression. This book wants every qualified person to

- Learn Bayesian Linear Regression,
- Know it intuitively as well as operationally,
- Know how to use and interpret it fluently and directly,
- Know "why,"
- Know how to install it themselves onto their own private computer using something as rudimentary as Excel (or more sophisticated methods if desired),
- Solve and explain difficult problems with it, and
- Presage solutions to analogous nonlinear problems.

Application and scale-up to large, real-world problems become easy at that juncture. The book strives to teach, demystify, and interpret Bayes, leaving out no steps along the way. People have complained that "Bayesian calculations are opaque and hard to understand both practically and conceptually." This book seeks to end that! The calculations in this book are possible, and almost instantaneous, in an Excel spreadsheet. (Large data sets may take more time or room than a spreadsheet offers, but hundreds of thousands of observations are definitely possible.) If you can program them in a spreadsheet, you can understand and intuit them!

The "Bible" in Bayesian econometrics is Zellner[107], and before that, Raiffa and Schlaiffer's[81] classic 1961 text. Alas, the Zellner[107] Bible is not fully integrated, self-standing, tightly focused on Bayes, and painstakingly derived step by step from first principles. It is our objective here to present all the conceptual and algebraic steps Zellner[107] and Raiffa and Schlaiffer[81] may have omitted from or included in their monographs, all the while being laser-focused on Bayes and the linear regression model. We have also chosen to include some salient results from classical statistics so that we can discern the precise relationship between classical and Bayes, a relationship that is subtle but worthy of explication.

One reviewer of this book said (approximately): "Zellner[107] published a book a bit like this in 1971, well before modern computation was available to him. Now that computation has come so far, do we really need methodology such as this from 1971?" The answer is a definitive YES! "Yes, we DO need a book like this; no one finished the definitive book last time!" We strive to make the method crystal clear and allow computer scientists, econometricians, engineers, etc. as well as theoreticians to understand and implement its solutions, in simple or complex problems. With the advent of powerful modern computation, statisticians and practitioners have become all-too-eager to initiate a simulation problem (which requires massive compute power), to put it on their parallel cloud computers where they can churn it, rather than "doing the math" carefully in advance. We strive herein to "do all the math" in advance to exploit all possible reductions and gain all possible insights *ex ante* so that practitioners don't have to rely on the computer for "results without insights." The philosophy of this monograph is to put forth the math scrupulously and completely so that there is literally "little to program" and "few errors to make" and so there is an archetype that tells them what to expect from simulation-type formulations.

What does Bayes substitute in place of computer-intensive simulation? Figure 1.3 presages. We do all the calculus and matrix algebra analytically in advance (with homage to Newton as caricatured in the figure.) We get all the math correct and complete before we do any programming. Once we do that, we obtain the following:

- **Fully Developed Mathematics.** Matrix manipulations, done once and done correctly, substitute for massive (or small) simulations that don't *ex*

Figure 1.3: Calculus or Computing?

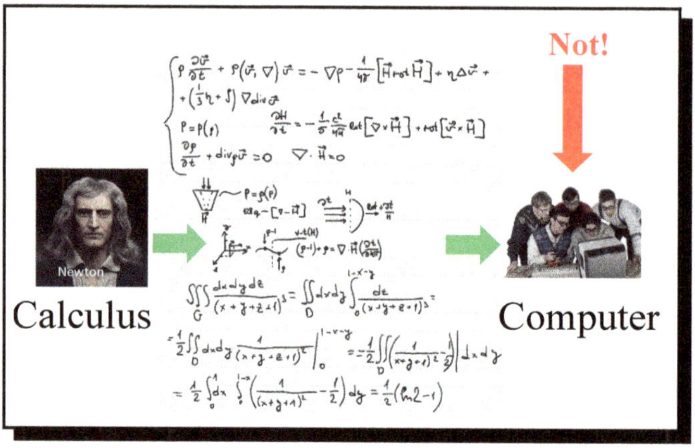

ante mine the Bayesian math to get the insight. They just calculate; numbers are thought to contain or substitute for insight. The conjugacy calculus derived painstakingly here does all the hard work and gets the right answer more simply and reliably thereafter. Programming it is easy, and insights are more obvious, and every bit as deep.

- **Knowledge.** We want readers to know what they are doing and why. They will, in this book, witness (and replicate) all the so-called "tedious calculations" that makes Bayes go. One reviewer said, "This book makes things seem easy. You just follow the calculus and matrix algebra, you can program it, and all is well." That is the intent, and that is the finest review this book could possibly imagine. Yes, and in spades. Readers will know how "these Bayesian problems" work, and that will help them know what their simulations in more complex, nonconjugate environments are really doing. True knowledge comes from the math and structure, not solely the data. What better way to analyze data?

- **Easy Computation.** Another reviewer of this book, a computer science person, said, "Excel is outdated; it is for engineers, not statisticians. The farthest downward statisticians should go on the 'evolutionary scale' is Python or R. Nesbitt, statisticians use 'systems' like Stata to do statistics." *Au contraire!* If we could, we would want the computations derived in this book on an abacus. In this monograph, we strive to make the computations subservient to the math, method, and data itself. We want calculations to be easily, directly, and, in fact, trivially understandable, transparent, and checkable. Thereafter, analysts can scale them up to large recurring database/analytics jobs. And we want all the notions underlying the equations to be fully demystified, understood, and transparent. This book strives to achieve that. Nothing would make us happier than every person working a regression problem by programming the equations developed in this book into his or her personal Excel spreadsheet and solving the problem there first. That is what we do in class. They can thereafter use 'systems' armed with full knowledge of what they compute and how straightforward it is.

- **Fast.** Much of the computational work that would otherwise have to be done by simulation has been done ex ante in Bayesian Linear Regression by matrix calculations, i.e., much of the work has been "calculized" into closed forms rather than calculated repeatedly in a simulation environment. Doing the hard work *ex ante* using calculus and matrices reduces the amount of work the computer has to do at the end and renders the calculation faster. This is not a theorem or a universal truth, but we have observed dramatic speed savings empirically.[7]
- Do you get **deeper insight** from the calculus-in-advance approach? You bet you do, and definitively so. Integrations are done exactly using calculus, not approximately using simulation.
- Is the **computing easier and faster** the further you go with calculus-in-advance? Yes; AI itself does that. All of the network math is long since done in advance. It is simply replicated in the AI calculation.
- Can you **solve bigger problems** with more dependent and independent variables? Yes.
- Will you **know exactly what you are doing and why?** Yes.
- Would you have an **archetype for harder problems** only solvable with simulation? Yes.
- Would you **steer clear from optimization (e.g., maximum likelihood) and render it a byproduct?** Yes; maximization is applied once you have probability distributions, and you will know why no optimization. John von Neumann said the only thing one should want to maximize is expected cardinal utility. Howard[45] has repeatedly asserted that maximizing anything else does not lead to good decision making.

The Bayesian problems put forth in the first part of this textbook are initially aimed at probability distributions over a single variable, i.e., univariate probability distributions. This is called "linear regression" or "simple linear regression." From Wiki:[8] *In statistics, linear regression is a linear approach for modeling the relationship between a scalar response and one or more explanatory variables (also known as dependent and independent variables).* The response is a scalar. This model is not particularly limiting. Such univariate probability distributions are conditional on one or more other variables. Prof. Howard has shown

[7] We observe this time and again in fields like economics. If one can develop a closed form that would otherwise have to be simulated or optimized, the computation time in the final solution is materially lower. Empirically speaking, "calculizing" a closed form saves computational-intensity.

[8] See https://en.wikipedia.org/wiki/Linear_regression.

with the influence diagram approach he and Dr. Matheson conceived (Howard and Matheson[46] and Howard[45]) how complex, joint, mutually relevant (i.e., dependent) distributions can be built up from simpler univariate distributions. Working with univariate distributions is therefore not as limiting as one might think at first blush. This textbook initially focuses on probabilistic projections based on conditional univariate probability distributions. Following Howard and Matheson[46], when we want to develop joint projections of two or more variables, we can simply "chain" conditionals together using relevance "arcs" in the implicit influence diagram among the variables that underlie the joint.

Would that the world were as simple as a single response variable. The book generalizes to the case in which the dependent variables, the response variables, can be two or more in number and are joint, i.e., mutually dependent. We include Chapters 15 through 17 that focus on direct estimation of a joint, multivariate probability distribution and show that it is an extension—quite a complicated extension at that—of the simple univariate Bayesian concepts. This model is called "multivariate linear regression," and it is articulated in the same Wiki reference.[9] Extensions of univariate concepts to direct estimation of joint distributions are straightforward but add complexity that partially obscures the simple and intuitive interpretations available in the univariate case. Happily, the extension to direct estimation of joint distributions gives results that are not only easily programmable in Excel but are also obvious extensions of the univariate case.

Bayesian Linear Regression does not need a lick of optimization embedded within it. There is no maximum likelihood, although the modes (the maximum points) of the prior, posterior, and classical distributions will emerge "like magic" and become obvious along the way and easy to visualize. In Bayes, there is no minimization of a "loss function." There is no optimization present or necessary. Bayesian Linear Regression is an application of pure probability. Bayes theorem. Marginal densities. Joint densities. Probabilistic prediction. Matrix operations. Happily, as such, it gets completely rid of such ideas as the maximum (mode) of a likelihood function or the mode of any density, rendering it a byline. Bayesian statistics is imminently "Jaynesian" in its approach, a tribute to the great Edwin Jaynes[49]. Bayesian statistics is a probability problem, not an optimization problem. We believe that any optimization we do should involve nothing other than the maximization of the expected values of von Neumann's[100] cardinal utility functions. Why would one maximize anything else?

Why do we make such a point of eliminating optimization from our considerations? The great John von Neumann's four axioms of rational behavior,[10] as articulated in the classic Von Neumann and Morgenstern[100] and beautifully re-articulated and motivated by Howard and Abbas[45] and before them Raiffa[80]

[9] See https://en.wikipedia.org/wiki/Linear_regression

[10] See https://en.wikipedia.org/wiki/Expected_utility_hypothesis for a summary of the von Neumann axioms—completeness, transitivity, continuity, and independence.

tell us that in decision-making, we want to make decisions that maximize "expected cardinal utility." Under von Neumann's construct, probability is explicitly distinguished from utility or value. Bayesian statistics and Bayesian Linear Regression seek to develop and articulate the probability dimension of the expected utility calculation in Decision Analysis, not the utility or value dimension. In developing expected utility, John von Neumann did not instruct us that we need only consider the mode of the underlying distribution or even only points near the mode, nor did he want us to craft an arbitrary quadratic loss function or some other arbitrary loss or penalty notion. He did not argue for special consideration of "predicted compared to observed." He instructed us to get the entire probability distribution over outcomes and thereafter make decisions that maximize expected utility under that probability distribution. The mode is but one infinitesimally tiny piece of that probability distribution von Neumann[100] instructed us to assemble. The idea of focusing solely on the mode would not appeal to von Neumann and in fact would catalyze "tissue rejection" because the mode is but "one tiny point" on an entire probability distribution, and the utility function will value each and every point on that entire distribution. Von Neumann instructed us to use the entire probability distribution, not just the maximum or the mode. We intend to sidestep many of the conceptual nuances associated with maximum likelihood or similar estimation procedures. When we move to prediction, the predicted items enter a decision-analytic approach just as any other probability distribution would. Jaynes[49] is the quintessential Bayesian, and so, by following his lead, are we. We search for Bayesian purity here.

This is not a book for people who are willing to take any detail of the development or the computation for granted, to turn it over to someone else, and to trust that they have done it correctly. Such people can simply skip to the sections that summarize the key results and type them into Excel, R, or Python. This is a book for people who want to know the applied mathematics and concepts underlying Bayesian Linear Regression and want to track through each and every probabilistic detail. It is for people who want to know marginal probability density functions (pdfs). Joint pdfs. Convolutions and conjugacies. This is intended as a complete work on the mathematics underlying Bayesian Linear Regression. This is a book for people who do not want to take any detail for granted. It is for show me" people. It is for "step-by-step" people. It is for people who need to interpret not only the results but also their genesis in terms of fundamental probability.

An important objective of this textbook is to put the mathematical developments together into a single integrated volume, with common notation, with no step left unstated. Virtually no exercises are "left to the reader."[11] Each and every argument, motivation, and step is articulated herein. Much of the historical development associated with Bayesian Linear Regression has been put forth in the literature in somewhat of a piecemeal fashion, save perhaps for Zellner[107]

[11] This is not entirely true. We leave simulations to the reader. We do not leave analytics to the reader.

and Press[78]. The objective here is to put the methodology all into one place and to interconnect it.

Bayesian statistics and Bayesian Linear Regression can be profound aids for "big data" people as well as small data people. Big data isn't necessarily immune from Bayesian or data quality issues. Big data doesn't mean that the data and only the data will suffice (recall the "mantra.") Some big data sets are fraught with challenges, e.g., measuring the same thing in the same way repeatedly over and over and over and over again, or ignoring or failing to identify relevances or redundancies.[12] Big data sets may inadvertently or unknowingly gather temperature data in Fahrenheit and Celsius in different ways (multicollinearity). Data sets can have adverse selection problems—drawing data from infants in order to draw inferences about elderly people! (We have seen this phenomenon more than once!) Ignoring failures or other extreme occurrences because they pose risk, liability, or "bad PR" to the data gatherer or the decision maker. Including data observations that contribute to the point one is trying to prove, systematically excluding other observations. It may be risky to gather and present such data. (For example, a physician may be loath to enter certain information into a patient chart who has experienced a bad outcome because he or she fears liability. There may be attribution and legal or economic loss attendant with entering such data.) Or a physician may enter a symptom or cause of death that "pays more" in order to augment revenue to the hospital. Just because a data set is "big" does not mean it is suitable for a regression-based predictive task. Bayes allows one to "punt" on whether a data base may be "good" or "bad." It allows one to update a point of view based on whatever data one might have, good or bad, lots or little. It works with one "poor" data point as well as with 10,000 "good" data points. Bayes is almost completely immune to the size of the data set, and by design. This is not a trivial point. The age-old mantra that "a large enough data set washes away the prior" may or may not be correct. The specific problem context and data gathered will determine that. Bayes "plays the data as it lies."

Bayesian calculus developed herein might be helpful for Bayesian networks[13] and other probabilistic constructs. That is in part why we have put the method forth in so much detail. We have not seen Bayesian math presented in a sequential, cogent, and complete fashion in the past, and to do so is one of our objectives here. We want a book that a person can understand and simply write his or her Bayesian statistics computer program directly from it without confusion. If it is useful for other constructs, so much the better.

In proceeding through this textbook, one might interpret it as advocating

[12] If the observations are exactly the same over many, many observations, the data matrix is singular or nearly so, meaning the prior is everything. It is well known that if the data are exactly the same for every observation, the data matrix is singular and its inverse is infinite, i.e., the variance is infinite.

[13] See https://en.wikipedia.org/wiki/Bayesian_network. The term Bayesian Belief networks is sometimes used.

that Bayesian Linear Regression is the be-all, end-all method for applying data to decisions under uncertainty (in a Decision Analysis context or a forecasting context). Bayesian Linear Regression provides a probability distribution over various distinctions that can be used in the von Neumann[100] framework. There are other methods as well, including subjective estimation of probabilities, discrete rather than continuous probability distributions with discrete alternatives, influence diagrams (which systematize Bayes theorem among multiple distinctions each characterized by a discrete probability distribution), Bayesian networks, and so forth. With such methods, there is a risk of pure subjectivity, bias, or incomplete observation or conceptualization during probability assessment, and there is no alternative but to accept it without data or experimentation. Kahneman and Tversky[51] have been very clear about that in their pathbreaking work. Methods that rely on probability can use the methods of Bayesian Linear Regression put forth here when there is data available. As an example, one can discretize a distribution developed using Bayesian Linear Regression for direct input to a decision tree or influence diagram. The true beauty of Bayesian Linear Regression is that it allows a systematic, methodologically correct procedure for inferring probability distributions from observed data, which we have dubbed "spinning straw into gold." Those inferred probability distributions can be used to solve decision problems. We suspect, for example, that the Bayesian probability distributions developed here can comprise nodes in a Bayesian network and the "training" thereof, although we do not seek to prove such assertions here.

In statistics classes and applications, regression can be presented as the be-all and end-all. It is the "way to solve probability problems." Happily, Bayesian Linear Regression opens our eyes to the fact that regression, properly accomplished, **actually generates a probability distribution (or more than one) over key distinctions as its fundamental output,** and those probability distributions are the grist for von Neumann[100] decision-making. This is not to be minimized or trivialized. Developing an authentic, authoritative probability distribution over various distinctions lies at the heart of Bayesian Linear Regression and solving decision problems. That is an important contribution of this textbook. Chapter 12 of this textbook illustrates this notion via an example drawn from Clemen[20]. In the words of Decision Analysis, Bayesian Linear Regression is a way of encoding a probability distribution using data.

This textbook will be as notable for what it omits as for what it includes. As best we can, we omit tangents to the Bayesian statistical method and solution. Such tangents obfuscate rather than clarify Bayesian statistical methods and results. Some readers may be offput by what we omit or trivialize. By such omission, we are in effect asserting that they are not central or important to Bayesian Linear Regression.

The need to use probability distributions to make decisions eliminates the need for us to get entrained in the discussion of "frequentist" versus "subjectivist" viewpoints of probability and a number of related topics. People before us have used frequentism as a frame of reference for discussing Bayes, rather than the reverse. We are not persuaded by the literally omnipresent focus on frequentism, ostensibly a nod to the bastions of classical regression. In fact, we (and others) reject the frequentist solution as the best reference point; we prefer Bayes as the appropriate reference point and develop frequentism as a special case of Bayes, a "corner solution." There is insight to be gained from doing so. Bayesian Linear Regression is a problem in pure probability, no matter where one gets the requisite probability distributions. In our view, adherence to frequentism does not help to understand and deploy Bayesian statistics; it is often quite the opposite.

Wiki defines frequentism as:[14] *defines an event's probability as the limit of its relative frequency in many trials (the long-run probability).* See Poirier[75]. Would that this could ever be true for real-world problems! As many monographs in the field note, frequentism never occurs in the real world. Sorry! Data sets are finite, and often in real-world decision problems, they are almost microscopically small. Bayesian methods don't require frequentist concepts. In golf, you don't get to experimentally hit your ball over a 140 yard water hazard 1000 times to establish probability of splashing it. You hit it once, for real. With Bayes, you take the data where it lies. You play the prior where it lies. There is no data on whether the San Francisco Giants will win the World Series this coming year, and there never will be, and yet you have to bet or not bet today. There is no data on the price of natural gas in Germany 1, 2, or 5 years from now, and there never will be. We have to combine judgment and data to get authentic probability distributions needed for decision making.

This textbook is unapologetically Bayesian. Priors come from any darn place you can get them! Even a "sports bar" is not out of the question; in fact, any tavern might do! (Or any operating room!) If you have to make a decision, you get your probabilities any way you can, and you fold them into your decision precisely the way von Neumann[100] told you to do it. With Bayes, the intensity of the prior and the size of the data set both impact your ultimate probabilistic view in a coherent way. This textbook is very mindful of Decision Analysis, especially as defined by Ronald Howard[45], the science that uses probability distributions together with alternatives and values to make good decisions. Bayesian statisticians endeavor to contribute authentic, accurate probabilities toward making good decisions. That is what Bayesian Linear Regression is all about, and that is where it shines.

One of the other crucial insights about Bayes is how it deals with "small" data sets. In real-world decision problems, data sets are almost universally never large. We have practiced Decision Analysis in the real world for decades, and literally never do we see an important decision supported by "enough data." In fact, the data needed to support decision-making is almost invariably parsimoniously small, often minuscule.[15] In the world of real decision-making, there

[14] See https://en.wikipedia.org/wiki/Frequentist_probability.

[15] Sometimes there is and can be no data. Where is the data on gold price in 2030? Where is the

simply are not very many observations, data points, or time or money to get them. For many or most decisions, no data has been gathered at all, and there is precious little time to gather it. As such, classical regression is not much help, no matter how much we might wish the situation were different. Classicists wring their hands about the paucity of data, limiting the number of coefficients in their models, or even fabricating data. But Bayes deals squarely with small data sets. Bayes provides a way to support better decision-making if all you have is a small, even a tiny, data set. What if you have only one slow, unreliable, Howard and Abbas[45] "Acme Rain Detector?"[16] You are certainly not alone; virtually every other decision-maker is in the same predicament. You and they routinely have to deal with very small data sets, almost microscopically so in reality. "Big data" is easy; real world decision makers know that it usually doesn't exist. It is rare when such data are developed carefully and completely from a decision-focused perspective. On the contrary, quite often big data bases are gathered, and people try to "mine them for insights" ex post facto by postulating and testing postulates, in effect groping through the data set to glean whatever from it they can.

Decision makers move forward with small or imperfect data because they HAVE to; there is no alternative. Decision makers will go forward with a paucity of data because they HAVE to. In today's era of "big data," precious little attention has been paid to the fact that for the really important uncertain variables, precious little data is available or gathered, or sometimes no data at all. Real-world decision makers have to deal with small data sets and know exactly how much they contribute to the decision, and Bayes helps them tremendously. When a decision is looming, the answer cannot be "Sorry, J.B., we ain't got data yet." This is not a trivial point. The big, tough decisions of the day involve substantial uncertainty, and there may be little or no data that helps us encode that uncertainty into a probability distribution. Uncertainties are large (wide variances), yet decisions have to be made today in the face of an often uncertain or risky situation. Small data is the rule, not the exception, in real-world decision-making. The Bayesian methods herein allow one to use whatever data are available to infer an authentic probability distribution, frequently with just a small handful of data points and maybe now and then with a large set of data points. Importantly, they may motivate the decision-maker to go and gather more data and to carefully structure a process to do so.

And let us turn to the notion of "data gathering." Data gathering is itself a **decision** (or, more properly, a spate of decisions), and often very messy and uncontrolled decisions at that. Howard and Abbas'[45] "Acme Rain Detector" represents a decision to incur costs to gather a little bit of impending rain data

---

data on the San Francisco 49ers winning the Superbowl in 2025?

[16]The Acme Rain Detector is a device that "gathers data" and does so inaccurately, just like real data gathering, when you pay it to do so. You put your money into the Acme Rain Detector, and it tells you whether it is going to be rainy or sunny at a party at which future decisions and rewards are dependent on the weather. The Acme Rain Detector generates data regarding the weather next week, and it has an "error rate." You pay your money and you get an imperfect reading.

with full recognition that the data is "imperfect." In the real world, data is generally gathered by junior people in deadly boring, repetitive jobs. Data are sometimes automatically gathered by machines, meters, or recorders (like doctors). Even when a sensor breaks or malfunctions or the process is changed to a "new method" or a new technology, the data keeps coming right on in, literally continuously, even though the method of gathering it has changed materially. So-called heteroskedasticity has occurred systematically, and yet it is often neither discerned nor considered. Questionnaires are whipped out, sent out, answered in a rush, sent back, and compiled. Quality control can be absent because it is extremely difficult to supervise or oversee data gathering. Adverse selection, which is virtually ubiquitous, is ignored. Disparate definitions are used, often without the requisite level of clarity, and definitions change over time. Contractors and vendors are hired to gather data and ship it to customers. Who knows what they are doing or how they are doing it? Sensitive considerations such as race, gender, or religion can be systematically omitted from personal data sets and thereby omitted as variables in regression functions. Data gathering is anything but a scientific activity. It is an intrinsically random activity, literally "sausage making." Bayesian and classical statistics anticipate and attempt to quantify this randomness using a simple probabilistic model of uncertainty surrounding data gathering put forth and articulated by Fisher and summarized by Salsburg[87] and others.

The contributions of Raiffa and Schlaiffer[81] were seminal; they were the first of which we are aware. Their references would encompass work that preceded theirs. The contributions of Arnold Zellner[107] are truly pivotal. He put a good bit of the content herein together initially, in pieces, in 1971, for all of us to understand and decode. This is a person whose work in statistics and economics was, in the author's view, worthy of the Nobel Prize. This was a person who was colossally generous to the author with his time and insight. The contributions of Edwin Jaynes[49] are pivotal. He put this all together in pieces for all of us to understand and decode. His clarity of exposition and philosophy on Bayesian analysis and issues were second to none. To a significant degree, we have compiled and integrated Zellner's and Jaynes' work, as well as others', into a single textbook.

1.2 This Textbook Isn't Solely Focused on Decision Analysis/Decision Theory

This textbook focuses squarely on Bayesian statistics (not Decision Analysis) as the central theme, which works well in every usual and customary problem setting in which one would otherwise use classical linear regression. While Bayes is an extension of classical statistics, it doesn't ignore its applications or applicability. It doesn't put its nose in the air, eschewing the heavy lifting of usual or typical regression problems. It works perfectly well in all the same contexts as classical. This textbook should fit as well into statistics courses as it does into

Decision Analysis courses; in fact, it should fit better. Classical statistics should never eschew Decision Analysis or Bayesian statistics.

1.3 No Step or Concept Omitted

We have another objective in mind here as well. This textbook strives to simplify, sequentialize, and articulate each and every step, from fundamental assumptions all the way to the results of Bayesian Linear Regression, to answer the "who, what, when, where, why, and how" questions so famously articulated by journalism schools. Regression techniques are so frequently used that we believe it is time to regularize and popularize them. There is no reason for analysts to use methods that are not fully understood, from soup to nuts. This will not only help current practitioners, but it may also assist practitioners of nonlinear Bayesian analysis to make the proper calculations in the proper order prior to applying a simulation or other solution. Linear becomes an archetype for nonlinear.

This textbook presumes substantial familiarity with matrices, matrix algebra, and integral calculus (multivariate and univariate). We provide appendices that highlight some of the key results from matrix and linear algebra, from which we draw liberally, and we have a number of key integral calculus prerequisite results in the appendices as well.

We present and belabor manipulations, even manipulations as pedestrian as integral substitutions or as sophisticated as nested Kronecker products and vectorization of outer products. We believe they are an important guide to understanding and observing EXACTLY what the underlying probability densities and transformations of variables truly are. Integral substitutions guide us, for example, between χ^2 distributions and inverse gamma distributions and ensure that we do not make errors by confusing among distributions. That is so crucial to the proper interpretation and use of Bayesian and classical statistical methods. What REALLY are these joint and marginal densities? What are they, and what are they not?

This textbook deals almost exclusively with the linear regression case. Nonlinear cases in practice are not amenable to analytic solutions. (Some linear situations are usually not amenable to closed forms. We point some of them out as we proceed through the discussion.) Such applications are relegated to simulation, outside the scope of this book for solutions and insight. This book systematically avoids simulation and simulation approaches for the important reason that the analytical results developed here shed profound insight on what the analyst is truly doing and assuming when he or she performs regression analysis, linear or nonlinear. This textbook has implications for nonlinear solutions, even though it does not address them directly. Our mission is didactic, and the linear case is the best environment to achieve it. Not to worry, there is plenty of challenge in the linear case.[17,18]

1.4 Organization

Chapter 2 of this textbook details the classical statistics approach. Before beginning with Bayes, we endeavor to put forth the fundamental mathematics and concepts underlying classical linear regression, which we call classical statistics. Classical statistics is, de facto, a "course prerequisite" for Bayesian statistics. It is difficult to understand Bayesian statistics until the principles of classical statistics are clear and mastered. This section, like classical statistics, does not end with an obvious, direct probability density function over model coefficients (as the later Bayesian sections will), but rather ends with the mean and variance-covariance estimates from classical statistics. We will develop analogous results, quite different in subtle but material ways, but nonetheless a direct substitute for and extension of classical statistical methods. In our section on classical statistics, we omit subjects like hypothesis testing, confidence intervals, and other considerations that we believe are superseded by Bayesian analysis. Pure probability is plenty sufficient!

Chapter 3 pivots to Bayes theorem in the univariate case. The section presumes knowledge of probability, the inferential notation of Jaynes[49], and calculus. This section presents the posterior as the product of prior probabilities times likelihood functions using the elemental principles of probability. It is the intellectual heart of Bayesian Linear Regression. Very simple, direct, and consistent with the theorem put forth by Thomas Bayes[5] in 1763 and extended by Pierre-Simon, Marquis de Laplace (Laplace), and others over the subsequent 250 years or so.[19] Very specific functional forms for the prior distribution, the likelihood function, and the posterior distribution are put forth. Our prior and posterior distributions in Bayes are identical in form, but with different parameterizations. As such, we can do all the mathematical developments using the common form, this "template," and merely apply the specific parameters for the prior and posterior to the template. We have heard all the criticisms of conjugate pairs as being limiting. To which we say, "Yeah, but they are still the state

---

[17] The author bemoans the fact that analysts jump too quickly to "simulation rather than thought." Today, it is much too easy to simulate, so much so that people are motivated to sprint right away to their computers and start sampling too early in the conceptual process. This book strives to avoid that, searching for exhaustive thought and analytical insight in advance of any simulation. There is nothing wrong with simulation and sampling intrinsically, but the sprint to do it from the outset without insight can overlook a lot of the insight one can "get for free" without plowing through sampling results. This textbook strives to get insight and thinking into a well-defined stack before crunching numbers.

[18] There are a number of "specialty topics" related to linear regression we have chosen to omit here, endeavoring instead to take a step-by-step, beginning-to-end, fundamental pass through basic Bayesian Linear Regression. We avoid specialty topics, particularly those related to "maximization," because Bayes is not based on maximization.

[19] It is interesting that probability emerged just over two centuries ago. It is still relatively young among the sciences.

of the art, not to mention blisteringly fast, interpretively invaluable, and such nice analogies that presage what should happen in simulations and more general Bayesian calculations." (We repeat; noncomputable forms seem to drag you right to the computer, which we strive to avoid until all the math and understanding are "in the can" and we have squeezed all the juice out of them we can. Our purpose is to put the intellectual basics into the can.)

Chapter 4 puts forth the specific structure of the prior, likelihood, and posterior. The functional form is similar for all, and this section uses that common form, which we term the "conjugate-archetype form" of equation to solve for all the parameters of the prior, the classical statistical solution, and the posterior using a common point of mathematical reference. This is a ponderous section and contains much of the mathematical essence of the (univariate) simple linear regression case. This section is de facto a "technical appendix," containing all the mathematical detail necessary for the prior-likelihood-posterior updating procedure implicit in the method. This is an important section for a course on Bayesian Linear Regression, not in reality a technical appendix at all, containing all relevant calculations and developments. Readers may be tempted to avoid it, but we caution against that because many of the key results and insights require facility with the mathematics.

Chapter 5 puts forth and examines the prior in greater detail than in Chapter 3, using the results derived for the conjugate-archetype form in Chapter 4. We have not been comprehensive herein with regard to any and all priors. We put forth the prior as a general construct and show effective ways to encode it. Later sections consider specific priors.

Chapter 6 uses the detailed representation and discussion of the prior, combines the likelihood, and creates the posterior. The chapter expounds on the mathematics of the Bayesian update and the Bayesian posterior itself. The chapter centers on the posterior and all its characteristics and properties.

In Chapter 2, and in fact in all of classical statistics, there is no notion of a "prior" anywhere present or mentioned—completely absent. (Or is it? As we shall see, it was there all the time, hiding in plain sight once we realize how to ferret it out.) Yet in Bayes in Chapter 3 and augmented in Chapter 6, the notion of a "prior" is intrinsic, front and center, unavoidable. You simply cannot operate without a prior; it is "required." Chapter 7 shows step by step how Bayesian Linear Regression can be specialized to classical regression if you render the prior invisible, ineffectual, inert, and informationless, i.e., if you totally "defang" the prior. Chapter 7 finds and applies a prior that is "invisible" or impotent. Chapter 7 puts forth precisely such a "defanged" prior, one that is intuitively appealing, and thereby shows how to specialize Bayesian Linear Regression to classical regression. In technical terminology, Chapter 7 defines what is called a "diffuse prior," ostensibly a prior that contains complete, abject, total, abstract, know-nothing, bone-numbing ignorance. The chapter then shows in detail how and why Bayesian analysis using a diffuse prior leads to the classical regression solution (with a minor alteration in a variance term).

Once we know the diffuse prior from Chapter 7, Chapter 8 inserts this diffuse prior in place of the prior that was used in Chapter 6 and derives the resulting posterior. It derives the posterior that results from applying the diffuse prior and studies it. The contribution of Chapter 8 is to show that the diffuse prior actually preserves the conjugacy of the prior-posterior pair and thereby allows all the previous results to go through unchanged. Profoundly important, Chapter 7 melds the diffuse prior with the likelihood function to create the posterior and thereby offers the classical regression solution as a pure probability distribution rather than seemingly arbitrary operations around a likelihood function. As a pure posterior probability distribution, this allows us to use it in von Neumann[100] decision contexts. That is profound and imminently conceptually helpful to gaining a true and correct understanding of classical statistics as well as Bayesian. We are chagrined that this interpretation of classical regression has been de-emphasized or overlooked in light of its conceptual importance, and we emphasize it in this textbook. Chapter 7 is clear:

- diffuse prior equals classical regression.
- diffuse prior equals abject, complete, total ex ante ignorance–zero information brought to the problem.
- diffuse prior is a construct, an abstraction, a point of reference, never a real-world reality.

The chapter amplifies those points in significant detail.

Chapter 9 summarizes the mathematical developments in the previous chapters for the univariate case and takes the time to systematically write the joint and marginal probability distributions of the Bayesian prior, the Bayesian posterior, and the classical regression posterior (classical regression solution). It is designed for programmers who are able to input these equations into as simple a system as Excel and make all the requisite calculations for classical and Bayesian statistics. "You mean Bayesian Linear Regression is so simple you can fully implement it in an Excel spreadsheet? With no macros or code?" Yep. How profound. Chapter 9 writes all the equations and constants that need to be programmed. It is a rather boring, mechanical chapter, but its purpose is to put all the equations one might want to program in one referencable location.

Chapter 10 addresses the special case in which the prior is inferred by classical regression from an "old," incomplete, inaccurate, or prefatory data base. (This occurs ubiquitously. Someone has an old, lousy database that is non-definitive. He or she wants to use that old data base to craft a prior to be updated by a much better, more accurate, more comprehensive, more consistent, systematically gathered data base.) An old, lousy data base seems like a good way, conceptually, to get a prior! Just grab that old, lower-quality, perhaps smaller or preliminary data set and see what regression on it might imply, interpreting the result as a prior. (Banish the thought of throwing away an old data base that wasn't all that great or that you paid too much to gather! Might as well squeeze some juice out of it!) Chapter 10 shows that the Bayesian updating calculation is exactly the same as if you merely append the old data to the new data and run a classical Bayesian analysis on the whole shootin' match! As we will prove, this means that you can initiate a statistical analysis with an initial set of data, and you can easily update it over time in increments as new data comes in, and you will get the selfsame answer as if you started over with the cumulative of all historically gathered data every time you get some new data. You carry along at each step the equivalent of a "sufficient statistic" for all the data you have seen up to then, and you add to that sufficient statistic based on the incremental data you are seeing now. The implications are profound. As new data come in over time, you do a very simple serial update and move forth with the entire historical data base! (This is not at all obvious with classical regression.) The proofs in this section tell us exactly how to run Bayesian (and classical) linear regression sequentially if new data comes in sequentially, which it frequently does. It also has mortal implications for some forms of empirical Bayes. This notion of serial updating is worth its weight in gold. Data literally never comes in all together in one huge mass. It comes in increments. The old refrain "We need more data" virtually ensures it.

Chapter 11 begins with an interesting and illustrative statistical analysis problem with only two independent variables and a single dependent variable, a score, and endeavors to project the probability distribution over that score as a function of those two variables. The problem in this chapter is to identify whether cognition (as measured by IQ) or test preparation (as measured by ex ante hours of study) is the bigger precursor for performance on an exam. The problem setting begins with two different people who have strong opinions about the subject, which they express via two quite different priors. The example is an idealized, simple first application of Bayes and offers great insight into how one might use the method.

Chapter 12 details and solves an interesting and illustrative Decision Analysis problem put forth by Clemen[20] in his Decision Analysis textbook using Bayesian statistical methods. (Clemen[20] used classical regression, which we generalize in this chapter to using Bayes.) Clemen's decision problem considers two distinct alternatives and one (uncertain) distinction. It is quite representative of the application and power of Bayesian statistical methods.

Chapter 13 details and solves an asset performance prediction problem from a Bayesian perspective. This problem, conducted by Drs. Nesbitt and Sparks in the 1990s in Canada, is where the work that catalyzed this textbook germinated. After all these many years, we thank the Canadian Strategic Highway Research (C-SHRP) Program for its seminal funding and motivation. They may scarcely remember and perhaps have never used the powerful tool we developed to aid them and extract true knowledge from their massive data set. The reader will be the judge if they should have when he or she reads Chapter 13.

Chapter 14 deals with the all-important problem of how one might go about assembling a subjective probability distribution, which can be used as a prior. To do so, we have drawn from Howard and Abbas'[45] "weight of the desk" problem. This is quite a nice application, drawn from Prof. Howard's famous Decision Analysis class at Stanford (which the author himself has taught since Dr. Howard's retirement), of how to get a prior over some distinction about which one might be highly uncertain and might have only limited, impressionistic knowledge. We have taken Howard's example and systematized, using Bayesian Linear Regression, how to get a continuous probability density function that represents one's prior. Once we have that, folding it into Bayes is straightforward.

Chapter 15, Chapter 16, and Chapter 17 put forth the substantially more complex mathematics of the multiple (multivariate) regression case. They are an extension of the methods for univariate Bayesian statistics built up painstakingly in the preceding chapters. This is a sophisticated and difficult extension, requiring sophisticated matrix mathematics, but it is entirely necessary for a complete volume on the topic. Suppose we wanted to infer a bivariate or multivariate model—a model of joint outputs. Chapter 15 begins the process by putting forth a multivariate likelihood function, a direct extension of the univariate likelihood function developed in Chapter 4. The addition of a second or more joint outputs adds significant complexity to the development.

Chapter 16 then posits a multivariate prior and develops all the key results for the prior—mean and variance-covariance of both the coefficients and the joint error terms. This chapter extends the sophisticated development in Chapter 4 related to the prior, the posterior, and the conjugate form. The extension to the multiparameter case is direct when we finish with it, but the subtleties and nuances along the way are rather intense.

Chapter 17 then multiplies likelihood in Chapter 15 times prior in Chapter 16 to obtain the posterior and studies the posterior in the same depth as we studied the prior in Chapter 4 in the simple regression case. It also specializes to the diffuse prior case–classical regression.

Chapter 18 summarizes the mathematical developments in the previous chapters for the multivariate case and takes the time to systematically write the joint and marginal probability distributions of the Bayesian prior, the Bayesian posterior, and the classical regression posterior (classical regression solution). It is designed for programmers who are able to input these equations into as simple a system as Excel and make all the requisite calculations for classical and Bayesian statistics. "You mean Bayesian Linear Regression is so simple you can fully implement it in an Excel spreadsheet? With no macros or code?" Yep. How profound. Chapter 18 writes all the equations and constants that need to be programmed.

Chapter 19 presents an extended example of the multivariate regression model in Chapter 15, Chapter 16, and Chapter 17. The topic of the example is the set of attributes that predict jointly the level of total crime and the level of violent crime in a city based on the characteristics of the city and its residents. Prospective contributory variables include per capita police funding, percent of people with high school diploma, percent of people who drop out of high school, percent of people currently away at college, and percent of people with a college degree. The example shows how to develop a joint density over these two highly correlated variables. Thereafter, we could impose a multiattribute utility function over them and calculate expected utility. We can also make marginal predictions as well as joint predictions.

One of the underlying reasons to pay the price in Chapter 15, Chapter 16, and Chapter 17 to extend from the univariate to the multivariate response case is because that is required to deal with the simultaneous equation regression case that motivated our initial consideration of the method a couple of decades ago. We were motivated by the need to infer price and income elasticities of demand in an economic problem, and that requires systems of regression equations. Chapter 20 addresses the question: What if you have a demand curve that is linear in coefficients and a supply curve that is linear in coefficients? The "magic crossing point" is the intersection of those linear curves—two simultaneous equations in two unknowns, price and quantity. What you observe in the real world is a series of "magic crossing points" (price-quantity pairs), and what you strive to do is identify the supply curve and identify the demand curve parameters based on the aforementioned observations. In economics, you are often looking to estimate the demand curve elasticity; it tends to be a "million dollar parameter" in real-world economic problems. We put forth the simultaneous equation method and mathematics in Chapter 20 and show how it must draw directly from the multivariate case in Chapter 15, Chapter 16, and Chapter 17. This section is not as satisfying because it cannot be fully solved in closed form. The ultimate solution would require simulation.

Chapter 21 contains an extended case study of simultaneous equation models as put forth in Chapter 20, illustrating key concepts with numerical calculations.

Chapter 22 concludes and integrates.

There are important technical appendices as well. Appendix A focuses on partitioning the variables in multivariate normal distributions so as to integrate one partition out and be left with the marginal distribution over the remaining unintegrated variables.

Appendix B summarizes the fundamentals of the multivariate normal distribution so that it can be used and integrated repeatedly in the text without reference or derivation.

Appendix C focuses on matrix and tensor derivatives. What is the derivative of a matrix Y that is full of variables y_{ij} with respect to another matrix X that is full of variables x_{ij}? Appendix C develops matrix derivatives that are needed in Chapter 15 for the multivariate case. This material is known, but in our experience, it is not as clearly or nicely motivated and presented as we would like.

Appendix D supports a specific result in Chapter 15 regarding the variance-covariance matrix in a multivariate (rather than univariate) setting.

Appendix E summarizes the fundamentals of univariate and multivariate change of variables problems.

1.5 Summary

Bayesian statistics is about developing probability distributions over key variables, melding data and judgment (or no judgment if desired).

1.6 Teaching

This textbook can serve as an effective supplement to education in Decision Analysis as well as for statistics or probability. (The author used material herein as he developed it as part of a Decision Analysis course at Stanford, MS&E 252, "Foundations of Decision Analysis," and another course in 2025 specifically about Bayesian Linear Regression.) More importantly, it facilitates education in Bayesian and classical statistics, illustrating in a cogent, easily understood, and highly focused context the power of Bayesian analysis and updating. The latter is the primary reason the author wrote it.

To reiterate, many practitioners of statistical analysis quickly embrace simple (and occasionally multiple) linear regression using simple or complex, pre-programmed tools like Stata (or even Excel regression), and the author worries that many practitioners are simply not armed to think deeply enough about what they are truly doing or calculating. Excel makes it so easy that there is no need to think. Practitioners have not been given the opportunity to learn and internalize Bayesian (or classical) statistics in a unified and consistent way in a course and to "grok"[20] its simplicity. We have not seen a book like this striving to do so.

The problem this textbook is trying to address is, tersely stated, that teachers and students need a book that

1. Starts by developing traditional classical linear regression;
2. As an extension, develops Bayesian Linear Regression with an informative prior from first principles;
3. Develops the posterior using that informed prior;
4. Specializes the informative prior so that you arrive at the classical regression solution as a posterior by using that specialized prior. We ask: What is the specific prior that gives you the classical regression solution?

[20]With all due credit to Robert A. Heinlein[40], who coined the word.

5. Examines the specific specializations that are required in order to do so;
6. Thereafter, makes predictions using prior, classical, and posterior information.
7. Includes sophisticated numerical examples programmed entirely in Excel. No complexity. No sophisticated coding.[21]
8. Pushes a single Bayesian construct (normal times inverse gamma) all the way through from assumptions to answers rather than appealing to tangential Bayesian situations using unrelated densities. All illustrations are in a Bayesian Linear Regression context.
9. Does not prematurely resort to simulation before working out all the mathematics all the way to the end.

Manipulating the prior so that you get the classical regression solution as the posterior is quite different from what we have seen before (e.g., fooling around with priors on their own in a vacuum to infer Jeffreys[50] or maximum entropy priors), yet it is absolutely pivotal to understand what classical regression truly is. It also allows you to make completely comparable predictions based on the

- prior only
- data only, appealing to the posterior that mimics classical regression
- posterior only, which melds prior and data

all correctly. People want a course and a textbook to project with the prior, project with the classical regression solution, and project with the posterior, all three being bona fide univariate or multivariate probability distributions. That helps them understand how much information is brought in via the prior versus how much information is in the posterior, the difference being attributable to the data. It helps them value the data and know how much more to gather.

We invariably hear practitioners utter the recurrent refrain, "We need more data." This textbook seeks to quell reflexive thinking like that, giving people the mathematical and probabilistic education to know what they are doing, why, and exactly how much true information is embedded in their data, be it small, medium, or large. In fact, applying the thinking embedded in the insightful "Acme Rain Detector" in Howard and Abbas[45], we can calculate the exact value of gathering more data—in fact, even one more tiny piece of additional data! (Howard, Abbas, and others have termed this the "value of imperfect information.") The Acme Rain Detector is a stroke of genius with regard to illustrating

[21]We have personally seen programming often become an intellectual "black hole," sucking facts, learning, and math inexorably into its abysm. We have seen learning all but stop when people program, especially when they program too early in the learning and analysis cycle. There is plenty of time to program effectually once you learn the Bayesian Linear Regression math. And your program will be simple, effectual, and grockable.

and motivating the value of imperfect information, and we bring that concept to the table.

This textbook is designed to help people understand precisely what they are doing when they choose regression. This textbook is intended to demystify classical regression and put it into a larger probabilistic context. It is intended to allow analysts to truly understand what they are doing. We recommend it for an upper-division undergraduate or first-year graduate student course in Bayesian Linear Regression. Students will need probability, integral calculus, and matrix algebra as prerequisites.

The truly nice part of this textbook is that all the material, all the lectures, and all the step-by-step developments are here. There are no exercises to be done by the professor as he or she prepares the course. Problems and exercises are omitted, but they exist literally everywhere there is a data base and a regression problem. The professor only needs to develop problem sets, exams, and analytical supplements for students to work on. The textbook is intended to be sequential and complete.

1.7 Literature

There are a number of monographs and texts, in whole or in part, that put forth Bayesian statistical methods and attendant issues. In our view, the references that lead the field in Bayesian Linear Regression are Zellner[107] and Press[78]. Key matrix results are contributed by Kollo and von Rosen[57]. Zellner[107] has put forth the definitive early text on univariate and multivariate Bayesian Linear Regression. Zellner[107] framed it for econometrics problems, and his pace-setting, now aging, book may not have gotten the parlay it might have if it were in the statistics realm. We have drawn inspiration from Zellner[107] and strived to reorganize his contributions in a statistical setting and to concentrate them as a textbook. Press[78] is an excellent, aging book on Bayesian statistics. Other Press references are somewhat subsidiary to this Bayesian monograph.

The two modern books that have risen to the fore are Bishop[10] and Murphy[69]. Bishop is quite general, like Gelman et al. al.[33], covering a host of topics but focusing more on computation than Bayesian theory. (The scope of the book is actually broader than the implications of the title.) This is an excellent book on the topic, even though it is almost two decades old. Murphy[69] is a just-published, excellent, comprehensive book. It contains a summary section on Bayesian Linear Regression, but it is neither step-by-step, didactic, nor illustrative on that topic. Our textbook is designed as a prelude to computation, i.e., it is intended to fill the needed niche of teaching Bayesian and classical linear regression with computation in Excel so that students and practitioners intrinsically understand.

We begin with the observation that many Bayesian books emphasize computation or data analysis—invariably simulation rather than analytic computation. Our focus is different—step-by-step, comprehensive logic developing Bayesian Linear Regression and its relation to classical linear regression in an analytic way. Computationally oriented books include Aitkin[2]; Bailer-Jones[4]; Bolstad[13], [12], [11]; Bozza, Taroni, and Biederman[15]; Chi[105]; Demidenko[24]; Downey[27]; Gatsonis[32]; Gelman, Carlin, Stern, Dunson, Vehtari and Rubin[33], which has become almost the "Bible" in the field; Ghosh, Delampady, and Samanta[35]; Koch[56]; Linden, Tose, and von Touissant[99]; Martin, Kumar, and Loo[66]; McElreath[67]; and Sivia and Skilling[92].

There are books centered on simple linear regression but (by choice) not the subsequent step, multivariate linear regression. Our focus is a step-by-step, comprehensive pass through Bayesian simple and multivariate linear regression and its relation to classical linear regression. Books relevant here include Berger[6]; Berry[9]; Gill[36]; Lee[61]; Lynch[65]; Press[77, 76, 79]; and Watanabe[103].

A number of books extend themselves beyond classical and Bayesian Linear Regression by introducing other conjugate pairs that have nothing to do with linear regression and other extensions. Our focus is a step-by-step, comprehensive development of Bayesian simple and multivariate linear regression and its relation to classical linear regression without any detours into attendant areas. Such books include Carlin and Louis[17]; Christensen, Johnson, Branscum, and Hanson[19]; Kaplan[52]; Robert[83]; Rowe[86]; Scutari and Denis[90]; Shemyakin and Kniazev[91]; Spiegelhalter, Abrams, and Myless[95]; and Turkman, Paulino, and Muller[98].

There are a number of books that adopt a less extensive approach than ours, including Florens[30], Schmitt[89], Stone[96], Slater[93], Robert and Marin[84], Reich and Ghosh[82], Kurt[60], Donovan and Mikey[26], Davidson-Pilon[16], and Hoff[43].

There are a few books that emphasize frequentism and/or the frequentist-Bayesian debate. These books are different from ours (which does not embrace frequentism as a frame of reference), including Klugman[54]; Koop[58]; and Wakefield[102].

Finally, there are books more focused on Bayesian philosophy, side issues, extensions, or collections of Bayesian and classical papers, including Agresti[1]; Bernardo, Bayarri, and Berger[8]; Bernardo, Bayarri, Berger, Dawid, Heckerman, Smith, and West[7]; Bovens[14]; Congdon[21]; Dey and Rao[25]; Fairfield and Charman[29]; Garnett[31]; Held[41]; Howson[47]; Lindley[62],[63]; Neal[70]; and Nielsen and Jensen[72].

Chapter 2

THE CLASSICAL REGRESSION MODEL

The probability of failure
doesn't matter to me.

Vinod Khosla

"*Linear regression is a linear approach to modeling the relationship between a scalar response (or dependent variable) and one or more explanatory variables (or independent variables). In linear regression, the relationships are modeled using linear predictor functions whose unknown model parameters are estimated from the data.*"[1] In its simplest form, with a single independent variable or "contributory variable" x and a single predicted or response variable y, the simple linear regression model is postulated:

$$y = \beta_0 + x\beta_1 + \varepsilon \quad ,$$

in which ε is an error term, a random variable. The error term accounts for the variability in y that cannot be explained by the assumed linear relationship between x and y. If ε were not present, that would mean that knowing x would provide enough information to deterministically calculate the value of y. The term ε is governed by a probability distribution, most often, by assumption, a univariate normal distribution with a mean of zero and standard deviation of σ, designated using Jaynesian inferential notation $\{\varepsilon|\sigma\}$. The σ parameter is not always exposed, sometimes using the abbreviated notation $\{\varepsilon\}$. We expose the parameter σ so as to deal with it explicitly. We emphasize that there are two evident and enumerated unknown parameters or coefficients in this equation,

[1] See https://en.wikipedia.org/wiki/Linear_regression.

namely β_0 and β_1, but there is a third parameter σ as well. If we specify a value of x, an independent variable, we need the coefficients β_0 and β_1 and the parameter of the error distribution σ if we are to predict the value of y. The two coefficients, β_0 and β_1, are obvious. We will deal with the specific functional representation of the error term after we generalize the linear term and deal with β_0 and β_1.

The linear model is immediately generalizable to a set of $p \geq 2$ independent or contributory variables.[2] By expanding the first two terms in the foregoing equation,

$$y = \beta_0 + \sum_{k=1}^{p-1} \beta_k x_k + \varepsilon \quad ,$$

in which p is the number of linear coefficients to be estimated $\beta_0, \beta_1, ...\beta_{p-1}$, and it goes about estimating those coefficients so that the response can be predicted as a function of a set of independent variables. This chapter and textbook will give a great deal of attention to carefully articulating this approach and solving it using classical methods in this section and Bayesian probabilistic methods in subsequent sections. The present chapter summarizes the classical regression approach, and subsequent chapters develop the Bayesian approach from first principles.

To estimate the coefficients, one collects a series of n observations of a dependent variable or response variable (designated y) together with a vector of $p - 1$ attendant independent variables (which we term contributory variables or explanatory variables, carefully not implying any causality) that prospectively have a contributing or perhaps (no guarantees) even a causal effect on that dependent variable. When we do so, we have for the first of the n observations:

$$\begin{aligned} y_1 &= \beta_0 + \beta_1 x_{1,1} + ... + \beta_{p-1} x_{1,p-1} + \varepsilon_1 \\ &= \beta_0 + \sum_{k=1}^{p-1} \beta_k x_{1,k} + \varepsilon_1 \quad . \end{aligned}$$

The observed value y_1 is modeled as a linear combination of the constant term β_0 and the coefficients β_1,...,β_{p-1} times the corresponding observed values x_{11},..., $x_{1,p-1}$ plus an error term ε_1. We have for the second of the n observations:

$$\begin{aligned} y_2 &= \beta_0 + \beta_1 x_{2,1} + ... + \beta_{p-1} x_{2,p-1} + \varepsilon_2 \\ &= \beta_0 + \sum_{k=1}^{p-1} \beta_k x_{2,k} + \varepsilon_2 \quad . \end{aligned}$$

[2]If p is 1, the model is just a constant. The method works, but few or none of the important results developed here are necessary. A constant model is usually too trivial to be useful.

We have for the ith of the n observations:

$$\begin{aligned} y_i &= \beta_0 + \beta_1 x_{i,1} + ... + +\beta_{p-1} x_{i,p-1} + \varepsilon_i \\ &= \beta_0 + \sum_{k=1}^{p-1} \beta_k x_{i,k} + \varepsilon_i \quad i = 1, ..., n \quad . \end{aligned}$$

We can write the foregoing equations in matrix form as follows:

$$\begin{bmatrix} y_1 \\ y_2 \\ \cdot \\ \cdot \\ \cdot \\ y_n \end{bmatrix} = \begin{bmatrix} 1 & x_{1,1} & x_{1,2} & \cdot & \cdot & \cdot & x_{1,p-1} \\ 1 & x_{2,1} & x_{2,2} & \cdot & \cdot & \cdot & x_{2,p-1} \\ \cdot & \cdot & \cdot & & & & \cdot \\ \cdot & \cdot & \cdot & & & & \cdot \\ \cdot & \cdot & \cdot & & & & \cdot \\ 1 & x_{n,1} & x_{n,2} & \cdot & \cdot & \cdot & x_{n,p-1} \end{bmatrix} \begin{bmatrix} \beta_0 \\ \beta_1 \\ \beta_2 \\ \cdot \\ \cdot \\ \cdot \\ \beta_{p-1} \end{bmatrix} + \begin{bmatrix} \varepsilon_1 \\ \varepsilon_2 \\ \cdot \\ \cdot \\ \cdot \\ \varepsilon_n \end{bmatrix} .$$

In this construction, there can be interpreted to be p contributory variables, including a constant or intercept term β_0, as well as $p - 1$ terms associated with independently observed variables. To wit, p counts the intercept as well as the observed variables. As such, for notational simplicity, we simply reindex the β terms $\beta_1, \beta_2, ...\beta_p$ and recognize that one of them is the intercept if we insert 1's into the observation data set for that coefficient. If we thus reindex, we can write the foregoing regression equations:

$$\begin{bmatrix} y_1 \\ y_2 \\ \cdot \\ \cdot \\ \cdot \\ y_n \end{bmatrix} = \begin{bmatrix} 1 & x_{1,2} & x_{1,3} & \cdot & \cdot & \cdot & x_{1,p} \\ 1 & x_{2,2} & x_{2,3} & \cdot & \cdot & \cdot & x_{2,p} \\ \cdot & \cdot & \cdot & & & & \cdot \\ \cdot & \cdot & \cdot & & & & \cdot \\ \cdot & \cdot & \cdot & & & & \cdot \\ 1 & x_{n,2} & x_{n,3} & \cdot & \cdot & \cdot & x_{n,p} \end{bmatrix} \begin{bmatrix} \beta_1 \\ \beta_2 \\ \beta_3 \\ \cdot \\ \cdot \\ \cdot \\ \beta_p \end{bmatrix} + \begin{bmatrix} \varepsilon_1 \\ \varepsilon_2 \\ \cdot \\ \cdot \\ \cdot \\ \varepsilon_n \end{bmatrix} .$$

This can be written in equation form:

$$y_i = \sum_{k=1}^{p} \beta_k x_{i,k} + \varepsilon_i \quad i = 1, ..., n \quad .$$

To reiterate, if we want a constant term, we must assume that all observations for the constant variable are 1. No confusion or obfuscation results. The series of observations of contributory variables and the dependent variable is organized as in Table 2.1.

The simple linear regression model that attempts to characterize these observations is the following:

$$y_1 - (x_{11}, ..., x_{1p})\,\beta = \varepsilon_1$$
$$y_2 - (x_{21}, ..., x_{2p})\,\beta = \varepsilon_2$$
$$...$$
$$y_n - (x_{n1}, ..., x_{np})\,\beta = \varepsilon_n \quad ,$$

Table 2.1: Observations

Observation	Dependent Variable	Independent Variables 1	2	3				p
1	y_1	1	x_{12}	x_{13}	.	.	.	x_{1p}
2	y_2	1	x_{22}	x_{23}	.	.	.	x_{2p}
3	y_3	1	x_{32}	x_{33}	.	.	.	x_{3p}
.	.	.	.	.	.	.	.	.
.	.	.	.	.	.	.	.	.
.	.	.	.	.	.	.	.	.
n	y_n	1	x_{n2}	x_{n3}	.	.	.	x_{np}

.

in which

$$\beta = \begin{bmatrix} \beta_1 \\ \beta_2 \\ . \\ . \\ . \\ \beta_p \end{bmatrix}$$

is a column vector of p unknown linear coefficients, the x vectors are successive row vectors extracted from the matrix of observations of independent variables, and the

$$\varepsilon = \begin{bmatrix} \varepsilon_1 \\ \varepsilon_2 \\ . \\ . \\ . \\ \varepsilon_n \end{bmatrix}$$

terms are "error" or, more properly, "predicted minus observed" error terms. Each entry is a random variable. By "linear," all we mean is that the model is linear in the sought-after coefficients of β. Statistical estimation techniques can be extended to nonlinear situations as well, but that is beyond the scope of this book. It will be convenient to organize the dependent and independent variable observations into a matrix of the independent variables and a vector of the dependent variable as follows:

$$y = \begin{bmatrix} y_1 \\ y_2 \\ . \\ . \\ . \\ y_n \end{bmatrix} ,$$

and

$$X = \begin{bmatrix} x_{11} & x_{12} & . & . & x_{1p} \\ x_{21} & x_{22} & . & . & x_{2p} \\ . & . & & & . \\ . & . & & & . \\ . & . & & & . \\ x_{n1} & x_{n2} & . & . & x_{np} \end{bmatrix} .$$

When we have done this, the error vector is easy to write

$$\varepsilon = y - X\beta \quad . \tag{2.1}$$

Consider again the issue of whether or not the model has an intercept term. If one wished to have an intercept term in one's model, suppose the first column of the X matrix was a column of 1's, i.e., the X matrix was

$$X = \begin{bmatrix} 1 & x_{12} & . & . & x_{1p} \\ 1 & x_{22} & . & . & x_{2p} \\ . & . & & & . \\ . & . & & & . \\ . & . & & & . \\ 1 & x_{n2} & . & . & x_{np} \end{bmatrix} .$$

In this particular case, the first model coefficient would always be multiplied by 1, meaning it would be a true intercept (or constant) term and would not vary by observation. The second, third,..., p^{th} model coefficients would always be multiplied by the appropriate observations, meaning the linear model coefficients correspond to columns $2, 3, ..., p$. This is far more convenient than carrying around the constant as a separate entity. The implication is that if there is a constant term, then p is, in practice, 2 or greater. However, there is no restriction on p, except that it must be greater than or equal to 1. If it is 1, the model trivializes to a constant, not a particularly sophisticated model but nonetheless valid. We reiterate that the variable p includes the constant plus all model linear coefficients.

In this model, the difference between

- the ith observed dependent variable, namely the ith element of the vector y, and
- the ith calculated or predicted dependent variable, namely the ith element of the vector $X\beta$

for each observation, the random variable ε_i, is assumed to be governed by a univariate normal distribution with a mean of 0 and a standard deviation of σ,[3] namely

$$N(0,\sigma) = \frac{1}{\sigma\sqrt{2\pi}} e^{-\frac{{\varepsilon_i}^2}{2\sigma^2}} \quad .$$

Because each of the normal distributions is assumed to be probabilistically independent (the error distributions are assumed to be independent, identically distributed, mutually irrelevant univariate normal distributions, each with a standard deviation σ and a zero mean),[4] the joint distribution over the n observed errors over the entire set of n observations is simply the product of the independent normal distributions, namely:

$$\begin{aligned} & f(\varepsilon_1, ..., \varepsilon_n) \\ & = \frac{1}{\sigma\sqrt{2\pi}} e^{-\frac{{\varepsilon_1}^2}{2\sigma^2}} \frac{1}{\sigma\sqrt{2\pi}} e^{-\frac{{\varepsilon_2}^2}{2\sigma^2}} ... \frac{1}{\sigma\sqrt{2\pi}} e^{-\frac{{\varepsilon_n}^2}{2\sigma^2}} \\ & = \frac{1}{\sigma^n (2\pi)^{\frac{n}{2}}} e^{-\frac{1}{2\sigma^2}\left({\varepsilon_1}^2 + {\varepsilon_2}^2 + ... + {\varepsilon_n}^2\right)} \quad . \end{aligned}$$

This is the joint probability of observing the errors $\varepsilon_1, ..., \varepsilon_n$. Observing this, people have chosen to use matrix notation, which is very convenient because it allows us to extend to an arbitrary number of variables (i.e., model coefficients) and an arbitrary number of observations. Using this notation, it is clear that the sum of squares in the exponent term in the previous equation is:

$${\varepsilon_1}^2 + {\varepsilon_2}^2 + ... + {\varepsilon_n}^2 = \varepsilon^T \varepsilon \quad .$$

Thus, the joint probability distribution over the error terms is more conveniently written in matrix form as follows:

$$f(\varepsilon_1, ..., \varepsilon_n) = \frac{1}{\sigma^n (2\pi)^{\frac{n}{2}}} e^{-\frac{1}{2\sigma^2}\varepsilon^T \varepsilon} \quad .$$

Using this notation and substituting from equation (2.1), we write the joint probability density function over the error terms to be:

[3]This is termed the homoskedastic or constant variance assumption. What that means is that every observation is governed by the exact same error function. Homoskedastic literally means "same variance."

[4]More sophisticated models of errors, e.g., heteroskedastic models, require a more sophisticated and structural error model. We do not deal with them herein.

$$\{y, X|\beta, \sigma\} = \frac{1}{\sigma^n (2\pi)^{\frac{n}{2}}} e^{-\frac{1}{2\sigma^2}(y-X\beta)^T(y-X\beta)} \quad . \tag{2.2}$$

By Bayesians and classicists alike, this equation is interpreted as the joint probability distribution over the observed values y and X given that the model parameters are β and σ. It represents the joint probability density function over the observations given the model coefficients. It is called the "**likelihood function.**" It is the probability that we see the observations y and X, assuming that the individual error terms are normally distributed and independent.

This is the classic Sir Ronald Aylmer Fisher[5] assumption—that the observed data had to have been generated from a purely random process. The likelihood function has to be a pure probability distribution over the data X, y. It cannot be systematic or non-random. The Fisher assumption is profound and a major contribution to thought about data and the process from which it has emanated. **Data must have been generated from a random process.** Period.[6] Not a systematic process like a game theory solution or an economic solution, but instead a purely random process characterized by a probability distribution (perhaps unobservable).[7] Howard's[45] "Acme Rain Detector" is very much this way. It has "sensitivity and specificity," which characterize a probability distribution over how detector readings are generated, i.e., "inaccurate," probabilistically-governed data. Howard[45] takes a bunch of readings (gathers data) from an imperfect machine, the readings governed by a probability distribution, just like real world data gathering, just like Fisher postulated for observed data.

The critical step at this stage is to write the exponent as a quadratic. We do that by using the matrix equivalent of completing the square. We take that approach so that we do not have to maximize the likelihood function; the maximum point is automatically evident as the top of a multivariate quadratic. We are going to remain averse in this book to maximizing anything or focusing on the mode of any probability distribution or function.[8] We begin with the exponent in equation (2.2):

$$Exp = (y - X\beta)^T (y - X\beta) \quad .$$

We are going to generalize an age-old algebraic operation—completing the square—in order to simplify the foregoing expression. To do so, we are going to postulate an as-yet unknown vector b, and then we are going to select the

[5] See https://en.wikipedia.org/wiki/Ronald_Fisher.

[6] We address this with an enlightening reference later.

[7] We have reservations about using these types of statistical methods with regard to game theory or economic (e.g., price) parameters. Those parameters resulted from systematic human action, not pure Fisherian randomness. As such, game theory and economic systems, e.g., the characterization of prices, may not be amenable to classical or Bayesian Linear Regression methods.

[8] The only thing we ever want to maximize is expected cardinal utility as laid out by the great John von Neumann[100].

particular value of that vector b that completes the square and allows us to write the foregoing expression as a pure matrix quadratic plus a constant. The vector or matrix completing the square defined here is a bit of a slant on univariate completing the square in high school algebra, so we will proceed through the steps in detail. We will be completing the square in the matrix sense defined here throughout this book, and we mean to cement the process here. We add and subtract our as yet unknown vector b from the coefficient vector β to write:

$$Exp = [y - X(\beta - b + b)]^T [y - X(\beta - b + b)] \quad .$$

For convenience of algebraic manipulation, we temporarily define

$$z = \beta - b$$

so that we can develop a matrix quadratic in z. Substitution into the foregoing equation yields:

$$Exp = [y - X(z + b)]^T [y - X(z + b)] \quad .$$

We expand this expression to form terms that are constant with respect to z, linear in z, and quadratic in z:

$$\begin{aligned} Exp &= [y - X(z + b)]^T [y - X(z + b)] \\ &= [(y - Xb) - Xz]^T [(y - Xb) - Xz] \\ &= \left[(y - Xb)^T - z^T X^T\right] [(y - Xb) - Xz] \\ &= (y - Xb)^T (y - Xb) - 2z^T X^T (y - Xb) + z^T X^T X z \quad . \end{aligned} \tag{2.3}$$

Notice in the last line of equation (2.3) that the leftmost term does not contain z at all. The rightmost term contains terms that are quadratic in z. The middle term is linear in z. If we made a judicious choice of b, the unknown vector, we could eliminate the middle term altogether, the linear term in z (or more technically, the linear term in z^T). That would allow us to write this exponent as a constant term plus a quadratic term with no linear term. To get that to occur, we could select b to zero out the middle term in the final row of the equation, the term that is linear in z^T as follows:

$$\begin{aligned} & X^T (y - Xb) = 0 \\ & \Rightarrow X^T y - X^T X b = 0 \\ & \Rightarrow b = \left(X^T X\right)^{-1} X^T y \equiv \bar{\beta} \quad , \end{aligned}$$

assuming that the indicated inverse exists. (We will address the assumed existence of the inverse immediately below.) The notation $\bar{\beta}$ has long been used in classical regression for the mean values of the coefficients. The "bar" notation is used by some to represent the mean value of a density, with the bar notation signifying the mean. We will honor the long-standing convention in classical regression to put the bar over key terms from standard regression, particularly the mean value of the model coefficients. If we set $b = \bar{\beta}$ to the classical regression estimate thus derived, set $z = \beta - \bar{\beta}$, and substitute both into the final line of equation (2.3), that equation reduces to:

(2.4) $$Exp = R + \left(\beta - \bar{\beta}\right)^T X^T X \left(\beta - \bar{\beta}\right) \quad ,$$

in which

(2.5) $$\bar{\beta} = \left(X^T X\right)^{-1} X^T y \quad ,$$

and

(2.6) $$R \equiv \left(y - X\bar{\beta}\right)^T \left(y - X\bar{\beta}\right) \quad .$$

The assumption that $X^T X$ is nonsingular and therefore invertible is important. This is a crucially important assumption for classical regression (but not necessarily for Bayes). It must be true, and the assumption is strictly dependent upon the observed data, or the classical method simply does not work. The matrix $X^T X$ is the "square" of the X matrix. It is both symmetric (by construction and no matter what the data) and positive semi-definite. It is invertible if and only if it is positive definite, so the assumption that $X^T X$ is nonsingular is an assumption that it is symmetric, positive definite (positive semi-definite is not strong enough), has all strictly positive eigenvalues, and is invertible. It is also tantamount to an assumption that the number of observations n is at least as large as the number of parameters p. Lacking that, $X^T X$ will be singular.[9]

When we get to the Bayesian method, there is no analogous restriction on the data matrix X. The matrix $X^T X$ has to be symmetric, which it is by construction, but it need not be nonsingular (as long as there is an informative prior that is not diffuse). With an informative prior, there will be a nonsingular, symmetric, positive definite matrix Q, which is related to the reciprocal of the variance-covariance matrix of the prior, that is added to the matrix $X^T X$. The requirement is that the sum of the prior matrix, which is symmetric, positive definite, and nonsingular and the data matrix, namely $Q + X^T X$, must be symmetric, positive definite, and nonsingular. That occurs in practical problems in our experience almost universally because both constituents of the sum are symmetric, positive definite, and nonsingular in real world problems. To wit, the only requirement we will have is that the prior have a positive definite, symmetric variance-covariance matrix and that the sum of the data matrix plus the prior variance-covariance matrix be nonsingular. There are no particular requirements

[9] If we had a purely deterministic process, $X^T X$ would be singular. Every row would be identical. The inverse, which, as we will see, leads to the variance-covariance matrix and therefore to the variances, will be infinite. That means zero informational value in the data, as we will see.

on the data. The only requirement is on the sum of the data plus prior. With Bayes, there will be no requirement that $X^T X$ be nonsingular. There will be no requirement that the number of observations n has to be at least as large as the number of parameters p. In fact, the only requirement is that $n \geq 1$, meaning that you have at least one observation.[10] Trivially, with no observations, you have no data. The assumption of nonsingularity of the $X^T X$ is specific to classical regression, not to Bayes. The assumptions detailed here are all a manifestation of the classical regression approach. There are a few "tune-ups" required with respect to these general statements, and we will note them during the development. All such revisions of traditional requirements (or non-requirements) are very satisfying, as they should be.

The requirement that $X^T X$ be nonsingular is, in the vernacular, actually a requirement that the data matrix X have a large number of rows n and that there is a lot of variation down the rows of each column. That enhances the probability of nonsingularity but does not guarantee it. It is de facto a repetition of the Fisher assumption that the data is generated as a random process, and you observe a lot of samples from that random process. In practice, if that occurs, the assumption that the matrix $X^T X$ is nonsingular is virtually always true.

Using no statistics, optimization, or anything other than simple algebraic completing of the square, we see the classical regression solution for the coefficients surfacing directly from completing the square. Completing the square replaces and obviates all need for optimization, making finding the maximum point of a quadratic trivial. Completing the square gives us a maximum, but it also gives the rest of the distribution as well. (Our assertion may not be precisely true that we ignore optimization, but it is true that optimization comes out of the formulation as a byproduct, and we do not dwell upon it because we want the whole distribution, not just the mode.) We can now write the likelihood function as follows:

$$
\begin{aligned}
\{y, X|\beta, \sigma\} &= \frac{1}{\sigma^n (2\pi)^{\frac{n}{2}}} e^{-\frac{1}{2\sigma^2}\left[R+(\beta-\bar{\beta})^T X^T X(\beta-\bar{\beta})\right]} \\
&= \frac{1}{\sigma^n (2\pi)^{\frac{n}{2}}} e^{-\frac{R}{2\sigma^2}} e^{-\frac{1}{2}(\beta-\bar{\beta})^T \left(\frac{X^T X}{\sigma^2}\right)(\beta-\bar{\beta})} \quad .
\end{aligned} \tag{2.7}
$$

This is the likelihood function, originally defined in equation (2.2), that has now in equation (2.7) been converted to a quadratic term plus a constant term in the exponent. It is becoming clear why. The quadratic term is related to a multivariate normal distribution. The constant term is related to a different univariate density, a special type of gamma density to be developed in detail here in Chapter 2.[11] We will explore this in great depth throughout this chapter so as to become

[10] Conceptually, one could update a prior with a single data estimate. In such a case the matrix $X^T X$ would be the product of a $1 \times p$ matrix X^T and a $p \times 1$ matrix X, which is a rank 1 $p \times p$ matrix.

[11] See https://en.wikipedia.org/wiki/Chi-square_distribution.

extremely precise.

There is something else important to observe. The distribution in equation (2.7) is the product of a multivariate normal distribution over the beta terms and a gamma-like distribution over the σ term.[12] This structure implies that the β terms are probabilistically independent from the σ term. Independence will continue to be evident as we use this distribution, and that fact, long known in statistics, offers great simplification and ease of interpretation.

By rendering the exponent a quadratic, finding the maximum point becomes trivial. And because of the convexity of a quadratic, we know there is a unique maximum, or "mode." The fact that completing the square trivially finds the maximum is irrelevant to the classical and Bayesian Linear Regression approaches throughout this textbook. It is an "innocent byproduct" of expressing a matrix quadratic function in the fashion that classical and Bayesian Linear Regression can do.

We think of the likelihood function as the probability that a random process governed by n independent normal distributions with mean zero and identical variance generates the observations embodied in the matrix X and the vector y. We term it the likelihood function because it quantifies the likelihood that the random process represented by the n independent normal distributions with mean zero and identical standard deviation governed by the parameters β and σ generates the data X and y actually observed. In probabilistic parlance, the likelihood function is the probability of X, y conditional on β, σ, which was written in equation (2.7) using Jaynesian[49] inferential notation[13] $\{X, y|\beta, \sigma\}$. So many classical statistics treatises chose not to sufficiently hammer home this critically important interpretation of the likelihood function. The likelihood function is the conditional probability of the observations given the model parameters. Equation (2.7) expresses the likelihood function in a very convenient and precise form. It will be convenient to represent the integrating constant in equation (2.7) as a single symbol

$$c_2 = \frac{1}{(2\pi)^{\frac{n}{2}}}$$

and write equation (2.7) as follows:

$$\{y, X|\beta, \sigma\} = c_2 \sigma^{-n} e^{-\frac{1}{2\sigma^2}R} e^{-\frac{1}{2}(\beta-\bar{\beta})^T \left(\frac{X^T X}{\sigma^2}\right)(\beta-\bar{\beta})} \quad .$$

We will return to this functional form repeatedly throughout the univariate chapters of the book, namely Chapter 3 through Chapter 10.

[12] It will actually be an "inverse gamma" distribution, a functional form that will become clear when we develop and emphasize it.

[13] Jaynes[49] has emphasized the value and clarity of inferential notation, which we use exclusively herein. His contribution to inferential notation adds profound clarity to probability theory and calculations.

There is a wonderful treatise on the probabilistic model of data development put forth by Fisher from a lay perspective written by Salsburg[87]. Salsburg makes very clear the notion, discovered by the revered statistician Sir Ronald Aylmer Fisher, that the data must come from a truly random process. Fisher's contribution to classical statistics was profound, and this assumption lay at the heart of the profundity. In the vernacular, "The data are generated by a random process."

2.1 Properties of the Classical Statistics Solution

The classical statistics model can be written in a matrix notation.[14] with its key assumptions as follows:

$y = X\beta + \varepsilon$ linear model

$< \varepsilon | X >= 0$ zero mean

$\varepsilon \sim N(0; \Sigma = \sigma^2 I)$

$< \varepsilon\varepsilon^T | X >= \sigma^2 I$

X is of rank p, which ensures that $X^T X$ is nonsingular.

(The $\sim$ notation in the third bullet indicates the variable to the left of the $\sim$ as the random variable and the expression to the right of the $\sim$ as the probability distribution that governs it.) The first statement is simply the linear model, the model put forth at the beginning of this section. The second statement is the statement about the error terms, which have a zero mean. The third statement is that all the error terms have a zero mean and a common variance, termed "homoskedasticity," which literally means "same variance."[15] The fourth statement reinforces that the errors are independent (uncorrelated) with common variance (homoskedastic). The final statement says that the X matrix is of rank p. Recall that the error term was

$$\varepsilon = y - X\beta \quad ,$$

which can be rewritten in the form

$$y = X\beta + \varepsilon \quad ,$$

and

[14] As with other probability textbooks, we use the notations $E(x)$ and $< x >$ interchangeably to represent the expected value of x. We use the bracketed notation$< x >$ when the $E(x)$ notation becomes cumbersome or when we want to strongly emphasize the notion of expected value.

[15] Models that do not have the "same variance" property are called heteroskedastic. Hierarchical Bayes models have that property. In order to be tractable, heteroskedastic models should have a well-defined relationship among all the variances that reside within them. We will not deal with heteroskedasticity herein because it is a logical extension of our homoskedastic model and can require simulation rather than offering a closed-form solution.

$$\bar{\beta} = \left(X^T X\right)^{-1} X^T y$$
$$= \left(X^T X\right)^{-1} X^T \left(X\beta + \varepsilon\right)$$
$$= \left(X^T X\right)^{-1} X^T X\beta + \left(X^T X\right)^{-1} X^T \varepsilon$$
$$= \beta + \left(X^T X\right)^{-1} X^T \varepsilon \quad .$$

2.2 Proof that Estimate of the Mean Is Unbiased

It is convenient when the estimate of the mean values of the coefficients (called the "estimator" of the mean) is what is called "unbiased." What does that mean? An estimator of a given parameter is said to be "unbiased" if its expected value is equal to the true value of the parameter. In other words, an estimator is unbiased if it produces parameter estimates that are on average correct.[16] If we look at the estimator of the mean $\bar{\beta}$, we might ask what would happen if we took the expectation of this estimate of the mean? We can think of expectation as a representation of the long-run average value of a random variable, i.e., the infinitely large sample representation of the long-run average value. We would like to know that the large sample representation trends toward the true and correct value of the parameter. Toward what value does the estimate trend? If we take the expectations of both sides and work through the requisite steps, we obtain the following series of calculations:

$$E\left(\bar{\beta}\right) = E\left[\beta + \left(X^T X\right)^{-1} X^T \varepsilon\right]$$
$$= E\left[\beta\right] + E\left[\left(X^T X\right)^{-1} X^T \varepsilon\right]$$
$$= E\left[\beta\right]$$
$$= \beta \quad .$$

This means that the expected value of the classical mean estimator trends toward the true and correct value and not toward some value other than the true and correct one. This is what is meant by "unbiased." Conceptually, the estimator of the parameter trends toward the true and correct value of the parameter. That is deemed an important and attractive property of the estimator. As we shall see, with Bayesian statistics, this is a non-issue. The biased and unbiased estimator issue goes away.

2.3 Residual Sum of Squared Errors

The first term of equation (2.4), rewritten as equation (2.6), is called the residual sum of squared errors. It is the square of the difference between the predicted

[16] See https://www.statlect.com/fundamentals-of-statistics/point-estimation.

values of y using the estimated model coefficients and the observed values of y. We begin by manipulating the residual sum of squared errors, substituting the expression for the estimated value of the coefficients in equation (2.5):

$$\begin{aligned}R &= (y - X\bar{\beta})^T (y - X\bar{\beta}) \\ &= \left[y - X(X^T X)^{-1} X^T y\right]^T \left[y - X(X^T X)^{-1} X^T y\right] \\ &= \left\{\left[I - X(X^T X)^{-1} X^T\right] y\right\}^T \left[I - X(X^T X)^{-1} X^T\right] y \\ &= y^T \left[I - X(X^T X)^{-1} X^T\right]^T \left[I - X(X^T X)^{-1} X^T\right] y \quad .\end{aligned}$$

The matrix embedded in the middle is written

$$M_X = I - X(X^T X)^{-1} X^T \quad ,$$

so the foregoing equation is written:

$$R = y^T M_X^T M_X y \quad .$$

We need to calculate the transpose of M_X:

$$\begin{aligned}M_X^T &= \left[I - X(X^T X)^{-1} X^T\right]^T \\ &= I^T - \left[X(X^T X)^{-1} X^T\right]^T \\ &= I - X\left[(X^T X)^{-1}\right]^T X^T \\ &= I - X(X^T X)^{-1} X^T \\ &= M_X \quad .\end{aligned}$$

Thus the matrix M_X is its own transpose, meaning it is symmetric. Thus:

$$R = y^T M_X M_X y \quad .$$

We now simplify the matrix product in the middle:

$$\begin{aligned}M_X M_X &= \left[I - X(X^T X)^{-1} X^T\right] \left[I - X(X^T X)^{-1} X^T\right] \\ &= \left[I - X(X^T X)^{-1} X^T\right] - X(X^T X)^{-1} X^T \left[I - X(X^T X)^{-1} X^T\right] \\ &= I - X(X^T X)^{-1} X^T - X(X^T X)^{-1} X^T + X(X^T X)^{-1} X^T X (X^T X)^{-1} X^T \\ &= I - X(X^T X)^{-1} X^T - X(X^T X)^{-1} X^T + X(X^T X)^{-1} X^T \\ &= I - X(X^T X)^{-1} X^T = M_X \quad .\end{aligned}$$

The technical term for a matrix being the square of itself is that the matrix is "idempotent." Therefore:

$$R = y^T M_X y \quad .$$

We are later going to develop a number of properties of the matrix M_X, properties that allow us to develop what is termed an "unbiased" estimate of σ^2. Keep in mind that R is calculated directly from the data X and y and the estimated model coefficients. Because the y and vectors are uncertain, the R term is uncertain too. Let us write an expression for the uncertainties in y:

$$\begin{aligned} y &= X\beta + \varepsilon \\ y^T &= \beta^T X^T + \varepsilon^T \quad . \end{aligned}$$

Let us substitute in y and M_X to find out what the probability distribution over R is:

$$\begin{aligned}
R &= y^T M_X y \\
&= \left(\beta^T X^T + \varepsilon^T\right)\left[I - X\left(X^T X\right)^{-1} X^T\right](X\beta + \varepsilon) \\
&= \left(\beta^T X^T + \varepsilon^T\right)\left[I\,(X\beta + \varepsilon) - X\left(X^T X\right)^{-1} X^T (X\beta + \varepsilon)\right] \\
&= \left(\beta^T X^T + \varepsilon^T\right)\left\{X\beta + \varepsilon - \left[X\left(X^T X\right)^{-1} X^T X\beta + X\left(X^T X\right)^{-1} X^T \varepsilon\right]\right\} \\
&= \left(\beta^T X^T + \varepsilon^T\right)\left(X\beta + \varepsilon - X\beta - X\left(X^T X\right)^{-1} X^T \varepsilon\right) \\
&= \left(\beta^T X^T + \varepsilon^T\right)\left[I - X\left(X^T X\right)^{-1} X^T\right]\varepsilon \\
&= \left[\left(\beta^T X^T + \varepsilon^T\right) I - \left(\beta^T X^T + \varepsilon^T\right) X\left(X^T X\right)^{-1} X^T\right]\varepsilon \\
&= \left[\beta^T X^T + \varepsilon^T - \left(\beta^T X^T + \varepsilon^T\right) X\left(X^T X\right)^{-1} X^T\right]\varepsilon \\
&= \left\{\beta^T X^T + \varepsilon^T - \left[\beta^T X^T X\left(X^T X\right)^{-1} X^T + \varepsilon^T X\left(X^T X\right)^{-1} X^T\right]\right\}\varepsilon \\
&= \left[\beta^T X^T + \varepsilon^T - \beta^T X^T - \varepsilon^T X\left(X^T X\right)^{-1} X^T\right]\varepsilon \\
&= \varepsilon^T\left[I - X\left(X^T X\right)^{-1} X^T\right]\varepsilon \\
&= \varepsilon^T M_X \varepsilon \quad .
\end{aligned}$$

Using this relationship, we can write the following expression for R:

$$R = \varepsilon^T M_X \varepsilon = \begin{bmatrix} \varepsilon_1 & \varepsilon_2 & \dots & \varepsilon_n \end{bmatrix} \begin{bmatrix} m_{11} & m_{12} & \dots & m_{1n} \\ m_{21} & m_{22} & & m_{2n} \\ \dots & & & \dots \\ m_{n1} & m_{n2} & \dots & m_{nn} \end{bmatrix} \begin{bmatrix} \varepsilon_1 \\ \varepsilon_2 \\ \dots \\ \varepsilon_n \end{bmatrix}$$

$$= \sum_{i=1}^{n} \varepsilon_i \sum_{j=1}^{n} \varepsilon_j m_{ij} = \sum_{i=1}^{n} \sum_{j=1}^{n} \varepsilon_i \varepsilon_j m_{ij} \quad .$$

If we take the expectation of R, an expression for the true and correct variance σ^2 will emerge. We will see that it will be an unbiased estimate of the variance. The expectation of R is:

$$\begin{aligned} E(R) &= E(\varepsilon^T M_X \varepsilon) \\ &= E\left[\sum_{i=1}^{n} \sum_{j=1}^{n} \varepsilon_i \varepsilon_j m_{ij}\right] \\ &= \sum_{i=1}^{n} \sum_{j=1}^{n} E\left(\varepsilon_i \varepsilon_j\right) m_{ij} \quad . \end{aligned}$$

The expectation can be moved inside the summation because the $m_{i,j}$ terms are merely constants that are a function of the observations. Because the error distributions are independent and normal, expectations of products of error distributions for which $i \neq j$ are zero. The only ones that are nonzero occur when $i = j$. Thus:

$$E(R) = \sum_{i=1}^{n} E(\varepsilon_i \varepsilon_i) m_{ii} \quad .$$

We know that the variance $E(\varepsilon_i \varepsilon_i)$ is σ^2 so that we have:

$$\begin{aligned} E(R) &= \sum_{i=1}^{n} \sigma^2 m_{ii} \\ &= \sigma^2 \sum_{i=1}^{n} m_{ii} \\ &= \sigma^2 Trace(M_X) \\ &= \sigma^2 Trace\left[I - X\left(X^T X\right)^{-1} X^T\right] \\ &= \sigma^2 \left\{Trace\left(I\right) - Trace\left[X\left(X^T X\right)^{-1} X^T\right]\right\} \quad . \end{aligned}$$

The first trace is an identity matrix of length n. Thus its value is n, and we write:

$$E(R) = \sigma^2 \left\{n - Trace\left[X\left(X^T X\right)^{-1} X^T\right]\right\} \quad .$$

We are going to have to work a bit to get the second trace. There is a result from linear algebra whose proof we shall sketch here. If A and B are two conformable (multipliable) matrices (in this forthcoming sketch, square matrices), then:

$$[AB]_{ij} = \sum_{k=1}^{n} a_{ik} b_{kj} \Rightarrow [AB]_{ii} = \sum_{k=1}^{n} a_{ik} b_{ki} \quad .$$

The first half of the following equation defines the trace, and the second half reverses the order of summation to prove the result for square matrices:

$$\begin{aligned} Trace\,(AB) &= \sum_{i=1}^{n} \sum_{k=1}^{n} a_{ik} b_{ki} \\ &= \sum_{k=1}^{n} \sum_{i=1}^{n} b_{ki} a_{ik} = Trace\,(BA) \quad . \end{aligned}$$

This generalizes to the product of three matrices that are conformable and multipliable, in which it is known that:

$$Trace\,(ABC) = Trace\,(CAB) = Trace\,(BCA) \quad .$$

In short, one can either bring C to the beginning or A to the end. This result is proven in a number of linear algebra texts and summarized above. We can directly apply this by moving the leftmost matrix to the end, as in the theorem, and writing:

$$\begin{aligned} & Trace\left[X\left(X^T X\right)^{-1} X^T\right] \\ &= Trace\left[\left(X^T X\right)^{-1} X^T X\right] \\ &= Trace\,[I] \quad . \end{aligned}$$

This is just the trace of the identity matrix, but we need to very carefully consider its rank. We know from the original characterization of the data matrix X that X is $n \times p$, and therefore X^T is $p \times n$. That means that $X^T X$ is $p \times p$. When we multiply a p times p matrix by its inverse, which is also p times p, we get an identity matrix that is $p \times p$. The trace of a $p \times p$ identity matrix is the sum of 1's p times, or p. That is:

$$Trace\left[\left(X^T X\right)^{-1} X^T X\right] = p \quad .$$

Therefore, we have the result

$$E(R) = \sigma^2 (n - p) \quad .$$

We can divide both sides of this by $n - p$ to write

$$E\left(\frac{R}{n - p}\right) = \sigma^2 \quad .$$

This is valid because $n-p$ is a multiplicative constant with regard to the expectations we have taken. This equation tells us that

$$\frac{R}{n-p} \tag{2.8}$$

is an unbiased estimate of the variance σ^2. The residual sum of squared errors R divided by $n-p$ is an unbiased estimate of the variance σ^2. This is one of the more important results in classical statistics. We notice from the definition of R that it is calculated directly from the X and y observations and nothing else. The numerator term is the sum of squared errors. The denominator is $n-p$, so, intuitively, this term represents the average of the squared errors of all the observations. To be clear, the unbiased estimate of the variance of the error term is just the residual sum of squared errors divided by the number of observations n minus the number of parameters (including intercept) p. Intuitively, the estimate of the error term is the average squared error over a sum with $n-p$ terms in it, the number of "extra" terms above and beyond the p terms being estimated, i.e., the magnitude of overdetermination.[17] The denominator $n-p$ is going to be called the "**degrees of freedom**." It is the number of observations minus the number of parameters, or conceptually, the degree to which the data overdetermines the equation. Classical statisticians use the notation:

$$\nu_C = n - p$$

and therefore the unbiased estimate of the error term is:

$$\sigma^2 = \frac{R}{\nu_C} \triangleq s_C^2 \quad . \tag{2.9}$$

It is common to use the Greek letter ν to characterize degrees of freedom, and we will subscript it using a C to depict the "classical" regression solution.

2.4 Variance-Covariance Matrix

Let us turn to the variance-covariance matrix.[18] We want to develop a true and correct (unbiased) estimate of the variance-covariance matrix. This notion is equally convenient to the mean value above, and the reason becomes clear as we articulate the proof. The variance-covariance matrix is the expectation of the square of the estimator we are considering minus its mean value. From the previous proof, the mean value is β and the estimator is $\bar{\beta}$. Consider the difference:

$$\bar{\beta} - \beta \quad .$$

[17] It is well known, but we will not repeat it here, that a pure maximum likelihood approach gives a biased estimator of the variance term.

[18] Others have termed it simply the covariance matrix, but we prefer the term variance-covariance matrix.

We showed immediately above that:

$$\bar{\beta} = \beta + \left(X^T X\right)^{-1} X^T \varepsilon \quad .$$

Therefore:

$$\begin{aligned}
&\left(\beta - \bar{\beta}\right) = -\left(X^T X\right)^{-1} X^T \varepsilon \\
&\Rightarrow \left(\beta - \bar{\beta}\right)^T = -\varepsilon^T X \left(X^T X\right)^{-1} \\
&\Rightarrow \left(\beta - \bar{\beta}\right)\left(\beta - \bar{\beta}\right)^T = \left(X^T X\right)^{-1} X^T \varepsilon \varepsilon^T X \left(X^T X\right)^{-1} \quad .
\end{aligned}$$

We need to take the expectation of this, for that is what the variance-covariance matrix over the model coefficients is. Thus:

$$\begin{aligned}
&E\left[\left(\beta - \bar{\beta}\right)\left(\beta - \bar{\beta}\right)^T\right] \\
&= E\left[\left(X^T X\right)^{-1} X^T \varepsilon \varepsilon^T X \left(X^T X\right)^{-1}\right] \\
&= \left(X^T X\right)^{-1} X^T E\left[\varepsilon \varepsilon^T\right] X \left(X^T X\right)^{-1} \quad .
\end{aligned}$$

The expectation of the error term inside the brackets is just σ^2 times the identity matrix because it is comprised of independent normals all with zero mean and all with the same variance term. It is as if there were the same independent normal on every element of the diagonal; the correlations are zero because of independence. Thus:

$$\begin{aligned}
&E\left[\left(\beta - \bar{\beta}\right)\left(\beta - \bar{\beta}\right)^T\right] \\
&= \left(X^T X\right)^{-1} X^T E\left[\varepsilon \varepsilon^T\right] X \left(X^T X\right)^{-1} \\
&= \left(X^T X\right)^{-1} X^T \sigma^2 I X \left(X^T X\right)^{-1} \\
&= \sigma^2 \left(X^T X\right)^{-1} X^T X \left(X^T X\right)^{-1} \\
&= \sigma^2 \left(X^T X\right)^{-1} \quad .
\end{aligned}$$

We know what the estimate of the variance-covariance matrix trends toward in the long term. For future reference, we write the variance-covariance matrix for the classical regression case:

$$VCV = \sigma^2 \left(X^T X\right)^{-1} \quad .$$

Substituting for σ^2 from equation (2.9), we write:

$$VCV = \frac{R}{\nu_C} \left(X^T X\right)^{-1} \quad .$$

2.5 This Monograph Omits Discussion of Consistency

We are not going to address consistency (asymptotic behavior as the number of observations becomes large) here. As presented in Wiki[19] regarding a consistent estimator, *"In practice one constructs an estimator as a function of an available sample of size n, and then imagines being able to keep collecting data and expanding the sample ad infinitum. In this way one would obtain a sequence of estimates indexed by n, and consistency is a property of what occurs as the sample size "grows to infinity." If the sequence of estimates can be mathematically shown to converge in probability to the true value..., it is called a consistent estimator..."*

2.6 Theorems about the Idempotent Matrix M_X

Let us state and prove some theorems about the foregoing idempotent matrix. We have already proven that the matrix M_X is symmetric and idempotent. An idempotent matrix has the property that it is its own square. Thus:

$$M_X M_X = M_X \quad .$$

2.6.1 Eigenvalues of M_X Are 0's and 1's Only

The expression for the eigenvalues λ of the matrix M_X is

$$M_X \xi = \lambda \xi \quad .$$

If we left multiply by M_X, we obtain

$$\begin{aligned}
&M_X M_X \xi = \lambda M_X \xi \\
&\Rightarrow M_X \xi = \lambda M_X \xi \\
&\Rightarrow \lambda \xi = \lambda^2 \xi \\
&\Rightarrow \lambda = \lambda^2 \\
&\Rightarrow \lambda = 0, 1 \quad ,
\end{aligned}$$

meaning that all eigenvalues of an idempotent matrix are either 0 or 1.

2.6.2 Every Column of X Is an Eigenvector of M_X with Eigenvalue 0

We know that the matrix M_X is

$$M_X = I - X\left(X^T X\right)^{-1} X^T \quad .$$

[19] https://en.wikipedia.org/wiki/Consistent_estimator

If we right multiply by X, we prove that $M_X X = 0$

$$\begin{aligned} M_X X &= \left[I - X\left(X^T X\right)^{-1} X^T\right] X \\ &= X - X\left(X^T X\right)^{-1} X^T X \\ &= X - X = 0 \quad . \end{aligned}$$

The matrix M_X is orthogonal to every column of X. That means that every column of X is an eigenvector of M_X with eigenvalue 0. There are p such columns, and because the X matrix is n times p, it cannot have a rank bigger than p. We will show in a subsequent section that the rank is generally exactly p.[20] There are p such eigenvectors, namely the columns of X, with eigenvalue 0. The matrix M_X has p eigenvalues that are zero.

2.6.3 The Trace of the Projection Matrix $P_X = I - M_X$ is p

We define the projection matrix

$$P_X = X\left(X^T X\right)^{-1} X^T \quad .$$

If we take the trace of both sides, we can write

$$Trace(P_X) = Trace\left[X\left(X^T X\right)^{-1} X^T\right] \quad .$$

If we use the trace theorem, and we move the leftmost term in the product on the right to the right hand side

$$\begin{aligned} Trace(P_X) &= Trace\left[\left(X^T X\right)^{-1} X^T X\right] \\ &= Trace[I] \quad . \end{aligned}$$

Because $X^T X$ is $p \times p$, the identity matrix I is $p \times p$ and therefore the trace at the right of the equation is p, i.e.,

$$Trace(P_X) = p \quad .$$

We will need this result below.

2.6.4 The Projection Matrix P_X Is Idempotent

Before moving on, we notice that the projection matrix itself is idempotent (i.e., is its own square):

[20] This section does not consider cases in which the rank of the data matrix is less than p, the number of coefficients. That would occur only if the matrix $X^T X$ is singular. Extensions to that case are possible.

$$\begin{aligned} P_X P_X &= \left[X\left(X^T X\right)^{-1} X^T\right]\left[X\left(X^T X\right)^{-1} X^T\right] \\ &= X\left(X^T X\right)^{-1} X^T X\left(X^T X\right)^{-1} X^T \\ &= X\left(X^T X\right)^{-1} X^T = P_X \quad . \end{aligned}$$

We know further from Wiki[21] that for an idempotent matrix, the trace of the matrix is equal to the rank of the matrix. We shall be proving and using that shortly.

2.6.5 Every Column of X Is an Eigenvector of P_X with Eigenvalue 1

We right multiply the projection matrix P_X by X:

$$P_X X = X\left(X^T X\right)^{-1} X^T X = X \quad .$$

That shows that the p eigenvectors of M_X that have eigenvalue 0, namely the p columns of X, are eigenvectors of P_X that have eigenvalue 1.

2.6.6 The Trace of the Matrix $M_X = I - P_X$ Is $n - p$

We derived the trace of the projection matrix to be

$$Trace(P_X) = p \quad .$$

Beginning with the relationship

$$M_X = I - P_X \quad ,$$

suppose we take the trace of both sides to obtain the relationship

$$\begin{aligned} Trace(M_X) &= Trace(I - P_X) \\ &= Trace(I) - Trace(P_X) \quad . \end{aligned}$$

We recognize that $Trace(I)$ is n because I is an $n \times n$ matrix here. We showed in the previous section that $Trace(P_X)$ is p. Therefore we have the following

$$Trace(M_X) = n - p \quad .$$

2.6.7 The Rank of the Matrix $M_X = I - P_X$ is $n - p$

If we can show that the rank of M_X is $n - p$, we know that its eigenvalues are exactly $n - p$ 1's and exactly p 0's. For every number that the rank is less than

[21] See https://en.wikipedia.org/wiki/Trace_(linear_algebra.)

n, that is a number where the eigenvalue is zero. There are p of them. To do so, we are going to need some basic results from linear algebra.

Begin by considering an $m \times n$ matrix A that has column rank r. That means that of the n columns, exactly r are linearly independent and the remaining $n-r$ are linearly dependent on the first r columns. There is a subset $r \leq n$ of the columns that forms the basis for the rest of the columns. Recognizing what column rank r means, every column of A can be written as a linear combination of the first r columns. That is, for all $j = 1, \ldots, n$, the jth column of A can be written as a linear combination of the first r columns of A (having put the r linearly independent columns into the first r positions), and the weights are represented by the f_{kj} terms

$$a_j = \sum_{k=1}^{r} f_{kj} a_k \quad j = 1, ..., n \quad .$$

This means that there is a matrix F and a matrix C such that

$$A_{m \times n} = C_{m \times r} F_{r \times n} \quad .$$

C is an $m \times r$ matrix containing the linearly independent basis, and F is an $r \times n$ matrix that contains the specific multipliers of the basis matrix that yields the columns of A. This is called the rank factorization of A. For the types of matrices here (square matrices), we only need to consider cases in which m and n are equal, i.e., square matrices. In that case, the rank factorization simplifies to

$$A_{n \times n} = C_{n \times r} F_{r \times n} \quad .$$

The matrices A we consider here are all idempotent, meaning that A is its own square

$$A^2 = A \quad .$$

Substituting the rank factorization on both sides for A, we can write

$$CFCF = CF \quad .$$

This implies that the last three terms on the left hand side of the equation must be equal to the last single term on the right

$$FCF = F = IF \quad .$$

This in turn implies that the first two terms on the left hand side of the equation must be equal to the identity matrix on the right

$$FC = I \quad .$$

If we examine the ranges of the foregoing matrices, we see the following

$$F_{r \times n} C_{n \times r} = I_{r \times r} \quad .$$

Let us take the trace of both sides

$$Trace\,(A) = Trace\,(CF) \quad .$$

We appeal to the trace of products theorem above

$$\begin{aligned} Trace\,(A) &= Trace\,(CF) \\ &= Trace\,(FC) \quad . \end{aligned}$$

We continue the logic through to completion, using the identity matrix result above

$$\begin{aligned} Trace\,(A) &= Trace\,(CF) \\ &= Trace\,(FC) \\ &= Trace\,(I_{r \times r}) \\ &= r \\ &= Rank(A) \quad . \end{aligned}$$

We have just proven that the trace of a square idempotent matrix is the rank of that matrix. This allows us to prove the sought after result for eigenvalues that are 1. Begin with an identity matrix of rank n

$$I_{n \times n} \quad .$$

We add and subtract P_X, which we know from above is idempotent.

$$\begin{aligned} &= I_{n \times n} - P_X + P_X \\ &= (I_{n \times n} - P_X) + P_X \quad . \end{aligned}$$

Take the trace of both sides

$$\begin{aligned} Trace\,(I_{n \times n}) &= Trace\,[(I_{n \times n} - P_X) + P_X] \\ &= Trace\,(P_X) + Trace\,(I_{n \times n} - P_X) \quad . \end{aligned}$$

Substituting

$$\begin{aligned} n &= p + Trace\,(I_{n \times n} - P_X) \\ &\Rightarrow Trace\,(I_{n \times n} - P_X) = n - p \end{aligned}$$

$$\Rightarrow Rank\,(I_{n\times n} - P_X) = n - p$$

in which we have in the final term relied on the fact that the rank and trace of the projection matrix P_X are equal and are equal to p. The reason we needed to prove this is to ascertain that the rank of the matrix M_X is EXACTLY $n-p$. We wanted to ascertain that there are exactly p eigenvalues of 0 and exactly $n-p$ eigenvalues of 1, and so we have.

2.6.8 Spectral Decomposition of M_X

We enumerate what the eigenvectors and eigenvalues of M_X are. We begin again with the square of the idempotent matrix developed previously

$$\begin{aligned}
&M_X M_X = M_X \\
&M_X\left[I - X\left(X^TX\right)^{-1}X^T\right] = M_X \\
&M_X - M_X X\left(X^TX\right)^{-1}X^T = M_X \\
&M_X X\left(X^TX\right)^{-1}X^T = 0 \\
&M_X P_X = 0 \quad .
\end{aligned}$$

This means that the columns of the projection matrix P_X are the eigenvectors with eigenvalue 0 of the matrix M_X. Because the rank of P_X is p, there are only p eigenvectors that span the columns of P_X. Every column of P_X can be written as a linear combination of just p eigenvectors.

We can make the foregoing manipulation in "reverse order." We begin again with

$$M_X M_X = 1 * M_X \quad .$$

This indicates that the columns of the matrix M_X are the eigenvectors with eigenvalue 1 of the matrix M_X. Because the rank of M_X is $n-p$, there are only $n-p$ eigenvectors that span the columns of M_X. Every column of M_X can be written as a linear combination of $n-p$ eigenvectors.

All this development has of course assumed that the $p \times p$ matrix X^TX is nonsingular; we have been working with its inverse. That means (see Horn[44]) we have assumed it is positive definite. It is also by construction definitely symmetric.

2.6.9 Illustration of Key Linear Algebra Theorems Through Use of an Example Rather Than Proofs

If a matrix A is symmetric, then linear algebra theorems tell us several things

- All the eigenvalues are real

- Eigenvectors corresponding to distinct eigenvalues are orthonormal
- A is orthonormally diagonalizeable, i.e., there exists an orthonormal matrix P such that $\Lambda = P^{-1}AP$ where Λ is a diagonal matrix, i.e., $A = P\Lambda P^{-1}$
- $P^{-1} = P^T$
- $|P| = 1$
- $|A| = |D| = \prod_{i=1}^{p} d_{ii}$

Eyes glass over when such theorems are put forth; they seem so abstract. Rather than abstract proofs, let us run through an example calculation to see how this works and what it really means. We have found that such an example is more intuitive and suitable to illustrate the key matrix theorems. Surely the example does not prove any theorems; it is designed to be illustrative rather than definitive. (The general proofs are left to linear algebra texts and can be readily found.) However, it illustrates the key theorems and their implications quite nicely. We give references to the proofs in the literature.

Begin with the following symmetric matrix

$$A = \begin{bmatrix} 4 & -2 & 2 \\ -2 & 7 & -4 \\ 2 & -4 & 7 \end{bmatrix} .$$

The fundamental equation for the eigenvalues is

$$\begin{aligned} & A\xi^i = \lambda_i \xi^i \\ & \Rightarrow A\xi^i = \lambda I \xi^i \\ & \Rightarrow (\lambda_i I - A)\,\xi^i = 0 \quad . \end{aligned}$$

We drop the subscript on the λ terms for notational simplicity. We want nonzero eigenvectors, so the ξ_i term cannot be 0. Therefore, the matrix at the left must be singular. We get an equation for the λ by finding the determinant of the matrix on the left and setting it to zero. To begin the calculation, the matrix to get the characteristic equation is

$$\lambda I - A = \begin{bmatrix} \lambda - 4 & 2 & -2 \\ 2 & \lambda - 7 & 4 \\ -2 & 4 & \lambda - 7 \end{bmatrix} .$$

The determinant of this matrix is

$$|\lambda I - A| = (\lambda - 4)\begin{vmatrix} \lambda - 7 & 4 \\ 4 & \lambda - 7 \end{vmatrix} - 2\begin{vmatrix} 2 & -2 \\ 4 & \lambda - 7 \end{vmatrix} - 2\begin{vmatrix} 2 & -2 \\ \lambda - 7 & 4 \end{vmatrix}$$
$$= (\lambda - 4)\,[(\lambda - 7)\,(\lambda - 7) - 16] - 2\,[2\,(\lambda - 7) + 8] - 2\,[8 + 2\,(\lambda - 7)]$$

$$= \lambda^3 - 18\lambda^2 + 81\lambda - 108$$
$$= (\lambda - 3)^2 (\lambda - 12) \quad .$$

Observing the structure of the characteristic equation, there are two eigenvalues: 3 is an eigenvalue of multiplicity 2 and 12 is an eigenvalue of multiplicity 1. We first want to substitute the specific eigenvalue $\lambda = 12$ into the characteristic equation to write

$$12 \times I - A = \begin{bmatrix} 8 & 2 & -2 \\ 2 & 5 & 4 \\ -2 & 4 & 5 \end{bmatrix} \quad .$$

We want to solve the system

$$(12I - A)\,\xi = \begin{bmatrix} 8 & 2 & -2 \\ 2 & 5 & 4 \\ -2 & 4 & 5 \end{bmatrix} \begin{bmatrix} \xi_1 \\ \xi_2 \\ \xi_3 \end{bmatrix} = \begin{bmatrix} 0 \\ 0 \\ 0 \end{bmatrix} \quad .$$

We would normally create a 3×4 tableau with the 0 vector at the right, and we would perform Gauss-Jordan[22] pivoting starting on the 3×4 matrix at the left. The 0 vector in the rightmost column is superfluous. No matter what pivots you engage on the left, the rightmost column will remain zero. Therefore, it is sufficient to Gauss-Jordan pivot the matrix on the left to its simplest, most reduced form:

$$\begin{bmatrix} 1 & 0 & -\frac{1}{2} \\ 0 & 1 & 1 \\ 0 & 0 & 0 \end{bmatrix} \begin{bmatrix} \xi_1 \\ \xi_2 \\ \xi_3 \end{bmatrix} = \begin{bmatrix} 0 \\ 0 \\ 0 \end{bmatrix} \quad .$$

We can solve

$$\begin{bmatrix} 1 & 0 & -\frac{1}{2} \\ 0 & 1 & 1 \\ 0 & 0 & 0 \end{bmatrix} \begin{bmatrix} \xi_1 \\ \xi_2 \\ \xi_3 \end{bmatrix} = \begin{bmatrix} \xi_1 - \frac{1}{2}\xi_3 \\ \xi_2 + \xi_3 \\ 0 \end{bmatrix} = \begin{bmatrix} 0 \\ 0 \\ 0 \end{bmatrix} \quad .$$

This means that

$$\xi_1 - \frac{1}{2}\xi_3 = 0 \Rightarrow \xi_1 = \frac{1}{2}\xi_3$$
$$\xi_2 + \xi_3 = 0 \Rightarrow \xi_2 = -\xi_3 \quad ,$$

which implies that the eigenvector is

[22] See https://people.richland.edu/james/lecture/m116/matrices/pivot.html for an illustration of Gauss-Jordan pivoting. We will not perform specific Gauss-Jordan pivoting steps herein.

$$\xi = \begin{bmatrix} \frac{1}{2}\xi_3 \\ -\xi_3 \\ \xi_3 \end{bmatrix} = const \begin{bmatrix} \frac{1}{2} \\ -1 \\ 1 \end{bmatrix} \quad .$$

The eigenvector of interest corresponding to the eigenvalue 12 is

$$\xi^1 = \begin{bmatrix} \frac{1}{2} \\ -1 \\ 1 \end{bmatrix} \quad .$$

It is easy to show that

$$A\xi^1 = 12\xi^1 = \begin{bmatrix} 6 \\ -12 \\ 12 \end{bmatrix} \quad .$$

So far, so good. We now need to deal with the two eigenvalues that are 3. We begin by substituting 3 into the characteristic equation

$$3I - A = \begin{bmatrix} -1 & 2 & -2 \\ 2 & -4 & 4 \\ -2 & 4 & -4 \end{bmatrix} \quad .$$

We want to solve the system

$$(3I - A)\,\xi = \begin{bmatrix} -1 & 2 & -2 \\ 2 & -4 & 4 \\ -2 & 4 & -4 \end{bmatrix} \begin{bmatrix} \xi_1 \\ \xi_2 \\ \xi_3 \end{bmatrix} = \begin{bmatrix} 0 \\ 0 \\ 0 \end{bmatrix} \quad .$$

We perform the reduction on the matrix on the left hand side to write

$$\begin{bmatrix} 1 & -2 & 2 \\ 0 & 0 & 0 \\ 0 & 0 & 0 \end{bmatrix} \begin{bmatrix} \xi_1 \\ \xi_2 \\ \xi_3 \end{bmatrix} = \begin{bmatrix} 0 \\ 0 \\ 0 \end{bmatrix} \quad .$$

This implies that

$$\begin{bmatrix} 1 & -2 & 2 \\ 0 & 0 & 0 \\ 0 & 0 & 0 \end{bmatrix} \begin{bmatrix} \xi_1 \\ \xi_2 \\ \xi_3 \end{bmatrix} = \begin{bmatrix} \xi_1 - 2\xi_2 + 2\xi_3 \\ 0 \\ 0 \end{bmatrix} = \begin{bmatrix} 0 \\ 0 \\ 0 \end{bmatrix} \quad .$$

This means that

$$\xi_1 - 2\xi_2 + 2\xi_3 = 0 \Rightarrow \xi_1 = 2\xi_2 - 2\xi_3 \quad ,$$

and in turn that the eigenvector has the form

$$\xi = \begin{bmatrix} \xi_1 \\ \xi_2 \\ \xi_3 \end{bmatrix} = \begin{bmatrix} 2\xi_2 - 2\xi_3 \\ \xi_2 \\ \xi_3 \end{bmatrix} = \xi_2 \begin{bmatrix} 2 \\ 1 \\ 0 \end{bmatrix} + \xi_3 \begin{bmatrix} -2 \\ 0 \\ 1 \end{bmatrix} .$$

This clearly points us toward the two eigenvectors as being

$$\xi^2 = \begin{bmatrix} 2 \\ 1 \\ 0 \end{bmatrix} \quad \xi^3 = \begin{bmatrix} -2 \\ 0 \\ 1 \end{bmatrix} .$$

The matrix comprised of the three eigenvectors determined thus far is thus

$$\begin{bmatrix} \frac{1}{2} & 2 & -2 \\ -1 & 1 & 0 \\ 1 & 0 & 1 \end{bmatrix} .$$

The inner product between columns 1 and 2 is 0. The inner product of column 1 and column 3 is 0. However, the inner product of columns 2 and 3 is not zero. It means that the columns of the matrix are not yet mutually orthogonal. We can make them mutually orthogonal by applying the Gram-Schmidt procedure, which is worth doing here so that we can illustrate the final result that a symmetric matrix A can always be diagonalized in an orthonormal fashion.

Let us perform Gram Schmidt orthonormalization.[23] The projection of a vector v onto a vector u is given by the equation

$$Proj_u(v) = \frac{\langle v, u \rangle}{\langle u, u \rangle} u \quad .$$

in which the $\langle . \rangle$ notation represents the inner product. If there is no projection of the vector v onto the vector u, then v and u are orthogonal. If there is a projection of v onto u, Gram-Schmidt wants to remove that projection. The process with the columns of the foregoing matrix designated v_1, v_2, v_3 is the following. We begin with the first column

$$u_1 = v_1 \quad .$$

We rescale the first column so that its size is 1:

$$e_1 = \frac{u_1}{|u_1|} = \frac{u_1}{\sqrt{\langle u_1, u_1 \rangle}} \quad .$$

This vector e_1 will be the first column of the orthonormal matrix. Following is the requisite calculation

[23] See https://en.wikipedia.org/wiki/Gram-Schmidt_process.

$$u_1 = \begin{bmatrix} \frac{1}{2} \\ -1 \\ 1 \end{bmatrix} \Rightarrow \langle u_1, u_1 \rangle = \frac{9}{4} \Rightarrow |u_1| = \frac{3}{2} \Rightarrow e_1 = \begin{bmatrix} \frac{1}{3} \\ -\frac{2}{3} \\ \frac{2}{3} \end{bmatrix} \quad .$$

We can check out the eigenfunction relationship

$$Au_1 = \begin{bmatrix} 4 \\ -8 \\ 8 \end{bmatrix} \quad ,$$

which is 12 times the newly developed ortho-eigenvector. The second step is to remove the projection of u_1 onto v_2. We want to get that portion out of v_2 by deploying the second step of the Gram-Schmidt process

$$\begin{aligned} u_2 &= v_2 - Proj_{u_1}(v_2) \\ &= v_2 - \frac{\langle v_2, u_1 \rangle}{\langle u_1, u_1 \rangle} u_1 \quad . \end{aligned}$$

Substituting the vectors and numbers

$$u_1 = \begin{bmatrix} \frac{1}{2} \\ -1 \\ 1 \end{bmatrix} \quad v_2 = \begin{bmatrix} 2 \\ 1 \\ 0 \end{bmatrix} \Rightarrow \langle u_1, v_2 \rangle = 0 \quad \langle u_1, u_1 \rangle = \frac{9}{4} \quad .$$

The second orthonormal vector is thus

$$u_2 = \begin{bmatrix} 2 \\ 1 \\ 0 \end{bmatrix} \Rightarrow \langle u_2, u_2 \rangle = 5 \Rightarrow |u_2| = \sqrt{5} \Rightarrow e_2 = \begin{bmatrix} \frac{2\sqrt{5}}{5} \\ \frac{\sqrt{5}}{5} \\ 0 \end{bmatrix} \quad .$$

We move on to the third vector v_3. The Gram-Schmidt step is

$$\begin{aligned} u_3 &= v_3 - Proj_{u_1}(v_3) - Proj_{u_2}(v_3) \\ &= v_3 - \frac{\langle v_3, u_1 \rangle}{\langle u_1, u_1 \rangle} u_1 - \frac{\langle v_3, u_2 \rangle}{\langle u_2, u_2 \rangle} u_2 \quad . \end{aligned}$$

The relevant vectors are

$$u_1 = \begin{bmatrix} \frac{1}{2} \\ -1 \\ 1 \end{bmatrix} \quad u_2 = \begin{bmatrix} 2 \\ 1 \\ 0 \end{bmatrix} \quad v_3 = \begin{bmatrix} -2 \\ 0 \\ 1 \end{bmatrix} \quad .$$

The relevant inner products, ordered according to the foregoing fractions, are

$$\langle u_1, u_1 \rangle = \frac{9}{4} \quad \langle u_2, u_2 \rangle = 5$$

$$\langle u_1, v_3\rangle = 0 \quad \langle u_2, v_3\rangle = -4 \quad .$$

The calculated value of u_3 is

$$u_3 = \begin{bmatrix} -2 \\ 0 \\ 1 \end{bmatrix} - \frac{0}{\frac{9}{4}} \begin{bmatrix} \frac{1}{2} \\ -1 \\ 1 \end{bmatrix} - \frac{-4}{5} \begin{bmatrix} 2 \\ 1 \\ 0 \end{bmatrix} = \begin{bmatrix} -\frac{2}{5} \\ \frac{4}{5} \\ 1 \end{bmatrix} \quad .$$

Renormalizing, we have

$$\langle u_3, u_3\rangle = \begin{bmatrix} -\frac{2}{5} \\ \frac{4}{5} \\ 1 \end{bmatrix} = \frac{9}{5} \Rightarrow |u_3| = \frac{3\sqrt{5}}{5} \Rightarrow e_3 = \begin{bmatrix} -\frac{2}{15}\sqrt{5} \\ \frac{4}{15}\sqrt{5} \\ \frac{1}{3}\sqrt{5} \end{bmatrix} \quad .$$

So the orthonormal matrix is the concatenation of the three orthonormal vectors e^1, e^2, e^3

$$[e_1, e_2, e_3] = \begin{bmatrix} \frac{1}{3} \\ -\frac{2}{3} \\ \frac{2}{3} \end{bmatrix}, \begin{bmatrix} \frac{2\sqrt{5}}{5} \\ \frac{\sqrt{5}}{5} \\ 0 \end{bmatrix}, \begin{bmatrix} -\frac{2}{15}\sqrt{5} \\ \frac{4}{15}\sqrt{5} \\ \frac{1}{3}\sqrt{5} \end{bmatrix} \quad .$$

Thus the orthonormal matrix and its transpose are

$$E = \begin{bmatrix} \frac{1}{3} & \frac{2\sqrt{5}}{5} & -\frac{2\sqrt{5}}{15} \\ -\frac{2}{3} & \frac{\sqrt{5}}{5} & \frac{4\sqrt{5}}{15} \\ \frac{2}{3} & 0 & \frac{\sqrt{5}}{3} \end{bmatrix} \quad E^T = \begin{bmatrix} \frac{1}{3} & -\frac{2}{3} & \frac{2}{3} \\ \frac{2\sqrt{5}}{5} & \frac{\sqrt{5}}{5} & 0 \\ -\frac{2\sqrt{5}}{15} & \frac{4\sqrt{5}}{15} & \frac{\sqrt{5}}{3} \end{bmatrix} \quad .$$

We verify several fundamental relationships. First

$$AE = \begin{bmatrix} 4 & -2 & 2 \\ -2 & 7 & -4 \\ 2 & -4 & 7 \end{bmatrix} \begin{bmatrix} \frac{1}{3} & \frac{2\sqrt{5}}{5} & -\frac{2\sqrt{5}}{15} \\ -\frac{2}{3} & \frac{\sqrt{5}}{5} & \frac{4\sqrt{5}}{15} \\ \frac{2}{3} & 0 & \frac{\sqrt{5}}{3} \end{bmatrix} = \begin{bmatrix} 4 & \frac{6\sqrt{5}}{5} & -\frac{2\sqrt{5}}{5} \\ -8 & \frac{3\sqrt{5}}{5} & \frac{4\sqrt{5}}{5} \\ 8 & 0 & \sqrt{5} \end{bmatrix} \quad .$$

We see that we get the same result with the diagonal eigenvalue matrix

$$E\Lambda = \begin{bmatrix} \frac{1}{3} & \frac{2\sqrt{5}}{5} & -\frac{2\sqrt{5}}{15} \\ -\frac{2}{3} & \frac{\sqrt{5}}{5} & \frac{4\sqrt{5}}{15} \\ \frac{2}{3} & 0 & \frac{\sqrt{5}}{3} \end{bmatrix} \begin{bmatrix} 12 & 0 & 0 \\ 0 & 3 & 0 \\ 0 & 0 & 3 \end{bmatrix} = \begin{bmatrix} 4 & \frac{6\sqrt{5}}{5} & -\frac{2\sqrt{5}}{5} \\ -8 & \frac{3\sqrt{5}}{5} & \frac{4\sqrt{5}}{5} \\ 8 & 0 & \sqrt{5} \end{bmatrix} \quad .$$

This means that

$$AE = E\Lambda \Rightarrow A = E\Lambda E^T$$
$$AE = E\Lambda \Rightarrow \Lambda = E^T AE \quad .$$

Let us evaluate the determinant of the matrix E

$$
\begin{aligned}
&\det\begin{vmatrix} \frac{1}{3} & \frac{2\sqrt{5}}{5} & -\frac{2\sqrt{5}}{15} \\ -\frac{2}{3} & \frac{\sqrt{5}}{5} & \frac{4\sqrt{5}}{15} \\ \frac{2}{3} & 0 & \frac{\sqrt{5}}{3} \end{vmatrix} \\
&= -\frac{2\sqrt{5}}{5}\begin{vmatrix} -\frac{2}{3} & \frac{4\sqrt{5}}{15} \\ \frac{2}{3} & \frac{\sqrt{5}}{3} \end{vmatrix} + \frac{\sqrt{5}}{5}\begin{vmatrix} \frac{1}{3} & -\frac{2\sqrt{5}}{15} \\ \frac{2}{3} & \frac{\sqrt{5}}{3} \end{vmatrix} \\
&= -\frac{2\sqrt{5}}{5}\left[\left(-\frac{2}{3}\right)\frac{\sqrt{5}}{3} - \frac{4\sqrt{5}}{15}\frac{2}{3}\right] + \frac{\sqrt{5}}{5}\left[\frac{1}{3}\frac{\sqrt{5}}{3} - -\frac{2\sqrt{5}}{15}\frac{2}{3}\right] \\
&= \frac{2\sqrt{5}}{5}\frac{18\sqrt{5}}{45} + \frac{\sqrt{5}}{5}\frac{9\sqrt{5}}{45} \\
&= \frac{36}{45} + \frac{9}{45} = 1 \quad .
\end{aligned}
$$

Clearly the matrix E is orthonormal.

But the example in this section presages something more. Because $X^T X$ is positive definite, all the eigenvalues are nonnegative. That means that the square root of the diagonal eigenvalue matrix exists and is well defined. Therefore

$$
A = E\Lambda^{\frac{1}{2}}\Lambda^{\frac{1}{2}}E^T = \left(E\Lambda^{\frac{1}{2}}\right)\left(E\Lambda^{\frac{1}{2}}\right)^T \quad .
$$

This means that A can be written as the square of a matrix, not unlike a Cholesky factorization. If we define

$$
B^T = E\Lambda^{\frac{1}{2}} \Rightarrow B = \Lambda^{\frac{1}{2}}E^T \quad .
$$

then A can be written as the square of a matrix

$$
A = B^T B \quad .
$$

A simple Cholesky decomposition[24] of the symmetric, positive definite matrix A shows that it can be written as the square of a triangular matrix.

Following are some key results and observations, and they all represent theorems from linear algebra for matrices. Of course, the example does not prove the theorems. It illustrates the theorems. Rather than merely quoting theorems, we see value in making calculations to illustrate them and thereafter referencing the theorems so that we really know what the theorems mean. Mere recitation of theorems sometimes lacks intuition.

1. The matrix A is diagonalizeable, i.e., there exists a matrix E whose determinant is 1 and whose columns are mutually orthogonal, for which

[24]See https://en.wikipedia.org/wiki/Cholesky_decomposition.

$A = E\Lambda E^{-1} = E\Lambda E^T$, and the elements of the diagonal Λ are the eigenvalues (single or multiple) of A.

2. E is nonsingular.
3. E is not necessarily symmetric. It is not symmetric in this example so cannot possibly be symmetric in general.
4. E^T and E^{-1} are the same. The transpose of E is the inverse of E. This is a general theorem for symmetric matrices–the inverse is the transpose.
5. Every column of E is orthogonal to every other column of E. There is always a mutually orthogonal subspace.
6. The size of every column of E is 1, by construction. E is an orthonormal diagonalization matrix.
7. The determinant of E is 1.
8. $A = E\Lambda E^T$ with $|E| = 1$
9. $|A| = |\Lambda| = \prod_{i=1}^{p} \lambda_i$
10. If, in addition, A is positive definite, all the eigenvalues are strictly positive and real.
11. M_X and P_X are easily shown to be symmetric, i.e., $A = A^T$.

2.6.10 The Standard Eigenvalue-Eigenvector Equation for the Matrix M_X

We now have the background to set up the standard eigenvalue matrix equation for a matrix M_X

$$M_X E = E\Lambda \quad .$$

We know therefore from the matrix results in the previous section

$$M_X = E\Lambda E^T \quad .$$

The eigenvalue matrix has been shown in the previous section to be that in Figure 2.1. The E matrix is orthonormal and nonsingular, meaning the columns of E form an orthonormal basis (length 1, all perpendicular). We are now well acquainted with the standard matrix properties we will use throughout this book.

2.7 Distribution Over the Error Term

This section puts forth the probability distribution over the error or variance term. We want to know what the estimate of σ^2 is, and this section will derive what its form is. Return to the term developed previously that characterizes the sum of squared errors,

$$R = \varepsilon^T M_X \varepsilon = \sum_{j=1}^{n} \varepsilon_i \varepsilon_i m_{ii} \quad .$$

Figure 2.1: The Diagonal Matrix of Eigenvalues

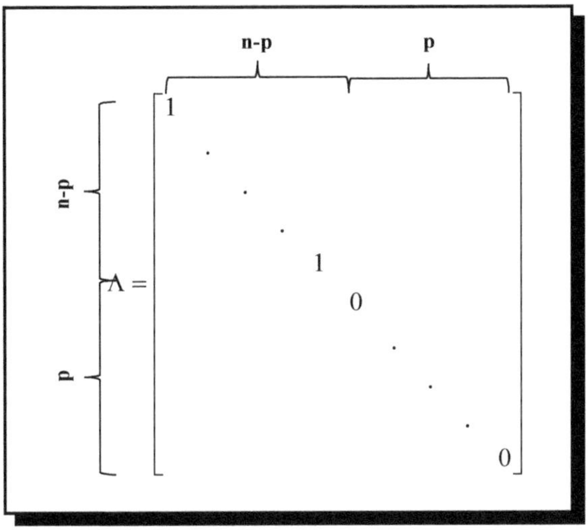

One might think that this is the sum of squared errors of n normal terms. That would be wrong. The right way to think about this is to write

$$\varepsilon^T M_X \varepsilon = \varepsilon^T E \Lambda E^T \varepsilon \quad .$$

Keep in mind, the columns of E that correspond to the 0 eigenvalues are the matrix X. It is useful to diagram the matrix

$$\Lambda = \begin{bmatrix} 1 & & & & & \\ & \cdots & & & & \\ & & 1 & & & \\ & & & 0 & & \\ & & & & \cdots & \\ & & & & & 0 \end{bmatrix} \quad ,$$

keeping in mind that the 1's occur on the first $n - p$ rows and $n - p$ columns, and the 0's occur in the final p rows and p columns. There are $n - p$ eigenvalues of 1, and p eigenvalues of 0, and this is the diagonal matrix that displays them. If we make the variable transformation:

$$y = E^T \varepsilon \quad ,$$

then the distribution over the variable y is a χ^2 distribution with degrees of freedom $n - p$

$$y^T \Lambda y = y^T \begin{bmatrix} 1 & & & & & \\ & \cdots & & & & \\ & & 1 & & & \\ & & & 0 & & \\ & & & & \cdots & \\ & & & & & 0 \end{bmatrix} y = \sum_{i=1}^{n-p} {y_i}^2 \quad .$$

This is a χ^2 distribution[25] with degrees of freedom $n-p$. To wit, the χ^2 distribution is obtained as the sum of the squares of the $n - p$ independent, zero-mean, unit-variance normal variables. It is the sum of variances, i.e., squares of errors, of $n - p$ normal distributions. The next section explains specifically what the χ^2 distribution looks like. This section simply establishes that we will be looking for a χ^2 distribution over the error term. This section does not give the actual χ^2 distribution. Indeed, the orthonormal diagonalization matrix is unknown here, a matrix that transforms the independent error functions into a y variable that is χ^2 distributed. It is not quite yet important to know the specific form of the χ^2 distribution, but rather to know that a χ^2 distribution is going to characterize the error term.

2.8 Technical Sidebar–Gamma and χ^2 Distribution from Fundamentals

A χ^2 distribution is a special case of a more general family of distributions.[26] Discussion of the χ^2 distribution exploits the fact that it is a gamma distribution, so we take a brief side journey in this chapter to articulate the fundamentals of the gamma distribution.[27] The probability density function for a gamma distribution is given in the most general terms as:

$$f(x) = \frac{1}{B^a \Gamma(a)} x^{a-1} e^{-\frac{x}{B}}$$

in which the parameters a and B are both greater than 0. For comparison, the χ^2 distribution has the form

$$f\left(x\right) = \frac{1}{2^{\frac{k}{2}} \Gamma\left(\frac{k}{2}\right)} x^{\frac{k}{2}-1} e^{-\frac{x}{2}} \quad .$$

[25] See https://en.wikipedia.org/wiki/Chi-squared_distribution, which tells us that *in probability theory and statistics, the chi-squared distribution with k degrees of freedom is the distribution of a sum of the squares of k independent standard normal random variables.*

[26] The χ^2 distribution will not be relevant to the Bayesian results in the rest of this textbook, but we present it here for completeness, comparison, and contrast.

[27] See https://en.wikipedia.org/wiki/Gamma_distribution.

The χ^2 distribution, which is the sum of squares of k standard normal distributions (k is termed the "degrees of freedom"), is a special case of a gamma distribution in which

$$a = \frac{k}{2}$$
$$B = 2 \quad .$$

The gamma distribution formula requires the evaluation of the gamma function defined as

$$\Gamma(a) \equiv \int_0^\infty x^{a-1} e^{-x} dx \quad .$$

The gamma distribution with a fixed value of $a = 2$ and various values of B appears in Figure 2.2. For any given value of a, the parameter B serves mostly to stretch or shrink the distribution in the x direction. The t transform (i.e., moment-generating function) of the general gamma function is by definition

Figure 2.2: Gamma Distribution with $a = 2$ and Various Values of B

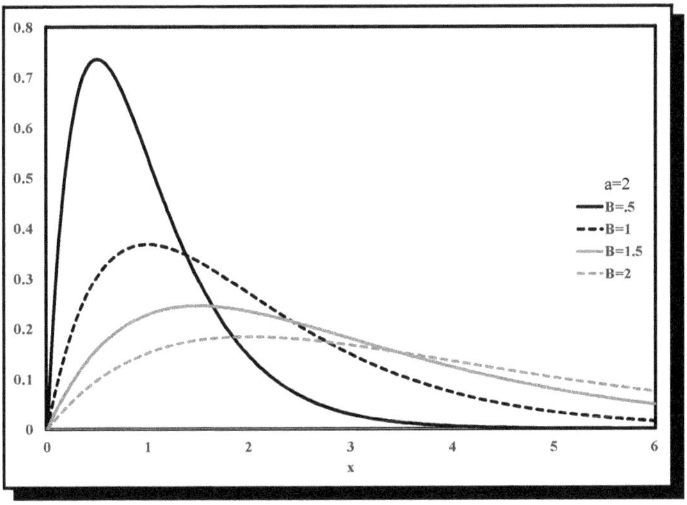

$$F(t) = \int_0^\infty f(x) e^{xt} dx$$
$$= \frac{1}{B^a \Gamma(a)} \int_0^\infty x^{a-1} e^{-\frac{x}{B}} e^{xt} dx$$

$$= \frac{1}{B^a\Gamma(a)} \int_0^\infty x^{a-1} e^{x\left(t-\frac{1}{B}\right)} dx \quad .$$

We make the substitution

$$\begin{aligned}
y &= x\left(-t+\frac{1}{B}\right) \Rightarrow \frac{B}{1-Bt}y = x \\
&\Rightarrow \frac{B}{1-Bt}dy = dx \\
&\Rightarrow x^{a-1} = \left(\frac{B}{1-Bt}\right)^{a-1} y^{a-1} \quad ,
\end{aligned}$$

to obtain

$$\begin{aligned}
F(t) &= \frac{1}{B^a\Gamma(a)} \int_0^\infty x^{a-1} e^{x\left(t-\frac{1}{B}\right)} dx \\
&= \frac{1}{B^a\Gamma(a)} \left(\frac{B}{1-Bt}\right)^a \int_0^\infty y^{a-1} e^{-y} dy \\
&= \frac{1}{\Gamma(a)} \left(\frac{1}{1-Bt}\right)^a \int_0^\infty y^{a-1} e^{-y} dy \\
&= \left(\frac{1}{1-Bt}\right)^a = (1-Bt)^{-a} \quad ,
\end{aligned}$$

in which the reduction of the second to last equation into the last equation makes use of the definition of the gamma function. We can calculate the mean and variance of this distribution by differentiating twice.

$$\begin{aligned}
&F(t) = (1-Bt)^{-a} \\
&F'(t) = aB(1-Bt)^{-a-1} \Rightarrow F'(0) = aB \\
&F''(t) = aBB\,(a+1)\,(1-Bt)^{-a-2} \Rightarrow F''(0) = aBB\,(a+1) = aaBB + aBB \\
&Var = aaBB + aBB - aBaB = aB^2 \quad .
\end{aligned}$$

We now want to look at one special case of the gamma function. Suppose that it has the parameters

$$\begin{aligned}
B &= 2\sigma^2 \\
a &= \frac{1}{2}
\end{aligned}$$

$$f(x) = \frac{1}{(2\sigma^2)^{\frac{1}{2}}\Gamma\left(\frac{1}{2}\right)} x^{-\frac{1}{2}} e^{-\frac{x}{2\sigma^2}} \quad .$$

When this occurs, the t transform (moment generating function) is

$$F(t) = \left(1 - 2\sigma^2 t\right)^{-\frac{1}{2}} \quad .$$

The moment-generating function for the sum of independent random variables is known to be the product of the individual moment-generating functions. If we raise this to the $(n-p)^{th}$ power, representing the sum of $n-p$ identical gamma distributions of this form, we write

$$H(t) = F(t)^{n-p} = \left(1 - 2\sigma^2 t\right)^{-\frac{(n-p)}{2}} \quad .$$

This is in the form

$$H(t) = (1 - Bt)^{-a} \quad ,$$

in which

$$B = 2\sigma^2$$
$$a = \frac{n-p}{2} \quad .$$

This is a gamma distribution with the foregoing parameters. Its specific functional form is

$$\begin{aligned} f(x) &= \frac{1}{(2\sigma^2)^{\frac{n-p}{2}}\Gamma\left(\frac{n-p}{2}\right)} x^{\frac{n-p}{2}-1} e^{-\frac{x}{2\sigma^2}} \\ &= \frac{1}{2^{\frac{n-p}{2}}\sigma^{n-p}\Gamma\left(\frac{n-p}{2}\right)} x^{\frac{n-p}{2}-1} e^{-\frac{x}{2\sigma^2}} \quad . \end{aligned}$$

This is a gamma distribution with degrees of freedom $n-p$. The Bayesian result is going to be quite different, in content as well as philosophy. This will be a profound insight. This χ^2 density from classical regression will not pertain in the Bayesian context. We can eliminate it from consideration in the Bayesian extension. Notwithstanding, it is useful to review how it was developed so that we will know what it is. The reason we show it here is to show the distribution over variance-covariance implied by classical regression.

How does this relate to the sum of squares of unit normals? Recall from earlier that a χ^2 distribution is the sum of squares of k standard normal random variables. We discussed the specializations in the gamma distribution that yield the χ^2 distribution. Therefore, the square of a single normal with mean 0 and standard deviation σ is a χ^2 distribution with degrees of freedom 1.

$$f(x) = \frac{1}{\sigma 2^{\frac{1}{2}} \Gamma\left(\frac{1}{2}\right)} x^{\frac{1}{2}-1} e^{-\frac{x}{2\sigma^2}} \quad .$$

We discussed previously that the specific parameters of the gamma distribution to create this χ^2 distribution are

$$a = \frac{1}{2}$$
$$B = 2\sigma^2 \quad .$$

The fact that a has the value $\frac{1}{2}$ points to a χ^2 distribution. The t transformation of this is

$$F(t) = \left(1 - 2\sigma^2 t\right)^{-\frac{1}{2}} \quad .$$

If we add $n - p$ such squares, the t transform of the density is F raised to the $n - p$ power

$$H(t) = \left(1 - 2\sigma^2 t\right)^{-\frac{(n-p)}{2}} \quad .$$

The density that gives this is

$$f(x) = \frac{1}{2^{\frac{(n-p)}{2}} \sigma^{n-p} \Gamma\left(\frac{n-p}{2}\right)} x^{\frac{(n-p)}{2}-1} e^{-\frac{x}{2\sigma^2}} \quad . \tag{2.10}$$

This is a gamma distribution, actually a χ^2 distribution, with degrees of freedom $n - p$. We have shown previously that the distribution over

$$\varepsilon^T M_X \varepsilon$$

is a χ^2 distribution with degrees of freedom $n - p$. The t transform of the distribution in equation (2.10)

$$H(t) = \left(1 - 2\sigma^2 t\right)^{-\frac{(n-p)}{2}} \quad .$$

The first and second derivatives are derived by differentiation

$$\begin{aligned}
&H(t) = \left(1 - 2\sigma^2 t\right)^{-\frac{n-p}{2}} \Rightarrow H(0) = 1 \\
&H'(t) = (n-p)\,\sigma^2 \left(1 - 2\sigma^2 t\right)^{-\frac{n-p}{2}-1} \Rightarrow H'(0) = (n-p)\,\sigma^2 \\
&H''(t) = (n-p)\,(n-p+2)\,\sigma^4 \left(1 - 2\sigma^2 t\right)^{-\frac{n-p}{2}-2} \\
&\Rightarrow H''(0) = (n-p)\,(n-p+2)\,\sigma^4 \quad .
\end{aligned}$$

This means that the mean $[H'(0)]$ and variance $[H''(0) - H'(0)^2]$ are the following

$$mean = (n-p)\,\sigma^2$$
$$var = (n-p)\,(n-p+2)\,\sigma^4 - (n-p)\,\sigma^2\,(n-p)\,\sigma^2 = 2\,(n-p)\,\sigma^4 \quad .$$

Using the estimate of the variance we developed previously,

$$\sigma^2 = \frac{\left(y - X\bar{\beta}\right)^T \left(y - X\bar{\beta}\right)}{n-p} = \frac{R}{n-p} \quad .$$

The estimated mean and variance of the error term are therefore

$$\text{mean} = R = \left(y - X\bar{\beta}\right)^T \left(y - X\bar{\beta}\right)$$
$$\text{var} = \frac{2}{n-p}\left[\left(y - X\bar{\beta}\right)^T \left(y - X\bar{\beta}\right)\right]^2 \quad .$$

We reiterate that this is not the variance we will see in the Bayesian case.

2.9 Only Cursory Attention to R-Squared

Gelman et al.[34] address the question of whether there is a meaningful "R-squared" in the Bayesian context and have proposed one. The reader is referred there for a discussion and extended definition. For the classical regression case, Wiki[28] gives a general equation in the case of classical linear regression. Bayesian Linear Regression doesn't really need it.

For the classical regression case, if we define the sample vector

$$c = \begin{bmatrix} corr\,(x_1 y) \\ . \\ . \\ . \\ corr\,(x_p y) \end{bmatrix}$$

and write the sample correlation matrix over the independent variables as in equation (2.11), then the square of the p-variate Pearson correlation coefficient, the R^2 value, is written as in equation

$$R_{xx} = \begin{bmatrix} corr\,(x_1 x_1) & . \; . \; . & corr\,(x_1 x_p) \\ . & & . \\ . & & . \\ . & & . \\ corr\,(x_p x_1) & . \; . \; . & corr\,(x_p x_p) \end{bmatrix} \tag{2.11}$$

and

[28] See https://en.wikipedia.org/wiki/Coefficient_of_multiple_correlation.

$$R^2 = c^T R_{xx}^{-1} c \quad .$$

2.10 Summary of Classical Regression Results

We conclude by defining and rewriting a few terms in classical regression. First, we define a term called "degrees of freedom" in the classical solution

$$\nu_C = n - p = \text{degrees of freedom} \Rightarrow n = \nu_C + p \quad .$$

If we divide the residual sum of squared errors by the degrees of freedom, we get the estimate of the variance σ^2 discussed throughout this section

$$s_C^2 = \frac{R}{\nu_C} \quad .$$

This allows us to write the residual sum of squared errors as follows

$$R = \nu_C s_C^2 \quad .$$

We can write the variance-covariance matrix of the coefficients as follows

$$VCV = s_C^2 (X^T X)^{-1} \quad .$$

The subscript C is used to denote classical statistics. The term s_C^2 is the sum of the squares of the errors divided by the degrees of freedom, which we can think of as the number of extra terms we have in our set of observations. It is clearly the average or mean error term, precisely the estimate in which we are interested. The degrees of freedom indicate how "overdetermined" the set of equations is. It is the number of equations (observations) one has above and beyond the bare minimum. The term $s_C^2 = \frac{R}{\nu_C}$ is the best estimate of the variance σ^2. It is the best guess, based on the data, what the variance of the error measure is. Using this definition and referring to the equation (2.7) with which we began this section, we can write the likelihood function.

$$\begin{aligned}
(2.12) \quad \{y, X|\beta, \sigma\} &= \frac{1}{\sigma^{\nu_C+p}(2\pi)^{\frac{\nu_C+p}{2}}} e^{-\frac{1}{2\sigma^2}\nu_C s_C^2} e^{-\frac{1}{2}(\beta-\bar{\beta})^T \left(\frac{X^T X}{\sigma^2}\right)(\beta-\bar{\beta})} \\
&= const \times \sigma^{-(\nu_C+p)} e^{-\frac{1}{2\sigma^2}\nu_C s_C^2} e^{-\frac{1}{2}(\beta-\bar{\beta})^T \left(\frac{X^T X}{\sigma^2}\right)(\beta-\bar{\beta})} \\
&= const \times \sigma^{-n} e^{-\frac{1}{2\sigma^2}\nu_C s_C^2} e^{-\frac{1}{2}(\beta-\bar{\beta})^T \left(\frac{X^T X}{\sigma^2}\right)(\beta-\bar{\beta})} \quad ,
\end{aligned}$$

in which we have used the relationship $n = \nu_C + p$. Importantly, this likelihood function is the product of σ to a negative power, and in the exponent we have a constant plus a quadratic term. This functional form is going to be central to the linear Bayesian model.

Let us examine and emphasize the specific form of equation (2.12). See Figure 2.3. Notice it contains the term σ raised to a negative power. The exponent is the sum of a scalar and a matrix quadratic centered on a mean value, namely the mean value $\bar{\beta}$. This is a very characteristic form, and we are going to exploit it in defining our Bayesian prior and getting the resulting posterior.

Figure 2.3: The Classical Likelihood Function

$$\{X, y|\beta, \sigma\} = \text{const} * \sigma^{-n} e^{-\frac{1}{2\sigma^2}\left[\nu_C s_C^2 + (\beta-\bar{\beta})^T X^T X(\beta-\bar{\beta})\right]}$$

Negative power
Scalar
Mean
Matrix quadratic

We cannot overemphasize what the likelihood function distribution is and what it is not. It is a distribution over X and y conditional on the parameters β and σ. It is decidedly $\{X, y|\beta, \sigma\}$, not $\{\beta, \sigma|X, y\}$. It is so tempting, so tauntingly delicious, to just go ahead and assume the conditioning goes the other way! Yet it does not. It is the probability over the observations given the model coefficients. The sought-after result is the opposite direction of conditioning, namely the conditional probability distribution over the model coefficients given the data. That, conceptually, is precisely what we seek with classical or Bayesian statistics. One looks at the data, and then infers the model coefficients from the data. We should emphasize that the maximum of this likelihood function is trivial. One does not need a collection of Karush-Kuhn-Tucker (KKT) optimization conditions[29] or difficult algebraic manipulations to infer the mode. The equation is quadratic in β. The maximum of a quadratic occurs, trivially, at $\bar{\beta}$.

In this summary, we have omitted concepts such as confidence intervals, hypothesis testing, and others. The reason is that Bayes (forthcoming) will deal with full and complete joint probability distributions over results. Armed with those full and complete joint distributions, it will become more clear what such tests are and what they might or might not add.

The other point we wish to reiterate, as Bayesian statisticians and Decision Analysts, is that the maximum point, the mode, of this distribution is not nearly enough for decision-making. It is literally a set of measure zero, a single point, one minuscule point on a complex, probably density function. Concepts such as von Neumann-Morgenstern[100] expected utility and maximization thereof require the entire distribution, not just the mode. Such is a key objective of Bayes: to obtain the entire distribution so that people can make decisions. And decisions are crafted based on maximizing expected cardinal utility.

[29] The Karush-Kuhn-Tucker conditions are the first order necessary conditions for local optimization. They are exquisitely motivated and developed in Luenberger.[64]

We now have the conceptual underpinning of classical linear regression. We now turn to Bayes.

Chapter 3

BAYESIAN REINTERPRETATION—APPLICATION OF BAYES THEOREM

> ***The most important questions of life are indeed, for the most part, really only problems of probability.***
>
> ***Pierre-Simon Laplace***

The idea underlying Bayes theorem is to multiply the likelihood function in equation (2.12) as summarized in Figure 2.3 (a probability distribution in its own right over the observations X and y given the model coefficients β and σ) by a marginal distribution over the model coefficients β and σ, what is called a "prior," to obtain a consequent joint distribution over the observations and the coefficients. Once we obtain the joint distribution, we can thereafter "reverse the conditionality" to obtain the desired Bayesian result. To be specific, according to Bayes theorem, we establish the joint distribution by multiplying the conditional distribution (likelihood function) in equation (2.12) by a marginal distribution over the model coefficients (the so-called prior):

$$\{X, y, \beta, \sigma\} = \{X, y \,|\beta, \sigma\}\{\beta, \sigma\} \quad .$$

Bayes theorem allows the reversal of the order of conditionality

$$\{X, y, \beta, \sigma\} = \{\beta, \sigma \,|X, y\}\{X, y\} = \{\beta, \sigma \,|X, y\}c_1 \quad , \tag{3.1}$$

in which c_1 is the constant

$$c_1 = \int \{X, y, \beta, \sigma\} d\beta d\sigma \quad .$$

On the right, equation (3.1) gives us the conditional probability distribution over the coefficients β, σ, i.e., conditional on the observations X, y. Intuitively, this is precisely what we are seeking—the probability distribution over the model coefficients β, σ conditional on the specific observations X, y, i.e., "after" seeing X, y, i.e., "after" seeing our data. Shouldn't the whole point be to infer the probability distribution over the coefficients β, σ after receiving the data X, y? Decidedly so!

After the substitution, the former equation becomes

$$\{X, y, \beta, \sigma\} = \{\beta, \sigma | X, y\} \, c_1 = \{X, y | \beta, \sigma\} \, \{\beta, \sigma\} \quad .$$

This is the quintessential statement of Bayes theorem for use in Bayesian Linear Regression.

3.1 The Prior—A Joint Density over Model Coefficients and Standard Deviation Term

The Savage[88] axioms of subjective probability, as reinterpreted and formalized by Stanford Professor Ronald Howard[45] in his famous graduate course entitled "Foundations of Decision Analysis" and articulated by de Finetti[23] allow the analyst or one or more of his appointed "experts" to begin with a subjective assessment, a subjective quantification, of the prior.[1] For example, a physician might well make a prior judgment about what affliction a patient might have before he or she gathers any data regarding that patient. An engineer might have a prior judgment about what is wrong with a particular machine. A plant foreman might have a prior judgment on what is wrong with a flotation process. An analyst may have a small, preliminary, and poor data set. We will see in Chapter 11 a very realistic example of how people can have dramatically different prior opinions and how Bayes can be used to update those judgments based on gathered data to get a justifiable answer. An economist has a pretty good empirical idea of what a price elasticity should be, and he or she may have a rather narrow range of uncertainty. Conversely, a probability assessor may have a pretty "wide" prior, i.e., a great deal of uncertainty. An analysis of observations that occur chronologically "before" the particular observations X, y are taken might be used to form a prior distribution. It is not yet important where we obtain the prior but rather to assert strongly that it exists formally, systematically, and explicitly. (Chapter 14 indicates how to formulate a prior using probability assessment procedures from Decision Analysis and is focused squarely on this point.) It is not appropriate to "gloss over" it, thereby treating it implicitly, obscurely, and, as we

[1] We will talk about the prior and quantification thereof in Chapter 14, which is related to the probability wheel.

believe, improperly, as in the classical regression formulation. We have seen in the previous chapter that classical statistics avoids the issue of a prior entirely, and often sanctimoniously so. Bayes does not.

The remainder of this section puts forth the notion of a prior distribution over the model coefficients and the standard deviation term. If one is to understand the Bayesian approach, it is central to understand the mathematics and philosophy of the prior as presented herein.

3.2 A Specific Functional Form (Conjugate Form) for the Prior

We want to exploit the specific functional structure of the likelihood function in equation (2.12) and Figure 2.3 to the maximum extent possible. That is, we want to postulate a functional form for the prior probability distribution over the coefficients β, σ in an exactly parallel form to that of the likelihood function in equation (2.12) and Figure 2.3 such that when we multiply prior times likelihood, we get a final form that is identical in functional form to the prior. We say that the prior is "conjugate"[2] to the posterior. We say that the prior and posterior are "conjugate forms," meaning they are from the same family of probability distributions. We want the prior and the posterior to have the same mathematical form, and this will not occur unless we have a specific form of likelihood function consonant with the prior. To be specific, suppose we postulate that the prior has precisely the same algebraic form as the likelihood function in equation (2.12) and Figure 2.3, namely

$$\{\beta, \sigma\} = const * \sigma^{-m} e^{-\frac{1}{2\sigma^2}\left[M + (\beta - \beta_0)^T Q (\beta - \beta_0)\right]} \quad . \tag{3.2}$$

We label and emphasize the specific parameters in this prior equation as shown in Figure 3.1, quite obviously parallel in form to the likelihood function in Figure 2.3, in which there is a constant of integration whose exact analytical expression we shall develop in the next chapter but do not need to know at this point in the development. Notice in the prior in Figure 3.1 there are four obvious parameters, namely m, M, β_0, and Q, and one not-so-obvious parameter, namely p.

Figure 3.1: Intrinsic Form of the Prior That Parallels The Likelihood Equation

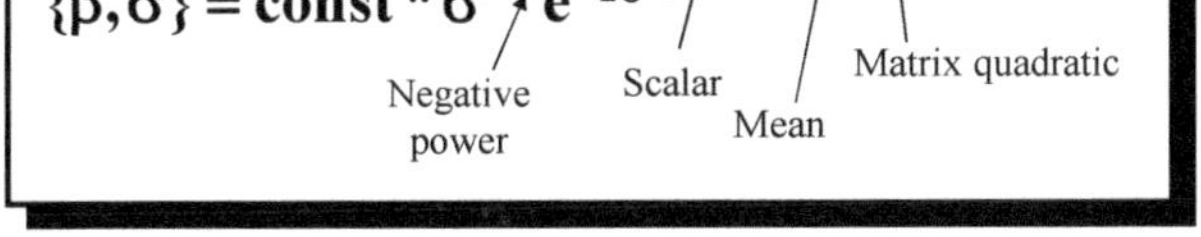

We enumerate the key terms for easy reference:

[2] See https://en.wikipedia.org/wiki/Conjugate_prior.

- m = a constant scalar power on the σ term.
- M = an additive scalar constant in the exponent.
- β_0 = a p-vector in the quadratic portion of the exponent.
- Q = a symmetric, $p \times p$, positive definite matrix in the quadratic portion of the exponent (which is called a precision matrix).[3] The precision matrix is the inverse of the variance-covariance matrix.
- p = number of parameters in one's linear model INCLUDING the constant if there is one. This is no different from the classical statistics p.

Intuitively, there are five parameters we must subjectively specify to comprise our prior; p is specified automatically because it represents the number of parameters under consideration:

- The number of model coefficients: p
- The constant scalar power on the σ term: m
- The additive scalar constant in the exponent: M
- The vector of parameter means (of length p) in the quadratic portion of the exponent: β_0
- The $p \times p$ symmetric, positive definite matrix in the quadratic portion of the exponent: Q

We emphasize that this collection of data comprises a "sufficient statistic" to fully specify the prior. If we put numerical values in for each of these parameters, we have a complete characterization of the prior. We should also mention that the number p is de facto given once we select the number of coefficients in our linear model. We define the degrees of freedom in the prior as follows:

$$\nu_0 = m - p - 1 \quad . \tag{3.3}$$

This will be interpreted and justified shortly. We shall see that this seemingly arbitrary definition of degrees of freedom is not, in fact, arbitrary at all.[4] We will see in Chapter 10 that our Bayesian approach allows data priors. We will also see why there is a 1 in equation (3.3) and where it comes from. (People could wonder why the 1 in the equation is subtracted and where it comes from. We shall see.) Under this definition, the coefficient m in the prior is given by the equation

[3]See https://en.wikipedia.org/wiki/Precision_(statistics) for the definition.

[4]This definition of degrees of freedom will prove to be quite the same as it is in classical regression. It is precisely the degrees of freedom one would get if one began with an old, perhaps lousy, data set comprised of m observations, analyzed it using classical statistics, and used the result from such classical statistics as a prior, reasoning that it is what we know going into the actual Bayesian analysis with a new data set.

(3.4) $$m = \nu_0 + p + 1 \quad .$$

We have defined the degrees of freedom parameter ν_0, and we have thereby specified the m parameter in the prior. To reiterate, at least for now, this definition of degrees of freedom is arbitrary, but it will soon be clear that it is not arbitrary at all.

If we inspect the prior closely, we see that the M term occupies an analogous position in the prior that the mean squared error parameter R occupied in classical regression. Therefore, following the notion in classical regression that the expectation of the error term is the quotient of M divided by the degrees of freedom, let us make the following definition of the term ${s_0}^2$ and see where it leads us and what it might mean in the succeeding calculations.

$${s_0}^2 = \frac{M}{\nu_0} \quad .$$

We will indeed see this term showing up as the mean of the error parameter. This definition means that

(3.5) $$M = \nu_0 {s_0}^2$$

so that the prior can be written

$$\{\beta, \sigma\} = const * \sigma^{-(\nu_0+p+1)} e^{-\frac{1}{2\sigma^2}\left[\nu_0 {s_0}^2 + (\beta-\beta_0)^T Q(\beta-\beta_0)\right]} \quad .$$

This could not be more parallel or analogous in form to the likelihood function in equation (2.12) and Figure 2.3. It is a directly analogous form, and that is precisely why we are choosing it to help develop our conjugate prior-posterior-likelihood relationship.

Intuitively, any prior knowledge by any expert should be embedded in the numerical values of m, M, β_0, and Q that are specified, and of course the number of coefficients p. They comprise judgments regarding what the model parameters should be, based on experience, observation, scientific insight, a previously existing data set, prior analysis, just plain guesses, etc. They are inputs to the statistical analysis by outside engineering-scientific experts or perhaps by a pre-existing or "old" data set.[5]

There have been so-called "purists" in the statistical and econometric realms who object to bringing any prior judgment, opinion, or extraneous data at all to the problem. They argue from the crenellated ivory tower of academic purity that statistics should be pure and free of bias or opinion. "Statistics should consider the data and only the data," they argue. "We don't want anything but the data

[5] We are going to see how to ignore or "defang" this prior, rendering it effectively impotent at delivering any a priori information to the problem. Stay tuned. Bayes does not force you to incorporate a "potent" prior.

determining the answer!" The very argument itself is a non sequitur. These are the same people who would impose ranges on certain coefficients or relationships between certain coefficients. Bayes is a way to do so formally, and it is a way to quantify carefully how confident or uncertain your prior is. It is a way to render any such judgments completely transparent and not buried in nuances of distorted statistical analysis or data alteration, rejection, or fabrication. And of course, there are other examples. We observe adverse selection in data sets, which markedly biases results (e.g., a doctor that fails to chart an "outlier" in a procedure because of fear of criticism, liability, censure, malpractice, suspension, etc.). Could a hospital misclassify or misrepresent a treatment or cause of death because they get paid more by insurance companies or the government for certain treatments or diseases? Could hospitals overreport COVID deaths because of the increased payments for COVID treatment and death? Could a highway engineer fail to report pavement or structural design or maintenance flaws that materially contributed to premature deterioration, fatalities, or property damage and thereby perhaps incriminate himself or herself or expose his State to liability? Is it possible for an aircraft company to suppress data about a misdesigned guidance and control system when there are two control-system based fatal crashes? What about feature selection when working with large data sets? Or model selection itself in the first place? Choosing different models for the same dataset could produce markedly different results and conclusions. Is it OK to "throw out" outlier data because it just doesn't "look right" or support some preconceived result? And of course, what about underfunding of data gathering or just plain incompetency or indolence? Data could be sullied by any or all of these phenomena.

We must also recognize that the data is highly uncertain. No one who has supervised, viewed, or participated in data gathering would argue that it is not a vaguely defined, often unsupervised, error-prone, uncertain, highly subjective process, literally "sausage making." Who would advocate, as Dr. Dale Jorgenson quipped, that "Man may have invented models, but God invented the data!" When we ponder this, such an assertion is downright preposterous on its face; deities have no particular hand in blessing data (nor in condemning judgment). Quite the contrary, data gathering is a messy and uncertain process, often downright ugly in fact, frequently outsourced to a low-cost vendor or gathered tangentially as part of some other process, often in a low-labor-cost venue. Data itself is often at the tail end of a substantial judgment process; the data itself embodies judgment! Let no one be fooled; data is laced with judgment, inconsistency, uncertainty, and subjectivity. Even if gathered automatically by machines or meters (or even Prof. Howard's Acme Rain Detectors), the reliability of the machines, meters, and underlying physical or economic processes comes into question. In addition, the error rates of meters can change, knowingly or unknowingly, throwing the whole data gathering process into question (heteroskedasticity). Sir Ronald Aylmer Fisher himself was very clear about that! Data is generated from a random process, quite often a random process with an extremely wide variance! That said, we endeavor to understand what such uncertainty might imply. No rational individual would argue that the data is "divine." Bayes is the best way to use data, as accurate or inaccurate as it might be.

3.3 The Bayesian Posterior with This Conjugate Prior

As we switch from the prior in the previous section to the posterior in this section, isn't Bayes Theorem disarmingly simple? We get the Bayesian posterior by multiplying the prior (in Figure 3.1) times the likelihood function (in Figure 2.3) as follows:

$$
\begin{aligned}
(3.6)\qquad & \{X, y\,|\beta,\sigma\}\{\beta,\sigma\} = \{\beta,\sigma\,|X,y\} \times const \\
&= \left\{ const \times \sigma^{-n} e^{-\frac{1}{2\sigma^2}\left[\nu_C s_C^2 + \left(\beta-\bar{\beta}\right)^T X^T X \left(\beta-\bar{\beta}\right)\right]} \right\} \\
&\times \left\{ const \times \sigma^{-m} e^{-\frac{1}{2\sigma^2}\left[M + (\beta-\beta_0)^T Q (\beta-\beta_0)\right]} \right\} \\
&= const \times \sigma^{-(n+m)} \\
&\times e^{-\frac{1}{2\sigma^2}\left[\nu_C s_C^2 + M + \left(\beta-\bar{\beta}\right)^T X^T X \left(\beta-\bar{\beta}\right) + (\beta-\beta_0)^T Q (\beta-\beta_0)\right]} \quad .
\end{aligned}
$$

There is no "weighting" or any other "sleight of hand" in combining the prior and the likelihood functions! It is a simple multiplication—prior times likelihood—implied directly by what Thomas Bayes and his descendents instructed us.[6]

If we relate the form of equation (3.6) to the form in Figure 3.1, we begin to see how all the various parameters in equation (3.6) can be regrouped to compute a posterior in the same form as Figure 3.1. An important observation at the outset is that the term m, a constant scalar power on the σ term in the prior, has been replaced with $m + n$ in the posterior. To wit, we have simply added n to the prior exponent of σ as given in equation (3.4) to obtain the exponent $m + n$, the exponent of σ in the posterior, as follows:

$$m + n = \nu_0 + p + 1 + n \quad .$$

This suggests that we subtract $p + 1$ from both sides to write

$$m + n - p - 1 = \nu_0 + n \quad .$$

For the posterior, we would expect the degrees of freedom to be larger than the prior by the amount n, the number of data points in our data base, which

[6]This is such an important point. How many times have students or colleagues asked me how they should just "weight" the prior versus the data to arrive at the posterior? You don't "weight" anything; Bayes Theorem says no such thing! The question is a non sequitur when you recognize what Bayes theorem says. It says to multiply prior times likelihood. No weighting, summing, weighted sum, or anything like that! Just simple prior times likelihood.

represents the amount of data above and beyond the prior. The reason we would expect this is that the prior within equation (3.6) has an exponent of m, but the posterior has an exponent of $m+n$. The foregoing equation suggests we define the posterior degrees of freedom, denoted ν, to be the left hand term:

$$\nu = m + n - p - 1 = \nu_0 + n \quad . \tag{3.7}$$

The analogy to the prior degrees of freedom is clear. It is as if the prior was based on m observations in equation (3.4) but now it is based on $m+n$ observations in equation (3.7), all else equal. That is the correct interpretation, and it will be borne out as we develop the mathematics further. It would seem intuitive given that the prior was based on m observations and we bring an additional n observations via data gathering.

Now that we have inferred the degrees of freedom for the posterior, we return to the entire posterior in equation (3.6), which seems to contain a rather foreboding exponential term. We can dramatically simplify it by completing the square of the matrices in the two exponents in a matrix sense, as we did previously in the classical regression development in Chapter 2. Again, there will be no need for optimization, loss functions, or similar concepts; just the simple algebraic completion of the square suffices.

We begin by adding and subtracting an as yet unknown vector β^* within the brackets as a first step toward completing the square:

$$\begin{aligned} Exp = R + M &+ \left(\beta - \bar{\beta} - \beta^* + \beta^*\right)^T \left(X^T X\right) \left(\beta - \bar{\beta} - \beta^* + \beta^*\right) \\ &+ \left(\beta - \beta_0 - \beta^* + \beta^*\right)^T Q \left(\beta - \beta_0 - \beta^* + \beta^*\right) \quad . \end{aligned}$$

We regroup terms as follows

$$\begin{aligned} Exp = R + M &+ \left[(\beta - \beta^*) + (\beta^* - \bar{\beta})\right]^T \left(X^T X\right) \left[(\beta - \beta^*) + (\beta^* - \bar{\beta})\right] \\ &+ \left[(\beta - \beta^*) + (\beta^* - \beta_0)\right]^T Q \left[(\beta - \beta^*) + (\beta^* - \beta_0)\right] \quad . \end{aligned}$$

For notational simplicity that facilitates forthcoming algebra, we define

$$z = \beta - \beta^* \quad ,$$

so that we can write

$$\begin{aligned} Exp = R + M &+ \left[z + (\beta^* - \bar{\beta})\right]^T \left(X^T X\right) \left[z + (\beta^* - \bar{\beta})\right] \\ &+ \left[z + (\beta^* - \beta_0)\right]^T Q \left[z + (\beta^* - \beta_0)\right] \quad . \end{aligned}$$

We expand the two quadratic terms

$$Exp = R + M$$

$$+ z^T \left(X^T X\right) z + 2(\beta^* - \bar{\beta})^T \left(X^T X\right) z + (\beta^* - \bar{\beta})^T \left(X^T X\right) (\beta^* - \bar{\beta})$$
$$+ z^T Q z + 2(\beta^* - \beta_0)^T Q z + (\beta^* - \beta_0)^T Q \left(\beta^* - \beta_0\right) \quad .$$

We then regroup into second order quadratic in z, a first order linear term in z, and a constant term independent of z as follows

$$\begin{aligned} Exp &= R + M \\ &+ (\beta^* - \bar{\beta})^T \left(X^T X\right) (\beta^* - \bar{\beta}) + (\beta^* - \beta_0)^T Q \left(\beta^* - \beta_0\right) \\ &+ 2 \left[(\beta^* - \bar{\beta})^T \left(X^T X\right) + (\beta^* - \beta_0)^T Q \right] z \\ &+ z^T (X^T X + Q) z \quad . \end{aligned}$$

We would like to choose β^* so as to eliminate the linear (middle) term in z, the standard completion of the square approach. We do so by taking the transpose of the middle term in brackets, the term that is linear in z.[7] If we examine every individual term in the overall summation in the foregoing equation, we will see that they are all scalars. We will recall further that because Q is symmetric, $Q^T = Q$. Furthermore, $X^T X$ is symmetric, meaning it is equal to its transpose. Therefore, we can transpose every term in the exponent (which only affects the term linear in z) to write

$$\begin{aligned} Exp &= R + M \\ &+ (\beta^* - \bar{\beta})^T \left(X^T X\right) (\beta^* - \bar{\beta}) + (\beta^* - \beta_0)^T Q \left(\beta^* - \beta_0\right) \\ &+ 2 z^T \left[\left(X^T X\right) (\beta^* - \bar{\beta}) + Q \left(\beta^* - \beta_0\right) \right] \\ &+ z^T (X^T X + Q) z \quad . \end{aligned} \tag{3.8}$$

Once we have transposed, we can find β^* that drives the bracketed term to zero

$$\begin{aligned} &\left(X^T X\right) (\beta^* - \bar{\beta}) + Q \left(\beta^* - \beta_0\right) = 0 \\ &\Rightarrow \left(X^T X + Q\right) \beta^* - \left[\left(X^T X\right) \bar{\beta} + Q \beta_0 \right] = 0 \\ &\Rightarrow \beta^* = \left(X^T X + Q\right)^{-1} \left[\left(X^T X\right) \bar{\beta} + Q \beta_0 \right] \quad . \end{aligned}$$

If we substitute the expression for the classical statistics solution from equation (2.5), namely $\bar{\beta}$, we obtain the expression for the posterior mean

$$\beta^* = \left(X^T X + Q\right)^{-1} \left(X^T y + Q \beta_0\right) \quad . \tag{3.9}$$

The term β^* in equation (3.9) is going to be the mean value of the Bayesian posterior.[8] It embodies a very special, very characteristic form—a matrix weighted

[7] Because the term is a scalar, taking the transpose has no material effect on the answer.

[8] Note that $X^T X$ does not have to be invertible in order for $X^T X + Q$ to be invertible and therefore that the posterior exist. That means that we do not have to have a classical regression solution in order to do a Bayesian update. More to come on this issue.

average of the prior and likelihood. The annotated equation in Figure 3.2 emphasizes. The arrows in the figure show intuitively that if there were no prior ($Q = 0$), we would have the classical statistics solution. If there were no data ($X^T X = 0$), we would have only the prior to rely on. Since we have both, we have a carefully constructed matrix-weighted average of both.

Figure 3.2: The Posterior Is a Specific "Mix" of Classical and Prior

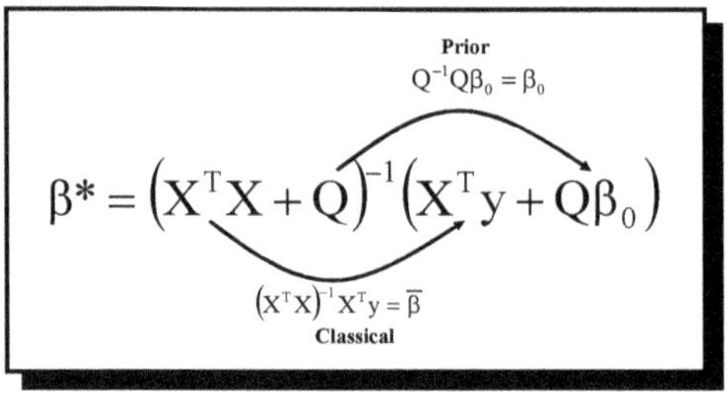

The "precision matrix" is a multiple of the matrix in the quadratic term in z in equation (3.8), namely:

$$H = X^T X + Q \quad . \tag{3.10}$$

This suggests that we think about the differences in two key terms from the prior in Figure 3.1 to the thus-calculated posterior

- β_0, a vector in the quadratic portion of the exponent in the prior, has been replaced with β^* from equation (3.9), the mean of the posterior, which appears in the quadratic portion of the exponent in the posterior.
- Q, a matrix in the quadratic portion of the exponent in the prior (which was called the precision matrix) has been replaced with the matrix $X^T X + Q$ in equation (3.10), which is the new precision matrix for the posterior.

If we substitute the expression for β^* into the third and fourth terms of the exponent in equation (3.8), recognizing that the linear term in z is zero by construction, we can rewrite that exponent

$$\begin{aligned} Exp &= R + M \\ &+ \left(\beta^* - \bar{\beta}\right)^T \left(X^T X\right)\left(\beta^* - \bar{\beta}\right) + \left(\beta^* - \beta_0\right)^T Q \left(\beta^* - \beta_0\right) \\ &+ \left(\beta - \beta^*\right)^T \left(X^T X + Q\right)\left(\beta - \beta^*\right) \quad . \end{aligned}$$

We are going to define the constant term as follows

$$\begin{aligned} A &= R + M \\ &+ \left(\beta^* - \bar{\beta}\right)^T \left(X^T X\right)\left(\beta^* - \bar{\beta}\right) + \left(\beta^* - \beta_0\right)^T Q \left(\beta^* - \beta_0\right) \quad . \end{aligned} \tag{3.11}$$

We know that the matrix $X^T X$ is positive definite, so the third term must be positive for all nonzero values of $\beta^* - \bar{\beta}$. We know that the matrix Q is positive definite, so the fourth term must be positive for all nonzero values of $\beta^* - \beta_0$. Because the first two terms are themselves both positive, all the individual terms are positive, and the sum A is positive. The foregoing equation allows us to write the exponent in the Bayesian posterior in a very simple form.

$$Exp = A + (\beta - \beta^*)^T \left(X^T X + Q\right)(\beta - \beta^*) \quad ,$$

in which A is a positive scalar. Substituting this exponent back into the equation for the Bayesian posterior, we write the probability distribution that is the posterior as follows

$$\begin{aligned} \{\beta, \sigma \,|X, y\} &= const \times \sigma^{-(m+n)} \\ &\quad \times e^{-\frac{1}{2\sigma^2}\left[A + (\beta-\beta^*)^T\left(X^T X + Q\right)(\beta-\beta^*)\right]} \\ &= const \times \sigma^{-(m+n)} \\ &\quad \times e^{-\frac{A}{2\sigma^2}} e^{-\frac{1}{2\sigma^2}(\beta-\beta^*)^T\left(X^T X + Q\right)(\beta-\beta^*)} \quad . \end{aligned} \tag{3.12}$$

This posterior density has exactly the same algebraic form as the prior density in equation (3.2) and Figure 3.1, which implies exactly the conjugate relationship between prior and posterior alluded to previously.

It is very useful and insightful to manipulate and simplify this equation, concentrating initially on A as we do so. It is because the A term is going to look like the residual sum of squared error term R in classical regression. It is going to have a similar, indeed analogous, role here as the constant term in the exponent in the posterior. We begin by expanding each of the two quadratic terms within the definition of the constant A in equation (3.11).

$$\begin{aligned} A &= R + M \\ &\quad + \beta^{*T}\left(X^T X\right)\beta^* - 2\beta^{*T}\left(X^T X\right)\bar{\beta} + \bar{\beta}^T\left(X^T X\right)\bar{\beta} \\ &\quad + \beta^{*T} Q \beta^* - 2\beta^{*T} Q \beta_0 + {\beta_0}^T Q \beta_0 \quad . \end{aligned}$$

Then we group the first and second two pairs of terms

$$\begin{aligned} A &= R + M \\ &\quad + \beta^{*T}\left(X^T X\right)\beta^* + \beta^{*T} Q \beta^* \\ &\quad - 2\beta^{*T}\left(X^T X\right)\bar{\beta} - 2\beta^{*T} Q \beta_0 \\ &\quad + \bar{\beta}^T\left(X^T X\right)\bar{\beta} + {\beta_0}^T Q \beta_0 \\ &= R + M \\ &\quad + \beta^{*T}\left(X^T X + Q\right)\beta^* \end{aligned}$$

$$
\begin{aligned}
&- 2\beta^{*T}\left[\left(X^TX\right)\bar{\beta} + Q\beta_0\right] \\
&+ \bar{\beta}^T\left(X^TX\right)\bar{\beta} + {\beta_0}^TQ\beta_0 \quad .
\end{aligned}
$$

Thereafter, we substitute the classical regression mean $\bar{\beta}$ in equation (2.5) into the final two terms, allowing a great deal of simplification

$$
\begin{aligned}
A &= R + M \\
&+ \beta^{*T}\left(X^TX + Q\right)\left(X^TX + Q\right)^{-1}\left(X^Ty + Q\beta_0\right) \\
&- 2\beta^{*T}\left[\left(X^TX\right)\left(X^TX\right)^{-1}X^Ty + Q\beta_0\right] \\
&+ \bar{\beta}^T\left(X^TX\right)\left(X^TX\right)^{-1}X^Ty + {\beta_0}^TQ\beta_0 \\
&= R + M \\
&- \beta^{*T}\left(X^Ty + Q\beta_0\right) \\
&+ \bar{\beta}^TX^Ty + {\beta_0}^TQ\beta_0 \quad .
\end{aligned}
$$

We can simplify the last two terms into a final, very useful and insightful form

$$
A = R + M + \left(\bar{\beta} - \beta^*\right)^TX^Ty + \left(\beta_0 - \beta^*\right)^TQ\beta_0 \quad . \tag{3.13}
$$

This is going to be an important interim result—proper computation of the constant A in the quadratic posterior. This term A is going to represent, for the posterior, the equivalent of the residual sum of squared errors term R for the classical regression case. We will be emphasizing this interpretation of this posterior constant A repeatedly as our developments proceed herein in Chapter 7. This suggests that we think about the changes in the rightmost two terms from the prior in Figure 3.1 to create the posterior A constant. In particular, M, a scalar in the first exponential term in the prior, has been replaced with A from equation (3.13) appearing in the first exponential term in the posterior. The transformation from prior to posterior is complete. We want to define the s^2 term analogously to what we did previously:

$$
\nu s^2 = A \quad ,
$$

which implies that

$$
s^2 = \frac{A}{\nu} \quad .
$$

We thus have all the terms we need to define and calculate the posterior in equation (3.12).

To conclude this section, it is convenient to think of gathering data as an "experiment."[9] We represent this idea as in Figure 3.3. In the figure, we see the

[9]Dr. Howard[45] has thought this through in detail and has presented a protracted example, an

most basic, traditional concept of classical statistics. One sticks the assembled data into a "regression calculator," and out the back door emerges a regression solution for the coefficients of your model. The more data, the better the answer! The better the predictive capability! That is the perception, which may or may not be correct but is certainly not a "theorem."

Figure 3.3: Classical Statistics Is an "Experiment" to Gather Data

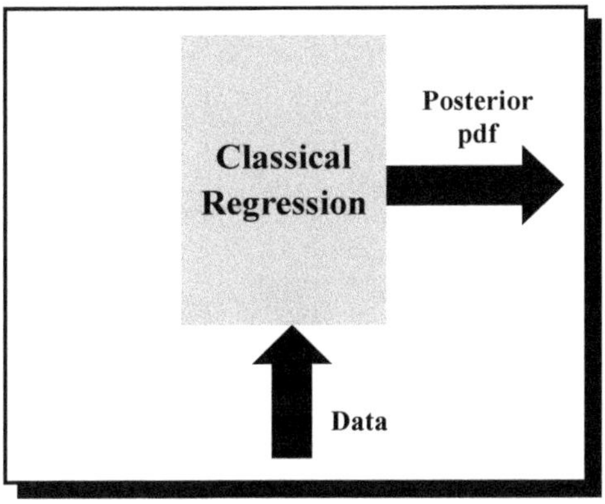

In the real world, a plethora of questions invariably arise. What if the data are problematic? What if the database is intrinsically small in size, say, for example, data based on destructive testing of an asset such as a nuclear warhead aboard a missile? What if the experiment gives you nothing new, nothing to be concerned about, delivering literally the same measurements every time with undiscernible variation across estimates? What if Howard and Abbas'[45] Acme Rain Detector kept saying the exact same thing irrespective of whether it was going to rain or not, like the proverbial stopped clock that gives the same reading all the time but indeed accurate data twice a day (if you use American time; only once if you use European time). What if there is a sea change in underlying technology (such as directional drilling and hydraulic fracturing, which occurred in the oil industry, literally revolutionizing costs and recoveries) that occurs part way through a protracted, serial monitoring and data gathering process? What good is historical data in such an environment? What if you bring to the table a lot of probabilistic judgment? Your people have pretty good intuition about what the answer is, but they have not gathered the data to verify it, and the data promises to trickle in over time. What if you have significant judgment to bring to the problem because you lack any historical data at all? Bayes brings judgment

Acme Rain Detector, that is quite analogous. Dr. Howard is gathering successive readings on the Acme Rain Detector as to whether it will rain tomorrow, each reading being a piece of "data." Gathering data is quite like an "experiment," getting more and more measurements of some phenomenon. Theoretically, the more experimental results one has, the better predictor one has. Everyone knows this is not true, except in an idealized setting. However, it seems to make classical statisticians feel better to think so because their theory holds that everything known or knowable is embedded in the assembled data, and the only way to get more information is to get more data.

to the problem in the form of a prior, upon which there are no material restrictions. Bayes melds experimental data with that prior, combining it with the prior to compute a posterior. We represent the Bayesian Linear Regression process as in Figure 3.4. On the left, we see a box that contains either old data or prior judgment within it. That old data or prior judgment are delivered to the Bayesian calculation in the form of a prior probability distribution over model parameters. What is our best judgment or best evidence of what the model coefficients are before we ever process a lick of new data? What do we know, and what is our confidence in knowing it, before we begin the data-gathering experiment?

Figure 3.4: Bayes Appends Data to Judgment (or to "Old" Data As a Substitute for Judgment)

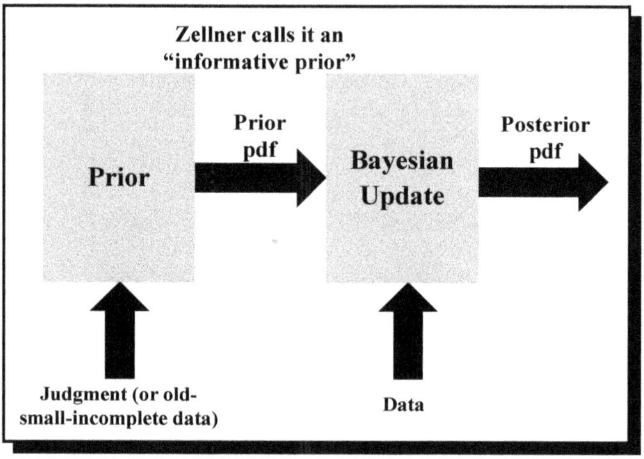

We see in the left-hand box in Figure 3.4 the process whereby old data or prior judgment is expressed as a probability distribution over model parameters. That probability distribution is called the Bayesian prior. Zellner[107] has called it an "informative prior," to emphasize that it is real, cogent knowledge that you bring to the problem before you ever gather or process any data. It is real, "true blue" knowledge. The words of Prof. Howard ring in our ears: "Is there any such thing as a prior that is not informative?" No! However, an abstract model of abject, total ignorance expressed in the form of a diffuse or "uninformative" prior is developed and explicated later herein. The Bayesian process represented by the box at the right is simply the process of combining the prior at the left with the data (the likelihood function) at the bottom and grinding them together to get the posterior associated with the arrow at the far right. This is a very standard Bayesian framework, and it is actually centered on the box at the right in the figure.

It will become clear in Chapter 7 that under very specific assumptions called a "diffuse prior," as indicated in Figure 3.5, there is an important equivalency, an isomorphism, between classical statistics and Bayesian Linear Regression. We discuss the diffuse prior in detail in Chapter 7, both in terms of its intuitive reasonableness as well as its validity in real-world applications.

Figure 3.5: Classical Is a Special Case of Bayes

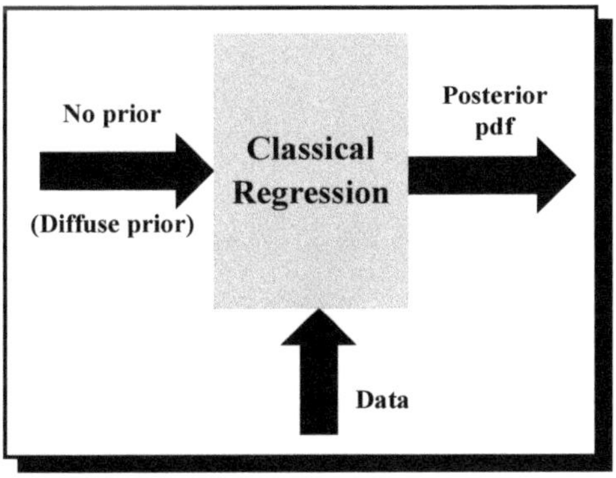

Bayes is an extension and clarification of classical regression and is therefore more powerful and intuitively satisfying. It is so difficult for many practitioners to understand what classical regression is truly doing, and by contrast, it is disarmingly simple to understand what Bayesian Linear Regression is really doing. People in undergrad statistics courses and practitioners struggle with the **interpretation** of regression results and concepts, and I did. The interpretation of Bayesian results is so disarmingly simple. It is easy to understand what you are getting out of Bayesian Linear Regression—a probability distribution over variables of interest. Bayes thereby intrinsically develops a probability distribution for use in von Neumann[100] expected utility calculations, the fundamental fabric of decision-making. There is nothing arbitrary or capricious in Bayes. You know exactly what you are doing and why. We like to think of Bayes as a generalization of classical statistical logic and, as such, as more powerful and interpretable. We shudder to say: Bayes renders classical obsolete and a "corner solution," precisely as Einstein's special and general relativity rendered Newtonian physics obsolete and a "corner solution."

Chapter 4

THE CONJUGATE-ARCHETYPE FUNCTIONAL FORM OF POSTERIOR AND PRIOR DISTRIBUTIONS

All knowledge degenerates into probability.

David Hume

Absolutely no optimization has been done thus far other than implicitly developing the mode of a pdf as "fallout" from completing the square and creating a quadratic within that distribution. We aim to keep it that way. We have done nothing except rely on pure probability. We have done no more than develop parallel forms for the various probability density functions by completing the square. We have developed a conjugate prior-posterior pair based on the form of a likelihood function. The specific conjugate functional forms developed thus far are shown below:

Likelihood:

$$\{X, y\,|\beta, \sigma\} = const \times \sigma^{-n} e^{-\frac{1}{2\sigma^2}\left[\nu_C s_C^2 + (\beta-\bar{\beta})^T (X^T X)(\beta-\bar{\beta})\right]} \quad .$$

This functional form of the likelihood function was derived as equation (2.12) in Chapter 2, was depicted in Figure 2.3, and is copied here for easy reference and comparison.

Prior:

$$\{\beta, \sigma\} = const \times \sigma^{-m} e^{-\frac{1}{2\sigma^2}\left[M + (\beta - \beta_0)^T Q (\beta - \beta_0)\right]} \quad .$$

This functional form of the prior was derived as equation (3.2) in Chapter 3, was depicted in Figure 3.1, and is copied here for easy reference and comparison.

Posterior:

$$\{\beta, \sigma \,| X, y\} = const \times \sigma^{-(n+m)} e^{-\frac{1}{2\sigma^2}\left[A + (\beta - \beta^*)^T \left(X^T X + Q\right)(\beta - \beta^*)\right]} \quad .$$

This functional form of the posterior was derived as equation (3.6) and is copied here for easy reference and comparison. It is, of course, the product of the prior times and the likelihood immediately above.

The very notion of conjugacy ensures that the functional forms of the prior and the posterior are identical. The prior and posterior densities immediately above exhibit exact identity of form. That is the whole point of conjugate densities—identical functional forms for the prior and posterior. We will be taking advantage of that identical form and, in fact, dwelling on it in this section.

Keep in mind that the likelihood function immediately above is **profoundly different** from the prior and the posterior (which appear to be the same in form). The three might appear identical at first blush, but they are not, not at all. The likelihood function is a probability density function over entirely different variables (X and y) than the prior and posterior (β and σ). We must be very clear: the likelihood function is a distribution over the data X and y. By contrast, the prior and posterior are distributions over the coefficients β and σ. The subtle difference between the likelihood function and the prior/posterior in terms of their arguments is going to be crucially important and repeatedly highlighted in this section.[1]

The prior and the posterior are identical in form—both have a very specific, identical functional form—which we call the "conjugate-archetype form:"

[1]This is no different from conjugate systems such as binomial-beta, Poisson-gamma, multinomial-Dirichlet, normal-inverse gamma, or exponential-gamma. The likelihood functions—binomial, Poisson, multinomial, normal, and exponential, respectively—are defined over fundamentally different arguments than are the priors and posteriors—beta, gamma, Dirichlet, inverse gamma, and gamma, respectively. This is true for all conjugacy pairs. Do analogous Bayesian concepts work for these other conjugate systems? Yes. Do they work for systems for which the prior and posterior are not conjugate at all (different functional forms between prior and posterior) and, in fact, might be complex functional forms or even computer programs? Yes. There are numerous examples of non-conjugate, nonlinear models, including, for example, Nesbitt[71].

$$\begin{aligned}\{\beta,\sigma\} &= const \times \sigma^{-r} e^{-\frac{1}{2\sigma^2}\left[g+(\beta-\mu)^T H(\beta-\mu)\right]} \\ &= const \times \sigma^{-r} e^{-\frac{g}{2\sigma^2}} e^{-\frac{1}{2}(\beta-\mu)^T\left(H\sigma^{-2}\right)(\beta-\mu)} \end{aligned} \tag{4.1}$$

in which the parameters are

- p = the number of model coefficients
- g = a scalar constant in the exponent
- μ = mean vector of length p
- $H = p \times p$ symmetric, positive definite matrix in the exponent
- r = a scalar exponent of the σ term .

We use the terminology "conjugate archetype" to emphasize that the prior and posterior are identical in functional form, the only difference being the specific settings of the foregoing parameters. Because the only difference between the prior and posterior is the parameter settings in the conjugate-archetype functional form of distribution, the archetype can be analyzed using generic parameters such as the foregoing. Thereafter, the specific parameters that characterize the prior can be substituted for the generic parameters, and Voila! We have the complete, detailed, full prior. Ditto for the posterior. In particular, by analyzing the archetype form using the foregoing generic parameters, we are able thereafter to substitute specific parameters of the prior and the posterior and garner the answers without additional work.

To represent our prior using this conjugate-archetype form, we simply substitute the specific prior parameters for the conjugate-archetype parameters:

- p = the number of model coefficients
- $g \leftarrow$ the scalar constant M
- $\mu \leftarrow \beta_0$
- $H \leftarrow Q$
- $r \leftarrow m = \nu_0 + p + 1$, a scalar .

To represent our posterior using this conjugate-archetype form, we simply substitute the specific posterior parameters for the conjugate-archetype parameters:

- p = the number of model coefficients
- $g \leftarrow$ the scalar constant A
- $\mu \leftarrow \beta^*$
- $H \leftarrow X^T X + Q$
- $r \leftarrow m + n = \nu_0 + p + 1 + n$, a scalar.

Our prior and posterior are in very good hands using this functional form (because the prior and posterior are both identically-formed probability distributions over β and σ). Such is the benefit of conjugacy.

It would seem apparent that we have to do some work before we can "connect the dots" from this Bayesian case to the classical case developed in Chapter 2. There is no notion of a prior in the classical case in Chapter 2, and yet a prior is central in the Bayesian case. To presage the Bayesian-to-classical connection, we are going to have to specialize the general "informative" prior from this section, in effect rendering it impotent and invisible. We defer discussion of that issue until Chapter 7, concentrating in this chapter on the fully general conjugate-archetype functional form itself, then moving to the prior, and then moving to the posterior. Once we have completed that, we will move on in Chapter 7 to "connect the dots" from the Bayesian case to the classical statistical case and derive a true and correct probabilistic representation of the classical case. In particular, we will make very specific assumptions about the prior, thereafter calculate the posterior subject to those assumptions, and note that the resulting posterior contains the same results we got in the classical case in Chapter 2. Our objective in this chapter is to develop results that pertain to the general prior and the general posterior by simple substitution of parameter values into the conjugate-archetype functional form.

4.1 Constant of Integration of the Conjugate-Archetype Distribution in Equation 4.1

Equation (4.1) contains an as yet uncalculated constant of integration within it. In this section, we will determine what that constant is. Unlike many developments, we mean to carry the integrating constant explicitly throughout our development. We identify the constant by integrating the density and setting the integral times the constant to 1. That tells us exactly what the constant is. We begin by writing a general expression for the multivariate normal distribution,[2] which will prove convenient for integrating the conjugate-archetype distribution in equation (4.1):

$$N\left(x;\mu,\Sigma\right)=f\left(x\right)=\frac{1}{\left(2\pi\right)^{\frac{p}{2}}\left|\Sigma\right|^{\frac{1}{2}}}e^{-\frac{1}{2}\left(x-\mu\right)^{T}\Sigma^{-1}\left(x-\mu\right)}\quad. \tag{4.2}$$

In this formulation, Σ is the variance-covariance matrix, and it is symmetric and nonnegative-definite. It is convenient to assume that the Σ matrix, the variance-covariance matrix, is strictly positive definite, which avoids degenerate cases that would otherwise emerge in classical and Bayesian Linear Regression. Our subsequent developments will assume that the Σ matrix, the variance-covariance matrix, is symmetric and strictly positive-definite. Practical applications invariably have positive, definite variance-covariance matrices. The inverse of the Σ matrix, namely Σ^{-1}, is called the "precision matrix." People have wanted to

[2]See https://en.wikipedia.org/wiki/Multivariate_normal_distribution.

eliminate the matrix inverse from the exponent, defining the matrix in the exponent as a single positive definite matrix called the precision matrix rather than the inverse of a matrix. We want to consider the special case in which the matrix in the conjugate archetype in equation (4.1) is exactly equal to the inverse variance-covariance matrix that appears in the multivariate normal distribution immediately above, namely:

$$\Sigma^{-1} = H\sigma^{-2} \quad .$$

We take the inverse of both sides:

$$\Sigma = H^{-1}\sigma^{2} \quad .$$

We take the determinant of both sides, recognizing that the determinant of a constant times a square $p \times p$ matrix is the constant raised to the p power times the determinant of the square matrix and the determinant of the inverse is the reciprocal of the determinant

$$|\Sigma| = \left|H^{-1}\sigma^{2}\right| = \sigma^{2p}\left|H^{-1}\right| = \frac{\sigma^{2p}}{|H|}$$
$$\Rightarrow |\Sigma|^{\frac{1}{2}} = \frac{\sigma^{p}}{|H|^{\frac{1}{2}}} \quad .$$

With this definition of the variance-covariance matrix, we are able to substitute for Σ into the standard multivariate normal density from equation (4.2) and rewrite it as follows:

$$N\left(x;\mu,H^{-1}\sigma^{2}\right) = f\left(x\right) = \frac{|H|^{\frac{1}{2}}}{(2\pi)^{\frac{p}{2}}\sigma^{p}} e^{-\frac{1}{2}(x-\mu)^{T}\left(H\sigma^{-2}\right)(x-\mu)} \quad ,$$

which we can manipulate into the form

$$e^{-\frac{1}{2}(x-\mu)^{T}\left(H\sigma^{-2}\right)(x-\mu)} = \frac{N\left(x;\mu,H^{-1}\sigma^{2}\right)}{|H|^{\frac{1}{2}}(2\pi)^{-\frac{p}{2}}\sigma^{-p}} \quad . \tag{4.3}$$

Substituting equation (4.3) for the rightmost exponential term in the conjugate-archetype form in equation (4.1), we obtain

$$\begin{aligned}\{\beta,\sigma\} &= const \times \sigma^{-r} e^{-\frac{g}{2\sigma^{2}}} e^{-\frac{1}{2}(\beta-\mu)^{T}\left(H\sigma^{-2}\right)(\beta-\mu)} \\ &= \frac{const}{|H|^{\frac{1}{2}}(2\pi)^{-\frac{p}{2}}} \sigma^{-(r-p)} e^{-\frac{g}{2\sigma^{2}}} N\left(\beta;\mu,H^{-1}\sigma^{2}\right) \quad .\end{aligned}$$

Using this representation of the conjugate-archetype form, we can easily derive the constant of integration by integrating the joint density over all variables, $d\sigma$

and $d\beta$, and noting that it must integrate to 1:[3]

$$
\begin{aligned}
1 &= \int_0^\infty d\sigma \int_{-\infty}^\infty d\beta \{\beta, \sigma\} \\
&= \int_0^\infty d\sigma \int_{-\infty}^\infty d\beta \frac{const}{|H|^{\frac{1}{2}}(2\pi)^{-\frac{p}{2}}} \sigma^{-(r-p)} e^{-\frac{g}{2\sigma^2}} N\left(\beta; \mu, H^{-1}\sigma^2\right) \\
&= \frac{const}{|H|^{\frac{1}{2}}(2\pi)^{-\frac{p}{2}}} \int_0^\infty d\sigma \sigma^{-(r-p)} e^{-\frac{g}{2\sigma^2}} \int_{-\infty}^\infty d\beta N\left(\beta; \mu, H^{-1}\sigma^2\right) \quad .
\end{aligned}
$$

We have manipulated this so that the last term on the right is the integral of a multivariate normal distribution over β alone. This integral alone will integrate to 1. Hence the equation becomes

$$
1 = \frac{const}{|H|^{\frac{1}{2}}(2\pi)^{-\frac{p}{2}}} \int_0^\infty d\sigma \sigma^{-(r-p)} e^{-\frac{g}{2\sigma^2}} \quad .
$$

In order to compute the integral, we want the exponent at the right to become a simple scalar term $-h$, and we therefore make the following substitution in the integral:

$$
\begin{aligned}
& \frac{g}{2\sigma^2} = h \\
& \Rightarrow \left(\frac{g}{2}\right)^{-1} \sigma^2 = h^{-1} \\
& \Rightarrow \sigma^2 = \left(\frac{g}{2}\right) h^{-1} \\
& \Rightarrow \sigma = \left(\frac{g}{2}\right)^{\frac{1}{2}} h^{-\frac{1}{2}} \\
& \Rightarrow d\sigma = -\frac{1}{2}\left(\frac{g}{2}\right)^{\frac{1}{2}} h^{-\frac{3}{2}} dh \quad .
\end{aligned}
$$

Using the limits of the integral on the variable σ, we establish the limits of integration on the new variable h

$$
\begin{aligned}
\sigma = 0 &\Rightarrow h = \infty \\
\sigma = \infty &\Rightarrow h = 0 \quad .
\end{aligned}
$$

The power of σ term can be written:

[3] We use the convention, where possible, to presage the first integral we are going to take by associating its differential with the innermost integral sign. No confusion should result.

$$\sigma^{-(r-p)} = \left(\frac{g}{2}\right)^{\frac{-(r-p)}{2}} h^{\frac{(r-p)}{2}} \quad .$$

These substitutions allow us to write:

$$1 = \frac{const}{|H|^{\frac{1}{2}}(2\pi)^{-\frac{p}{2}}} \int_{\infty}^{0} -\frac{1}{2}\left(\frac{g}{2}\right)^{\frac{1}{2}} h^{-\frac{3}{2}} \left(\frac{g}{2}\right)^{\frac{-(r-p)}{2}} h^{\frac{(r-p)}{2}} e^{-h} dh \quad .$$

We know that we can reverse the limits on the integral if we take the negative of the integrand (eliminating the minus sign in the integrand):

$$1 = \frac{const}{|H|^{\frac{1}{2}}(2\pi)^{-\frac{p}{2}}} \frac{1}{2}\left(\frac{g}{2}\right)^{-\frac{(r-p-1)}{2}} \int_{0}^{\infty} h^{\left[\frac{(r-p-1)}{2}-1\right]} e^{-h} dh \quad .$$

We use the definition of the gamma function[4]

$$\Gamma(n) = \int_{0}^{\infty} x^{n-1} e^{-x} dx \quad ,$$

which implies for positive integers n (which we see frequently hereafter because we are counting observations or variables, ending up with arguments that are integers), that

$$\Gamma(n) = (n-1)! \quad .$$

We substitute for the integral in the previous equation:

$$1 = \frac{const}{|H|^{\frac{1}{2}}(2\pi)^{-\frac{p}{2}}} \frac{1}{2}\left(\frac{g}{2}\right)^{-\frac{r-p-1}{2}} \Gamma\left(\frac{r-p-1}{2}\right)$$

$$\Rightarrow const = \frac{2|H|^{\frac{1}{2}}}{(2\pi)^{\frac{p}{2}}\Gamma\left(\frac{r-p-1}{2}\right)} \left(\frac{g}{2}\right)^{\frac{r-p-1}{2}} \quad .$$

We notice that the term $r-p-1$ appears as the numerator of the exponent of $\frac{g}{2}$, and it also appears as the numerator of the argument of the Γ function in the denominator of the foregoing equation. This motivates us to define the parameter ν as follows:

(4.4) $$\nu = r - p - 1 \quad .$$

There is a name for the ν parameter defined in equation (4.4), namely "degrees of freedom." We see by this development that the proper definition of degrees of

[4]See https://en.wikipedia.org/wiki/Gamma_function.

freedom "falls naturally out of the math." There it is, one of the key exponents. This definition allows us to write the constant using the ν parameter as follows:

$$const = \frac{2|H|^{\frac{1}{2}}}{(2\pi)^{\frac{p}{2}}\Gamma\left(\frac{\nu}{2}\right)}\left(\frac{g}{2}\right)^{\frac{\nu}{2}} \quad .$$

If we substitute into equation (4.1), we can write the density with the constant expressed explicitly

$$\{\beta,\sigma\} = \frac{2|H|^{\frac{1}{2}}}{(2\pi)^{\frac{p}{2}}\Gamma\left(\frac{\nu}{2}\right)}\left(\frac{g}{2}\right)^{\frac{\nu}{2}}\sigma^{-r}e^{-\frac{g}{2\sigma^2}} \times e^{-\frac{1}{2}(\beta-\mu)^T\left(H\sigma^{-2}\right)(\beta-\mu)} \quad .$$

We can express the parameter r of the conjugate-archetype distribution by reversing equation (4.4) to write:

$$r = \nu + p + 1 \quad . \tag{4.5}$$

In preparation for future analysis, we will define the variable

$$s^2 = \frac{g}{\nu} \quad . \tag{4.6}$$

This will prove to be a convenient definition as well. This implies that

$$g = \nu s^2 \quad . \tag{4.7}$$

This will allow us to write the distribution in the following form, which properly integrates to 1.

$$\{\beta,\sigma\} = \frac{2|H|^{\frac{1}{2}}}{(2\pi)^{\frac{p}{2}}\Gamma\left(\frac{\nu}{2}\right)}\left(\frac{\nu s^2}{2}\right)^{\frac{\nu}{2}}\sigma^{-(\nu+p+1)}e^{-\frac{\nu s^2}{2\sigma^2}} \times e^{-\frac{1}{2}(\beta-\mu)^T\left(H\sigma^{-2}\right)(\beta-\mu)} \quad . \tag{4.8}$$

If we substitute the exponential term in equation (4.3), we can rewrite this density in the form

$$\{\beta,\sigma\} = \frac{2\left(\frac{\nu s^2}{2}\right)^{\frac{\nu}{2}}}{\Gamma\left(\frac{\nu}{2}\right)}\sigma^{-(\nu+1)}e^{-\frac{\nu s^2}{2\sigma^2}}N\left(\beta;\mu,H^{-1}\sigma^2\right) \quad . \tag{4.9}$$

We can see the multivariate normal form at the right, and we can see the gamma-like form at the left.

It is time to talk about what this distribution truly is. Even though the well-known χ^2 form is a special case of the gamma density, this equation does not contain a χ^2 density and should not be interpreted as a χ^2 density. It should be interpreted for what it is—an inverse gamma density. It is inverse because the variable σ appears in the denominator of all the key terms in the distribution. It is therefore de facto a distribution over $1/\sigma$. We will give a great deal of attention to this in the forthcoming discussion. The Bayesian probability distribution derived here does not contain the χ^2 distribution at all. The χ^2 distribution has been eliminated in Bayes and replaced by an inverse gamma distribution. We will be deriving and giving a great deal of attention to this inverse gamma distribution in forthcoming discussions. This is truly a probability distribution, and its arguments are β and σ, not X and y. Keep in mind that this is a joint probability distribution over the (linear) model coefficients β and the variance variable σ. Henceforth, we will use the expressions in equations (4.8) and (4.9) for the conjugate-archetype distribution. We will continue to develop the density in equation (4.9) in the next section.

4.2 Analyzing and Re-expressing the Joint Distribution

We begin with the complete expression for the distribution over the coefficients and the error term in equation (4.8).[5] In order to explore the structure of this distribution, we begin by knowing that this distribution in equation (4.8) must integrate to unity:

$$1 = \int_{-\infty}^{\infty} d\beta \int_{0}^{\infty} d\sigma \, \{\beta, \sigma\}$$

$$= \frac{2|H|^{\frac{1}{2}}(2\pi)^{-\frac{p}{2}}\left(\frac{\nu s^2}{2}\right)^{\frac{\nu}{2}}}{\Gamma\left(\frac{\nu}{2}\right)} \int_{-\infty}^{\infty} d\beta \int_{0}^{\infty} d\sigma \sigma^{-(\nu+p+1)} e^{-\frac{\nu s^2}{2\sigma^2}} e^{-\frac{1}{2\sigma^2}(\beta-\mu)^T H(\beta-\mu)} .$$

In this integral, we want to represent the distribution over the quantity $z = \sigma^2$ (the variance) rather than over the quantity σ (the standard deviation). We want the variance rather than the standard deviation as the argument of the probability distribution inside the integral; this is akin to a variable transformation in probability. We want to make the substitution $z = \sigma^2$ and note its implications for the integrand:

$$\sigma^2 = z \Rightarrow \sigma = z^{\frac{1}{2}} \Rightarrow d\sigma = \frac{1}{2} z^{-\frac{1}{2}} dz \Rightarrow \sigma^{-(\nu+p+1)} = z^{-\frac{\nu+p+1}{2}} .$$

Making these substitutions, we rewrite the foregoing density in a convenient form

[5] We seek simple way to transform the distribution over σ into a more recognizable and identifiable form. We will make the transformation formally later in this section.

$$1 = \frac{2|H|^{\frac{1}{2}}(2\pi)^{-\frac{p}{2}}\left(\frac{\nu s^2}{2}\right)^{\frac{\nu}{2}}}{\Gamma\left(\frac{\nu}{2}\right)}$$
$$\times \int_{-\infty}^{\infty} d\beta \int_{0}^{\infty} \left(\frac{1}{2}z^{-\frac{1}{2}}dz\right)\left(z^{-\frac{\nu+p+1}{2}}\right) e^{-\frac{\nu s^2}{2z}} e^{-\frac{1}{2z}(\beta-\mu)^T H(\beta-\mu)}$$
$$= \frac{|H|^{\frac{1}{2}}(2\pi)^{-\frac{p}{2}}\left(\frac{\nu s^2}{2}\right)^{\frac{\nu}{2}}}{\Gamma\left(\frac{\nu}{2}\right)} \int_{-\infty}^{\infty} d\beta \int_{0}^{\infty} \left[z^{-\frac{\nu+p+2}{2}} e^{-\frac{\nu s^2}{2z}} e^{-\frac{1}{2z}(\beta-\mu)^T H(\beta-\mu)}\right] dz \quad .$$

We extract and combine the leading constant and the bracketed term to write the joint density that uses the variance variable z rather than the standard deviation variable σ, thereby reconstituting the new joint density

$$\{\beta, z\} = \frac{|H|^{\frac{1}{2}}(2\pi)^{-\frac{p}{2}}\left(\frac{\nu s^2}{2}\right)^{\frac{\nu}{2}}}{\Gamma\left(\frac{\nu}{2}\right)} z^{-\frac{\nu+p}{2}+1} e^{-\frac{\nu s^2}{2z}} e^{-\frac{1}{2z}(\beta-\mu)^T H(\beta-\mu)} \quad . \tag{4.10}$$

We want to write this in the form of a normal times another distribution. We begin with the multivariate normal distribution put forth in equation (4.2) and thereafter to make the following substitutions so as to derive a form of normal distribution that is implied by the rightmost exponential term within our distribution in equation (4.10):

$$\sigma^{-2}H = \Sigma^{-1} \Rightarrow H^{-1}\sigma^2 = \Sigma \Rightarrow |H|^{-1}\sigma^{2p} = |\Sigma| \Rightarrow |H|^{\frac{1}{2}}\sigma^{-p} = |\Sigma|^{-\frac{1}{2}} \quad .$$

Making these substitutions, we write a specific multivariate normal distribution as follows

$$N\left(\beta; \mu, H^{-1}\sigma^2\right) = \left(\frac{1}{2\pi}\right)^{\frac{p}{2}} |H|^{\frac{1}{2}}\sigma^{-p} e^{-\frac{1}{2\sigma^2}(\beta-\mu)^T H(\beta-\mu)} \quad .$$

We see the rightmost exponential term in equation (4.10) herein, and we wish to substitute it. Rearranging, we isolate the rightmost exponential term in equation (4.10) so that we can thereafter substitute it:

$$e^{-\frac{1}{2\sigma^2}(\beta-\mu)^T H(\beta-\mu)} = \frac{N\left(\beta; \mu, H^{-1}\sigma^2\right)\sigma^p(2\pi)^{\frac{p}{2}}}{|H|^{\frac{1}{2}}} \quad .$$

We substitute z in place of σ^2

$$z = \sigma^2 \Rightarrow z^{\frac{1}{2}} = \sigma \Rightarrow z^{\frac{p}{2}} = \sigma^p \quad .$$

Making these substitutions, we can write the exponential term using the same variables as in equation (4.10)

$$e^{-\frac{1}{2z}(\beta-\mu)^T H(\beta-\mu)} = \frac{N\left(\beta;\mu,H^{-1}z\right) z^{\frac{p}{2}}(2\pi)^{\frac{p}{2}}}{|H|^{\frac{1}{2}}} \quad .$$

Thereafter, we can substitute the exponential term into the joint distribution in equation (4.10) using z rather than σ^2.

$$\begin{aligned}\{\beta,z\} &= \frac{|H|^{\frac{1}{2}}(2\pi)^{-\frac{p}{2}}\left(\frac{\nu s^2}{2}\right)^{\frac{\nu}{2}}}{\Gamma\left(\frac{\nu}{2}\right)} z^{-\left(\frac{\nu+p}{2}+1\right)} e^{-\frac{\nu s^2}{2z}} \frac{N\left(\beta;\mu,H^{-1}z\right) z^{\frac{p}{2}}(2\pi)^{\frac{p}{2}}}{|H|^{\frac{1}{2}}} \\ &= \frac{\left(\frac{\nu s^2}{2}\right)^{\frac{\nu}{2}}}{\Gamma\left(\frac{\nu}{2}\right)} z^{-\frac{\nu}{2}-1} e^{-\frac{\nu s^2}{2z}} N\left(\beta;\mu,H^{-1}z\right) \quad .\end{aligned}$$

It is clear that there is an embedded normal distribution here, and there is a second distribution that is comprised of all the other terms besides the normal. The normal will integrate to one, so the distribution comprised of all the other terms must integrate to one as well. All the remaining terms other than the normal comprise what we will observe is a standard inverse gamma distribution:[6]

(4.11) $$IG\left(z,\frac{\nu s^2}{2},\frac{\nu}{2}\right) = \frac{\left(\frac{\nu s^2}{2}\right)^{\frac{\nu}{2}}}{\Gamma\left(\frac{\nu}{2}\right)} z^{-\left(\frac{\nu}{2}+1\right)} e^{-\frac{\nu s^2}{2z}} \quad .$$

It is useful to explicitly write the product rule using the variance z

(4.12) $$\{\beta,z\} = IG\left(z,\frac{\nu s^2}{2},\frac{\nu}{2}\right) N\left(\beta;\mu,H^{-1}z\right) \quad .$$

This is the famous result quoted by Zellner[107], stated but rarely derived from first principles. We will derive in Sections 4.4-4.5 below the mean and variance of this inverse gamma distribution. For later reference, the mean value of the coefficients β is

(4.13) $$Mean = \mu$$

and the H matrix is

(4.14) $$\text{H matrix} = H \quad .$$

[6] See https://en.wikipedia.org/wiki/Inverse-gamma_distribution.

4.3 The Marginal Distribution over the Model Coefficients β

What is the marginal distribution over the model coefficients β, taking full account of all probabilistic dependencies between coefficients? What is the mean and variance-covariance matrix of that marginal distribution? What is the functional form or "family" that this marginal distribution belongs to? This question is going to persist throughout the Bayesian statistical development of the conjugate prior and posterior, just as it has throughout classical statistics. We need to calculate it carefully and completely. To calculate the marginal distribution, we merely integrate over the joint density in equation (4.8) to eliminate the σ term.

$$\begin{aligned}
\{\beta\} &= \int_0^\infty \{\beta,\sigma\} d\sigma \\
&= \int_0^\infty d\sigma \frac{2|H|^{\frac{1}{2}}(2\pi)^{-\frac{p}{2}}\left(\frac{\nu s^2}{2}\right)^{\frac{\nu}{2}}}{\Gamma\left(\frac{\nu}{2}\right)} \sigma^{-(\nu+p+1)} e^{-\frac{\nu s^2}{2\sigma^2}} e^{-\frac{1}{2\sigma^2}(\beta-\mu)^T H(\beta-\mu)} \\
&= \frac{2|H|^{\frac{1}{2}}(2\pi)^{-\frac{p}{2}}\left(\frac{\nu s^2}{2}\right)^{\frac{\nu}{2}}}{\Gamma\left(\frac{\nu}{2}\right)} \int_0^\infty d\sigma \sigma^{-(\nu+p+1)} e^{-\frac{\nu s^2}{2\sigma^2}\left[1+(\beta-\mu)^T \frac{H}{\nu s^2}(\beta-\mu)\right]} \quad .
\end{aligned}$$

We define the variable h, a variable to be used for substitution in the integral, as follows

$$\begin{aligned}
h &= \frac{\nu s^2}{2\sigma^2}\left[1+(\beta-\mu)^T\left(\frac{H}{\nu s^2}\right)(\beta-\mu)\right] \\
&= \sigma^{-2}\frac{\nu s^2}{2}\left[1+(\beta-\mu)^T\left(\frac{H}{\nu s^2}\right)(\beta-\mu)\right] \quad .
\end{aligned}$$

Note that when $\sigma = 0$, the value of h is ∞. When $\sigma = \infty$, the value of h is zero. Thus, the integral over h reverses the limits of integration relative to those of σ. When we reverse this integral, we have to multiply the integrand by -1, and that removes a negative sign that emerges in the integrand itself. We will see this pattern emerge throughout all the integrals in this section, so we should be on the lookout for it when getting our signs and integrals correct. We will note this when it occurs in forthcoming integrals. We can manipulate this so as to express σ as a function of h:

$$\sigma = h^{-\frac{1}{2}}\left(\frac{\nu s^2}{2}\right)^{\frac{1}{2}}\left[1+(\beta-\mu)^T\left(\frac{H}{\nu s^2}\right)(\beta-\mu)\right]^{\frac{1}{2}} \quad .$$

We take the total differential in preparation for substituting in the integral

$$d\sigma = -\frac{1}{2}h^{-\frac{3}{2}}dh\left(\frac{\nu s^2}{2}\right)^{\frac{1}{2}}\left[1+(\beta-\mu)^T\left(\frac{H}{\nu s^2}\right)(\beta-\mu)\right]^{\frac{1}{2}} \quad .$$

We will be eliminating this negative sign below when we reverse the order of integration. We raise the σ term to the correct power in preparation for substituting into the integral

$$\sigma^{-(\nu+p+1)} = h^{\frac{\nu+p+1}{2}}\left(\frac{\nu s^2}{2}\right)^{-\frac{\nu+p+1}{2}}\left[1+(\beta-\mu)^T\left(\frac{H}{\nu s^2}\right)(\beta-\mu)\right]^{-\frac{\nu+p+1}{2}} \quad .$$

We take the product of the previous two equations and multiply them by the substituted exponential

$$\begin{aligned}
d\sigma\sigma^{-(\nu+p+1)}e^{-h} &= -\frac{1}{2}h^{-\frac{3}{2}}dh\left(\frac{\nu s^2}{2}\right)^{\frac{1}{2}}\left[1+(\beta-\mu)^T\frac{H}{\nu s^2}(\beta-\mu)\right]^{\frac{1}{2}} \\
&\times h^{\frac{\nu+p+1}{2}}\left(\frac{\nu s^2}{2}\right)^{-\frac{\nu+p+1}{2}}\left[1+(\beta-\mu)^T\frac{H}{\nu s^2}(\beta-\mu)\right]^{-\frac{\nu+p+1}{2}} e^{-h} \\
&= -\frac{1}{2}dhh^{\frac{\nu+p}{2}-1}\left(\frac{\nu s^2}{2}\right)^{\frac{-\nu-p}{2}}\left[1+(\beta-\mu)^T\frac{H}{\nu s^2}(\beta-\mu)\right]^{\frac{-\nu-p}{2}} e^{-h} \quad .
\end{aligned}$$

If we pull the constants outside the integral and simplify, we obtain:

$$\begin{aligned}
\{\beta\} &= \frac{2|H|^{\frac{1}{2}}(2\pi)^{-\frac{p}{2}}\left(\frac{\nu s^2}{2}\right)^{\frac{\nu}{2}}}{\Gamma\left(\frac{\nu}{2}\right)} \\
&\times \int_{\infty}^{0} -\frac{1}{2}dhh^{\frac{\nu+p}{2}-1}\left(\frac{\nu s^2}{2}\right)^{\frac{-\nu-p}{2}}\left[1+(\beta-\mu)^T\frac{H}{\nu s^2}(\beta-\mu)\right]^{\frac{-\nu-p}{2}} e^{-h} \\
&= \frac{2|H|^{\frac{1}{2}}(2\pi)^{-\frac{p}{2}}\left(\frac{\nu s^2}{2}\right)^{-\frac{p}{2}}\left[1+(\beta-\mu)^T\frac{H}{\nu s^2}(\beta-\mu)\right]^{\frac{-\nu-p}{2}}}{\Gamma\left(\frac{\nu}{2}\right)} \\
&\times \int_{\infty}^{0} -\frac{1}{2}dhh^{\frac{\nu+p}{2}-1}e^{-h} \\
&= \frac{|H|^{\frac{1}{2}}(2\pi)^{-\frac{p}{2}}\left(\frac{\nu s^2}{2}\right)^{-\frac{p}{2}}\left[1+(\beta-\mu)^T\frac{H}{\nu s^2}(\beta-\mu)\right]^{\frac{-\nu-p}{2}}}{\Gamma\left(\frac{\nu}{2}\right)} \\
&\times \int_{0}^{\infty} dhh^{\frac{\nu+p}{2}-1}e^{-h}
\end{aligned}$$

$$= \frac{|H|^{\frac{1}{2}} 2^{-\frac{p}{2}} \pi^{-\frac{p}{2}} \left(\nu s^2\right)^{-\frac{p}{2}} \left(\frac{1}{2}\right)^{-\frac{p}{2}} \left[1 + (\beta - \mu)^T \frac{H}{\nu s^2} (\beta - \mu)\right]^{\frac{-\nu-p}{2}}}{\Gamma\left(\frac{\nu}{2}\right)}$$

$$\times \Gamma\left(\frac{\nu + p}{2}\right)$$

$$= \frac{|H|^{\frac{1}{2}} \Gamma\left(\frac{\nu+p}{2}\right)}{\left(\pi \nu s^2\right)^{\frac{p}{2}} \Gamma\left(\frac{\nu}{2}\right)} \left[1 + (\beta - \mu)^T \frac{H}{\nu s^2} (\beta - \mu)\right]^{-\frac{\nu+p}{2}} .$$

The final marginal joint distribution over the model coefficients is:

$$\{\beta\} = \frac{|H|^{\frac{1}{2}} \Gamma\left(\frac{\nu+p}{2}\right)}{\left(\pi \nu s^2\right)^{\frac{p}{2}} \Gamma\left(\frac{\nu}{2}\right)} \left[1 + (\beta - \mu)^T \frac{H}{\nu s^2} (\beta - \mu)\right]^{-\frac{\nu+p}{2}} . \tag{4.15}$$

Is this multivariate density a function of any family of distributions? You bet it is! The multivariate Student's t distribution is known to have the form:[7]

$$T_n(r; \mu, \Omega, \nu) = \frac{\Gamma\left(\frac{\nu+n}{2}\right) |\Omega|^{-\frac{1}{2}}}{\Gamma\left(\frac{\nu}{2}\right) (\pi \nu)^{\frac{n}{2}}} \left[1 + (r - \mu)^T \frac{\Omega^{-1}}{\nu} (r - \mu)\right]^{-\frac{\nu+n}{2}} , \tag{4.16}$$

for which the variance is

$$\text{variance} = \frac{\nu}{\nu - 2} \Omega . \tag{4.17}$$

If we set $n = p$ in this definition and we set

$$\frac{H}{\nu s^2} = \frac{\Omega^{-1}}{\nu} ,$$

we can simplify as follows

$$\begin{aligned}
&\frac{H}{s^2} = \Omega^{-1} \\
&\Rightarrow \left(\frac{H}{s^2}\right)^{-1} = \Omega \\
&\Rightarrow H^{-1} s^2 = \Omega \\
&\Rightarrow |H|^{-1} s^{2p} = |\Omega| \\
&\Rightarrow |H|^{\frac{1}{2}} \left(s^2\right)^{-\frac{p}{2}} = |\Omega|^{-\frac{1}{2}} .
\end{aligned}$$

The fourth line of the foregoing equation uses the result that the determinant of

[7] See https://en.wikipedia.org/wiki/Multivariate_t-distribution and Kotz and Natarajah[59].

a square matrix times a constant is the constant raised to the pth power times the determinant of the matrix. We then substitute into the standard form of the Student's t distribution (see Kotz and Natarajah)[59]:

$$T_p\left(r;\mu,H,\nu\right)=\frac{\Gamma\left(\frac{\nu+p}{2}\right)|H|^{\frac{1}{2}}}{\Gamma\left(\frac{\nu}{2}\right)\left(\pi\nu s^2\right)^{\frac{p}{2}}}\left[1+\left(r-\mu\right)^T\frac{H}{\nu s^2}\left(r-\mu\right)\right]^{-\frac{\nu+p}{2}} .$$

Ours is a multivariate Student's t distribution over the elements of the vector r. Kotz and Natarajah[59] contains a comprehensive discussion and analysis of multivariate Student's t distributions.

We will, of course, want to calculate the mean and variance-covariance matrix for this marginal distribution over the model coefficients, which are

$$Mean=\mu \quad ,$$

and

$$VCV=\frac{\nu s^2}{\nu-2}H^{-1} \quad ,$$

and we will want to calculate the marginal distributions over every one of the individual model coefficients. We will return to the original formulation of the joint density function in equation (4.8) and integrate to obtain all the sought-after terms.

In practice, Student's t distributions[8] are more difficult to work with than normal or other distributions. They do not integrate analytically. They have no moment generation function, i.e., no Laplace transform or t transform. They do not readily admit of closed-form analytical solutions. One does not want to summarily assume that such a distribution is easily analytically tractable and amenable to easy analytics. We will push the closed-form solutions as far as we can. Such analytical difficulties are presaged by noting that there is no moment generation function for the Student's t distribution and that the characteristic function is a modified Bessel function of the second kind. We need not consider such issues here, which are tangential to our purpose. The Student's t distribution is an ultimate goal for us.

4.4 Mean of Marginal Distribution Over Model Coefficients

We obtain the mean of the foregoing marginal distribution over model coefficients, designated $<\beta>$, by integrating the original conjugate-archetype joint density in equation (4.9), multiplied by β, over both β and σ:

[8] See https://en.wikipedia.org/wiki/Student's_t-distribution and Kotz and Natarajah[59].

$$
\begin{aligned}
<\beta> &= \int d\beta \int d\sigma \beta\{\beta, \sigma\} \\
&= \int_0^\infty d\sigma \int_{-\infty}^\infty d\beta\beta \frac{2\left(\frac{\nu s^2}{2}\right)^{\frac{\nu}{2}}}{\Gamma\left(\frac{\nu}{2}\right)} \sigma^{-(\nu+1)} e^{-\frac{\nu s^2}{2\sigma^2}} N\left(\beta; \mu, H^{-1}\sigma^2\right) \\
&= \frac{2\left(\frac{\nu s^2}{2}\right)^{\frac{\nu}{2}}}{\Gamma\left(\frac{\nu}{2}\right)} \int_0^\infty d\sigma\sigma^{-(\nu+1)} e^{-\frac{\nu s^2}{2\sigma^2}} \int_{-\infty}^\infty d\beta\beta N\left(\beta; \mu, H^{-1}\sigma^2\right) \quad .
\end{aligned}
$$

We see that we have isolated the mean of a multivariate normal in the final expression on the right. That term trivially integrates to the mean value μ.

$$
\begin{aligned}
<\beta> &= \int d\beta \int d\sigma \beta\{\beta, \sigma\} \\
&= \frac{2\left(\frac{\nu s^2}{2}\right)^{\frac{\nu}{2}}}{\Gamma\left(\frac{\nu}{2}\right)} \int_0^\infty d\sigma\sigma^{-(\nu+1)} e^{-\frac{\nu s^2}{2\sigma^2}} \mu \\
&= \mu \frac{2\left(\frac{\nu s^2}{2}\right)^{\frac{\nu}{2}}}{\Gamma\left(\frac{\nu}{2}\right)} \int_0^\infty d\sigma\sigma^{-(\nu+1)} e^{-\frac{\nu s^2}{2\sigma^2}} \quad .
\end{aligned}
$$

We prepare to make the substitutions in the integral by defining a substitution variable h:

$$
\begin{aligned}
h &= \frac{\nu s^2}{2\sigma^2} \Rightarrow \sigma^2 = \left(\frac{\nu s^2}{2}\right) h^{-1} \Rightarrow \sigma = \left(\frac{\nu s^2}{2}\right)^{\frac{1}{2}} h^{-\frac{1}{2}} \\
d\sigma &= -\frac{1}{2}\left(\frac{\nu s^2}{2}\right)^{\frac{1}{2}} h^{-\frac{3}{2}} dh \\
\sigma &= \left(\frac{\nu s^2}{2}\right)^{\frac{1}{2}} h^{-\frac{1}{2}} \\
\sigma^{-(\nu+1)} &= \left(\frac{\nu s^2}{2}\right)^{-\frac{\nu+1}{2}} h^{\frac{\nu+1}{2}} \quad .
\end{aligned}
$$

Making the substitutions in the integral, reversing the limits of integration, and eliminating the minus sign in the integrand as we have done repeatedly in the foregoing developments, we have

$$< \beta >= \mu \frac{2\left(\frac{\nu s^2}{2}\right)^{\frac{\nu}{2}}}{\Gamma\left(\frac{\nu}{2}\right)} \int_0^\infty \frac{1}{2}\left(\frac{\nu s^2}{2}\right)^{\frac{1}{2}} h^{-\frac{3}{2}} dh \left(\frac{\nu s^2}{2}\right)^{-\frac{\nu+1}{2}} h^{\frac{\nu+1}{2}} e^{-h}$$

$$= \mu \frac{1}{\Gamma\left(\frac{\nu}{2}\right)} \int_0^\infty dh h^{\frac{\nu}{2}-1} e^{-h}$$

$$= \mu \quad ,$$

recognizing that the final integral is the same gamma function that is in the denominator. We write the final answer

$$< \beta >= \mu \quad . \tag{4.18}$$

The mean values of the coefficients are the mean values μ of the embedded normal, a most intuitive and satisfying result.

4.5 The Variance-Covariance Matrix of the Conjugate-archetype Density

The variance-covariance[9] term requires a few more steps but is very similar in approach to computing the mean. We begin with the definition of the variance-covariance matrix of the conjugate-archetype density in equation (4.9) above. We have to insert into the integrand the "outer product" quadratic term $(\beta - \mu)(\beta - \mu)^T$:

$$VCV = \int_0^\infty d\sigma \int_{-\infty}^\infty d\beta \, (\beta - \mu) \, (\beta - \mu)^T \{\beta, \sigma\}$$

$$= \int_0^\infty d\sigma \int_{-\infty}^\infty d\beta \, (\beta - \mu) \, (\beta - \mu)^T \frac{2\left(\frac{\nu s^2}{2}\right)^{\frac{\nu}{2}}}{\Gamma\left(\frac{\nu}{2}\right)} \sigma^{-(\nu+1)} e^{-\frac{\nu s^2}{2\sigma^2}} N\left(\beta; \mu, H^{-1}\sigma^2\right)$$

$$= \int_0^\infty d\sigma \frac{2\left(\frac{\nu s^2}{2}\right)^{\frac{\nu}{2}}}{\Gamma\left(\frac{\nu}{2}\right)} \sigma^{-(\nu+1)} e^{-\frac{\nu s^2}{2\sigma^2}} \int_0^\infty d\beta \, (\beta - \mu) \, (\beta - \mu)^T N\left(\beta; \mu, H^{-1}\sigma^2\right) \quad .$$

As before, we have manipulated this equation so that the final term on the right is the integral equation that defines the variance-covariance matrix of the multivariate normal density. It integrates into the variance-covariance matrix, which

[9]Often in the literature, we see the terminology "covariance matrix." We prefer the term variance-covariance matrix because the variance is going to have a special place in our subsequent analysis when we examine marginal distributions.

we see in the argument that the normal distribution is $H^{-1}\sigma^2$. Therefore, we can write:

$$
\begin{aligned}
VCV &= \int_0^\infty d\sigma \frac{2\left(\frac{\nu s^2}{2}\right)^{\frac{\nu}{2}}}{\Gamma\left(\frac{\nu}{2}\right)} \sigma^{-(\nu+1)} e^{-\frac{\nu s^2}{2\sigma^2}} H^{-1}\sigma^2 \\
&= H^{-1} \frac{2\left(\frac{\nu s^2}{2}\right)^{\frac{\nu}{2}}}{\Gamma\left(\frac{\nu}{2}\right)} \int_0^\infty d\sigma \sigma^{-(\nu-1)} e^{-\frac{\nu s^2}{2\sigma^2}} \quad .
\end{aligned}
$$

We need to make substitutions in the final integral to express it in closed form. In particular, if we substitute

$$
\begin{aligned}
h &= \frac{\nu s^2}{2\sigma^2} \Rightarrow \sigma^2 = \left(\frac{\nu s^2}{2}\right) h^{-1} \Rightarrow \sigma = \left(\frac{\nu s^2}{2}\right)^{\frac{1}{2}} h^{-\frac{1}{2}} \\
d\sigma &= -\frac{1}{2}\left(\frac{\nu s^2}{2}\right)^{\frac{1}{2}} h^{-\frac{3}{2}} dh \\
\sigma^{-(\nu-1)} &= \left(\frac{\nu s^2}{2}\right)^{-\frac{\nu-1}{2}} h^{\frac{\nu-1}{2}} \quad ,
\end{aligned}
$$

we note that the limits of the integral are reversed (∞ to 0), so we can reverse those limits and eradicate the minus sign in the differential $d\sigma$ to write

$$
\begin{aligned}
VCV &= H^{-1} \frac{2\left(\frac{\nu s^2}{2}\right)^{\frac{\nu}{2}}}{\Gamma\left(\frac{\nu}{2}\right)} \int_0^\infty \frac{1}{2}\left(\frac{\nu s^2}{2}\right)^{\frac{1}{2}} h^{-\frac{3}{2}} dh \left(\frac{\nu s^2}{2}\right)^{-\frac{\nu-1}{2}} h^{\frac{\nu-1}{2}} e^{-h} \\
&= H^{-1} \frac{\left(\frac{\nu s^2}{2}\right)}{\Gamma\left(\frac{\nu}{2}\right)} \int_0^\infty dh h^{\frac{\nu-2}{2}-1} e^{-h} \\
&= H^{-1} \frac{\left(\frac{\nu s^2}{2}\right)}{\Gamma\left(\frac{\nu}{2}\right)} \Gamma\left(\frac{\nu}{2}-1\right) \quad .
\end{aligned}
$$

We need to use the recursion relationship for the gamma function introduced previously

$$
\Gamma(a) = (a-1)\Gamma(a-1) \quad ,
$$

to show in the denominator that

$$
\Gamma\left(\frac{\nu}{2}\right) = \left(\frac{\nu}{2}-1\right)\Gamma\left(\frac{\nu}{2}-1\right) \quad ,
$$

and thus that

$$\begin{aligned}
VCV &= H^{-1}\frac{\left(\frac{\nu s^2}{2}\right)}{\Gamma\left(\frac{\nu}{2}\right)}\Gamma\left(\frac{\nu}{2}-1\right)\\
&= H^{-1}\frac{\left(\frac{\nu s^2}{2}\right)}{\left(\frac{\nu}{2}-1\right)\Gamma\left(\frac{\nu}{2}-1\right)}\Gamma\left(\frac{\nu}{2}-1\right)\\
&= H^{-1}s^2\frac{\nu}{\nu-2} \quad .
\end{aligned}$$

We write the final answer

$$VCV = H^{-1}s^2\frac{\nu}{\nu-2} = H^{-1}\frac{g}{\nu-2} \quad . \tag{4.19}$$

The precision matrix is the inverse of the VCV matrix.[10] This allows us to relate the matrix H to the variance covariance matrix

$$H = VCV^{-1}\frac{\nu s^2}{\nu-2} = VCV^{-1}\frac{g}{\nu-2} \quad . \tag{4.20}$$

In the words of Raiffa and Schlaiffer[81], the variance-covariance matrix is proportional to the inverse of the matrix H.[11] The precision matrix is the matrix H that appears in the quadratic term in the multivariate normal distribution. We are going to see that the multiplicative constant that multiplied H^{-1} is extremely significant; it will be the mean of the marginal distribution over σ^2.

We will see in Chapter 15 now nicely and directly this calculation generalizes to the case in which we want to consider multivariate rather than univariate probabilistic models.

4.6 The Marginal Density Over the Standard Deviation Term σ

The previous development considered the marginal (joint) density over the model coefficients β. There is another element of the density that merits attention as well, namely the marginal density over the standard deviation term σ. This section develops the mean and variance of the marginal probability distribution over the standard deviation term, beginning with the joint density in equation (4.9) and integrating out the model parameter term β. It is straightforward to integrate out the model coefficients β from equation (4.9) to obtain the marginal distribution over σ.

[10] See https://en.wikipedia.org/wiki/Precision_(statistics.)

[11] Technically, the precision matrix is the inverse of the variance-covariance matrix. That is, Precision $= VCV^{-1} = H \times \frac{\nu-2}{g}$. We sometimes refer to H as the precision matrix when that technically isn't correct. No confusion should result.

$$\begin{aligned}
\{\sigma\} &= \int_{-\infty}^{\infty} d\beta\,\{\beta,\sigma\} = \int_{-\infty}^{\infty} d\beta \frac{2\left(\frac{\nu s^2}{2}\right)^{\frac{\nu}{2}}}{\Gamma\left(\frac{\nu}{2}\right)}\sigma^{-(\nu+1)}e^{-\frac{\nu s^2}{2\sigma^2}} N\left(\beta;\mu,H^{-1}\sigma^2\right) \\
&= \frac{2\left(\frac{\nu s^2}{2}\right)^{\frac{\nu}{2}}}{\Gamma\left(\frac{\nu}{2}\right)}\sigma^{-(\nu+1)}e^{-\frac{\nu s^2}{2\sigma^2}} \int_{-\infty}^{\infty} d\beta N\left(\beta;\mu,H^{-1}\sigma^2\right) \\
&= \frac{2\left(\frac{\nu s^2}{2}\right)^{\frac{\nu}{2}}}{\Gamma\left(\frac{\nu}{2}\right)}\sigma^{-(\nu+1)}e^{-\frac{\nu s^2}{2\sigma^2}} \quad .
\end{aligned}$$

This means that the marginal density over the standard deviation term σ looks like a gamma density function governed by the equation

$$\{\sigma\} = \frac{2\left(\frac{\nu s^2}{2}\right)^{\frac{\nu}{2}}}{\Gamma\left(\frac{\nu}{2}\right)}\sigma^{-(\nu+1)}e^{-\frac{\nu s^2}{2\sigma^2}} \quad . \tag{4.21}$$

We will relate this to an inverse gamma distribution after we calculate its mean and variance analytically and directly.[12] This expression is exactly consistent with the density on the right hand side of equation (4.9), and we will shortly show that it is an inverse gamma function in the variable σ^2.

4.7 Mean of the Marginal Distribution Over Square of Standard Deviation Term

We shall calculate the mean and the variance of the foregoing distribution in standard fashion. However, we are not going to focus on the standard deviation σ per se. Often in statistics classes, we are asked to find the mean and standard deviation of a variable, σ, that is governed by a distribution over $\sigma > 0$. Instead, we want to develop the corresponding distribution over the square of the standard deviation, i.e., σ^2. Thereafter, we want to calculate the mean value of that distribution over the square and the variance of that distribution over the square. We use the terms "mean of the variance" and "variance of the variance" to articulate what we are computing in this section. The conversion of the distribution from σ to σ^2 will occur within the integral substitutions used in calculating the key results. We will note the form of those distributions, which are related to inverse gamma distributions, as we proceed. It is worth emphasizing that the standard deviation is defined as the square root of the variance, and conversely, the variance is the square of the standard deviation. We are calculating the average value

[12] This will be a very important discussion, and it anticipates the discussion in Chapter 15 for the multivariate case, which is an extension from an inverse gamma distribution to an inverse Wishart distribution. Understanding this as an inverse gamma distribution helps us move intellectually to the multivariate case and the Wishart, which are more difficult.

of the variance and the variance of the variance. When we focus on the expectation and variance of σ^2, the math is slightly simpler, and the insight is much deeper. In this section, then, we will calculate the mean value of the variance, namely $\langle \sigma^2 \rangle$, directly by integration. We will not resort to the inverse gamma interpretation quite yet but instead perform the integration calculation directly without substituting an intermediate form. Substitution to an intermediate form adds no real insight that we have not already obtained by unearthing the inverse gamma distribution. The expected value of σ^2 is

$$
\begin{aligned}
<\sigma^2> &\equiv \int_0^\infty d\sigma \sigma^2 \{\sigma\} \\
&= \int_0^\infty d\sigma \sigma^2 \frac{2\left(\frac{\nu s^2}{2}\right)^{\frac{\nu}{2}}}{\Gamma\left(\frac{\nu}{2}\right)} \sigma^{-(\nu+1)} e^{-\frac{\nu s^2}{2\sigma^2}} \\
&= \frac{2\left(\frac{\nu s^2}{2}\right)^{\frac{\nu}{2}}}{\Gamma\left(\frac{\nu}{2}\right)} \int_0^\infty d\sigma \sigma^{-(\nu-1)} e^{-\frac{\nu s^2}{2\sigma^2}} \quad .
\end{aligned}
$$

We want to make the same substitution we have made previously throughout this section, namely

$$
\begin{aligned}
h &= \frac{\nu s^2}{2\sigma^2} \Rightarrow \sigma^2 = \left(\frac{\nu s^2}{2}\right) h^{-1} \Rightarrow \sigma = \left(\frac{\nu s^2}{2}\right)^{\frac{1}{2}} h^{-\frac{1}{2}} \\
d\sigma &= -\frac{1}{2}\left(\frac{\nu s^2}{2}\right)^{\frac{1}{2}} h^{-\frac{3}{2}} dh \\
\sigma^{-(\nu-1)} &= \left(\frac{\nu s^2}{2}\right)^{-\frac{\nu-1}{2}} h^{\frac{\nu-1}{2}} \quad ,
\end{aligned}
$$

we note that the limits of the integral are reversed (∞ to 0), so we can reverse those limits and eradicate the minus sign in the differential $d\sigma$ to write:

$$
\begin{aligned}
<\sigma^2> &= \frac{\left(\frac{\nu s^2}{2}\right)}{\Gamma\left(\frac{\nu}{2}\right)} \int_0^\infty dh h^{\left(\frac{\nu}{2}-1\right)-1} e^{-h} \\
&= \frac{\left(\frac{\nu s^2}{2}\right)}{\Gamma\left(\frac{\nu}{2}\right)} \Gamma\left(\frac{\nu}{2}-1\right) = \frac{\left(\frac{\nu s^2}{2}\right)}{\left(\frac{\nu}{2}-1\right)\Gamma\left(\frac{\nu}{2}-1\right)} \Gamma\left(\frac{\nu}{2}-1\right) \\
&= s^2 \frac{\nu}{\nu-2} \quad .
\end{aligned}
$$

We rewrite this result, which is the mean of the inverse gamma over σ^2, for future reference

$$<\sigma^2>=s^2\frac{\nu}{\nu-2} \quad . \tag{4.22}$$

The variance-covariance matrix in equation (4.19) is the inverse of the H matrix times this mean value of σ^2. We will see the direct analogy of this result emerge when, in Chapter 15, we generalize to joint probability distributions with two or more outputs. The generalization involves calculating the mean of a Wishart distribution as a multivariate extension of an inverse gamma distribution.

4.8 Variance of the Marginal Distribution Over Standard Deviation Term

The second moment of the distribution over the variance is by definition

$$<\left(\sigma^2\right)^2>=\int\limits_0^\infty \sigma^4 d\sigma \frac{2\left(\frac{\nu s^2}{2}\right)^{\frac{\nu}{2}}}{\Gamma\left(\frac{\nu}{2}\right)}\sigma^{-(\nu+1)}e^{-\frac{\nu s^2}{2\sigma^2}}$$

$$=\frac{2\left(\frac{\nu s^2}{2}\right)^{\frac{\nu}{2}}}{\Gamma\left(\frac{\nu}{2}\right)}\int\limits_0^\infty d\sigma\sigma^{-(\nu-3)}e^{-\frac{\nu s^2}{2\sigma^2}} \quad .$$

We need to make the same substitution we have made previously throughout this chapter, which reverses the limits on the integral. When we re-reverse them, we eradicate the negative sign on the $d\sigma$ term. We make the same substitution:

$$h=\frac{\nu s^2}{2\sigma^2}\Rightarrow\sigma^2=\left(\frac{\nu s^2}{2}\right)h^{-1}\Rightarrow\sigma=\left(\frac{\nu s^2}{2}\right)^{\frac{1}{2}}h^{-\frac{1}{2}}$$

$$d\sigma=-\frac{1}{2}\left(\frac{\nu s^2}{2}\right)^{\frac{1}{2}}h^{-\frac{3}{2}}dh$$

$$\sigma^{-(\nu-3)}=\left(\frac{\nu s^2}{2}\right)^{-\frac{\nu-3}{2}}h^{\frac{\nu-3}{2}} \quad .$$

With this substitution, we obtain the result

$$<\left(\sigma^2\right)^2>=\frac{\left(\frac{\nu s^2}{2}\right)^2}{\Gamma\left(\frac{\nu}{2}\right)}\int\limits_0^\infty dhh^{\left(\frac{\nu}{2}-2\right)-1}e^{-h}$$

$$=\frac{\left(\frac{\nu s^2}{2}\right)^2}{\Gamma\left(\frac{\nu}{2}\right)}\Gamma\left(\frac{\nu}{2}-2\right)=\frac{\left(\frac{\nu s^2}{2}\right)^2}{\left(\frac{\nu}{2}-1\right)\Gamma\left(\frac{\nu}{2}-1\right)}\Gamma\left(\frac{\nu}{2}-2\right)$$

$$= \frac{\left(\frac{\nu s^2}{2}\right)^2}{\left(\frac{\nu}{2}-1\right)\left(\frac{\nu}{2}-2\right)\Gamma\left(\frac{\nu}{2}-2\right)}\Gamma\left(\frac{\nu}{2}-2\right) = \left(s^2\right)^2 \frac{\nu}{\nu-2}\frac{\nu}{\nu-4} \quad .$$

The variance of σ^2 is thus

$$\begin{aligned}
\left\langle \left(\sigma^2\right)^2 \right\rangle - \left\langle \sigma^2 \right\rangle^2 &= \left(s^2\right)^2 \frac{\nu}{\nu-2}\frac{\nu}{\nu-4} - s^2\frac{\nu}{\nu-2}s^2\frac{\nu}{\nu-2} \\
&= \left(s^2\right)^2 \frac{\nu^2}{\nu-2}\left(\frac{1}{\nu-4} - \frac{1}{\nu-2}\right) \\
&= \left(s^2\right)^2 \frac{\nu^2}{(\nu-2)}\left[\frac{2}{(\nu-2)(\nu-4)}\right] \\
&= \frac{2}{\nu-4}\left(\frac{\nu s^2}{\nu-2}\right)^2 \\
&= \frac{2}{\nu-4}\left\langle \sigma^2 \right\rangle^2 \quad .
\end{aligned}$$

We rewrite this for ease of future reference

$$\begin{aligned}
< \left(\sigma^2\right)^2 > - < \sigma^2 >^2 &= \frac{2}{\nu-4}\left(s^2\frac{\nu}{\nu-2}\right)^2 \\
&= \frac{2}{\nu-4} < \sigma^2 >^2 \quad .
\end{aligned} \tag{4.23}$$

This is precisely the variance derived for the inverse gamma. It is not the variance derived in Chapter 2 for the classical regression case, which was based on a χ^2 distribution. In this section, under Bayes, the error distribution has become an inverse gamma distribution rather than a χ^2 distribution, which is discussed in great detail shortly. Note that this equation holds for degrees of freedom strictly larger than 4. The practical implications of this will become clear when we analyze the prior and the posterior. For now, whenever the degrees of freedom is 4 or smaller, the solution is undefined and the method does not work. The generalization to the multivariate case in Chapter 15 is not at all trivial, requiring many pages of matrix algebra and calculus.

4.9 Sidebar: What Kind of Distribution Is the Marginal Density Over σ in Equation 4.21?

Equation (4.21), reproduced here, is the following

$$\{\sigma\} = \frac{2\left(\frac{\nu s^2}{2}\right)^{\frac{\nu}{2}}}{\Gamma\left(\frac{\nu}{2}\right)}\sigma^{-(\nu+1)}e^{-\frac{\nu s^2}{2\sigma^2}} \quad .$$

Does this imply an inverse gamma distribution, reproduced from Wiki:[13]

$$f(x) = \frac{\beta^{\alpha}}{\Gamma(\alpha)} x^{-(\alpha+1)} e^{-\frac{\beta}{x}} \quad .$$

Is our distribution really an inverse gamma distribution? Yes; inspection makes it clear that it is an inverse gamma function in σ^2. It is clear, comparing the forms of the foregoing two equations, what type of density this is. We develop the answer to the second moment of σ in two ways: (1.) a simple change of variables problem using integrals, and (2.) a change of variables problem using the Jacobian. To begin the former, we set the integral equal to 1 and begin substituting to examine the embedded density.

$$1 = \int_0^{\infty} \frac{2\left(\frac{\nu s^2}{2}\right)^{\frac{\nu}{2}}}{\Gamma\left(\frac{\nu}{2}\right)} \sigma^{-(\nu+1)} e^{-\frac{\nu s^2}{2\sigma^2}} d\sigma \quad .$$

Let us make the following substitution

$$\sigma^2 = x \quad .$$

This leads to a series of substitution results

$$\begin{aligned} \sigma &= x^{\frac{1}{2}} \\ d\sigma &= \frac{1}{2} x^{-\frac{1}{2}} dx \\ \sigma^{-(\nu+1)} &= x^{-\frac{\nu+1}{2}} \\ \sigma = 0 &\Rightarrow x = 0 \\ \sigma = \infty &\Rightarrow x = \infty \quad . \end{aligned}$$

We now make all these substitutions in the integral and simplify

$$\begin{aligned} 1 &= \int_0^{\infty} \frac{2\left(\frac{\nu s^2}{2}\right)^{\frac{\nu}{2}}}{\Gamma\left(\frac{\nu}{2}\right)} x^{-\frac{\nu+1}{2}} e^{-\frac{\nu s^2}{2x}} \left(\frac{1}{2} x^{-\frac{1}{2}} dx\right) \\ &= \int_0^{\infty} \frac{\left(\frac{\nu s^2}{2}\right)^{\frac{\nu}{2}}}{\Gamma\left(\frac{\nu}{2}\right)} x^{-\frac{\nu}{2}-1} e^{-\frac{\nu s^2}{2x}} dx \quad . \end{aligned}$$

We are nearing the insight. Suppose we define

$$\alpha = \frac{\nu}{2}$$

[13] See https://en.wikipedia.org/wiki/Inverse-gamma_distribution.

$$B = \alpha s^2 = \frac{\nu s^2}{2} \quad .$$

Substituting into the foregoing integral gives the following equation

$$1 = \int_0^\infty \frac{B^\alpha}{\Gamma(\alpha)} x^{-(\alpha+1)} e^{-\frac{B}{x}} dx \quad .$$

What is the integrand? **It is an inverse gamma distribution in x.** It is not a χ^2 distribution at all. If we examine the equation for the inverse gamma distribution in x in Wiki,[14] we see exactly the equation in this integrand. This tells us that the distribution we have been working with is truly an inverse gamma distribution in the variable x. But what is the variable x? It is σ^2! It is the variance term! We have been working with an inverse gamma distribution in the error term variance variable σ^2. This inverse gamma differs conceptually from the χ^2 distribution we saw with classical statistics. The old notion that the σ term should be governed by a χ^2 distribution is no longer pertinent. Bayes is different and more general, even when it emulates classical statistics. With Bayes, we do all the math and examine the result at the end, not making arbitrary assumptions or assertions along the way. We plot the inverse gamma distribution in Figure 4.1. We might be tempted to say that the χ^2 assumption in classical regression is "wrong." We would not say that, but we would say that the Bayesian development herein is preferable, more general, and much more lucid about what you actually bring to the problem with classical statistics, and what you actually bring to the problem is not as satisfying. Lucidity about exactly what we bring to the problem in classical as well as Bayesian statistics is profoundly important.

We know from Wiki that the mean of this distribution is

$$\mu = \frac{B}{\alpha - 1}$$

and the variance is

$$V = \frac{B^2}{(\alpha-1)^2(\alpha-2)} = \frac{\mu^2}{\alpha - 2} \quad .$$

We can solve for the parameters α and B by manipulating these two equations to write:

$$V = \frac{\mu^2}{\alpha - 2}$$
$$\Rightarrow \alpha = \frac{\mu^2}{V} + 2 = \frac{\mu^2 + 2V}{V} = t^2 + 2$$

and

[14] See https://en.wikipedia.org/wiki/Inverse-gamma_distribution.

Figure 4.1: Inverse Gamma Distribution with $B = 1$ and Various Values of α

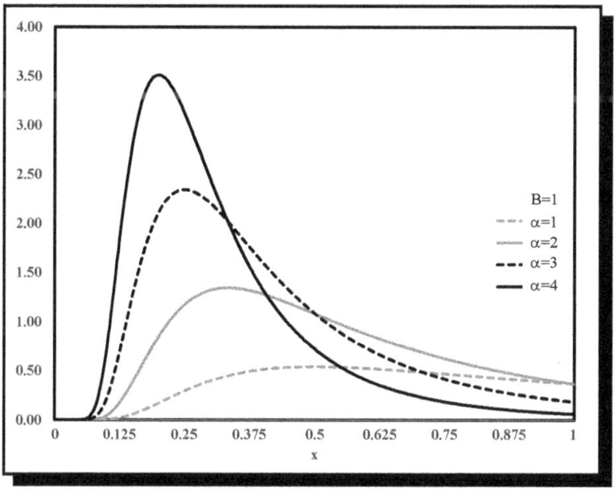

$$\mu = \frac{B}{\alpha - 1}$$
$$\Rightarrow B = \mu \frac{\mu^2 + V}{V} = \mu \left(t^2 + 1\right) \quad .$$

The second approach to determining what type of density this is appeals to the single variable change of variables theorem.[15] If we have a density over σ, namely

$$f(\sigma)$$

and we want a density over a variable y that is a monotonic function of σ, i.e.,

$$y = g(\sigma) \Rightarrow \sigma = g^{-1}(y) \triangleq h(y) \quad ,$$

then that new density is

$$\phi(y) = \left| \frac{\partial h(y)}{\partial y} \right| f[h(y)] \quad .$$

Applying this theorem to the marginal density over σ we developed in equation (4.21), we want to consider a variable

$$y = g(\sigma) = \sigma^2 \quad ,$$

which implies the following

[15]See https://en.wikipedia.org/wiki/Probability_density_function.

$$\sigma = y^{\frac{1}{2}} \Rightarrow g^{-1}(y) = h(y) = y^{\frac{1}{2}} \Rightarrow \frac{\partial h(y)}{\partial y} = \frac{1}{2} y^{-\frac{1}{2}} \quad .$$

The density we want to transform, the marginal density over σ that we derived, is that in equation (4.21). If we make the foregoing change of variables, we see that the density over y is computed to be

$$\phi(y) = \left| \frac{\partial h(y)}{\partial y} \right| f[h(y)] = \frac{\left(\frac{\nu s^2}{2}\right)^{\frac{\nu}{2}}}{\Gamma\left(\frac{\nu}{2}\right)} \left(\frac{1}{y}\right)^{\frac{\nu}{2}+1} e^{-\frac{\nu s^2}{2}\frac{1}{y}} \quad .$$

The inverse gamma function[16] is written

$$\phi(x) = \frac{B^{\alpha}}{\Gamma(\alpha)} \left(\frac{1}{x}\right)^{\alpha+1} e^{-\frac{B}{x}} \quad .$$

Matching up terms, we see that our derived distribution is indeed an inverse gamma distribution in x with parameters

$$\alpha = \frac{\nu}{2}$$
$$B = \frac{\nu s^2}{2} \quad .$$

Wiki[17] gives the mean and variance of the inverse gamma distribution to be respectively

$$Mean = \frac{B}{\alpha - 1} = \frac{\nu s^2}{\nu - 2} \tag{4.24}$$

and

$$Variance = \left(\frac{B}{\alpha - 1}\right)^2 \frac{1}{\alpha - 2} = Mean^2 \frac{2}{\nu - 4} \quad . \tag{4.25}$$

4.9.1 Sidebar–Relationship to the t-statistic

Let us define the mean to be M and the standard deviation to be S. We can write the variance equation, equation (4.25), as follows

$$S^2 = M^2 \frac{2}{\nu - 4} \quad .$$

We solve for ν

[16] See https://en.wikipedia.org/wiki/Inverse-gamma_distribution.
[17] See https://en.wikipedia.org/wiki/Inverse-gamma_distribution.

$$\nu = 2\frac{M^2}{S^2} + 4 \quad .$$

We consider that the t-statistic t, the mean over the standard deviation, allows us to express the ν term strictly in terms of t, the ratio of the mean to the standard deviation

$$\nu = 2t^2 + 4 \quad .$$

We move to the first equation, equation (4.24), and write

$$M = s^2\frac{\nu}{\nu - 2} \quad .$$

We can solve for s^2 as follows

$$s^2 = M\frac{\nu - 2}{\nu} \quad . \tag{4.26}$$

We can write an expression for the term on the right of M on the right

$$\nu = 2t^2 + 4 \Rightarrow \nu - 2 = 2t^2 + 2 \Rightarrow \frac{\nu - 2}{\nu} = \frac{t^2 + 1}{t^2 + 2} \quad .$$

We can thus write an expression for s^2

$$s^2 = M\frac{t^2 + 1}{t^2 + 2} \quad .$$

This allows us to write the expression for νs^2 as follows

$$\nu s^2 = \left(2t^2 + 4\right) M\frac{t^2 + 1}{t^2 + 2} = 2M\left(t^2 + 1\right) \quad .$$

In order to subjectively specify the inverse gamma function, all we would need is the mean M and the t-statistic.

4.9.2 Sidebar–Reciprocal of the Inverse Gamma

It is also valuable to discover what this inverse gamma distribution implies about its own reciprocal based on the same parameters α and β. In particular, if we begin with the inverse gamma distribution

$$f(x) = \frac{B^\alpha}{\Gamma(\alpha)}x^{-(\alpha+1)}e^{-\frac{B}{x}} \quad .$$

We know that this distribution integrates to unity

$$1=\int_0^\infty \frac{B^\alpha}{\Gamma(\alpha)}x^{-(\alpha+1)}e^{-\frac{B}{x}}dx \quad .$$

Making the substitutions

$$z=\frac{1}{x}=x^{-1}\Rightarrow x=z^{-1}\Rightarrow dx=-z^{-2}dz\Rightarrow x^{-(\alpha+1)}=z^{(\alpha+1)} \quad ,$$

we can write the foregoing integral

$$1=\int_0^\infty \frac{B^\alpha}{\Gamma(\alpha)}z^{\alpha-1}e^{-\beta z}dz \quad .$$

If we examine the integrand, we see the distribution is a regular gamma distribution

$$f(z)=\frac{B^\alpha}{\Gamma(\alpha)}z^{\alpha-1}e^{-\beta z}$$

which is what we would expect. This distribution is, of course, not a direct "division" of the inverse gamma. It is more complex than that, as probability invariably is.

4.10 Back to the Inverse Gamma

We will develop these results from first principles below. Using these foregoing three expressions, we are able to write the inverse gamma distribution in $z=\sigma^2$ as follows

$$IG\left(z,\frac{\nu s^2}{2},\frac{\nu}{2}\right)=\frac{\left(\frac{\nu s^2}{2}\right)^{\frac{\nu}{2}}}{\Gamma\left(\frac{\nu}{2}\right)}\left(\frac{1}{z}\right)^{\frac{\nu}{2}+1}e^{-\frac{\nu s^2}{2z}} \quad . \tag{4.27}$$

The foregoing mean and variance over σ^2 continue to hold. In the forthcoming developments in this section, we will not dwell on this distribution, which is in fact an inverse gamma distribution over σ^2. There is no need to. We will continue to initiate all integrations with the density as written without trying to name it, relate it directly to an inverse gamma, or take any shortcuts. When we get to the multiple regression case in Chapter 15, we will not be able to do so quite so easily. For now, it is much more conceptually elegant to integrate the distributions in their most elemental forms.

4.11 Summary: Mean and Variance-Covariance Matrix of the Conjugate-Archetype Distribution

The mean and variance-covariance matrix of β and σ^2 are thus summarized as follows

$$\text{Mean} = \begin{bmatrix} \mu \\ s^2 \frac{\nu}{\nu-2} \end{bmatrix}$$

and

$$\text{Variance-Covariance} = \begin{bmatrix} H^{-1} s^2 \frac{\nu}{\nu-2} & 0 \\ 0 & \frac{2}{\nu-4}\left(s^2 \frac{\nu}{\nu-2}\right)^2 \end{bmatrix} ,$$

in which the degrees of freedom ν is defined by equation (4.4) and the variance term s^2 is given by equation (4.6). To be clear, we are considering the square, σ^2, not σ, in our consideration of the error term here.

We can see that the covariance terms between all of the β terms and the σ term are calculated to be zero just like classical statistics. And what does zero covariance signal? It signals that in the conjugate-archetype distribution, β and σ are probabilistically independent, i.e., mutually irrelevant. The zero covariances in the variance-covariance matrix signal probabilistic independence between every one of the β elements and the σ term. We shall see this independence persist in the multivariate extension of these results in Chapter 15.

4.12 The Marginal Distributions Over Individual Model Coefficients

What is the marginal distribution over each of the individual model coefficients? This is one of the most important parameters of the density that we must understand. It may be difficult to develop a closed form for it, but it is a closed form we really want. The closed form developed here allows great ease in programming, visualization, and interpretation, ultimately allowing marginal distributions over individual coefficients (or subsets of coefficients). The closed form developed here will involve the partitioning of the precision matrix into a subset of the coefficients we want to keep and a complementary subset of the coefficients that we want to integrate out as we form our marginal density.

We begin by partitioning the vector of model coefficients into two parts. Beginning with equation (4.8), we define the partitioned vector and matrix

$$\beta - \mu = \begin{bmatrix} \beta_1 - \mu_1 \\ \beta_2 - \mu_2 \end{bmatrix} = \begin{bmatrix} \xi_1 \\ \xi_2 \end{bmatrix} ,$$

and

$$H = \begin{bmatrix} H_{11} & H_{12} \\ H_{21} & H_{22} \end{bmatrix} \quad .$$

We think of the partition as being of length L from index 1 to index L, and therefore the second partition is of length $p - L$ between index $L + 1$ and index p. That means that the vector μ is broken into two subvectors μ_1 of length L from $1, ..., L$ and μ_2 of length $p - L$ from $L + 1, ..., p$. The reason we do so is that we want to derive the marginal distribution by integrating out all the variables in the first partition of length L to leave the marginal distribution over the variables in the second partition of length $p - L$. To do so, we will have to group the variables in the first partition of length L and group the variables in the second partition of length $p - L$. Using this notation, we can write the quadratic term in equation (4.8) using the partitioned matrix and vector by first writing the right-hand product

$$\begin{aligned} H\xi &= \begin{bmatrix} H_{11} & H_{12} \\ H_{21} & H_{22} \end{bmatrix} \begin{bmatrix} \xi_1 \\ \xi_2 \end{bmatrix} = \begin{bmatrix} H_{11}\xi_1 + H_{12}\xi_2 \\ H_{21}\xi_1 + H_{22}\xi_2 \end{bmatrix} \\ \xi^T H\xi &= \begin{bmatrix} \xi_1{}^T & \xi_2{}^T \end{bmatrix} \begin{bmatrix} H_{11}\xi_1 + H_{12}\xi_2 \\ H_{21}\xi_1 + H_{22}\xi_2 \end{bmatrix} \\ &= \xi_1{}^T H_{11}\xi_1 + \xi_1{}^T H_{12}\xi_2 + \xi_2{}^T H_{21}\xi_1 + \xi_2{}^T H_{22}\xi_2 \quad . \end{aligned}$$

We add and subtract a term (in the second line) that allows us to factor this expression for the exponent conveniently

$$\begin{aligned} &(\beta - \mu)^T H (\beta - \mu) \\ &= \xi_1{}^T H_{11}\xi_1 + \xi_1{}^T H_{12}\xi_2 + \xi_2{}^T H_{21}\xi_1 + \xi_2{}^T H_{22}\xi_2 \\ &- \xi_2{}^T H_{21} H_{11}{}^{-1} H_{12}\xi_2 + \xi_2{}^T H_{21} H_{11}{}^{-1} H_{12}\xi_2 \quad . \end{aligned}$$

Regrouping

$$\begin{aligned} &(\beta - \mu)^T H (\beta - \mu) \\ &= \xi_1{}^T H_{11}\xi_1 + \xi_1{}^T H_{12}\xi_2 + \xi_2{}^T H_{21}\xi_1 + \xi_2{}^T H_{21} H_{11}{}^{-1} H_{12}\xi_2 \\ &+ \xi_2{}^T \left(H_{22} - H_{21} H_{11}{}^{-1} H_{12}\right) \xi_2 \quad . \end{aligned}$$

We extract the middle line in the equation and write it independently

$$T = \xi_1{}^T H_{11}\xi_1 + \xi_1{}^T H_{12}\xi_2 + \xi_2{}^T H_{21}\xi_1 + \xi_2{}^T H_{21} H_{11}{}^{-1} H_{12}\xi_2 \quad .$$

We transpose the two H_{21} matrices to the form H_{12}^T in the third and fourth terms to write

$$T = \xi_1{}^T H_{11}\xi_1 + \xi_1{}^T H_{12}\xi_2 + \xi_2{}^T H_{12}{}^T \xi_1 + \xi_2{}^T H_{12}{}^T H_{11}{}^{-1} H_{12}\xi_2 \quad .$$

We then insert three carefully placed terms that express the identity matrix as H_{11} times its inverse

$$
\begin{aligned}
T &= {\xi_1}^T H_{11} \xi_1 + {\xi_1}^T \left(H_{11} {H_{11}}^{-1}\right) H_{12} \xi_2 \\
&+ {\xi_2}^T {H_{12}}^T \left({H_{11}}^{-1} H_{11}\right) \xi_1 + {\xi_2}^T {H_{12}}^T {H_{11}}^{-1} \left(H_{11} {H_{11}}^{-1}\right) H_{12} \xi_2 \quad .
\end{aligned}
$$

We then group the first two terms and the second two terms

$$
\begin{aligned}
T &= \left({\xi_1}^T H_{11} \xi_1 + {\xi_1}^T H_{11} {H_{11}}^{-1} H_{12} \xi_2\right) \\
&+ \left({\xi_2}^T {H_{12}}^T {H_{11}}^{-1} H_{11} \xi_1 + {\xi_2}^T {H_{12}}^T {H_{11}}^{-1} H_{11} {H_{11}}^{-1} H_{12} \xi_2\right) \quad .
\end{aligned}
$$

Each of the terms is factorable

$$
\begin{aligned}
T &= {\xi_1}^T H_{11} \left(\xi_1 + {H_{11}}^{-1} H_{12} \xi_2\right) \\
&+ {\xi_2}^T {H_{12}}^T {H_{11}}^{-1} H_{11} \left(\xi_1 + {H_{11}}^{-1} H_{12} \xi_2\right) \quad .
\end{aligned}
$$

The left hand terms can be grouped conveniently as well

$$
T = \left(\xi_1 + {H_{11}}^{-1} H_{12} \xi_2\right)^T H_{11} \left(\xi_1 + {H_{11}}^{-1} H_{12} \xi_2\right) \quad .
$$

This term substitutes back into the original quadratic to create the following expression

$$
\begin{aligned}
(\beta - \mu)^T H (\beta - \mu) &= {\xi_2}^T \left(H_{22} - H_{21} {H_{11}}^{-1} H_{12}\right) \xi_2 \\
&+ \left(\xi_1 + {H_{11}}^{-1} H_{12} \xi_2\right)^T H_{11} \left(\xi_1 + {H_{11}}^{-1} H_{12} \xi_2\right) \quad .
\end{aligned}
$$

This implies that the probability density function in equation (4.8) can be written

$$
\begin{aligned}
\{\beta, \sigma\} &= \frac{2|H|^{\frac{1}{2}} (2\pi)^{-\frac{p}{2}} \left(\frac{\nu s^2}{2}\right)^{\frac{\nu}{2}}}{\Gamma\left(\frac{\nu}{2}\right)} \sigma^{-(\nu+p+1)} e^{-\frac{\nu s^2}{2\sigma^2}} e^{-\frac{1}{2\sigma^2}(\beta-\mu)^T H (\beta-\mu)} \\
&= \frac{2|H|^{\frac{1}{2}} (2\pi)^{-\frac{p}{2}} \left(\frac{\nu s^2}{2}\right)^{\frac{\nu}{2}}}{\Gamma\left(\frac{\nu}{2}\right)} \sigma^{-(\nu+p+1)} e^{-\frac{\nu s^2}{2\sigma^2}} \\
&\times e^{-\frac{1}{2\sigma^2}\left[\xi_1 + {H_{11}}^{-1} H_{12} \xi_2\right]^T H_{11} \left[\xi_1 + {H_{11}}^{-1} H_{12} \xi_2\right]} \\
&\times e^{-\frac{1}{2\sigma^2} {\xi_2}^T \left(H_{22} - H_{21} {H_{11}}^{-1} H_{12}\right) \xi_2} \quad .
\end{aligned}
$$

We reinsert

$$
\begin{aligned}
\xi_1 &= \beta_1 - \mu_1 \\
\xi_2 &= \beta_2 - \mu_2 \quad .
\end{aligned}
$$

This allows us to write

$$\begin{aligned}
&\xi_1 - {H_{11}}^{-1} H_{12}\xi_2 \\
&= (\beta_1 - \mu_1) + {H_{11}}^{-1} H_{12} (\beta_2 - \mu_2) \\
&= \beta_1 - \left[\mu_1 - {H_{11}}^{-1} H_{12} (\beta_2 - \mu_2)\right] \quad .
\end{aligned}$$

The term in brackets, a vector of length L, can be defined

$$\mu_{12} = \mu_1 - {H_{11}}^{-1} H_{12} (\beta_2 - \mu_2) \quad .$$

We use the index 12 in μ_{12} to emphasize that it is a function of μ_1 and μ_2. This definition allows us to write the joint density in what will be seen to be a marginal times a conditional density

$$\begin{aligned}
\{\beta, \sigma\} = & \frac{2|H|^{\frac{1}{2}} (2\pi)^{-\frac{p}{2}} \left(\frac{\nu s^2}{2}\right)^{\frac{\nu}{2}}}{\Gamma\left(\frac{\nu}{2}\right)} \sigma^{-(\nu+p+1)} e^{-\frac{\nu s^2}{2\sigma^2}} \\
& \times e^{-\frac{1}{2\sigma^2}(\beta_1-\mu_{12})^T H_{11} (\beta_1-\mu_{12})} \\
& \times e^{-\frac{1}{2\sigma^2}(\beta_2-\mu_2)^T \left(H_{22}-H_{21}{H_{11}}^{-1}H_{12}\right)(\beta_2-\mu_2)} \quad .
\end{aligned}$$

Note that we have the L partition in the first of the last two exponentials, and we have the $p - L$ partition in the second of the last two exponentials. It should be clear what we have in mind; we are going to want to integrate out the partition of length L so as to be left with the marginal distribution over a remaining partition of length $p - L$. To wit, we want to integrate out the model coefficients in the first partition of the matrix of length L. (This corresponds to integrating out the second-to-last exponential term in the immediately foregoing equation.) We want to integrate out the second-to-last term in that equation containing the simple precision matrix H_{11}. This will leave behind a marginal distribution over the coefficients in the partition of length $p - L$. We are going to integrate out the L partition by multiplying and dividing by the appropriate terms to get the correct multivariate integrating constant for the L partition so that we can easily integrate out that term. Prior to manipulating the second to last term, the L term, so that it aligns with the proper multivariate normal form so that the integral becomes 1, we rewrite the general multivariate normal.

$$f(\beta) = \frac{1}{(2\pi)^{\frac{L}{2}} |\Sigma|^{\frac{1}{2}}} e^{-\frac{1}{2}(\beta-\mu)^T \Sigma^{-1} (x-\mu)} \quad .$$

In order to represent our second to last term as a multivariate normal, we must make the following substitutions

$$\Sigma^{-1} = \sigma^{-2} H_{11}$$

$$\Rightarrow \Sigma = \sigma^2 {H_{11}}^{-1}$$
$$\Rightarrow |\Sigma| = \left|\sigma^2 {H_{11}}^{-1}\right| = \sigma^{2L} \left|{H_{11}}^{-1}\right|$$
$$\Rightarrow |\Sigma|^{\frac{1}{2}} = \sigma^L \left|{H_{11}}^{-1}\right|^{\frac{1}{2}} \quad .$$

These substitutions enable us to write the normal distribution

$$f(\beta) = \frac{1}{(2\pi)^{\frac{L}{2}} \sigma^L \left|{H_{11}}^{-1}\right|^{\frac{1}{2}}} e^{-\frac{1}{2\sigma^2}(\beta_1-\mu_{12})^T H_{11}(\beta_1-\mu_{12})} \quad .$$

Knowing this functional form for the normal, we then multiply and divide the foregoing density by the three terms in the denominator so that we can accumulate a complete normal distribution within our expression:

$$\begin{aligned}
\{\beta,\sigma\} &= \frac{2|H|^{\frac{1}{2}}(2\pi)^{-\frac{p}{2}}\left(\frac{\nu s^2}{2}\right)^{\frac{\nu}{2}}}{\Gamma\left(\frac{\nu}{2}\right)} \sigma^{-(\nu+p+1)} e^{-\frac{\nu s^2}{2\sigma^2}} \\
&\times \frac{(2\pi)^{\frac{L}{2}} \sigma^L \left|{H_{11}}^{-1}\right|^{\frac{1}{2}}}{(2\pi)^{\frac{L}{2}} \sigma^L \left|{H_{11}}^{-1}\right|^{\frac{1}{2}}} e^{-\frac{1}{2\sigma^2}(\beta_1-\mu_{12})^T H_{11}(\beta_1-\mu_{12})} \\
&\times e^{-\frac{1}{2\sigma^2}(\beta_2-\mu_2)^T\left(H_{22}-H_{21}{H_{11}}^{-1}H_{12}\right)(\beta_2-\mu_2)} \\
&= \frac{2|H|^{\frac{1}{2}}(2\pi)^{-\frac{p}{2}}\left(\frac{\nu s^2}{2}\right)^{\frac{\nu}{2}}(2\pi)^{\frac{L}{2}}\sigma^L\left|{H_{11}}^{-1}\right|^{\frac{1}{2}}}{\Gamma\left(\frac{\nu}{2}\right)} \sigma^{-(\nu+p+1)} e^{-\frac{\nu s^2}{2\sigma^2}} \\
&\times \frac{1}{(2\pi)^{\frac{L}{2}} \sigma^L \left|{H_{11}}^{-1}\right|^{\frac{1}{2}}} e^{-\frac{1}{2\sigma^2}(\beta_1-\mu_{12})^T H_{11}(\beta_1-\mu_{12})} \\
&\times e^{-\frac{1}{2\sigma^2}(\beta_2-\mu_2)^T\left(H_{22}-H_{21}{H_{11}}^{-1}H_{12}\right)(\beta_2-\mu_2)} \quad .
\end{aligned}$$

The second-to-last term in the last equation is a standard multivariate normal distribution with L parameters whose mean is μ_{12}, whose precision matrix is H_{11}, and whose integrating constant has been properly accumulated immediately preceding the exponential term. This will integrate over the variable β_1 to be 1 because it is a standard multivariate normal. Such integration leaves behind the remaining marginal density over the $p-L$ vector β_2, which after such integration is:

$$\begin{aligned}
\{\beta_2,\sigma\} &= \frac{2|H|^{\frac{1}{2}}\left(\frac{\nu s^2}{2}\right)^{\frac{\nu}{2}}}{|H_{11}|^{\frac{1}{2}}(2\pi)^{\frac{p-L}{2}}\Gamma\left(\frac{\nu}{2}\right)} \sigma^{-(\nu+p+1-L)} e^{-\frac{\nu s^2}{2\sigma^2}} \\
&\times e^{-\frac{1}{2\sigma^2}(\beta_2-\mu_2)^T\left(H_{22}-H_{21}{H_{11}}^{-1}H_{12}\right)(\beta_2-\mu_2)} \quad .
\end{aligned}$$

This simplifies to the final form of the joint density

$$\{\beta_2, \sigma\} = \frac{2|H|^{\frac{1}{2}}\left(\frac{\nu s^2}{2}\right)^{\frac{\nu}{2}}}{|H_{11}|^{\frac{1}{2}}(2\pi)^{\frac{p-L}{2}}\Gamma\left(\frac{\nu}{2}\right)}\sigma^{-(\nu+p+1-L)}e^{-\frac{\nu s^2}{2\sigma^2}} \times e^{-\frac{1}{2}(\beta_2-\mu_2)^T\frac{H_{22}-H_{21}H_{11}^{-1}H_{12}}{\sigma^2}(\beta_2-\mu_2)} \quad .$$

We see the quotient of two determinants in the constant of integration. We want to simplify those, and we can do so by appealing to the matrix determinant term in equation (A.1) Appendix A to the H matrix as follows

$$|H| = |H_{11}|\left|H_{22} - H_{21}H_{11}^{-1}H_{12}\right|$$
$$\Rightarrow \frac{|H|}{|H_{11}|} = \left|H_{22} - H_{21}H_{11}^{-1}H_{12}\right| \quad .$$

This simplifies the joint distribution to the form

$$\{\beta_2, \sigma\} = \frac{2\left|H_{22} - H_{21}H_{11}^{-1}H_{12}\right|^{\frac{1}{2}}}{(2\pi)^{\frac{p-L}{2}}\Gamma\left(\frac{\nu}{2}\right)}\left(\frac{\nu s^2}{2}\right)^{\frac{\nu}{2}}\sigma^{-(\nu+p+1-L)}e^{-\frac{\nu s^2}{2\sigma^2}} \times e^{-\frac{1}{2}(\beta_2-\mu_2)^T\left[\left(H_{22}-H_{21}H_{11}^{-1}H_{12}\right)\sigma^{-2}\right](\beta_2-\mu_2)} \quad . \tag{4.28}$$

In order to recharacterize the underlying distributions, we begin by setting the integral of the first form of the joint distribution in equation (4.28) to unity

$$1 = \frac{2\left|H_{22} - H_{21}H_{11}^{-1}H_{12}\right|^{\frac{1}{2}}\left(\frac{\nu s^2}{2}\right)^{\frac{\nu}{2}}}{(2\pi)^{\frac{p-L}{2}}\Gamma\left(\frac{\nu}{2}\right)} \times \int_{-\infty}^{\infty} d\beta \int_{0}^{\infty} d\sigma\sigma^{-(\nu+p+1-L)}e^{-\frac{\nu s^2}{2\sigma^2}}e^{-\frac{1}{2}(\beta_2-\mu_2)^T\frac{H_{22}-H_{21}H_{11}^{-1}H_{12}}{\sigma^2}(\beta_2-\mu_2)} \quad .$$

We then prepare to make the following substitutions so as to change the integrand into its true form

$$z = \sigma^2 \Rightarrow \sigma = z^{\frac{1}{2}} \Rightarrow d\sigma = \frac{1}{2}z^{-\frac{1}{2}}dz \Rightarrow \sigma^{-(\nu+p+1-L)} = z^{-\frac{1}{2}(\nu+p+1-L)} \quad .$$

This changes the integral as follows

$$1 = \frac{2\left|H_{22} - H_{21}H_{11}^{-1}H_{12}\right|^{\frac{1}{2}}\left(\frac{\nu s^2}{2}\right)^{\frac{\nu}{2}}}{(2\pi)^{\frac{p-L}{2}}\Gamma\left(\frac{\nu}{2}\right)} \times \int_{-\infty}^{\infty} d\beta \int_{0}^{\infty}\left\{\begin{array}{l} d\sigma \\ \sigma^{-(\nu+p+1-L)} \\ e^{-\frac{\nu s^2}{2\sigma^2}}e^{-\frac{1}{2}(\beta_2-\mu_2)^T\frac{H_{22}-H_{21}H_{11}^{-1}H_{12}}{\sigma^2}(\beta_2-\mu_2)} \end{array}\right\}$$

$$= \frac{2\left|H_{22} - H_{21}{H_{11}}^{-1}H_{12}\right|^{\frac{1}{2}}\left(\frac{\nu s^2}{2}\right)^{\frac{\nu}{2}}}{(2\pi)^{\frac{p-L}{2}}\Gamma\left(\frac{\nu}{2}\right)}$$

$$\times \int_{-\infty}^{\infty} d\beta \int_{0}^{\infty} \left\{ \begin{array}{l} \frac{1}{2}z^{-\frac{1}{2}}dz \\ z^{-\frac{1}{2}(\nu+p+1-L)} \\ e^{-\frac{\nu s^2}{2z}} e^{-\frac{1}{2}(\beta_2-\mu_2)^T \frac{H_{22}-H_{21}{H_{11}}^{-1}H_{12}}{z}(\beta_2-\mu_2)} \end{array} \right\}$$

$$= \frac{\left|H_{22} - H_{21}{H_{11}}^{-1}H_{12}\right|^{\frac{1}{2}}\left(\frac{\nu s^2}{2}\right)^{\frac{\nu}{2}}}{(2\pi)^{\frac{p-L}{2}}\Gamma\left(\frac{\nu}{2}\right)}$$

$$\times \int_{-\infty}^{\infty} d\beta \int_{0}^{\infty} dz z^{-\frac{\nu+p-L}{2}-1} e^{-\frac{\nu s^2}{2z}} e^{-\frac{1}{2}(\beta_2-\mu_2)^T \frac{H_{22}-H_{21}{H_{11}}^{-1}H_{12}}{z}(\beta_2-\mu_2)} \quad .$$

The transformed distribution is that inside the integral, which is

$$\{\beta_2, z\} = \frac{\left|H_{22} - H_{21}{H_{11}}^{-1}H_{12}\right|^{\frac{1}{2}}\left(\frac{\nu s^2}{2}\right)^{\frac{\nu}{2}}}{(2\pi)^{\frac{p-L}{2}}\Gamma\left(\frac{\nu}{2}\right)} \times z^{-\frac{\nu+p-L}{2}-1} e^{-\frac{\nu s^2}{2z}} e^{-\frac{1}{2}(\beta_2-\mu_2)^T \frac{H_{22}-H_{21}{H_{11}}^{-1}H_{12}}{z}(\beta_2-\mu_2)} \quad . \tag{4.29}$$

Now we ponder the variance-covariance term in the exponential term on the right in this equation

$$\Sigma^{-1} = \frac{H_{22} - H_{21}{H_{11}}^{-1}H_{12}}{z}$$

$$\Rightarrow \Sigma = z\left(H_{22} - H_{21}{H_{11}}^{-1}H_{12}\right)^{-1}$$

$$\Rightarrow |\Sigma| = z^{p-L}\left|H_{22} - H_{21}{H_{11}}^{-1}H_{12}\right|^{-1}$$

$$\Rightarrow |\Sigma|^{\frac{1}{2}} = z^{\frac{p-L}{2}}\left|H_{22} - H_{21}{H_{11}}^{-1}H_{12}\right|^{-\frac{1}{2}} \quad .$$

We think of the following $p - L$ variate normal distribution

$$N\left(\beta_2; p-L, \mu_2, \Sigma\right) = \frac{1}{(2\pi)^{\frac{p-L}{2}}|\Sigma|^{\frac{1}{2}}} e^{-\frac{1}{2}(\beta_2-\mu_2)^T\Sigma^{-1}(\beta_2-\mu_2)} \quad .$$

We can reorganize this to isolate the exponential term

$$e^{-\frac{1}{2}(\beta_2-\mu_2)^T\Sigma^{-1}(\beta_2-\mu_2)} = (2\pi)^{\frac{p-L}{2}}|\Sigma|^{\frac{1}{2}} N\left(\beta_2; p-L, \mu_2, \Sigma\right) \quad .$$

We can then substitute these parameters into the density in equation (4.29) to write

$$e^{-\frac{1}{2}(\beta_2-\mu_2)^T \frac{H_{22}-H_{21}H_{11}{}^{-1}H_{12}}{z}(\beta_2-\mu_2)}$$
$$= (2\pi)^{\frac{p-L}{2}} z^{\frac{p-L}{2}} \left|H_{22} - H_{21}H_{11}{}^{-1}H_{12}\right|^{-\frac{1}{2}} N\left(\beta_2; p-L, \mu_2, \Sigma\right) \quad .$$

This substitutes into the revised distribution

$$\begin{aligned}
\{\beta_2, z\} &= \frac{\left|H_{22} - H_{21}H_{11}{}^{-1}H_{12}\right|^{\frac{1}{2}} \left(\frac{\nu s^2}{2}\right)^{\frac{\nu}{2}}}{(2\pi)^{\frac{p-L}{2}}\Gamma\left(\frac{\nu}{2}\right)} z^{-\frac{\nu+p-L}{2}-1} \\
&\times e^{-\frac{\nu s^2}{2z}} e^{-\frac{1}{2}(\beta_2-\mu_2)^T \frac{H_{22}-H_{21}H_{11}{}^{-1}H_{12}}{z}(\beta_2-\mu_2)} \\
&= \frac{\left|H_{22} - H_{21}H_{11}{}^{-1}H_{12}\right|^{\frac{1}{2}} \left(\frac{\nu s^2}{2}\right)^{\frac{\nu}{2}}}{(2\pi)^{\frac{p-L}{2}}\Gamma\left(\frac{\nu}{2}\right)} z^{-\frac{\nu+p-L}{2}-1} e^{-\frac{\nu s^2}{2z}} (2\pi)^{\frac{p-L}{2}} z^{\frac{p-L}{2}} \\
&\times \left|H_{22} - H_{21}H_{11}{}^{-1}H_{12}\right|^{-\frac{1}{2}} N\left(\beta_2; p-L, \mu_2, \Sigma\right) \\
&= \frac{\left(\frac{\nu s^2}{2}\right)^{\frac{\nu}{2}}}{\Gamma\left(\frac{\nu}{2}\right)} z^{-\frac{\nu}{2}-1} e^{-\frac{\nu s^2}{2z}} N\left[\beta_2; p-L, \mu_2, z\left(H_{22} - H_{21}H_{11}{}^{-1}H_{12}\right)^{-1}\right] \quad .
\end{aligned}$$

This allows us to write the density quite simply as follows

$$\begin{aligned}
\{\beta_2, z\} &= \frac{\left(\frac{\nu s^2}{2}\right)^{\frac{\nu}{2}}}{\Gamma\left(\frac{\nu}{2}\right)} z^{-\frac{\nu}{2}-1} e^{-\frac{\nu s^2}{2z}} \\
&\times N\left[\beta_2; p-L, \mu_2, z\left(H_{22} - H_{21}H_{11}{}^{-1}H_{12}\right)^{-1}\right] \\
&= IG\left(z, \frac{\nu s^2}{2}, \frac{\nu}{2}\right) \\
&\times N\left[\beta_2; p-L, \mu_2, z\left(H_{22} - H_{21}H_{11}{}^{-1}H_{12}\right)^{-1}\right] \quad .
\end{aligned} \tag{4.30}$$

Clearly this density contains the same inverse gamma distribution

$$IG\left(z, \frac{\nu s^2}{2}, \frac{\nu}{2}\right) = \frac{\left(\frac{\nu s^2}{2}\right)^{\frac{\nu}{2}}}{\Gamma\left(\frac{\nu}{2}\right)} z^{-\frac{\nu}{2}-1} e^{-\frac{\nu s^2}{2z}} \tag{4.31}$$

times a normal distribution, the same as the form of the coefficients determined previously. Notice that the inverse gamma distribution is identical in form to that in equation (4.27). We will soon simplify to the univariate case below, i.e., the case in which $p - L = 1$. For now, we are going to develop the marginal distribution over the model coefficients β_2 where the length of the vector can be 1 or higher.

4.13 The Marginal Distribution Over the Model Coefficients

We begin with equation (4.28) and combine the final two exponential terms to form an aggregate exponent to compute the marginal distribution over the coefficients

$$\{\beta_2, \sigma\} = \frac{2\left|H_{22} - H_{21}{H_{11}}^{-1}H_{12}\right|^{\frac{1}{2}}\left(\frac{\nu s^2}{2}\right)^{\frac{\nu}{2}}}{(2\pi)^{\frac{p-L}{2}}\Gamma\left(\frac{\nu}{2}\right)}\sigma^{-(\nu+p+1-L)} \times e^{-\frac{\nu s^2}{2\sigma^2}\left[1+(\beta_2-\mu_2)^T\frac{H_{22}-H_{21}{H_{11}}^{-1}H_{12}}{\nu s^2}(\beta_2-\mu_2)\right]} . \tag{4.32}$$

We are going to want to return to the joint density in equation (4.32) and integrate out the σ term to write the marginal density over β_2

$$\{\beta_2, \sigma\} = \frac{2\left|H_{22} - H_{21}{H_{11}}^{-1}H_{12}\right|^{\frac{1}{2}}\left(\frac{\nu s^2}{2}\right)^{\frac{\nu}{2}}}{(2\pi)^{\frac{p-L}{2}}\Gamma\left(\frac{\nu}{2}\right)}\sigma^{-(\nu+p+1-L)} \times e^{-\frac{\nu s^2}{2\sigma^2}\left[1+(\beta_2-\mu_2)^T\frac{H_{22}-H_{21}{H_{11}}^{-1}H_{12}}{\nu s^2}(\beta_2-\mu_2)\right]} .$$

We want to prepare the following substitutions

$$h = \frac{\nu s^2}{2\sigma^2}\left[1 + (\beta_2 - \mu_2)^T\frac{H_{22} - H_{21}{H_{11}}^{-1}H_{12}}{\nu s^2}(\beta_2 - \mu_2)\right]$$

$$\sigma = h^{-\frac{1}{2}}\left[\frac{\nu s^2}{2} + (\beta_2 - \mu_2)^T\frac{\nu s^2}{2}\frac{H_{22} - H_{21}{H_{11}}^{-1}H_{12}}{\nu s^2}(\beta_2 - \mu_2)\right]^{\frac{1}{2}}$$

$$d\sigma = -\frac{1}{2}h^{-\frac{3}{2}}dh\left[\frac{\nu s^2}{2} + (\beta_2 - \mu_2)^T\frac{\nu s^2}{2}\frac{H_{22} - H_{21}{H_{11}}^{-1}H_{12}}{\nu s^2}(\beta_2 - \mu_2)\right]^{\frac{1}{2}}$$

$$\sigma^{-\nu-p-1+L} = h^{\frac{\nu+p+1-L}{2}} \times \left[\frac{\nu s^2 + (\beta_2 - \mu_2)^T\left(H_{22} - H_{21}{H_{11}}^{-1}H_{12}\right)(\beta_2 - \mu_2)}{2}\right]^{\frac{-\nu-p-1+L}{2}} .$$

We substitute and simplify

$$\{\beta_2\} = \int_0^\infty \{\beta_2, \sigma\}\, d\sigma$$
$$= \frac{2\left|H_{22} - H_{21}{H_{11}}^{-1}H_{12}\right|^{\frac{1}{2}}\left(\frac{\nu s^2}{2}\right)^{\frac{\nu}{2}}}{(2\pi)^{\frac{p-L}{2}}\Gamma\left(\frac{\nu}{2}\right)}$$

$$\times \int_0^\infty \left\{ \begin{array}{l} \sigma^{-(\nu+p+1-L)} \\ d\sigma \\ e^{-\frac{\nu s^2}{2\sigma^2}\left[1+(\beta_2-\mu_2)^T \frac{H_{22}-H_{21}H_{11}{}^{-1}H_{12}}{\nu s^2}(\beta_2-\mu_2)\right]} \end{array} \right\}$$

$$= \frac{2\left|H_{22} - H_{21}H_{11}{}^{-1}H_{12}\right|^{\frac{1}{2}} \left(\frac{\nu s^2}{2}\right)^{\frac{\nu}{2}}}{(2\pi)^{\frac{p-L}{2}}\Gamma\left(\frac{\nu}{2}\right)}$$

$$\times \int_0^\infty \left\{ \begin{array}{l} h^{\frac{\nu+p+1-L}{2}} \left[\frac{\nu s^2+(\beta_2-\mu_2)^T\left(H_{22}-H_{21}H_{11}{}^{-1}H_{12}\right)(\beta_2-\mu_2)}{2}\right]^{\frac{-\nu-p-1+L}{2}} \\ -\frac{1}{2}h^{-\frac{3}{2}}dh\left[\frac{\nu s^2}{2} + (\beta_2-\mu_2)^T \frac{\nu s^2}{2}\frac{H_{22}-H_{21}H_{11}{}^{-1}H_{12}}{\nu s^2}(\beta_2-\mu_2)\right]^{\frac{1}{2}} \\ e^{-h} \end{array} \right\}$$

$$= -\frac{\left|H_{22} - H_{21}H_{11}{}^{-1}H_{12}\right|^{\frac{1}{2}} \left(\frac{\nu s^2}{2}\right)^{\frac{\nu}{2}}}{(2\pi)^{\frac{p-L}{2}}\Gamma\left(\frac{\nu}{2}\right)}$$

$$\times \int_\infty^0 \left\{ \begin{array}{l} h^{\frac{\nu+p+1-L-3}{2}} \\ \left[\frac{\nu s^2+(\beta_2-\mu_2)^T\left(H_{22}-H_{21}H_{11}{}^{-1}H_{12}\right)(\beta_2-\mu_2)}{2}\right]^{\frac{-\nu-p+L}{2}} \\ e^{-h}dh \end{array} \right\}$$

$$= \frac{\left|H_{22} - H_{21}H_{11}{}^{-1}H_{12}\right|^{\frac{1}{2}} \left(\frac{\nu s^2}{2}\right)^{\frac{\nu}{2}}}{(2\pi)^{\frac{p-L}{2}}\Gamma\left(\frac{\nu}{2}\right)} \int_0^\infty h^{\frac{\nu+p-L}{2}-1}e^{-h}dh$$

$$\times \left[\frac{\nu s^2 + (\beta_2-\mu_2)^T\left(H_{22} - H_{21}H_{11}{}^{-1}H_{12}\right)(\beta_2-\mu_2)}{2}\right]^{\frac{-\nu-p+L}{2}}$$

$$= \frac{\left|H_{22} - H_{21}H_{11}{}^{-1}H_{12}\right|^{\frac{1}{2}} \left(\frac{\nu s^2}{2}\right)^{\frac{\nu}{2}}\Gamma\left(\frac{\nu+p-L}{2}\right)}{(2\pi)^{\frac{p-L}{2}}\Gamma\left(\frac{\nu}{2}\right)}$$

$$\times \left[\frac{\nu s^2 + (\beta_2-\mu_2)^T\left(H_{22} - H_{21}H_{11}{}^{-1}H_{12}\right)(\beta_2-\mu_2)}{2}\right]^{\frac{-\nu-p+L}{2}} .$$

We can eliminate the powers of 2 from this expression

$$\{\beta_2\} = \frac{\left|H_{22} - H_{21}H_{11}{}^{-1}H_{12}\right|^{\frac{1}{2}} \left(\nu s^2\right)^{\frac{\nu}{2}}\Gamma\left(\frac{\nu+p-L}{2}\right)}{\pi^{\frac{p-L}{2}}\Gamma\left(\frac{\nu}{2}\right)}$$

$$\times \left[\nu s^2 + (\beta_2-\mu_2)^T\left(H_{22} - H_{21}H_{11}{}^{-1}H_{12}\right)(\beta_2-\mu_2)\right]^{\frac{-\nu-p+L}{2}} .$$

We can simplify terms containing νs^2 as follows to write the final form of the marginal density

$$(4.33)\quad \{\beta_2\} = \frac{\left|H_{22} - H_{21}{H_{11}}^{-1}H_{12}\right|^{\frac{1}{2}}\Gamma\left(\frac{\nu+p-L}{2}\right)}{\left(\pi\nu s^2\right)^{\frac{p-L}{2}}\Gamma\left(\frac{\nu}{2}\right)}$$
$$\times\left[1 + \left(\beta_2 - \mu_2\right)^T \frac{\left(H_{22} - H_{21}{H_{11}}^{-1}H_{12}\right)}{\nu s^2}\left(\beta_2 - \mu_2\right)\right]^{\frac{-\nu-p+L}{2}} .$$

We appeal to Wiki[18] for the functional form of the multivariate Student's t distribution

$$\{x\} = \frac{\Gamma\left(\frac{\nu+p}{2}\right)}{\Gamma\left(\frac{\nu}{2}\right)\left(\nu\pi\right)^{\frac{p}{2}}\left|\Sigma\right|^{\frac{1}{2}}}\left[1 + \frac{1}{\nu}(x-\mu)^T\Sigma^{-1}\left(x-\mu\right)\right]^{-\frac{\nu+p}{2}}$$

whose mean is μ and whose variance is

$$\text{variance} = \frac{\nu}{\nu-2}\Sigma \quad .$$

We want to adapt this multivariate Student's t distribution to our derived distribution. We do so in two steps, the first being

$$p \leftarrow p - L \quad .$$

After this substitution, the multivariate Student's t distribution is written

$$\{x\} = \frac{\Gamma\left(\frac{\nu+p-L}{2}\right)}{\Gamma\left(\frac{\nu}{2}\right)\left(\nu\pi\right)^{\frac{p-L}{2}}\left|\Sigma\right|^{\frac{1}{2}}}\left[1 + \frac{1}{\nu}(x-\mu)^T\Sigma^{-1}\left(x-\mu\right)\right]^{-\frac{\nu+p-L}{2}} .$$

We then define the following substitution and its implication

$$\begin{aligned}
&\frac{\Sigma^{-1}}{\nu} \leftarrow \frac{\left(H_{22} - H_{21}{H_{11}}^{-1}H_{12}\right)}{\nu s^2}\\
&\Rightarrow \Sigma^{-1} \leftarrow s^{-2}\left(H_{22} - H_{21}{H_{11}}^{-1}H_{12}\right)\\
&\Rightarrow \Sigma \leftarrow s^2\left(H_{22} - H_{21}{H_{11}}^{-1}H_{12}\right)^{-1}\\
&\Rightarrow \left|\Sigma\right| \leftarrow s^{2(p-L)}\left|H_{22} - H_{21}{H_{11}}^{-1}H_{12}\right|^{-1}\\
&\Rightarrow \left|\Sigma\right|^{\frac{1}{2}} \leftarrow s^{(p-L)}\left|H_{22} - H_{21}{H_{11}}^{-1}H_{12}\right|^{-\frac{1}{2}} \quad .
\end{aligned}$$

We then substitute into the density to write

[18] See https://en.wikipedia.org/wiki/Multivariate_t-distribution.

$$\{x\} = \frac{\left|H_{22} - H_{21}{H_{11}}^{-1}H_{12}\right|^{\frac{1}{2}}\Gamma\left(\frac{\nu+p-L}{2}\right)}{\left(\nu\pi s^2\right)^{\frac{p-L}{2}}\Gamma\left(\frac{\nu}{2}\right)}$$
$$\left[1 + (x-\mu)^T \frac{\left(H_{22} - H_{21}{H_{11}}^{-1}H_{12}\right)}{\nu s^2}(x-\mu)\right]^{-\frac{\nu+p-L}{2}}.$$

This is precisely identical to the multivariate Student's t distribution we have developed in equation (4.33) representing the marginal density over model coefficients. We know by substitution that the variance-covariance matrix is

$$\text{variance-covariance} = \frac{\nu}{\nu-2}\Sigma = \frac{\nu s^2}{\nu-2}\left(H_{22} - H_{21}H_{11}^{-1}H_{12}\right)^{-1}$$

4.14 The Important Special Case of $p - L = 1$, a Univariate Marginal

The special case in which $p-L = 1$, meaning that $L = p-1$, is the focus of this section. It is the case in which the marginal distribution is univariate. Following is the full density for that case including both the scalar single coefficient β_2 and σ:

$$\{\beta_2, \sigma\} = \frac{2\left|H_{22} - H_{21}{H_{11}}^{-1}H_{12}\right|^{\frac{1}{2}}\left(\frac{\nu s^2}{2}\right)^{\frac{\nu}{2}}}{(2\pi)^{\frac{1}{2}}\Gamma\left(\frac{\nu}{2}\right)}\sigma^{-(\nu+2)}e^{-\frac{\nu s^2}{2\sigma^2}}$$
$$\times\, e^{-\frac{1}{2}(\beta_2-\mu_2)^T\frac{H_{22}-H_{21}{H_{11}}^{-1}H_{12}}{\sigma^2}(\beta_2-\mu_2)} \tag{4.34}$$

in which we emphasize that β_2 is now a scalar and the distribution is univariate in one coefficient. We rewrite the marginal joint distribution for $p - L$ coefficients from the previous section

$$\{\beta_2\} = \frac{\left|H_{22} - H_{21}{H_{11}}^{-1}H_{12}\right|^{\frac{1}{2}}\Gamma\left(\frac{\nu+p-L}{2}\right)}{\left(\pi\nu s^2\right)^{\frac{p-L}{2}}\Gamma\left(\frac{\nu}{2}\right)}$$
$$\times\left[1 + (\beta_2-\mu_2)^T \frac{\left(H_{22} - H_{21}{H_{11}}^{-1}H_{12}\right)}{\nu s^2}(\beta_2-\mu_2)\right]^{\frac{-\nu-p+L}{2}}.$$

We then specialize this to the case in which $p-L = 1$, i.e., the marginal distribution over a single coefficient alone. This is the case in which we are looking for the marginal distribution over a single model coefficient. We want to integrate out the $p-1$ coefficients to leave a single coefficient and its marginal density. In this case, the mean of the marginal distribution for the ith coefficient is clearly the ith element of μ_2. In the case in which $L = p - 1$, the univariate marginal density becomes

(4.35)

$$\{\beta_2\} = \frac{\left|H_{22} - H_{21}{H_{11}}^{-1}H_{12}\right|^{\frac{1}{2}}\Gamma\left(\frac{\nu+1}{2}\right)}{(\pi\nu s^2)^{\frac{1}{2}}\Gamma\left(\frac{\nu}{2}\right)} \times \left[1 + (\beta_2 - \mu_2)^T \frac{\left(H_{22} - H_{21}{H_{11}}^{-1}H_{12}\right)}{\nu s^2}(\beta_2 - \mu_2)\right]^{-\frac{\nu+1}{2}} .$$

The numerator (precision matrix) is actually a scalar in this case because we have only preserved one row after integrating out $p-1$ rows. To wit,

$$H_{22} - H_{21}H_{11}^{-1}H_{12}$$

is a scalar. As a scalar, its inverse is its reciprocal. We are going to examine this in more detail below and justify this result in more detail. All the terms in the foregoing equation are easy to calculate, so the univariate Student's t distribution is analytically tractable and plottable and integrates to 1 by construction. There is no need for numerical methods or numerical integration in dealing with this marginal univariate density. We need to reiterate the critically important insight here. When a matrix is actually a scalar, the determinant of the matrix is the scalar itself. Therefore, when calculating the determinant of a scalar matrix, we simply write the scalar. This, of course, only pertains when we are considering 1×1 matrices, i.e., scalars. Equally important, the inverse of a scalar is simply the reciprocal of that scalar. That is going to be important here.

The foregoing results require precise indexation, and it is therefore insightful to specialize from the outset in the case in which the first partition is of length $p-1$ and the last partition is of length 1, and calculate each of the partitioned elements explicitly. This gives us the intuition we need to understand that we need only be concerned with the diagonals of the H matrix when we are considering the case in which $L = p-1$. This is a subtle yet central message that needs to be illustrated. The partitioned matrix H is assumed to have the form

$$H = \begin{bmatrix} A & b \\ b^T & c \end{bmatrix} ,$$

in which

$$\begin{aligned} A: &\quad (p-1) \times (p-1) \\ b: &\quad (p-1) \times 1 \\ c: &\quad scalar \quad . \end{aligned}$$

We appeal to Appendix A to write the defining equation for a partitioned matrix H and its partitioned inverse Z

$$\begin{bmatrix} Z_{11} & Z_{12} \\ Z_{21} & Z_{22} \end{bmatrix} \begin{bmatrix} H_{11} & H_{12} \\ H_{21} & H_{22} \end{bmatrix} = \begin{bmatrix} I & 0 \\ 0 & I \end{bmatrix} \quad .$$

Appendix A develops a general analytic expression for the partitioned inverse, which is:

$$\begin{aligned} Z_{11} &= \left(H_{11} - H_{12}{H_{22}}^{-1}H_{21}\right)^{-1} \\ Z_{12} &= -\left(H_{11} - H_{12}{H_{22}}^{-1}H_{21}\right)^{-1} H_{12}{H_{22}}^{-1} \\ Z_{21} &= -\left(H_{22} - H_{21}{H_{11}}^{-1}H_{12}\right)^{-1} H_{21}{H_{11}}^{-1} \\ Z_{22} &= \left(H_{22} - H_{21}{H_{11}}^{-1}H_{12}\right)^{-1} \end{aligned}$$

Let us suppose that the first partition is of length $p-1$ and the final partition is of length 1. That means that the matrices that comprise the foregoing calculations have the following sizes

$$\begin{array}{ll} H_{11} & (p-1) \times (p-1) \\ H_{12} & (p-1) \times 1 \\ H_{21} & 1 \times (p-1) \\ H_{22} & 1 \times 1 \quad . \end{array}$$

The equation for the number in the ii^{th} position (i.e., the position of the row and column that corresponds to the single row and column i, i partition) of H^{-1} based on the foregoing inverse structure is

$$\left[H^{-1}\right]_{ii} = \left(H_{22} - H_{21}{H_{11}}^{-1}H_{12}\right)^{-1} = z_{ii} \quad . \tag{4.36}$$

This is what is in the inverse matrix H^{-1} in the ii^{th} position. Because this is a scalar, the inverse of the matrix is simple–the reciprocal of the scalar in the ii^{th} position. Therefore, we can take the inverse of both sides to write

$$H_{22} - H_{21}{H_{11}}^{-1}H_{12} = \frac{1}{\left[H^{-1}\right]_{ii}} = \frac{1}{z_{ii}} \quad .$$

This is the inverse of what is in $[H^{-1}]_{ii}$. Using the simple partitioned matrix in this section, a partition that distinguishes the first $p-1$ rows from the last row, the inverse equation becomes

$$\begin{bmatrix} Z_{11} & Z_{12} \\ Z_{21} & Z_{22} \end{bmatrix} \begin{bmatrix} A & b \\ b^T & c \end{bmatrix} = \begin{bmatrix} I & 0 \\ 0 & 1 \end{bmatrix} \quad .$$

When we substitute the specific matrix considered in this discussion, we can write

$$\begin{bmatrix} Z_{11} & Z_{12} \\ Z_{21} & Z_{22} \end{bmatrix} = \begin{bmatrix} \left(A - c^{-1}bb^T\right)^{-1} & -\left(A - c^{-1}bb^T\right)^{-1}bc^{-1} \\ -\left(c - b^TA^{-1}b\right)^{-1}b^TA^{-1} & \left(c - b^TA^{-1}b\right)^{-1} \end{bmatrix} .$$

The entry in the lower right reinforces that $(c - b^TA^{-1}b)^{-1} = [H^{-1}]_{ii}$. Comparing the immediately preceding equation summarizing our specific example here and the previous equation characterizing the general analytic case drawn from Appendix A, the two lower right corners imply that

$$\begin{aligned} \left(H_{22} - H_{21}{H_{11}}^{-1}H_{12}\right)^{-1} &= \left[H^{-1}\right]_{ii} = z_{ii} = \left(c - b^TA^{-1}b\right)^{-1} \\ \Rightarrow H_{22} - H_{21}{H_{11}}^{-1}H_{12} &= \frac{1}{\left[H^{-1}\right]_{ii}} = \frac{1}{z_{ii}} = c - b^TA^{-1}b \quad . \end{aligned}$$

We see that this is the numerator on the quadratic term of the Student's t distribution in equation (4.35). And this is very easy to obtain. It is the reciprocal of the inverse of the H matrix. This means that

$$\begin{aligned} c - b^TA^{-1}b &= \frac{1}{\left[H^{-1}\right]_{ii}} = \frac{1}{z_{ii}} \\ &= H_{22} - H_{21}H_{11}^{-1}H_{12} \quad . \end{aligned}$$

This means that we calculate the inverse of the full H matrix in the normal fashion. Thereafter, the reciprocals of the diagonal terms are the numerators in the univariate Student's t distributions of each individual variable. First the univariate joint distribution for β_2 in equation (4.34) becomes, after substitution of z_{ii} for the matrix terms, the following

$$\text{(4.37)} \quad \{\beta_2, \sigma\} = \frac{2\left(\frac{\nu s^2}{2}\right)^{\frac{\nu}{2}}}{\left(2\pi z_{ii}\right)^{\frac{1}{2}}\Gamma\left(\frac{\nu}{2}\right)}\sigma^{-(\nu+2)}e^{-\frac{\nu s^2}{2\sigma^2}}e^{-\frac{1}{2}(\beta_2-\mu_2)^T\frac{1}{z_{ii}\sigma^2}(\beta_2-\mu_2)} \quad .$$

For later purposes that will be clear shortly, we define

$$\text{(4.38)} \quad P = \nu s^2 \quad ,$$

$$\text{(4.39)} \quad Q = \frac{1}{z_{ii}} \quad ,$$

$$\text{(4.40)} \quad k = \nu + 2 \quad .$$

The univariate normal distribution with the variance σ^2 multiplied by z_{ii} is,

$$N\left(\beta_2; \mu_2, z_{ii}\sigma^2\right) = \frac{1}{\sigma\sqrt{2\pi z_{ii}}}e^{-\frac{1}{2}(\beta_2-\mu_2)^T\frac{1}{z_{ii}\sigma^2}(\beta_2-\mu_2)}$$

in which the mean continues to be μ but the variance is changed to $\sigma^2 z_{ii}$. We rearrange this to write

$$e^{-\frac{1}{2}(\beta_2-\mu_2)^T\frac{1}{z_{ii}\sigma^2}(\beta_2-\mu_2)} = N\left(\beta_2;\mu_2,z_{ii}\sigma^2\right)\sigma\sqrt{2\pi z_{ii}} \quad .$$

When we substitute this into the density, we obtain the joint distribution

$$\{\beta_2,\sigma\} = \frac{2\left(\frac{\nu s^2}{2}\right)^{\frac{\nu}{2}}}{\Gamma\left(\frac{\nu}{2}\right)}\sigma^{-\nu-1}e^{-\frac{\nu s^2}{2\sigma^2}}N\left(\beta_2;\mu_2,z_{ii}\sigma^2\right) \quad .$$

To identify the underlying density using our "quick and dirty" substitution, we integrate this and set the answer to 1

$$1 = \frac{\left(\frac{\nu s^2}{2}\right)^{\frac{\nu}{2}}}{\Gamma\left(\frac{\nu}{2}\right)}\int \sigma^{-\nu-1}e^{-\frac{\nu s^2}{2\sigma^2}}N\left(\beta_2;\mu_2,z_{ii}\sigma^2\right) \quad .$$

If we conceive the following substitutions

$$z=\sigma^2 \Rightarrow \sigma = z^{\frac{1}{2}} \Rightarrow d\sigma = \frac{1}{2}z^{-\frac{1}{2}}dz \Rightarrow \sigma^{-\nu-1} = z^{\frac{-\nu-1}{2}}$$

we obtain the integral that clearly contains the key underlying densities

$$1 = \frac{\left(\frac{\nu s^2}{2}\right)^{\frac{\nu}{2}}}{\Gamma\left(\frac{\nu}{2}\right)}\int z^{\frac{-\nu}{2}-1}e^{-\frac{\nu s^2}{2z}}N\left(\beta_2;\mu_2,z_{ii}z\right) \quad .$$

This integral unveils what the joint distribution truly is

$$\{\beta_2,z\} = IG\left(z,\frac{\nu s^2}{2},\frac{\nu}{2}\right)N\left(\beta_2;\mu_2,z_{ii}z\right) \quad . \tag{4.41}$$

We see the same inverse gamma distribution, but it is now multiplied by a univariate rather than multivariate normal and it needs to have $q=1$. The multivariate marginal is given in equation (4.35) and reproduced here with $q=1$ for easy reference

$$\{\beta_2\} = \frac{\left|H_{22}-H_{21}{H_{11}}^{-1}H_{12}\right|^{\frac{1}{2}}\Gamma\left(\frac{\nu+1}{2}\right)}{(\pi\nu s^2)^{\frac{1}{2}}\Gamma\left(\frac{\nu}{2}\right)} \times\left[1+(\beta_2-\mu_2)^T\frac{\left(H_{22}-H_{21}{H_{11}}^{-1}H_{12}\right)}{\nu s^2}(\beta_2-\mu_2)\right]^{-\frac{\nu+1}{2}} \quad .$$

If we substitute the relationship in equation (4.36) into this equation, we show that the final univariate marginal is

$$(4.42)\ \{\beta_2\} = \frac{\Gamma\left(\frac{\nu+1}{2}\right)}{(\pi\nu s^2 z_{ii})^{\frac{1}{2}}\Gamma\left(\frac{\nu}{2}\right)}\left[1+(\beta_2-\mu_2)^T\frac{1}{z_{ii}\nu s^2}(\beta_2-\mu_2)\right]^{-\frac{\nu+1}{2}} .$$

There is no need to simulate to get the marginals; they are strictly analytic. This is why we don't simulate; analytical insight is lost when we resort to simulation.

It is worthwhile to summarize these calculations to get the precision matrix. We have borrowed a precision matrix generated from a data set introduced in Chapter 12, which appears in Table 4.1. We will be looking to generate the precision matrix (actually a scalar) for the univariate Student's t distribution for the final, rightmost variable in the matrix, namely CP.[19] In the top panel of the figure, we see the complete H matrix. In that H matrix at the top, the A matrix in the upper left corner is shaded gray, the b vector down the rightmost column is unshaded at the upper right, the b^T vector is at the lower left, and the c scalar is shaded gray at the lower right. It should be clear from the shading of the figure how it relates to the CP variable in the H matrix—the c scalar lies at the intersection of the CP row and the CP column.

Table 4.1: An Example H Matrix

H

	Int	Ad	P	CP
Int	36.0000	18,296.0000	3,400.9600	3,411.8000
Ad	18,296.0000	9,454,524.0000	1,728,295.1600	1,733,148.4400
P	3,400.9600	1,728,295.1600	321,558.0044	322,343.6064
CP	3,411.8000	1,733,148.4400	322,343.6064	323,576.3140

H^{-1}

	Int	Ad	P	CP
Int	69.950682630	-0.005567355	-0.319147816	-0.389810452
Ad	-0.005567355	0.000006523	0.000001359	0.000022409
P	-0.319147816	0.000001359	0.003811520	-0.000439171
CP	-0.389810452	0.000022409	-0.000439171	0.004430736

The inverse matrix H^{-1} appears in the lower panel of Table 4.1. We have shaded the intersection of the CP row and the CP column in the H^{-1} matrix for quick and easy reference. We have shown that the shaded element is going to be the scalar

$$(c - b^T A^{-1} b)^{-1} = \frac{1}{c - b^T A^{-1} b} = z_{4,4} = [H^{-1}]_{4,4} .$$

We have extracted the key submatrices from the matrix in Table 4.1 and arrayed them in Table 4.2. We see the matrix A, which is grayed at the top of Table 4.1 and reproduced directly in gray in Table 4.2. We see the b vector from the upper

[19] All other terms can be analyzed similarly by simply permuting rows and columns.

right in the unshaded areas of Table 4.2. We see the c scalar from the lower left of Table 4.1 transferred to the bottom of Table 4.2.

Table 4.2: The Components of the H Matrix

A	36.0000	18,296.0000	3,400.9600
	18,296.0000	9,454,524.0000	1,728,295.1600
	3,400.9600	1,728,295.1600	321,558.0044
b	3,411.8000		
	1,733,148.4400		
	322,343.6064		
b^T	3,411.8000	1,733,148.4400	322,343.6064
c	323,576.3140		

The requisite calculations are summarized in Table 4.3. It is worthwhile to walk through the figure in detail to understand the nuances of the univariate calculation. We see at the top the inverse of the matrix A from Table 4.2. It is crucially important to notice that A^{-1} is NOT equal to the first three rows and columns, i.e., the 3×3 submatrix, of the full inverse H^{-1}. In particular, one cannot assume that the subset of H^{-1} that is in the position where the A matrix was withdrawn is in fact the A^{-1} matrix. It is NOT. That is why this calculation is subtle. The second panel of Table 4.3 shows the first half of the matrix multiplication $b^T A^{-1}$, and the third panel shows the entire calculation $b^T A^{-1} b$. The next panel shows the calculation of $c - b^T A^{-1} b$. This value is the precision matrix for the univariate Student's t distribution over the CP term.

Table 4.3: The Inverse Matrix

A^{-1}	35.655662985	-0.003595855	-0.357785479
	-0.003595855	0.000006410	0.000003580
	-0.357785479	0.000003580	0.003767990
$b^T A^{-1}$	87.978707224	-0.005057585	0.099119104
$b^T A^{-1} b$	323,350.6179		
$c - b^T A^{-1} b$	225.696121635	Precision	
z_{44}	0.004430736		
$\frac{1}{z_{44}}$	225.6961216		

We move the $z_{4,4}$ term from the inverse of the original matrix H^{-1} in Table 4.1 into the shaded position in Table 4.3. This is the entry in the inverse of the original H matrix calculated in full generality. Its reciprocal is shown at the bottom of Table 4.3. Voila! It is exactly equal to the precision matrix term above! If we calculate the inverse of the H matrix in the normal fashion and extract the 4,4 entry, namely 0.004430736, and thereafter take its reciprocal, we have the correct and proper precision matrix (in reality, a precision scalar) for the univariate Student's t distribution of the fourth parameter CP. This is what the development in the text tells us: calculate the inverse of the original, multivariate precision matrix in the normal fashion, extract the diagonals, take their reciprocals, and you thereby have the precision matrices of the four univariate Student's t distributions, one for each variable. There are no shortcuts; you must make the calculation correctly, serially, and explicitly.

It is important to note that this simple technique does not work for bivariate or higher-order joint marginals. The reason is that their inverses are not their reciprocals. It only works for single-variable, univariate marginals. We have included this numerical example to show in detail what the subtle calculation is and how it must be made.

4.15 Derivation of General Univariate Student's t from First Principles

We need to take a detour to see EXACTLY how we systematically derive a univariate Student's t distribution such as that in equation (4.42) with the proper coefficients in a univariate setting. We do so in order to understand the correct and proper orientation of the terms in the univariate Student's t distributions derived as the marginal distributions previously, which we will use to display key Bayesian results. The conjugate-archetype distribution, with which we begin by integrating to 1, is

$$\{\beta, \sigma\} = c\sigma^{-k} e^{-\frac{P}{2\sigma^2}} e^{-\frac{1}{2\sigma^2}(\beta-\mu)^T Q(\beta-\mu)}$$

in which Q, P, μ, and β are all scalars. This is the starting point to develop the univariate Student's t distribution and its key attributes.

4.15.1 Integrating Constant

The first task is to calculate the constant of integration by integrating the foregoing joint density to 1

$$1 = \int_{-\infty}^{\infty} \int_{0}^{\infty} d\beta d\sigma \, \{\beta, \sigma\}$$

$$= \int_{-\infty}^{\infty} \int_{0}^{\infty} d\beta d\sigma c \sigma^{-k} e^{-\frac{P}{2\sigma^2}} e^{-\frac{1}{2\sigma^2}(\beta-\mu)^T Q(\beta-\mu)}$$

$$= \int_{-\infty}^{\infty} \int_{0}^{\infty} d\beta d\sigma c \sigma^{-k} e^{-\frac{P}{2\sigma^2}} e^{-\frac{1}{2\left(\frac{\sigma^2}{Q}\right)}(\beta-\mu)^T (\beta-\mu)} \quad .$$

This suggests the substitutions

$$S^2 = \frac{\sigma^2}{Q} \Rightarrow S = \frac{\sigma}{\sqrt{Q}} \Rightarrow \sigma = S\sqrt{Q} \Rightarrow \sigma^2 = QS^2$$
$$d\sigma = dS\sqrt{Q}$$
$$\sigma^{-k} = S^{-k} Q^{-\frac{k}{2}} \quad .$$

We substitute this into the integral and simplify

$$1 = \int_{-\infty}^{\infty} \int_{0}^{\infty} d\beta d\sigma c \sigma^{-k} e^{-\frac{P}{2\sigma^2}} e^{-\frac{1}{2\left(\frac{\sigma^2}{Q}\right)}(\beta-\mu)^T (\beta-\mu)}$$

$$= c \int_{-\infty}^{\infty} \int_{0}^{\infty} d\beta \left(dS\sqrt{Q}\right) \left(S^{-k} Q^{-\frac{k}{2}}\right) e^{-\frac{P}{2QS^2}} e^{-\frac{1}{2S^2}(\beta-\mu)^T (\beta-\mu)}$$

$$= c \int_{-\infty}^{\infty} \int_{0}^{\infty} d\beta \left(dS\sqrt{Q}\right) \left(S^{-k} Q^{-\frac{k}{2}}\right) e^{-\frac{P}{2QS^2}} \frac{S\sqrt{2\pi}}{S\sqrt{2\pi}} e^{-\frac{1}{2S^2}(\beta-\mu)^T (\beta-\mu)}$$

$$= c \int_{0}^{\infty} \left(dS\sqrt{Q}\right) \left(S^{-k} Q^{-\frac{k}{2}}\right) e^{-\frac{P}{2QS^2}} S\sqrt{2\pi} \int_{-\infty}^{\infty} d\beta \frac{1}{S\sqrt{2\pi}} e^{-\frac{1}{2S^2}(\beta-\mu)^T (\beta-\mu)} \quad .$$

The final term on the right is structured as a univariate normal that integrates to 1. Therefore

$$1 = c \int_{0}^{\infty} \left(dS\sqrt{Q}\right) \left(S^{-k} Q^{-\frac{k}{2}}\right) e^{-\frac{P}{2QS^2}} S\sqrt{2\pi}$$

$$= c\sqrt{2\pi} Q^{\frac{-k+1}{2}} \int_{0}^{\infty} dS S^{-k+1} e^{-\frac{P}{2QS^2}} \quad .$$

We establish the h substitutions as before

$$h = \frac{P}{2QS^2} \Rightarrow S^2 = h^{-1} \frac{P}{2Q} \Rightarrow S = h^{-\frac{1}{2}} \left(\frac{P}{2Q}\right)^{\frac{1}{2}}$$

$$\Rightarrow dS = -\frac{1}{2}h^{-\frac{3}{2}}dh\left(\frac{P}{2Q}\right)^{\frac{1}{2}}$$
$$\Rightarrow S^{(-k+1)} = h^{-\frac{1}{2}(-k+1)}\left(\frac{P}{2Q}\right)^{\frac{1}{2}(-k+1)} \quad .$$

Substitute into the integral and manipulate

$$1 = c\sqrt{2\pi}Q^{\frac{-k+1}{2}}\int_0^\infty dS S^{-k+1}e^{-\frac{P}{2QS^2}}$$
$$= c\sqrt{2\pi}Q^{\frac{-k+1}{2}}\int_\infty^0\left[-\frac{1}{2}h^{-\frac{3}{2}}dh\left(\frac{P}{2Q}\right)^{\frac{1}{2}}\right]\left[h^{-\frac{1}{2}(-k+1)}\left(\frac{P}{2Q}\right)^{\frac{1}{2}(-k+1)}\right]e^{-h}$$
$$= \frac{1}{2}c\sqrt{2\pi}Q^{\frac{-k+1}{2}}\left(\frac{P}{2Q}\right)^{-\frac{1}{2}k+\frac{2}{2}}\int_0^\infty dh h^{\left(\frac{k-2}{2}\right)-1}e^{-h}$$
$$= \frac{1}{2}c\sqrt{2\pi}\left(\frac{P}{2}\right)^{-\frac{k-2}{2}}Q^{-\frac{1}{2}}\Gamma\left(\frac{k-2}{2}\right)$$

in which we have used the gamma function. This gives us the constant

$$c = \frac{2\left(\frac{P}{2}\right)^{\frac{k-2}{2}}Q^{\frac{1}{2}}}{\sqrt{2\pi}\Gamma\left(\frac{k-2}{2}\right)} \quad .$$

The density, complete with the integrating constant, is thus shown to be

$$\text{(4.43)} \quad \{\beta,\sigma\} = \frac{2\sqrt{Q}}{\sqrt{2\pi}\Gamma\left(\frac{k-2}{2}\right)}\left(\frac{P}{2}\right)^{\frac{k-2}{2}}\sigma^{-k}e^{-\frac{P}{2\sigma^2}}e^{-\frac{1}{2\sigma^2}(\beta-\mu)^TQ(\beta-\mu)} \quad .$$

Comparing the two univariate Student's t distributions, we see that

$$\frac{Q}{P} = \frac{1}{z_{ii}\nu s^2} \Rightarrow \frac{P}{Q} = z_{ii}\nu s^2$$

and

$$\frac{\Gamma\left(\frac{k-1}{2}\right)}{\pi^{\frac{1}{2}}\Gamma\left(\frac{k-2}{2}\right)}\left(\frac{1}{z_{ii}\nu s^2}\right)^{\frac{1}{2}} = \frac{\Gamma\left(\frac{\nu+1}{2}\right)}{(\pi z_{ii}\nu s^2)^{\frac{1}{2}}\Gamma\left(\frac{\nu}{2}\right)}$$
$$\Rightarrow \frac{\Gamma\left(\frac{k-1}{2}\right)}{\Gamma\left(\frac{k-2}{2}\right)} = \frac{\Gamma\left(\frac{\nu+1}{2}\right)}{\Gamma\left(\frac{\nu}{2}\right)} \Rightarrow k = \nu + 2 \quad .$$

4.15.2 Underlying Univariate Student's t Distribution

Let us now integrate out the σ term to calculate the underlying univariate Student's t distribution given this joint distribution

$$\begin{aligned}
&\int_0^\infty d\sigma \{\beta,\sigma\} \\
&= \int_0^\infty d\sigma \frac{2\sqrt{Q}}{\sqrt{2\pi}\Gamma\left(\frac{k-2}{2}\right)}\left(\frac{P}{2}\right)^{\frac{k-2}{2}} \sigma^{-k} e^{-\frac{P}{2\sigma^2}\left[1+(\beta-\mu)^T\frac{Q}{P}(\beta-\mu)\right]} \quad .
\end{aligned}$$

We make the following substitutions

$$\begin{aligned}
h &= \frac{P}{2\sigma^2}\left[1+(\beta-\mu)^T\frac{Q}{P}(\beta-\mu)\right] \\
\Rightarrow \sigma &= \left(\frac{P}{2}\right)^{\frac{1}{2}} h^{-\frac{1}{2}}\left[1+(\beta-\mu)^T\frac{Q}{P}(\beta-\mu)\right]^{\frac{1}{2}} \\
d\sigma &= -\frac{1}{2}\left(\frac{P}{2}\right)^{\frac{1}{2}} h^{-\frac{3}{2}}dh\left[1+(\beta-\mu)^T\frac{Q}{P}(\beta-\mu)\right]^{\frac{1}{2}} \\
\sigma^{-k} &= \left(\frac{P}{2}\right)^{-\frac{k}{2}} h^{\frac{k}{2}}\left[1+(\beta-\mu)^T\frac{Q}{P}(\beta-\mu)\right]^{-\frac{k}{2}} \quad .
\end{aligned}$$

We compute the marginal distribution over the coefficients β by integrating over σ

$$\begin{aligned}
\{\beta\} &= \int_0^\infty d\sigma \{\beta,\sigma\} \\
&= \frac{\Gamma\left(\frac{k-1}{2}\right)}{\sqrt{\pi}\Gamma\left(\frac{k-2}{2}\right)}\left(\frac{Q}{P}\right)^{\frac{1}{2}}\left[1+(\beta-\mu)^T\frac{Q}{P}(\beta-\mu)\right]^{-\frac{k-1}{2}} \quad .
\end{aligned} \tag{4.44}$$

4.15.3 The Mean

We return to equation (4.43) to calculate the mean value of the coefficients

$$\begin{aligned}
&\int_0^\infty d\sigma \int_{-\infty}^\infty d\beta\beta \{\beta,\sigma\} = \langle\beta\rangle \\
&= \int_0^\infty d\sigma \int_{-\infty}^\infty d\beta\beta \frac{2\sqrt{Q}}{\sqrt{2\pi}\Gamma\left(\frac{k-2}{2}\right)}\left(\frac{P}{2}\right)^{\frac{k-2}{2}} \sigma^{-k} e^{-\frac{P}{2\sigma^2}} e^{-\frac{1}{2\sigma^2}(\beta-\mu)^T Q(\beta-\mu)} \quad .
\end{aligned}$$

The mean value of the coefficients is calculated as follows

$$\begin{aligned}
\langle\beta\rangle &= \int_0^\infty d\sigma \int_{-\infty}^\infty d\beta\beta\left\{\beta,\sigma\right\} \\
&= \int_0^\infty d\sigma \int_{-\infty}^\infty d\beta\beta \frac{2\sqrt{Q}}{\sqrt{2\pi}\Gamma\left(\frac{k-2}{2}\right)}\left(\frac{P}{2}\right)^{\frac{k-2}{2}} \sigma^{-k} e^{-\frac{P}{2\sigma^2}} e^{-\frac{1}{2}(\beta-\mu)^T \frac{Q}{\sigma^2}(\beta-\mu)} \quad .
\end{aligned}$$

We make the following definition in the first equation and write its implications

$$\begin{aligned}
&\frac{Q}{\sigma^2} = \frac{1}{S^2} \\
&\Rightarrow \sigma^2 = S^2 Q \\
&\Rightarrow \sigma = S Q^{\frac{1}{2}} \\
&\Rightarrow \sigma^{-k} = S^{-k} Q^{-\frac{k}{2}} \\
&\Rightarrow d\sigma = dS\sqrt{Q} \\
&\Rightarrow \frac{P}{2\sigma^2} = \frac{P}{2QS^2} \quad .
\end{aligned}$$

We make the foregoing substitutions and simplify

$$\begin{aligned}
\langle\beta\rangle &= \frac{2\sqrt{Q}}{\sqrt{2\pi}\Gamma\left(\frac{k-2}{2}\right)}\left(\frac{P}{2}\right)^{\frac{k-2}{2}} \\
&\times \int_0^\infty \left[dS\sqrt{Q}\right]\left[S^{-k}Q^{-\frac{k}{2}}\right] e^{-\frac{P}{2QS^2}} \int_{-\infty}^\infty d\beta\beta e^{-\frac{1}{2S^2}(\beta-\mu)^T(\beta-\mu)} \\
&= \frac{2\sqrt{Q}}{\sqrt{2\pi}\Gamma\left(\frac{k-2}{2}\right)}\left(\frac{P}{2}\right)^{\frac{k-2}{2}} \\
&\times \int_0^\infty dS S^{-k} Q^{\frac{-k+1}{2}} e^{-\frac{P}{2QS^2}} \int_{-\infty}^\infty d\beta\beta e^{-\frac{1}{2S^2}(\beta-\mu)^T(\beta-\mu)} \quad .
\end{aligned}$$

The normal distribution, from Wiki,[20] is written with S in place of σ

$$\begin{aligned}
N\left(\beta;\mu,S\right) &= \frac{1}{S\sqrt{2\pi}} e^{-\frac{1}{2S^2}(\beta-\mu)^T(\beta-\mu)} \\
\Rightarrow e^{-\frac{1}{2S^2}(\beta-\mu)^T(\beta-\mu)} &= S\sqrt{2\pi} N\left(\beta;\mu,S\right) \quad .
\end{aligned}$$

We can substitute and simplify

[20] https://en.wikipedia.org/wiki/Normal_distribution.

$$
\begin{aligned}
\langle\beta\rangle &= \frac{2\sqrt{Q}}{\sqrt{2\pi}\Gamma\left(\frac{k-2}{2}\right)}\left(\frac{P}{2}\right)^{\frac{k-2}{2}} \\
&\times \int_0^\infty dS S^{-k} Q^{\frac{-k+1}{2}} e^{-\frac{P}{2QS^2}} \int_{-\infty}^{\infty} d\beta \beta S \sqrt{2\pi} N(\beta;\mu,S) \\
&= \frac{\sqrt{2\pi}2\sqrt{Q}}{\sqrt{2\pi}\Gamma\left(\frac{k-2}{2}\right)}\left(\frac{P}{2}\right)^{\frac{k-2}{2}} Q^{\frac{-k+1}{2}} \\
&\times \int_0^\infty dS S^{-k+1} e^{-\frac{P}{2QS^2}} \int_{-\infty}^{\infty} d\beta \beta N(\beta;\mu,S) \\
&= \frac{2}{\Gamma\left(\frac{k-2}{2}\right)}\left(\frac{P}{2}\right)^{\frac{k-2}{2}} Q^{\frac{-k+2}{2}} \int_0^\infty dS S^{-k+1} e^{-\frac{P}{2QS^2}} \mu \\
&= \mu \frac{2}{\Gamma\left(\frac{k-2}{2}\right)}\left(\frac{P}{2Q}\right)^{\frac{k-2}{2}} \int_0^\infty dS S^{-k+1} e^{-\frac{P}{2QS^2}} \quad .
\end{aligned}
$$

Let us make the substitution

$$
\begin{aligned}
h &= \frac{P}{2QS^2} \\
&\Rightarrow S^2 = \frac{P}{2Q} h^{-1} \\
&\Rightarrow S = h^{-\frac{1}{2}}\left(\frac{P}{2Q}\right)^{\frac{1}{2}} \\
&\Rightarrow dS = -\frac{1}{2} h^{-\frac{3}{2}} dh \left(\frac{P}{2Q}\right)^{\frac{1}{2}} \\
&\Rightarrow S^{-k+1} = h^{\frac{k-1}{2}}\left(\frac{P}{2Q}\right)^{\frac{-k+1}{2}} \quad .
\end{aligned}
$$

When we do so, we get to the salient result

$$
\begin{aligned}
\langle\beta\rangle &= \mu \frac{2}{\Gamma\left(\frac{k-2}{2}\right)}\left(\frac{P}{2Q}\right)^{\frac{k-2}{2}} \\
&\times \int_\infty^0 \left[-\frac{1}{2} h^{-\frac{3}{2}} dh \left(\frac{P}{2Q}\right)^{\frac{1}{2}}\right]\left[h^{\frac{k-1}{2}}\left(\frac{P}{2Q}\right)^{\frac{-k+1}{2}}\right] e^{-h}
\end{aligned}
$$

$$= \mu \frac{1}{\Gamma\left(\frac{k-2}{2}\right)} \int_0^\infty dh h^{\frac{k-2}{2}-1} e^{-h}$$

$$= \mu \quad . \tag{4.45}$$

We have proven that the mean of the coefficients under the bivariate distribution in equation (4.43) is μ.

4.15.4 The Variance

Turning now to the variance of that same distribution in equation (4.43)

$$\text{variance} = \frac{2\sqrt{Q}}{\sqrt{2\pi}\Gamma\left(\frac{k-2}{2}\right)} \left(\frac{P}{2}\right)^{\frac{k-2}{2}}$$
$$\times \int_{-\infty}^{\infty} \int_0^{\infty} d\beta d\sigma (\beta - \mu)^T (\beta - \mu) \sigma^{-k} e^{-\frac{P}{2\sigma^2}} e^{-\frac{1}{2\sigma^2}(\beta-\mu)^T Q (\beta-\mu)} \quad .$$

This suggests the same substitution we performed earlier

$$S^2 = \frac{\sigma^2}{Q} \Rightarrow S = \frac{\sigma}{\sqrt{Q}} \Rightarrow \sigma = S\sqrt{Q} \Rightarrow \sigma^2 = QS^2$$
$$d\sigma = dS\sqrt{Q}$$
$$\sigma^{-k} = S^{-k} Q^{-\frac{k}{2}} \quad .$$

This creates simplification of the variance calculation

$$\text{variance} = \frac{2\sqrt{Q}}{\sqrt{2\pi}\Gamma\left(\frac{k-2}{2}\right)} \left(\frac{P}{2}\right)^{\frac{k-2}{2}} Q^{\frac{-k+1}{2}}$$
$$\times \int_{-\infty}^{\infty} \int_0^{\infty} d\beta dS (\beta - \mu)^T (\beta - \mu) S^{-k} e^{-\frac{P}{2QS^2}} \frac{S\sqrt{2\pi}}{S\sqrt{2\pi}} e^{-\frac{1}{2S^2}(\beta-\mu)^T(\beta-\mu)}$$
$$= \frac{2\sqrt{Q}}{\sqrt{2\pi}\Gamma\left(\frac{k-2}{2}\right)} \left(\frac{P}{2}\right)^{\frac{k-2}{2}} Q^{\frac{-k+1}{2}} \sqrt{2\pi}$$
$$\times \int_{-\infty}^{\infty} \int_0^{\infty} d\beta dS (\beta - \mu)^T (\beta - \mu) S^{-k+1} e^{-\frac{P}{2QS^2}} \frac{1}{S\sqrt{2\pi}} e^{-\frac{1}{2S^2}(\beta-\mu)^T(\beta-\mu)}$$
$$= \frac{2\sqrt{Q}}{\sqrt{2\pi}\Gamma\left(\frac{k-2}{2}\right)} \left(\frac{P}{2}\right)^{\frac{k-2}{2}} Q^{\frac{-k+1}{2}} \sqrt{2\pi}$$
$$\times \int_0^{\infty} dS S^{-k+1} e^{-\frac{P}{2QS^2}} \int_{-\infty}^{\infty} d\beta (\beta - \mu)^T (\beta - \mu) \frac{1}{S\sqrt{2\pi}} e^{-\frac{1}{2S^2}(\beta-\mu)^T(\beta-\mu)} \quad .$$

The final term on the right is the variance of a normal distribution, namely S^2. Substituting that variance and rearranging gives the equation

$$\text{variance} = \frac{2Q^{\frac{-k+2}{2}}}{\Gamma\left(\frac{k-2}{2}\right)}\left(\frac{P}{2}\right)^{\frac{k-2}{2}}\int_0^\infty dSS^{-k+3}e^{-\frac{P}{2QS^2}} \quad .$$

Prepare the same substitutions as above

$$h = \frac{P}{2QS^2} \Rightarrow S^2 = h^{-1}\frac{P}{2Q} \Rightarrow S = h^{-\frac{1}{2}}\left(\frac{P}{2Q}\right)^{\frac{1}{2}}$$
$$dS = -\frac{1}{2}h^{-\frac{3}{2}}dh\left(\frac{P}{2Q}\right)^{\frac{1}{2}}$$
$$S^{(-k+3)} = h^{\frac{k-3}{2}}\left(\frac{P}{2Q}\right)^{\frac{-k+3}{2}} \quad .$$

This leads to the following answer for the integral

$$\int_0^\infty dSS^{(-k+3)}e^{-\frac{P}{2QS^2}} = \int_\infty^0 -\frac{1}{2}h^{-\frac{3}{2}}dh\left(\frac{P}{2Q}\right)^{\frac{1}{2}}h^{\frac{k-3}{2}}\left(\frac{P}{2Q}\right)^{\frac{-k+3}{2}}e^{-h}$$
$$= \frac{1}{2}\left(\frac{P}{2Q}\right)^{-\left(\frac{k-4}{2}\right)}\int_0^\infty dhh^{\left(\frac{k-4}{2}\right)-1}e^{-h}$$
$$= \frac{1}{2}\left(\frac{P}{2Q}\right)^{-\left(\frac{k}{2}-2\right)}\Gamma\left(\frac{k}{2}-2\right) \quad .$$

Substitute and simplify

$$\text{variance} = \frac{2Q^{\frac{-k+2}{2}}}{\Gamma\left(\frac{k-2}{2}\right)}\left(\frac{P}{2}\right)^{\frac{k-2}{2}}\int_0^\infty dSS^{-k+3}e^{-\frac{P}{2QS^2}} \tag{4.46}$$
$$= \frac{2Q^{\frac{-k+2}{2}}}{\Gamma\left(\frac{k-2}{2}\right)}\left(\frac{P}{2}\right)^{\frac{k-2}{2}}\frac{1}{2}\left(\frac{P}{2Q}\right)^{-\left(\frac{k}{2}-2\right)}\Gamma\left(\frac{k}{2}-2\right)$$
$$= Q^{\frac{-k+2}{2}}\left(\frac{P}{2}\right)^{\frac{k-2}{2}}\left(\frac{P}{2Q}\right)^{-\left(\frac{k}{2}-2\right)}\frac{\Gamma\left(\frac{k}{2}-2\right)}{\Gamma\left(\frac{k}{2}-1\right)}$$
$$= \left(\frac{P}{2Q}\right)\frac{\Gamma\left(\frac{k}{2}-2\right)}{\Gamma\left(\frac{k}{2}-1\right)} = \left(\frac{P}{2Q}\right)\frac{\Gamma\left(\frac{k}{2}-2\right)}{\left(\frac{k}{2}-2\right)\Gamma\left(\frac{k}{2}-2\right)}$$
$$= \frac{1}{k-4}\left(\frac{P}{Q}\right) \quad .$$

We have just developed a simple expression for the variance of the univariate Student's t distribution, and it applies directly to our univariate marginal distributions. We are going to examine our univariate Student's t distributions and associate terms with P, Q, and k in this development.

4.15.5 Sidebar–Variable Change to Hotelling t^2 and Fisher-Snedecor F Distributions

This section develops change-of-variable transformations so that one can derive a distribution over t^2 and can derive a Fisher-Snedecor F distribution, both of which might be used to define ranges over which to examine the posterior (or the prior). We perform the analysis in the univariate case here, which can be extended to the multivariate case if desired. That is an exercise left to the reader.

We begin with the joint distribution over $\{\beta, \sigma\}$ from equation (4.43) and the marginal distribution over $\{\beta\}$ from equation (4.44) that derives from it. We rewrite equation (4.44) to begin our variable transformation

$$\{\beta\} = \frac{\Gamma\left(\frac{k-1}{2}\right)}{\sqrt{\pi}\Gamma\left(\frac{k-2}{2}\right)}\left(\frac{Q}{P}\right)^{\frac{1}{2}}\left[1+\frac{Q}{P}(\beta-\mu)^2\right]^{-\frac{k-1}{2}} .$$

There is a well known gamma function relationship[21]

$$\Gamma\left(\frac{1}{2}\right) = \sqrt{\pi}$$

that can be substituted to write the Student's t distribution in the form

$$\{\beta\} = \frac{\Gamma\left(\frac{k-1}{2}\right)}{\Gamma\left(\frac{1}{2}\right)\Gamma\left(\frac{k-2}{2}\right)}\left(\frac{Q}{P}\right)^{\frac{1}{2}}\left[1+\frac{Q}{P}(\beta-\mu)^2\right]^{-\frac{k-1}{2}} .$$

We can use the beta function[22] to simplify the notation so that we can write

$$\{\beta\} = \frac{1}{B\left(\frac{k-2}{2},\frac{1}{2}\right)}\left(\frac{Q}{P}\right)^{\frac{1}{2}}\left[1+\frac{Q}{P}(\beta-\mu)^2\right]^{-\frac{k-1}{2}}$$

in which

$$B\left(\frac{k-2}{2},\frac{1}{2}\right) = \frac{\Gamma\left(\frac{k-2}{2}\right)\Gamma\left(\frac{1}{2}\right)}{\Gamma\left(\frac{k-1}{2}\right)} .$$

The mean and variance were shown above to be

$$M = \mu$$
$$V = \frac{1}{(k-4)}\frac{1}{\frac{Q}{P}} .$$

This implies that

[21] See https://en.wikipedia.org/wiki/Gamma_function.
[22] See https://en.wikipedia.org/wiki/Beta_function.

$$\frac{Q}{P} = \frac{1}{V(k-4)} \quad .$$

We can substitute this to rewrite the marginal distribution over β using the parameters M and V as follows

$$\{\beta\} = \frac{1}{B\left(\frac{k-2}{2}, \frac{1}{2}\right)} \left(\frac{1}{V(k-4)}\right)^{\frac{1}{2}} \left[1 + \frac{(\beta - M)^2}{V(k-4)}\right]^{-\frac{k-1}{2}} \quad .$$

We endeavor to preserve the variable names M for the mean and V for the variance in our expressions so that we understand the final distributions in terms of the mean and variance of the underlying Student's t distribution. We want to have consistency between our degrees of freedom and the definition of k. That means

$$\frac{k-1}{2} = \frac{\nu+1}{2} \Rightarrow k = \nu + 2 \quad .$$

We can substitute for k and write

$$\{\beta\} = \frac{1}{B\left(\frac{\nu}{2}, \frac{1}{2}\right)} \left[\frac{1}{V(\nu-2)}\right]^{\frac{1}{2}} \left[1 + \frac{(\beta - \mu)^2}{V(\nu-2)}\right]^{-\frac{\nu+1}{2}} \quad .$$

To begin our variable transformations, we note that this density must integrate to 1 as follows

$$\begin{aligned} 1 &= \int_{-\infty}^{\infty} \{\beta\}\, d\beta \\ &= \frac{1}{B\left(\frac{\nu}{2}, \frac{1}{2}\right)} \left[\frac{1}{V(\nu-2)}\right]^{\frac{1}{2}} \int_{-\infty}^{\infty} \left[1 + \frac{(\beta - M)^2}{V(\nu-2)}\right]^{-\frac{\nu+1}{2}} d\beta \quad . \end{aligned} \tag{4.47}$$

4.15.5.1 Sidebar–Variable Change to Hotelling t^2 Distribution

We begin with the distribution in Equation (4.47) and make the substitution

$$t^2 = \frac{(\beta - M)^2}{V} \Rightarrow t = \frac{\beta - M}{V^{\frac{1}{2}}} \quad .$$

This substitution uses the notion of the number of standard deviations away from the mean the distribution is spanning. We rearrange to write

$$\beta = M + tV^{\frac{1}{2}}$$

$$d\beta = V^{\frac{1}{2}} dt \quad .$$

Substitution into the distribution in Equation (4.47), we can write

$$1 = \int_{-\infty}^{\infty} \{\beta\} \, d\beta = \frac{1}{B\left(\frac{\nu}{2}, \frac{1}{2}\right)} \left[\frac{1}{V(\nu-2)}\right]^{\frac{1}{2}} \int_{-\infty}^{\infty} \left(1 + \frac{t^2}{\nu-2}\right)^{-\frac{\nu+1}{2}} V^{\frac{1}{2}} dt$$

$$= \frac{1}{B\left(\frac{\nu}{2}, \frac{1}{2}\right)\sqrt{\nu-2}} \int_{-\infty}^{\infty} \left(1 + \frac{t^2}{\nu-2}\right)^{-\frac{\nu+1}{2}} dt \quad .$$

The density over the number of standard deviations we are away from the mean, namely t, thus becomes

$$\{t\} = \frac{1}{B\left(\frac{\nu}{2}, \frac{1}{2}\right)\sqrt{\nu-2}} \left(1 + \frac{t^2}{\nu-2}\right)^{-\frac{\nu+1}{2}} \quad . \tag{4.48}$$

We plot this t distribution in Figure 4.2 using the value $\nu = 20$. We can easily find a t-value such that the probability under both tails from that t-value outward is say five percent. Notice that this distribution is only a function of the degrees of freedom parameter ν. With this distribution, the Hotelling t^2 distribution, one can specify bounds in terms of t, and determine their bounds and probabilities in the original Student's t distribution in equation (4.44). You can posit ranges for t and infer ranges in the original distribution in equation (4.44).

Figure 4.2: t-Statistic Plot with ν = 20

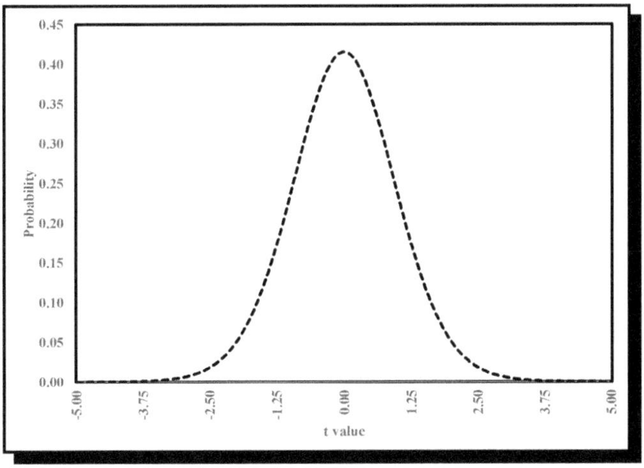

Note that the mean of this distribution in equation (4.48) is zero. We need to calculate the variance. We do this by noting that the variance is $\langle t^2 \rangle$. We write the expression for the mean of the square of t

$$\langle t^2 \rangle = \int_{-\infty}^{\infty} t^2 \{t\}\, dt = \frac{\Gamma\left(\frac{\nu+1}{2}\right)}{\Gamma\left(\frac{\nu}{2}\right)\Gamma\left(\frac{1}{2}\right)\sqrt{\nu-2}} \int_{-\infty}^{\infty} t^2 \left(1 + \frac{t^2}{\nu-2}\right)^{-\frac{\nu+1}{2}} dt \quad .$$

Consider the variable transformation

$$\frac{t^2}{\nu-2} = \frac{x^2}{\nu} \Rightarrow t^2 = \frac{\nu-2}{\nu}x^2 \Rightarrow t = \frac{\sqrt{\nu-2}}{\sqrt{\nu}}x \Rightarrow dt = \frac{\sqrt{\nu-2}}{\sqrt{\nu}}dx \quad .$$

If we substitute this transformation in, we can write

$$\begin{aligned}
\langle t^2 \rangle &= \frac{\Gamma\left(\frac{\nu+1}{2}\right)}{\Gamma\left(\frac{\nu}{2}\right)\Gamma\left(\frac{1}{2}\right)\sqrt{\nu-2}} \int_{-\infty}^{\infty} \frac{\nu-2}{\nu}x^2 \left(1 + \frac{x^2}{\nu}\right)^{-\frac{\nu+1}{2}} \frac{\sqrt{\nu-2}}{\sqrt{\nu}}dx \\
&= \frac{\nu-2}{\nu} \int_{-\infty}^{\infty} \frac{\Gamma\left(\frac{\nu+1}{2}\right)}{\Gamma\left(\frac{\nu}{2}\right)\Gamma\left(\frac{1}{2}\right)\sqrt{\nu}} x^2 \left(1 + \frac{x^2}{\nu}\right)^{-\frac{\nu+1}{2}} dx \\
&= \frac{\nu-2}{\nu}\frac{\nu}{\nu-2} = 1 \quad .
\end{aligned}$$

This distribution has a mean of 0 and a variance (and standard deviation) of 1 no matter what the value of ν. However, it is useful to establish a t-statistic bound for any given value of ν. Figure 4.3 illustrates different shapes for the t-statistics from lower (with higher peaks) to higher (with lower peaks).

Figure 4.3: Three Hotelling t^2 Plots with ν = 5, 20, 100

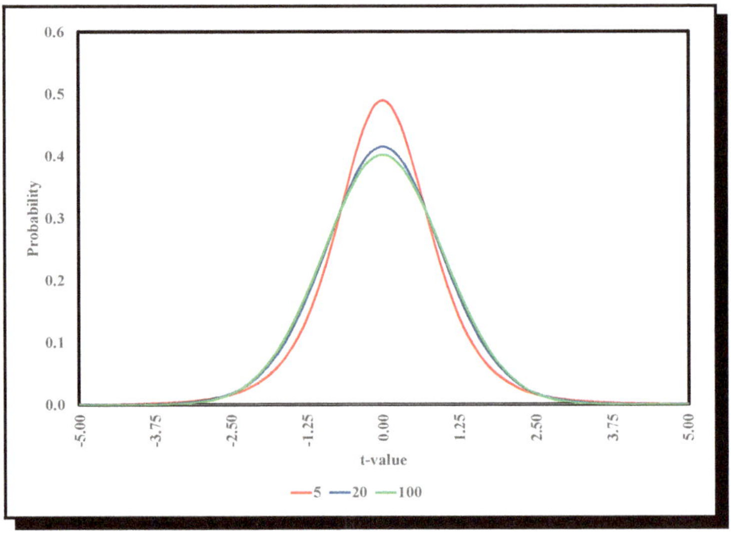

4.15.5.2 Sidebar–Hotelling T^2 Distribution—Total Substitution

This section makes a slightly different substitution:

$$T^2 = \frac{(\beta - M)^2}{(\nu - 2)\,V} \Rightarrow T^{2\frac{1}{2}} = \frac{(\beta - M)^{2\frac{1}{2}}}{(\nu - 2)^{\frac{1}{2}} V^{\frac{1}{2}}} \Rightarrow T = \frac{\beta - M}{(\nu - 2)^{\frac{1}{2}} V^{\frac{1}{2}}}$$

$$\beta = M + T(\nu - 2)^{\frac{1}{2}} V^{\frac{1}{2}}$$

$$d\beta = (\nu - 2)^{\frac{1}{2}} V^{\frac{1}{2}} dT \quad .$$

If we make these substitutions in Equation (4.48), we obtain the equation

$$1 = \int_{-\infty}^{\infty} \{\beta\}\, d\beta = \frac{1}{B\left(\frac{\nu}{2}, \frac{1}{2}\right)} \left[\frac{1}{V\,(\nu - 2)}\right]^{\frac{1}{2}} \int_{-\infty}^{\infty} \left(1 + T^2\right)^{-\frac{\nu+1}{2}} (\nu - 2)^{\frac{1}{2}} V^{\frac{1}{2}} dT$$

$$= \frac{1}{B\left(\frac{\nu}{2}, \frac{1}{2}\right)} \int_{-\infty}^{\infty} \left(1 + T^2\right)^{-\frac{\nu+1}{2}} dT \quad .$$

This implies that the density is

$$\{T\} = \frac{1}{B\left(\frac{\nu}{2}, \frac{1}{2}\right)} \left(1 + T^2\right)^{-\frac{\nu+1}{2}} \quad .$$

The mean of this distribution is 0. We need to determine the variance of this distribution. To do so, consider the integral of this distribution for $\langle T^2 \rangle$

$$\langle T^2 \rangle = \int_{-\infty}^{\infty} T^2 \frac{1}{B\left(\frac{\nu}{2}, \frac{1}{2}\right)} \left(1 + T^2\right)^{-\frac{\nu+1}{2}} dT \quad .$$

Making the same substitutions as we saw previously,

$$\frac{T^2}{\nu - 2} = \frac{x^2}{\nu} \Rightarrow T^2 = \frac{\nu - 2}{\nu} x^2 \Rightarrow T = \frac{\sqrt{\nu - 2}}{\sqrt{\nu}} x \Rightarrow dT = \frac{\sqrt{\nu - 2}}{\sqrt{\nu}} dx \quad .$$

to determine $\langle T^2 \rangle$, we can write

$$\langle T^2 \rangle = \int_{-\infty}^{\infty} \frac{1}{B\left(\frac{\nu}{2}, \frac{1}{2}\right)} \frac{x^2}{\nu} \left(1 + \frac{x^2}{\nu}\right)^{-\frac{\nu+1}{2}} \nu^{-\frac{1}{2}} dx$$

$$= \frac{1}{\nu} \int_{-\infty}^{\infty} x^2 \frac{1}{B\left(\frac{\nu}{2}, \frac{1}{2}\right) \nu^{\frac{1}{2}}} \left(1 + \frac{x^2}{\nu}\right)^{-\frac{\nu+1}{2}} dx$$

$$= \frac{1}{\nu} \frac{\nu}{\nu - 2} = \frac{1}{\nu - 2} \quad .$$

We have plotted the T^2 distribution in Figure 4.4.

Figure 4.4: Full T^2 Plot with $\nu = 20$

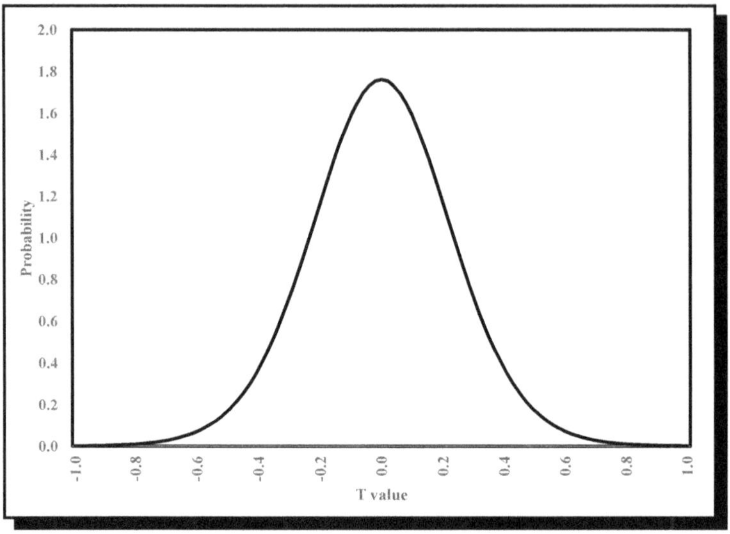

4.15.5.3 Sidebar–Substitution to Create Fisher-Snedecor *F* Distribution

We need to design a substitution in Equation (4.48) so that we end up with an F-distribution. Consider the substitution

$$F = \frac{t^2}{\nu - 2} \Rightarrow t = F^{\frac{1}{2}}\sqrt{\nu - 2} \Rightarrow dt = \frac{1}{2}F^{-\frac{1}{2}}\sqrt{\nu - 2}dF \quad .$$

If we substitute this into in Equation (4.48), we obtain

$$\begin{aligned} 1 &= \frac{2}{B\left(\frac{\nu}{2}, \frac{1}{2}\right)\sqrt{\nu - 2}} \int_0^\infty (1 + F)^{-\frac{\nu+1}{2}} \frac{1}{2} F^{-\frac{1}{2}} \sqrt{\nu - 2} dF \\ &= \frac{1}{B\left(\frac{\nu}{2}, \frac{1}{2}\right)} \int_0^\infty F^{-\frac{1}{2}} (1 + F)^{-\frac{\nu+1}{2}} dF \quad . \end{aligned}$$

The F-distribution is the integrand itself.

$$\{F\} = \frac{1}{B\left(\frac{\nu}{2}, \frac{1}{2}\right)} F^{-\frac{1}{2}} (1 + F)^{-\frac{\nu+1}{2}} \quad .$$

Let us start by setting this integral to 1 so that we can make some substitutions

$$1 = \int \{F\} \, dF = \frac{1}{B\left(\frac{1}{2}, \frac{\nu}{2}\right)} \int\limits_0^\infty F^{-\frac{1}{2}} (1+F)^{-\frac{\nu+1}{2}} \, dF \quad .$$

Consider the following substitution

$$F = \frac{x}{d_2} \Rightarrow dF = dx \frac{1}{d_2} \Rightarrow F^{-\frac{1}{2}} = x^{-\frac{1}{2}} \left(\frac{1}{d_2}\right)^{-\frac{1}{2}} = x^{-\frac{1}{2}} {d_2}^{\frac{1}{2}} \quad .$$

Making the substitution

$$1 = \int \{F\} \, dF = \frac{1}{B\left(\frac{1}{2}, \frac{\nu}{2}\right)} \int\limits_0^\infty \left[x^{-\frac{1}{2}} \left(\frac{1}{d_2}\right)^{-\frac{1}{2}}\right] \left(1 + \frac{x}{d_2}\right)^{-\frac{\nu+1}{2}} dx \frac{1}{d_2}$$

$$= \frac{1}{B\left(\frac{1}{2}, \frac{\nu}{2}\right)} \int\limits_0^\infty x^{-\frac{1}{2}} \nu^{\frac{1}{2}} \frac{1}{\nu} \left(1 + \frac{x}{\nu}\right)^{-\frac{\nu+1}{2}} dx$$

$$= \frac{1}{B\left(\frac{1}{2}, \frac{\nu}{2}\right)} \left(\frac{1}{\nu}\right)^{\frac{1}{2}} \int\limits_0^\infty x^{-\frac{1}{2}} \left(1 + \frac{x}{\nu}\right)^{-\frac{\nu+1}{2}} dx \quad .$$

The embedded density is

$$\{x\} = \frac{1}{B\left(\frac{1}{2}, \frac{\nu}{2}\right)} \left(\frac{1}{\nu}\right)^{\frac{1}{2}} x^{-\frac{1}{2}} \left(1 + \frac{x}{\nu}\right)^{-\frac{\nu+1}{2}} \quad . \tag{4.49}$$

Referring to Wiki for the Fisher-Snedecor distribution[23], we see the general definition written as follows

$$\{x\} = \frac{1}{B\left(\frac{d_1}{2}, \frac{d_2}{2}\right)} \left(\frac{d_1}{d_2}\right)^{\frac{d_1}{2}} x^{\frac{d_1}{2}-1} \left(1 + \frac{d_1}{d_2} x\right)^{-\frac{d_1+d_2}{2}}$$

$$d_1 : numerator \quad d_2 : denominator \quad .$$

If we set the numerator parameter d_1 of this distribution to 1 and we set the denominator parameter d_2 to ν, we obtain precisely the distribution we developed in equation (4.49). All the results in Wiki referenced in the preceding footnote are thus directly available to us. In particular, we see in that referenced Wiki section that the mean of the Fisher-Snedecor distribution we have developed is

$$\frac{\nu}{\nu - 2}$$

and its variance is

[23] See https://en.wikipedia.org/wiki/F-distribution.

$$\frac{2\nu^2(\nu-1)}{(\nu-2)^2(\nu-4)} \quad .$$

The plot of the F-distribution for this case appears in Figure 4.5. It is easy to see how one might develop a right-side (tail) F test for this distribution. In particular, we can determine a value of F^* such that the probability of falling somewhere higher than F^* in that right hand side tail is say 0.05.

Figure 4.5: F-Distribution Plot with $\nu = 20$

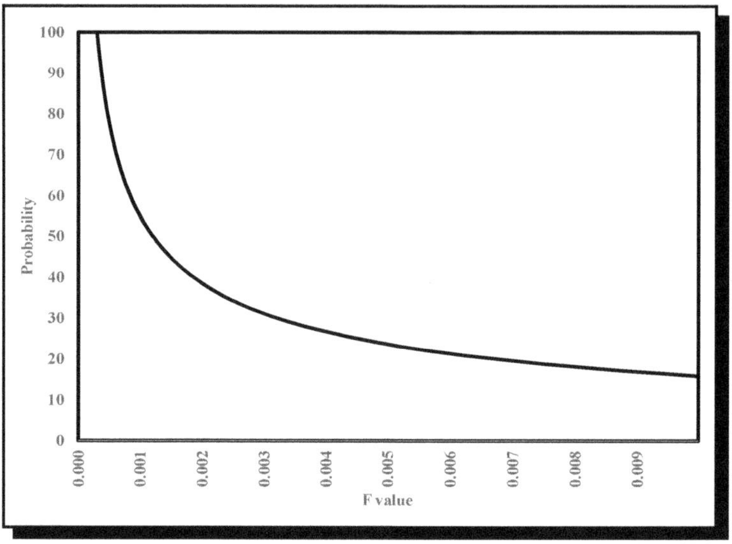

4.15.6 Interim Summary

We now know exactly how a normal times an inverse gamma distribution integrates into a Student's t. Here is the summary:

- The underlying joint distribution is that in equation (4.43).
- The underlying univariate Student's t distribution is that in equation (4.44).
- The mean of the underlying univariate Student's t distribution is shown in equation (4.45) to be μ.
- The variance of the underlying univariate Student's t distribution is shown in equation (4.46) to be

$$\text{variance} = \frac{1}{k-4}\frac{P}{Q}$$

4.15.7 Application to Our Own Univariate Student's t Distribution

The development thus far in this section allows us to go back to our specific univariate Student's t and understand means, variances, and so forth intrinsically. The distribution we have derived in equation (4.42) is rewritten

$$(4.50)\quad \{\beta_2\} = \frac{\Gamma\left(\frac{\nu+1}{2}\right)}{\left(\pi\nu s^2 z_{ii}\right)^{\frac{1}{2}}\Gamma\left(\frac{\nu}{2}\right)}\left[1+(\beta_2-\mu_2)^T\frac{1}{z_{ii}\nu s^2}(\beta_2-\mu_2)\right]^{-\frac{\nu+1}{2}} \quad ,$$

and our newly derived univariate Student's t is written

$$(4.51)\qquad \{\beta\} = \frac{\Gamma\left(\frac{k-1}{2}\right)}{\sqrt{\pi}\Gamma\left(\frac{k-2}{2}\right)}\left(\frac{Q}{P}\right)^{\frac{1}{2}}\left[1+(\beta-\mu)^T\frac{Q}{P}(\beta-\mu)\right]^{-\frac{k-1}{2}} \quad .$$

Suppose we set

$$\begin{aligned} k &= \nu+2 \\ Q &= \frac{1}{z_{ii}} \\ P &= \nu s^2 \quad . \end{aligned}$$

We will be able to establish the equivalency between our specific univariate Student's t as expressed in equation (4.50) and the just-derived, general, parametric one in equation (4.51). In particular, if we make the foregoing substitutions into the generic form in equation (4.51), we obtain our specific form in equation (4.50). Keep in mind, the matrix c indicated here reduces to a single row and column and is actually a scalar. That is why we can eliminate the determinant in this section–the requisite matrix is a scalar, a 1 by 1 matrix. We want to check out the variance, which we do as follows

$$\begin{aligned} \text{variance} &= \frac{1}{k-4}\left(\frac{P}{Q}\right) = \frac{1}{\nu-2}\left(\frac{\nu s^2}{H_{22}-H_{21}{H_{11}}^{-1}H_{12}}\right) \\ &= \frac{\nu s^2}{\nu-2}\frac{1}{\left(H_{22}-H_{21}{H_{11}}^{-1}H_{12}\right)} \\ &= \frac{\nu s^2}{\nu-2}\frac{1}{(c-b^TA^{-1}b)} \\ &= \frac{\nu s^2}{\nu-2}[H^{-1}]_{ii} \\ &= \frac{\nu s^2}{\nu-2}z_{ii} \quad . \end{aligned}$$

The precision matrix is defined to be $\frac{Q}{P}$, which is

$$\begin{aligned}\text{precision} &= \frac{H_{22} - H_{21}{H_{11}}^{-1}H_{12}}{\nu s^2} \\ &= \frac{c - b^T A^{-1} b}{\nu s^2} \\ &= \frac{1}{\nu s^2 [H^{-1}]_{ii}} \\ &= \frac{1}{\nu s^2 z_{ii}} \quad .\end{aligned}$$

implying that

$$\text{precision} = \frac{1}{\nu - 2} \times \frac{1}{\text{variance}} \quad .$$

The process of obtaining the variance of the univariate marginal distribution of the ith coefficient, the precision matrix of that marginal distribution, and in fact the marginal distribution itself is now evident:

1. Calculate the inverse of the entire H matrix, i.e., H^{-1}.
2. Get the ii^{th} diagonal element of the H^{-1} matrix. Designate it z_{ii}

$$\begin{aligned}z_{ii} &= \left[H^{-1}\right]_{ii} \\ &= \left(H_{22} - H_{21}{H_{11}}^{-1}H_{12}\right)^{-1} \\ &= \left(c - b^T A^{-1} b\right)^{-1} \quad .\end{aligned}$$

3. Using these definitions, the univariate Student's t density for the ith coefficient as given by equation (4.44) can be written:

$$\{\beta_i\} = \frac{\Gamma\left(\frac{\nu+1}{2}\right)}{\Gamma\left(\frac{\nu}{2}\right)} \frac{1}{(\pi z_{ii} \nu s^2)^{\frac{1}{2}}} \times \left[1 + (\beta_i - \mu_i) \frac{1}{\nu s^2 z_{ii}} (\beta_i - \mu_i)\right]^{-\frac{\nu+1}{2}} \quad . \tag{4.52}$$

4. This density will integrate to 1 because the constant is correctly calculated. (The proper constant ensures that this integrates to 1.)
5. The variance of the ith variable is the ith diagonal element of the H^{-1} matrix z_{ii} multiplied by the factor $\frac{\nu s^2}{\nu - 2}$ to obtain the variance

$$\text{variance} = \frac{\nu s^2}{\nu - 2} z_{ii} \quad .$$

6. The precision matrix of the ith variable requires the ith diagonal element of the H^{-1} matrix z_{ii} used in the following formula

(4.53) $$\text{precision} = \frac{1}{\nu s^2 z_{i,i}} \quad .$$

which we see in the expression for the univariate density immediately above.

7. It is obvious that the mean is μ_i.

Using this series of steps, we can calculate and plot every marginal density for every coefficient. This will indicate what we really know about each of the coefficients, expressed as a marginal probability of the value of each coefficient.

4.16 The Predictive Distribution

4.16.1 Development of the Predictive Distribution

The final item of business required in this section is to develop a predictive distribution. What does this mean? This means that after we find the coefficients, how can we predict, for a given set of observations, what our model projection will be? This prediction is the simulation problem. In particular, we postulate a matrix D with q rows and p columns that contains row vectors of the values of the independent variables whose value d we want to predict. Specifically, we want to postulate that

$$d = D\beta + w \quad .$$

where w is a q-vector of random variables, each governed by an independent, identically distributed error term with mean zero and variance σ^2 and where the model coefficients β are distributed according to the conjugate-archetype distribution.[24] What we are after is the joint probability distribution, which is q-dimensional, over the vector d? This is not just a simple "change of variables" problem. There is also a "predicted versus observed" univariate distribution at work here, just as there was when the data were generated. We will see how to "peel them apart" later in this section so that we can identify the change of variables problem per se.

In the vernacular of Bayes theorem, what we want to obtain is the joint density over the predicted values d and D and the model coefficients β and σ given the observed data X and y. It is the joint density of the vector d to be predicted, the coefficients given the data we have seen, and the independent variables whose values we wish to predict.

[24]This is to be carefully distinguished from a simple "change of variables" problem $d = D\beta$. That would be incorrect; it would miss the measurement error term σ^2 altogether, i.e., the difference between predicted and actual embedded in the data. It would de facto miss the measurement error between calculated $D\beta$ and actual d, i.e., miss the key term that governs uncertainty in the observed data.

$$\{d, D, \beta, \sigma \,|X, y\} \quad .$$

We decompose this joint distribution into two parts using conditional probability relationships

$$\{d, D, \beta, \sigma \,|X, y\} = \{d, D \,|\beta, \sigma, X, y\}\{\beta, \sigma \,|X, y\} \quad .$$

This is very straightforward and very parallel in form to the original likelihood function in Chapter 2 in the classical regression case. The term furthest to the right is simply our derived distribution over model coefficients in equation (4.8). It is the joint probability distribution over the model coefficients given the observations.

The first term on the right-hand side is the likelihood of observing the vector d and the matrix D, exactly analogous to the original likelihood function observing y and X in equation (2.2), given the model parameters and the original observations. This is no more than the likelihood function of the q predictions to be made. To see why, refer to equation (2.2) and make the substitutions:

$$\begin{aligned} n &\leftarrow q \\ y &\leftarrow d \\ X &\leftarrow D \end{aligned}$$

to obtain the likelihood function, which is very standard in form, identical in form to equation (2.2)

$$\{d, D \,|\beta, \sigma, X, y\} = (2\pi)^{-\frac{q}{2}}\sigma^{-q}e^{-\frac{1}{2\sigma^2}(d-D\beta)^T(d-D\beta)}$$

in which q is the number of observations we want to predict, i.e., the number of rows in the matrix D. We simply multiply the probability density function over coefficients in the distribution in equation (4.8) by this function to obtain the predictions that would be made under the prior and do the same with the predictions that would be made under the posterior. Multiplying this by the prior, we obtain the following functional form for the predictive distribution over d:

$$\begin{aligned} \{d, D, \beta, \sigma \,|X, y\} &= (2\pi)^{-\frac{q}{2}}\sigma^{-q}e^{-\frac{1}{2\sigma^2}(d-D\beta)^T I(d-D\beta)} \\ &\times \frac{2|H|^{\frac{1}{2}}(2\pi)^{-\frac{p}{2}}\left(\frac{\nu s^2}{2}\right)^{\frac{\nu}{2}}}{\Gamma\left(\frac{\nu}{2}\right)}\sigma^{-(\nu+p+1)} \\ &\times e^{-\frac{\nu s^2}{2\sigma^2}}e^{-\frac{1}{2\sigma^2}(\beta-\mu)^T H(\beta-\mu)} \\ &= (2\pi)^{-\frac{q+p}{2}}\sigma^{-(\nu+p+q+1)}\frac{2|H|^{\frac{1}{2}}\left(\frac{\nu s^2}{2}\right)^{\frac{\nu}{2}}}{\Gamma\left(\frac{\nu}{2}\right)}e^{-\frac{\nu s^2}{2\sigma^2}} \end{aligned}$$

$$\times\, e^{-\frac{1}{2\sigma^2}\left[(d-D\beta)^T I(d-D\beta)+(\beta-\mu)^T H(\beta-\mu)\right]} \quad . \tag{4.54}$$

We wish to complete the square on the exponent so we can integrate out the β and σ terms. We focus on the exponent, ignoring the multiplicative factor $-\frac{1}{2\sigma^2}$, which we can re-append at the end. We want to write the exponent as a constant, linear, and quadratic term in β

$$\begin{aligned}
&\text{exponent} = (d - D\beta)^T I\,(d - D\beta) + (\beta - \mu)^T H\,(\beta - \mu) \\
&= \left(d^T - \beta^T D^T\right) I\,(d - D\beta) + \left(\beta^T - \mu^T\right) H\,(\beta - \mu) \\
&= d^T d - 2\beta^T D^T d + \beta^T D^T D\beta + \beta^T H\beta - 2\beta^T H\mu + \mu^T H\mu \\
&= \left(d^T d + \mu^T H\mu\right) - 2\beta^T \left(D^T d + H\mu\right) + \beta^T \left(D^T D + H\right)\beta \quad .
\end{aligned}$$

We then add and subtract an unknown (as yet) constant λ from β

$$\begin{aligned}
\text{exponent} = {} & d^T d - 2\left[(\beta - \lambda)^T + \lambda^T\right] D^T d \\
& + \left[(\beta - \lambda)^T + \lambda^T\right] D^T D\left[(\beta - \lambda) + \lambda\right] \\
& + \left[(\beta - \lambda)^T + \lambda^T\right] H\left[(\beta - \lambda) + \lambda\right] \\
& - 2\left[(\beta - \lambda)^T + \lambda^T\right] H\mu + \mu^T H\mu \quad .
\end{aligned}$$

For notational simplicity, we define

$$z = \beta - \lambda \quad .$$

This allows us to write

$$\begin{aligned}
&\text{exponent} = d^T d - 2\left(z^T + \lambda^T\right) D^T d + \left(z^T + \lambda^T\right) D^T D\,(z + \lambda) \\
&+ \left(z^T + \lambda^T\right) H\,(z + \lambda) - 2\left(z^T + \lambda^T\right) H\mu + \mu^T H\mu \quad .
\end{aligned}$$

We expand out all the terms

$$\begin{aligned}
&\text{exponent} = d^T d \\
&- 2z^T D^T d - 2\lambda^T D^T d \\
&+ z^T D^T Dz + 2z^T D^T D\lambda + \lambda^T D^T D\lambda \\
&+ z^T Hz + 2z^T H\lambda + \lambda^T H\lambda \\
&- 2z^T H\mu - 2\mu^T H\lambda \\
&+ \mu^T H\mu \quad .
\end{aligned}$$

We rearrange into terms that are constant with respect to z, linear in z, and quadratic in z

$$\begin{aligned}\text{exponent} &= \lambda^T\left(D^T D + H\right)\lambda - 2\left(d^T D + \mu^T H\right)\lambda + \mu^T H\mu + d^T d \\ &+ 2z^T\left[\left(D^T D + H\right)\lambda - \left(H\mu + D^T d\right)\right] \\ &+ z^T\left(D^T D + H\right)z \quad .\end{aligned}$$

We want to select λ so as to eliminate the term that is linear in z^T, i.e., to zero out the term in brackets that is left multiplied by z^T

$$\begin{aligned}&\left(D^T D + H\right)\lambda - \left(H\mu + D^T d\right) = 0 \\ &\lambda = \left(D^T D + H\right)^{-1}\left(H\mu + D^T d\right) \\ &\Rightarrow \lambda^T = \left(\mu^T H + d^T D\right)\left(D^T D + H\right)^{-1} \quad .\end{aligned}$$

For notational convenience, we are going to redefine the inverse matrix

$$\begin{aligned}V &= \left(D^T D + H\right)^{-1} \\ \Rightarrow V^{-1} &= D^T D + H \quad .\end{aligned} \tag{4.55}$$

We can use this to rewrite the definition of λ

$$\begin{aligned}&\lambda = V\left(H\mu + D^T d\right) \\ &\Rightarrow \lambda^T = \left(\mu^T H + d^T D\right)V \quad .\end{aligned}$$

We substitute this value of λ, which eliminates the term that is linear in z by construction

$$\begin{aligned}\text{exponent} &= \left(\mu^T H + d^T D\right)\left(D^T D + H\right)^{-1}\left(D^T D + H\right) \\ &\times \left(D^T D + H\right)^{-1}\left(H\mu + D^T d\right) \\ &- 2\left(d^T D + \mu^T H\right)\left(D^T D + H\right)^{-1}\left(H\mu + D^T d\right) \\ &+ \mu^T H\mu + d^T d \\ &+ z^T\left(D^T D + H\right)z \quad .\end{aligned}$$

This simplifies to

$$\begin{aligned}\text{exponent} &= \mu^T H\mu + d^T d - \left(d^T D + \mu^T H\right)\left(D^T D + H\right)^{-1}\left(H\mu + D^T d\right) \\ &+ z^T\left(D^T D + H\right)z \quad .\end{aligned}$$

This equation motivates us to define

$$Y_0 = \mu^T H\mu + d^T d - \left(d^T D + \mu^T H\right)\left(D^T D + H\right)^{-1}\left(H\mu + D^T d\right) \quad ,$$

so that the exponent can be written compactly

$$\begin{aligned}\text{exponent} &= Y_0 + z^T\left(D^T D + H\right)z \\ &= Y_0 + (\beta-\lambda)^T\left(D^T D + H\right)(\beta-\lambda) \quad .\end{aligned}$$

We can substitute this expression into the density in equation (4.54) to write it in notationally simpler fashion

$$\begin{aligned}\{d, D, \beta, \sigma\,|X, y\} &= \frac{2(2\pi)^{-\frac{q+p}{2}}|H|^{\frac{1}{2}}\left(\frac{\nu s^2}{2}\right)^{\frac{\nu}{2}}}{\Gamma\left(\frac{\nu}{2}\right)} \\ &\quad\times \sigma^{-(\nu+p+q+1)}e^{-\frac{\nu s^2}{2\sigma^2}}e^{-\frac{1}{2\sigma^2}\left[Y_0+(\beta-\lambda)^T\left(D^T D+H\right)(\beta-\lambda)\right]} \\ &= \frac{2(2\pi)^{-\frac{q+p}{2}}|H|^{\frac{1}{2}}\left(\frac{\nu s^2}{2}\right)^{\frac{\nu}{2}}}{\Gamma\left(\frac{\nu}{2}\right)} \\ &\quad\times \sigma^{-(\nu+p+q+1)} \times e^{-\frac{1}{2\sigma^2}\left[\nu s^2+Y_0+(\beta-\lambda)^T\left(D^T D+H\right)(\beta-\lambda)\right]} \quad .\end{aligned}$$

Consider a p-variate normal density of the form

$$\begin{aligned}N(\beta;\lambda,\Sigma) &= \frac{1}{(2\pi)^{\frac{p}{2}}|\Sigma|^{\frac{1}{2}}}e^{-\frac{1}{2}(\beta-\lambda)^T\Sigma^{-1}(\beta-\lambda)} \\ \Rightarrow e^{-\frac{1}{2}(\beta-\lambda)^T\Sigma^{-1}(\beta-\lambda)} &= (2\pi)^{\frac{p}{2}}|\Sigma|^{\frac{1}{2}}N(\beta;\lambda,\Sigma) \quad .\end{aligned}$$

If we let

$$\begin{aligned}\Sigma^{-1} &= \left(D^T D + H\right)\sigma^{-2} \\ \Rightarrow \Sigma &= \left(D^T D + H\right)^{-1}\sigma^2 \\ \Rightarrow |\Sigma| &= \left|D^T D + H\right|^{-1}\sigma^{2p} \\ \Rightarrow |\Sigma|^{\frac{1}{2}} &= \left|D^T D + H\right|^{-\frac{1}{2}}\sigma^p \quad ,\end{aligned}$$

we can substitute into the definition for the normal

$$\begin{aligned}e^{-\frac{1}{2}(\beta-\lambda)^T\left(\frac{D^T D+H}{\sigma^2}\right)(\beta-\lambda)} &= (2\pi)^{\frac{p}{2}}\left|D^T D + H\right|^{-\frac{1}{2}}\sigma^p \\ &\quad\times N\left[\beta;\lambda,\left(D^T D + H\right)^{-1}\sigma^2\right] \quad .\end{aligned}$$

We can thereafter substitute into the density to write

$$\begin{aligned}\{d, D, \beta, \sigma|X, y\} &= \frac{2(2\pi)^{-\frac{q+p}{2}}|H|^{\frac{1}{2}}\left(\frac{\nu s^2}{2}\right)^{\frac{\nu}{2}}}{\Gamma\left(\frac{\nu}{2}\right)} \\ &\quad\times \sigma^{-(\nu+p+q+1)}e^{-\frac{1}{2\sigma^2}\left(\nu s^2+Y_0\right)}(2\pi)^{\frac{p}{2}}\left|D^T D + H\right|^{-\frac{1}{2}}\sigma^p\end{aligned}$$

$$\times N\left[\beta; \lambda, \left(D^T D + H\right)^{-1} \sigma^2\right]$$

$$= \frac{2(2\pi)^{-\frac{q}{2}} |H|^{\frac{1}{2}} \left(\frac{\nu s^2}{2}\right)^{\frac{\nu}{2}}}{|D^T D + H|^{\frac{1}{2}} \Gamma\left(\frac{\nu}{2}\right)}$$

$$\times \sigma^{-(\nu+q+1)} e^{-\frac{1}{2\sigma^2}\left(\nu s^2 + Y_0\right)}$$

$$\times N\left[\beta; \lambda, \left(D^T D + H\right)^{-1} \sigma^2\right] \quad .$$

We integrate out β by dropping the trailing normal through integration

$$\{d, D, \sigma | X, y\} = \frac{2(2\pi)^{-\frac{q}{2}} |H|^{\frac{1}{2}} \left(\frac{\nu s^2}{2}\right)^{\frac{\nu}{2}}}{|D^T D + H|^{\frac{1}{2}} \Gamma\left(\frac{\nu}{2}\right)} \times \sigma^{-(\nu+q+1)} e^{-\frac{1}{2\sigma^2}\left(\nu s^2 + Y_0\right)} \quad .$$

The exponent in the second-to-final term motivates us to define

$$\begin{aligned} Y &= \nu s^2 + Y_0 \\ &= \nu s^2 + \mu^T H \mu + d^T d - \left(d^T D + \mu^T H\right) \left(D^T D + H\right)^{-1} \left(H\mu + D^T d\right) \end{aligned}$$

and thereby change the density to

$$\{d, D, \sigma \,| X, y\} = \frac{2(2\pi)^{-\frac{q}{2}} |H|^{\frac{1}{2}} \left(\frac{\nu s^2}{2}\right)^{\frac{\nu}{2}}}{|D^T D + H|^{\frac{1}{2}} \Gamma\left(\frac{\nu}{2}\right)} \sigma^{-(\nu+q+1)} e^{-\frac{1}{2\sigma^2} Y} \quad . \tag{4.56}$$

We need to further simplify Y, which we rewrite

$$\begin{aligned} Y &= \nu s^2 + Y_0 \\ &= \nu s^2 + d^T d + \mu^T H \mu - \left(d^T D + \mu^T H^T\right) \left(D^T D + H\right)^{-1} \left(D^T d + H\mu\right) \quad . \end{aligned}$$

For convenience, we continue the definition from equation (4.55)

$$\begin{aligned} V &= \left(D^T D + H\right)^{-1} \\ V^{-1} &= D^T D + H \quad . \end{aligned}$$

We return to the notation in equation (4.55) and write Y in the following form

$$Y = \nu s^2 + d^T d + \mu^T H \mu - \left(d^T D + \mu^T H^T\right) V \left(D^T d + H\mu\right) \quad .$$

We now want to express Y as a quadratic in d without a linear term, and we know we can complete the square to do so. In particular, we want to add and subtract a constant η (as yet unknown) from d, the intent being to complete the square so as to express Y as a quadratic in d without a linear term.

$$\begin{aligned}
&d = [(d-\eta)+\eta] \Rightarrow d^T = \left[(d-\eta)^T + \eta^T\right] \\
&z = d - \eta \\
&d = z + \eta \Rightarrow d^T = z^T + \eta^T \quad .
\end{aligned}$$

With this substitution, we can write

$$\begin{aligned}
Y &= \nu s^2 + \mu^T H \mu \\
&+ \left(z^T + \eta^T\right)(z+\eta) \\
&- \left[\left(z^T+\eta^T\right) D + \mu^T H\right] V \left[H\mu + D^T (z+\eta)\right] \quad .
\end{aligned}$$

We next expand out all the products and begin grouping and simplifying

$$\begin{aligned}
Y &= \nu s^2 + \mu^T H \mu \\
&+ z^T z + 2z^T \eta + \eta^T \eta \\
&- \left(z^T + \eta^T\right) DV \left[H\mu + D^T (z+\eta)\right] \\
&- \mu^T HV \left[H\mu + D^T (z+\eta)\right] \\
&= \nu s^2 + \mu^T H \mu \\
&+ z^T z + 2z^T \eta + \eta^T \eta \\
&- z^T DVH\mu - z^T DVD^T (z+\eta) - \eta^T DVH\mu - \eta^T DVD^T (z+\eta) \\
&- \mu^T HVH\mu - \mu^T HVD^T z - \mu^T HVD^T \eta \\
&= \nu s^2 + \mu^T H \mu \\
&+ z^T z + 2z^T \eta + \eta^T \eta \\
&- z^T DVH\mu - z^T DVD^T z - 2z^T DVD^T \eta - \eta^T DVH\mu - \eta^T DVD^T \eta \\
&- \mu^T HVH\mu - \mu^T HVD^T z - \mu^T HVD^T \eta \quad .
\end{aligned}$$

We reorganize these terms into constant, linear, and quadratic terms in z and thereafter group and simplify

$$\begin{aligned}
Y &= \nu s^2 + \mu^T (H - HVH)\mu + \eta^T \left(I - DVD^T\right)\eta - 2\mu^T HVD^T \eta \\
&+ 2z^T \left[\left(I - DVD^T\right)\eta - DVH\mu\right] \\
&+ z^T \left(I - DVD^T\right) z \quad .
\end{aligned}$$

In order to complete the square, we want to zero out the term that is linear in z, the middle term. To do so, we solve for η to eliminate the middle term

$$\left(I - DVD^T\right)\eta - DVH\mu = 0 \quad .$$

We solve for η

$$\begin{aligned} \eta &= \left(I - DVD^T\right)^{-1} DVH\mu \\ \Rightarrow \eta^T &= \mu^T HVD^T \left(I - DVD^T\right)^{-1} \quad . \end{aligned} \tag{4.57}$$

We substitute into the equation for Y and begin grouping and simplifying

$$\begin{aligned} Y &= \nu s^2 + \mu^T \left(H - HVH\right)\mu + \eta^T \left(I - DVD^T\right)\eta - 2\mu^T HVD^T \eta \\ &+ z^T \left(I - DVD^T\right) z \\ &= \nu s^2 + \mu^T \left(H - HVH\right)\mu + \mu^T HVD^T \left(I - DVD^T\right)^{-1} \left(I - DVD^T\right)\eta \\ &- 2\mu^T HVD^T \eta \\ &+ z^T \left(I - DVD^T\right) z \\ &= \nu s^2 + \mu^T \left(H - HVH\right)\mu + \mu^T HVD^T \eta - 2\mu^T HVD^T \eta \\ &+ z^T \left(I - DVD^T\right) z \\ &= \nu s^2 + \mu^T \left(H - HVH\right)\mu - \mu^T HVD^T \eta \\ &+ z^T \left(I - DVD^T\right) z \\ &= \nu s^2 + \mu^T H\mu - \mu^T HV \left(H\mu + D^T \eta\right) \\ &+ z^T \left(I - DVD^T\right) z \quad . \end{aligned}$$

We will return to this equation shortly after we have developed a key insight and result related to the mean of the predicted value d. We need to develop an alternative expression for the inverse in the last term in the foregoing equation $(I - DVD^T)$. We use the definition of the matrix V in equation (4.55) to write

$$I - DVD^T = I - D\left(D^T D + H\right)^{-1} D^T \quad .$$

We postulate and prove that the inverse of this matrix is:

$$I + DH^{-1}D^T \quad .$$

We prove that the second matrix is the inverse of the first by simply multiplying and verifying that we get the identity matrix

$$\begin{aligned} &\left[I - D\left(D^T D + H\right)^{-1} D^T\right]\left[I + DH^{-1}D^T\right] \\ &= I\left[I + DH^{-1}D^T\right] - D\left(D^T D + H\right)^{-1} D^T \left[I + DH^{-1}D^T\right] \\ &= I + DH^{-1}D^T - D\left(D^T D + H\right)^{-1} D^T \\ &- D\left(D^T D + H\right)^{-1} D^T D H^{-1} D^T \\ &= I + DH^{-1}D^T - D\left(D^T D + H\right)^{-1} D^T \end{aligned}$$

$$
\begin{aligned}
&- D\left(D^T D + H\right)^{-1}\left[\left(D^T D + H\right) - H\right] H^{-1} D^T \\
&= I + D H^{-1} D^T - D\left(D^T D + H\right)^{-1} D^T \\
&- D\left(D^T D + H\right)^{-1}\left(D^T D + H\right) H^{-1} D^T \\
&\qquad + D\left(D^T D + H\right)^{-1} H H^{-1} D^T \\
&= I + D H^{-1} D^T - D\left(D^T D + H\right)^{-1} D^T \\
&- D H^{-1} D^T + D\left(D^T D + H\right)^{-1} D^T \\
&= I \quad .
\end{aligned}
$$

Thus we have verified by simple matrix multiplication that

$$
\begin{aligned}
&\left[I - D\left(D^T D + H\right)^{-1} D^T\right]^{-1} = \left[I + D H^{-1} D^T\right] \\
&\Rightarrow \left[I - D\left(D^T D + H\right)^{-1} D^T\right] = \left[I + D H^{-1} D^T\right]^{-1} \quad .
\end{aligned}
\tag{4.58}
$$

Using this inverse, we can rewrite the expression for η in equation (4.57)

$$
\begin{aligned}
\eta &= \left[I - D\left(D^T D + H\right)^{-1} D^T\right]^{-1} D\left(D^T D + H\right)^{-1} H\mu \\
&= \left[I + D H^{-1} D^T\right] D\left(D^T D + H\right)^{-1} H\mu \quad .
\end{aligned}
$$

This allows us to write a highly simplified expression for the vector η, which we now derive

$$
\begin{aligned}
\eta &= \left[I + D H^{-1} D^T\right] D\left(D^T D + H\right)^{-1} H\mu \\
&= \left[D + D H^{-1} D^T D\right]\left(D^T D + H\right)^{-1} H\mu \\
&= \left[D + D H^{-1}\left(D^T D + H - H\right)\right]\left(D^T D + H\right)^{-1} H\mu \\
&= \left[D + D H^{-1}\left(D^T D + H\right) - D H^{-1}(H)\right]\left(D^T D + H\right)^{-1} H\mu \\
&= \left[D + D H^{-1}\left(D^T D + H\right) - D\right]\left(D^T D + H\right)^{-1} H\mu \\
&= D H^{-1}\left(D^T D + H\right)\left(D^T D + H\right)^{-1} H\mu \\
&= D H^{-1} H\mu \\
&= D\mu \quad .
\end{aligned}
$$

The mean of d is simply the matrix D times the mean of μ, which is intuitively what we would expect (but not straightforward to prove). Armed with this new notion of the mean, we return to the exponent Y and substitute for η

$$
\begin{aligned}
Y &= \nu s^2 + \mu^T H\mu - \mu^T H V\left(H\mu + D^T D\mu\right) \\
&+ z^T\left(I - D V D^T\right) z
\end{aligned}
$$

$$
\begin{aligned}
&= \nu s^2 + \mu^T H \mu - \mu^T H V \left(H + D^T D\right) \mu \\
&+ z^T \left(I - D V D^T\right) z \\
&= \nu s^2 + \mu^T H \mu - \mu^T H \mu \\
&+ z^T \left(I - D V D^T\right) z \\
&= \nu s^2 + z^T \left(I - D V D^T\right) z \quad .
\end{aligned}
$$

This is simpler than one might have thought–a quadratic plus a simple constant. We need to simplify the quadratic matrix in the final term $I - DVD^T$. We begin by substituting the definition of V from equation (4.55) to write

$$
I - DVD^T = I - D\left(D^T D + H\right)^{-1} D^T \quad .
$$

We then use the second inverse relationship from equation (4.58) to write

$$
\begin{aligned}
I - DVD^T &= I - D\left(D^T D + H\right)^{-1} D^T \\
&= \left(I + D^T H^{-1} D\right)^{-1} \quad .
\end{aligned}
$$

If we make this matrix substitution, together with the definition of z

$$
\begin{aligned}
z &= d - \eta \\
&= d - D\mu
\end{aligned}
$$

into the quadratic term at the end of the foregoing expression for Y, we can write

$$
\begin{aligned}
Y &= \nu s^2 + z^T \left[I - D\left(D^T D + H\right)^{-1} D^T\right] z \\
&= \nu s^2 + (d - D\mu)^T \left(I + D H^{-1} D^T\right)^{-1} (d - D\mu) \\
&= \nu s^2 \left[1 + (d - D\mu)^T \frac{\left(I + D H^{-1} D^T\right)^{-1}}{\nu s^2} (d - D\mu)\right] \quad .
\end{aligned}
$$

Substituting into the density in equation (4.56), we write

$$
\begin{aligned}
\{d, D, \sigma | X, y\} &= \frac{2(2\pi)^{-\frac{q}{2}} |H|^{\frac{1}{2}} \left(\frac{\nu s^2}{2}\right)^{\frac{\nu}{2}}}{|D^T D + H|^{\frac{1}{2}} \Gamma\left(\frac{\nu}{2}\right)} \sigma^{-(\nu+q+1)} e^{-\frac{\nu s^2}{2\sigma^2}} \\
&\times e^{-\frac{1}{2}(d - D\mu)^T \frac{\left(I + DH^{-1}D^T\right)^{-1}}{\sigma^2} (d - D\mu)} \quad . \qquad (4.59)
\end{aligned}
$$

The matrix determinant lemma,[25] which addresses the determinant of the sum of an invertible matrix and a dyadic product, is expressed as follows for H being an invertible $n \times n$ matrix and U and V being $n \times m$ matrices

[25] See https://en.wikipedia.org/wiki/Matrix_determinant_lemma.

$$\left|H + UV^T\right| = \left|I + V^T H^{-1} U\right| |H| \quad .$$

Wanting to match the first term on the right in the matrix determinant lemma with our precision matrix, we would like to set $V^T = D \Rightarrow V = D^T$ and $U = D^T \Rightarrow U^T = D$ and write the matrix determinant lemma in the corresponding form

$$\left|H + D^T D\right| = \left|I + DH^{-1} D^T\right| |H| \quad .$$

This can be rearranged to

$$\frac{|H|}{|D^T D + H|} = \frac{1}{|I + DH^{-1} D^T|} \quad .$$

This means that the foregoing density can be rearranged to the form

$$\begin{aligned}\{d, D, \sigma | X, y\} = &\frac{2(2\pi)^{-\frac{q}{2}} \left(\frac{\nu s^2}{2}\right)^{\frac{\nu}{2}}}{|I + DH^{-1} D^T|^{\frac{1}{2}} \Gamma\left(\frac{\nu}{2}\right)} \sigma^{-(\nu+q+1)} e^{-\frac{\nu s^2}{2\sigma^2}} \\ &\times e^{-\frac{1}{2}(d - D\mu)^T \frac{\left(I + DH^{-1} D^T\right)^{-1}}{\sigma^2} (d - D\mu)} \quad .\end{aligned}$$

If we assume that

$$\begin{aligned}&\Omega^{-1} = \sigma^{-2}\left(I + DH^{-1} D^T\right)^{-1} \\ &\Rightarrow \Omega = \sigma^2 \left(I + DH^{-1} D^T\right) \\ &\Rightarrow I + DH^{-1} D^T = \Omega \sigma^{-2} \\ &\Rightarrow \left|I + DH^{-1} D^T\right| = |\Omega| \, \sigma^{-2q} \\ &\Rightarrow \left|I + DH^{-1} D^T\right|^{\frac{1}{2}} = |\Omega|^{\frac{1}{2}} \sigma^{-q}\end{aligned}$$

then we can write the foregoing distribution

$$\begin{aligned}\{d, D, \sigma | X, y\} &= \frac{2(2\pi)^{-\frac{q}{2}} \left(\frac{\nu s^2}{2}\right)^{\frac{\nu}{2}}}{|\Omega|^{\frac{1}{2}} \sigma^{-q} \Gamma\left(\frac{\nu}{2}\right)} \sigma^{-(\nu+q+1)} e^{-\frac{\nu s^2}{2\sigma^2}} \\ &\quad \times e^{-\frac{1}{2}(d - D\mu)^T \Omega^{-1} (d - D\mu)} \\ &= \frac{2(2\pi)^{-\frac{q}{2}} \left(\frac{\nu s^2}{2}\right)^{\frac{\nu}{2}}}{|\Omega|^{\frac{1}{2}} \Gamma\left(\frac{\nu}{2}\right)} \sigma^{-(\nu+1)} e^{-\frac{\nu s^2}{2\sigma^2}} \\ &\quad \times e^{-\frac{1}{2}(d - D\mu)^T \Omega^{-1} (d - D\mu)}\end{aligned}$$

$$= \frac{2(2\pi)^{-\frac{q}{2}}\left(\frac{\nu s^2}{2}\right)^{\frac{\nu}{2}}}{|\Omega|^{\frac{1}{2}}\Gamma\left(\frac{\nu}{2}\right)}\sigma^{-(\nu+1)}e^{-\frac{\nu s^2}{2\sigma^2}} \times e^{-\frac{1}{2}(d-D\mu)^T\Omega^{-1}(d-D\mu)} \quad .$$

Rewriting the last expression and substituting key relationships for Ω from above, we write the joint density over d and σ

$$\{d, D, \sigma|X, y\} = \frac{2(2\pi)^{-\frac{q}{2}}\left(\frac{\nu s^2}{2}\right)^{\frac{\nu}{2}}}{|I + DH^{-1}D^T|^{\frac{1}{2}}\Gamma\left(\frac{\nu}{2}\right)}\sigma^{-(\nu+q+1)}e^{-\frac{\nu s^2}{2\sigma^2}} \times e^{-\frac{1}{2\sigma^2}(d-D\mu)^T\left(I+DH^{-1}D^T\right)^{-1}(d-D\mu)} \quad . \tag{4.60}$$

If we refer to Wiki,[26] the multivariate normal distribution for a $q \times q$ variance-covariance matrix $\Omega = I + DH^{-1}D^T$ with mean value $D\mu$ is

$$N(x; D\mu, \Omega) = \frac{1}{(2\pi)^{\frac{q}{2}}|\Omega|^{\frac{1}{2}}}e^{-\frac{1}{2}(x-D\mu)^T\Omega^{-1}(x-D\mu)} \quad . \tag{4.61}$$

This can be rearranged to the form

$$e^{-\frac{1}{2}(d-D\mu)^T\Omega^{-1}(d-D\mu)} = (2\pi)^{\frac{q}{2}}|\Omega|^{\frac{1}{2}}N(d; D\mu, \Omega) \quad .$$

Substituting into the density above, we can simplify the predictive density to the form

$$\{d, D, \sigma|X, y\} = \frac{2\left(\frac{\nu s^2}{2}\right)^{\frac{\nu}{2}}}{\Gamma\left(\frac{\nu}{2}\right)}\sigma^{-(\nu+1)}e^{-\frac{\nu s^2}{2\sigma^2}} \times N\left[d; D\mu, \left(I + DH^{-1}D^T\right)\sigma^2\right] \quad .$$

We seem to have come right back to our old friend, the product of a normal times an inverse gamma. We want to do the same integral change of variables problem as we did with the model coefficients to assess what this distribution actually is. We start with the integral over the joint distribution and set it to 1:

$$1 = \int_{-\infty}^{\infty} dd \int_{0}^{\infty} d\sigma \{d, D, \sigma|X, y\} =$$

$$= \frac{2\left(\frac{\nu s^2}{2}\right)^{\frac{\nu}{2}}}{\Gamma\left(\frac{\nu}{2}\right)}\int_{-\infty}^{\infty} dd \int_{0}^{\infty} d\sigma\sigma^{-(\nu+1)}e^{-\frac{\nu s^2}{2\sigma^2}}N\left[d; D\mu, \left(I + DH^{-1}D^T\right)\sigma^2\right] \quad .$$

[26] See https://en.wikipedia.org/wiki/Multivariate_normal_distribution.

We then make the following substitutions

$$z = \sigma^2 \Rightarrow \sigma = z^{\frac{1}{2}} \Rightarrow d\sigma = \frac{1}{2} z^{-\frac{1}{2}} dz \Rightarrow \sigma^{-(\nu+1)} = z^{-\frac{\nu+1}{2}} \quad .$$

The substitutions change the integrand as indicated in the following equations

$$1 = \frac{2\left(\frac{\nu s^2}{2}\right)^{\frac{\nu}{2}}}{\Gamma\left(\frac{\nu}{2}\right)} \int_{-\infty}^{\infty} dd \int_{0}^{\infty} \frac{1}{2} z^{-\frac{1}{2}} dz z^{-\frac{\nu+1}{2}} e^{-\frac{\nu s^2}{2z}} N\left[d; D\mu, \left(I + DH^{-1}D^T\right) z\right]$$

$$= \frac{\left(\frac{\nu s^2}{2}\right)^{\frac{\nu}{2}}}{\Gamma\left(\frac{\nu}{2}\right)} \int_{-\infty}^{\infty} dd \int_{0}^{\infty} dz z^{-\left(\frac{\nu}{2}+1\right)} e^{-\frac{\nu s^2}{2z}} N\left[d; D\mu, \left(I + DH^{-1}D^T\right) z\right] \quad .$$

This implies that the underlying distribution over d and z is

$$(4.62) \quad \{d, z\} = \frac{\left(\frac{\nu s^2}{2}\right)^{\frac{\nu}{2}}}{\Gamma\left(\frac{\nu}{2}\right)} z^{-\left(\frac{\nu}{2}+1\right)} e^{-\frac{\nu s^2}{2z}} N\left[d; D\mu, \left(I + DH^{-1}D^T\right) z\right] \quad .$$

If we integrate out the variable d, we are left with the integral over z

$$(4.63) \qquad IG(z; \frac{\nu s^2}{2}, \frac{\nu}{2}) = \{z\} = \frac{\left(\frac{\nu s^2}{2}\right)^{\frac{\nu}{2}}}{\Gamma\left(\frac{\nu}{2}\right)} z^{-\left(\frac{\nu}{2}+1\right)} e^{-\frac{\nu s^2}{2z}} \quad .$$

This is an inverse gamma distribution, and it is precisely the same inverse gamma distribution we saw with the coefficients above. This is highly enlightening. The inverse gamma is exactly the same as the inverse gamma in the joint distribution over coefficients and error term. This is a consequence of the independence between σ^2 and d.

Just as we have seen before, the joint density is the product of an inverse gamma in equation (4.63) times a normal distribution in equation (4.61), and all the mathematics we have done before applies directly.

$$(4.64) \qquad \{d, z\} = IG(z; \frac{\nu s^2}{2}, \frac{\nu}{2}) N\left[d; D\mu, \left(I + DH^{-1}D^T\right) z\right] \quad .$$

We begin with the density in equation (4.59) and regroup terms so that we can integrate over the σ terms to create the marginal distribution over the prediction:

$$\{d, D, \sigma | X, y\} = \frac{2(2\pi)^{-\frac{q}{2}} |H|^{\frac{1}{2}} \left(\frac{\nu s^2}{2}\right)^{\frac{\nu}{2}}}{|D^T D + H|^{\frac{1}{2}} \Gamma\left(\frac{\nu}{2}\right)}$$
$$\times \sigma^{-(\nu+q+1)} e^{-\frac{\nu s^2}{2\sigma^2}} e^{-\frac{1}{2}(d - D\mu)^T \frac{\left(I + DH^{-1}D^T\right)^{-1}}{\sigma^2} (d - D\mu)}$$

$$= \frac{2(2\pi)^{-\frac{q}{2}}|H|^{\frac{1}{2}}\left(\frac{\nu s^2}{2}\right)^{\frac{\nu}{2}}}{|D^T D + H|^{\frac{1}{2}}\Gamma\left(\frac{\nu}{2}\right)}\sigma^{-(\nu+q+1)}$$

$$\times\, e^{-\frac{\nu s^2+(d-D\mu)^T\left(I+DH^{-1}D^T\right)^{-1}(d-D\mu)}{2\sigma^2}} \quad .$$

In order to simplify future algebra, we define

$$Z = \frac{\nu s^2 + (d - D\mu)^T \left(I + DH^{-1}D^T\right)^{-1}(d - D\mu)}{2} = \quad .$$

We substitute this into the joint density to write

$$\{d, D, \sigma|X, y\} = \frac{2(2\pi)^{-\frac{q}{2}}|H|^{\frac{1}{2}}\left(\frac{\nu s^2}{2}\right)^{\frac{\nu}{2}}}{|D^T D + H|^{\frac{1}{2}}\Gamma\left(\frac{\nu}{2}\right)}\sigma^{-(\nu+q+1)}e^{-\frac{Z}{\sigma^2}} \quad .$$

We would like to know what the marginal distribution over the prediction vector d is. We therefore integrate it over σ as follows

$$\int_0^\infty \{d, D, \sigma|X, y\}\, d\sigma = \{d, D|X, y\}$$

$$= \int_0^\infty \frac{2(2\pi)^{-\frac{q}{2}}|H|^{\frac{1}{2}}\left(\frac{\nu s^2}{2}\right)^{\frac{\nu}{2}}}{|D^T D + H|^{\frac{1}{2}}\Gamma\left(\frac{\nu}{2}\right)}\sigma^{-(\nu+q+1)}e^{-\frac{Z}{\sigma^2}}\, d\sigma \quad .$$

In order to perform the integration, we define a substitution variable h as follows

$$h = \frac{Z}{\sigma^2} \Rightarrow \sigma^2 = Zh^{-1} \Rightarrow \sigma = Z^{\frac{1}{2}}h^{-\frac{1}{2}}$$

$$\Rightarrow d\sigma = -\frac{1}{2}Z^{\frac{1}{2}}h^{-\frac{3}{2}}dh \Rightarrow \sigma^{-(\nu+q+1)} = Z^{\frac{-\nu-q-1}{2}}h^{\frac{\nu+q+1}{2}} \quad .$$

Substituting into the foregoing integral to establish the marginal over d, we have

$$\{d, D|X, y\}$$

$$= \int_\infty^0 \frac{2(2\pi)^{-\frac{q}{2}}|H|^{\frac{1}{2}}\left(\frac{\nu s^2}{2}\right)^{\frac{\nu}{2}}}{|D^T D + H|^{\frac{1}{2}}\Gamma\left(\frac{\nu}{2}\right)}\left(Z^{\frac{-\nu-q-1}{2}}h^{\frac{\nu+q+1}{2}}\right)e^{-h}\left(-\frac{1}{2}Z^{\frac{1}{2}}h^{-\frac{3}{2}}dh\right)$$

$$= \frac{(2\pi)^{-\frac{q}{2}}|H|^{\frac{1}{2}}\left(\frac{\nu s^2}{2}\right)^{\frac{\nu}{2}}}{|D^T D + H|^{\frac{1}{2}}\Gamma\left(\frac{\nu}{2}\right)}Z^{-\frac{\nu+q}{2}}\int_0^\infty h^{\frac{\nu+q}{2}-1}e^{-h}dh$$

$$= \frac{(2\pi)^{-\frac{q}{2}}|H|^{\frac{1}{2}}\left(\frac{\nu s^2}{2}\right)^{\frac{\nu}{2}}\Gamma\left(\frac{\nu+q}{2}\right)}{|D^TD+H|^{\frac{1}{2}}\Gamma\left(\frac{\nu}{2}\right)} Z^{-\frac{\nu+q}{2}} \quad .$$

We need to express the Z term so that we can substitute and simplify

$$\begin{aligned}
Z^{-\frac{\nu+q}{2}} &= \left[\frac{\nu s^2 + (d-D\mu)^T\left(I+DH^{-1}D^T\right)^{-1}(d-D\mu)}{2}\right]^{-\frac{\nu+q}{2}} \\
&= \left[\nu s^2 + (d-D\mu)^T\frac{\nu s^2}{\nu s^2}\left(I+DH^{-1}D^T\right)^{-1}(d-D\mu)\right]^{-\frac{\nu+q}{2}} 2^{\frac{\nu+q}{2}} \\
&= \left\{\nu s^2\left[1+(d-D\mu)^T\frac{\left(I+DH^{-1}D^T\right)^{-1}}{\nu s^2}(d-D\mu)\right]\right\}^{-\frac{\nu+q}{2}} 2^{\frac{\nu+q}{2}} \\
&= \left[1+(d-D\mu)^T\frac{\left(I+DH^{-1}D^T\right)^{-1}}{\nu s^2}(d-D\mu)\right]^{-\frac{\nu+q}{2}}\left(\frac{\nu s^2}{2}\right)^{-\frac{\nu+q}{2}} \quad .
\end{aligned}$$

Substituting into the marginal density yields the expression

$$\begin{aligned}
\{d, D|X, y\} &= \frac{(2\pi)^{-\frac{q}{2}}|H|^{\frac{1}{2}}\left(\frac{\nu s^2}{2}\right)^{\frac{\nu}{2}}\Gamma\left(\frac{\nu+q}{2}\right)}{|D^TD+H|^{\frac{1}{2}}\Gamma\left(\frac{\nu}{2}\right)}\left(\frac{\nu s^2}{2}\right)^{-\frac{\nu+q}{2}} \\
&\quad \times\left[1+(d-D\mu)^T\frac{\left(I+DH^{-1}D^T\right)^{-1}}{\nu s^2}(d-D\mu)\right]^{-\frac{\nu+q}{2}} \\
&= \frac{|H|^{\frac{1}{2}}}{|D^TD+H|^{\frac{1}{2}}}\frac{\left(\pi\nu s^2\right)^{-\frac{q}{2}}\Gamma\left(\frac{\nu+q}{2}\right)}{\Gamma\left(\frac{\nu}{2}\right)} \\
&\quad \times\left[1+(d-D\mu)^T\frac{\left(I+DH^{-1}D^T\right)^{-1}}{\nu s^2}(d-D\mu)\right]^{-\frac{\nu+q}{2}} \quad .
\end{aligned}$$

We can apply the matrix determinant lemma,[27]

$$\begin{aligned}
&\left|D^TD+H\right| = |H|\left|I+DH^{-1}D^T\right| \\
&\Rightarrow \frac{|H|}{|D^TD+H|} = \frac{1}{|I+DH^{-1}D^T|} \quad .
\end{aligned}$$

This allows us to simplify the quotient of determinants at the left to write

$$\{d, D|X, y\} = \frac{\Gamma\left(\frac{\nu+q}{2}\right)}{\left(\pi\nu s^2\right)^{\frac{q}{2}}|I+DH^{-1}D^T|^{\frac{1}{2}}\Gamma\left(\frac{\nu}{2}\right)}$$

[27]See https://en.wikipedia.org/wiki/Matrix_determinant_lemma.

$$\times \left[1 + (d - D\mu)^T \frac{\left(I + DH^{-1}D^T\right)^{-1}}{\nu s^2}(d - D\mu)\right]^{-\frac{\nu+q}{2}} \quad . \tag{4.65}$$

This is the crucial finding and most general result of this section, the general joint predictive distribution. It is a q-variate Student's t distribution. The mean of this distribution is

$$D\mu \quad . \tag{4.66}$$

The variance-covariance matrix is easily discerned from the standard multivariate t distribution to be:

$$VCV = \left(I + DH^{-1}D^T\right)\frac{\nu s^2}{\nu - 2} \quad . \tag{4.67}$$

We will not consider in this monograph the case of predicting jointly a subset of the rows of D. It is not a particularly interesting problem in practice, only technically. The reader might want to take that problem on, for it includes all the central elements of the distributions and predictions.

4.16.2 Special Case of the Predictive Distribution with a Single Row

As a frequent special case, the matrix D will just be a single row of observations which we shall denote by $D = \delta$ in which the number of rows of δ is $q = 1$, and what we want is the corresponding probability distribution over what the model would project for that single row of observations over the scalar random variable d. When D is a single row of observations, $q = 1$, we write D as the row vector (single row) δ. It is dimension $1 \times p$. In the special case in which D is a single row vector δ, equation (4.60) specializes to the product form

$$\begin{aligned}\{d, \delta, \sigma | X, y\} = & \frac{2(2\pi)^{-\frac{1}{2}}\left(\frac{\nu s^2}{2}\right)^{\frac{\nu}{2}}}{|1 + \delta H^{-1}\delta^T|^{\frac{1}{2}}\Gamma\left(\frac{\nu}{2}\right)}\sigma^{-(\nu+2)}e^{-\frac{\nu s^2}{2\sigma^2}} \\ & \times e^{-\frac{1}{2\sigma^2}(d-\delta\mu)^T\left(1+\delta H^{-1}\delta^T\right)^{-1}(d-\delta\mu)} \quad .\end{aligned} \tag{4.68}$$

This implies from equation (4.62) that the underlying distribution over d and $z = \sigma^2$ is

$$\{d, z\} = \frac{\left(\frac{\nu s^2}{2}\right)^{\frac{\nu}{2}}}{\Gamma\left(\frac{\nu}{2}\right)} z^{-\left(\frac{\nu}{2}+1\right)}e^{-\frac{\nu s^2}{2z}}N\left[d; \delta\mu, \left(1 + \delta H^{-1}\delta^T\right)z\right] \quad .$$

The inverse gamma that emerges is that in equation (4.63), specialized to $D = \delta$

$$IG(z; \frac{\nu z}{2}, \frac{\nu}{2}) = \{z\} = \frac{\left(\frac{\nu z}{2}\right)^{\frac{\nu}{2}}}{\Gamma\left(\frac{\nu}{2}\right)} z^{-\left(\frac{\nu}{2}+1\right)} e^{-\frac{\nu z}{2\sigma^2}} \quad .$$

This specializes to the product form in equation (4.64) specialized to the case of $D = \delta$ and $q = 1$, which is written as follows

$$\{d, z\} = IG(z; \frac{\nu s^2}{2}, \frac{\nu}{2}) N\left[d; \delta\mu, \left(1 + \delta H^{-1}\delta^T\right) z\right] \quad .$$

The marginal distribution is derived from equation (4.65) by setting $D = \delta$ and $q = 1$

$$\{d, \delta|X, y\} = \frac{\Gamma\left(\frac{\nu+1}{2}\right)}{(\pi\nu s^2)^{\frac{1}{2}}|1 + \delta H^{-1}\delta^T|^{\frac{1}{2}}\Gamma\left(\frac{\nu}{2}\right)} \times \left[1 + (d - \delta\mu)^T \frac{\left(1 + \delta H^{-1}\delta^T\right)^{-1}}{\nu s^2} (d - \delta\mu)\right]^{-\frac{\nu+1}{2}} \quad .$$

The mean of this distribution is derived from equation (4.66)

(4.69) $$Mean = \delta\mu \quad .$$

The variance-covariance matrix in equation (4.67) specializes to the form

(4.70) $$VCV = \left(1 + \delta H^{-1}\delta^T\right) \frac{\nu s^2}{\nu - 2} \quad .$$

When the number of independent variables is $p = 5$, the H^{-1} matrix is written

$$H^{-1} = \begin{bmatrix} \zeta_{11} & \zeta_{12} & \zeta_{13} & \zeta_{14} & \zeta_{15} \\ \zeta_{21} & \zeta_{22} & \zeta_{23} & \zeta_{24} & \zeta_{25} \\ \zeta_{31} & \zeta_{32} & \zeta_{33} & \zeta_{34} & \zeta_{35} \\ \zeta_{41} & \zeta_{42} & \zeta_{43} & \zeta_{44} & \zeta_{45} \\ \zeta_{51} & \zeta_{52} & \zeta_{53} & \zeta_{54} & \zeta_{55} \end{bmatrix} \quad .$$

If we assume that the δ matrix is

$$\delta = \begin{bmatrix} \delta_{11} & \delta_{12} & \delta_{13} & \delta_{14} & \delta_{15} \end{bmatrix} \quad ,$$

then

$$\begin{bmatrix} \delta_1 & \delta_2 & \delta_3 & \delta_4 & \delta_5 \end{bmatrix} \begin{bmatrix} \zeta_{11} & \zeta_{12} & \zeta_{13} & \zeta_{14} & \zeta_{15} \\ \zeta_{21} & \zeta_{22} & \zeta_{23} & \zeta_{24} & \zeta_{25} \\ \zeta_{31} & \zeta_{32} & \zeta_{33} & \zeta_{34} & \zeta_{35} \\ \zeta_{41} & \zeta_{42} & \zeta_{43} & \zeta_{44} & \zeta_{45} \\ \zeta_{51} & \zeta_{52} & \zeta_{53} & \zeta_{54} & \zeta_{55} \end{bmatrix} \begin{bmatrix} \delta_1 \\ \delta_2 \\ \delta_4 \\ \delta_5 \end{bmatrix}$$

$$= \begin{bmatrix} \sum_{i=1}^{5} \delta_i \zeta_{i1} & \sum_{i=1}^{5} \delta_i \zeta_{i2} & \sum_{i=1}^{5} \delta_i \zeta_{i3} & \sum_{i=1}^{5} \delta_i \zeta_{i4} & \sum_{i=1}^{5} \delta_i \zeta_{i5} \end{bmatrix} \begin{bmatrix} \delta_1 \\ \delta_2 \\ \delta_3 \\ \delta_4 \\ \delta_5 \end{bmatrix}$$

$$= \sum_{i=1}^{5} \sum_{j=1}^{5} \delta_i \delta_j \zeta_{ij}$$

and

$$\begin{bmatrix} \delta_1 & \delta_2 & \delta_3 & \delta_4 & \delta_5 \end{bmatrix} \begin{bmatrix} \mu_1 \\ \mu_2 \\ \mu_3 \\ \mu_4 \\ \mu_5 \end{bmatrix} = \sum_{i=1}^{5} \delta_i \mu_i \quad .$$

We see that this has reduced to a univariate distribution over a single row, i.e., $q = 1$.

4.16.3 Predictions for a Single Row and a Single Column (3) with $\delta = e_3$ with $q = 1$–Prior

We need to put in place the capability to forecast the value of a single coefficient, say coefficient 3. We do that using the special case in which

$$\delta = e_3 = \begin{bmatrix} 0 & 0 & 1 & 0 & 0 \end{bmatrix} \quad ,$$

in which we want to explore the marginal distribution over the third coefficient. In this situation

$$\begin{bmatrix} 0 & 0 & 1 & 0 & 0 \end{bmatrix} \begin{bmatrix} \zeta_{11} & \zeta_{12} & \zeta_{13} & \zeta_{14} & \zeta_{15} \\ \zeta_{21} & \zeta_{22} & \zeta_{23} & \zeta_{24} & \zeta_{25} \\ \zeta_{31} & \zeta_{32} & \zeta_{33} & \zeta_{34} & \zeta_{35} \\ \zeta_{41} & \zeta_{42} & \zeta_{43} & \zeta_{44} & \zeta_{45} \\ \zeta_{51} & \zeta_{52} & \zeta_{53} & \zeta_{54} & \zeta_{55} \end{bmatrix} \begin{bmatrix} 0 \\ 0 \\ 1 \\ 0 \\ 0 \end{bmatrix}$$

$$= \begin{bmatrix} \zeta_{31} & \zeta_{32} & \zeta_{33} & \zeta_{34} & \zeta_{35} \end{bmatrix} \begin{bmatrix} 0 \\ 0 \\ 1 \\ 0 \\ 0 \end{bmatrix} = \zeta_{33}$$

and

$$\begin{bmatrix} 0 & 0 & 1 & 0 & 0 \end{bmatrix} \begin{bmatrix} \mu_1 \\ \mu_2 \\ \mu_3 \\ \mu_4 \\ \mu_5 \end{bmatrix} = \mu_3 \quad .$$

This means that the univariate projected density is inferred from equation (4.65). In the special case in which D is a single row vector e_3, equation (4.60) specializes to the product form

$$\{d, e_3, \sigma | X, y\} = \frac{2(2\pi)^{-\frac{1}{2}} \left(\frac{\nu s^2}{2}\right)^{\frac{\nu}{2}}}{|1 + [H^{-1}]_{33}|^{\frac{1}{2}} \Gamma\left(\frac{\nu}{2}\right)} \sigma^{-(\nu+2)} e^{-\frac{\nu s^2}{2\sigma^2}} \times e^{-\frac{1}{2\sigma^2}(d-\mu_3)^T \left(1+[H^{-1}]_{33}\right)^{-1}(d-\mu_3)} \quad . \tag{4.71}$$

This implies from equation (4.62) that the underlying distribution over d and $z = \sigma^2$ is

$$\{d, z\} = \frac{\left(\frac{\nu s^2}{2}\right)^{\frac{\nu}{2}}}{\Gamma\left(\frac{\nu}{2}\right)} z^{-\left(\frac{\nu}{2}+1\right)} e^{-\frac{\nu s^2}{2z}} N\left[d; \mu_3, \left(1 + \left[H^{-1}\right]_{33}\right) z\right] \quad . \tag{4.72}$$

The inverse gamma that emerges is that in equation (4.63), specialized to $D = e_3$

$$IG(z; \frac{\nu z}{2}, \frac{\nu}{2}) = \{z\} = \frac{\left(\frac{\nu z}{2}\right)^{\frac{\nu}{2}}}{\Gamma\left(\frac{\nu}{2}\right)} z^{-\left(\frac{\nu}{2}+1\right)} e^{-\frac{\nu z}{2\sigma^2}} \quad . \tag{4.73}$$

This specializes to the product form in equation (4.64) specialized to the case of $D = e_3$ and $q = 1$, which is written as follows

$$\{d, z\} = IG(z; \frac{\nu s^2}{2}, \frac{\nu}{2}) N\left[d; \mu_3, \left(1 + \left[H^{-1}\right]_{33}\right) z\right] \quad . \tag{4.74}$$

The marginal distribution is derived from equation (4.65) by setting $D = e_3$ and $q = 1$

$$\{d, e_3 | X, y\} = \frac{\Gamma\left(\frac{\nu+1}{2}\right)}{(\pi \nu s^2)^{\frac{1}{2}} |1 + [H^{-1}]_{33}|^{\frac{1}{2}} \Gamma\left(\frac{\nu}{2}\right)} \times \left[1 + (d - \mu_3)^T \frac{\left(1 + \left[H^{-1}\right]_{33}\right)^{-1}}{\nu s^2} (d - \mu_3)\right]^{-\frac{\nu+1}{2}} \quad . \tag{4.75}$$

The mean of this distribution is derived from equation (4.66)

$$\text{Mean} = \mu_3 \quad . \tag{4.76}$$

The variance-covariance matrix in equation (4.67) specializes to the form

$$VCV = \left(1 + \left[H^{-1}\right]_{33}\right)\frac{\nu s^2}{\nu - 2} \quad . \tag{4.77}$$

This term looks very much like the marginal distribution over the coefficient β_3 except for the leading 1 term in the numerator. If we omit that 1 term, we obtain precisely the marginal distribution over the coefficient β_3. When we omit the 1 term, the marginal distribution over the third coefficient below becomes absolutely identical with equation (4.52)

$$\{d|e_3\} = \frac{\Gamma\left(\frac{\nu+1}{2}\right)}{\Gamma\left(\frac{\nu}{2}\right)\{\pi\nu s^2[H^{-1}]_{33}\}^{\frac{1}{2}}}\left[1 + \frac{(d-\mu_3)^2}{[H^{-1}]_{33}\nu s^2}\right]^{-\frac{\nu+1}{2}} \quad . \tag{4.78}$$

Its mean is

$$\text{mean} = \mu_3 \quad . \tag{4.79}$$

Its variance is

$$\text{variance} = \left[H^{-1}\right]_{33}\frac{\nu s^2}{\nu - 2} \quad . \tag{4.80}$$

We have developed precisely the probability distribution over the third model coefficient. So we can indeed develop marginal probabilities over individual model coefficients by simply predicting that coefficient using a unit vector with a 1 in the three position and eliminating the $1+$ term in the projective distribution, which arises because of the specific model of prediction $D\beta + w$, which contains the w component. This insight is going to prove useful when me move to more complex joint forms.

4.17 We Will Array These Results Together with the Prior, Posterior, and Classical in Chapter 9

These results, all of which have common functional forms for the Archetype, Prior, Posterior, and Classical, will be arrayed in Chapter 9 for easy access by analysts and programmers.

Chapter 5

THE PRIOR

When asked what he meant by a miracle: Oh, anything with a probability of less than 20%.

Enrico Fermi

The prior in Bayesian analysis has become a topic of great discussion (and we believe, alas, confusion). In crafting a prior, we are striving in this chapter to create one that is simple and easy to quantify at its heart. Zellner[107] sought the "brass ring" of a perfect, and perfectly general, prior, one that could express any prior information no matter how subtle or complex. We do not see the need for that in practice. Subjective priors are to a high degree "impressionistic." We just need to approximate them so that they contain the essence of our prior beliefs. Prior information is not often four significant figures of accuracy nor scrupulously consistent with one specific density functional form! Priors are impressionistic. Priors are "conventional wisdom." Most prior information is simpler, probabilistically independent, and impressionistic than sophisticated functional forms people might select. We seek a way to quantify impressionistic, simple prior knowledge within a conjugate-archetype form for the prior density. We want the knowledgeable but impressionistic prior used by the analyst, not the fully general, fully nuanced prior that no doubt over-expresses the prior information that is truly held. Our approximate approach herein does not decry highly detailed, nuanced, accurate prior information. It approximates it so that the Bayesian Linear Regression calculations can be made without work or hesitation. Notwithstanding the approximate approach taken in this chapter, our prior is fully general within the confines of the conjugate-archetype density functional form.

5.1 Summary of Parameters of the Prior

In Chapter 4, we went to great lengths to develop all the salient aspects of the prior distribution and the posterior distribution. Keep in mind that in the conjugate setting in this textbook, the prior and posterior both have the identical functional form, the same mathematical expression–the selfsame equation. That is what conjugacy means. As such, the only distinction between the prior and the posterior is the specific numerical settings of constants of the functional form that characterize the prior versus the specific numerical settings of constants of the functional form that characterize the posterior. That is why in Chapter 4 we only need to analyze the functional form of the probability distribution once. This section concentrates on the specific parameters that characterize the prior. What are they? How do we get them? What underpinning do they have? For the prior, we have a very specific parametrization, one that can be substituted directly into the appropriate terms in the conjugate-archetype density results in Chapter 4 immediately preceding, as stated at the beginning of that section.

(5.1)
$$\begin{aligned} g &\leftarrow M \\ \mu &\leftarrow \beta_0 \\ H &\leftarrow Q \\ r &\leftarrow m \\ \nu &\leftarrow \nu_0 \\ s^2 &\leftarrow \frac{M}{\nu_0} \end{aligned}$$

These four parameters are a "sufficient statistic" for the prior, all the numbers (except for the number of independent parameters p). If we make these substitutions from equation (5.1) into all the various relationships in Chapter 4, we will have a bona fide prior characterized by these parameters. In this section, we are going to make the foregoing assignments for the indicated variables into all dimensions of the conjugate-archetype density developed in Chapter 4 so as to properly and fully constitute the (conjugate) prior. We define the degrees of freedom for the prior precisely as in equation (4.4) and make the substitutions from equation (5.1) to write

(5.2)
$$\nu_0 \stackrel{\Delta}{=} m - p - 1 \quad .$$

This implies, by substitution from equation (5.1) into equation (4.5), that

(5.3)
$$m = \nu_0 + p + 1 \quad .$$

This definition of "degrees of freedom" and the constant m may appear to be somewhat arbitrary, but we will show immediately below in this section that this is far from arbitrary. If the prior, whose form is summarized in Figure 3.1, were given by an old data set with n_0 elements, the degrees of freedom would have been, as we showed with classical regression:

$$\nu_C = n_0 - p \quad .$$

Under the foregoing definition of ν_0 and m, that would mean that

$$\begin{aligned} m &= \nu_C + p + 1 \\ &= n_0 - p + p + 1 \\ &= n_0 + 1 \end{aligned}$$

and therefore the exponent on the σ term in Figure 3.1 would have been $-(n_0 + 1)$. We shall see that the seemingly extra term -1 as compared with the likelihood function itself emanates from the fact that the likelihood function is a probability distribution over the data given the parameters $\{X, y|\beta, \sigma\}$, while the prior and the posterior are probability distributions over the parameters $\{\beta, \sigma\}$, unconditionally for the prior and conditional on the data X, y for the posterior. The prior and the likelihood functions are probabilities over different things. This -1 term is also going to become obvious in an intuitively appealing way, as we shall see in Chapter 7 when we consider the diffuse prior.[1] We substitute from equation (5.1) into equation (4.7) to develop the expression for M

$$M = g = \nu_0 {s_0}^2 \quad . \tag{5.4}$$

We define the variance term of the prior s_0^2 to be precisely as in equation (4.6) with the substitutions from equation (5.1) and compute s_0^2 terms as follows:

$${s_0}^2 = \frac{M}{\nu_0} \quad . \tag{5.5}$$

The term M is going to look like the residual sum of squared errors $R = \nu_C s_C^2$ from the classical regression case.

Using these definitions, we write the crucial results for the prior

- The mean value of model coefficients devolves directly from substituting equations (5.1) into equation (4.13) to write

$$< \beta >= \mu = \beta_0 \quad .$$

- The H matrix for this case is obtained by substituting equations (5.1) into equation (4.14) to write

[1] This same -1 or +1 factor, depending on whether you view it in the denominator or numerator, occurs in the Laplace Rule of Succession, and for exactly the same reason. The Laplace Rule of Succession was one of the earliest applications of the type of Bayesian thinking articulated in detail in this textbook. See https://en.wikipedia.org/wiki/Rule_of_Succession for a discussion of the Rule of Succession.

$$H = Q \quad . \tag{5.6}$$

5.2 Subjective Assessment of the Prior

Doesn't it seem alluring to subjectively estimate the two means and the two variances (or standard deviations) of the two distributions that comprise the prior in equation (5.1), namely ν, s^2, μ, and H, and be done with it! Good idea, and there is an intuitive, approximate way to do so that we will put forth in this section. In particular, we will estimate the two means and the two variances and use them to represent prior information that is brought by the analyst (or subject matter experts available to the analyst) and injected into the problem as the prior.[2] Suppose we have identified an "expert" who knows the following by virtue of his or her experience and knowledge, or perhaps extracted from an earlier data base:

- $<\beta>$, the mean values of the model coefficients,
- VCV, the variance-covariance matrix of the model coefficients,
- $<\sigma^2>$, the mean value of the variance term, for which the expert is willing to provide his or her estimate,
- V_{S^2}, the variance of the variance term, Quite often, it is easier for the expert to think about the standard deviation of the variance term, which is $SD = \sqrt{V_{S^2}}$ and to consider it relative to the mean term $<\sigma^2>$.

To summarize, the statistician asks the subject-matter expert for his or her estimate of these four terms. The statistician asks: "Please give me your view of

- $<\beta>$, the expected value of the coefficients;
- VCV, the variance-covariance matrix of the coefficients;
- $<\sigma^2>$, the expected value of the error term; and
- $\sqrt{V_{S^2}}$, the standard deviation of the error term."

[2] Often there is no prior data available, yet people have an "intuition," an "inkling", about the mean values of the coefficients and about the level of uncertainty and interdependence they have in those coefficients. We are unaware of any analyst anywhere at any time who doesn't have and doesn't impose his or her judgment, i.e., who doesn't have a clue, an inkling. A priori information is not immediately visible or intuitive to most practitioners of classical regression because there is no provision for prior information. In classical regression, the analyst brings NO information–zip, zero, nada–to the problem because he or she is not ALLOWED to do so. He or she only brings the data–zero judgment. In Bayes, analysts and experts must think about and offer ex ante judgment and introduce it accurately because Bayes requires it. (We will put forth a way to sidestep that, but we leave that to later in this book.) In classical statistics, analysts and experts have to impose their judgment some other, less transparent way!

We are thereby implicitly given the t-statistic of the inverse gamma distribution, which is the mean over the standard deviation. The t-statistic is

$$t = \frac{< \sigma^2 >}{\sqrt{V_{S^2}}}$$

and is implied directly by the expert's assessment. The expert volunteers a vector of means $< \beta >$, a scalar $< \sigma^2 >$, a matrix VCV, and a scalar $\sqrt{V_S^2}$, respectively. We shall see that there are very insightful and clever ways to get the requisite prior parameters from such estimates, but for now, simply assume they have been provided by some subject matter expert.

Hopefully, this is not merely "dumb guessing." Such estimates should be provided by people who have a notion, an idea, or an intuition about what these coefficients might be, albeit with uncertainty. We shall introduce ways to elicit and represent such information later in this textbook, in Chapters 10 through 14. Once we are given these parameters of the prior, we can back-substitute to derive the posterior coefficients per se in a way to made clear shortly in this section.

To begin, suppose the expert gives you his or her estimate of the mean values of the coefficients, namely $< \beta >$. The statistician directly sets the mean value of the coefficients prior to the expert's judgment as follows:

$$\beta_0 \leftarrow < \beta > \quad . \tag{5.7}$$

Suppose the expert estimates the mean value of the error term $< \sigma^2 >$ and the standard deviation of the error term $\sqrt{V_{S^2}}$. In particular, the expert offers the estimates

$$s_0^2 \frac{\nu_0}{\nu_0 - 2} \leftarrow < \sigma^2 > \tag{5.8}$$

for the mean value and

$$\sqrt{\frac{2}{\nu_0 - 4}} \langle \sigma^2 \rangle \leftarrow \sqrt{V_{S^2}}$$

for the standard deviation, the square root of the variance. In so doing, the expert is thinking about the mean per se of σ^2 and the degree of uncertainty he has about it expressed as to how wide is the distribution relative to its mean. He then squares the standard deviation term to get the variance of the error term

$$\frac{2}{\nu_0 - 4} \langle \sigma^2 \rangle^2 \leftarrow V_{S^2} \quad . \tag{5.9}$$

We can use equation (5.8) and equation (5.9) to solve for ν_0 and s_0^2. We rewrite equation (5.9) to solve for ν_0:

$$\nu_0 = 2\frac{\langle\sigma^2\rangle^2}{V_{S^2}} + 4 = 2\left(t^2+2\right)$$
$$\Rightarrow \nu_0 - 2 = 2\frac{\langle\sigma^2\rangle^2 + V_{S^2}}{V_{S^2}} = 2t^2 + 2 \quad .$$

It is not necessary that ν_0 be an integer. It can be a positive real number. We can substitute this equation into equation (5.8) and simplify to isolate s_0^2

$$\text{(5.10)} \qquad \begin{aligned} s_0^2 &= \langle\sigma^2\rangle \frac{\langle\sigma^2\rangle^2 + V_{S^2}}{\langle\sigma^2\rangle^2 + 2V_{S^2}} \\ &= \langle\sigma^2\rangle \frac{t^2+1}{t^2+2} \quad . \end{aligned}$$

We have thus calculated the prior ν_0 and the prior s_0^2 based on what the expert has told us. We can compute $\nu_0 s_0^2$ by multiplying the former two equations, reverifying that

$$\frac{\nu_0 s_0^2}{\nu_0 - 2} = \frac{2\langle\sigma^2\rangle \frac{\langle\sigma^2\rangle^2 + V_{S2}}{V_{S2}}}{2\frac{\langle\sigma^2\rangle^2 + V_{S2}}{V_{S2}}} = \langle\sigma^2\rangle \quad .$$

This can be expressed as in equation (4.26) with the appropriate substitutions for the prior. We reorganize this equation to write

$$s_0^2 = \langle\sigma^2\rangle \frac{\nu_0 - 2}{\nu_0} \quad .$$

We have previously derived the relationship for the prior variance-covariance matrix

$$VCV \leftarrow Q^{-1} s_0^2 \frac{\nu_0}{\nu_0 - 2} = Q^{-1} < \sigma^2 > \quad .$$

Inverting this gives the expression for the precision matrix Q based on the matrix VCV that the expert has provided to us

$$\text{(5.11)} \qquad Q = VCV^{-1} \langle\sigma^2\rangle \quad .$$

When the variance-covariance matrix is diagonal (zeroes on all the off diagonals), the Q matrix is also diagonal. In that case, all covariances are zero by construction, and all correlation coefficients are zero. This means that the prior distribution becomes a sequence of independent distributions over the model coefficients. The marginal distribution over every model coefficient is independent of the marginal distribution over every other model coefficient. This is a strong simplification, perhaps an oversimplification, but usually consistent with the judgment of subject matter experts. It really facilitates using our particular conjugate prior. Real world experts often have sketchy ideas at best about covariance or dependence among their ex ante estimates (unless they have an old data set and have run a regression analysis on it that might give them insight about the prior that is represented by that old data set.)

When we accumulate the foregoing four equations all in one place (i.e., equations (5.7), (5.9), (5.10), and (5.11)), we arrive at the inputs to the prior that we need, derived directly from the estimates the expert provided to us:

$$\begin{aligned}
\beta_0 &=< \beta > \\
s_0^2 &=< \sigma^2 > \frac{\left(< \sigma^2 >\right)^2 + V_{S^2}}{\left(< \sigma^2 >\right)^2 + 2V_{S^2}} \\
Q &= VCV^{-1} < \sigma^2 > \\
\nu_0 &= 2\frac{\left(< \sigma^2 >\right)^2}{V_{S^2}} + 4 = \frac{2}{V_{S^2}}\left[\left(< \sigma^2 >\right)^2 + 2V_{S^2}\right] \quad .
\end{aligned}$$

We can take the second and the fourth equation and calculate the $\nu_0 s_0^2$ term

$$\nu_0 s_0^2 = 2\left(\frac{\langle\sigma^2\rangle^2}{V_{S^2}} + 1\right)\langle\sigma^2\rangle \quad .$$

Equation (3.4) gave us the definition of m in the prior, and equation (3.5) gave us the definition of M in the prior. Using the terms just developed and substituting into those equations, two of the fundamental parameters of the prior are shown to be

$$\begin{aligned}
m &= \nu_0 + p + 1 \\
&= \frac{2 < \sigma^2 >^2}{V_{S^2}} + p + 5 = 2t^2 + p + 5 \\
M &= \nu_0 {s_0}^2 \\
&= 2 < \sigma^2 > \left(1 + \frac{< \sigma^2 >^2}{V_{S^2}}\right) = 2 < \sigma^2 > \left(t^2 + 1\right) \quad .
\end{aligned}$$

The mean term β_0 is obvious from the previous equation.

Let us summarize what we have learned in this section. The expert gives us the following four parameter values and implicitly the value of p

- $< \beta >$
- $< \sigma^2 >$
- VCV
- V_{S^2}

- p

and from them we derive the specific parameters of the prior density

- ν_0
- ${s_0}^2$
- Q
- β_0
- p .

5.3 p-vector Multivariate Density–Prior

We now go to work characterizing the parameters of the prior. We know that the prior has the functional form identical with equation (4.8) but with parameter substitutions from equation (5.1). Making such substitutions allows us to write the prior joint density over β and σ in the following form:[3]

$$\{\beta, \sigma\} = \frac{2|Q|^{\frac{1}{2}}}{(2\pi)^{\frac{p}{2}}\Gamma\left(\frac{\nu_0}{2}\right)} \left(\frac{\nu_0 s_0^2}{2}\right)^{\frac{\nu_0}{2}} \sigma^{-(\nu_0+p+1)} e^{-\frac{\nu_0 s_0^2}{2\sigma^2}} \times e^{-\frac{1}{2}(\beta-\beta_0)^T\left(Q\sigma^{-2}\right)(\beta-\beta_0)} \quad . \tag{5.12}$$

Note that this is precisely in the form of equation (4.8), the conjugate-archetype equation functional form for the conjugate prior and posterior, but after the parameter substitutions inserted from equation (5.1) to actually represent the prior. We notice that there are exactly five parameters, which collectively we might term the "sufficient statistics," for the prior

- p = number of model parameters
- ν_0 = degrees of freedom
- ${s_0}^2$ = variance term
- Q = precision matrix
- β_0 = mean of the coefficients.

In Chapter 4, we changed the foregoing joint density over β and σ in (4.8) into a joint density over β and $z = \sigma^2$ to form the joint density in equation (4.10). We thereafter substitute the relations in equation (5.1) for the prior to write

[3]Keep in mind, because the prior and posterior are conjugate, we can use the same functional form for both, but with different parameter settings.

(5.13)
$$\{\beta, z\} = \frac{|Q|^{\frac{1}{2}}(2\pi)^{-\frac{p}{2}}\left(\frac{\nu_0 s_0^2}{2}\right)^{\frac{\nu_0}{2}}}{\Gamma\left(\frac{\nu_0}{2}\right)} \times z^{-\frac{\nu_0+p}{2}+1} e^{-\frac{\nu_0 s_0^2}{2z}} e^{-\frac{1}{2z}(\beta-\beta_0)^T Q(\beta-\beta_0)} \quad .$$

We will be using this density for the prior frequently in the forthcoming developments. There will see an inverse gamma distribution embedded in this prior now that we have expressed the joint density using the variance term $z = \sigma^2$ rather than the standard deviation term σ. We manipulate the prior density by substituting into equation (4.11) using equation (5.1) to obtain for the prior

(5.14)
$$IG\left(z; \frac{\nu_0 s_0^2}{2}, \frac{\nu_0}{2}\right) = \frac{\left(\frac{\nu_0 s_0^2}{2}\right)^{\frac{\nu_0}{2}}}{\Gamma\left(\frac{\nu_0}{2}\right)} z^{-\left(\frac{\nu_0}{2}+1\right)} e^{-\frac{\nu_0 s_0^2}{2z}} \quad .$$

Its mean and variance are derived by substituting from equation (5.1) into the equation (4.22) for the mean and into equation (4.23) for the variance respectively to write

(5.15)
$$< \sigma^2 >= \frac{\nu_0 s_0^2}{\nu_0 - 2} = \text{Mean}$$

and

(5.16)
$$< \left(\sigma^2\right)^2 > - < \sigma^2 >^2 = \frac{2}{\nu_0 - 4}\left(\frac{\nu_0 s_0^2}{\nu_0 - 2}\right)^2 = \text{Variance} \quad .$$

The "product rule" in terms of the variance term z is determined by substituting into equation (4.12) from equation (5.1) to write the following, a form so dwelt upon by Zellner[107]

(5.17)
$$\{\beta, z\} = IG\left(z; \frac{\nu_0 s_0^2}{2}, \frac{\nu_0}{2}\right) N\left(\beta; \beta_0, Q^{-1}z\right) \quad .$$

The remainder of this section provides a comprehensive discussion of this prior and its various marginals.

5.4 p-vector Marginal Density–Prior

This section integrates the σ term (or more conveniently the z term) out of equation (5.13) to create the marginal density over β for the prior. Recall that the joint marginal over the model coefficients, taken directly from equation (4.15) for the conjugate-archetype distribution but with all parameters substituted as in equation (5.1) to represent the prior, is the following:

$$(5.18)\quad \{\beta\} = \frac{|Q|^{\frac{1}{2}}\Gamma\left(\frac{\nu_0+p}{2}\right)}{(\pi\nu_0 s_0^2)^{\frac{p}{2}}\Gamma\left(\frac{\nu_0}{2}\right)}\left[1+(\beta-\beta_0)^T\left(\frac{Q}{\nu_0 s_0^2}\right)(\beta-\beta_0)\right]^{-\frac{\nu_0+p}{2}} \quad .$$

The mean of the coefficients of this marginal distribution for the prior, a p-vector, is obtained by substituting from equation (5.1) into equation (4.18) to write

$$(5.19)\quad <\beta>=\beta_0 \quad ,$$

and the variance-covariance matrix of the coefficients, obtained by substituting from equation (5.1) into equation (4.19), is

$$(5.20)\quad VCV = Q^{-1}s_0^2\frac{\nu_0}{\nu_0-2} \quad .$$

The H matrix, obtained by substituting from equation (5.1) into equation (4.20), is

$$(5.21)\quad H = Q \quad .$$

This completes the characterization of the marginal distribution over the p-vector of model coefficients.

5.5 Marginal Distribution for $(p-L)$-subvector of β–Prior

We may not want to carry all p of the model coefficients for subsequent analysis or prediction. We may want to consider a subset of only $p-L$ of the p elements of β, eliminating L of them by integration to form the marginal. What does the joint and marginal distribution for the prior look like for that subset of $p-L$ model coefficients? We are interested in the joint density over β_2 (assumed to be a $p-L$ vector of β) and σ. We obtain the partially integrated marginal distribution for the prior by substituting from equation (5.1) into equation (4.28) and writing as in equation (5.22)

$$(5.22)\quad \{\beta_2,\sigma\} = \frac{2\left|Q_{22}-Q_{21}{Q_{11}}^{-1}Q_{12}\right|^{\frac{1}{2}}}{(2\pi)^{\frac{p-L}{2}}\Gamma\left(\frac{\nu_0}{2}\right)}\left(\frac{\nu_0 s_0^2}{2}\right)^{\frac{\nu_0}{2}}\sigma^{-(\nu_0+p+1-L)}$$
$$\times e^{-\frac{\nu_0 s_0^2}{2\sigma^2}}$$
$$\times e^{-\frac{1}{2}(\beta_2-\beta_{02})^T\left[\left(Q_{22}-Q_{21}{Q_{11}}^{-1}Q_{12}\right)\sigma^{-2}\right](\beta_2-\beta_{02})} \quad .$$

In this equation, β_2 represents the subset of length $p-L$ of the p-vector β we want to consider. The structure of the equations becomes clear once we recognize that the length of β_2 is $p-L$. The remaining subset will be integrated away to create the marginal distribution over the $p-L$ subset.

$$\{\beta_2, z\} = \frac{\left|Q_{22} - Q_{21}{Q_{11}}^{-1}Q_{12}\right|^{\frac{1}{2}} \left(\frac{\nu_0 s_0^2}{2}\right)^{\frac{\nu_0}{2}}}{(2\pi)^{\frac{p-L}{2}} \Gamma\left(\frac{\nu_0}{2}\right)}$$
$$\times\, z^{-\frac{\nu_0+p-L}{2}-1} e^{-\frac{\nu_0 s_0^2}{z}} e^{-\frac{1}{2}(\beta_2-\beta_{02})^T \frac{Q_{22}-Q_{21}{Q_{11}}^{-1}Q_{12}}{z}(\beta_2-\beta_{02})} \quad .$$

As shown in Chapter 4, the product rule density over the $p - L$-vector β_2 and σ for the prior is obtained by substituting from equation (5.1) into equation (4.30) to write the product rule density in equation (5.23)

$$\text{(5.23)} \qquad \{\beta_2, z\} = IG\left(z, \frac{\nu_0 s_0^2}{2}, \frac{\nu_0}{2}\right)$$
$$\times\, N\left[\beta_2; p - L, \beta_{02}, z\left(Q_{22} - Q_{21}{Q_{11}}^{-1}Q_{12}\right)^{-1}\right] \quad .$$

The inverse gamma density over $z = \sigma^2$ is obtained by substituting from equation (5.1) into equation (4.31) to write equation (5.24)

$$\text{(5.24)} \qquad IG\left(z, \frac{\nu_0 s_0^2}{2}, \frac{\nu_0}{2}\right) = \frac{\left(\frac{\nu_0 s_0^2}{2}\right)^{\frac{\nu_0}{2}}}{\Gamma\left(\frac{\nu_0}{2}\right)} z^{-\frac{\nu_0}{2}-1} e^{-\frac{\nu_0 s_0^2}{2z}} \quad .$$

This is the same inverse gamma developed for the original p-vector β in equation (5.14). Finally, the marginal distribution over the $p - L$-vector β_2 is obtained by substituting from equation (5.1) into equation (4.33), which yields the following

$$\text{(5.25)} \quad \{\beta_2\} = \frac{\left|Q_{22} - Q_{21}{Q_{11}}^{-1}Q_{12}\right|^{\frac{1}{2}} \Gamma\left(\frac{\nu_0+p-L}{2}\right)}{\left(\pi\nu_0 s_0^2\right)^{\frac{p-L}{2}} \Gamma\left(\frac{\nu_0}{2}\right)}$$
$$\times \left[1 + (\beta_2 - \beta_{02})^T \frac{\left(Q_{22} - Q_{21}{Q_{11}}^{-1}Q_{12}\right)}{\nu_0 s_0^2} (\beta_2 - \beta_{02})\right]^{\frac{-\nu_0-p+L}{2}} \quad .$$

5.6 The Special Case in Which $p - L = 1$, a Univariate Subvector of β–Prior

This section deals with the case in which the marginal distribution is over one single element of β, a scalar. To be clear, in this development, β_2 will be a scalar. In effect, we eliminate $p - 1$ of the elements of β and consider only a single element β_2 thereof. This implies we have boiled β down to a univariate marginal distribution for the prior. If we substitute the parameters from equation (5.1) into equation (4.34), we obtain

$$\{\beta_2, \sigma\} = \frac{2\left|Q_{22} - Q_{21}{Q_{11}}^{-1}Q_{12}\right|^{\frac{1}{2}} \left(\frac{\nu_0 s_0^2}{2}\right)^{\frac{\nu_0}{2}}}{(2\pi)^{\frac{1}{2}}\Gamma\left(\frac{\nu_0}{2}\right)} \sigma^{-(\nu_0+2)} e^{-\frac{\nu_0 s_0^2}{2\sigma^2}} \times e^{-\frac{1}{2\sigma^2}(\beta_2-\beta_{02})^T\left(Q_{22}-Q_{21}{Q_{11}}^{-1}Q_{12}\right)(\beta_2-\beta_{02})} \quad . \tag{5.26}$$

If we substitute the parameters from equation (5.1) into equation (4.30), making sure to set $q = 1$, we obtain the prior joint distribution over the scalar β_2 and z in equation (5.27)

$$\{\beta_2, z\} = IG\left(z, \frac{\nu_0 s_0^2}{2}, \frac{\nu_0}{2}\right) \times N\left[\beta_2; \beta_{02}, z\left(Q_{22} - Q_{21}{Q_{11}}^{-1}Q_{12}\right)^{-1}\right] \quad . \tag{5.27}$$

If we substitute the parameters from equation (5.1) into equation (4.31), we obtain equation (5.28), precisely and exactly the same inverse gamma distribution as we saw in the p-dimensional density in equation (5.24):

$$IG\left(z, \frac{\nu_0 s_0^2}{2}, \frac{\nu_0}{2}\right) = \frac{\left(\frac{\nu_0 s_0^2}{2}\right)^{\frac{\nu_0}{2}}}{\Gamma\left(\frac{\nu_0}{2}\right)} z^{-\frac{\nu_0}{2}-1} e^{-\frac{\nu_0 s_0^2}{2z}} \quad . \tag{5.28}$$

If we substitute the parameters from equation (5.1) into equation (4.35), we obtain equation (5.29), which is the marginal distribution over the $p - L = 1$ coefficients β_2:

$$\{\beta_2\} = \frac{\left|Q_{22} - Q_{21}{Q_{11}}^{-1}Q_{12}\right|^{\frac{1}{2}} \Gamma\left(\frac{\nu_0+1}{2}\right)}{(\pi\nu_0 s_0^2)^{\frac{1}{2}}\Gamma\left(\frac{\nu_0}{2}\right)} \times \left[1 + (\beta_2 - \beta_{02})^T \frac{\left(Q_{22} - Q_{21}{Q_{11}}^{-1}Q_{12}\right)}{\nu_0 s_0^2} (\beta_2 - \beta_{02})\right]^{-\frac{\nu_0+1}{2}} \quad . \tag{5.29}$$

This is clearly a univariate Student's t distribution. We will simplify notation to show this more clearly. If we substitute the parameters from equation (5.1) into equation (4.36), we obtain equation (5.30)

$$\left[Q^{-1}\right]_{ii} = \left(Q_{22} - Q_{21}{Q_{11}}^{-1}Q_{12}\right)^{-1} = z_{ii} \quad . \tag{5.30}$$

If we substitute the parameters from equation (5.1) into equation (4.37), we obtain the following equation, which is the full density over β_2 and σ expressed using the diagonal term z_{ii} for notational simplicity and clarity

$$\{\beta_2, \sigma\} = \frac{2\left(\frac{\nu_0 s_0^2}{2}\right)^{\frac{\nu_0}{2}}}{(2\pi z_{ii})^{\frac{1}{2}}\Gamma\left(\frac{\nu_0}{2}\right)} \sigma^{-(\nu_0+2)} e^{-\frac{\nu_0 s_0^2}{2\sigma^2}}$$
$$\times\, e^{-\frac{1}{2}(\beta_2-\beta_{02})^T \frac{1}{z_{ii}\sigma^2}(\beta_2-\beta_{02})} \quad .$$

If we substitute the parameters from equation (5.1) into equation (4.41), we obtain equation (5.31), which is the full product rule density over β_2 and $z = \sigma^2$:

$$\{\beta_2, z\} = IG\left(z, \frac{\nu_0 s_0^2}{2}, \frac{\nu_0}{2}\right) N(\beta_2; z) \quad . \tag{5.31}$$

If we substitute the parameters from equation (5.1) into equation (4.42), we obtain the following, which is the marginal density over the sought after $p - L = 1$ subset of β, namely β_2:

$$\{\beta_2\} = \frac{\Gamma\left(\frac{\nu_0+1}{2}\right)}{(\pi\nu_0 s_0^2 z_{ii})^{\frac{1}{2}}\Gamma\left(\frac{\nu_0}{2}\right)}$$
$$\times\left[1 + (\beta_2 - \beta_{02})^T \frac{1}{\nu_0 s_0^2 z_{ii}}(\beta_2 - \beta_{02})\right]^{-\frac{\nu_0+1}{2}} \quad . \tag{5.32}$$

5.7 The Generic Univariate Student's t Distribution–Prior

We rewrite the univariate Student's t distribution from equation (4.43) for easy reference:

$$\{\beta, \sigma\} = \frac{2\sqrt{Q}}{\sqrt{2\pi}\Gamma\left(\frac{k-2}{2}\right)}\left(\frac{P}{2}\right)^{\frac{k-2}{2}} \sigma^{-k} e^{-\frac{P}{2\sigma^2}} e^{-\frac{1}{2\sigma^2}(\beta-\mu)^T Q(\beta-\mu)} \quad .$$

For this generic univariate Student's t distribution, we define for the prior from equation (4.38) with the substitutions from equation (5.1)

$$P = \nu_0 s_0^2 \quad ,$$

For the generic univariate Student's t distribution, we define for the prior from equation (4.39) with the substitutions from equation (5.1)

$$Q = \frac{1}{z_{ii}} \quad .$$

For the generic univariate Student's t distribution, we define for the prior from equation (4.40) with the substitutions from equation (5.1)

$$k = \nu_0 + 2 \quad .$$

For the generic univariate Student's t distribution, we define for the prior from equation (4.44) with the substitutions from equation (5.1)

$$\{\beta\} = \int_0^\infty d\sigma \, \{\beta, \sigma\} = \frac{\Gamma\left(\frac{k-1}{2}\right)}{\sqrt{\pi}\Gamma\left(\frac{k-2}{2}\right)} \left(\frac{Q}{P}\right)^{\frac{1}{2}} \left[1 + (\beta-\mu)^T \frac{Q}{P} (\beta-\mu)\right]^{-\frac{k-1}{2}}$$

$$= \frac{\Gamma\left(\frac{\nu_0+1}{2}\right)}{\sqrt{\pi}\Gamma\left(\frac{\nu_0}{2}\right)} \left(\frac{1}{z_{ii}\nu_0 s_0^2}\right)^{\frac{1}{2}} \left[1 + (\beta-\mu)^T \frac{1}{z_{ii}\nu_0 s_0^2} (\beta-\mu)\right]^{-\frac{\nu_0+1}{2}} \quad .$$

For the generic univariate Student's t distribution, we define for the prior from equation (4.45) with the substitutions from equation (5.1)

$$\langle\beta\rangle = \mu = \beta_0 \quad .$$

For the generic univariate Student's t distribution, we define for the prior from equation (4.46) with the substitutions from equation (5.1)

$$\text{variance} = \frac{1}{k-4}\left(\frac{P}{Q}\right) = \frac{\nu_0 s_0^2 z_{ii}}{\nu_0 - 2} \quad .$$

It is instructive for people who might want to calculate or program the marginal distributions over the coefficients to enumerate step by step the calculations one has to make to move from the joint distribution over the coefficients β to the p individual coefficients β_i

1. Calculate the inverse of the entire Q matrix, i.e.,

$$Q^{-1} \quad .$$

2. Get the ii^{th} diagonal element of the Q^{-1} matrix. Designate it z_{ii}

$$z_{ii} = \left[Q^{-1}\right]_{ii} = \left(Q_{22} - Q_{21}{Q_{11}}^{-1}Q_{12}\right)^{-1} \quad .$$

3. Using these definitions, the univariate Student's t density in equation (5.32) becomes

$$\{\beta_i\} = \frac{\Gamma\left(\frac{\nu_0+1}{2}\right)}{\Gamma\left(\frac{\nu_0}{2}\right)} \frac{1}{\left(\pi z_{ii}\nu_0 s_0^2\right)^{\frac{1}{2}}} \left[1 + \left(\beta_i - [\beta_0]_i\right) \frac{1}{\nu_0 s_0^2 z_{ii}} \left(\beta_i - [\beta_0]_i\right)\right]^{-\frac{\nu_0+1}{2}} \quad . \tag{5.33}$$

4. This density will integrate to 1 because the constant is correctly calculated. We carry it correctly so that the distributions can be easily programmed and plotted without rescaling. (The constant ensures that this integrates to 1.)

5. The mean of this distribution is

$$[\beta_0]_i \quad . \tag{5.34}$$

6. The variance of the ith variable is the ith diagonal element of the Q^{-1} matrix z_{ii} multiplied by the factor, i.e.,

$$\frac{\nu_0 s_0^2}{\nu_0 - 2}$$

to obtain the variance of the univariate marginal

$$Variance = \frac{\nu_0 s_0^2}{\nu_0 - 2} z_{ii} \quad . \tag{5.35}$$

We summarize the vector of mean values of the $(p+1)$-vector β appended to the scalar σ and the variance-covariance of rank $(p,1) \times (p,1)$ in the following two equations respectively:

$$\text{Mean} = \begin{bmatrix} \beta_0 \\ s_0^2 \frac{\nu_0}{\nu_0 - 2} \end{bmatrix}$$

and

$$\text{Variance-Covariance} = \begin{bmatrix} Q^{-1} s_0^2 \frac{\nu_0}{\nu_0-2} & 0 \\ 0 & \frac{2}{\nu_0 - 4}\left(s_0^2 \frac{\nu_0}{\nu_0 - 2}\right)^2 \end{bmatrix} \quad .$$

5.8 The Full Predictive Distribution–Prior

The multivariate predictive distribution (based on the matrix D whose q rows contain the predictions to be made and whose columns numbering p represent the number of coefficients in the model) has been calculated in equation (4.60) after substituting all the prior parameters from equation (5.1)

$$\{d, D, \sigma | X, y\} = \frac{2(2\pi)^{-\frac{q}{2}} \left(\frac{\nu_0 s_0^2}{2}\right)^{\frac{\nu_0}{2}}}{|I + DQ^{-1}D^T|^{\frac{1}{2}} \Gamma\left(\frac{\nu_0}{2}\right)} \sigma^{-(\nu_0 + q + 1)} e^{-\frac{\nu_0 s_0^2}{2\sigma^2}} \times e^{-\frac{1}{2\sigma^2}(d - D\beta_0)^T \left(I + DQ^{-1}D^T\right)^{-1}(d - D\beta_0)} \quad . \tag{5.36}$$

Keep in mind, the predicted variable *d* is a q-vector. We reiterate that the foregoing is not just a rote change of variables problem. It also considers the Fisherian assumption about data being generated by a random process. The multivariate predictive distribution (based on the matrix D whose q rows contain the predictions to be made and whose columns numbering p represent the number of coefficients in the model) has been calculated in equation (4.62) with the variance term z rather than σ, substituting all the prior parameters from equation (5.1)

$$\{d, z\} = \frac{\left(\frac{\nu_0 s_0^2}{2}\right)^{\frac{\nu_0}{2}}}{\Gamma\left(\frac{\nu_0}{2}\right)} z^{-\left(\frac{\nu_0}{2}+1\right)} e^{-\frac{\nu_0 s_0^2}{2z}} \times N\left[d; D\beta_0, \left(I + DQ^{-1}D^T\right) z\right] \quad . \tag{5.37}$$

The inverse gamma that emerges has been calculated in product form using the inverse gamma form developed in equation (4.63). Substituting all the prior parameters from equation (5.1), we write

$$IG\left(z; \frac{\nu_0 s_0^2}{2}, \frac{\nu_0}{2}\right) = \frac{\left(\frac{\nu_0 s_0^2}{2}\right)^{\frac{\nu_0}{2}}}{\Gamma\left(\frac{\nu_0}{2}\right)} z^{-\left(\frac{\nu_0}{2}+1\right)} e^{-\frac{\nu_0 s_0^2}{2z}} \quad . \tag{5.38}$$

The product form density developed in equation (4.64), after substituting all the prior parameters from equation (5.1), is

$$\{d, z\} = IG(z; \frac{\nu_0 s_0^2}{2}, \frac{\nu_0}{2}) N\left[d; D\beta_0, \left(I + DQ^{-1}D^T\right) z\right] \quad . \tag{5.39}$$

The marginal distribution over the predictive parameter d, a q-vector, calculated in equation (4.65), after substituting all the prior parameters from equation (5.1), becomes

$$\{d\} = \frac{\Gamma\left(\frac{\nu_0+q}{2}\right)}{|I + DQ^{-1}D^T|^{\frac{1}{2}} (\pi\nu_0 s_0^2)^{\frac{q}{2}} \Gamma\left(\frac{\nu_0}{2}\right)} \times \left[1 + (d - D\beta_0)^T \frac{\left(I + DQ^{-1}D^T\right)^{-1}}{\nu_0 s_0^2} (d - D\beta_0)\right]^{-\frac{\nu_0+q}{2}} \quad . \tag{5.40}$$

The mean of this distribution, which appeared in equation (4.66), has received substitutions from equation (5.1) to represent the prior mean

$$D\beta_0 \quad . \tag{5.41}$$

The variance-covariance matrix, which appeared in equation (4.67), has received substitutions from equation (5.1) to represent the prior variance-covariance matrix

$$\left(I + DQ^{-1}D^T\right) \frac{\nu_0 s_0^2}{\nu_0 - 2} \quad . \tag{5.42}$$

5.9 Predictions for a Single Row D, denoted δ, with $q = 1$–Prior

In the case in which the D matrix consists of a single row δ, which is of length p, the univariate predictive distribution over the (now scalar) d for the prior is obtained by setting $q = 1$, simplifying the D term to be δ, and making other substitutions from equation (5.36) to write

$$\{d, \delta, \sigma|X, y\} = \frac{2(2\pi)^{-\frac{1}{2}}\left(\frac{\nu_0 s_0^2}{2}\right)^{\frac{\nu_0}{2}}}{|1+\delta Q^{-1}\delta^T|^{\frac{1}{2}}\Gamma\left(\frac{\nu_0}{2}\right)}\sigma^{-(\nu_0+2)}e^{-\frac{\nu_0 s_0^2}{2\sigma^2}} \times e^{-\frac{1}{2\sigma^2}(d-\delta\beta_0)^T\left(1+\delta Q^{-1}\delta^T\right)^{-1}(d-\delta\beta_0)} .$$

The univariate predictive distribution whose columns are p in number and represent the coefficients in the model has been calculated in equation (5.37) with the variance term z rather than σ. We substitute the prior parameters from equation (5.1) to write

$$\{d, z\} = \frac{\left(\frac{\nu_0 s_0^2}{2}\right)^{\frac{\nu_0}{2}}}{\Gamma\left(\frac{\nu_0}{2}\right)} z^{-\left(\frac{\nu_0}{2}+1\right)}e^{-\frac{\nu_0 s_0^2}{2z}} N\left[d; \delta\beta_0, \left(1+\delta Q^{-1}\delta^T\right)z\right] .$$

The inverse gamma that emerges has been calculated in product form uses the inverse gamma form developed in equation (5.38). We substitute all the prior parameters from equation (5.1) to write

$$IG(z; \frac{\nu_0 s_0^2}{2}, \frac{\nu_0}{2}) = \{z\} = \frac{\left(\frac{\nu_0 s_0^2}{2}\right)^{\frac{\nu_0}{2}}}{\Gamma\left(\frac{\nu_0}{2}\right)} z^{-\left(\frac{\nu_0}{2}+1\right)}e^{-\frac{\nu_0 s_0^2}{2z}} .$$

The product form density developed in equation (5.39) is converted to the prior by substituting the prior parameters from equation (5.1) as follows

$$\{d, z\} = IG(z; \frac{\nu_0 s_0^2}{2}, \frac{\nu_0}{2}) N\left[d; \delta\beta_0, \left(1+\delta Q^{-1}\delta^T\right)z\right] .$$

The marginal distribution over the predictive parameter d has been calculated in equation (5.40) after substituting all the prior parameters from equation (5.1). We write

$$(5.43) \qquad \{d\} = \frac{\Gamma\left(\frac{\nu_0+1}{2}\right)}{|1+\delta Q^{-1}\delta^T|^{\frac{1}{2}}(\pi\nu_0 s_0^2)^{\frac{1}{2}}\Gamma\left(\frac{\nu_0}{2}\right)} \times \left[1 + (d-\delta\beta_0)^T\frac{\left(1+\delta Q^{-1}\delta^T\right)^{-1}}{\nu_0 s_0^2}(d-\delta\beta_0)\right]^{-\frac{\nu_0+1}{2}} .$$

The mean of this distribution, which appeared in equation (5.41) and has received substitutions from equation (5.1) to characterize the prior mean, is the following

$$\delta\beta_0 \quad . \tag{5.44}$$

The variance-covariance matrix, which appeared in equation (5.42) and has received substitutions from equation (5.1) to characterize the prior variance-covariance, is

$$\left(1+\delta Q^{-1}\delta^T\right)\frac{\nu_0 s_0^2}{\nu_0-2} \quad . \tag{5.45}$$

5.10 Predictions for a Single Row and a Single Column (3) with $\delta = e_3$ with $q = 1$–Prior

In the case in which the δ matrix consists of a single row e_3, which is of length p and contains all zeroes with a 1 in position 3, the univariate predictive distribution over the (scalar) d for the prior is obtained from equation (4.71) by modifying δ to e_3 and simplifying to write

$$\begin{aligned}\{d, e_3, \sigma | X, y\} = &\frac{2(2\pi)^{-\frac{1}{2}}\left(\frac{\nu_0 s_0^2}{2}\right)^{\frac{\nu_0}{2}}}{|1+[Q^{-1}]_{33}|^{\frac{1}{2}}\Gamma\left(\frac{\nu_0}{2}\right)}\sigma^{-(\nu_0+2)}e^{-\frac{\nu_0 s_0^2}{2\sigma^2}}\\ &\times e^{-\frac{1}{2\sigma^2}(d-\beta_{03})^T\left(1+[Q^{-1}]_{33}\right)^{-1}(d-\beta_{03})} \quad .\end{aligned}$$

The univariate predictive distribution, whose columns number p and represent the number of coefficients in the model, has been calculated in equation (4.72) using the variance term z rather than the standard deviation term σ. Substituting all the prior parameters from equation (5.1) leads to the equation

$$\{d, z\} = \frac{\left(\frac{\nu_0 s_0^2}{2}\right)^{\frac{\nu_0}{2}}}{\Gamma\left(\frac{\nu_0}{2}\right)} z^{-\left(\frac{\nu_0}{2}+1\right)} e^{-\frac{\nu_0 s_0^2}{2z}} N\left[d; \beta_{03}, \left(1+\left[Q^{-1}\right]_{33}\right)z\right] \quad .$$

The inverse gamma distribution that emerges has been thus calculated in product form, and it is in the same form as the as the inverse gamma form in equation (4.73). Substituting all the prior parameters from equation (5.1), we write

$$IG(z; \frac{\nu_0 s_0^2}{2}, \frac{\nu_0}{2}) = \{z\} = \frac{\left(\frac{\nu_0 s_0^2}{2}\right)^{\frac{\nu_0}{2}}}{\Gamma\left(\frac{\nu_0}{2}\right)} z^{-\left(\frac{\nu_0}{2}+1\right)} e^{-\frac{\nu_0 s_0^2}{2z}} \quad .$$

This is the selfsame inverse gamma distribution we have seen throughout this prior section. The product form density developed in equation (4.74) is converted to the prior by substituting all the prior parameters from equation (5.1) to write

$$\{d, z\} = IG(z; \frac{\nu_0 s_0^2}{2}, \frac{\nu_0}{2}) N\left[d; \beta_{03}, \left(1 + \left[Q^{-1}\right]_{33}\right) z\right] \quad .$$

The marginal distribution over the predictive parameter d has been calculated in equation (4.75). We substitute the prior parameters from equation (5.1) to write

$$\{d\} = \frac{\Gamma\left(\frac{\nu_0+1}{2}\right)}{|1 + [Q^{-1}]_{33}|^{\frac{1}{2}} (\pi \nu_0 s_0^2)^{\frac{1}{2}} \Gamma\left(\frac{\nu_0}{2}\right)}$$
$$\times \left[1 + (d - e_3 \beta_{03})^T \frac{\left(1 + \left[Q^{-1}\right]_{33}\right)^{-1}}{\nu_0 s_0^2} (d - e_3 \beta_{03})\right]^{-\frac{\nu_0+1}{2}} \quad .$$

The mean of this distribution, which appeared in equation (4.76), has received substitutions from equation (5.1) to become the following

$$\beta_{03} \quad .$$

Its variance-covariance matrix, which appeared in equation (4.77), has received substitutions from equation (5.1) to become

$$\left(1 + \left[Q^{-1}\right]_{33}\right) \frac{\nu_0 s_0^2}{\nu_0 - 2} \quad .$$

If we want to use e_3 in place of δ, we are going to be predicting the third coefficient of the coefficient vector β. The density over this third term is

$$\{d_3\} = \frac{\Gamma\left(\frac{\nu_0+1}{2}\right)}{|1 + [Q^{-1}]_{33}|^{\frac{1}{2}} (\pi \nu_0 s_0^2)^{\frac{1}{2}} \Gamma\left(\frac{\nu_0}{2}\right)}$$
$$\times \left[1 + (d_3 - \beta_{03})^T \frac{1}{(1 + [Q^{-1}]_{33}) \nu_0 s_0^2} (d_3 - \beta_{03})\right]^{-\frac{\nu_0+1}{2}} \quad .$$

Eliminating the 1 so that we are executing a mere change of variables problem here (and not taking account of the stochastic nature of the data process), we can write the density over coefficient 3 to be

$$\{d_3\} = \frac{\Gamma\left(\frac{\nu_0+1}{2}\right)}{|[Q^{-1}]_{33}|^{\frac{1}{2}} (\pi \nu_0 s_0^2)^{\frac{1}{2}} \Gamma\left(\frac{\nu_0}{2}\right)} \tag{5.46}$$
$$\times \left[1 + (d_3 - \beta_{03})^T \frac{1}{[Q^{-1}]_{33} \nu_0 s_0^2} (d_3 - \beta_{03})\right]^{-\frac{\nu_0+1}{2}} \quad .$$

Compare this with equation (5.33); it is identical. Happily so. We have equated the third coefficient done intrinsically to the prediction distribution for the third coefficient. This will become very useful the further we do into this book. The mean of this univariate density is of course

$$Mean = \beta_{03} \quad . \tag{5.47}$$

Compare this with equation (5.34); it is identical. The variance is

$$\left(1 + \left[Q^{-1}\right]_{33}\right) \frac{\nu_0 s_0^2}{\nu_0 - 2} \quad .$$

If we eliminate the predicted versus actual term represented by the 1+ term, the variance is

$$\left[Q^{-1}\right]_{33} \frac{\nu_0 s_0^2}{\nu_0 - 2} \quad . \tag{5.48}$$

Compare this with equation (5.35); it is identical. This process of eliminating the 1 term, which represents the error distribution of predicted relative to actual, is going to be important again later in this monograph when we seek marginal distributions for more complex outputs than a single univariate output to our sought after probability distribution.

5.11 We Will Array These Results Together with the Archetype, Posterior, and Classical in Chapter 9

These results, all of which have common functional forms for the Archetype, Prior, Posterior, and Classical will be arrayed in Chapter 9 for easy access by analysts and programmers.

5.12 Ways to Get a Prior

The equations for the prior are now complete. We have a fully mathematical expression for the prior, for predictions using the prior, and a number of the marginals and moments in the prior. This distribution can assuredly characterize a wide range of prior beliefs and thereby anchor Bayesian statistical analysis. We have seen one simple way to get a prior. There are others, as we summarize here.

We have already seen one way to get a prior–just guess, a rudimentary guess at that–but a guess that has the merit of reflecting in a rudimentary way what an expert might impressionistically (or analytically) think ex ante. We do not mean to be pejorative about just guessing. Guesses are far better in a mean value sense than assuming complete, abject, bone-numbing, legally-dead-above-the-neck ignorance or agnosticism. Everyone knows more than that, even anecdotally. Because the analyst or his designated experts have a priori estimates, the honest analyst will put a big variance around his or her prior estimate to reflect relative ignorance (or smaller if experienced and sophisticated). In order to provide a prior estimate using these simple estimates, one needs to estimate

- the coefficient mean and variance-covariance.
- the variance term mean and variance term variance.

This is actually far more useful and far less arbitrary than might be implied here. However, there are other ways as well.

What if the prior is derived from a pre-existing data set, a "lousy," "incomplete," "old," "small," or "questionable" data set? What if it is a subset of a larger data set gathered for a different purpose? What if it is "small" because the data gathering person ran out of data gathering money when the decision had to be made. Many real-world decisions do not wait for data! Can we put together a prior from an old, pre-existing "lousy" data set? Of course we can! And we shall address that later after we have worked the Bayesian method all the way through to its logical bottom line. We shall do so carefully in Chapter 10 because this is such an important topic. Our finding there is rather profound. We show that a "data prior" is actually more closely integrated into the Bayesian result than one might think a priori.

Can we find clever ways to assess the prior directly from expert assessment? Again, the answer is a definitive yes, and we shall see one such method in Chapter 14. The McFadden[68] quantal choice work also gives a huge clue and head start. This section has dealt with the mathematics of the prior, not the origins of the prior.

We have put forth only the simplest, most subjective possible prior here so that the analyst can easily "get right to work." Unlike Zellner[107], we are not in search of a completely general (which we might term infinitely general) prior as we often do in Decision Analysis. Zellner[107] sought a fully general prior, one that could be "anything." We are looking for a characterization of prior judgment in its simplest possible, and most subjectively estimable, form. We want to make it easy to initiate regression analysis with approximate, impressionistic prior information. In one of the final conversations we had with Dr. Zellner, he admitted that a simple, direct characterization of the prior is far better than "the nothingburger that classical regression gives you." Our prior is not "correct" or "robust." It is simple, approximate, useful, illustrative, and elucidating. We can proceed toward general priors outside this book, and that will erode the prior-posterior conjugacy that is so elucidating and so directly developed from classical regression over the last century.

Chapter 6

CALCULATING THE POSTERIOR JOINT DENSITY FROM THE PRIOR AND THE LIKELIHOOD

> ***A problem of statistical inference or, more simply, a statistics problem is a problem in which data that have been generated in accordance with some unknown probability distribution must be analyzed and some type of inference about the unknown distribution must be made.***
>
> *Morris H. DeGroot*

We begin this discussion of the posterior by appealing to the key results from Chapter 3 for easy use in this section. We need to rewrite and appeal to the definitions of the key quantities from Chapter 3 and summarize the crucial results for the posterior:

- Model coefficients. In Chapter 3, we designated the posterior mean value of the coefficients as β^*, which we derived in equation (3.9). We are going to be able to make the substitution $\mu \leftarrow \beta^*$ in the equations developed for the conjugate-archetype in Chapter 4 to characterize the prior in full mathematical detail. The mean of the coefficients under the posterior is rewritten here for reference:

$$<\beta>=\beta^* = \left(X^TX+Q\right)^{-1}\left(X^Ty+Q\beta_0\right) \quad .$$

- The sum of squared errors term A was derived in equation (3.13) in Chapter 3. This is going to allow us to make the substitution $g \leftarrow A$ in the equations developed for the conjugate-archetype in Chapter 4 to develop the posterior. The A constant is rewritten here for reference:

$$A = R + M + \left(\bar{\beta}-\beta^*\right)^T X^Ty + \left(\beta_0-\beta^*\right)^T Q\beta_0 \quad .$$

- The degrees of freedom for the posterior was developed in Chapter 3 in equation (3.7) and is rewritten here for reference:

$$\nu = m+n-p-1 = \nu_0+n \quad .$$

  This allows us to make the substitution in the equations in Chapter 4 that $r \leftarrow m+n$ to develop the proper degrees of freedom in the posterior.

- The H matrix is that in equation (3.10) in Chapter 3 , which is rewritten here for reference

$$H = X^TX+Q \quad .$$

We will be developing variance-covariance, mean, and density values for the posterior here in Chapter 6 using the foregoing substitutions using the conjugate-archetype from Chapter 4. As with the prior in Chapter 5, there is no point actually working through all the calculus from first principles. It is preferable, and more clear, to substitute into Chapter 4 the key substitutions of this chapter.

To summarize, for the posterior, we have the following specific parametrization, which can be substituted directly into the appropriate terms in the conjugate-archetype density results in Chapter 4, stated in the beginning of that chapter.

$$\begin{aligned} g &\leftarrow A \\ \mu &\leftarrow \beta^* \\ r &\leftarrow m+n \\ \nu &\leftarrow \nu \\ s^2 &\leftarrow \frac{A}{\nu} \end{aligned} \tag{6.1}$$

If we make these substitutions from equation (6.1) into all the various relationships in Chapter 4, we will have a bona fide posterior characterized by these parameters, i.e., we will have the bona fide posterior developed in Chapter 3. This chapter makes the foregoing assignments for the indicated variables into the conjugate-archetype density developed in Chapter 4 so as to properly constitute the (conjugate) posterior.

We begin by characterizing the degrees of freedom for the posterior beginning with equation (4.4) and thereafter making the substitutions from equation (6.1) to write

$$\nu \triangleq m + n - p - 1 \quad . \tag{6.2}$$

This implies, by substitution from equation (6.1) into equation (4.5), that

$$m + n = \nu + p + 1 \quad . \tag{6.3}$$

As noted in Chapter 4, this might again appear to be somewhat arbitrary, but we will re-verify in this section that it is far from arbitrary. We substitute from equation (6.1) into equation (4.7) to develop the expression for A:

$$\begin{aligned} A &= \nu s^2 \\ &= R + M + \left(\bar{\beta} - \beta^*\right)^T X^T y + (\beta_0 - \beta^*)^T Q \beta_0 \quad . \end{aligned} \tag{6.4}$$

This allows us to develop the s^2 term for the posterior, precisely the same as equation (4.6) with the substitutions from equation (6.1), as follows

$$s^2 = \frac{A}{\nu} \quad . \tag{6.5}$$

A is going to be analogous to the residual sum of squared errors $R = \nu_C s_C^2$ from the classical regression case. Using these definitions, we write the crucial results for the posterior

- The mean value of model coefficients devolves directly from substituting equations (6.1) into equation (4.13) to write

$$< \beta >= \mu = \beta^* \quad . \tag{6.6}$$

- The H matrix for this case is obtained by substituting equations (6.1) into equation (4.14) to write

$$H = X^T X + Q \quad . \tag{6.7}$$

  We reiterate that $X^T X$ need not be non-singular for these equations to work.

6.1 p-vector Multivariate Density–Posterior

We know that the posterior has the functional form derived from equation (4.8) but with parameter substitutions from equation (6.1). Such substitutions allow us to write the posterior joint density over β and σ in the following form:[1]

$$\{\beta,\sigma\} = \frac{2\left|X^TX+Q\right|^{\frac{1}{2}}}{(2\pi)^{\frac{p}{2}}\Gamma\left(\frac{\nu}{2}\right)}\left(\frac{\nu s^2}{2}\right)^{\frac{\nu}{2}}\sigma^{-(\nu+p+1)}e^{-\frac{\nu s^2}{2\sigma^2}} \times e^{-\frac{1}{2}(\beta-\beta^*)^T\left[\left(X^TX+Q\right)\sigma^{-2}\right](\beta-\beta^*)} \quad . \tag{6.8}$$

Note that this is precisely in the form of equation (4.8), the conjugate-archetype equation functional form for the conjugate prior and posterior, but with the parameter substitutions inserted from equation (6.1) to create the posterior. We reiterate that there are exactly five parameters, which collectively we term the "sufficient statistics," for the posterior

- p = number of model parameters
- ν = degrees of freedom
- s^2 = variance term
- X^TX+Q = precision matrix
- β^* = mean of the coefficients .

We change the foregoing joint density over β and σ into a joint density over β and $z=\sigma^2$ by substitution into equation (4.10) using the relations in equation (6.1), and thereafter writing

$$\{\beta,z\} = \frac{\left|X^TX+Q\right|^{\frac{1}{2}}(2\pi)^{-\frac{p}{2}}\left(\frac{\nu s^2}{2}\right)^{\frac{\nu}{2}}}{\Gamma\left(\frac{\nu}{2}\right)}z^{-\frac{\nu+p}{2}+1} \times e^{-\frac{\nu s^2}{2z}}e^{-\frac{1}{2z}(\beta-\beta^*)^T\left(X^TX+Q\right)(\beta-\beta^*)} \quad . \tag{6.9}$$

We will be using this density frequently in the forthcoming developments. There will be an inverse gamma distribution for the posterior embedded here, which becomes evident when we express the distribution using the variance using $z=\sigma^2$. We obtain the inverse gamma by substituting into equation (4.11) using equation (6.1) to obtain

$$IG\left(z;\frac{\nu s^2}{2},\frac{\nu}{2}\right) = \frac{\left(\frac{\nu s^2}{2}\right)^{\frac{\nu}{2}}}{\Gamma\left(\frac{\nu}{2}\right)}z^{-\left(\frac{\nu}{2}+1\right)}e^{-\frac{\nu s^2}{2z}} \quad . \tag{6.10}$$

Its mean and variance are derived by substituting from equation (6.1) into the equation (4.22) for the mean and into equation (4.23) for the variance respectively to write the mean and variance-covariance for the posterior

[1] Keep in mind, because the posterior and posterior are conjugate, we can use the same functional form for both, but with different parameter values.

(6.11)
$$<\sigma^2>=\frac{\nu s^2}{\nu-2}=\text{mean}$$

and

(6.12)
$$<\left(\sigma^2\right)^2>-<\sigma^2>^2=\frac{2}{\nu-4}\left(\frac{\nu s^2}{\nu-2}\right)^2=\text{variance} \quad .$$

The "product rule" in terms of the variance term z is determined by substituting into equation (4.12) from equation (6.1) to obtain the following

(6.13)
$$\{\beta,z\}=IG\left(z;\frac{\nu s^2}{2},\frac{\nu}{2}\right)N\left\{\beta;\beta^*,\left[\left(X^TX+Q\right)^{-1}z\right]\right\} \quad .$$

This section provides a complete discussion of the posterior.

6.2 p-vector Marginal Density–Posterior

This section integrates the σ term (or equivalently the z term) out to create the marginal density over β. Recall that the joint marginal over the model coefficients, taken directly from equation (4.15) for the conjugate-archetype distribution but with all parameters substituted as in equation (6.1), is the following:

(6.14)
$$\{\beta\}=\frac{\left|X^TX+Q\right|^{\frac{1}{2}}\Gamma\left(\frac{\nu+p}{2}\right)}{\left(\pi\nu s^2\right)^{\frac{p}{2}}\Gamma\left(\frac{\nu}{2}\right)}$$
$$\left[1+(\beta-\beta^*)^T\frac{X^TX+Q)}{\nu s^2}(\beta-\beta^*)\right]^{-\frac{\nu+p}{2}} \quad .$$

The mean of the coefficients of this marginal distribution, obtained by substituting from equation (6.1) into equation (4.18), is

(6.15)
$$<\beta>=\beta^* \quad .$$

The variance-covariance matrix of the coefficients, obtained by substituting from equation (6.1) into equation (4.19), is

(6.16)
$$VCV=\left(X^TX+Q\right)^{-1}s^2\frac{\nu}{\nu-2} \quad .$$

The H matrix, obtained by substituting from equation (6.1) into equation (4.20), is

(6.17)
$$H=X^TX+Q \quad .$$

We summarize the vector of mean values of the $(p+1)$-vector β appended to the scalar σ and the variance-covariance of rank $(p,1)\times(p,1)$ in the following two equations respectively:

$$\text{Mean} = \begin{bmatrix} \beta^* \\ s^2\frac{\nu}{\nu-2} \end{bmatrix}$$

and

$$\text{Variance-Covariance} = \begin{bmatrix} \left(X^TX+Q\right)^{-1}s^2\frac{\nu}{\nu-2} & 0 \\ 0 & \frac{2}{\nu-4}\left(s^2\frac{\nu}{\nu-2}\right)^2 \end{bmatrix} \quad .$$

This completes the characterization of the marginal distribution over the model coefficients.

6.3 Marginal Distribution for the $(p-L)$-subvector of the p-vector β–Posterior

We may not want to carry all p of the model coefficients. We may want to consider a subset of $p-L$ of the p elements of β. What does the joint and marginal distribution look like for that subset of $p-L$ model coefficients? We are interested in the joint density over β_2 (which is assumed to be a $p-L$ vector subset of the p-vector β) and σ. For notational convenience, we write

$$\Omega = X^TX+Q \quad .$$

We are going to use the same partitioning of the matrix Ω as we used for the matrix H previously. The partitioning will be obvious; no confusion will result. We have an L partition and a $p-L$ partition. (We will be able to move away from this definition of Ω, which is merely made purely for notational convenience during the development.)

We obtain the (partially integrated) marginal distribution over β_2 and σ by substituting from equation (6.1) into equation (4.28), the answer appearing in equation (6.18)

$$\{\beta_2,\sigma\} = \frac{2\left|\Omega_{22}-\Omega_{21}{\Omega_{11}}^{-1}\Omega_{12}\right|^{\frac{1}{2}}}{(2\pi)^{\frac{p-L}{2}}\Gamma\left(\frac{\nu}{2}\right)}\left(\frac{\nu s^2}{2}\right)^{\frac{\nu}{2}}\sigma^{-(\nu+p+1-L)}e^{-\frac{\nu s^2}{2\sigma^2}} \times e^{-\frac{1}{2}(\beta_2-\beta_2^*)^T\left[\left(\Omega_{22}-\Omega_{21}{\Omega_{11}}^{-1}\Omega_{12}\right)\sigma^{-2}\right](\beta_2-\beta_2^*)} \quad . \tag{6.18}$$

Use of the partitioned matrix Ω is apparent, obvious, and precisely analogous to the partitioning of the matrix H in Chapter 4 and the matrix Q in Chapter 5. In this equation, β_2 represents the subset of the p-vector β we want to have "left over" after the integration of L variables. The term β_2^* represents the $p-L$ subvector of the p-vector β^* corresponding to β_2. The remaining subset will be integrated away to create the marginal distribution over the $p-L$ subset. We want to convert this density over to the form that uses the variance term $z=\sigma^2$ in place of the standard deviation term σ, which we accomplish by substituting (6.1) into equation (4.29), the answer appearing in equation (6.19)

$$\{\beta_2, z\} = \frac{\left|\Omega_{22}-\Omega_{21}{\Omega_{11}}^{-1}\Omega_{12}\right|^{\frac{1}{2}}\left(\frac{\nu s^2}{2}\right)^{\frac{\nu}{2}}}{(2\pi)^{\frac{p-L}{2}}\Gamma\left(\frac{\nu}{2}\right)} \times z^{-\frac{\nu+p-L}{2}-1} e^{-\frac{\nu s^2}{2z}} e^{-\frac{1}{2}(\beta_2-\beta_2^*)^T\frac{\Omega_{22}-\Omega_{21}{\Omega_{11}}^{-1}\Omega_{12}}{z}(\beta_2-\beta_2^*)} \quad . \tag{6.19}$$

As shown in Chapter 4, the product rule density over β_2 and σ is obtained by substituting from equation (6.1) into equation (4.30) to write the product rule density in equation (6.20)

$$\{\beta_2, z\} = IG\left(z, \frac{\nu s^2}{2}, \frac{\nu}{2}\right) \times N\left[\beta_2; p-L, \beta_2^*, z\left(\Omega_{22}-\Omega_{21}{\Omega_{11}}^{-1}\Omega_{12}\right)^{-1}\right] \quad . \tag{6.20}$$

The inverse gamma density over z is obtained by substituting from equation (6.1) into equation (4.31) to write equation (6.21)

$$IG\left(z, \frac{\nu s^2}{2}, \frac{\nu}{2}\right) = \frac{\left(\frac{\nu s^2}{2}\right)^{\frac{\nu}{2}}}{\Gamma\left(\frac{\nu}{2}\right)} z^{-\frac{\nu}{2}-1} e^{-\frac{\nu s^2}{2z}} \quad . \tag{6.21}$$

Finally, the marginal distribution over the $p-L$-vector β_2 is obtained by substituting from equation (6.1) into equation (4.33), as follows:

$$\{\beta_2\} = \frac{\left|\Omega_{22}-\Omega_{21}{\Omega_{11}}^{-1}\Omega_{12}\right|^{\frac{1}{2}}\Gamma\left(\frac{\nu+p-L}{2}\right)}{(\pi\nu s^2)^{\frac{p-L}{2}}\Gamma\left(\frac{\nu}{2}\right)} \times \left[1+(\beta_2-\beta_2^*)^T\frac{\left(\Omega_{22}-\Omega_{21}{\Omega_{11}}^{-1}\Omega_{12}\right)}{\nu s^2}(\beta_2-\beta_2^*)\right]^{\frac{-\nu-p+L}{2}} \quad . \tag{6.22}$$

6.4 When $p-L=1$-subvector of β–Posterior

If we substitute the parameters from equation (6.1) into equation (4.34), we obtain the posterior univariate full density over the scalar β_2 and σ in equation (6.23):

$$\{\beta_2, \sigma\} = \frac{2\left|\Omega_{22}-\Omega_{21}{\Omega_{11}}^{-1}\Omega_{12}\right|^{\frac{1}{2}}\left(\frac{\nu s^2}{2}\right)^{\frac{\nu}{2}}}{(2\pi)^{\frac{1}{2}}\Gamma\left(\frac{\nu}{2}\right)}\sigma^{-(\nu+2)} e^{-\frac{\nu s^2}{2\sigma^2}}$$

$$\times\, e^{-\frac{1}{2}(\beta_2-\beta_2^*)^T\frac{\Omega_{22}-\Omega_{21}\Omega_{11}{}^{-1}\Omega_{12}}{\sigma^2}(\beta_2-\beta_2^*)} \quad . \tag{6.23}$$

If we substitute the parameters from equation (6.1) into equation (4.30), making sure to set $q = 1$, we obtain the following equation over β_2 and $z = \sigma^2$

$$\{\beta_2, z\} = IG\left(z, \frac{\nu s^2}{2}, \frac{\nu}{2}\right) N\left[\beta_2; \beta_2^*, z\left(\Omega_{22} - \Omega_{21}\Omega_{11}{}^{-1}\Omega_{12}\right)^{-1}\right] \quad .$$

If we substitute the parameters from equation (6.1) into equation (4.31), we obtain equation (6.24)

$$IG\left(z, \frac{\nu s^2}{2}, \frac{\nu}{2}\right) = \frac{\left(\frac{\nu s^2}{2}\right)^{\frac{\nu}{2}}}{\Gamma\left(\frac{\nu}{2}\right)} z^{-\frac{\nu}{2}-1} e^{-\frac{\nu s^2}{2z}} \quad . \tag{6.24}$$

If we substitute the parameters from equation (6.1) into equation (4.35), we obtain equation (6.25)

$$\begin{aligned}\{\beta_2\} = {} & \frac{\left|\Omega_{22} - \Omega_{21}\Omega_{11}{}^{-1}\Omega_{12}\right|^{\frac{1}{2}}\Gamma\left(\frac{\nu+1}{2}\right)}{(\pi\nu s^2)^{\frac{1}{2}}\Gamma\left(\frac{\nu}{2}\right)} \\ & \times \left[1 + (\beta_2 - \beta_2^*)^T\frac{\Omega_{22} - \Omega_{21}\Omega_{11}{}^{-1}\Omega_{12}}{\nu s^2}(\beta_2 - \beta_2^*)\right]^{-\frac{\nu+1}{2}} \quad .\end{aligned} \tag{6.25}$$

If we substitute the parameters from equation (6.1) into equation (4.36), we obtain equation (6.26)

$$\left[\Omega^{-1}\right]_{ii} = \left(\Omega_{22} - \Omega_{21}\Omega_{11}{}^{-1}\Omega_{12}\right)^{-1} = z_{ii} \quad . \tag{6.26}$$

If we substitute the parameters from equation (6.1) into equation (4.37), we obtain equation (6.27)

$$\{\beta_2, \sigma\} = \frac{2\left(\frac{\nu s^2}{2}\right)^{\frac{\nu}{2}}}{(2\pi z_{ii})^{\frac{1}{2}}\Gamma\left(\frac{\nu}{2}\right)}\sigma^{-(\nu+2)} e^{-\frac{\nu s^2}{2\sigma^2}} e^{-\frac{1}{2}(\beta_2-\beta_2^*)^T\frac{1}{z_{ii}\sigma^2}(\beta_2-\beta_2^*)} \quad . \tag{6.27}$$

If we substitute the parameters from equation (6.1) into equation (4.41), we obtain equation (6.28)

$$\{\beta_2, z\} = IG\left(z, \frac{\nu s^2}{2}, \frac{\nu}{2}\right) N\left(\beta_2; z\right) \quad . \tag{6.28}$$

If we substitute the parameters from equation (6.1) into equation (4.42), we obtain equation (6.29)

$$(6.29)\quad \{\beta_2\} = \frac{\Gamma\left(\frac{\nu+1}{2}\right)}{(\pi\nu s^2 z_{ii})^{\frac{1}{2}}\Gamma\left(\frac{\nu}{2}\right)}\left[1+(\beta_2-\beta_2^*)^T\frac{1}{z_{ii}\nu s^2}(\beta_2-\beta_2^*)\right]^{-\frac{\nu+1}{2}} \quad .$$

6.5 The Generic Univariate Student's t Distribution

If we substitute the parameters from equation (6.1) into equation (4.43), we obtain equation (6.30)

$$(6.30)\quad \{\beta,\sigma\} = \frac{2\sqrt{Q}}{\sqrt{2\pi}\Gamma\left(\frac{k-2}{2}\right)}\left(\frac{P}{2}\right)^{\frac{k-2}{2}}\sigma^{-k}e^{-\frac{P}{2\sigma^2}}e^{-\frac{1}{2\sigma^2}(\beta-\mu)^TQ(\beta-\mu)} \quad .$$

For this generic univariate Student's t distribution, we define for the posterior from equation (4.38) with the substitutions from equation (6.1)

$$P = \nu s^2 \quad .$$

For the generic univariate Student's t distribution, we define for the posterior from equation (4.39) with the substitutions from equation (6.1)

$$Q = \frac{1}{z_{ii}} \quad .$$

For the generic univariate Student's t distribution, we define for the posterior from equation (4.40) with the substitutions from equation (6.1)

$$k = \nu + 2 \quad .$$

For the generic univariate Student's t distribution, we define for the posterior from equation (4.44) with the substitutions from equation (6.1)

$$\{\beta\} = \int_0^\infty d\sigma\,\{\beta,\sigma\} = \frac{\Gamma\left(\frac{k-1}{2}\right)}{\sqrt{\pi}\Gamma\left(\frac{k-2}{2}\right)}\left(\frac{Q}{P}\right)^{\frac{1}{2}}\left[1+(\beta-\beta^*)^T\frac{Q}{P}(\beta-\beta^*)\right]^{-\frac{k-1}{2}}$$
$$= \frac{\Gamma\left(\frac{\nu+1}{2}\right)}{\sqrt{\pi}\Gamma\left(\frac{\nu}{2}\right)}\left(\frac{1}{z_{ii}\nu s^2}\right)^{\frac{1}{2}}\left[1+(\beta-\beta^*)^T\frac{1}{z_{ii}\nu s^2}(\beta-\beta^*)\right]^{-\frac{\nu+1}{2}} \quad .$$

For the generic univariate Student's t distribution, we define for the posterior from equation (4.45) with the substitutions from equation (6.1)

$$\langle\beta\rangle = \mu = \beta^* \quad .$$

For the generic univariate Student's t distribution, we define for the posterior from equation (4.46) with the substitutions from equation (6.1)

$$\text{variance} = \frac{1}{k-4}\left(\frac{P}{Q}\right) = \frac{\nu s^2 z_{ii}}{\nu - 2} \quad .$$

It is instructive for people who might want to calculate or program the marginal distributions over the coefficients to enumerate step by step the calculations one has to make to move from the joint distribution over the coefficients β to the p individual coefficients β_i

1. Calculate the inverse of the entire $\left(X^T X + Q\right)$ matrix, i.e.,

$$\left(X^T X + Q\right)^{-1} \quad .$$

2. Get the ii^{th} diagonal element of the $\left(X^T X + Q\right)^{-1}$ matrix. Designate it z_{ii}

$$z_{ii} = \left[\left(X^T X + Q\right)^{-1}\right]_{ii} = \left(\Omega_{22} - \Omega_{21}{\Omega_{11}}^{-1}\Omega_{12}\right)^{-1} \quad .$$

3. Using these definitions, the univariate Student's t density in equation (6.29) becomes

$$\{\beta_i\} = \frac{\Gamma\left(\frac{\nu+1}{2}\right)}{\Gamma\left(\frac{\nu}{2}\right)} \frac{1}{\left(\pi z_{ii} \nu s^2\right)^{\frac{1}{2}}} \times \left[1 + \left(\beta_i - [\beta^*]_i\right) \frac{1}{\nu s^2 z_{ii}} \left(\beta_i - [\beta^*]_i\right)\right]^{-\frac{\nu+1}{2}} \quad . \tag{6.31}$$

4. This density will integrate to 1 because the constant is correctly calculated. (The constant ensures that this integrates to 1.)

5. The mean of this distribution is

$$[\beta^*]_i \quad . \tag{6.32}$$

6. The variance of the ith variable is the ith diagonal element of the $\left(X^T X + Q\right)^{-1}$ matrix z_{ii} multiplied by the factor, i.e.,

$$\frac{\nu s^2}{\nu - 2}$$

to obtain the variance of the univariate marginal

$$\frac{\nu s^2}{\nu - 2} z_{ii} \quad . \tag{6.33}$$

6.6 The Full q-Dimensional Predictive Distribution–Posterior

The multivariate predictive distribution (based on the matrix D whose q rows contain the predictions to be made and whose columns numbering p represent the number of coefficients in the model) has been calculated in equation (4.60). Substituting all the posterior parameters from equation (6.1), we can write

$$\{d, D, \sigma|X, y\} = \frac{2(2\pi)^{-\frac{q}{2}}\left(\frac{\nu s^2}{2}\right)^{\frac{\nu}{2}}}{\left|I + D(X^TX+Q)^{-1}D^T\right|^{\frac{1}{2}}\Gamma\left(\frac{\nu}{2}\right)}\sigma^{-(\nu+q+1)}e^{-\frac{\nu s^2}{2\sigma^2}} \times e^{-\frac{1}{2\sigma^2}(d-D\beta^*)^T\left[I+D\left(X^TX+Q\right)^{-1}D^T\right]^{-1}(d-D\beta^*)} \quad . \tag{6.34}$$

The multivariate predictive distribution (based on the matrix D whose q rows contain the predictions to be made and whose columns numbering p represent the number of coefficients in the model) has been calculated in equation (4.62) with the variance term z rather than σ. Substituting all the posterior parameters from equation (6.1), we write

$$\{d, z\} = \frac{\left(\frac{\nu s^2}{2}\right)^{\frac{\nu}{2}}}{\Gamma\left(\frac{\nu}{2}\right)}z^{-\left(\frac{\nu}{2}+1\right)}e^{-\frac{\nu s^2}{2z}} N\left\{d; D\beta^*, \left[I + D\left(X^TX+Q\right)^{-1}D^T\right]z\right\} \quad . \tag{6.35}$$

The inverse gamma that emerges has been calculated in product form uses the inverse gamma form developed in equation (4.63), substituting all the posterior parameters from equation (6.1)

$$IG(z; \frac{\nu s^2}{2}, \frac{\nu}{2}) = \{z\} = \frac{\left(\frac{\nu s^2}{2}\right)^{\frac{\nu}{2}}}{\Gamma\left(\frac{\nu}{2}\right)}z^{-\left(\frac{\nu}{2}+1\right)}e^{-\frac{\nu s^2}{2z}} \quad . \tag{6.36}$$

The product form density developed in equation (4.64), substituting all the posterior parameters from equation (6.1), is

$$\{d, z\} = IG(z; \frac{\nu s^2}{2}, \frac{\nu}{2})N\left[d; D\beta^*, \left(I + DH^{-1}D^T\right)z\right] \quad . \tag{6.37}$$

The marginal distribution over the predictive parameter d has been calculated in equation (4.65), substituting all the posterior parameters from equation (6.1)

(6.38)

$$\{d\} = \frac{\Gamma\left(\frac{\nu+q}{2}\right)}{\left|I + D(X^TX+Q)^{-1}D^T\right|^{\frac{1}{2}} (\pi\nu s^2)^{\frac{q}{2}} \Gamma\left(\frac{\nu}{2}\right)}$$

$$\left\{1 + (d - D\beta^*)^T \frac{\left[I + D(X^TX+Q)^{-1}D^T\right]^{-1}}{\nu s^2} (d - D\beta^*)\right\}^{-\frac{\nu+q}{2}} .$$

The mean of this distribution, which appeared in equation (4.66), has received substitutions from equation (6.1) to write

(6.39) $$D\beta^* \quad .$$

and its variance-covariance matrix, which appeared in equation (4.67), has received substitutions from equation (6.1) to write

(6.40) $$\left[I + D(X^TX+Q)^{-1}D^T\right] \frac{\nu s^2}{\nu - 2} \quad .$$

6.7 Predictions for a Single Row D, denoted δ, with q=1–Posterior

In the case in which the D matrix consists of a single row δ, which is of length p, the univariate predictive distribution over the (scalar) d for the posterior is obtained by setting $q = 1$, simplifying the D to δ, and substituting other terms from equation (6.38) as follows:

$$\{d, \delta, \sigma | X, y\} = \frac{2(2\pi)^{-\frac{1}{2}} \left(\frac{\nu s^2}{2}\right)^{\frac{\nu}{2}}}{\left|1 + \delta(X^TX+Q)^{-1}\delta^T\right|^{\frac{1}{2}} \Gamma\left(\frac{\nu}{2}\right)} \sigma^{-(\nu+2)} e^{-\frac{\nu s^2}{2\sigma^2}}$$

$$\times\, e^{-\frac{1}{2\sigma^2}(d-\delta\beta^*)^T\left[1+\delta(X^TX+Q)^{-1}\delta^T\right]^{-1}(d-\delta\beta^*)} \quad .$$

The univariate predictive distribution whose columns numbering p represent the number of coefficients in the model) has been calculated in equation (6.35) with the variance term z rather than σ, substituting all the posterior parameters from equation (6.1)

$$\{d, z\} = \frac{\left(\frac{\nu s^2}{2}\right)^{\frac{\nu}{2}}}{\Gamma\left(\frac{\nu}{2}\right)} z^{-\left(\frac{\nu}{2}+1\right)} e^{-\frac{\nu s^2}{2z}} N\left\{d; \delta\beta^*, \left[1 + \delta(X^TX+Q)^{-1}\delta^T\right] z\right\} \quad .$$

The inverse gamma that emerges has been calculated in product form uses the inverse gamma form developed in equation (6.36), substituting all the posterior parameters from equation (6.1)

$$IG(z; \frac{\nu s^2}{2}, \frac{\nu}{2}) = \{z\} = \frac{\left(\frac{\nu s^2}{2}\right)^{\frac{\nu}{2}}}{\Gamma\left(\frac{\nu}{2}\right)} z^{-\left(\frac{\nu}{2}+1\right)} e^{-\frac{\nu s^2}{2z}} \quad .$$

The product form density developed in equation (6.37), substituting all the posterior parameters from equation (6.1), is

$$\{d, z\} = IG\left(z; \frac{\nu s^2}{2}, \frac{\nu}{2}\right) N\left\{d; \delta\beta^*, \left[1 + \delta\left(X^T X + Q\right)^{-1} \delta^T\right] z\right\} \quad .$$

The marginal distribution over the predictive parameter d has been calculated in equation (6.38), substituting all the posterior parameters from equation (6.1)

$$(6.41) \quad \{d\} = \frac{\Gamma\left(\frac{\nu+1}{2}\right)}{\left|1 + \delta(X^T X + Q)^{-1} \delta^T\right|^{\frac{1}{2}} (\pi \nu s^2)^{\frac{1}{2}} \Gamma\left(\frac{\nu}{2}\right)} \times$$
$$\left\{1 + (d - \delta\beta^*)^T \frac{\left[1 + \delta\left(X^T X + Q\right)^{-1} \delta^T\right]^{-1}}{\nu s^2} (d - \delta\beta^*)\right\}^{-\frac{\nu+1}{2}} \quad .$$

The mean of this distribution, which appeared in equation (6.39), has received substitutions from equation (6.1) that imply the following

$$(6.42) \qquad \text{mean} = \delta\beta^* \quad .$$

Its variance-covariance matrix, which appeared in equation (6.40), has received substitutions from equation (6.1) that imply the following

$$(6.43) \qquad \text{variance} = \left[1 + \delta\left(X^T X + Q\right)^{-1} \delta^T\right] \frac{\nu s^2}{\nu - 2} \quad .$$

6.8 Predictions for a Single Row and a Single Column (3) Assuming $\delta = e_3$ with $q = 1$–Posterior

In the case in which the δ matrix consists of a single row e_3, which is of length p and has all zeroes except for a 1 in the 3 position, the univariate predictive distribution over the (scalar) d for the posterior is obtained by setting $q = 1$ and simplifying the $D = \delta = e_3$ and other terms in equation (6.38) to write

$$\{d, e_3, \sigma | X, y\} = \frac{2(2\pi)^{-\frac{1}{2}}\left(\frac{\nu s^2}{2}\right)^{\frac{\nu}{2}}}{\left|1+\left[(X^TX+Q)^{-1}\right]_{33}\right|^{\frac{1}{2}}\Gamma\left(\frac{\nu}{2}\right)}\sigma^{-(\nu+2)}e^{-\frac{\nu s^2}{2\sigma^2}} \times e^{-\frac{1}{2\sigma^2}(d-\beta_3^*)^T\left(1+\left[(X^TX+Q)^{-1}\right]_{33}\right)^{-1}(d-\beta_3^*)} \quad .$$

The univariate predictive distribution whose columns numbering p represent the number of coefficients in the model) has been calculated in equation (6.35) using the variance term z rather than σ, substituting all the posterior parameters from equation (6.1)

$$\{d, z\} = \frac{\left(\frac{\nu s^2}{2}\right)^{\frac{\nu}{2}}}{\Gamma\left(\frac{\nu}{2}\right)}z^{-\left(\frac{\nu}{2}+1\right)}e^{-\frac{\nu s^2}{2z}}N\left[d; \beta_3^*, \left(1+\left[(X^TX+Q)^{-1}\right]_{33}\right)z\right] \quad .$$

The inverse gamma distribution has been calculated in product form uses the inverse gamma form developed in equation (6.36). Substituting all the posterior parameters from equation (6.1), we write

$$IG(z; \frac{\nu s^2}{2}, \frac{\nu}{2}) = \{z\} = \frac{\left(\frac{\nu s^2}{2}\right)^{\frac{\nu}{2}}}{\Gamma\left(\frac{\nu}{2}\right)}z^{-\left(\frac{\nu}{2}+1\right)}e^{-\frac{\nu s^2}{2z}} \quad .$$

The product form density developed in equation (6.37), substituting all the posterior parameters from equation (6.1), is

$$\{d, z\} = IG(z; \frac{\nu s^2}{2}, \frac{\nu}{2})N\left[d; \beta_3^*, \left(1+\left[(X^TX+Q)^{-1}\right]_{33}\right)z\right] \quad .$$

The marginal distribution over the predictive parameter d has been calculated in equation (6.38). Substituting all the posterior parameters from equation (6.1), we can write

$$\{d\} = \frac{\Gamma\left(\frac{\nu+1}{2}\right)}{\left|1+\left[(X^TX+Q)^{-1}\right]_{33}\right|^{\frac{1}{2}}(\pi\nu s^2)^{\frac{1}{2}}\Gamma\left(\frac{\nu}{2}\right)} \times \left[1+(d-\beta_3^*)^T\frac{\left(1+\left[(X^TX+Q)^{-1}\right]_{33}\right)^{-1}}{\nu s^2}(d-\beta_3^*)\right]^{-\frac{\nu+1}{2}} \quad .$$

The mean of this distribution, which appeared in equation (6.39), has received substitutions from equation (6.1) to become the following

$$\text{mean} = \beta_3^* \quad ,$$

and its variance-covariance matrix, which appeared in equation (6.40), has received substitutions from equation (6.1) to become the following

$$\text{variance} = \left(1 + \left[\left(X^T X + Q\right)^{-1}\right]_{33}\right) \frac{\nu s^2}{\nu - 2} \quad .$$

If we want to use e_3 in place of δ, we are going to be predicting the third coefficient. The density was just shown to be

$$\{d_3\} = \frac{\Gamma\left(\frac{\nu+1}{2}\right)}{\left|1 + \left[\left(X^T X + Q\right)^{-1}\right]_{33}\right|^{\frac{1}{2}} (\pi \nu s^2)^{\frac{1}{2}} \Gamma\left(\frac{\nu}{2}\right)}$$
$$\times \left[1 + (d_3 - \beta_3^*)^T \frac{1}{\left(1 + \left[\left(X^T X + Q\right)^{-1}\right]_{33}\right) \nu s^2} (d_3 - \beta_3^*)\right]^{-\frac{\nu+1}{2}} \quad .$$

Eliminating the 1 so that we are doing a mere change of variables problem and eliminate the predicted minus actual issues in the data, we can write the density over coefficient 3 to be

$$\text{(6.44)} \qquad \{d_3\} = \frac{\Gamma\left(\frac{\nu+1}{2}\right)}{\left|\left[\left(X^T X + Q\right)^{-1}\right]_{33}\right|^{\frac{1}{2}} (\pi \nu s^2)^{\frac{1}{2}} \Gamma\left(\frac{\nu}{2}\right)}$$
$$\times \left[1 + (d_3 - \beta_3^*)^T \frac{1}{\left[\left(X^T X + Q\right)^{-1}\right]_{33} \nu s^2} (d_3 - \beta_3^*)\right]^{-\frac{\nu+1}{2}} \quad .$$

Compare this with equation (6.31); it is identical. The mean of this univariate density is of course

$$\text{(6.45)} \qquad \text{mean} = \beta_3^* \quad .$$

Compare this with equation (6.32); it is identical. The variance is

$$\text{variance} = \left(1 + \left[\left(X^T X + Q\right)^{-1}\right]_{33}\right) \frac{\nu s^2}{\nu - 2} \quad .$$

If we eliminate the predicted versus actual term, the variance is

$$\text{(6.46)} \qquad \text{variance} = \left[\left(X^T X + Q\right)^{-1}\right]_{33} \frac{\nu s^2}{\nu - 2} \quad .$$

Compare this with equation (6.33); it is identical. This process of eliminating the 1 term, which represents the error distribution of predicted relative to actual, is going to be important again later in this monograph when we seek marginal distributions for more complex outputs than a single univariate.

6.9 We Will Array These Results Together with the Archetype, Prior, and Posterior in Chapter 9

These results, all of which have common functional forms for the Archetype, Prior, Posterior, and Classical will be arrayed in Chapter 9 for easy access by analysts and programmers.

Chapter 7

THE "KNOW NOTHING" (DIFFUSE) PRIOR AND ITS EQUIVALENCE TO CLASSICAL REGRESSION

When a coincidence seems amazing, that's because the human mind isn't wired to naturally comprehend probability and statistics.

Neil deGrasse Tyson

Recalling Chapter 2, which explicates classical linear regression, there is no notion of a "prior" whatsoever embedded within classical regression. The notion of a prior is completely absent. Or is it? The notion of a "prior" does not exist at all, at least on the surface. Yet in Chapter 3 and augmented in Chapter 6, the notion of a "prior" is front and center, implicitly with classical statistics and explicitly here. You cannot move forward without a prior. This suggests that Bayesian statistics can be specialized to classical statistics if you somehow render the prior invisible, ineffectual, inert, impotent, i.e., if you "defang" the prior. That is, you search for an "invisible" or impotent prior, one that is non-discernible. If you find that prior, you should be able to specialize Bayesian statistics into classical statistics, and you will have something substantive to say the specific nature of that now-invisible prior. This chapter shows precisely how.

There is profound depth to this "defanging" of the prior, as we will see in this section. Based in part on past work of Jeffreys[50] and the maximum entropy people, we can actually find and impose a prior that replicates the classical statistical approach in the posterior. A defanged prior that replicates classical regression with the posterior. And what this prior is and means is rather astonishing and enlightening. The meaning of that prior, called a "diffuse prior" is complete, total, abject, blockheaded, uncomprehending ignorance, and it is the quintessential maximum entropy prior. You know nothing about the problem when it comes time to do the statistical analysis, and the Jeffreys prior defines rather precisely what "nothing" means. "No clue," in the vernacular. With a Jeffreys prior, there is absolutely zero knowledge in the prior. The sole and complete knowledge is that which is embedded in the data. This chapter is going to make such concepts crystal clear.

To get the ball rolling in this section, we begin by considering the case in which the precision matrix in the prior Q is a diagonal matrix, the case in which the p model coefficients are all probabilistically independent, i.e., the case with all zero covariances. To wit, there will be zero covariance and only individual coefficient variances in the prior:

$$Q = \begin{bmatrix} \omega_1 & 0 & 0 & . & . & . & 0 \\ 0 & \omega_2 & 0 & . & . & . & 0 \\ 0 & 0 & \omega_3 & . & . & . & 0 \\ . & . & . & . & . & . & . \\ . & . & . & . & . & . & . \\ . & . & . & . & . & . & . \\ 0 & 0 & 0 & . & . & . & \omega_p \end{bmatrix} . \tag{7.1}$$

We discussed this situation in Chapter 5. The inverse of the precision matrix, denoted Q^{-1}, contains the reciprocals of the diagonals of Q:

$$Q^{-1} = \begin{bmatrix} \frac{1}{\omega_1} & 0 & 0 & . & . & . & 0 \\ 0 & \frac{1}{\omega_2} & 0 & . & . & . & 0 \\ 0 & 0 & \frac{1}{\omega_3} & . & . & . & 0 \\ . & . & . & . & . & . & . \\ . & . & . & . & . & . & . \\ . & . & . & . & . & . & . \\ 0 & 0 & 0 & . & . & . & \frac{1}{\omega_p} \end{bmatrix} .$$

This means, based on equation (4.19), the variance-covariance matrix is

$$VCV = \frac{\nu_0 s_0^2}{\nu_0 - 2} \begin{bmatrix} \frac{1}{\omega_1} & 0 & 0 & . & . & . & 0 \\ 0 & \frac{1}{\omega_2} & 0 & . & . & . & 0 \\ 0 & 0 & \frac{1}{\omega_3} & . & . & . & 0 \\ . & . & . & . & . & . & . \\ . & . & . & . & . & . & . \\ . & . & . & . & . & . & . \\ 0 & 0 & 0 & . & . & . & \frac{1}{\omega_p} \end{bmatrix} . \tag{7.2}$$

In this section, we will be examining the case in which $\omega_1, \omega_2, \ldots \omega_p$ all approach 0 in the limit from the positive orthant. We see clearly that this approaches a case in which the variance-covariance matrix has all zero covariances and all the variances of the prior over model coefficients approach infinity. We will be thinking about what infinite variance (or more properly, approaching infinite variance in the limit) means shortly, and it will be key to our development.

Using the foregoing notion of the variance-covariance matrix, the classical regression solution will now be derived as a limiting special case of the Bayesian posterior with a very specific prior, one that is developed from the diagonal matrices just introduced. In particular, if we assume abject, unequivocal, idealistic, blissfully negligent ignorance within the prior, we obtain the classical regression solution from the Bayesian technique (in a very exacting sense to be developed in this section). One would have to work overtime to be as ignorant as what we will term in this section a "diffuse prior" would imply. That is not a criticism at all; that is a statement that a diffuse prior is a specific, explicit model of ignorance, a lack of any and all ex ante knowledge, as pointed out by Prof. Shachter. As a deadly serious matter, the diffuse prior is a "model" of ignorance. It is actually a rather careful representation of what complete and total ignorance (lack of any and all information) is. It would be impossible in the real world to be this ignorant, but it is well to understand the model of utter ignorance, i.e., "zero information," to know what that might really mean. This level of ignorance will never really be achieved in the real world.

When we make the specialization from Bayes to classical in this section, we do not get precisely and exactly the same answer as one obtains under the traditional classical statistics model in Chapter 2, and that is a terrifically enlightening finding. In particular, the variances of the inverse gamma distribution over the square of the σ term will be different in form from classical statistics as it was characterized in Chapter 2, all other terms being identical. The Bayesian specialization of the classical statistics case is far more compelling than the classical statistics case is on its own merits. When we reflect that the classical likelihood function is the sole focus of classical statistics and that it represents the Fisherian probability distribution over the data given the parameters of the model (rather than a distribution over the parameters conditional on the observed data), it should be no surprise that there is a subtle difference.

In this section, we are going to begin with the posterior distribution developed in Chapter 3 and augmented in Chapter 6. In particular, we are going to start with the posterior joint density expressed in equation (6.8), and we are going to make a sequence of changes to the prior that drive the posterior closer and closer with each step and ultimately all the way to the point at which the posterior is identical to the classical regression solution. To wit, change the prior until the posterior becomes the classical regression solution from Chapter 2! What do you have to do to the prior in order to make the posterior become exactly identical with the classical regression solution? That is the approach we are going to take sequentially in this chapter. The changes to the prior are going to be extremely enlightening.

We begin with the following equations for the posterior:

- Equation (6.6) for the mean values of the coefficients.
- Equation (6.11) for the mean value of the variance term.
- Equation (6.16) for the variance-covariance matrix of the coefficients.
- Equation (6.12) for the variance of the variance term.
- Equation (6.4) for the A term.
- Equation (6.2) for the ν term.
- Equation (6.5) for the s^2 term.

7.1 First Step–Precision Matrix $Q = 0$

As an initial step, suppose we wish to assume that the precision matrix becomes closer and closer to $Q = 0$. The way to make this happen is to select the precision matrix in equation (7.1) and thereafter let all the diagonal terms ω_1, ω_2,... ω_p approach 0 in the limit from the positive orthant. That means that the precision matrix is always diagonal and approaches the zero matrix in a very specific way. The diagonal is always infinitesimally strictly positive, and the off-diagonal terms are always exactly 0. And there is never a whit of covariance. No knowledge whatsoever. No variable gives any other conditional information about another variable. Values of variables to not inform each other.

That would simultaneously mean that we are assuming that the variance-covariance matrix of the coefficients in equation (7.2) has its diagonals becoming infinite as the ω_1, ω_2,... ω_p coefficients are all approaching 0 in the limit from the positive orthant. That would also mean that we are assuming zero covariances among the coefficients. There are no correlations; every coefficient is independent of every other coefficient. In the limit, the variance-covariance matrix is diagonal, and the diagonal terms approach infinity—increasingly large positive numbers on the diagonal but zeroes off the diagonal.

Using this diagonal construct, we de facto are assuming zero covariance, i.e., zero conditional knowledge of any variable based on any other variable. Zero knowledge about any variable no matter what the value of any other variable is. Abject conditional ignorance. No knowledge about any variable gives any knowledge about any other variable. Following this assumption, a zero-precision matrix is constructed as the limit of a diagonal matrix and implies an infinite variance-covariance matrix with zero covariances. Infinite variance means total, abject, complete, utter, blockheaded ignorance, i.e., a uniform distribution from minus infinity to plus infinity over each model coefficient β_i and the supposition there is no inter-dependency among coefficients. Conceptually, if the variance of a random variable approaches infinity, what does it matter what the mean is? The distribution ranges from minus infinity to plus infinity, centered around any mean value one might posit! The value of the mean becomes irrelevant if the variance is infinite. This is about as agnostic and ignorant as one can get. This is a prior in which you have absolutely zero prior knowledge about any of the coefficients—uniformly distributed between $-a$ and a as a goes to ∞, i.e., uniformly distributed between $-\infty$ and ∞. See Figure 7.1. This is a mighty convenient and intuitively appealing model of ignorance. The variance of all the model coefficients goes to infinity. Who cares what the mean is when this massive, colossal degree of uncertainty exists?

Figure 7.1: Uniform Distribution between $-a$ and a

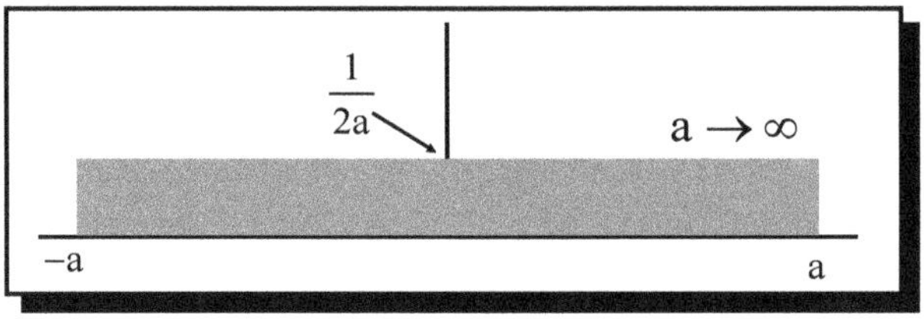

Under this assumption of $Q = 0$, the model coefficients within the posterior are transformed to the following:

- Mean Values

$$\begin{aligned}\langle\beta\rangle &= \beta^* \\ &= \left(X^T X + Q\right)^{-1}\left(X^T y + Q\beta_0\right) \\ &= \left(X^T X + 0\right)^{-1}\left(X^T y + 0\beta_0\right) \\ &= \left(X^T X\right)^{-1} X^T y \\ &= \bar{\beta} \quad .\end{aligned}$$

$$< \sigma^2 >= s^2 \frac{\nu}{\nu - 2} \quad .$$

- Variances

$$VCV = \begin{bmatrix} \frac{\nu s^2}{\nu-2}\left(X^T X + 0\right)^{-1} & 0 \\ 0 & \frac{2}{\nu-4}\left(s^2 \frac{\nu}{\nu-2}\right)^2 \end{bmatrix}$$

$$= \begin{bmatrix} \frac{\nu s^2}{\nu-2}\left(X^T X\right)^{-1} & 0 \\ 0 & \frac{2}{\nu-4}\left(s^2 \frac{\nu}{\nu-2}\right)^2 \quad . \end{bmatrix}$$

By setting $Q = 0$, we already have the classical regression solution with regard to the mean, but we do not have the full classical statistics variance-covariance solution yet. We need to update the other key terms to reflect this assumption

$$\begin{aligned} A &= R + M + \left(\bar{\beta} - \bar{\beta}\right)^T X^T y + (\beta_0 - \beta^*)^T 0 \beta_0 \\ &= R + M \end{aligned}$$

$$\nu = m + n - p - 1 = \nu_0 + n \quad .$$

$$s^2 = \frac{A}{\nu} \quad .$$

7.2 Second Step–Degrees of Freedom Set Equal to Classical

To move closer to the classical solution, let us examine the degrees of freedom of the prior, the classical regression solution and the posterior. The degrees of freedom of the classical regression solution was shown in Chapter 2, equation (2.8), to be

$$\nu_C = n - p \quad .$$

The assumed degrees of freedom of the prior in Chapter 5, equation (5.2) was shown to be

$$\nu_0 = m - p - 1 \quad .$$

The assumed degrees of freedom of the posterior in Chapter 3, equation (3.7) was shown to be

$$\nu = m + n - p - 1 \quad .$$

If we want a Bayesian solution that is equal in every dimension to the classical regression solution, we would need to have the degrees of freedom in the classical regression case be equal to the degrees of freedom in the Bayesian posterior as follows:

$$\begin{aligned} &\nu = m + n - p - 1 = \nu_C = n - p \\ &\Rightarrow m - 1 = 0 \Rightarrow m = 1 \quad . \end{aligned}$$

If we set $m = 1$ (arbitrarily but temporarily for now and proven to be valid from first principles in the next section), we will have reduced the posterior degrees of freedom to the classical regression degrees of freedom. This will alter the four coefficient results as follows

- Mean Values

$$< \beta >= \left(X^T X\right)^{-1} X^T y = \bar{\beta} \quad .$$

$$< \sigma^2 >= s^2 \frac{\nu_C}{\nu_C - 2} \quad .$$

- Variances

$$VCV = \begin{bmatrix} \frac{\nu_C s^2}{\nu_C - 2} \left(X^T X\right)^{-1} & 0 \\ 0 & \frac{2}{\nu_C - 4} \left(s^2 \frac{\nu_C}{\nu_C - 2}\right)^2 \end{bmatrix} \quad .$$

We need to update the other key term

$$A = R + M \quad .$$

This A term with the specialization that $Q = 0$, $\beta^* = \bar{\beta}$, and $\nu = \nu_C$ reduces to

$$A = \nu_C s_C^2 + \nu_0 s_0^2 \quad .$$

Furthermore,

$$\nu_C = n - p \quad ,$$

and

$$s^2 = \frac{A}{\nu_C} \quad .$$

The s^2 term with the specialization that $Q = 0$, $\beta^* = \bar{\beta}$, and $\nu = \nu_C$ reduces to

$$s^2 = \frac{\nu_C s_C^2 + \nu_0 s_0^2}{\nu_C} \quad .$$

We have, up until now, arbitrarily assumed that $m = 1$. We now turn our attention toward establishing that $m = 1$ is in fact correct and justifiable. Relevant to this discussion, there has been discussion in the literature of the "Jeffreys prior."[1] We will relate our derived prior to maximum entropy thinking herein.

It is worthwhile here to consider the notion of maximum entropy in a prefatory fashion.[2] *The uniform distribution on the interval [a,b] is the maximum*

[1]The Jeffreys prior is thought to be the "minimal prior information" prior. All the machinations about what constitutes minimum information in a general setting seem beyond our purpose here. We will simply develop an intuitively appealing minimal information prior directly, and we will note how it pertains to maximum entropy advocates' more theoretical developments. See https://en.wikipedia.org/wiki/Jeffreys_prior for a prefatory discussion with references.

[2]From https://en.wikipedia.org/wiki/Maximum_entropy_probability_distribution.

entropy distribution among all continuous distributions that are supported in the interval [a,b], and thus the probability density is 0 outside of the interval. This uniform density can be related to Laplace's principle of indifference, sometimes called the principle of insufficient reason. More generally, if we are given a subdivision $a = a_0 < a_1 < ... < a_k = b$ *of the interval [a,b] and probabilities* $p_1,...,p_k$ *that add up to one, then we can consider the class of all continuous distributions such that* $\{a_{j-1} \leq X < a_j\} = p_j \; j = 1, ..., k$ *The density of the maximum entropy distribution for this class is constant on each of the intervals* $[a_{j-1}, a_j]$*. The uniform distribution on the finite set* $x_1, ..., x_n$ *(which assigns a probability of* $1/n$ *to each of these values) is the maximum entropy distribution among all discrete distributions supported on this set.*

Wiki[3] clarifies what the Laplace Principle of Indifference is: *The principle of indifference (also called the principle of insufficient reason) is a rule for assigning epistemic probabilities. The principle of indifference states that in the absence of any relevant evidence, agents should distribute their credence (or "degrees of belief") equally among all the possible outcomes under consideration.*

7.3 Is There Any Fundamental Justification for $m = 1$?

Is there any justification for selecting $m = 1$ in the diffuse prior? Is such an assumption arbitrary, an appellation just to "make the math work?" Yes, there is indeed fundamental justification, which this section derives from the notion of a diffuse prior related to the σ parameter. How might we regard a distribution over the logarithm of σ that is uniformly distributed between $-ln(a)$ and $ln(a)$ for some large (limiting) value of a, i.e., for some value of a that can be passed to the infinite limit? (Clearly this is an assumption of a more and more uniform distribution between $-\infty$ and ∞ that is being approached. This is about as diffuse as the distribution over $ln(\sigma)$ can get.) Under such an assumption, the logarithm of σ would be uniformly distributed between $-\infty$ and ∞ in the limit. Would this be a diffuse prior? Of course it would. It would be a diffuse prior, an "informationless" prior, over $ln(\sigma)$ and therefore a diffuse prior over σ itself. Abject ignorance of $ln(\sigma)$. What would it imply for the prior over σ itself? Let us develop the answer to those questions.

We want to consider the transformed variable $x = ln(\sigma)$, which is governed by the density $f(x)$, which we want to assume is uniform between $ln(1/a)$ and $ln(a)$. This suggests that the "diffuse prior" over $ln(\sigma)$ is that in the plot in Figure 7.2. The logarithm of σ is uniformly distributed between $ln(1/a)$ to $ln(a)$, which is the same as uniformly distributed between $-ln(a)$ to $ln(a)$. The uniform density over x is therefore written in standard form:

[3] https://en.wikipedia.org/wiki/Principle_of_Indifference.

$$f(x) = \begin{cases} \frac{1}{\ln(a)-\ln(1/a)} & \ln(1/a) \le x \le \ln(a) \\ 0 & otherwise \end{cases} \quad .$$

Figure 7.2: Uniform Distribution between $ln(1/a)$ and $ln(a)$

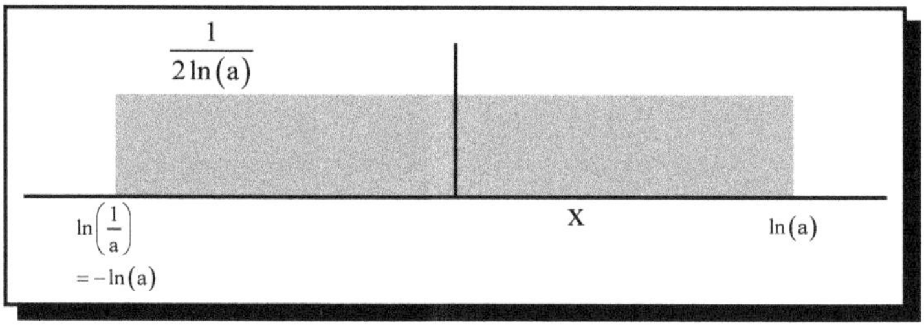

Let us now integrate this density over x across its entire range, which of course must integrate to 1:

$$1 = \int_{\ln\left(\frac{1}{a}\right)}^{\ln(a)} \frac{1}{2\ln(a)} dx \quad .$$

We consider a variable change from x to $\sigma = e^x$

$$\sigma = e^x \Rightarrow x = \ln(\sigma) \quad .$$

If we take the total differential of the right hand equation, we account for the change from dx to $d\sigma$

$$dx = \frac{1}{\sigma} d\sigma \quad .$$

We have to adjust the lower and upper limits of integration to be consistent with the new variable σ. We know for the lower limit that

$$x = \ln\left(\frac{1}{a}\right) \Rightarrow \sigma = \frac{1}{a} \quad ,$$

and we know for the upper limit that

$$x = \ln(a) \Rightarrow \sigma = a \quad .$$

We therefore know that the integral has transformed to

$$1 = \int_{\frac{1}{a}}^{a} \frac{1}{2\ln(a)} \frac{1}{\sigma} d\sigma$$

We know that the density over σ is the integrand save for the $d\sigma$ term, namely

$$\{\sigma\} = \frac{1}{2\ln(a)} \frac{1}{\sigma} = c_0\sigma^{-1}$$

This is profound. A diffuse prior over the logarithm of σ yields the following functional form for the prior over σ itself

$$\{\sigma\} \propto \frac{1}{\sigma} = \sigma^{-1} \quad .$$

This is a powerful result. We have shown that $m = 1$ is indeed a uniform (and therefore diffuse) prior over $ln(\sigma)$ between $ln(\frac{1}{\alpha})$ and $ln(\alpha)$, which is between $-ln(\alpha)$ and $ln(\alpha)$. You are abjectly ignorant about $ln(\sigma)$–uniformly ignorant between $-\infty$ and ∞ in the limit as a becomes large. That diffuse prior over σ has the functional form of the reciprocal of σ.[4]

If we return to Figure 3.1, we have argued that to create a diffuse prior, M must be zero and Q must be zero, leaving the prior to be characterized solely by the term σ^{-m}. And now, we have just shown that the prior is truly diffuse when $m = 1$. We have done so by showing that the density $\{\sigma\}$ is a constant times σ^{-1}, and this is consistent with the density over $ln(\sigma)$ approaching a uniform distribution between $-\infty$ and ∞. We have shown specifically why $m = 1$ creates a diffuse prior over $ln(\sigma)$. We are now ready for the final assumption that will fully equate the posterior to the classical regression solution. This value of $m = 1$ not only represents a diffuse prior over $ln(\sigma)$ but at the same time allows the degrees of freedom for the Bayesian diffuse posterior to be the same as the degrees of freedom for classical regression developed in Chapter 2.

7.4 Third Step–Set the A Value in the Posterior Equal to the R Value in the Classical Regression Case

There is only one additional term in the foregoing development that is different from the classical regression case, and that is the A term. We must have it be equal to its classical regression counterpart $\nu_C s_C^2 = R$. To effect this, it must occur that

$$A = \nu_C s_C^2 + \nu_0 s_0^2 = \nu_C s_C^2 = R \quad .$$

This will occur only if

[4]This result will generalize in Chapter 15 to the more sophisticated multivariate case in which the inverse gamma distribution is replaced by a multivariate Wishart distribution.

$$\nu_0 s_0^2 = 0 \quad .$$

The expression $\nu_0 s_0^2$ is the expression for the M term in the prior. The situation we want occurs in the prior only if the M term is 0, i.e., the case in which the exponential term is

$$e^{\frac{M}{2\sigma^2}} = 1 \quad ,$$

and thereby disappears from the prior altogether (i.e., becomes 1). Only if this occurs can the prior be what we derived in the section immediately above, namely

$$const \times \sigma^{-1} \quad .$$

This would imply the classical regression result that

$$s_C^2 = \frac{R}{\nu_C} \quad .$$

The interpretation seems clear when we realize what $M = \nu_0 s_0^2 = 0$ truly means. If we refer back to equation (5.2) for the prior, $m = 1$ means that the degrees of freedom $\nu_0 = -p$, a negative number. The only way M could be zero with a negative number as part of its product is for s_0^2, the variance term, to be zero, i.e., $< \sigma^2 >= 0$. This means that there is no information about the error term in the prior other than the diffuse assumption about $ln(\sigma)$ discussed above. There is no measurement or measurement error at all in the prior; we are absolutely sure, without the slightest doubt, that $ln(\sigma)$ is uniformly distributed between $-\infty$ and ∞. The exponential term containing M is gone altogether. The prior has a mean error of zero and a variance in error of zero. Under the assumption that $M = 0$, which we have just developed in this section, we arrive at the final answer we have been seeking: the parameter settings that equalize the posterior to the classical regression solution:

Mean Values

$$< \beta >= \left(X^T X\right)^{-1} X^T y = \bar{\beta} \quad .$$

$$< \sigma^2 >= \frac{\nu_C {s_C}^2}{\nu_C - 2} \quad .$$

Variances

$$VCV = \begin{bmatrix} \frac{\nu_C {s_C}^2}{\nu_C - 2}\left(X^T X\right)^{-1} & 0 \\ 0 & \frac{2}{\nu_C - 4}\left(\frac{\nu_C {s_C}^2}{\nu_C - 2}\right)^2 \end{bmatrix} \quad .$$

If we set the prior over both σ and β to be the reciprocal of σ (only), we have a diffuse prior over both σ and β. To summarize, this reflects the assumptions made thus far in this chapter that

- $Q = 0$, which means infinite variance and zero covariance about every model coefficient. That is quintessentially ignorant!
- The degrees of freedom are the same as the classical regression case, meaning infinite variance in $ln(\sigma)$. That is quintessentially ignorant!
- $A = R \Rightarrow s_0^2 = 0$, meaning there is no uncertainty about the infinite variance in $ln(\sigma)$. That is quintessentially ignorant!

and it implies that $m = 1$ and that the diffuse prior on σ is the reciprocal of σ.

We are about to see that, unlike the classical regression case, the posterior developed here is a bona fide probability distribution over σ and β GIVEN the observed data X and y, and not the reverse. This is exactly the interpretation we want. The simple addition of -1 to the exponent in the σ term in classical regression (and revision of the constant of integration) turns the whole likelihood function around 180 degrees. The Bayesian posterior now becomes $\{\beta, \sigma|X, y\}$, which is precisely what we want, and not the likelihood function $\{X, y|\beta, \sigma\}$, which we already have from our first-principles development of the likelihood function in Chapter 2. The entire prior, a totally diffuse prior, becomes:

$$\{\beta, \sigma\} = const \times \sigma^{-1} e^{-\frac{1}{2\sigma^2}\left[0+(\beta-\beta_0)^T 0(\beta-\beta_0)\right]} = const \times \sigma^{-1} \quad .$$

This means that $m = 1$, $Q = 0$, and $\nu_0 s_0^2 = 0$, and the degrees of freedom being the same as the classical regression case gives us the situation in which $ln(\sigma)$ is uniformly distributed from minus infinity to plus infinity, and we have the following composite diffuse prior over all the prior variables β and σ:

$$\{\beta, \sigma\} = const \times \sigma^{-1} \quad .$$

This shows specifically why setting $m = 1$ is not arbitrary at all. On the contrary, it comes from a respectable diffuse prior, extending the notion not only to the model coefficients but also to the error term. It clearly represents abject ignorance about the magnitude of σ but implies that we know that its sign is strictly positive. We have no idea ex ante what σ is save that it is positive.

7.5 Summary of Assumptions to Equate Classical with Bayes

We summarize the key assumptions that comprise the diffuse prior:

- $Q = 0$ this means an infinite variance on the β coefficients with zero covariances or correlations.

- $M = 0$, this means that zero information is delivered about the error term in the prior.
- $m = 1$, this means a uniform distribution on the logarithm of the standard deviation term from $-\infty$ to ∞, abject ignorance about the log of the standard deviation.

We have under these assumptions for our posterior the following expression for the joint density

$$\{\beta, \sigma | X, y\} = \frac{2|X^T X|^{\frac{1}{2}} (2\pi)^{-\frac{p}{2}} \left(\frac{\nu_C s_C^2}{2}\right)^{\frac{\nu_C}{2}}}{\Gamma\left(\frac{\nu_C}{2}\right)} \times \sigma^{-(\nu_C+p+1)} e^{-\frac{\nu_C s_C^2}{2\sigma^2}} e^{-\frac{1}{2\sigma^2}(\beta-\bar{\beta})^T (X^T X)(\beta-\bar{\beta})} .$$

Voila; this is the posterior for the diffuse prior. And this is precisely where the Bayesian reinterpretation of classical statistics starts to pay **colossally large dividends**, especially the conceptual dimension. We know exactly what zero prior knowledge means. We know exactly what zero knowledge coming from anywhere other than the data truly means. There is no equivocation in Bayes.

- We have, under the assumptions of classical statistics only, an explicit, unabashed joint probability distribution over β and σ given y and X.
- We can calculate posterior marginals over β and separately over σ.
- We can observe that the conditionals between β and σ are zero, meaning total probabilistic independence.
- We can calculate the means and variance-covariance of β.
- We can calculate the mean and variance of σ^2.
- We can predict linear functions of β, a hugely important capability.
- We have the full, actual posterior density itself! And we can contrive tests based on those fully densities. No estimators. No approximations. Why transform variables? We will know exactly why.

And we do not have to resort back to arcane calculations or interpretations related to likelihood functions to do so. The joint probability distribution over model coefficients is right before our very eyes. It is there for the taking! We have arrived at an explicitly probabilistic model that mirrors the classical regression solution. Is that ever valuable from a decision making and forecasting perspective!

7.6 Is a Diffuse Prior Real?

Is a diffuse prior real or practical? Representative of anything in the real world? In the real world, we would argue that **there is no such thing as a noninfor-**

mative prior. A diffuse prior. Most assuredly, the analyst, the subject matter expert, the scientist, and even the dilettante always have at least an inkling. (An inkling might be a prior probability distribution over a variable with a huge variance around some assumed mean. We will see that assumption in the one of the future examples in Chapter 13.) He or she is just not very sure! A diffuse prior is an abstract "model" of ignorance, an abstract model of zero information, a limiting case, a corner solution. A diffuse prior is an abstraction, an abstract model of absolute, total, abject, unqualified ignorance, exactly as Prof. Shachter has suggested. One would have to convince this author that anyone in the real world is actually as ignorant about anything as the diffuse prior would suggest. The author finds it hard to dismiss the question Prof. Howard asked: "Is there anything other than an informative prior?" Heck, even the author knows SOMETHING about ballet! His prior over key parameters of ballet doesn't satisfy the abstract, idealized, corner solution model of complete, abject ignorance articulated by Zellner[107] and analyzed by Jeffreys in the 1930s.[5] Notwithstanding its empirical validity as an important point of reference, Zellner's model of ignorance is a revealing argument about what true ignorance might really mean and how data begins to reshape and overcome such ignorance. The author does not believe that a sentient being in reality can ever be as ignorant (or dumb) as a diffuse prior implies. How could that occur? Prof. Ross Shachter asked, "Is the diffuse prior a model of ignorance?" Very insightful question, and the answer is yes. "Variance approaching infinity" is a model, an abstraction, of what true ignorance might be.

The author met with Dr. Zellner at the University of Chicago. A wonderful, collaborative, and professional person, his help was instrumental here. When the author worked with Dr. Zellner, he found him to be as much a Bayesian as Dr. Howard, Dr. Raiffa, and the author himself. And he had come to that state directly out of the classical statistical space that so rabidly advocated a diffuse prior. We sensed that Dr. Zellner himself had "tissue rejection" about the diffuse prior. He kept saying, "We know the approximate range of the price elasticity of demand. We know that some values are simply nonsensical. We cannot be agnostic about that ex ante. We know there might be some peaks or valleys in it. We know where 'outlier' values might begin. Goodness, Dale, just by saying THAT, we have dynamited the diffuse prior irretrievably and irrevocably. And we know an awful lot more, like that a certain elasticity should lie somewhere between -0.05 and -0.2. Outside that range, it just isn't credible or consistent with observed behavior in the market." It just doesn't seem right. That implied we needed what

[5] All this analysis, such as Jeffrey's, striving to carefully define what abject ignorance is, doesn't motivate the Bayesian. The Bayesian knows that people bring serious, cogent knowledge to any statistical problem, albeit often highly uncertain, and they want to incorporate it into the final answer. With all due humility on their part, they know they have very wide probability distributions over things that they don't understand well. This is true in hearts, spades, clubs, and diamonds for senior management. They have to make decisions based almost entirely on their prior experiences. They seek data to update those priors, but data comes slowly and in small quantities; data analysis comes slowly; and decisions need to be made right now.

he termed an "informative prior" to distinguish it from an uninformative or diffuse prior. Dr. Zellner would agree with the author and Dr. Howard on this. Dr. Howard objected to the notion of "informative prior," arguing that all priors are informative, even the diffuse prior, and thus the nomenclature/definition is a non sequitur. The author accepts that, but we insist on keeping the abstract model of abject ignorance clear in our minds so that we have some clarity on what degree of information we might truly be bringing to the problem as we emulate classical regression. Plus, it gives us a coherent way to judge what classical regression is really assuming.

What Drs. Howard and Dr. Shachter were saying is that a diffuse prior is intrinsically itself informative, i.e., revealing of information. It is a statement that the amount of information a parson knows is absolute zero, zero degrees Kelvin, zip, zero, nada. That is informative in its own way. It says that all that is known or knowable is resident within the data. The Dale Jorgenson statement! The statement is intrinsic in the structure of classical regression, not an "ah hah!" finding of classical regression.

Christian Ostberg, an editor and collaborator on this manuscript, tells the anecdote: "I was once in a job interview for a technical job, and the interviewer asked, 'What are the assumptions of linear regression?' The answer he was looking for was linearity, homoscedasticity, independence, mode of likelihood, etc. Should I have said, 'Well, linear regression assumes abject ignorance of model parameters?' I probably wouldn't have gotten the job (I didn't get it anyway), but I suspect it would have made them think!" The answer, of course, is yes; it does assume abject ex ante ignorance, as we have now seen. Unbounded uniform distributions on every model coefficient and unbounded uniform distribution on the logarithm of standard deviation! Classical linear regression assumes that everything that is known or knowable is in the data. There is zero degrees of freedom, de facto, in the prior. When you really think about it, how can such a statement possibly be true? It simply cannot, but it is an interesting "abstraction" for comparison with the posterior and the prior.

We should mention that the mathematical developments in this section work with a diffuse prior, and we are about to see in Chapter 8 that they work just as well with an "old data" prior or a purely subjective, informative prior as in Chapter 6. Any of the three is admissible. The real contribution of this section is to put the classical regression solution into exactly the same probabilistic context and into exactly the same probabilistic form, the conjugate-archetype form, as the prior and Bayesian posterior. This facilitates EASY comparison. In the forthcoming sections, when we refer to the classical regression solution, we are explicitly referring to the Bayesian Linear Regression solution with a diffuse prior.

During the early COVID ruminations of the past several years, we have heard the refrain, "We follow the science!" by statisticians who apparently embrace nothing but purely classical statistical methods and data. What these people are really saying is that using ONLY a diffuse prior and using specifically gathered data ONLY is "the science." Scientists like Dr. Fauci continually equate "their specifically gathered data" with "the science," i.e., deploying a diffuse prior was allegedly intrinsic to "the science." That is extremely hard to buy. There are clinicians who have made empirical observations of dozens of treatments and dozens of resulting outcomes. They have observed and mentally processed a lot of data, albeit perhaps imperfectly. Does "We follow the science" mean that we must reject completely what clinicians have been observing for a century in a clinical setting, accepting nothing but formal, approved, overseen, blue ribbon, government approved data? People who argue that they follow the science seem to be answering yes to that question, and we should not be comfortable with the answer "yes" when important decisions need to be made today. Such people reject any and all information that exists outside the data gathered from their experimental or clinical situations. No judgment or empirical observations are allowed to enter because they allegedly pollute the "science" embedded in the gathered data. This is mighty hard to accept.

With Bayes, we do not have to take such a dogmatic or narrow view. We can allow judgment based on empirical observations to enter via the prior, and we can measure the effect of alternative priors on the posterior. The example in Chapter 11 illustrates exactly that. Man, oh man, is that powerful for decision-making, particularly decision-making that needs to be done NOW before definitive data (or perhaps any data at all) are gathered or observations made? That is one of the profound contributions of Bayes—allowing prior knowledge to be formally used today until definitive, carefully gathered, and ostensibly "accurate" data have been gathered, and such data can logically and systematically replace some or all of the initial prior judgment. Good decisions today; better decisions tomorrow. Maybe changes tomorrow in decisions that were made today based on the best, but uncertain, information we had today. In the real world, we often cannot wait to make today's decisions.[6]

Returning to the mantra about classical regression "Everything that is known or knowable is embedded in the data," **the statement simply cannot possibly be true**. In fact, it is intellectually preposterous on its face. It means, "You have to embrace the diffuse prior. Period; unconditionally. If you don't, aspects of the prior will enter the solution, and your data won't be the sole source of information in your solution. You are not allowed to have a prior other than the diffuse one that is completely ignorant, totally and blissfully so." This is intellectually bothersome, and happily, Bayes allows you to reject the statement in favor of bringing whatever judgment you might have to the analysis and making it explicit and apparent in the prior.

While a diffuse, totally ignorant prior is an interesting "corner solution" and

[6]Chernow[18] wrote about the medical practice of applying leeches to the body of George Washington to draw blood containing "ill humors" from his body, "ill humors" thought to be causing disease. This was the leading medical practice of the day, and it has since been supplanted by better knowledge. But it was state of the art in Washington's day when medical decisions had to be made, and there were no data (and no probability and statistics!) to supplant it.

point of reference, it does not appear to the author to be intrinsically valuable in its own right. It is indeed difficult to believe that everything that is known or knowable about a problem resides totally within whatever data you have gathered. This is preposterous on its face in the real world. We invariably use the diffuse prior as a point of reference, but we invariably enter a Zellnerian[107] informative prior for our final analysis. Paying tribute to old science simply for the sake of paying tribute to old results isn't our objective here. Our objective is to move the ball down the field, articulating the newer science of Bayes, getting better forecasts, and helping people make better decisions. That is what happened in physics—general relativity (Bayes) replaced Newtonian physics (classical regression) and reinterpreted the older science as an interesting special case, a corner solution.

Chapter 8

CLASSICAL REGRESSION–A POSTERIOR WITH A DIFFUSE PRIOR

"I can live with doubt and uncertainty and not knowing. I think it is much more interesting to live not knowing than to have answers that might be wrong."

Richard P. Feynman

We will be developing all the posterior results for the classical regression case (which we call posterior with diffuse prior) using key parameter substitutions:

$$
\begin{aligned}
g &\leftarrow R \\
\mu &\leftarrow \bar{\beta} \\
H &\leftarrow X^T X \\
r &\leftarrow \nu_C + p + 1 \\
\nu &\leftarrow \nu_C \\
s_C^2 &\leftarrow \frac{R}{\nu_C} \quad .
\end{aligned}
\tag{8.1}
$$

The r term merits discussion. Recall that ν_C represents the classical regression degrees of freedom developed in Chapter 2, and we **absolutely must** preserve that in the classical regression case. In Chapter 2, we developed an expression for the degrees of freedom that pertains in the classical regression case as over the past century, a value we must preserve:

$$\nu_C = n - p \quad .$$

This implies, because r is the exponent on σ in the ultimate joint distribution, that the r parameter will become the degrees of freedom plus p plus 1, just as for the other cases:

$$r = \nu_C + p + 1 \quad .$$

We will repeat and justify this value of r throughout this section, but we need to put it forth from the outset to facilitate the required substitutions. If we make the foregoing substitutions from equation (8.1) into all the various relationships in Chapter 4, we will have a bona fide classical regression case characterized by these parameters, i.e., we will have the bona fide classical regression case developed in Chapter 7. This chapter makes the foregoing substitutions for the indicated variables into the conjugate-archetype density developed in Chapter 4 so as to properly constitute the (conjugate) classical.

We begin by re-characterizing the degrees of freedom for the classical regression case to be that in equation (8.2), rewritten here for convenient reference

$$\nu_C = n - p \quad . \tag{8.2}$$

This implies, by substitution from equation (8.1) into equation (4.5), that

$$r = \nu_C + p + 1 \quad . \tag{8.3}$$

As noted in Chapter 4, this definition of degrees of freedom might again appear to be somewhat arbitrary, but we will verify again in this section that it is far from arbitrary. It truly must be the classical regression degrees of freedom. We substitute from equation (8.1) into equation (4.7) to develop the expression for R

$$R = g = \nu_C s_C^2 = \left(y - X\bar{\beta}\right)^T \left(y - X\bar{\beta}\right) \quad . \tag{8.4}$$

This allows us to develop the s_C^2 term for the classical, precisely the same as equation (4.6) with the substitutions from equation (8.1), as follows

$$s_C^2 = \frac{R}{\nu_C} \quad . \tag{8.5}$$

R is going to be identical to the residual sum of squared errors $R = \nu_C s_C^2$ from the classical regression case in Chapter 2. Using these definitions, we write the crucial results for the classical

- The mean value of model coefficients devolves directly from substituting equations (8.1) into equation (4.13) to write

$$<\beta>=\mu=\bar{\beta}=\left(X^TX\right)^{-1}X^Ty \quad . \tag{8.6}$$

- The H matrix for this case is obtained by substituting equations (8.1) into equation (4.14) to write

$$H=X^TX \quad . \tag{8.7}$$

8.1 p-vector Multivariate Density–Classical

We know that the classical regression case has the functional form derived from equation (4.8) but with parameter substitutions from equation (8.1). Making that substitution allows us to write the classical joint density over β and σ in the following form:

$$\begin{aligned}\{\beta,\sigma\} &= \frac{2\left|X^TX\right|^{\frac{1}{2}}}{(2\pi)^{\frac{p}{2}}\Gamma\left(\frac{\nu_C}{2}\right)}\left(\frac{\nu_C s_C^2}{2}\right)^{\frac{\nu_C}{2}}\sigma^{-(\nu_C+p+1)}e^{-\frac{\nu_C s_C^2}{2\sigma^2}}\\ &\times e^{-\frac{1}{2}(\beta-\bar{\beta})^T\left[(X^TX)\sigma^{-2}\right](\beta-\bar{\beta})} \quad .\end{aligned} \tag{8.8}$$

Note that this is precisely in the form of equation (4.8), the conjugate-archetype equation functional form for the conjugate prior and posterior, but with the parameter substitutions inserted from equation (8.1) to create the classical regression solution as the posterior. We notice that there are exactly five parameters, which collectively we might term the "sufficient statistics," for the classical regression solution:

- p = number of model parameters
- ν_C = degrees of freedom
- s_C^2 = variance term
- X^TX = precision matrix
- $\bar{\beta}$ = mean of the coefficients .

We change the foregoing joint density over β and σ in equation (8.8) into a joint density over β and the variance term $z=\sigma^2$ by substitution into equation (4.10) using the relationships in equation (8.1) to write

$$\begin{aligned}\{\beta,z\} &= \frac{\left|X^TX\right|^{\frac{1}{2}}(2\pi)^{-\frac{p}{2}}\left(\frac{\nu_C s_C^2}{2}\right)^{\frac{\nu_C}{2}}}{\Gamma\left(\frac{\nu_C}{2}\right)}z^{-\frac{\nu_C+p}{2}+1}e^{-\frac{\nu_C s_C^2}{2z}}\\ &\times e^{-\frac{1}{2z}(\beta-\bar{\beta})^T(X^TX)(\beta-\bar{\beta})} \quad .\end{aligned} \tag{8.9}$$

We will be using this classical regression joint density repeatedly in the forthcoming developments in this section (and beyond). There will be an inverse gamma distribution for the classical case that again emerges now that we have expressed the distribution of the variance $z = \sigma^2$. Fantastic! We know how to deal with inverse gammas by now. We obtain the inverse gamma by substituting into equation (4.11) using equation (8.1) to obtain

$$(8.10) \qquad IG\left(z; \frac{\nu_C s_C^2}{2}, \frac{\nu_C}{2}\right) = \frac{\left(\frac{\nu_C s_C^2}{2}\right)^{\frac{\nu_C}{2}}}{\Gamma\left(\frac{\nu_C}{2}\right)} z^{-\left(\frac{\nu_C}{2}+1\right)} e^{-\frac{\nu_C s_C^2}{2z}} \quad .$$

Its mean and variance are derived by substituting from equation (8.1) into the equation (4.22) for the mean and into equation (4.23) for the variance respectively

$$(8.11) \qquad <\sigma^2>= \frac{\nu_C s_C^2}{\nu_C - 2} = \text{mean}$$

and

$$(8.12) \qquad <\left(\sigma^2\right)^2> - <\sigma^2>^2= \frac{2}{\nu_C - 4}\left(\frac{\nu_C s_C^2}{\nu_C - 2}\right)^2 = \text{variance} \quad .$$

The "product rule" in terms of the variance term z is determined by substituting into equation (4.12) from equation (8.1) to obtain the following

$$(8.13) \qquad \{\beta, z\} = IG\left(z; \frac{\nu_C s_C^2}{2}, \frac{\nu_C}{2}\right) N\left\{\beta; \bar{\beta}, \left[\left(X^T X\right)^{-1} z\right]\right\} \quad .$$

This section provides a complete characterization of the classical posterior. This is a truly important section. It puts forth the classical regression solution as a "plain vanilla" posterior for the diffuse prior, just like any posterior derived from any conjugate-archetype prior. And it centers classical regression squarely on the conceptual sensibility of the diffuse prior.

8.2 p-vector Marginal Density–Classical

This section integrates the σ term (or equivalently the z term) out to create the marginal density over β. The joint marginal density over the model coefficients, taken directly from equation (4.15) for the conjugate-archetype distribution but with all parameters substituted as in equation (8.1), is the following:

$$(8.14) \quad \{\beta\} = \frac{\left|X^T X\right|^{\frac{1}{2}} \Gamma\left(\frac{\nu_C + p}{2}\right)}{\left(\pi \nu_C s_C^2\right)^{\frac{p}{2}} \Gamma\left(\frac{\nu_C}{2}\right)} \left[1 + \left(\beta - \bar{\beta}\right)^T \frac{X^T X}{\nu_C s_C^2}\left(\beta - \bar{\beta}\right)\right]^{-\frac{\nu_C + p}{2}} \quad .$$

The mean of the coefficients of this marginal distribution, obtained by substituting equation (8.1) into equation (4.18), is that in equation (8.6) and the variance-covariance matrix of the coefficients, obtained by substituting equation (8.1) into equation (4.19), is

$$VCV = \left(X^T X\right)^{-1} s_C^2 \frac{\nu_C}{\nu_C - 2} \quad . \tag{8.15}$$

The H matrix, obtained by substituting from equation (8.1) into equation (4.20), is

$$H = X^T X \quad . \tag{8.16}$$

This completes the characterization of the marginal distribution over the model coefficients.

We summarize the vector of mean values of the $(p+1)$-vector β appended to the scalar σ and the variance-covariance of rank $(p, 1) \times (p, 1)$ in the following two equations respectively:

$$\text{Mean} = \begin{bmatrix} \bar{\beta} \\ s_C^2 \frac{\nu_C}{\nu_C - 2} \end{bmatrix}$$

and

$$\text{Var-Cov} = \begin{bmatrix} \left(X^T X\right)^{-1} s_C^2 \frac{\nu_C}{\nu_C-2} & 0 \\ 0 & \frac{2}{\nu_C-4}\left(s_C^2 \frac{\nu_C}{\nu_C-2}\right)^2 \end{bmatrix} \quad . \tag{8.17}$$

8.3 Marginal Distribution for $(p - L)$-subvector β_2 of p-vector β–Classical

We may not want to carry all p of the model coefficients for future analysis or prediction. We may want to consider a subset of $p - L$ of the p elements of β, integrating out L of the coefficients. What does the joint and marginal distribution look like for that subset of $p - L$ model coefficients? We are interested in the joint density over β_2 (assumed to be a $p - L$ vector subset of β) and σ.

For notational convenience, we define

$$\Psi = X^T X \quad .$$

We are going to use the same partitioning of the matrix Ψ as we used for the matrix H previously in Chapter 4. The partitioning in the forthcoming development will be obvious. It will be the same as the previous partitioning of H, Q, and Ω in Chapters 4, 5, and 6 respectively. Using this partitioned Ψ matrix, we obtain the partially integrated marginal distribution by substituting from equation (8.1) into equation (4.29), is the following:

$$\{\beta_2, \sigma\} = \frac{2\left|\Psi_{22} - \Psi_{21}\Psi_{11}{}^{-1}\Psi_{12}\right|^{\frac{1}{2}}}{(2\pi)^{\frac{p-L}{2}}\Gamma\left(\frac{\nu_C}{2}\right)} \times \left(\frac{\nu_C s_C^2}{2}\right)^{\frac{\nu_C}{2}} \sigma^{-(\nu_C+p+1-L)} e^{-\frac{\nu_C s_C^2}{2\sigma^2}} \times e^{-\frac{1}{2}(\beta_2-\bar{\beta}_2)^T\left[(\Psi_{22}-\Psi_{21}\Psi_{11}{}^{-1}\Psi_{12})\sigma^{-2}\right](\beta_2-\bar{\beta}_2)} \tag{8.18}$$

In this equation, β_2 is the subset of the p-vector β we want to consider. The term $\bar{\beta}_2$ represents the $p - L$ subvector of the p-vector $\bar{\beta}$ corresponding to β_2. The remaining subset will be integrated away to create the marginal distribution over the $p - L$ subset. As shown in Chapter 4, the product rule density over β_2 and $z = \sigma^2$ is obtained by substituting from equation (8.1) into equation (4.30) to write the product rule density as follows:

$$\{\beta_2, z\} = IG\left(z, \frac{\nu_C s_C^2}{2}, \frac{\nu_C}{2}\right) N\left[\beta_2; p - L, \bar{\beta}_2, z\left(\Psi_{22} - \Psi_{21}\Psi_{11}{}^{-1}\Psi_{12}\right)^{-1}\right] . \tag{8.19}$$

The inverse gamma density over z is obtained by substituting from equation (8.1) into equation (4.31) as follows:

$$IG\left(z, \frac{\nu_C s_C^2}{2}, \frac{\nu_C}{2}\right) = \frac{\left(\frac{\nu_C s_C^2}{2}\right)^{\frac{\nu_C}{2}}}{\Gamma\left(\frac{\nu_C}{2}\right)} z^{-\frac{\nu_C}{2}-1} e^{-\frac{\nu_C s_C^2}{2z}} . \tag{8.20}$$

Finally, the marginal distribution over the $p - L$-vector β_2 is obtained by substituting from equation (8.1) into equation (4.33), which yields the following:

$$\{\beta_2\} = \frac{\left|\Psi_{22} - \Psi_{21}\Psi_{11}{}^{-1}\Psi_{12}\right|^{\frac{1}{2}}\Gamma\left(\frac{\nu_C+p-L}{2}\right)}{(\pi\nu_C s_C^2)^{\frac{p-L}{2}}\Gamma\left(\frac{\nu_C}{2}\right)} \times \left[1 + (\beta_2 - \bar{\beta}_2)^T \frac{(\Psi_{22} - \Psi_{21}\Psi_{11}{}^{-1}\Psi_{12})}{\nu_C s_C^2} (\beta_2 - \bar{\beta}_2)\right]^{\frac{-\nu_C-p+L}{2}} . \tag{8.21}$$

8.4 The Special Case in Which $p - L = 1$-subvector of p-vector β–Classical

We now move to the case in which $p-L = 1$, i.e., we are looking for a univariate marginal distribution β_2 over a single coefficient. In this section, β_2 represents a scalar variable and we are pursuing a univariate distribution over the coefficient. If we substitute the parameters from equation (8.1) into equation (4.34) for this case in which we have reduced to a single, scalar marginal distribution with

$p - L = 1$, we obtain the classical regression equation (8.22)

$$\{\beta_2, \sigma\} = \frac{2\left|\Psi_{22} - \Psi_{21}{\Psi_{11}}^{-1}\Psi_{12}\right|^{\frac{1}{2}} \left(\frac{\nu_C s_C^2}{2}\right)^{\frac{\nu_C}{2}}}{(2\pi)^{\frac{1}{2}}\Gamma\left(\frac{\nu_C}{2}\right)} \sigma^{-(\nu_C+2)} e^{-\frac{\nu_C s_C^2}{2\sigma^2}} \times e^{-\frac{1}{2}(\beta_2-\bar{\beta}_2)^T \frac{\Psi_{22}-\Psi_{21}{\Psi_{11}}^{-1}\Psi_{12}}{\sigma^2}(\beta_2-\bar{\beta}_2)} \quad . \tag{8.22}$$

If we substitute the parameters from equation (8.1) into equation (4.30), making sure to set $q = 1$, we obtain the classical product rule equation (8.23)

$$\begin{aligned} \{\beta_2, z\} &= IG\left(z, \frac{\nu_C s_C^2}{2}, \frac{\nu_C}{2}\right) \\ & N\left[\beta_2; \bar{\beta}_2, z\left(\Psi_{22} - \Psi_{21}{\Psi_{11}}^{-1}\Psi_{12}\right)^{-1}\right] \quad . \end{aligned} \tag{8.23}$$

If we substitute the parameters from equation (8.1) into equation (4.31) representing the inverse gamma distribution, we obtain equation (8.24) for the inverse gamma distribution here

$$IG\left(z, \frac{\nu_C s_C^2}{2}, \frac{\nu_C}{2}\right) = \frac{\left(\frac{\nu_C s_C^2}{2}\right)^{\frac{\nu_C}{2}}}{\Gamma\left(\frac{\nu_C}{2}\right)} z^{-\frac{\nu_C}{2}-1} e^{-\frac{\nu_C s_C^2}{2z}} \quad . \tag{8.24}$$

If we substitute the parameters from equation (8.1) into equation (4.35), we obtain equation (8.25), which represents the marginal distribution over the single coefficient

$$\{\beta_2\} = \frac{\left|\Psi_{22} - \Psi_{21}{\Psi_{11}}^{-1}\Psi_{12}\right|^{\frac{1}{2}}\Gamma\left(\frac{\nu_C+1}{2}\right)}{(\pi\nu_C s_C^2)^{\frac{1}{2}}\Gamma\left(\frac{\nu_C}{2}\right)} \times \left[1 + (\beta_2 - \bar{\beta}_2)^T \frac{\Psi_{22} - \Psi_{21}{\Psi_{11}}^{-1}\Psi_{12}}{\nu_C s_C^2}(\beta_2 - \bar{\beta}_2)\right]^{-\frac{\nu_C+1}{2}} \quad . \tag{8.25}$$

If we substitute the parameters from equation (8.1) into equation (4.36), we obtain equation (8.26), which is the diagonal of the inverse precision matrix that allows a simple expression for the marginal

$$\left[\Psi^{-1}\right]_{ii} = \left[\left(X^T X\right)^{-1}\right]_{ii} = \left(\Psi_{22} - \Psi_{21}{\Psi_{11}}^{-1}\Psi_{12}\right)^{-1} = z_{ii} \quad . \tag{8.26}$$

If we substitute the parameters from equation (8.1) into equation (4.37), we obtain equation (8.27), which integrates the diagonal term into the appropriate position

$$\{\beta_2, \sigma\} = \frac{2\left(\frac{\nu_C s_C^2}{2}\right)^{\frac{\nu_C}{2}}}{(2\pi z_{ii})^{\frac{1}{2}}\Gamma\left(\frac{\nu_C}{2}\right)} \sigma^{-(\nu_C+2)}$$

$$e^{-\frac{\nu_C s_C^2}{2\sigma^2}} e^{-\frac{1}{2}(\beta_2-\bar{\beta}_2)^T \frac{1}{z_{ii}\sigma^2}(\beta_2-\bar{\beta}_2)} \quad . \tag{8.27}$$

If we substitute the parameters from equation (8.1) into equation (4.41), we obtain equation (8.28), which characterizes the product rule for this case

$$\{\beta_2, z\} = IG\left(z, \frac{\nu_C s_C^2}{2}, \frac{\nu_C}{2}\right) N(\beta_2; z) \quad . \tag{8.28}$$

If we substitute the parameters from equation (8.1) into equation (4.42), we obtain the following equation, which integrates the z term out to give the marginal distribution over a single coefficient.

$$\{\beta_2\} = \frac{\Gamma\left(\frac{\nu_C+1}{2}\right)}{(\pi \nu_C s_C^2 z_{ii})^{\frac{1}{2}}\Gamma\left(\frac{\nu_C}{2}\right)} \tag{8.29}$$

$$\times \left[1 + (\beta_2 - \bar{\beta}_2)^T \frac{1}{z_{ii}\nu_C s_C^2}(\beta_2 - \bar{\beta}_2)\right]^{-\frac{\nu_C+1}{2}} \quad .$$

8.5 The Generic Univariate Student's t Distribution

If we substitute the parameters from equation (8.1) into equation (4.43), we obtain equation (8.30)

$$\{\beta, \sigma\} = \frac{2\sqrt{Q}}{\sqrt{2\pi}\Gamma\left(\frac{k-2}{2}\right)}\left(\frac{P}{2}\right)^{\frac{k-2}{2}} \sigma^{-k} e^{-\frac{P}{2\sigma^2}} e^{-\frac{1}{2\sigma^2}(\beta-\mu)^T Q(\beta-\mu)} \quad . \tag{8.30}$$

For this generic univariate Student's t distribution, we define for the classical from equation (4.38) with the substitutions from equation (8.1)

$$P = \nu_C s_C^2 \quad .$$

For the generic univariate Student's t distribution, we define for the classical from equation (4.39) with the substitutions from equation (8.1)

$$Q = \frac{1}{z_{ii}} \quad .$$

For the generic univariate Student's t distribution, we define for the classical from equation (4.40) with the substitutions from equation (8.1)

$$k = \nu_C + 2 \quad .$$

For the generic univariate Student's t distribution, we define for the classical from equation (4.44) with the substitutions from equation (8.1)

$$\{\beta\} = \int\limits_0^\infty d\sigma\,\{\beta,\sigma\} = \frac{\Gamma\left(\frac{k-1}{2}\right)}{\sqrt{\pi}\Gamma\left(\frac{k-2}{2}\right)}\left(\frac{Q}{P}\right)^{\frac{1}{2}}\left[1+(\beta-\mu)^T\frac{Q}{P}(\beta-\mu)\right]^{-\frac{k-1}{2}}$$

$$= \frac{\Gamma\left(\frac{\nu_C+1}{2}\right)}{\sqrt{\pi}\Gamma\left(\frac{\nu_C}{2}\right)}\left(\frac{1}{z_{ii}\nu_C s_C^2}\right)^{\frac{1}{2}}\left[1+(\beta-\mu)^T\frac{1}{z_{ii}\nu_C s_C^2}(\beta-\mu)\right]^{-\frac{\nu_C+1}{2}} \quad .$$

For the generic univariate Student's t distribution, we define for the classical from equation (4.45) with the substitutions from equation (8.1)

$$\langle\beta\rangle = \mu = \bar{\beta} \quad .$$

For the generic univariate Student's t distribution, we define for the classical case from equation (4.46) with the substitutions from equation (8.1)

$$\text{variance} = \frac{1}{k-4}\left(\frac{P}{Q}\right) = \frac{\nu_C s_C^2 z_{ii}}{\nu_C - 2} \quad .$$

It is instructive for people who might want to calculate or program the marginal distributions over the coefficients to enumerate step by step the calculations one has to make to move from the joint distribution over the coefficients β to the p individual coefficients β_i

1. Calculate the inverse of the entire $\left(X^TX\right)$ matrix, i.e.,

$$\left(X^TX\right)^{-1} \quad .$$

2. Get the ii^{th} diagonal element of the $\left(X^TX\right)^{-1}$ matrix. Designate it z_{ii}

$$z_{ii} = \left[\left(X^TX\right)^{-1}\right]_{ii} = \left(\Psi_{22} - \Psi_{21}\Psi_{11}{}^{-1}\Psi_{12}\right)^{-1} \quad .$$

3. Using these definitions, the univariate Student's t density in equation (8.29) becomes

$$\{\beta_i\} = \frac{\Gamma\left(\frac{\nu_C+1}{2}\right)}{\Gamma\left(\frac{\nu_C}{2}\right)}\frac{1}{\left(\pi z_{ii}\nu_C s_C^2\right)^{\frac{1}{2}}} \tag{8.31}$$
$$\times\left[1+\left(\beta_i - \left[\bar{\beta}\right]_i\right)\frac{1}{\nu_C s_C^2 z_{ii}}\left(\beta_i - \left[\bar{\beta}\right]_i\right)\right]^{-\frac{\nu_C+1}{2}} \quad .$$

4. This density will integrate to 1 because the constant has been developed completely and correctly.

5. The mean of this distribution is

$$[\bar{\beta}]_i \quad . \tag{8.32}$$

6. The variance of the ith variable is the ith diagonal element of the $\left(X^T X\right)^{-1}$ matrix z_{ii} multiplied by the factor

$$\frac{\nu_C s_C^2}{\nu_C - 2}$$

to obtain the variance of the univariate marginal

$$\frac{\nu_C s_C^2}{\nu_C - 2} z_{ii} \quad . \tag{8.33}$$

8.6 The Full Predictive Distribution–Classical

The multivariate predictive distribution (based on the matrix D whose q rows contain the predictions to be made and whose columns that are p in number representing the coefficients in the model) has been calculated in the following equation. We substitute the classical parameters from equation (8.1) to write

$$\begin{aligned} \{d, D, \sigma | X, y\} = & \frac{2(2\pi)^{-\frac{q}{2}} \left(\frac{\nu_C s_C^2}{2}\right)^{\frac{\nu_C}{2}}}{\left|I + D(X^T X)^{-1} D^T\right|^{\frac{1}{2}} \Gamma\left(\frac{\nu_C}{2}\right)} \\ & \times \sigma^{-(\nu_C + q + 1)} e^{-\frac{\nu_C s_C^2}{2\sigma^2}} \\ & \times e^{-\frac{1}{2\sigma^2}(d - D\bar{\beta})^T \left[I + D\left(X^T X\right)^{-1} D^T\right]^{-1} (d - D\bar{\beta})} \end{aligned} \tag{8.34}$$

The multivariate predictive distribution (based on the matrix D has been calculated in equation (4.62) using the variance term z rather than σ. We substitute all the classical parameters from equation (8.1) to write

$$\{d, z\} = \frac{\left(\frac{\nu_C s_C^2}{2}\right)^{\frac{\nu_C}{2}}}{\Gamma\left(\frac{\nu_C}{2}\right)} z^{-\left(\frac{\nu_C}{2} + 1\right)} e^{-\frac{\nu_C s_C^2}{2z}} N\left\{d; D\bar{\beta}, \left[I + D\left(X^T X\right)^{-1} D^T\right] z\right\} \quad .$$

The inverse gamma that emerges has been calculated in product form in equation (4.63). We substitute all the classical parameters from equation (8.1) in order to write

$$IG\left(z; \frac{\nu_C s_C^2}{2}, \frac{\nu_C}{2}\right) = \{z\} = \frac{\left(\frac{\nu_C s_C^2}{2}\right)^{\frac{\nu_C}{2}}}{\Gamma\left(\frac{\nu_C}{2}\right)} (z)^{-\left(\frac{\nu_C}{2} + 1\right)} e^{-\frac{\nu_C s_C^2}{2z}} \quad . \tag{8.35}$$

The product form density developed in equation (4.64) can receive substitutions of all the classical parameters from equation (8.1) to write

$$\{d, z\} = IG(z; \frac{\nu_C s_C^2}{2}, \frac{\nu_C}{2}) N\left[d; D\bar{\beta}, \left(I + DH^{-1}D^T\right) z\right] \quad . \tag{8.36}$$

The marginal distribution over the predictive parameter d has been calculated in equation (4.65). We can substitute all the classical parameters from equation (8.1) to write

$$\{d, D|X, y\} = \frac{\Gamma\left(\frac{\nu_C + q}{2}\right)}{(\pi \nu_C s_C^2)^{\frac{q}{2}} \left|I + D(X^T X)^{-1} D^T\right|^{\frac{1}{2}} \Gamma\left(\frac{\nu_C}{2}\right)}$$

$$\times \left[1 + \left(d - D\bar{\beta}\right)^T \frac{\left[I + D\left(X^T X\right)^{-1} D^T\right]^{-1}}{\nu_C s_C^2} \left(d - D\bar{\beta}\right)\right]^{-\frac{\nu_C + q}{2}} \quad . \tag{8.37}$$

The mean of this distribution appeared in equation (4.66), It has received substitutions from equation (8.1) to write

$$D\bar{\beta} \quad . \tag{8.38}$$

Its variance-covariance matrix appeared in equation (4.67). It has received substitutions from equation (8.1) to be written

$$\left[I + D\left(X^T X\right)^{-1} D^T\right] \frac{\nu_C s_C^2}{\nu_C - 2} \quad . \tag{8.39}$$

8.7 Predictions for a Single Row D, denoted δ, with $q = 1$–Classical

In the case in which the D matrix consists of a single row δ, which is of length p, the univariate predictive distribution over the (scalar) d for the classical regression case is obtained by setting $q = 1$ and simplifying the D and other terms in equation (8.34) to write

$$\{d, \delta, \sigma|X, y\} = \frac{2(2\pi)^{-\frac{1}{2}} \left(\frac{\nu_C s_C^2}{2}\right)^{\frac{\nu_C}{2}}}{\left|1 + \delta(X^T X)^{-1} \delta^T\right|^{\frac{1}{2}} \Gamma\left(\frac{\nu_C}{2}\right)} \sigma^{-(\nu_C + 2)} e^{-\frac{\nu_C s_C^2}{2\sigma^2}}$$

$$\times\, e^{-\frac{1}{2\sigma^2}\left(d - \delta\bar{\beta}\right)^T \left[1 + \delta\left(X^T X\right)^{-1} \delta^T\right]^{-1} \left(d - \delta\bar{\beta}\right)} \quad . \tag{8.40}$$

The univariate predictive distribution whose columns numbering p represent the number of coefficients in the model) has been calculated in equation (4.62) using the variance term z rather than σ. Substituting all the classical parameters from equation (8.1) allows us to write

$$\{d,z\} = \frac{\left(\frac{\nu_C s_C^2}{2}\right)^{\frac{\nu_C}{2}}}{\Gamma\left(\frac{\nu_C}{2}\right)} z^{-\left(\frac{\nu_C}{2}+1\right)} e^{-\frac{\nu_C s_C^2}{2z}} N\left\{d; \delta\bar{\beta}, \left[1+\delta\left(X^T X\right)^{-1}\delta^T\right] z\right\} \quad .$$

The inverse gamma that emerges has been calculated in product form using the inverse gamma form developed in equation (4.63). We substitute all the classical parameters from equation (8.1) and write

$$IG\left(z; \frac{\nu_C s_C^2}{2}, \frac{\nu_C}{2}\right) = \{z\} = \frac{\left(\frac{\nu_C s_C^2}{2}\right)^{\frac{\nu_C}{2}}}{\Gamma\left(\frac{\nu_C}{2}\right)} z^{-\left(\frac{\nu_C}{2}+1\right)} e^{-\frac{\nu_C s_C^2}{2z}} \quad .$$

The product form density was developed in equation (4.64). Substituting all the classical parameters from equation (8.1) yields the equation

$$\{d,z\} = IG\left(z; \frac{\nu_C s_C^2}{2}, \frac{\nu_C}{2}\right) N\left\{d; \delta\bar{\beta}, \left[1+\delta\left(X^T X\right)^{-1}\delta^T\right] z\right\} \quad .$$

The marginal distribution over the predictive parameter d has been calculated in equation (4.65). We substitute all the classical parameters from equation (8.1) to write

$$\{d\} = \frac{\Gamma\left(\frac{\nu_C+1}{2}\right)}{\left|1+\delta(X^T X)^{-1}\delta^T\right|^{\frac{1}{2}} (\pi \nu_C s_C^2)^{\frac{1}{2}} \Gamma\left(\frac{\nu_C}{2}\right)} \left\{1 + (d-\delta\bar{\beta})^T \frac{\left[1+\delta\left(X^T X\right)^{-1}\delta^T\right]^{-1}}{\nu_C s_C^2} (d-\delta\bar{\beta})\right\}^{-\frac{\nu_C+1}{2}} \quad . \tag{8.41}$$

The mean of this distribution, which appeared in equation (4.18), has received substitutions from equation (8.1) to become

$$\delta\bar{\beta} \quad . \tag{8.42}$$

and its variance-covariance matrix, which appeared in equation (4.67), has received substitutions from equation (8.1) to be written

$$\left[1+\delta\left(X^T X\right)^{-1}\delta^T\right] \frac{\nu_C s_C^2}{\nu_C - 2} \quad . \tag{8.43}$$

8.8 Predictions for a Single Row and a Single Column (3) with $\delta = e_3$ and $q = 1$–Classical

In the case in which the δ matrix consists of a single row e_3, which is of length p, the univariate predictive distribution over the (scalar) d for the classical is obtained by setting $q = 1$ and simplifying the D and other terms in equation (8.40) to write

$$\{d, e_3, \sigma|X, y\} = \frac{2(2\pi)^{-\frac{1}{2}} \left(\frac{\nu_C s_C^2}{2}\right)^{\frac{\nu_C}{2}}}{\left|1 + \left[(X^T X)^{-1}\right]_{33}\right|^{\frac{1}{2}} \Gamma\left(\frac{\nu_C}{2}\right)} \sigma^{-(\nu_C+2)} e^{-\frac{\nu_C s_C^2}{2\sigma^2}}$$
$$\times e^{-\frac{1}{2\sigma^2}(d-\bar{\beta}_3)^T \left(1+\left[(X^T X)^{-1}\right]_{33}\right)^{-1}(d-\bar{\beta}_3)} .$$

The univariate predictive distribution whose columns numbering p represent the number of coefficients in the model) has been calculated in equation (4.62) with the variance term z rather than σ. Substituting all the classical parameters from equation (8.1) gives the following

$$\{d, z\} = \frac{\left(\frac{\nu_C s_C^2}{2}\right)^{\frac{\nu_C}{2}}}{\Gamma\left(\frac{\nu_C}{2}\right)} z^{-\left(\frac{\nu_C}{2}+1\right)} e^{-\frac{\nu_C s_C^2}{2z}} N\left[d; \bar{\beta}_3, \left(1 + \left[(X^T X)^{-1}\right]_{33}\right) z\right] .$$

The inverse gamma that emerges has been calculated in product form uses the inverse gamma form developed in equation (4.63). Substituting all the classical parameters from equation (8.1) gives the equation

$$IG(z; \frac{\nu_C s_C^2}{2}, \frac{\nu_C}{2}) = \{z\} = \frac{\left(\frac{\nu_C s_C^2}{2}\right)^{\frac{\nu_C}{2}}}{\Gamma\left(\frac{\nu_C}{2}\right)} z^{-\left(\frac{\nu_C}{2}+1\right)} e^{-\frac{\nu_C s_C^2}{2z}} .$$

The product form density is developed in equation (4.64). We substitute all the classical parameters from equation (8.1) to write

$$\{d, z\} = IG(z; \frac{\nu_C s_C^2}{2}, \frac{\nu_C}{2}) N\left[d; \bar{\beta}_3, \left(1 + \left[(X^T X)^{-1}\right]_{33}\right) z\right] .$$

The marginal distribution over the predictive parameter d has been calculated in equation (4.65). Substituting all the classical parameters from equation (8.1) gives the following

$$(8.44) \quad \{d\} = \frac{\Gamma\left(\frac{\nu_C+1}{2}\right)}{\left|1 + \left[(X^T X)^{-1}\right]_{33}\right|^{\frac{1}{2}} (\pi \nu_C s_C^2)^{\frac{1}{2}} \Gamma\left(\frac{\nu_C}{2}\right)}$$

$$\left[1 + \left(d - e_3\bar{\beta}_3\right)^T \frac{\left(1 + \left[\left(X^T X\right)^{-1}\right]_{33}\right)^{-1}}{\nu_C s_C^2} \left(d - e_3\bar{\beta}_3\right)\right]^{-\frac{\nu_C+1}{2}} .$$

The mean of this distribution appeared in equation (4.66). When we substitute from equation (8.1), we write

$$\bar{\beta}_3 \quad .$$

We wrote its variance-covariance matrix in equation (4.67). When we substitute from equation (8.1), we write

$$\left(1 + \left[\left(X^T X\right)^{-1}\right]_{33}\right) \frac{\nu_C s_C^2}{\nu_C - 2} \quad .$$

If we use e_3 in place of δ, we are going to be predicting the third coefficient. The density is

$$\{d_3\} = \frac{\Gamma\left(\frac{\nu_C+1}{2}\right)}{\left|1 + \left[\left(X^T X\right)^{-1}\right]_{33}\right|^{\frac{1}{2}} (\pi\nu_C s_C^2)^{\frac{1}{2}} \Gamma\left(\frac{\nu_C}{2}\right)}$$

$$\times \left[1 + \left(d_3 - \bar{\beta}_3\right)^T \frac{1}{\left(1 + \left[\left(X^T X\right)^{-1}\right]_{33}\right) \nu_C s_C^2} \left(d_3 - \bar{\beta}_3\right)\right]^{-\frac{\nu_C+1}{2}} .$$

Eliminating the 1 so that we are doing a mere change of variables problem, we can write the density over coefficient 3 to be

$$\{d_3\} = \frac{\Gamma\left(\frac{\nu_C+1}{2}\right)}{\left|\left[\left(X^T X\right)^{-1}\right]_{33}\right|^{\frac{1}{2}} (\pi\nu_C s_C^2)^{\frac{1}{2}} \Gamma\left(\frac{\nu_C}{2}\right)}$$

$$\times \left[1 + \left(d_3 - \bar{\beta}_3\right)^T \frac{1}{\left[\left(X^T X\right)^{-1}\right]_{33} \nu_C s_C^2} \left(d_3 - \bar{\beta}_3\right)\right]^{-\frac{\nu_C+1}{2}} .$$

Compare this with equation (8.31); it is identical. The mean of this univariate density is, of course,

$$Mean = \bar{\beta}_3 \quad . \tag{8.45}$$

Compare this with equation (8.32); it is identical. The variance is

$$\left\{1 + \left[\left(X^T X\right)^{-1}\right]_{33}\right\} \frac{\nu_C s_C^2}{\nu_C - 2} \quad .$$

If we eliminate the predicted versus actual term, the variance is

(8.46)
$$\left[\left(X^T X\right)^{-1}\right]_{33} \frac{\nu_C s_C^2}{\nu_C - 2} \quad .$$

Compare this with equation (8.33); it is identical. This process of eliminating the 1 term, which represents the error distribution of predicted relative to actual, is going to be important again later in this monograph when we seek marginal distributions for more complex outputs than a single univariate.

Chapter 9

SUMMARY OF KEY EQUATIONS AND PARAMETERS FOR THE UNIVARIATE CASE

> ***"Il n'est pas certain que tout soit incertain." (Translation: It is not certain that everything is uncertain.)***
>
> ***Blaise Pascal, Pascal's Pensees***

We summarize the critical results from Chapter 4, which will be used for the prior, posterior, and classical regression solution. These are the results that are plotted and used as probability distributions for decision making and forecasting. These are the equations we will need to use and program.

9.1 General Terms

9.1.1 ν (Degrees of Freedom) term

Archetype: Equation (4.4)
Prior: Equation (5.2)
Posterior: Equation (6.2)
Classical: Equation (8.2)

9.1.2 r, (the exponent on σ in the joint density over β and σ, i.e., the exponent on σ is $-r$)

Archetype: Equation (4.5)
Prior: Equation (5.3)
Posterior: Equation (6.3)
Classical: Equation (8.3)

9.1.3 g, the sum of squared errors term, which is equal to νs^2 and successively g, M, A, and R in the four cases

Archetype: Equation (4.7)
Prior: Equation (5.4)
Posterior: Equation (6.4)
Classical: Equation (8.4)

9.1.4 s^2 term $\frac{g}{\nu}$ = mean of error distribution

Archetype: Equation (4.6)
Prior: Equation (5.5)
Posterior: Equation (6.5)
Classical: Equation (8.5)

9.1.5 Mean of coefficients, which is equal to $< \beta >= \mu$

Archetype: Equation (4.18)
Prior: Equation (5.19)
Posterior: Equation (6.15)
Classical: Equation (8.6)

9.1.6 H matrix of the exponential term in the joint density over β and σ, which has a different form in each case

Archetype: Equation (4.14)
Prior: Equation (5.6)
Posterior: Equation (6.7)
Classical: Equation (8.7)

9.2 Full p-vector Multivariate Density Terms over σ and β

9.2.1 Full Multivariate Joint Distribution over Both the p-vector of Coefficients β and the Scalar σ in Basic Form (Using σ Rather Than $z = \sigma^2$)

This is the joint density over coefficients and σ for the archetype, prior, posterior, and classical cases.

Archetype: Equation (4.8)
Prior: Equation (5.12)
Posterior: Equation (6.8)
Classical: Equation (8.8)

9.2.2 Full multivariate joint density over p-vector β and $z = \sigma^2$

This is the joint density over coefficients and $z = \sigma^2$, converted from the previous section.

Archetype: Equation (4.10)
Prior: Equation (5.13)
Posterior: Equation (6.9)
Classical: Equation (8.9)

9.2.3 Full joint density over p-vector β and $z = \sigma^2$ in Product Form, inverse gamma times normal

Archetype: Equation (4.12)
Prior: Equation (5.17)
Posterior: Equation (6.13)
Classical: Equation (8.13)

9.3 Full p-vector Marginal Terms over β Only, the σ Having Been Integrated Out

9.3.1 Full marginal distribution over the p-vector of coefficients β, the σ Having Been Integrated Out

Archetype: Equation (4.15)
Prior: Equation (5.18)
Posterior: Equation (6.14)
Classical: Equation (8.14)

9.3.2 Mean of full marginal distribution over the p-vector of coefficients, μ

Archetype: Equation (4.18)
Prior: Equation (5.19)
Posterior: Equation (6.15)
Classical: Equation (8.6)

9.3.3 Variance-covariance matrix of full marginal distribution over the p-vector of coefficients

Archetype: Equation (4.19)
Prior Equation (5.20)
Posterior: Equation (6.16)
Classical: Equation (8.15)

9.3.4 H matrix of full marginal distribution over the p-vector of coefficients

Archetype: Equation (4.20)
Prior: Equation (5.21)
Posterior: Equation (6.17)
Classical: Equation (8.16)

9.4 Marginal Terms Over $z = \sigma^2$

9.4.1 Inverse Gamma distribution over z

Archetype: Equation (4.11)
Prior: Equation (5.14)
Posterior: Equation (6.10)
Classical: Equation (8.10)

9.4.2 Mean of inverse gamma (IG) Distribution over the variance $z = \sigma^2$

Archetype: Equation (4.22)
Prior: Equation (5.15)
Posterior: Equation (6.11)
Classical: Equation (8.11)

9.4.3 Variance of inverse gamma (IG) Distribution over the variance σ^2

Archetype: Equation (4.23)
Prior: Equation (5.16)
Posterior: Equation (6.12)
Classical: Equation (8.12)

9.5 Results for the $p - L$-vector Subset of the p-vector β

9.5.1 Full joint density over the $p - L$ coefficients of β including both the $p - L$-vector of coefficients β_2 and σ

Archetype: Equation (4.28)
Prior: Equation (5.22)
Posterior: Equation (6.18)
Classical: Equation (8.18)

9.5.2 Product rule density over the $p - L$ coefficients of β, namely β_2, and σ

Archetype: Equation (4.30)
Prior: Equation (5.23)
Posterior: Equation (6.20)
Classical: Equation (8.19)

9.5.3 Inverse gamma (IG) density over the $p - L$ coefficients

Archetype: Equation (4.31)
Prior: Equation (5.28)
Posterior: Equation (6.24)
Classical: Equation (8.20)

9.5.4 Marginal distribution over the $p - L$ coefficients of β, namely β_2

Archetype: Equation (4.33)
Prior: Equation (5.25)
Posterior: Equation (6.22)
Classical: Equation (8.21)

9.6 Special Case in which $p - L = 1$–Univariate Distribution over a Single Element of the p-vector β

9.6.1 Full joint density over $p - L = 1$ coefficients including both the scalar β_2 and σ

Archetype: Equation (4.34)
Prior: Equation (5.26)
Posterior: Equation (6.23)
Classical: Equation (8.22)

9.6.2 Marginal univariate Student's t coefficient distribution over $p - L = 1$ coefficient β_2 after integrating out σ, expressed in partitioned matrix form

Archetype: Equation (4.35)
Prior: Equation (5.29)
Posterior: Equation (6.25)
Classical: Equation (8.25)

9.6.3 The ith diagonals of the H matrix, namely z_{ii}

Archetype: Equation (4.36)
Prior: Equation (5.30)
Posterior: Equation (6.26)
Classical: Equation (8.26)

9.6.4 Marginal univariate Student's t coefficient distribution over $p - L = 1$ coefficient β_2 after integrating out σ, expressed in z_{ii} form

Archetype: Equation (4.42)
Prior: Equation (5.32)
Posterior: Equation (6.29)
Classical: Equation (8.29)

9.6.5 Univariate product rule joint density over scalar β_2 and σ

Archetype: Equation (4.41)
Prior: Equation (5.31)
Posterior: Equation (6.28)
Classical: Equation (8.28)

9.7 Generic Univariate Student's t Distribution

9.7.1 Wiki univariate joint density over β_i and σ in generic form given the parameters P,Q,k

Archetype: Equation (4.52)
Prior: Equation (5.33)
Posterior: Equation (6.31)
Classical: Equation (8.31)

9.7.2 Generic P value

Archetype: Equation (4.38)

9.7.3 Generic Q value

Archetype: Equation (4.39)

9.7.4 Generic k value

Archetype: Equation (4.40)

9.7.5 Generic underlying univariate t, P,Q,k

Archetype: Equation (4.44)

9.7.6 Generic underlying mean, P,Q,k

Archetype: Equation (4.45)

9.7.7 Generic underlying variance, P,Q,k

Archetype: Equation (4.46)

9.8 Predictive Densities over q-Vector D

9.8.1 Univariate predictive joint density over q-vector d and σ

Archetype: Equation (4.60)
Prior: Equation (5.36)
Posterior: Equation (6.34)
Classical: Equation (8.34)

9.8.2 Univariate inverse gamma density over z

Archetype: Equation (4.63)
Prior: Equation (5.38)
Posterior: Equation (6.36)
Classical: Equation (8.35)

9.8.3 Univariate predictive product form density over q-vector d and z

Archetype: Equation (4.64)
Prior: Equation (5.39)
Posterior: Equation (6.37)
Classical: Equation (8.36)

9.9 Univariate Marginal Predictive Distributions over q-Vector D

9.9.1 Univariate Predictive Marginal over q-vector d

Archetype: Equation (4.65)
Prior: Equation (5.40)
Posterior: Equation (6.38)
Classical: Equation (8.37)

9.9.2 Mean of Marginal over q-vector d

Archetype: Equation (4.66)
Prior: Equation (5.41)
Posterior: Equation (6.39)
Classical: Equation (8.38)

9.9.3 Variance-Covariance matrix of marginal over q-vector d

Archetype: Equation (4.67)
Prior: Equation (5.42)
Posterior: Equation (6.40)
Classical: Equation (8.39)

9.10 Univariate Marginal Predictive Distributions over Scalar $\delta\beta$

9.10.1 Joint density for single row $\delta\beta$

Archetype: Equation (4.68)
Prior: Equation (5.43)
Posterior: Equation (6.41)
Classical: Equation (8.41)

9.10.2 Mean for single row $\delta\beta$

Archetype: Equation (4.69)
Prior: Equation (5.44)
Posterior: Equation (6.42)
Classical: Equation (8.42)

9.10.3 VCV for single row $\delta\beta$

Archetype: Equation (4.70)
Prior: Equation (5.45)
Posterior: Equation (6.43)
Classical: Equation (8.43)

9.11 Univariate Marginal Predictive Distribution over Scalar $e_3\beta$

9.11.1 Joint density for single row/single element of β, namely $e_3\beta$

Archetype: Equation (4.78)
Prior: Equation (5.46)
Posterior: Equation (6.44)
Classical: Equation (8.44)

9.11.2 Mean for single row/single row/single element of β, namely $e_3\beta$

Archetype: Equation (4.79)
Prior: Equation (5.47)
Posterior: Equation (6.45)
Classical: Equation (8.45)

9.11.3 Variance for single row/single row/single element of β, namely $e_3\beta$

Archetype: Equation (4.80)
Prior: Equation (5.48)
Posterior: Equation (6.46)
Classical: Equation (8.46)

Chapter 10

WHEN THE PRIOR IS CALCULATED FROM AN OLD DATA SET AND A NEW DATA SET IS COLLECTED TO UPDATE

> ***Why, sir, there is every probability that you will soon be able to tax it! (Said to William Gladstone, the Chancellor of the Exchequer, when he asked about the practical worth of electricity.)***
>
> ***Michael Faraday***

Quite often, in practice, data gathering is cumulative and ongoing. We will have an answer today based on data gathered through all past history through yesterday. We will gather additional data today, and we would dearly love at the end of the day today to perform a Bayesian update using only today's data and not have to graft the small amount of today's new data onto the huge, cumulative amount of historical (old) data and re-run the whole shootin' match from scratch every day![1] This section shows that we can in fact do precisely that, and it shows exactly how and how easy it is. Let us depict the data that is old with the

[1] This is not in general true with nonlinear or simulation-only "prior-times-likelihood" functional forms. They are not generally amenable to easy incremental updating.

subscript "0" and the data that is new and gathered on an ongoing basis with the subscript "N." This "updating" we are about to develop here will not hold for distributions that require simulation.

We think of the old and new data as represented in the matrix in Figure 10.1. On the left, we see a matrix containing the old data in the first n_0 rows and p columns (shaded gray), and we see the new data in the successive n_N rows and p columns below (unshaded). The data are arrayed as in the left-hand matrix in Figure 10.1. This is precisely the way the data would look had one gathered all the old and new data in one fell swoop. The figure is designed to show us how to calculate $X^T X$ for the matrix that contains all old plus all new data. Before we do so, we need to diagram the matrix $X^T y$ that we will need for our overall classical regression solution. See Figure 10.2.

Figure 10.1: The Old Data (0) and the New Data (N)–$X^T X$

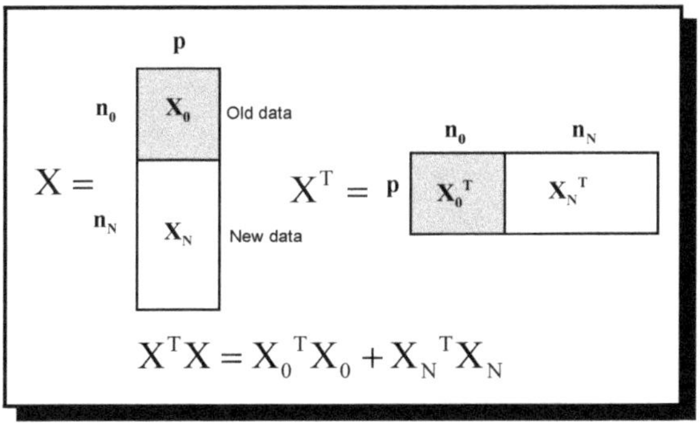

Figure 10.2: The Old Data (0), the New Data (N), X^T, and y

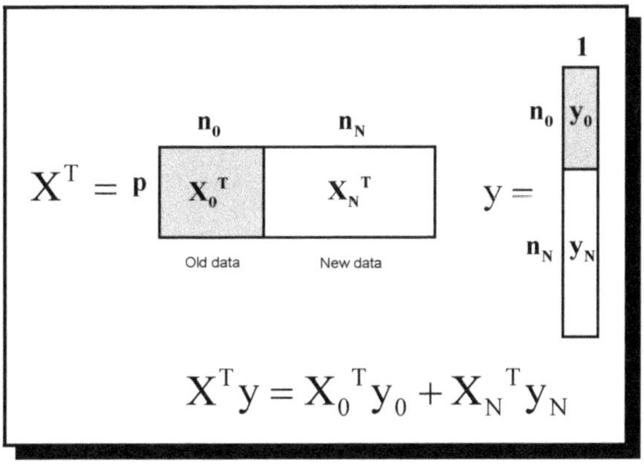

Throughout this section, the term "classical regression solution" will be understood to mean the Bayesian posterior assuming a diffuse prior. It is our characterization of classical regression as articulated in Chapter 7, not that in Chapter

2. We can safely walk away from the version in Chapter 2 in favor of a probabilistically rigorous development and interpretation of the statistical and prediction problems. When we say classical regression in this and subsequent sections, we mean Bayesian posterior with a diffuse prior, as developed in detail in Chapter 7. We have already shown that classical statistics and Bayesian statistics with a diffuse prior are virtually isomorphic and precisely equal in all the important dimensions. And we have eliminated the "tests" that have been endemic in classical statistics, substituting instead a complete multivariate posterior probability distribution.

10.1 The Classical Regression Solution (Bayes with Diffuse Prior) on the Whole Data Set, Old Combined with New

This first section gives the answer we really want—the answer that considers all data, old and new, in one big data set, and creates a posterior from that data, old plus new. We want to calculate the Bayesian posterior with diffuse prior for the whole data set, old plus new, and this section does so. To do so, we will need the matrix X^TX and the vector X^Ty for the whole data set, old plus new. That whole data set is comprised of all old and new data aggregated row by row, as shown in Figure 10.1 and Figure 10.2. The figures show us why the equations for the relevant terms needed for classical regression on this database are

$$X^TX = {X_0}^TX_0 + {X_N}^TX_N \quad , \tag{10.1}$$

and

$$X^Ty = {X_0}^Ty_0 + {X_N}^Ty_N \quad .$$

The inverse of the matrix containing all old plus new data is

$$\left(X^TX\right)^{-1} = \left({X_0}^TX_0 + {X_N}^TX_N\right)^{-1} \quad .$$

The classical statistics solution for the mean value of the coefficient vector using all old plus new data in aggregate has been shown in previous developments to be

$$\begin{aligned} \bar{\beta} &= \left(X^TX\right)^{-1}X^Ty = \left(X_0^TX_0 + X_N^TX_N\right)^{-1}\left(X_0^Ty_0 + X_N^Ty_N\right) \\ \Rightarrow \bar{\beta}^T &= \left(y_0^TX_0 + y_N^TX_N\right)\left(X_0^TX_0 + X_N^TX_N\right)^{-1} \quad . \end{aligned} \tag{10.2}$$

The residual sum of squared errors using the total data set, old plus new, is given by equation (8.4) using a script $\Re$ notation to denote the residual sum of squared errors over the composite of the old plus new data bases

$$\Re = \left(y - X\bar{\beta}\right)^T\left(y - X\bar{\beta}\right)$$

$$
\begin{aligned}
&= \left(y^T - \bar{\beta}^T X^T\right)\left(y - X\bar{\beta}\right) \\
&= y^T y - 2\bar{\beta}^T X^T y + \bar{\beta}^T X^T X \bar{\beta} \\
&= y^T y - \bar{\beta}^T X^T y \\
&= y^T y - y^T X \bar{\beta} \quad .
\end{aligned}
$$

We substitute the detailed equations for the old and new data sets into the last of the foregoing equations to write

$$
\begin{aligned}
\Re &= y^T y - y^T X \bar{\beta} \\
&= y_0^T y_0 + y_N^T y_N - \left(X_0^T y_0 + X_N^T y_N\right)^T \bar{\beta} \qquad (10.3) \\
&= y_0^T y_0 + y_N^T y_N - \bar{\beta}^T \left(X_0^T y_0 + X_N^T y_N\right) \quad .
\end{aligned}
$$

We will return to this equation after developing a few more terms when we analyze the old data set only and thereafter the new data set only. For a data set with $n_0 + n_N$ data elements in it, the classical regression and the Bayesian solution with the diffuse prior must end up with posterior degrees of freedom equal to the total number of data points minus the number of coefficients p:

$$
\nu = n_0 + n_N - p \quad . \qquad (10.4)
$$

The posterior with a diffuse prior and the aggregate combination of both data sets has an exponent on σ as derived in equation (8.3) of

$$
\nu + p + 1 \quad ,
$$

which means that if we add p to both sides of equation (10.4) and add 1, we obtain the following expression for the exponent of σ

$$
n_0 + n_N + 1 \quad .
$$

The similarity with the diffuse prior result in Chapter 7 is striking. The exponent is the total number of data points plus 1, just as it was in Chapter 7. We write the equation for the error term using the equation (8.5):

$$
s^2 = \frac{\Re}{\nu} \quad , \qquad (10.5)
$$

in which we are using the diffuse prior and therefore are borrowing results directly from Chapter 7. We summarize the posterior density of the combined analysis of the old plus new data sets

$$
\{\beta, \sigma\} = \frac{2\left|X^T X\right|^{\frac{1}{2}} (2\pi)^{-\frac{p}{2}} \left(\frac{\nu s^2}{2}\right)^{\frac{\nu}{2}}}{\Gamma\left(\frac{\nu}{2}\right)}
$$

$$\times\, \sigma^{-(\nu+p+1)} e^{-\frac{\nu s^2}{2\sigma^2}} e^{-\frac{1}{2\sigma^2}(\beta-\bar{\beta})^T (X^T X)(\beta-\bar{\beta})}$$
$$= \frac{2|X^T X|^{\frac{1}{2}} (2\pi)^{-\frac{p}{2}} \left(\frac{\Re}{2}\right)^{\frac{\nu}{2}}}{\Gamma\left(\frac{\nu}{2}\right)}$$
$$\times\, \sigma^{-(n_0+n_N+1)} e^{-\frac{\Re}{2\sigma^2}} e^{-\frac{1}{2\sigma^2}(\beta-\bar{\beta})^T (X^T X)(\beta-\bar{\beta})} \quad .$$

It is well to summarize the "sufficient statistics" that comprise this density

- p = number of model parameters
- ν = degrees of freedom, from equation (10.4)
- s^2 = variance term, from equation (10.5), based on the $\Re$ term in equation (10.3)
- $\Re = \nu s^2$
- $X^T X$ = precision matrix, from equation (10.1)
- $\bar{\beta}$ = mean, from equation, (10.2)

We are now finished with the Bayesian posterior using the diffuse prior (the classical regression solution) using the combined old and new data sets.

It is well to reassert what we have done here. We have performed a Bayesian Linear Regression with diffuse priors (classical regression) on the data set comprised of $n_0 + n_N$ old plus new data observations. It is as if we clumped all the old data points plus all the new data points into one large "old plus new" data set and performed classical regression analysis using that entire data set. The classical regression "answers" we get depend on the entirety of both data sets. Having this solution in hand, we are going to consider creating a prior using only the old data set and thereafter updating that prior using only the new data set. We will show that the latter gives an identical answer to what we just got. It is well to make the foregoing substitutions to write the joint density for the overall total data set, old plus new

$$\{\beta,\sigma\}$$
$$= \frac{2|X_0^T X_0 + X_N^T X_N|^{\frac{1}{2}}}{(2\pi)^{\frac{p}{2}} \Gamma\left(\frac{\nu}{2}\right)} \left(\frac{\nu s^2}{2}\right)^{\frac{\nu}{2}}$$
$$\times\, \sigma^{-(\nu+p+1)} e^{-\frac{\nu s^2}{2\sigma^2}} e^{-\frac{1}{2\sigma^2}(\beta-\bar{\beta})^T (X_0^T X_0 + X_N^T X_N)(\beta-\bar{\beta})}$$
$$= \frac{2|X_0^T X_0 + X_N^T X_N|^{\frac{1}{2}}}{(2\pi)^{\frac{p}{2}} \Gamma\left(\frac{\nu}{2}\right)} \left(\frac{\Re}{2}\right)^{\frac{\nu}{2}}$$
$$\times\, \sigma^{-(n_0+n_N+1)} e^{-\frac{\Re}{2\sigma^2}} e^{-\frac{1}{2\sigma^2}(\beta-\bar{\beta})^T (X_0^T X_0 + X_N^T X_N)(\beta-\bar{\beta})} \quad .$$

10.2 The Classical Regression Solution (Bayes with Diffuse Prior) on the Old Data Set Only

Our second task begins by conducting classical analysis of the old data set, and the old data set only. (By classical analysis, we mean Bayesian posterior with diffuse prior on the old data set only, in the sense of Chapter 7.) This produces a posterior on the old data set only. We thereafter use this posterior on the old data set only as an informative prior for a second analysis—Bayesian updating of that informative prior using the new data set only. To summarize

- Step 1: Conduct classical analysis of the old data set only, i.e., a Bayesian posterior using the diffuse prior and the old data set only in the sense of Chapter 8.
- Step 2: Interpret the posterior from Step 1 as an informative prior for a second Bayesian analysis in which the informative prior will be updated by the new data set only in the sense of Chapter 7.

We compute the posterior based on the diffuse prior and ONLY the old data set, i.e., the classical regression analysis (Bayesian posterior with diffuse prior) on the old data set only. Thereafter, we use that result as the prior to be updated by ONLY the new data set.

Conceptually, the classical analysis of the old data set represents the prior information to be later updated by the new data set in a Bayesian sense. The old data set represents the information we bring to the problem, i.e., the information that predates the new data set. It is what we know before applying the new data set, the quintessential notion of a prior. We will show that "prior from old data thereafter updated by new data" gives the identical answer as the Bayesian posterior using the diffuse prior on the aggregate of all old plus new data in total.

We begin by conducting a classical regression analysis on just the old data base using a diffuse prior. In order to calculate the classical regression solution for the old data base only, we will need the matrix $X_0^T X_0$ and the vector $X_0^T y_0$, in both cases containing all old data, as in Figure 10.1 and Figure 10.2. The classical statistics mean coefficient using only the old data base is

$$\bar{\beta}_0 = \left(X_0^T X_0\right)^{-1} X_0^T y_0 \Rightarrow \bar{\beta}_0^T = y_0^T X_0 \left(X_0^T X_0\right)^{-1} \quad . \tag{10.6}$$

Because we will need it later, we write

$$\left(X_0^T X_0\right)^{-1} \bar{\beta}_0 = X_0^T y_0 \Rightarrow \left(X_0^T X_0\right)^{-1} \bar{\beta}_0^T = y_0^T X_0 \quad .$$

The residual sum of squared errors from the classical statistics residual sum of squared errors using only the old data base is given by equation (8.4) to be

$$\begin{aligned} R_0 &= (y_0 - X_0\bar{\beta}_0)^T(y_0 - X_0\bar{\beta}_0) \\ &= y_0^T\,(y_0 - X_0\bar{\beta}_0) \\ &= {y_0}^T y_0 - {y_0}^T X_0\left({X_0}^T X_0\right)^{-1}{X_0}^T y_0 \quad . \end{aligned} \tag{10.7}$$

We need to rearrange terms

$${y_0}^T y_0 = R_0 + {y_0}^T X_0\left({X_0}^T X_0\right)^{-1}{X_0}^T y_0 \quad .$$

For a data set with n_0 data elements in it, the classical regression and the Bayesian solution with the diffuse prior must end up with a posterior degrees of freedom equal to the total number of old data points minus the number of coefficients p:

$$\nu_0 = n_0 - p \quad . \tag{10.8}$$

This diffuse prior analysis of the old data set leads to an exponent on the σ term in the density as derived in equation (5.3) that is

$$\nu_0 + p + 1 \quad .$$

Substituting equation (10.8) into the immediately preceding equation, we obtain the following exponent on the σ term in the density resulting from the old data base only

$$n_0 + 1 \quad .$$

This is precisely as we expect. The exponent on the σ term in the density is the number of old data points n_0 plus 1. We write the equation for the error term using equation (8.5)

$$s_0^2 = \frac{R_0}{\nu_0} \quad . \tag{10.9}$$

We summarize the posterior for the diffuse prior analysis (classical regression) on the old data set alone, which we might dub the "old data base posterior"

$$\begin{aligned} \{\beta, \sigma\} &= \frac{2\left|X_0^T X_0\right|^{\frac{1}{2}}(2\pi)^{-\frac{p}{2}}\left(\frac{\nu_0 s_0^2}{2}\right)^{\frac{\nu_0}{2}}}{\Gamma\left(\frac{\nu_0}{2}\right)} \\ &\quad \times \sigma^{-(\nu_0+p+1)} e^{-\frac{\nu_0 s_0^2}{2\sigma^2}} e^{-\frac{1}{2\sigma^2}(\beta-\bar{\beta}_0)^T(X_0^T X_0)(\beta-\bar{\beta}_0)} \\ &= \frac{2\left|X_0^T X_0\right|^{\frac{1}{2}}(2\pi)^{-\frac{p}{2}}\left(\frac{R_0}{2}\right)^{\frac{\nu_0}{2}}}{\Gamma\left(\frac{\nu_0}{2}\right)} \\ &\quad \times \sigma^{-(n_0+1)} e^{-\frac{R_0}{2\sigma^2}} e^{-\frac{1}{2\sigma^2}(\beta-\bar{\beta}_0)^T(X_0^T X_0)(\beta-\bar{\beta}_0)} \quad . \end{aligned} \tag{10.10}$$

It is well to summarize the "sufficient statistics" that comprise this density derived as the diffuse prior/classical regression solution of the old data set only

- p = number of model parameters
- ν_0 = degrees of freedom, from equation (10.8)
- s_0^2 = variance term, from equation (10.9), based on the R_0 term in equation (10.7)
- $X_0^T X_0$ = precision matrix
- $\bar{\beta}_0$ = mean, from equation (10.6) .

We are finished with the old data set. And what have we done with it? We have calculated the Bayesian posterior distribution, or what we have been terming the classical regression solution, on that old data set and that old data set alone. In particular, it is the posterior one would get if one used a diffuse prior and ONLY used the old data set. While it is a posterior on the old data set, we want it to be a prior for the new data set. That is, it is precisely the prior we want to use for a Bayesian update using the new data and the new data alone. These calculated parameters fully define a distribution in the form of the prior, and they have resulted from classical regression. We are going to make that connection to the functional form of the prior explicit in the coming section.

10.3 Equating the Old Data Posterior with a Prior in Standard Form

Let us return to the joint prior in standard form that we presented in equation (5.12) and rewrite it here for easy reference

$$\begin{aligned}\{\beta, \sigma\} &= \frac{2|Q|^{\frac{1}{2}}}{(2\pi)^{\frac{p}{2}}\Gamma\left(\frac{\nu_0}{2}\right)}\left(\frac{\nu_0 s_0^2}{2}\right)^{\frac{\nu_0}{2}} \sigma^{-(\nu_0+p+1)} e^{-\frac{\nu_0 s_0^2}{2\sigma^2}} \\ &\quad \times e^{-\frac{1}{2}(\beta-\beta_0)^T\left(Q\sigma^{-2}\right)(\beta-\beta_0)} \quad .\end{aligned} \tag{10.11}$$

We want to examine whether this joint prior in standard form can be term for term related to the joint prior we developed immediately above as equation (10.10) for the old data base only, rewritten here for easy reference

$$\begin{aligned}\{\beta, \sigma\} &= \frac{2\left|X_0^T X_0\right|^{\frac{1}{2}}(2\pi)^{-\frac{p}{2}}\left(\frac{\nu_0 s_0^2}{2}\right)^{\frac{\nu_0}{2}}}{\Gamma\left(\frac{\nu_0}{2}\right)} \\ &\quad \times \sigma^{-(\nu_0+p+1)} e^{-\frac{\nu_0 s_0^2}{2\sigma^2}} e^{-\frac{1}{2\sigma^2}(\beta-\bar{\beta}_0)^T\left(X_0^T X_0\right)(\beta-\bar{\beta}_0)} \\ &= \frac{2\left|X_0^T X_0\right|^{\frac{1}{2}}(2\pi)^{-\frac{p}{2}}\left(\frac{R_0}{2}\right)^{\frac{\nu_0}{2}}}{\Gamma\left(\frac{\nu_0}{2}\right)}\end{aligned} \tag{10.12}$$

$$\times\, \sigma^{-(n_0+1)} e^{-\frac{R_0}{2\sigma^2}} e^{-\frac{1}{2\sigma^2}(\beta-\bar{\beta}_0)^T (X_0^T X_0)(\beta-\bar{\beta}_0)} \quad .$$

We begin with the second equation, our equation, equation (10.12), and see what parameters we can substitute directly from it into identical positions in the first equation (10.11), which is the prior in standard form. It is obvious that if we define ν_0 as the degrees of freedom calculated by us in equation (10.12) when we performed classical regression on the old data set only, that same term and notation occur in exactly the same place in equation (10.11). We therefore make the substitution

$$\nu_0 \leftarrow \nu_0$$

the left arrow implying that the parameter on the right is the parameter calculated in equation (10.12), and the left arrow places it into the appropriate position in equation (10.11).

It is also obvious that if we define s_0^2 as the degrees of freedom calculated by us in equation (10.12) when we performed classical regression on the old data set only, that same term and notation occur in exactly the same place in equation (10.11). We therefore make the substitution

$$s_0^2 \leftarrow s_0^2$$

the left arrow implying that the parameter on the right is the parameter calculated in equation (10.12), and the left arrow places it into the appropriate position in equation (10.11).

We notice that the term $\nu_0 s_0^2$ occurs in exactly the same place in (10.12) as it does in equation (10.11), and that in equation (10.11), it occupies the position of the M term in the standard prior. Therefore we make the substitution

$$M \leftarrow \nu_0 s_0^2$$

the right hand side of the arrow again representing equation (10.12) and the right hand side representing equation (10.11). We notice that the term ν_0+p+1 occurs in exactly the same place in equation (10.12) as it does in equation (10.11), and that in equation (10.11), it occupies the position of the m term in the standard prior. Therefore we make the substitution

$$m = \nu_0 + p + 1 \leftarrow \nu_0 + p + 1$$

the right hand side of the arrow again representing equation (10.12) and the left hand side representing equation (10.11). We notice that the β_0 term in the standard prior occurs in exactly the same places and the same way in equation (10.11) that the classical regression mean on the old data base β_0 occurs in equation (10.12). Therefore, we make the substitution

$$\beta_0 \leftarrow \bar{\beta}_0 = \left(X_0^T X_0\right)^{-1} X_0^T y_0$$

the right hand side of the arrow again representing equation (10.12) and the left hand side representing equation (10.11). If we left multiply every term in this equation by Q

$$Q\beta_0 = Q\bar{\beta}_0 = \left(X_0^T X_0\right)\bar{\beta} = X_0^T y_0 \quad .$$

Finally of course, we have

$$p \leftarrow p$$

When we make all these substitutions into equation (10.11), we have the prior that we have developed from Bayesian Linear Regression with a diffuse prior (which we have called classical regression). It is

$$\{\beta, \sigma\} = \frac{2\left|X_0^T X_0\right|^{\frac{1}{2}}}{(2\pi)^{\frac{p}{2}}\Gamma\left(\frac{\nu_0}{2}\right)}\left(\frac{R_0}{2}\right)^{\frac{\nu_0}{2}} \times \sigma^{-(n_0+1)} e^{-\frac{R_0}{2\sigma^2}} e^{-\frac{1}{2\sigma^2}(\beta-\bar{\beta}_0)^T\left(X_0^T X_0\right)(\beta-\bar{\beta}_0)} \quad . \tag{10.13}$$

10.4 Technical Sidebar to Develop One Key Result That Will Be Needed later–Classical Regression Analysis of the New Data Only

This section is a technical sidebar that develops an algebraic result that will be convenient later. That is the only purpose of this section—to develop an equation that we will need later when we start with the "old data prior" and thereafter update with the new data. To initiate this technical aside, suppose we were conducting classical statistical analysis using ONLY the new data, i.e., completely ignoring all the old data. When we say classical statistics, of course we mean Bayesian statistics with a diffuse prior. If we analyzed only the new data, we would obtain the following results: First, we write the classical regression solution for the new data set (meaning Bayesian analysis with a diffuse prior) using only the new data set:

$$\bar{\beta}_N = \left(X_N^T X_N\right)^{-1} X_N^T y_N \Rightarrow \bar{\beta}_N^T = y_N^T X_N\left(X_N^T X_N\right)^{-1} \quad . \tag{10.14}$$

Thereafter the residual sum of squared errors would be

$$R_N = {y_N}^T y_N - {y_N}^T X_N\left({X_N}^T X_N\right)^{-1} {X_N}^T y_N \quad . \tag{10.15}$$

We need to rearrange terms

$$y_N{}^T y_N = R_N + y_N{}^T X_N \left(X_N{}^T X_N\right)^{-1} X_N{}^T y_N \quad .$$

Keep in mind that R_N is the residual sum of squared errors for the classical regression (Bayes with diffuse prior) solution to the new data set only (no influence of the old data set in the prior or anywhere else). The functional form of the likelihood function is evident from equation (2.7). If we subscript y, X, R, and $\bar{\beta}$ by N, we can write the likelihood function for the new data base in the following form

$$\{y_N, X_N|\beta,\sigma\} = \frac{1}{\sigma^{n_N}(2\pi)^{\frac{n_N}{2}}} e^{-\frac{R_N}{2\sigma^2}} e^{-\frac{(\beta-\bar{\beta}_N)^T\left(X_N^T X_N\right)(\beta-\bar{\beta}_N)}{2\sigma^2}} \tag{10.16}$$

in which the parameters for the new data base are

$$\begin{aligned}
& n_N \\
& R_N = \left(y_N - X_N\bar{\beta}_N\right)^T \left(y_N - X_N\bar{\beta}_N\right) \\
& X_N^T X_N \\
& \bar{\beta}_N = \left(X_N^T X_N\right)^{-1} X_N^T y_N \quad .
\end{aligned}$$

10.5 Start with the Solution to the Old Data Set as the Prior and Update Using the New Data Set

We have the prior. It is the posterior of the Bayesian analysis on the old data base using a diffuse prior. It appears in equation (10.13). Bayes tells us to multiply the prior by the likelihood function in equation (10.16), which is exactly what we will do here, to obtain the posterior:

$$\begin{aligned}
&\{y_N, X_N|\beta,\sigma\}\{\beta,\sigma\} \\
&= \frac{2\left|X_0^T X_0\right|^{\frac{1}{2}}}{(2\pi)^{\frac{p}{2}}\Gamma\left(\frac{\nu_0}{2}\right)}\left(\frac{R_0}{2}\right)^{\frac{\nu_0}{2}} \sigma^{-(n_0+1)} e^{-\frac{R_0}{2\sigma^2}} e^{-\frac{1}{2\sigma^2}(\beta-\bar{\beta}_0)^T\left(X_0^T X_0\right)(\beta-\bar{\beta}_0)} \\
&\times \frac{1}{\sigma^{n_N}(2\pi)^{\frac{n_N}{2}}} e^{-\frac{R_N}{2\sigma^2}} e^{-\frac{(\beta-\bar{\beta}_N)^T\left(X_N^T X_N\right)(\beta-\bar{\beta}_N)}{2\sigma^2}} \\
&= \frac{2\left|X_0^T X_0\right|^{\frac{1}{2}}}{(2\pi)^{\frac{p+n_N}{2}}\Gamma\left(\frac{\nu_0}{2}\right)}\left(\frac{R_0}{2}\right)^{\frac{\nu_0}{2}} \sigma^{-(n_0+n_N+1)} \\
&\times e^{-\frac{1}{2\sigma^2}\left[R_0+R_N+(\beta-\bar{\beta}_0)^T\left(X_0^T X_0\right)(\beta-\bar{\beta}_0)+(\beta-\bar{\beta}_N)^T\left(X_N^T X_N\right)(\beta-\bar{\beta}_N)\right]} \quad .
\end{aligned}$$

We rewrite the exponent in the foregoing equation for convenient future reference

(10.17)

$$exponent = R_0 + R_N$$
$$+ \left(\beta - \bar{\beta}_0\right)^T \left(X_0^T X_0\right) \left(\beta - \bar{\beta}_0\right) + \left(\beta - \bar{\beta}_N\right)^T \left(X_N^T X_N\right) \left(\beta - \bar{\beta}_N\right) \quad .$$

As before, we are relegated to complete the square in the exponent to develop a quadratic from it. To begin that, we define the following temporary notation to simplify calculations

$$Q = X_0^T X_0$$
$$Y = X_N^T X_N$$
$$\beta - \beta^* = z \Rightarrow \beta = z + \beta^* \quad .$$

We substitute this notation into the final two terms of the exponent in equation to write

$$\left(z + \beta^* - \bar{\beta}_0\right)^T Q \left(z + \beta^* - \bar{\beta}_0\right) + \left(z + \beta^* - \bar{\beta}_N\right)^T Y \left(z + \beta^* - \bar{\beta}_N\right) \quad .$$

Let us begin with the left hand of the two terms

$$\left(z + \beta^* - \bar{\beta}_0\right)^T Q \left(z + \beta^* - \bar{\beta}_0\right) \quad .$$

To help expand this left hand term, we temporarily define

$$a = \beta^* - \beta_0$$

which allows us to write the foregoing left hand term

$$\left(z + a\right)^T Q \left(z + a\right)$$
$$= z^T Q z + 2 z^T Q a + a^T Q a$$
$$= z^T Q z + 2 z^T Q \left(\beta^* - \bar{\beta}_0\right) + \left(\beta^* - \bar{\beta}_0\right)^T Q \left(\beta^* - \bar{\beta}_0\right) \quad .$$

Let us now shift to the right hand of the two terms

$$\left(z + \beta^* - \bar{\beta}_N\right)^T Y \left(z + \beta^* - \bar{\beta}_N\right) \quad .$$

We temporarily define

$$c = \beta^* - \bar{\beta}_N$$

which allows us to write the right hand term

$$\left(z + \beta^* - \bar{\beta}_N\right)^T Y \left(z + \beta^* - \bar{\beta}_N\right)$$
$$= \left(z + c\right)^T Y \left(z + c\right)$$

$$
\begin{aligned}
&= z^T Y z + 2z^T Y c + c^T Y c \\
&= z^T Y z + 2z^T Y \left(\beta^* - \bar{\beta}_N\right) + \left(\beta^* - \bar{\beta}_N\right)^T Y \left(\beta^* - \bar{\beta}_N\right) \quad .
\end{aligned}
$$

Now we combine the left hand term and the right hand term, in anticipation of completing the square, to obtain

$$
\begin{aligned}
&\left(z + \beta^* - \bar{\beta}_0\right)^T Q \left(z + \beta^* - \bar{\beta}_0\right) + \left(z + \beta^* - \bar{\beta}_N\right)^T Y \left(z + \beta^* - \bar{\beta}_N\right) \\
&= z^T \left(Q + Y\right) z \\
&+ 2z^T \left[Q \left(\beta^* - \bar{\beta}_0\right) + Y \left(\beta^* - \bar{\beta}_N\right)\right] \\
&+ \left(\beta^* - \bar{\beta}_0\right)^T Q \left(\beta^* - \bar{\beta}_0\right) + \left(\beta^* - \bar{\beta}_N\right)^T Y \left(\beta^* - \bar{\beta}_N\right) \quad .
\end{aligned}
$$

We complete the square by eliminating the middle term, the one that is multiplied by z^T:

$$
\begin{aligned}
0 &= \left[Q \left(\beta^* - \bar{\beta}_0\right) + Y \left(\beta^* - \bar{\beta}_N\right)\right] \\
&\Rightarrow \left(Q + Y\right) \beta^* = \left(Q\bar{\beta}_0 + Y\bar{\beta}_N\right) \\
&\Rightarrow \beta^* = \left(Q + Y\right)^{-1} \left(Q\bar{\beta}_0 + Y\bar{\beta}_N\right) \\
&\Rightarrow \beta^{*T} = \left(Q\bar{\beta}_0 + Y\bar{\beta}_N\right)^T \left(Q + Y\right)^{-1} \quad .
\end{aligned}
$$

It is worth substituting for Q and Y to get an explicit expression for the posterior mean.

$$
\beta^* = \left(Q + Y\right)^{-1} \left(Q\bar{\beta}_0 + Y\bar{\beta}_N\right) \quad .
$$

We substitute the foregoing expression for β^* into the sum of matrices to obtain a quadratic plus a constant

$$
\begin{aligned}
&\left(z + \beta^* - \bar{\beta}_0\right)^T Q \left(z + \beta^* - \bar{\beta}_0\right) + \left(z + \beta^* - \bar{\beta}_N\right)^T Y \left(z + \beta^* - \bar{\beta}_N\right) \\
&= z^T \left(Q + Y\right) z \\
&+ \left(\beta^* - \bar{\beta}_0\right)^T Q \left(\beta^* - \bar{\beta}_0\right) + \left(\beta^* - \bar{\beta}_N\right)^T Y \left(\beta^* - \bar{\beta}_N\right) \quad .
\end{aligned}
$$

Expanding the two quadratics

$$
\begin{aligned}
&z^T \left(Q + Y\right) z \\
&+ \left(\beta^* - \bar{\beta}_0\right)^T Q \left(\beta^* - \bar{\beta}_0\right) + \left(\beta^* - \bar{\beta}_N\right)^T Y \left(\beta^* - \bar{\beta}_N\right) \\
&= z^T \left(Q + Y\right) z \\
&+ \beta^{*T} Q \beta^* - 2\beta^{*T} Q \bar{\beta}_0 + \bar{\beta}_0^T Q \bar{\beta}_0 \\
&+ \beta^{*T} Y \beta^* - 2\beta^{*T} Y \bar{\beta}_N + \bar{\beta}_N^T Y \bar{\beta}_N \quad .
\end{aligned}
$$

We group terms

$$
\begin{aligned}
& z^T (Q+Y) z \\
& + \beta^{*T} Q \beta^* - 2\beta^{*T} Q \bar{\beta}_0 + \bar{\beta}_0^T Q \bar{\beta}_0 \\
& + \beta^{*T} Y \beta^* - 2\beta^{*T} Y \bar{\beta}_N + \bar{\beta}_N^T Y \bar{\beta}_N \\
& = z^T (Q+Y) z \\
& + \beta^{*T} (Q+Y) \beta^* - 2\beta^{*T} \left(Q\bar{\beta}_0 + Y\bar{\beta}_N\right) \\
& + \bar{\beta}_0^T Q \bar{\beta}_0 + \bar{\beta}_N^T Y \bar{\beta}_N \quad .
\end{aligned}
$$

Substitute for β^* in the third line of the final equality

$$
\begin{aligned}
& z^T (Q+Y) z + \beta^{*T} (Q+Y) \beta^* \\
& - 2\beta^{*T} \left(Q\bar{\beta}_0 + Y\bar{\beta}_N\right) + \bar{\beta}_0^T Q \bar{\beta}_0 + \bar{\beta}_N^T Y \bar{\beta}_N \\
& = z^T (Q+Y) z + \beta^{*T} (Q+Y) (Q+Y)^{-1} \left(Q\bar{\beta}_0 + Y\bar{\beta}_N\right) \\
& - 2\beta^{*T} \left(Q\bar{\beta}_0 + Y\bar{\beta}_N\right) + \bar{\beta}_0^T Q \bar{\beta}_0 + \bar{\beta}_N^T Y \bar{\beta}_N \\
& = z^T (Q+Y) z + \beta^{*T} \left(Q\bar{\beta}_0 + Y\bar{\beta}_N\right) \\
& - 2\beta^{*T} \left(Q\bar{\beta}_0 + Y\bar{\beta}_N\right) + \bar{\beta}_0^T Q \bar{\beta}_0 + \bar{\beta}_N^T Y \bar{\beta}_N \quad .
\end{aligned}
$$

Reduce the two middle terms and simplify to the final expression

$$
\begin{aligned}
& z^T (Q+Y) z + \beta^{*T} \left(Q\bar{\beta}_0 + Y\bar{\beta}_N\right) \\
& - 2\beta^{*T} \left(Q\bar{\beta}_0 + Y\bar{\beta}_N\right) + \bar{\beta}_0^T Q \bar{\beta}_0 + \bar{\beta}_N^T Y \bar{\beta}_N \\
& = z^T (Q+Y) z \\
& + \bar{\beta}_0^T Q \bar{\beta}_0 + \bar{\beta}_N^T Y \bar{\beta}_N - \beta^{*T} \left(Q\bar{\beta}_0 + Y\bar{\beta}_N\right) \\
& = z^T (Q+Y) z \\
& + \left(\bar{\beta}_0 - \beta^*\right)^T Q \bar{\beta}_0 + \left(\bar{\beta}_N - \beta^*\right)^T Y \bar{\beta}_N \quad .
\end{aligned}
$$

In summary, the final two terms in the exponent have reduced to

$$
\begin{aligned}
& z^T (Q+Y) z \\
& + \left(\bar{\beta}_0 - \beta^*\right)^T Q \bar{\beta}_0 + \left(\bar{\beta}_N - \beta^*\right)^T Y \bar{\beta}_N \quad .
\end{aligned}
$$

Substituting for z, we can write

$$
\begin{aligned}
& \left(\beta - \beta^*\right)^T (Q+Y) \left(\beta - \beta^*\right) \\
& + \left(\bar{\beta}_0 - \beta^*\right)^T Q \bar{\beta}_0 + \left(\bar{\beta}_N - \beta^*\right)^T Y \bar{\beta}_N \quad .
\end{aligned}
$$

Therefore, the final exponent in equation (10.17) contains as its final two terms the foregoing expression plus the other two terms, and the final exponent is written

$$exponent = (\beta - \beta^*)^T (Q + Y)(\beta - \beta^*) + R_0 + R_N$$
$$+ (\bar{\beta}_0 - \beta^*)^T Q\bar{\beta}_0 + (\bar{\beta}_N - \beta^*)^T Y\bar{\beta}_N \quad .$$

We now substitute Q, Y, and z to reinsert the original terms

$$exponent = (\beta - \beta^*)^T (Q + Y)(\beta - \beta^*) + R_0 + R_N$$
$$+ (\bar{\beta}_0 - \beta^*)^T Q\bar{\beta}_0 + (\bar{\beta}_N - \beta^*)^T Y\bar{\beta}_N$$
$$= (\beta - \beta^*)^T \left(X_0^T X_0 + X_N^T X_N\right)(\beta - \beta^*) + R_0 + R_N$$
$$+ (\bar{\beta}_0 - \beta^*)^T X_0^T X_0\bar{\beta}_0 + (\bar{\beta}_N - \beta^*)^T X_N^T X_N\bar{\beta}_N \quad .$$

We substitute for the regression means for both the old and the new data

$$exponent = (\beta - \beta^*)^T \left(X_0^T X_0 + X_N^T X_N\right)(\beta - \beta^*)$$
$$+ R_0 + (\bar{\beta}_0 - \beta^*)^T X_0^T y_0$$
$$+ R_N + (\bar{\beta}_N - \beta^*)^T X_N y_N \quad .$$

The exponent shows the quadratic term, and it shows the constant term, which is the A term. We rewrite just the constant term here for future reference:

$$A = R_0 + R_N + (\bar{\beta}_0 - \beta^*)^T X_0^T y_0 + (\bar{\beta}_N - \beta^*)^T X_N y_N \quad .$$

Let us now substitute for R_0 from equation (10.7) and for R_N from equation (10.15)

$$A = \left[y_0^T y_0 - y_0^T X_0 (X_0^T X_0)^{-1} X_0^T y_0\right]$$
$$+ \left[y_N^T y_N - y_N^T X_N (X_N^T X_N)^{-1} X_N^T y_N\right]$$
$$+ (\bar{\beta}_0 - \beta^*)^T X_0^T y_0 + (\bar{\beta}_N - \beta^*)^T X_N^T y_N \quad .$$

We now substitute for β_0^T from equation (10.6) and for β_N^T from equation (10.14) to write

$$A = y_0^T y_0 + y_N^T y_N - \bar{\beta}_0^T X_0^T y_0 - \bar{\beta}_N^T X_N^T y_N$$
$$+ (\bar{\beta}_0 - \beta^*)^T X_0^T y_0 + (\bar{\beta}_N - \beta^*)^T X_N^T y_N \quad .$$

We see that the term $\bar{\beta}_0^T X_0^T y_0$ occurs twice, and the term $\bar{\beta}_N^T X_N^T y_N$ occurs twice. The two occurrences of each cancel, so this equation reduces to the much simpler final form

$$A = y_0^T y_0 + y_N^T y_N - \beta^{*T} \left(X_0^T y_0 + X_N^T y_N\right) \quad . \tag{10.18}$$

We almost have the solution at hand and we can appeal to the equations in Chapter 3 characterizing the posterior.

- Equation (3.10) characterizes the posterior precision matrix

$$X_N^T X_N + X_0^T X_0 \quad .$$

Term by term comparison shows this to be identical to equation (10.1), which gave the precision matrix for the aggregated old plus new data.

- Equation (3.9) characterizes the posterior mean values of β

$$\begin{aligned}\beta^* &= \left(X_N^T X_N + X_0^T X_0\right)^{-1}\left(X_N^T y_N + X_0^T X_0 \beta_0\right) \\ &= \left(X_N^T X_N + X_0^T X_0\right)^{-1}\left(X_N^T y_N + X_0^T y_0\right) \quad .\end{aligned} \tag{10.19}$$

Term by term comparison shows this to be identical to equation (10.2), which gave the mean value of the coefficients for the aggregated old plus new data set. This means that the mean β^* (achieved by updating the old data set using the new data set) seen in the immediately preceding equation and $\bar{\beta}$ from equation (10.2) are absolutely identical. The single analysis of the old plus new data set is identical to the classical regression analysis of the old data set followed by a Bayesian update using the new data set as far as the mean values of coefficients is concerned. It is thus confirmed that the mean is the same, i.e., $\beta^* = \bar{\beta}$.

- Equation (3.7), which characterizes the posterior degrees of freedom, which is

$$\begin{aligned}\nu &= n_0 + n_1 - p \\ \nu + p + 1 &= n_0 + n_1 + 1 \quad .\end{aligned}$$

- Equation (10.18), which characterizes the posterior value of the residual sum of squared error term A will be shown to be equal to $\Re$.
- Equation (6.5), which characterizes the posterior value of the s^2 term

$$s^2 = \frac{A}{\nu} \quad .$$

We want to begin with these general posterior equations and insert into these equations the correct specification of the prior developed in Section 8.3.

We have the solution at hand. First, notice that β^* developed here in equation (10.19) is precisely identical to the mean developed for the aggregate of the old and new data based in equation (10.2). That means that the mean value of the coefficients of the classical statistical analysis that aggregates both the old and new data bases into a single large data base is exactly identical to the mean value of the coefficients if you first use the old data base to create a prior and then update that prior using the new data base. The Bayesian update procedure gives the same answer for the mean values of the coefficients as does the direct analysis of the entire data base.

And now comes the crucial insight. If we compare this value of A in equation (10.18) with the value of $\Re$ in equation (10.3), we see that they too are identical. We have therefore shown that the sum of squared errors term is identical for

- doing classical regression solely on the old data base and thereafter doing a Bayesian update of that old data base using the new data base
- directly aggregating the old plus new data bases into a single large data base and doing a classical regression on the whole thing.

The posterior density in either case is

$$\{\beta,\sigma\} = \frac{2\left|X_N^T X_N + X_0^T X_0\right|^{\frac{1}{2}} (2\pi)^{-\frac{p}{2}} \left(\frac{A}{2}\right)^{\frac{\nu}{2}}}{\Gamma\left(\frac{\nu}{2}\right)} \times \sigma^{-(\nu+p+1)} e^{-\frac{A}{2\sigma^2}} e^{-\frac{1}{2\sigma^2}(\beta-\bar{\beta})^T (X_N^T X_N + X_0^T X_0)(\beta-\bar{\beta})} \quad .$$

And so far, we have shown that the degrees of freedom, the exponent of σ, and the exponent of the error term are identical. We have shown that the posterior mean and the precision matrix are identical. Continuing down this path, for the Bayesian posterior, we have the specific parametrization:

$$\begin{aligned}
g &= A = {y_0}^T y_0 + {y_N}^T y_N \\
&\quad - \left({X_0}^T y_0 + {X_N}^T y_N\right)^T \left({X_0}^T X_0 + {X_N}^T X_N\right)^{-1} \left({X_0}^T y_0 + {X_N}^T y_N\right) \\
H &= {X_0}^T X_0 + {X_N}^T X_N \\
\mu &= \bar{\beta} = ({X_0}^T X_0 + {X_N}^T X_N)^{-1} ({X_0}^T y_0 + {X_N}^T y_N) \\
r &= n_0 + n_N + 1 \\
\nu &= \nu_0 + n_N \quad .
\end{aligned}$$

In summary, if you have an "old" data base and a "new" data base, the classical regression solution to the combined, aggregated data base is identical to the posterior of the Bayesian Linear Regression solution, with the old data serving as a prior and the new data serving to update the prior to the posterior. How could it be otherwise (although challenging to prove)? There is precisely and exactly the same amount of information resident within the two data sets collectively: the classical analysis of the old plus the new data base and the Bayesian analysis that starts with a prior from the old data base and updates it with the new data base. How could the answer be otherwise?[2]

[2]This is not a happy message for things like Empirical Bayes because is says that no matter

10.6 Serially Repeated Update

So what? We have merely shown that you can append together all the old data and all the new data. You can thereafter run a normal regression on the whole aggregate data base and get the same answer as you would have gotten with a mathematically sophisticated Bayesian update. Where is the value, therefore, in Bayes? You get the same answer if you just append together a mega-data set and start from scratch. Why all the complexity? Why would you care a whit about Bayesian updating? Why don't you just get the new numbers coming in, graft them to the end of the old numbers, do a full classical regression solution on the whole data set, and go drink coffee?

This is a rather naive viewpoint for a number of reasons that will be made clear herein. First, and frequently true in practice, what if your old data set was a terabyte? What if you had 2 million rows times 10 columns of old data, gathered over a significant period of time? Why would you want to keep running it over and over again every time new data comes in? What truly is a "sufficient statistic" for a prior based on this old data set? What, in fact, is the minimal set of numbers required to fully characterize this prior terabyte data set? Do you even need all the individual historical data gathered thus far, or are there results you have already calculated that fully characterize this historical data set? Is the posterior coming out of this historical data per se enough to fully represent it? We shall see that the answer to this last question is a definitive yes. How would you calculate the incremental value of new data?

To answer the question, if the prior were derived from a historical data set of length n_0, that prior would have the specific form

$$\{\beta, \sigma\} = \frac{2\left|X_0^T X_0\right|^{\frac{1}{2}} (2\pi)^{-\frac{p}{2}} \left(\frac{\nu_0 s_0{}^2}{2}\right)^{\frac{\nu_0}{2}}}{\Gamma\left(\frac{\nu_0}{2}\right)} \times \sigma^{-(\nu_0+p+1)} e^{-\frac{\nu_0 s_0{}^2}{2\sigma^2}} e^{-\frac{1}{2\sigma^2}(\beta-\beta_0)^T (X_0^T X_0)(\beta-\beta_0)} \quad .$$

The specific parameters it takes to fully characterize that prior are

- n_0 = number of data points that comprise the prior
- p = number of model coefficients. This is a fixed parameter of the problem
- ν_0 = degrees of freedom = $n_0 - p$
- s_0^2, which is derived from the relationship $R_0 = \frac{\nu_0 s_0^2}{\nu_0 - 2}$ in which R_0 is the residual sum of squared errors
- β_0 = mean of coefficients
- $X_0^T X_0$ = square of data matrix.

how you sub-group the data set, you get the SAME answer at the end. This of course has only been proven for the linear case herein.

What is pivotal to notice is that these six parameters comprise a "sufficient statistic" for the entire distribution itself and in fact for the entire data base itself. (The parameter n_0 is implicit in the degrees of freedom, but we have included it explicitly anyway even though it is redundant.) You don't NEED the massive historical data set; you can toss it away! All you need are the final five of the foregoing six parameters, sort of "running totals" for those terms. That is a full and complete characterization of the historical database, no matter how large it is. Equation (5.1) was very clear about the parametrization of the prior; it was very clear that the "sufficient statistic," i.e., the full parametrization, of the prior consists solely of p, ν_0, s_0^2, β_0, and $X_0^T X_0$.

So how does this work in real life? Figure 10.3 begins to show how. As shown, we begin with an "Initial Data Set" designated I_0. This Initial Data Set I_0 is comprised of old, initial data (or could be a purely subjective prior, which we shall ignore for now) that initiates the analysis. The initiating analysis on the Initial Data Set I_0 is a "Bayesian-with-Diffuse-Prior" calculation, which we term "Classical" for brevity in light of the equivalence proven previously. We see coming out of the right a black arrow with the designation a. This arrow represents the result of "Bayesian-with-Diffuse-Prior" analysis, which we have equated with "Classical Statistics," on the Initial Data Set I_0. Specifically, the information on the black arrow designated a is the full posterior density, complete with all its "sufficient statistics," calculated from the Initial Data Set I_0 using the "Bayesian-with-Diffuse-Prior" classical method. Make no mistake, it is the posterior density, complete with all its requisite parameters enumerated above. The figure indicates that the density at a is the result of "Bayesian-with-Diffuse-Prior" classical regression on the Initial Data Set I_0. All very intuitive. In summary, a is the sufficient statistic resulting from the diffuse prior and the data I_0.

Figure 10.3: Bayesian-with-Diffuse-Prior Classical Analysis of I_0

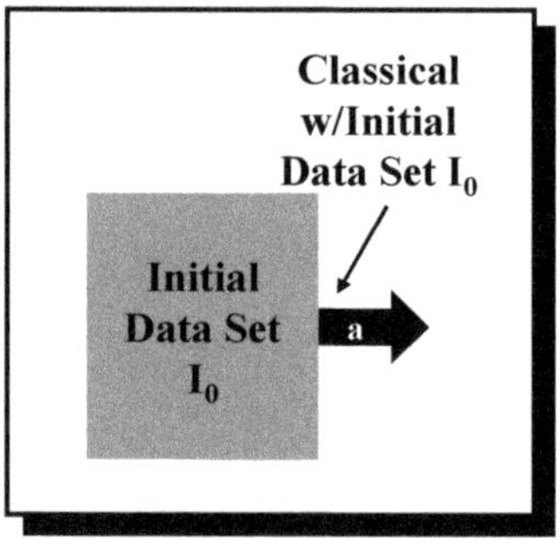

In preparation for further calculations represented by further figures, we append at the top of Figure 10.3 to form Figure 10.4 that the running cumulative total set of updates and calculations thus far is equivalent to the Bayesian-with-Diffuse-Prior or Classical calculation using the Initial Data Set I_0 but no additional data.

Figure 10.4: Bayesian-with-Diffuse-Prior Classical Posterior Onto Arrow a

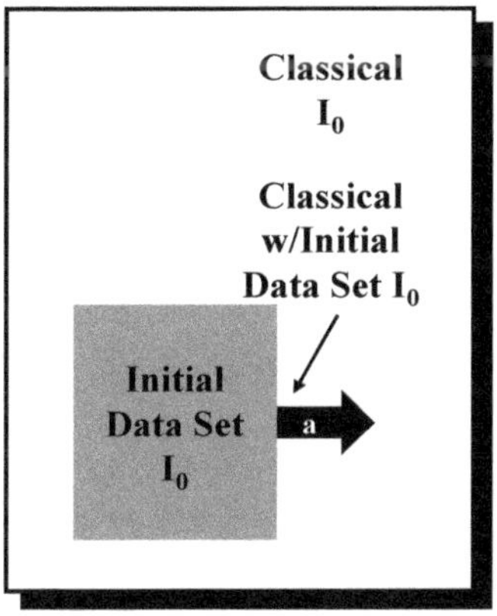

What do we do with the full posterior density, complete with all its "sufficient statistics," calculated from the Initial Data Set I_0? Figure 10.5 emphasizes that this is the prior that will be used for the next, subsequent Bayesian update. In particular, it will be the prior used for the first level of update, which will be "Data 1." Conceptually, the Initial Data Set I_0 was already in place, and Data 1 just came in today, and we want to incorporate Data 1 into our analysis immediately, right as the new data comes in.

Figure 10.5: Bayesian-with-Diffuse-Prior Classical As Prior into Data 1

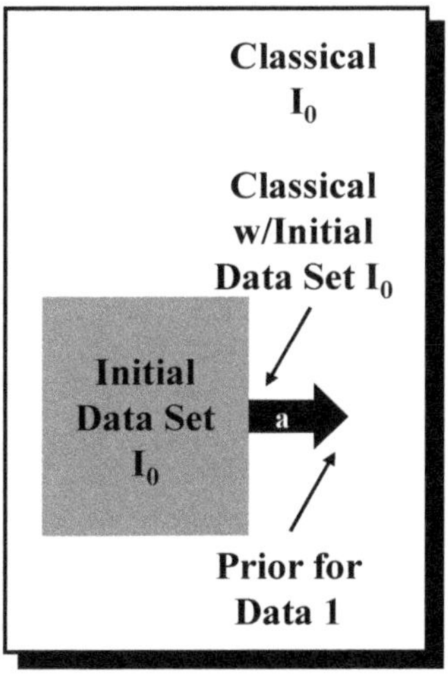

Figure 10.6 begins the process of updating the prior density a using the newly arriving data, Data 1. As indicated in the figure, the full density on the arrow at a serves as the Bayesian prior. The just-arrived data set is designated Data

1. The diagram clearly shows that the prior is the full probability distribution a and the data used for the Bayesian update is Data 1. The prior (or more properly, the sufficient statistic a characterizing the prior) contains all relevant information from the data set I_0. It is not necessary to carry along the Initial Data Set I_0 per se any further. The "sufficient statistic" is resident within the distribution at a. The Bayesian posterior thus created is represented by the black arrow exiting from the right and designated "b." To be clear, the Bayesian posterior was created using the prior a as a sufficient statistic, and the data in Data 1 is represented by the arrow b. In short, a updated by Data 1 leads to b. The Bayesian posterior is placed on the arrow b. Figure 10.6 makes this clear. To summarize, b represents the sufficient statistic for the posterior developed from the prior a and the data set Data1.

Figure 10.6: Bayesian-with-Diffuse-Prior Classical Analysis of I_0 Plus Data 1

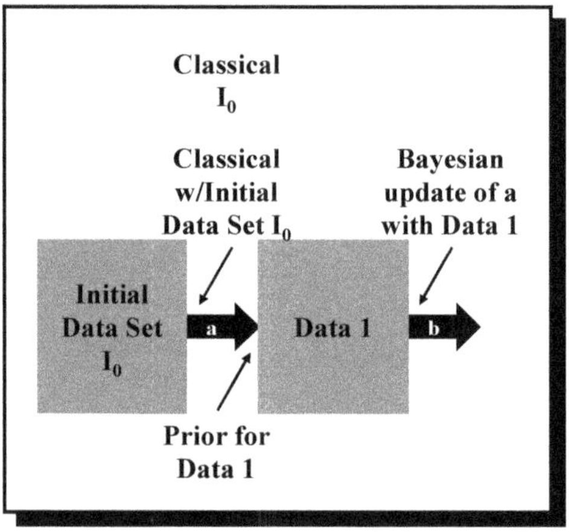

Proven in the previous section, and imminently insightful here, is that the posterior at arrow b is precisely and exactly what you would get if you ran the "Bayesian-with-Diffuse-Prior" analysis using as input the aggregate of the Initial Data Set I_0 PLUS the new data set 1. It is as if you never did a Bayesian update at all but rather ran classical statistics on the combined data sets I_0 plus Data 1 that you had received by time 1. You could merely graft the new data set Data 1 onto the end of the Initial Data set I_0, run classical regression, and obtain the posterior at b. That fact is emphasized in Figure 10.7. The heading at the top right indicates that the full posterior density at b is precisely and exactly equal to the posterior density you would have gotten if you had run classical analysis on the combined data set I_0 plus Data 1 without any appeal to Bayes. And yet with Bayes, you have calculated that posterior in two steps: classically on the Initial Data Set I_0 to give the posterior at a, followed by a Bayesian update using that posterior as a prior to be updated by the new data set Data 1. No difference at all at b.

Figure 10.7: Bayesian-with-Diffuse-Prior Classical Analysis of I_0 Plus Data 1 Onto Arrow b

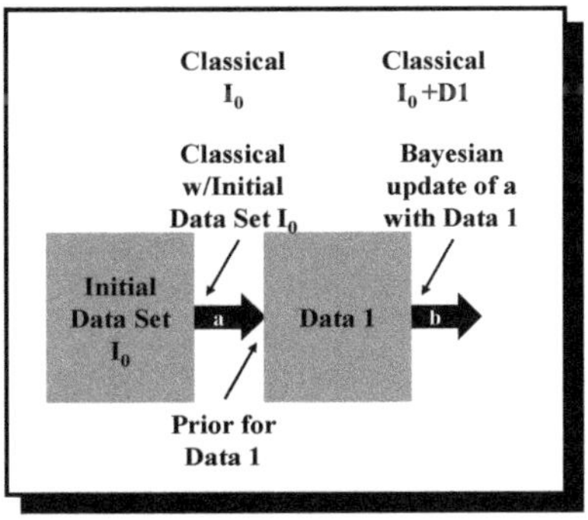

In light of all this, why would you even consider serial Bayesian updating? The reason is that the Bayesian update in the Data 1 box is trivial. You do not have to return to the Initial Data Set I_0 at all. You have everything that was resident within that data set embedded in the distribution at a. The parameters of the probability distribution at a are a "sufficient statistic" for the entire Initial Data Set I_0. If that Initial Data Set I_0 had a million rows and ten parameters, you would no longer need it at all once you reached point a. All you would need is the distribution at a calculated from it. The Bayesian update across the Data 1 box merely needs the new data in Data 1. Updating the precision matrix and all the other sufficient statistics is trivial and blisteringly fast, much faster than retreating to the entire historical data base I_0 plus Data 1 and doing a classical regression on the whole aggregate data set. No need!

What do we do with the full posterior density, complete with all its "sufficient statistics," calculated from the Initial Data Set I_0 plus Data 1 and associated with the black arrow at "b". Figure 10.8 emphasizes that this is the prior that will be used for subsequent Bayesian updates for the second and subsequent incoming data sets. In particular, the full distribution at b will serve as the prior used for the second Bayesian update, which will use the newly incoming set of data, "Data 2." Conceptually, the Initial Data Set I_0 and Data 1 have already been fully processed and embedded in the density at b, and Data 2 has just been reported to us. We want to incorporate Data 2 into our analysis as soon as it comes in. We see the caption immediately under the arrow coming out of the Data 1 box designated "Prior for Data 2." This indicates that the posterior at a will be used as the prior for a Bayesian update using the newly arriving data set Data 2, designated "Data 2" in this and subsequent diagrams. These are the data that are gathered today, new data above and beyond the Initial Data Set I_0 and Data 1. New data is coming in over time, and Data 2 represents the new data that has come in most recently. This is very typical of serial data-gathering activities. Data comes in "bundles" over a protracted period of time, and we want to be completely up to date every minute with respect to ALL the data to date that has ever come in. We want to update our posterior with every bundle of data that comes in.

Figure 10.8: Bayesian-with-Diffuse-Prior Classical Analysis of I_0 Plus Data 1 As Prior into Data 2

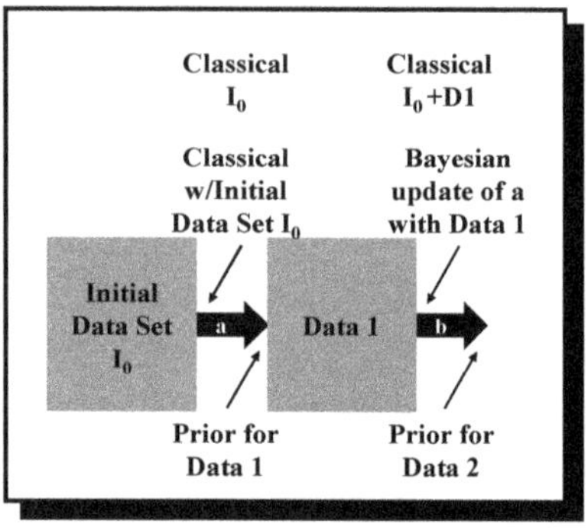

The pattern is becoming clear, as we see in the serial diagram in Figure 10.9. We see Figure 10.8 as the leftmost two blocks in Figure 10.9. Continuing the previous logic, we see the probability distribution at b serving as a prior to be updated by Data 2. We see the posterior from this update at c. We see the probability distribution at c serving as a prior to being updated by Data 3. We see the posterior from this update at d. We see the probability distribution at d serving as a prior to being updated by Data 4. We see the posterior from this update at e. And so forth it goes as long as data keeps coming in.

Figure 10.9: Serial Bayesian Updating

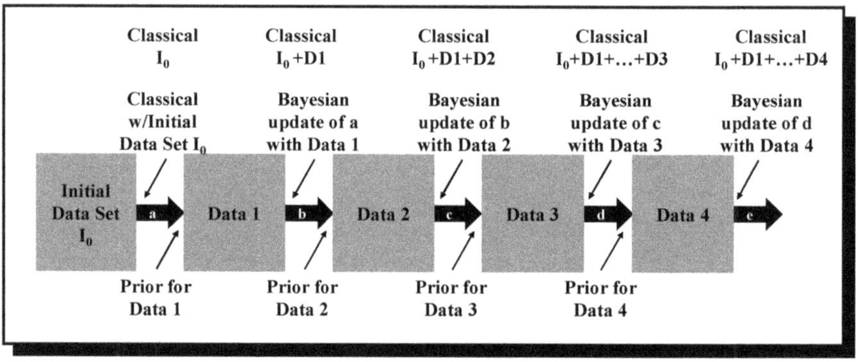

Importantly, we see the distribution at c as the "Bayesian-with-Diffuse-Prior" classical posterior for the data set that has been gathered up to that time, namely I_0 + Data 1. By similar inductive reasoning and proof, the distribution at c is the "Bayesian-with-Diffuse-Prior" Classical posterior for the data set that has been gathered up to that time but not beyond, namely I_0 + Data 1 + Data 2. The distribution at d is the "Bayesian-with-Diffuse-Prior" classical posterior for the data set that has been gathered up to that time but not beyond, namely I_0 + Data 1 + Data 2 + Data 3. And so forth it goes. We proved in the previous section that a Bayesian posterior is the Bayesian update of the entering prior using the most recent data set. The implications are actually quite profound. Every single step of the way is exactly consistent with a "Bayesian with diffuse prior" classical analysis of all the data gathered cumulatively to date. The step-by-step Bayesian update along the way gives us exactly this answer.

However, the Bayesian update all along the way gives you something much more valuable. It gives you the ability to calculate the incremental impact, or incremental value, of each additional data bundle that comes in. You see explicitly and overtly what the incremental value of each additional data bundle is! You can compare the posterior at each stage with the prior to see how much difference the bundle of data at that point in the sequence makes.

This is exactly akin to the Acme Rain Detector in Howard and Abbas[45], which shows the incremental value of each successive reading of an imperfect predictor or detector (not unlike imperfect reading of data at each step). Each successive reading of the Acme Rain Detector updates the aggregate sum of all the previous readings. This has been illustrated quite profoundly by Howard and Abbas[45]—the incremental value of each successive reading generally drops, not increases! The marginal value of successive data actually decreases because it does not change decisions. More data may be available, but generally, the marginal value of each new element of data decreases, not increases. More data may be better, but not at the margin! You have to pay for incremental new data at the margin, often about the same for each incremental burst of data at every point along the way, and the value of marginal new data tends to decrease.[3]

The second benefit of the serial Bayesian updating approach is that you have to process the Initial Data Set I_0 and each incremental data set (Data 1, Data 2, Data 3, and so forth) once and only once. Once you have processed the Initial Data Set I_0 once, you need never process it again. You merely save its "sufficient statistics" in the form of its posterior and thereafter use that posterior in the next subsequent Bayesian update. In fact, you need only use each successive "sufficient static" once as a prior in a subsequent data bundle. Once used, even the interim distributions can be discarded and folded as they are into the next successive interim posterior, which in turn becomes an interim prior in the next data bundle further downstream. The beauty of Bayes is that the update at each step of the diagram is computationally trivial. There is virtually no work or computation at all, and the interim data bundle Data i is generally very, very small relative to the aggregate size of all the data in the Initial Data Set I_0 plus all incoming bundles along the way up to and including $i - 1$. You know exactly

[3]The Acme Rain Detector in Howard and Abbas[45] really illustrates this phenomenon nicely. The value of new incremental data drops the more old data you have. If you knew the cost of new data acquisition, you could gather data until the marginal value of a unit of data dropped to be equal to the marginal cost of that new unit of data, at which point you would stop gathering more data, make the decision, and move on.

where you stand at every step of the way.

The results of this section help us challenge the ubiquitous old refrain: "We need more data." Using the successive Bayesian updates herein, we are able to carefully envision and quantify what each incremental bundle of data might be worth in terms of changing the posterior distribution or changing the decision. It is the posterior distribution we use to make decisions; it is NOT only the variance of the posterior distribution! If we change it and make a different decision, then the additional data has substantial, material value. If a bundle of incoming data changes the posterior to the degree that we would make a different decision, then the new bundle is valuable. If it does not change the decision, i.e., if we make the same decision under the new posterior, then the incoming bundle of data isn't worth anything. This notion of the value of imperfect information, as personified by Howard and Abbas[45] Acme Rain Detector, is profoundly important here. The old refrain, "We need more data!" can be questioned and guided. In fact, in practice, the "We need more data" refrain is often just flat wrong.

If new data does not lead to a different decision, then, according to decision theory, it has no value. This is a primal tenet of Decision Analysis, yet not a tenet of regression analysis. Howard and Abbas[45] cite numerous examples in which more information is worth nothing or next to nothing. The point of gathering data is not to achieve certainty or near certainty. The point of gathering data is to gather and process enough so that further data doesn't change any decision one might make. The decision drives the value, not the lack of uncertainty.

This result should not be particularly surprising in retrospect. The amount of data brought to the problem in the Bayesian case is no different from the amount of information that is brought to the problem in the classical case. And yet the Bayesian technique allows you to process it serially, a little bit at a time, and not have to return and reprocess the entire data set. But it had to be proven, as it has been here by induction. Classical doesn't necessarily admit of an updating process.

There is no hope of successive updates if classical regression is selected. Bayes gives you the benefit of incrementally processing each new data set rather than going back to the entire data set every time even one new piece of data is introduced.

10.7 A Note on "Empirical Bayes"

What if we took any subset m of the original n observations and regarded those m observations as "prior?" What if, thereafter, we regarded the remaining $n - m$ observations as the observed set? By so doing, we obtain what is called "empirical Bayes." This is what empirical Bayesians do in the simple linear regression case.

If we were to do so, the first m observations would comprise what we have called a data prior in the sense we have just analyzed in the previous sections of this chapter. The remaining $n - m$ observations would comprise the data. To wit, if we

- Used m of the observations to create a prior and its sufficient statistics
- Updated that prior using the remaining $n - m$ observations

we would get exactly the same answer as if we had simply run Bayesian Linear Regression using a diffuse prior (classical regression) using the entire data set containing all n observations. There would be no difference (or benefit) to using part of a data set to comprise a prior. This may not generally be true for cases other than the linear regression case may be a waste of time—it does not change the final answer, nor does it give the analyst any hint about any particular prior given the results of this section. It doesn't relieve the analyst from formulating a prior for the whole shootin' match, or choose a diffuse prior for the whole shootin' match and be done with it! In the linear regression case, the only prior that changes the answer, i.e., brings different information to the problem, is a purely subjective, external prior that is NOT based on a data set. If it IS based on a prior data set, one can and should simply append that prior data set to the observed data set and get exactly the same answer. We have proven that as soon as you try to comprise your prior from a second data set, you have de facto merely appended that data set to the observed data set. There is no information in a data set that resides outside that data set.

So what is this empirical Bayes attempting to do? We start with the definition in Wikipedia:[4] *Empirical Bayes methods are procedures for statistical inference in which the prior probability distribution is estimated from the data. This approach stands in contrast to standard Bayesian methods, for which the prior distribution is fixed before any data are observed.* We have proven earlier in this chapter that empirical Bayes can be a non sequitur in the simple linear regression case, unnecessary and unrevealing with the linear simple regression model. What is the point of it? One merely appends the old data to the new data and moves forward with correct analysis using the standard Bayesian or classical linear method. If there are two or more old data sets, we append them individually to the new, observed data set and get on with it.

[4]https://en.wikipedia.org/wiki/Empirical_Bayes_method.

Chapter 11

EXAMPLE–MIGHT IQ OR HOURS STUDIED-IN-ADVANCE MATTER MORE TO TEST SCORE?

You mean your statistics are facts, but my facts are just statistics?

Paul Askew

We are going to explore a very simple example in this section, one that is nicely illustrative of how one might use Bayesian statistics. Suppose we were going to gather a database to explore what is more important to a test score: IQ or cumulative hours spent studying in advance of the test. Is "being smart," expressed using IQ (assuming that the notion of IQ is an acceptable measure of cognitive ability), the most important predictor of test score? Is "cumulative hard work in advance" the most important predictor of test scores? Problems of this type have been around in many forms over many years, and there is likely no definitive answer yet. The present example is designed to illustrate how one might start to address such a question using Bayesian Linear Regression. We will use caricatures of conversations in this analysis, conversations people often have about these types of statistical problems, to render the discussion more active.[1]

It seems a physics professor (whom we shall abbreviate by the acronym PP) quipped, "You know, it doesn't matter much how many hours you study

[1] As we will see in this example, the definitions and implications of hours studied and IQ may not be as distinct as one might think.

for my physics test. Physics is hard. Only people with high IQs can possibly do well. People without high IQs, i.e., without high cognition, are pretty much condemned to low scores in physics, i.e., to mediocre performance." We have probably all heard variants of this utterance from both technical and lay people. The argument is generally unsupported, so our Bayesian statistical researcher (abbreviated by the acronym BSR) sets out to see if such a statement might be supported by observations of the performance of various physics students on the final exam given during the past several years.

PP expects that his observation of performance over a number of physics finals during the past few years would support his conjecture, thinking, "I have been teaching physics here for twenty years, and I have seen this phenomenon borne out repeatedly—high IQ students do well, and lower IQ students don't do as well no matter how much they study." BSR's job is not to challenge that statement but rather to convince PP to tell him ex ante how he thinks the study will come out, i.e., what he thinks the ultimate contribution of IQ versus ex ante study hours will turn out to be. In the lingo of Bayesian analysis, BSR's job is to get PP to provide him or her with an ex ante opinion in the form of a Bayesian prior. Give BSR a prior! This is a perfectly natural thing to do, done implicitly or explicitly in virtually every statistical analysis we have witnessed. The researcher invariably has a judgment about how the study will come out. Bayes allows him or her to input that ex ante judgment in the form of a prior and see what it might imply! It allows him or her to put in the prior judgment of anyone they might want and to see to what degree the data either conforms with or refutes that prior judgment.

The head of the Psychology department, a kind and gentle person for whom we use the acronym KGP, is weary of hearing the refrain from people in technical disciplines that "you have to be smarter to do what we do." KGP thinks such assertions are self-serving and generally unsupported. In contrast to PP's view, KGP believes that everyone who comes to her university is fully and equally capable of mastering the material and scoring well on key exams, the big differentiator being how many cumulative hours they are willing to invest in studying and preparation. She often echoes the quote from Thomas Edison: "Genius is 1 percent inspiration and 99 percent perspiration." She encourages students to study and study and study to command the material, and if they do so, she tells them they can expect to perform at the top of the class. She discounts the effect of IQ on performance for those who study assiduously. She is so adamant that she too agrees to give BSR her prior judgment on how much she weights IQ versus hours of study in predicting student performance on the final. She asks to be on the management team for the project and have oversight, to which PP and BSR gladly agree. There is no attempt to conceal or bias anything here.

Extrapolated to similar problems in the real world, PP, like many senior managers, wants to hire a data-gathering statistical person to get some relevant data on the problem, analyze it, and bring PP back the answer, which he expects to be exactly the same answer he thought he would get going in! PP's motives may not be entirely objective, but he agrees to participate with BSR and KGP openly and completely, confident that his judgment will be borne out by the gathered data! Has anyone ever heard this refrain before? Of course, it is rather ubiquitous: "The data will support what I am advocating!"

PP elects to start with a simple, small, easily obtainable database. The procedure in this example represents a pilot program designed to generate interesting results and garner interest in an expanded data analysis of the issue of IQ or preparation. Not wanting to spend a lot of money on the pilot program, PP and BSR have agreed to use a data set that PP has gathered from his previous four years of physics finals in his class. There were ten student observations for those classes, which included the test score, the IQ, and cumulative hours worked studying for the test as provided by those ten students themselves. A great place to start! How many times do we see pilot programs get started with small, simple, uncontrolled data sets such as this? Frequently. But how much adverse selection is embedded in that database? What if anything was omitted?

Make no mistake, this is a typical setting for statistics problems in industry, government, and academia. A would-be decision maker wants to catalyze a major study or make an imminent decision, so he or she starts with a small, self-gathered data set to establish preliminary values and gain preliminary insights. At such a preliminary stage, BSR cannot be sure of the veracity, accuracy, or even the definition of the variables observed in his data set. However, BSR wants to examine the analysis of this initial data set using Bayesian methods to see if cogent insights or decisions might emerge. Keep in mind that BSR won't know a whole lot about the database; it will have been gathered from PP's records with minimal resource expended. BSR (and PP) want to see what an easily gathered, easily available data set might imply. From small starts grow great ideas, PP and BSR reason. The data set BSR gathers from PP is that in Table 11.1. There are two independent variables: IQ and Hours Studied. PP has taken the former from the students' GRE scores and the hours spent studying by asking the students at the time of the test.[2]

11.1 The Classical Regression Solution

Both PP and KGP start out the time-tested way. They don't use Bayesian statistics. They never have! They never bothered to learn, so they know not what they don't know (perhaps)! They default to classical statistics. BSR begins in the standard fashion by calculating the matrix X^TX, which appears in Table 11.2. BSR then dutifully inverts it as in Table 11.3 because BSR knows the variance-covariance matrix and the classical regression solution will require this inverse.

BSR then assembles the key parameters of the classical regression analysis (which we know in this textbook is Bayesian analysis with a diffuse prior) and

[2]Of course, there may be omitted variables here. There are potentially other variables that may be contributory to high or low test scores. We are not going to address omitted variables here.

Table 11.1: Data Set for the IQ Versus Study Problem

Student	Test score	Intercept	IQ	Study Hours
1	100	1	125	30
2	95	1	104	40
3	92	1	110	25
4	90	1	105	20
5	85	1	100	20
6	80	1	100	20
7	78	1	95	15
8	75	1	95	10
9	72	1	85	0
10	65	1	90	5

Table 11.2: X^TX for the IQ Versus Study Problem

X^TX	Intercept	IQ	Study Hours
Intercept	10	1009	185
IQ	1009	102941	19585
Study Hours	185	19585	4675

Table 11.3: $(X^TX)^{-1}$ for the IQ Versus Study Problem

$(X^TX)^{-1}$	Intercept	IQ	Study Hours
Intercept	16.97799	-0.19013	0.12466
IQ	-0.19013	0.00218	-0.00160
Study Hours	0.12466	-0.00160	0.00197

arrays them in Table 11.4. When we say "classical regression" throughout this book, we specifically mean "Bayesian Linear Regression with diffuse prior." We will make some plots that better illustrate this solution shortly.

The variance-covariance matrix and the key t-statistics results for the classical regression solution are then calculated in Table 11.5,[3] which completes the traditional classical regression analysis. This is what comes out of Excel (almost) when you use the linear regression add-in. But we know that waiting in the wings is enough information to calculate the full and correct posterior distribution based on the diffuse prior—the classical regression solution.

Based on their t-statistics in Table 11.5, the IQ and Hours of Study distributions would seem to be rather broadly distributed around their means. We will see that this is indeed the case. The data set is sufficiently small and non-definitive that we just can't be too sure of these coefficients' values based simply on data assembled. If BSR didn't move on to Bayes, BSR might say that the re-

[3]The t-statistic is defined as the mean divided by the standard deviation.

Table 11.4: Key Calculated Parameters of the IQ Versus Study Problem

	$\bar{B}$	$X^T y$
Intercept	23.156141	832
IQ	0.509434	84955
Study hours	0.467134	16445
n	10	
p	3	
ν_C	7	
$R = \nu_C s_C^2$	105.1158061	
$s_C^2 = \frac{R}{\nu_C}$	15.01654373	
$\langle \sigma^2 \rangle = \frac{\nu_C s_C^2}{\nu_C - 2}$	21.02316122	
$\frac{2}{\nu_C - 4}\left(s_C^2 \frac{\nu_C}{\nu_C - 2}\right)^2$	294.6488719	
Std. Dev. Error	17.16533926	
t-statistic Error	1.224744871	

Table 11.5: The Variance-Covariance Matrix and t-statistics of the Classical Regression Result

VCV	Intercept	IQ	Study hours	
Intercept	356.931119	-3.997147	2.620719	
IQ	-3.997147	0.045769	-0.033564	
Study hours	2.620719	-0.033564	0.041398	
Key Results	$\bar{\beta}$	Variance	Std. Dev.	t-statistic
Intercept	23.156141	356.931119	18.892621	1.225671
IQ	0.509434	0.045769	0.213937	2.381238
Study hours	0.467134	0.041398	0.203466	2.295880

sults are non-definitive now and say that "we need more data!" The old refrain!

Happily, before we sing the old refrain, we can calculate the entire marginal probability distribution over each coefficient expressed analytically. We turn to the univariate marginal Student's t distribution for each of the three variables: intercept, IQ, and Hours Studied. We see in Table 11.6 at the top the calculation of $(X^TX)^{-1}$. In the middle of the table, we pull the diagonals off of the calculation of $(X^TX)^{-1}$. Finally, we multiply by $\nu_0 s_0^2$ to obtain the precision matrix of the univariate marginal distribution. The precision term, which occurs in the denominator of the univariate Student's t distribution, appears in the second to last position in Table 11.6. We note there is a different precision term for each of the three variables: intercept, IQ, and Hours of Study.

Table 11.6: Calculation of the Univariate Precision Term in the Univariate Student's t Distribution

		$-\frac{\nu_C+1}{2}$	-4
$(X^TX)^{-1}$	Intercept	IQ	Study hours
Intercept	16.977995	-0.190131	0.124659
IQ	-0.190131	0.002177	-0.001597
Study hours	0.124659	-0.001597	0.001969
	Intercept	IQ	Study hours
$z_{ii} = \left[(X^TX)^{-1}\right]_{ii}$	16.97799466	0.002177068	0.001969182
Precision	Intercept	IQ	Study hours
$z_{ii}\nu_C s_C{}^2$	1784.655595	0.22884428	0.206992164

We next turn to the integrating constant in the marginal Student's t distribution. That calculation is summarized in Table 11.7. We see in the first four lines the calculation of the logarithms of the components of the integrating constant. We then exponentiate to obtain the actual integrating constants for each of the three variables in the final line of numbers. The equation at the bottom is the equation characterizing the marginal univariate Student's t distribution. The integrating constant, which appears in the final line of Table 11.6, is the entire front term outside the braces in the equation for the marginal distribution at the bottom in the forthcoming Table 11.7.

The equation at the bottom of Table 11.7 using constant terms developed higher in Table 11.7 and precision terms developed in Table 11.6 can now be used to plot the marginal distributions for the intercept, IQ, and Hours of Study coefficients. We plot the marginal probability distribution under classical regression in Figure 11.1 for the intercept, in Figure 11.2 for the IQ term, and in Figure 11.3 for the Hours Studying term. We note that they are wide.

Table 11.7: Calculation of the Constant in the Univariate Student's t Distribution

$ln\Gamma\left(\frac{\nu_C+1}{2}\right)$	$ln\Gamma\left(\frac{\nu_C}{2}\right)$
1.791759469	1.200973602

π	$ln(\pi)$
3.141592654	1.144729886

$$\ln c = \ln\Gamma\left(\frac{\nu_C+1}{2}\right) - \ln\Gamma\left(\frac{\nu_C}{2}\right) - \frac{1}{2}\ln\left(z_{ii}\nu_C s_C^2\right) - \frac{1}{2}\ln\pi$$

Intercept	IQ	Study hours
-3.725069442	0.755777676	0.805958095

Constant c

Intercept	IQ	Study hours
0.024111426	2.12926676	2.238840494

$$\{\beta_i\} = \frac{\Gamma\left(\frac{\nu_C+1}{2}\right)}{\pi^{\frac{1}{2}}\Gamma\left(\frac{\nu_C}{2}\right)} \frac{1}{\left(z_{ii}\nu_C s_C^2\right)^{\frac{1}{2}}} \left\{1 + \left(\beta_i - \bar{\beta}_i\right) \frac{1}{z_{ii}\nu_C s_C^2} \left(\beta_i - \bar{\beta}_i\right)\right\}^{-\frac{\nu_C+1}{2}}$$

Figure 11.1: Marginal Probability Distribution Over the Intercept Term–Classical Regression

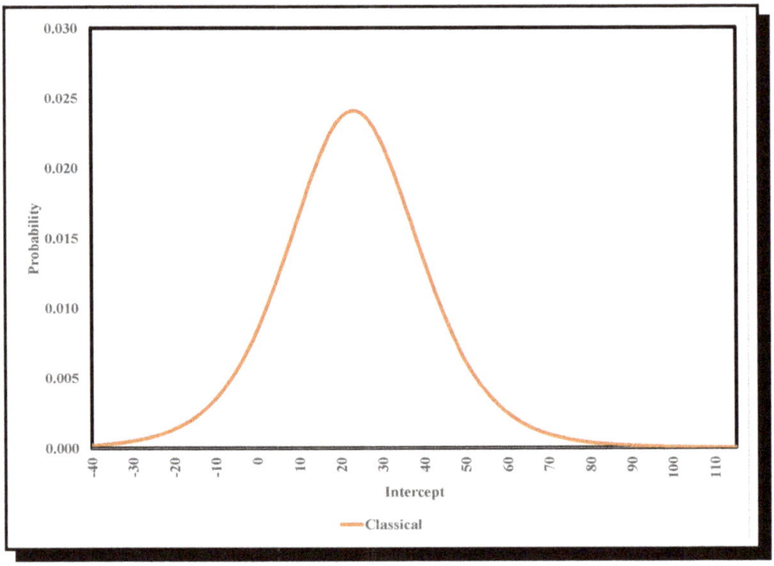

Figure 11.2: Marginal Probability Distribution Over the IQ Term–Classical Regression

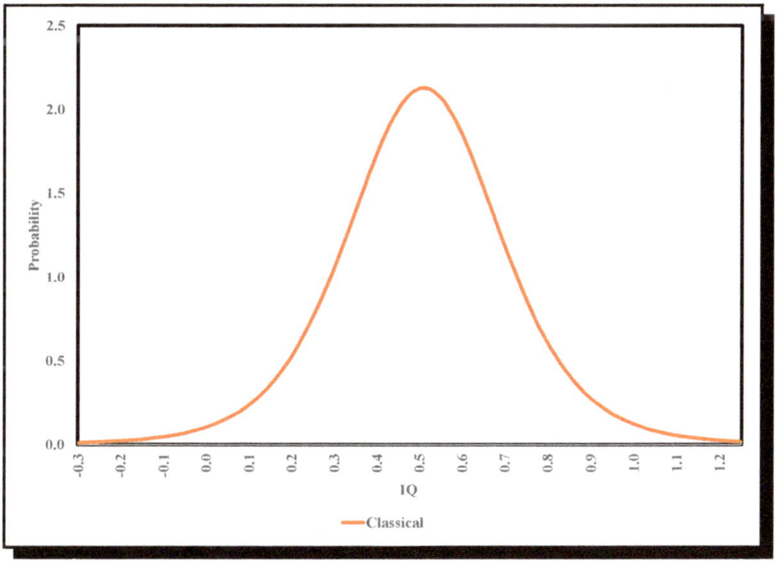

Figure 11.3: Marginal Probability Distribution Over the Hours of Study Term–Classical Regression

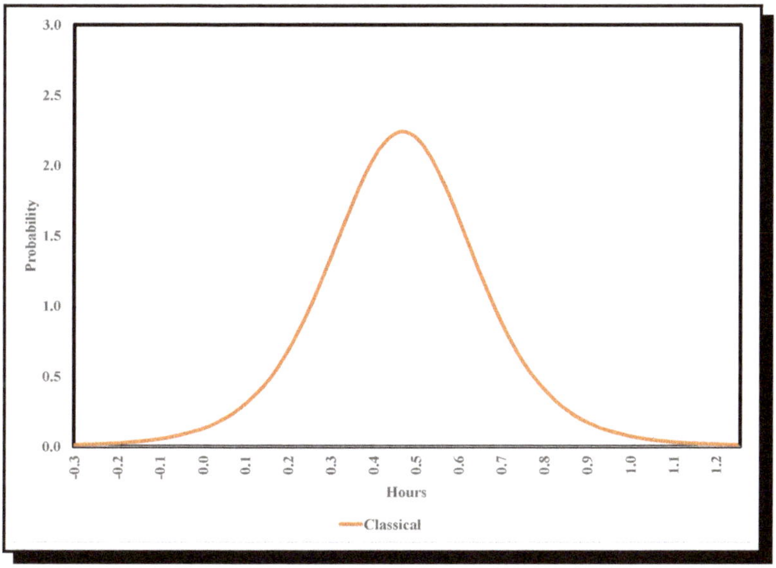

It is interesting to view the inverse gamma probability distribution, i.e., the distribution over the reciprocal of σ^2. See Figure 11.4. Notice that the distribution is pretty wide. It is seen to have a mean of 21.0232 and a standard deviation of 17.1653. This is a fairly wide distribution over reciprocal variance.

Figure 11.4: The Error Distribution Over the Inverse of σ^2

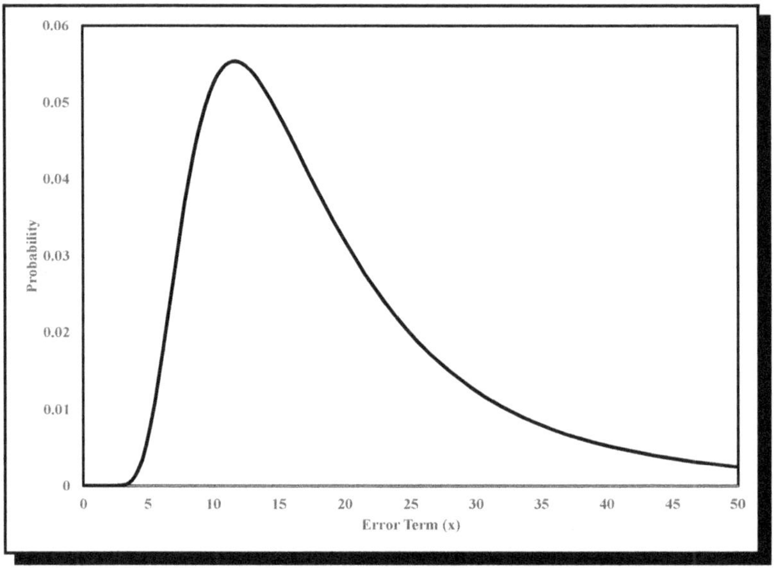

11.2 The Bayesian Regression Solution

Our physics professor, PP, is anxious to demonstrate that the only people who can master physics tests are high-IQ students and that they do well almost no matter how hard they work. They pick it all up in lectures, contacts with professors, office hours, homework, broad mathematical and physical knowledge, and contacts with other high-IQ students. As a corollary, he believes it is somewhat futile for lower-IQ students to study for hours on end for a physics test. They are, in his judgment, less likely to succeed as a result of such study. He has prepared a prior that quantifies this point of view. He has been thinking about the topic for a long time, and he is itching to put it forth. He is eager to see whether the data supports his prior beliefs. If not, he wants to know why.

11.2.1 The Prior

How does BSR assemble and quantify PP's prior BSR might begin by asking PP, "What are the mean values of the intercept, the IQ coefficient, and the Hours Studied coefficient?" PP might answer: "I think for one additional point of IQ, the score should increase by about 0.66. I think that for one additional hour of study, the score should increase by about 0.165. Therefore, my subjectively determined mean values of my coefficients for IQ and Hours are 0.66 and 0.165, respectively, the former being 4 times larger than the latter. Let's just go with 15 for the intercept because that seems to give a reasonable equation in the range where the data were gathered. I realize that my interpretation of the intercept is unclear, and I haven't thought much about omitted variables, but let's go with it." Those estimates of the mean are put into the model as raw inputs in gray, as shown at the left of Table 11.8.

Table 11.8: Prior Mean Value and Standard Deviation of the Coefficients

		0.1
	Prior Means	SD
Intercept	15	1.500000
IQ	0.66	0.066000
Study Hours	0.165	0.016500

BSR enters. "Hey, Prof., how certain of these coefficients are you?"

"Well, I guess I am plus or minus 10 percent on each of the coefficients," replies PP. BSR interprets that to mean that the standard deviation of each of the model coefficients is ten percent of the mean. We can see that calculation at the right in Table 11.8.

BSR asks him if there is any covariance. PP looks at her like she has just suggested Newtonian mechanics for a Hadron Collider problem. "For goodness sake, ma'am, if you specified an IQ of 100 and asked me for a conditional distribution on hours studied, I would give the same answer as if you specified an IQ of 110. I can't think of any dependencies or relevancies among the variables."

BSR verifies: "So we will assume that all covariances are zero in the prior?"

PP answers, "That's what I believe."

"Are you comfortable with that prior, based on your years of empirical observations?" BSR asks the professor one more time. "Your standard deviation is fairly small. Are you implying you are fairly certain of your opinion?"

"Yes, that seems to be quite a reasonable characterization of my prior point of view. I realize my prior is quite confident, perhaps even bordering on arrogant. But I have seen this phenomenon so many times, this is my prior belief."

We have seen how we might represent the PP in Table 11.8. We see the means, and we see the standard deviations calculated as 0.1 of the mean values at the right.

We next ask the professor if his opinions were based on a number of historical observations, and if so, what that number might have been. What is the error associated with these observations? He offers a pretty big mean observation error 20, and the standard deviation he infers is 0.5 of that. The standard deviation is significantly larger relative to the mean than it was for the coefficients. He is very uncomfortable with the concept of measurement error. We tell him we can assume complete, total, and abject ignorance, meaning literally infinite variance at the outset, meaning a diffuse prior only on the error term. But he says that his observations were all taken from the same kind of people in the same problem setting, and he is somewhat confident in a value of 20 test points for the measurement error term. Based on his ruminations, BSR reluctantly puts in as a prior the error term of 20 with a standard deviation of 10 for the measurement error term. PP is insistent that we not use a diffuse prior, or zero prior opinion, of the measurement error. He is not willing to attribute variability in his judgment to poor measurement methods, i.e., to intrinsic uncertainty in the measurement method. The darker gray entries in Table 11.9 show the 20 and the 10 values, and the table shows all the requisite calculations to get to the desired prior probability distribution. Keep in mind, the 20 value in line 1 and the 10 value in line 5 are subjective inputs from PP; the rest of the values, light gray or white, are calculated values.

Table 11.9: Computed and Estimated Values Relevant to the Prior

$< \sigma^2 >$	20
$< \sigma^2 >^2$	400
$V_{S^2} = \frac{2}{\nu_0 - 4}\left(\frac{\nu_0 s_0^2}{\nu_0 - 2}\right)^2$	100
	0.5
$\sqrt{V_{S^2}} = \langle\sigma^2\rangle \sqrt{\frac{2}{\nu_0 - 4}}$	10
$\nu_0 = 2\frac{<\sigma^2>^2}{V_{S^2}} + 4$	12
$s_0^2 = < \sigma^2 > \frac{\nu_0 - 2}{\nu_0}$	16.66666667
$\nu_0 s_0^2$	200
$\frac{\nu_0 s_0^2}{\nu_0 - 2}$	20

We see in the top line of Table 11.9 the professor's estimated measurement error of 20. We see the square of that number, which is needed in subsequent calculations, in the second line. If we move down to the next darker gray area in the table, we see the estimate of the standard deviation of the error term at 10. We calculate in the third line of the table the square of the standard deviation, which is the variance of the error term.

We see in the gray area in the fifth line of the table the calculated value of the degrees of freedom term ν_0. Notice that the calculated degrees of freedom here are 12. We might think of this as the number of observations PP has implicitly made. BSR tells the professor that the number of implicit observations is fewer than one might have thought given the strength of his convictions. The number of student observations embedded in his prior is actually less than PP himself might have thought. This could occur, for example, if PP had seen exactly the same thing happen 50 times but only reported it once in aggregate.

Once we have the degrees of freedom ν_0, we are able to use the formula $s_0^2 = \frac{R}{\nu_C}$ given previously to calculate the mean error term s_0^2. Recall this is the mean value of σ^2.

Armed with the degrees of freedom ν_0 and the error term s_0^2, we are able to calculate $\nu_0 s_0^2$ and, of course, the quotient $\frac{\nu_0 s_0^2}{\nu_0 - 2}$, which we quickly see is the same as $\langle\sigma^2\rangle$. In the last line, we see the variance-covariance matrix that our physics professor has estimated. It is calculated directly from the assumed a priori standard deviations of the uncertainties in the coefficients.

We know that subjective estimation of a measurement error by a so-called subject matter expert is problematic. People don't think about or even recognize things such as that. PP has no idea what BSR might be talking about! However, we let PP specify what he thinks the measurement error is ex ante in his prior.

Using the terms calculated in the foregoing equations and tables, we are able in Table 11.10 to calculate the terms that are relevant to the variance-covariance matrix of the prior. We see in the table the variance-covariance matrix itself, its inverse, and the precision matrix Q calculated from it.

Table 11.10: Calculations Derived from Variance-Covariance Matrix of Prior

VCV	Intercept	IQ	Study hours
Intercept	2.25	0	0
IQ	0	0.004356	0
Study hours	0	0	0.00027225
VCV^{-1}	Intercept	IQ	Study hours
Intercept	0.4444	0	0
IQ	0	229.5684	0
Study hours	0	0	3673.0946
$Q = VCV^{-1}\frac{\nu_0 s_0{}^2}{\nu_0 - 2}$	Intercept	IQ	Study hours
Intercept	8.8889	0	0
IQ	0	4591.3682	0
Study hours	0	0	73461.8916

We next do some preliminary work to prepare the prior to be in exactly the right form in Table 11.11. Notice in the table that we have calculated the values of m and M that are convenient for updating the prior to form the posterior. We have everything we need to do a Bayesian update from the prior to the posterior.

Table 11.11: Summary Calculations Related to the Prior

$m = \nu_0 + p + 1$	16		
$M = \nu_0 s_0^2$	200		
$\frac{\nu_0}{\nu_0 - 2} s_0^2$	20		
$\frac{2}{\nu_0 - 4}\left(s_0^2 \frac{\nu_0}{\nu_0 - 2}\right)^2$	100		
	$Q\beta_0$		
Intercept	133.33		
IQ	3030.30		
Study hours	12121.21		
	Intercept	IQ	Study hours
Diagonals VCV	2.2500	0.0044	0.0003
Std. Dev.	1.5000	0.0660	0.0165
Means	15.0000	0.6600	0.1650
t-statistics	10.0000	10.0000	10.0000

We use the foregoing information for PP's prior to calculate the marginal univariate probability distributions for the intercept term in Figure 11.5, the IQ term in Figure 11.6, and the Hours of Study term in Figure 11.7. We notice immediately how "narrow PP's prior seems to be." As he now tells BSR, he has what Decision Analysis calls a "central tendency" in his prior probability distribution.[4] We have the classical regression solution—its complete multivariate density and all its marginal densities. We reiterate that the classical regression solution is the Bayesian posterior with a diffuse prior.

11.2.2 The Posterior

How does BSR compute the posterior based on PP's prior, updating that prior with the data? Such Bayesian updating has been the topic of this book. This section summarizes and discusses that calculation. See Table 11.12, which summarizes calculations for the posterior. The top section of the table is the calculation of $X^T X + Q$ from the classical solution and the prior. The next section

[4] Howard and Abbas[45] define and discuss the central tendency in probability estimation. A central tendency means an unrealistically low variance in his prior probability estimates. He is confident that he is right. This is all there is to it; he is confident he is right.

Figure 11.5: Marginal Probability Distribution Over the Intercept Term–Prior

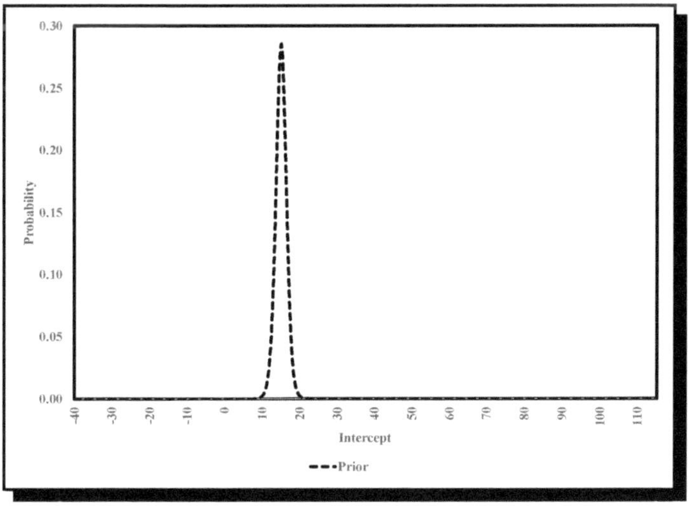

Figure 11.6: Marginal Probability Distribution Over the IQ Term–Prior

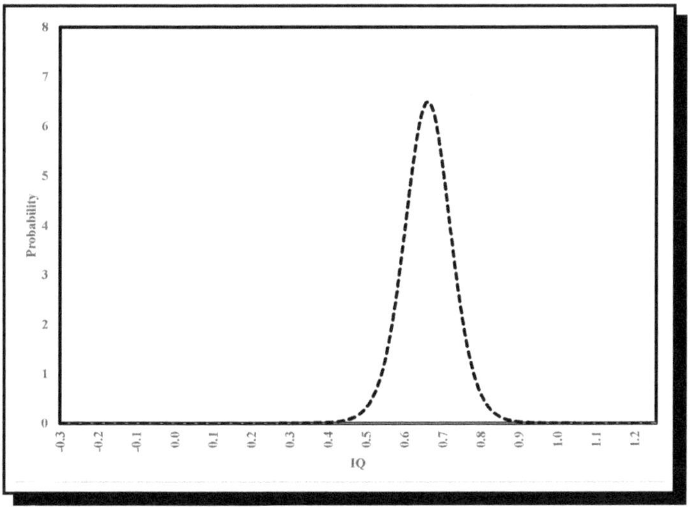

Figure 11.7: Marginal Probability Distribution Over the Hours of Study Term–Prior

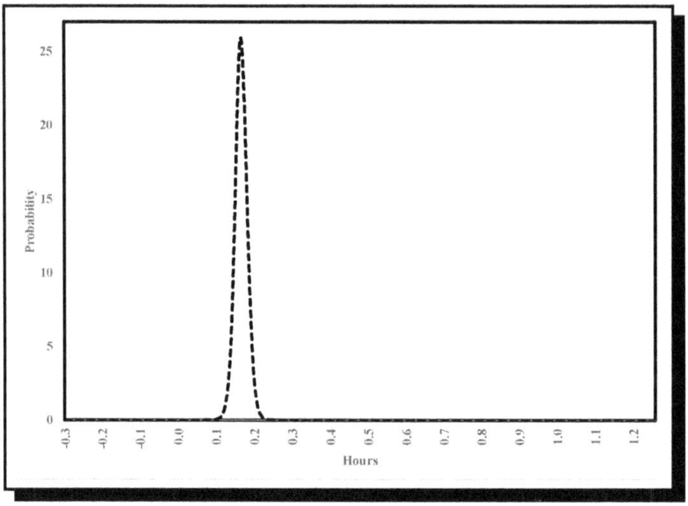

Table 11.12: Calculations Related to β^* in Posterior

X^TX+Q	Intercept	IQ	Study hours
Intercept	18.88888889	1009	185
IQ	1009	107532.3682	19585
Study hours	185	19585	78136.89164

$\left(X^TX+Q\right)^{-1}$	Intercept	IQ	Study hours
Intercept	0.106143609	-0.00099565	-1.75024E-06
IQ	-0.00099565	1.90838E-05	-2.42601E-06
Study hours	-1.75024E-06	-2.42601E-06	1.34103E-05

X^Ty	$Q\beta_0$	$X^Ty+Q\beta_0$
832	133.3333333	965.3333333
84955	3030.30303	87985.30303
16445	12121.21212	28566.21212

	β^*
Intercept	14.8113823
IQ	0.648656296
Study hours	0.16793825

$\nu=\nu_0+n$	22

calculates the inverse $\left(X^T X + Q\right)^{-1}$. The third section of the table calculates the key components of β^*. The calculation of β^* appears in the fourth section, and the calculation of the posterior degrees of freedom appears in the final line.

Using these terms in Table 11.12 related to the calculation of the mean β^*, we turn to the calculation of the constant A, which we have interpreted as the residual sum of squared errors in the posterior. Its four components are clear in equation (6.4). See Table 11.13. The top line of the table shows the calculation of the $\bar{\beta}-\beta^*$ terms, the difference between the classical regression and Bayesian posterior terms. We have dubbed this the "third term" in the four-term expression for the constant A. The next line of the table shows the calculation of the β_0 and β^* terms, the difference between the prior and Bayesian posterior terms. We have dubbed this the "fourth" term in the four-term expression for the constant A. The next line of the table shows all four terms that comprise the constant A, which is calculated at the lower left of the table.

Table 11.13: Calculations Related to A in Posterior

$\bar{\beta}-\beta^*$		$\left(\bar{\beta}-\beta^*\right)^T$		
8.344758		8.344758	-0.139222	0.299195
-0.139222		$\left(\bar{\beta}-\beta^*\right)^T X^T y$		
0.299195		35.473981		
		Third		
$\beta_0-\beta^*$				
0.188618		$(\beta_0-\beta^*)^T$		
0.011344		0.188618	0.011344	-0.002938
-0.002938		$(\beta_0-\beta^*)^T Q\beta_0$		
		23.908737		
		Fourth		
R	105.1158			
M	200.0000			
Third	35.4740			
Fourth	23.9087			
$A=\nu s^2$	364.4985			

Armed with the constant A term and the foregoing calculations in the posterior, we are in a position to complete the calculation of all the terms that comprise the Bayesian posterior. See Table 11.14. We see the calculation of the s^2 term, the νs^2 term, the mean of the inverse gamma error distribution $\left\langle\sigma^2\right\rangle$, and the variance and standard deviation of the inverse gamma error term.

We can use all these quantities to calculate and display the marginal univariate distributions over the intercept term in Figure 11.8, the IQ term in Figure 11.9, and the hours spent studying term in Figure 11.10 for the posterior. No-

Table 11.14: Calculations Related to Variance-Covariance in Posterior

$s^2 = \frac{A}{\nu}$	16.56811474				
νs^2	364.4985244				
$\langle \sigma^2 \rangle = \frac{\nu s^2}{\nu - 2}$	18.22492622				
$\frac{2}{\nu-4}\left(\frac{\nu s^2}{\nu-2}\right)^2$	36.90532619			$SD = \sqrt{\frac{2}{\nu-4}} \langle \sigma^2 \rangle$	6.074975406
VCV	Intercept	IQ	Study hours		
Intercept	1.934459	-0.018146	-0.000032		
IQ	-0.018146	0.000348	-0.000044		
Study hours	-0.000032	-0.000044	0.000244		
Key Params	β^*	Variance	Std. Dev.	t-statistic	
Intercept	14.811382	1.934459	1.390848	10.649170	
IQ	0.648656	0.000348	0.018649	34.781597	
Study hours	0.167938	0.000244	0.015633	10.742316	

Figure 11.8: Marginal Probability Distribution Over the Intercept Term–Posterior

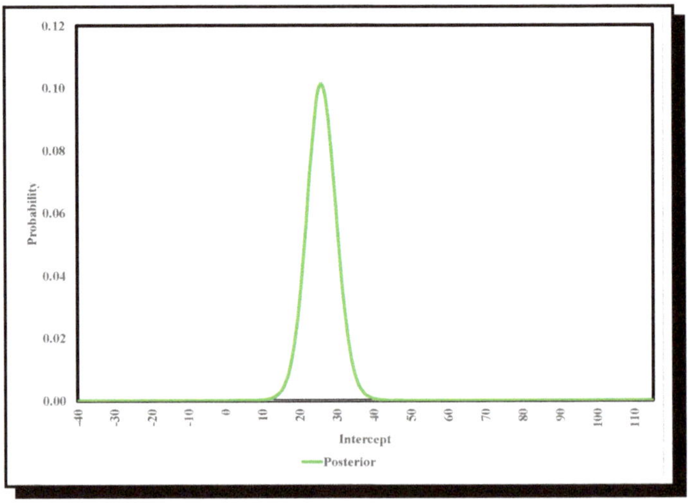

Figure 11.9: Marginal Probability Distribution Over the IQ Term–Posterior

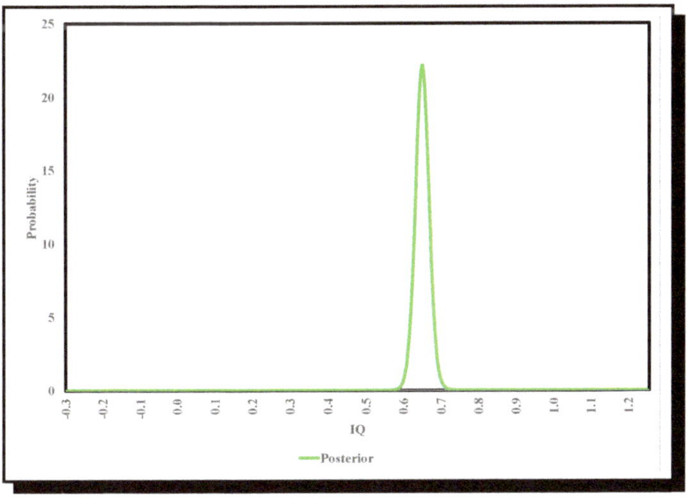

Figure 11.10: Marginal Probability Distribution Over the Hours of Study Term–Posterior

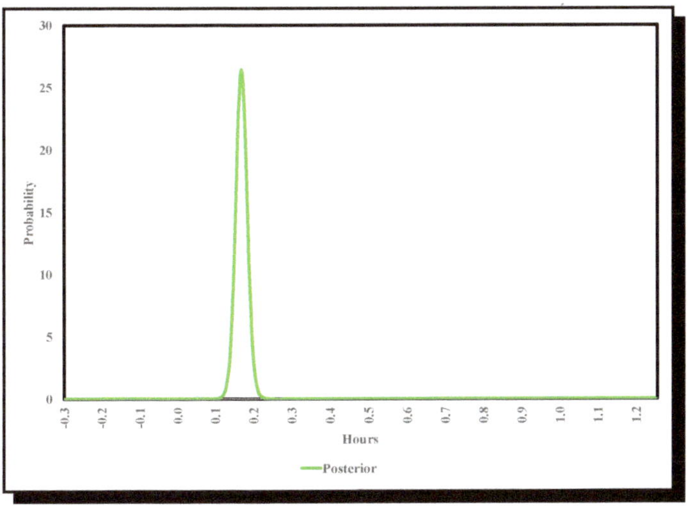

tice how narrow these distributions have become—narrower than the prior. That means that the data that we have used in a Bayesian sense to update the prior is quite consistent with PP's prior point of view. The data do not contradict but rather support PP's prior. That narrows his posterior relative to the prior.

11.2.3 Comparative Analysis of the Coefficient Estimates

It is insightful to compare BSR's coefficient estimates under the classical, PP's prior, and the posterior. Keep in mind that the prior is PP's prior estimate of the answer. If all three distributions—prior, classical, and posterior—are fairly similar, we can have increasing confidence that the data does not contradict PP's prior judgment. If they are not all similar or if the data is too small, we worry that the data contradicts prior judgment. In such a situation, we would think that more data, more observations, or better judgment are required.

Such is not the case here if we examine the classical, prior, and posterior marginal probability distributions over the intercept term in Figure 11.11. It is striking how the posterior is almost identical to the prior. The data have not changed PP's prior much. That is not the material insight, however. The material insight is that the small data set has not refuted PP's prior. The data set is, of course, very small, and it doesn't have the size or substance to refute the prior. We cannot say that the data supports his prior, but we can certainly say that the data does not refute it. This is an interesting insight where the intercept term is concerned. The data are not particularly material to changing the prior, refuting it, or modifying it very much. We would conjecture that the data is consistent with the prior, but that is too strong. In the vernacular, "Wow, PP's prior point of view seems to be borne out by the data for the intercept term."

Let us see if the same is true with the IQ term. We examine the classical, prior, and posterior marginal probability distributions over the IQ term in Figure 11.12. It is striking again how the posterior mean is almost identical to the prior mean. However, it is even more striking that the posterior variance is materially smaller than the prior variance. The data has narrowed BSR's estimate of the posterior relative to the prior. The data have not changed the mean value of PP's prior one iota as we move to the posterior! That is not the salient insight, however. The salient insight is that the small data set has not refuted PP's prior either through size or substance. We might conjecture that the data supports PP's prior, but we would certainly assert that the data does not refute it. This is an interesting insight where the IQ term is concerned. The IQ data are not particularly material to the prior, and, importantly, they do not refute it. In the vernacular, "Wow, PP's prior point of view is not contradicted by the data for the IQ term."

We examine the classical, prior, and posterior marginal probability distributions over the hours of the study term in Figure 11.13. It is striking yet again how the posterior is literally identical to the prior, precisely as it was for the intercept term. The data have not changed PP's prior one iota where the hours

Figure 11.11: Marginal Probability Distributions Over Intercept Term–Classical, Prior, and Posterior

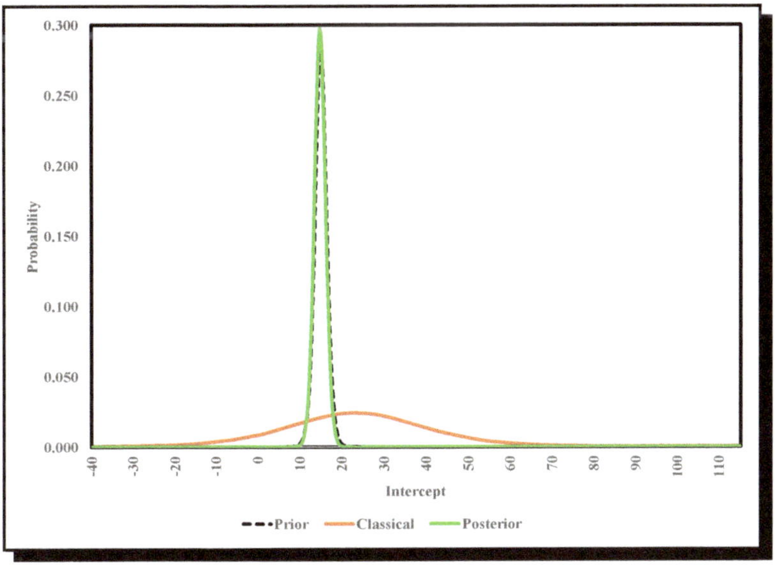

Figure 11.12: Marginal Probability Distributions Over IQ Term–Classical, Prior, and Posterior

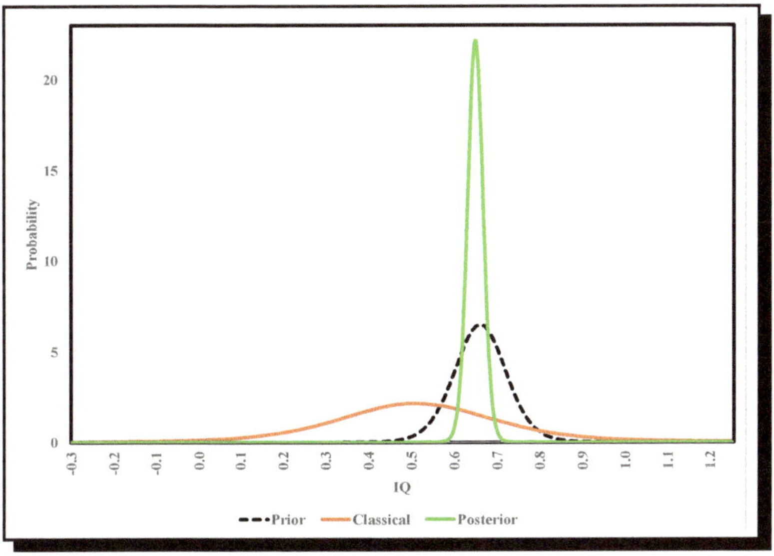

of preparation or intercept term are concerned. The material insight is that the small data set has not refuted PP's prior. We may not be able to say that the data supports his prior, but we can certainly say that the data does not refute it. This is a brilliant insight as far as the hours of preparation are concerned. The data are not particularly material to the prior, not refuting it in any way and being rather insignificant in content here. In the vernacular, "Wow, PP's prior point of view seems to be borne out by the data for the Hours Studied term."

Figure 11.13: Marginal Probability Distributions Over Hours of Study Term–Classical, Prior, and Posterior

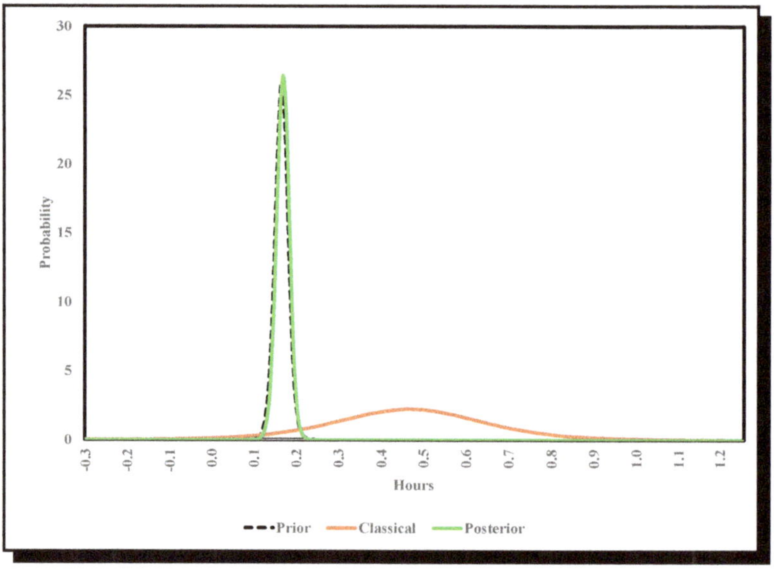

My goodness! The ten data points seem to support PP's prior view to a significant degree. They do not refute it. The ten observations are, in effect, "more of the same" with regard to the key parameters of the model. PP's prior is looking to be a pretty credible estimate of the three coefficients. Thus far, it seems that PP's views will be borne out in the future.

It is instructive to examine the prior, posterior, and classical inverse gamma distributions. See Figure 11.14. We see a narrowing from the prior through the classical and to the posterior.

11.2.4 Predictions for Test Performance of Eleanor and Samuel

It is tricky to discern definitive conclusions based on marginal or joint distributions over coefficients. It is much more insightful to postulate a couple of students and predict their performance based on their assumed IQ and hours of study.

Figure 11.14: Inverse Gamma Distributions–Classical, Prior, and Posterior

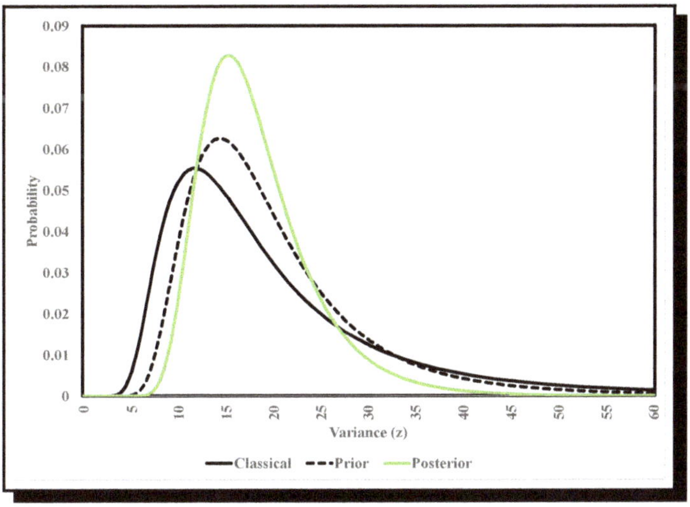

Armed with the foregoing probability distributions over coefficients (in fact, armed with the entire joint distribution over all coefficients), we now make a prediction of test scores for two different students. One is PP's favorite student from two years ago. Eleanor Rigby has an IQ of 120. Her score on the GRE is at the top of the charts. She has aced most of the physics courses in the department, in spite of playing in the local symphony orchestra and running Iron Man triathalons. Realistically, Eleanor doesn't have much time to study for tests. She has only studied for 18 hours for this particular test. Contrast that with Samuel Adams, who has a more modest IQ of 95. He gets by in the physics program by studying for hours on end. He always says to people, "They may be smarter than me, but they won't outwork me." He can be found in a dark corner of the physics laboratory at an old, broken-down desk, away from the crowd, studying essentially full time. He has studied for 35 hours for this particular test. The two students are represented in Table 11.15. It is difficult to discern definitive conclusions based on marginal or joint distributions over coefficients. It is much more insightful to postulate a couple of students and predict their performance based on assumed IQ and hours of study.

Table 11.15: Two Students–Eleanor and Samuel

	Intercept	IQ	Hours
Eleanor	1	120	18
Samuel	1	95	35

What would we predict for Eleanor's performance? For Samuel? Absolutely? Relatively? Can we easily tell from the distributions over the individual coefficients themselves? Alas, it just isn't clear. These predictions will be more clear. What would we expect PP's prior to predict for the performance of Eleanor and Samuel? What would we expect the classical regression solution (the data) to predict for Eleanor and Samuel? What would we expect the posterior to predict for Eleanor and Samuel? These are subtle questions, but with our predictive distribution, we are positioned to answer them easily. They lie at the heart of what the study endeavors to know.

If we stick to classical regression, ignoring the prior of PP altogether, the predicted performances of Eleanor and Samuel are shown in Figure 11.15. It is significant that based solely on the collected data—a mere ten observations—Eleanor is predicted to perform better than Samuel, say by approximately five points. And because the Eleanor curve is slightly higher at its peak, we can say that there is slightly less uncertainty in Eleanor's curve. The ten observations would seem to be consistent with and supportive of PP's prior view. We will examine that further when we update PP's prior views using the ten observations and later when we update KGP's views with the ten observations.

Figure 11.15: Predicted Performance for Eleanor and Samuel–Classical Regression

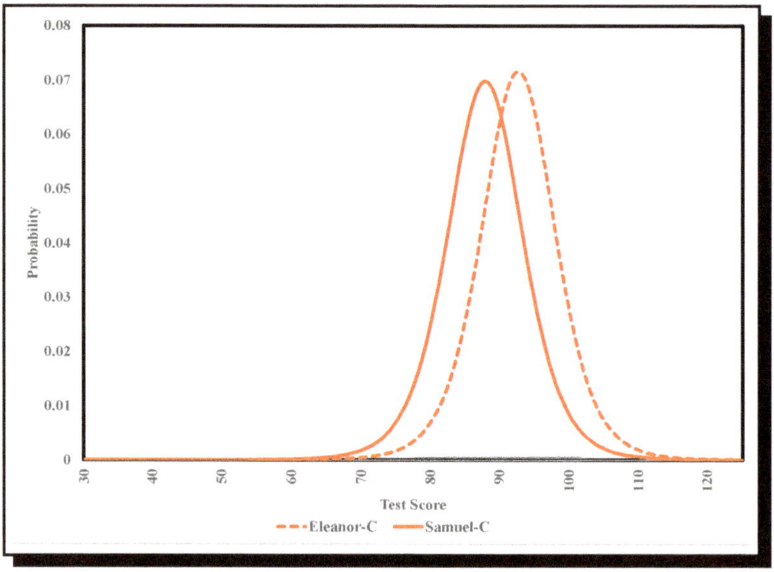

If we predict Eleanor's and Samuel's performances based on the prior, which represents PP's views going into the analysis before any data are collected, we see the situation in Figure 11.16. As indicated in the figure, PP foresees about a 20-point expected difference in performance between the two, with Eleanor having the advantage. He foresees slightly more uncertainty in his prediction of Eleanor's performance; her performance curve is "fatter." He is sure Samuel is going to do poorly, but he is not sure how far Eleanor will knock it out of the park! There is substantial uncertainty in his forecast of the performance of the two. He is not as "arrogant" in his views as he might have thought a priori based on the shape and width of his prior. The perception that he has used central tendency to obviate objective updating does not seem to be indicated here.

Figure 11.16: Predicted Performance for Eleanor and Samuel–Prior

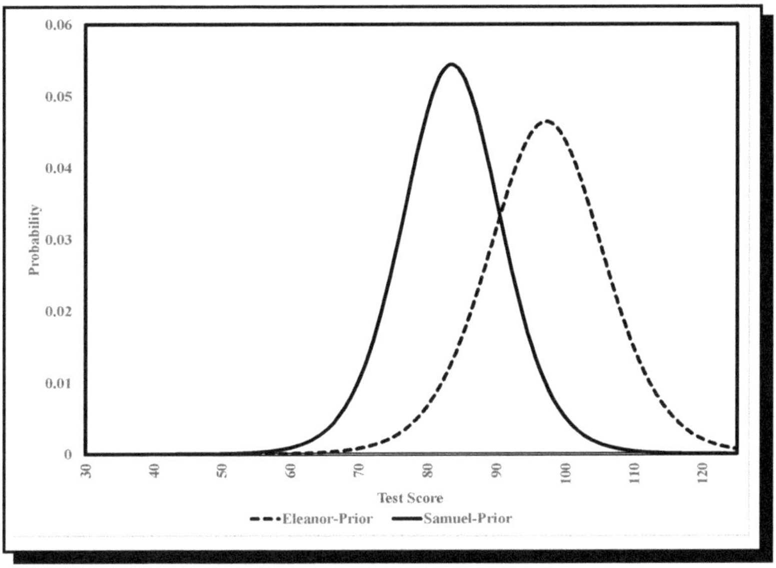

And now the tour de force—we predict Eleanor's and Samuel's performances based on the posterior, which represents the updating of PP's views going into the analysis before any data are collected with the data that actually are collected by PP. This leads to the posterior situation in Figure 11.17. As indicated in the figure, PP foresees about a 20-point expected difference in performance between the two, with Eleanor having the advantage. The two curves are significantly narrower than the prior or the classical, meaning that the prior and the classical are mutually reinforcing and lead to less uncertainty in the posterior. Data that confirms or does not refute a prior leads to lower posterior uncertainty. It stands to reason, and it tells BSR that the data and PP's prior are saying much the same thing. The data narrows the prior into a slimmer posterior! He is quite sure that Eleanor will have a higher score than Samuel because his posterior curves do not overlap very much. Interestingly, the magnitude of uncertainty in the two curves is comparable. We often see this empirically, but there are no rigid theorems that we know of. In light of the posterior, PP is not as "arrogant" in his views as one might have thought a priori. His experience, knowledge, and past observations seem to be consistent with the data.

Figure 11.17: Predicted Performance for Eleanor and Samuel–Posterior

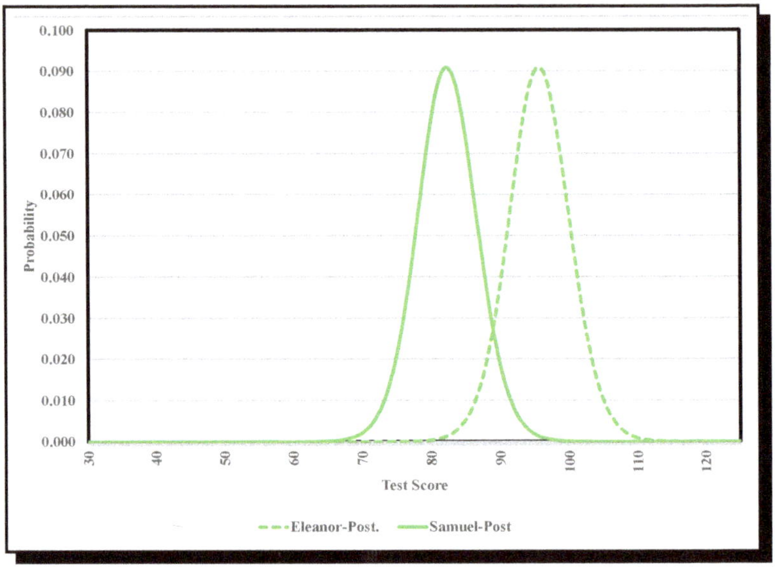

How does Eleanor's predicted performance vary between the classical, the prior, and the posterior situations? Does it matter whether we use the prior, the classical, or the posterior in analyzing Eleanor? The three densities are plotted in Figure 11.18. Our initial impression and answer to the former question are "not too much difference." Notice that the posterior falls between the prior and the classical. That is intuitively what one might expect, but it is not a "theorem" except in the univariate case. It is serendipitous but occurs frequently in real-world problems. We see that the prior gives the "fattest" distribution, followed by the classical (data) and finally the posterior. This type of structure indicates that the data are reinforcing the prior. The data are reinforcing the prior notion PP advanced that IQ matters more than hours of study. Notice that all three are clustered fairly nicely together, meaning that the data and the prior are mutually reinforcing. That is why the posterior variance is decreasing. The data that PP gave us supports the view that PP started with. Decisions one might make using any of these three marginal distributions might be expected to be the same. You would have to run the numbers with the three distributions to see, but they are rather strikingly close, with the prior being a slight bit more uncertain (higher variance).

Let us observe how Samuel's predicted performance varies between the classical, the prior, and the posterior situation. Refer to Figure 11.19, which plots all three on a common graph. Notice that the posterior continues to fall between the prior and the classical. Again, this is intuitively what one might expect, but it is not a "theorem" except in the univariate case. We see that the prior

Figure 11.18: Predicted Performance for Eleanor–Classical, Prior, and Posterior

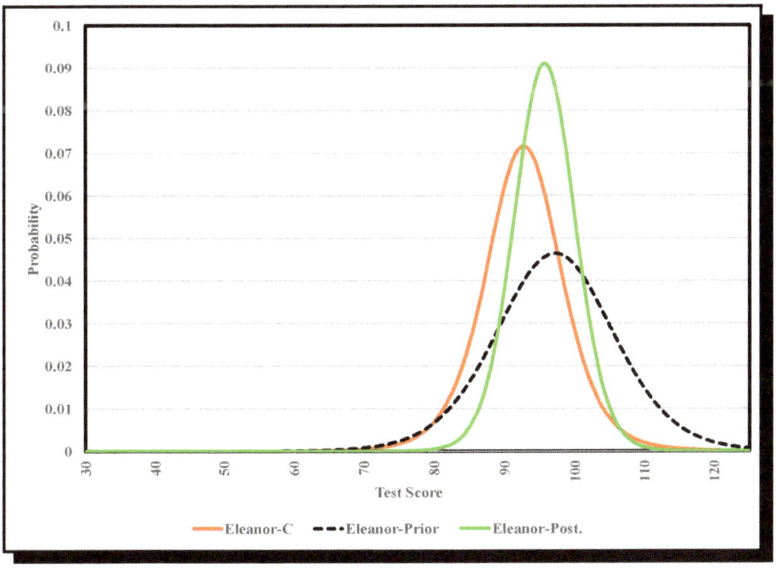

continues to give the "fattest" distribution, followed by the classical (data) and finally the posterior. This type of structure continues to indicate that the data are reinforcing the prior. The data are reinforcing the prior notion that PP had that IQ matters more than hours of study. Notice that all three are clustered fairly nicely together, meaning that the data and the prior are mutually reinforcing. That is why the posterior variance is decreasing. The data that PP gave us continues to support the view that PP started with.

It seems that PP's prior views are consistent with the gathered data, and when you combine the two, his variance in the answer drops. He becomes more certain. The data he provided has, in fact, reinforced his prior judgment. Is that just a coincidence? Is there adverse selection in the data? It would be time to question that at this point in the analysis, but we continue past that.

11.2.5 What Difference Would KGP's Prior Make in All This Analysis?

The kind and gentle psychologist KGP had a markedly different prior view than PP, and she was concerned with the results given PP's prior and given the data he provided. In particular, KGP believes, and believes she has observed, that students who are "hard workers" benefit markedly from their hard work and studying and that in fact they can catch and pass the high IQ students by virtue of diligent studying and attention to results, tools, and techniques. She believes that physics students who work hard can catch and pass high IQ physics students who may not choose to work so hard. She wants to know if the ten observations

Figure 11.19: Predicted Performance for Samuel–Classical, Prior, and Posterior

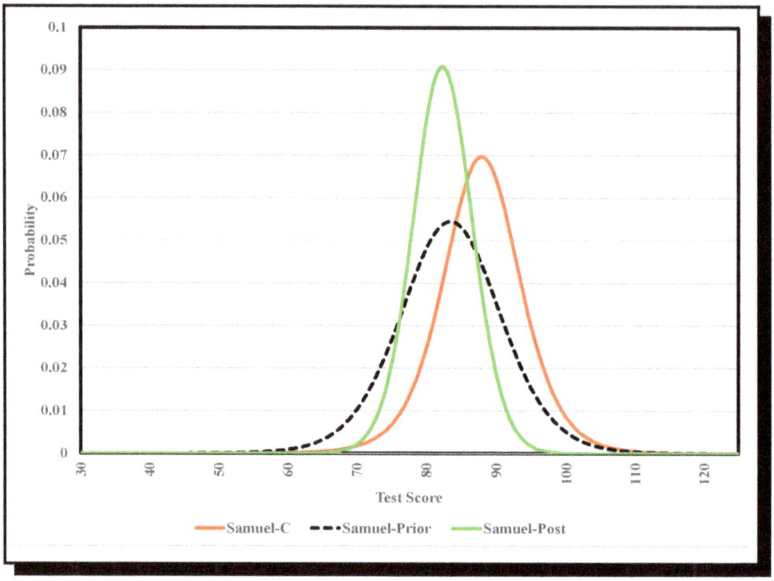

support that point of view or refute it. BSR (and Bayesian Linear Regression) are going to show her how.

How does BSR quantify KGP's prior? BSR might begin by asking KGP "What are the mean values of the intercept, the IQ coefficient, and the Hours Studied coefficient?" KGP might answer the question: "I think for one additional point of IQ, the score should increase by about 0.165. I think that for one additional hour of study, the score should increase by about 0.66. I think that Hours Studied is four times as important as IQ. Therefore, my subjectively determined mean values of my coefficients for IQ and Hours are 0.66 and 0.165, respectively, the former being 4 times larger than the latter. Let's just go with 15 for the intercept because I don't know any better. I realize that my interpretation of the intercept is unclear, and I haven't thought much about omitted variables, but let's go with it. I have seen so many students over the years, and I perceive it is only 20 percent IQ and a full 80 percent ex ante studying effort, and I truly hope that is true. Otherwise, we may be wasting time studying and teaching here. Just let in the smart students and relax!" KGP's estimates of the mean are put into the model, raw inputs in darker gray, as shown at the left of Table 11.16.

We must emphasize that the classical regression solution is exactly as it was before. No change at all. There is no prior in classical regression save for the diffuse prior, and the posterior for the diffuse prior is the classical regression solution itself. We do not have to revisit the classical regression analysis at all for KGP's prior; there isn't a prior.

Table 11.16: Summary of KGP Prior

	Bayesian Prior Means		0.1
			SD
Intercept	15		1.500000
IQ	0.165		0.016500
Study Hours	0.66		0.066000
$< \sigma^2 >$	20		
$< \sigma^2 >^2$	400		
$V_{S^2} = \frac{2}{\nu_0 - 4}\left(\frac{\nu_0 s_0^2}{\nu_0 - 2}\right)^2$	100		
	0.5		
$\sqrt{V_{S^2}} = \langle \sigma^2 \rangle \sqrt{\frac{2}{\nu_0 - 4}}$	10		
$\nu_0 = 2\frac{<\sigma^2>^2}{V_{S^2}} + 4$	12		
$s_0^2 =< \sigma^2 > \frac{\nu_0 - 2}{\nu_0}$	16.66666667		
$\nu_0 s_0^2$	200		
$\frac{\nu_0 s_0^2}{\nu_0 - 2}$	20		
VCV	Intercept	IQ	Study hours
Intercept	2.25	0	0
IQ	0	0.00027225	0
Study hours	0	0	0.004356

Under KGP's prior expectations, KGP would expect Eleanor to score rather poorly on the test. Figure 11.20 depicts the situation. KGP believes that study hours will be more important in determining the test score than Eleanor's high IQ. KGP believes that Eleanor will perform more poorly than Samuel because she did not put in the hours of preparation that he did. Her prior discounted both Eleanor's and Samuel's IQs and put most of the weight on the hours of preparation. When we see both predictions discounted like this, we fear we might be operating outside the normal range of students under the assumed prior, i.e., the assumed prior may not be an accurate representation of what the subject truly believes.

Turning to the KGP posterior in Figure 11.21, the answer is similar in nature, but there is substantially more uncertainty in the posterior than there is in the prior. The green posterior curves are wider than the prior curves. Eleanor is still expected to do worse than Samuel. How could this occur—more, not less,

Figure 11.20: Predictive Distributions for Eleanor and Samuel Under KGP Prior

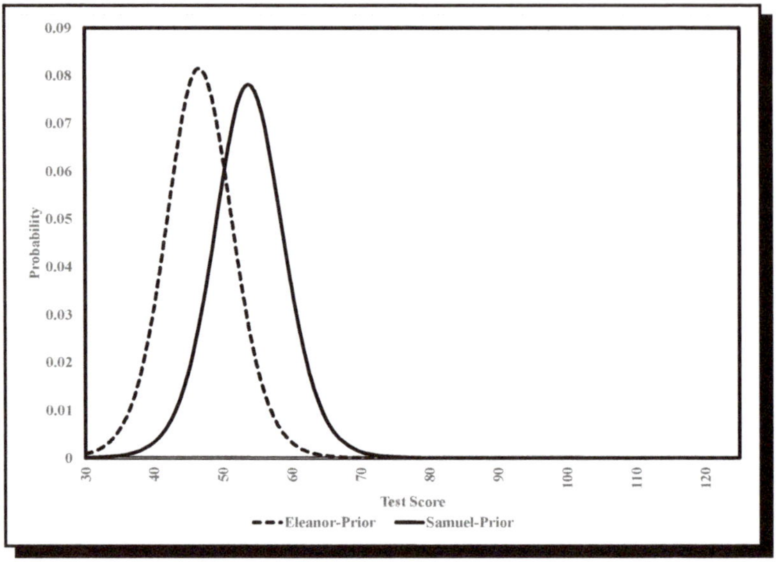

uncertainty after you have gathered data? More uncertainty after you have run Howard and Abbas'[45] Acme Rain Detector and gathered more readings? The reason is very simple, and the phenomenon is very real, as we shall see in the discussion to follow.

Figure 11.21: Predictive Distributions for Eleanor and Samuel Under KGP Posterior

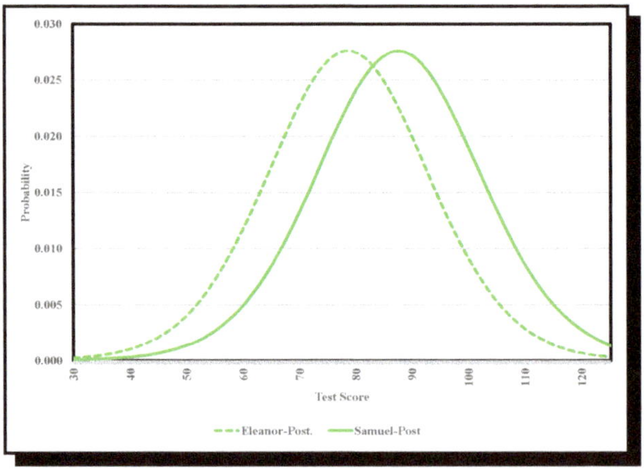

It is important to summarize that KGP's prior would expect Eleanor to score at a low level because Eleanor simply has not put in the advance study effort. In

fact, under KGP's prior expectations, she would expect quite a low grade for Eleanor on the test, lower than we might think is credible. (This can be a sign that the prior is biased toward a point of view and not an authentic representation of what the subject truly believes.) We see above that the classical solution and the data project a much higher score for Eleanor. The prior has a tight variance, and the classical regression solution has a wider variance, and they are obviously directly contradictory. Notice that when this happens, the variance of the posterior is understandably higher than the variance of either the prior or the classical regression solution. The data that comes in is sharply at odds with the prior, and this logically INCREASES the degree of uncertainty embedded in the posterior. The degree of uncertainty in the posterior becomes much higher. You have two points of view that are "fighting it out," and that increases, not decreases, uncertainty. This is very characteristic of situations in which the prior is centered around a wrong perception and/or exhibits too much central tendency, both of which apparently occur here.

We will amplify, beginning with Figure 11.22. If we focus on the classical regression case at the right, classical regression suggests that Eleanor will score well on the test because the data suggests that IQ is a more important precursor than KGP's prior beliefs. The graph shows us that KGP's prior is not consistent with the small amount of data that has been gathered. This could be because the data was offered by PP, who has a dramatically different perception from KGP, and he might have implicitly or explicitly incorporated bias or adverse selection in the students he selected for the data base. Nonetheless, it seems that the data support the notion that IQ is a more important contributing factor to test scoring than KGP believed in the past. This does not indict either PP or KGP but rather suggests further scrutiny of the data and the underlying sampling process. We would want to identify or ferret out any adverse selection therein.

The situation in Figure 11.22 also shows the mean situated between the mean of the prior and the mean of the classical regression solution. The posterior mean is "indecisive" in the vernacular, unable to decide whether to favor the mean of the prior or the mean of the data. One cannot yet tell whether the prior or the data should have more predominance in the posterior mean. This represents a highly uncertain situation and argues for more and better data to be gathered. Neither the data nor the prior are definitive, and we are left with a lot of uncertainty resident within the posterior. Most would consider this an unacceptable degree of uncertainty. If one had to make a decision, it would be a decision in the face of a great deal of uncertainty. Howard and Abbas'[45] Acme Rain Detector would swing into action if this were the situation in Howard and Abbas'[45] party problem, gathering more readings. There wouldn't be enough readings (enough data) to get an authentic probability distribution thus far.

Under KGP's prior expectations, KGP would expect Samuel to score somewhat poorly on the test but better than Eleanor because Samuel has done almost twice as much preparation as Eleanor. Figure 11.23 shows the situation with Samuel. To reiterate, KGP believes that study hours will be more important in

Figure 11.22: Predicted Performance for Eleanor Under KGP Prior

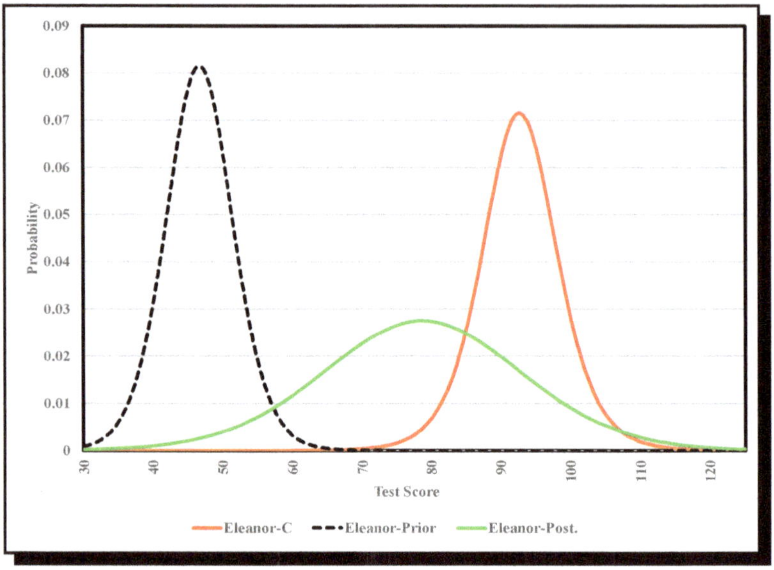

determining the test score than Eleanor's high IQ. If we focus on the classical regression case at the right, classical regression suggests that Eleanor will score well on the test because the data suggests that IQ is a more important precursor than KGP's prior beliefs. The graph shows us that KGP's prior is contradicted by the small amount of data that has been gathered. This could be because the data were offered by PP, who has a dramatically different perception from KGP, and he might have exhibited bias in the students he selected for the database. It might show that PP's prior work was just plain better than KGP's. Nonetheless, it seems that the data provided here actually supports the notion that IQ is a more important contributing factor to test scoring than KGP believed in the past. This type of result could occur with omitted variables, and that is a next step forward for BSR.

The situation in Figure 11.23 shows the mean NOT situated between the mean of the prior and the mean of the classical regression solution but in fact almost right on top of the classical regression solution. The data is dominating under KGP's prior for Samuel. Her prior is de facto being discounted. The posterior mean is decisive in the vernacular, able on its own to decide whether to favor the mean of the prior or the mean of the data. It is the data that wins. However, the dramatic increase in uncertainty in the posterior is still evident in the figure. The fact that the posterior is so different from the prior introduces a lot of uncertainty. This is likely to be an unacceptably high degree of uncertainty in the posterior, one that must be addressed by assembling a lot more data. One cannot generally allow this magnitude of uncertainty to persist. That said, we

Figure 11.23: Predicted Performance for Samuel Under KGP Prior

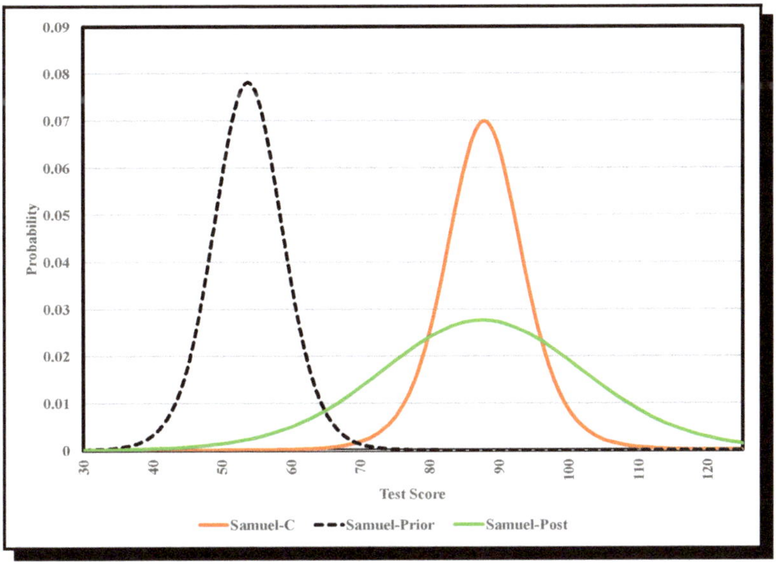

will want to be cognizant of the increased value of each incremental new data point relative to the cost of acquiring each new data point and cut off additional data gathering when the two become equal.

There is such an interesting message in Samuel's prediction graph. When you receive information that contradicts your prior, you INCREASE uncertainty relative to either the prior or the classical solution. The contradiction between your prior and the information coming in "flies in the face of your prior" and thereby increases your posterior level of uncertainty. Such an interesting but subtle concept. Data INCREASES uncertainty.

We want to emphasize that this is just a simple illustration of the data, method, and calculation and is not intended as a definitive analysis of intelligence or cumulative study effort before the fact. It is just a really convenient medium to conduct Bayesian Linear Regression analysis and prediction, and the type of thinking that goes into it.

Chapter 12

A DECISION ANALYSIS PROBLEM POSED BY CLEMEN[20] USING BAYESIAN PRIOR, POSTERIOR, AND CLASSICAL REGRESSION

When you know absolutely nothing about the topic, make your forecast by asking a carefully selected probability sample of 300 others who don't know the answer either.

Edgar Fiedler

Clemen[20] has posed an interesting problem:

"*In a business context, a common example is forecasting sales, and to come up with a conditional distribution for sales (Y), we might want to use explanatory variables such as prices (X_1), advertising (X_2), a competitor's price (X_3), or other important microeconomic variables. These variables could be either quantities that are uncertain (like the competitors' price) or variables that are under the control of the decision maker (e.g.., own price and advertising). Often it is possible to obtain data indicating the amount of sales that was associated with a particular price, amount of advertising, and so on. How to use such data to create a model of the relationships among the variables is our topic.*"

The use of data to understand the relationship of sales to own price, competitor price, and level of advertising is not trivial and certainly not unusual. Consider the influence diagram in Figure 12.1. The brute force approach would require obtaining enough data to estimate the conditional distribution of the particular variable of interest (Y, sales) for every possible combination of values for its conditioning or predecessor variables (X_1, our price; X_2, advertising; and X_3, competitor price). In addition, of course, we would need to know what the feasible values are for the decision variables (X_1, our price) and (X_2, advertising), and we would have to assess a distribution for the possible values for the uncertain variable (X_3, competitor price). That would require a rather substantial quantity of data. If we chose direct assessment by the analyst or an appointed expert, we would not know whether the analysis by that person was derived from some sort of observed data or simply subjectively assessed. This section puts forth a prior based on an *a priori* data set, an old, perhaps inaccurate, small data set, precisely the scenario anticipated in Chapter 10 (successive updating). This is, in fact, quite illustrative of the real-world situation. You have some "lousy" data, but you know you have to systematically gather and analyze better, more accurate data. The lousy data is better than nothing, and in fact, it represents your prior state of knowledge before you gather any new, additional, higher-quality, more focused data.

Figure 12.1: Influence Diagram for Clemen's Decision Problem

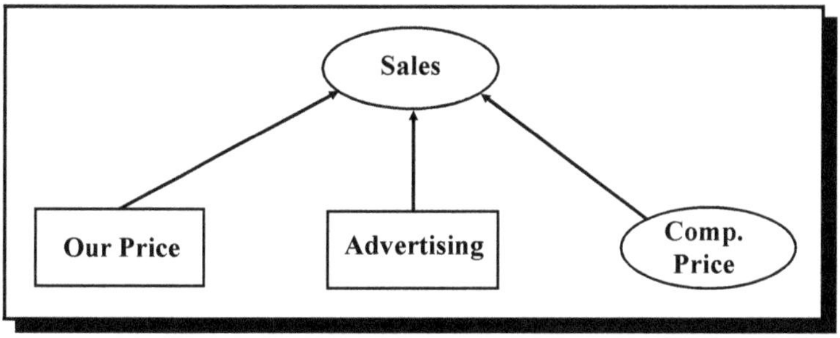

We should also note that this problem from Clemen[20] only contains three independent variables, two of which are alternatives and one of which is an uncertain variable (called a distinction by Howard and Abbas[45]). There are possibly one or more omitted variables. Clemen[20] did not attempt a comprehensive assessment of the number or types of variables that may affect sales but was rather attempting to be simpler and more didactic regarding how classical statistics might be used to build models that could support Decision Analysis. There may well be "omitted variables" here. We wager there are. We proceed through the simple example here without trying to expand or generalize. That does not meet our need or the expressed need of Clemen[20].

Suppose we have the observed data from Clemen[20] in Table 12.1. We see in the central area of the table the independent variable observations (own price, advertising, and competitor price), including the intercept as a fourth variable.

We reiterate from Chapter 2 that an intercept is a constant, a variable not associated with an independent observation. (In practice, people virtually always include an intercept term, whose interpretation is not always clear.) The shaded area is the matrix X. We see the correspondingly observed revenues in the y vector (revenue) in Table 12.1. We want to develop a probability distribution over revenue given our own price, advertising, and competitor prices.

Table 12.1: Data Set for the Clemen Problem

Observation	Constant	Advertising ($1000s)	Price ($)	Competition Price ($)	Sales ($1000s)
	Int	Ad	P	CP	S
1	1	366	90.99	96.95	10541
2	1	377	90.99	93.99	8891
3	1	387	94.99	90.99	5905
4	1	418	96.99	97.95	8251
5	1	434	92.99	97.95	11461
6	1	450	95.95	93.95	6924
7	1	457	93.95	90.99	7347
8	1	466	91.95	96.95	10972
9	1	467	96.95	94.99	7811
10	1	468	92.95	96.95	10559
11	1	468	97.99	98.95	9825
12	1	475	91.95	90.99	9130
13	1	479	99.95	91.95	5116
14	1	479	96.99	95.95	7830
15	1	481	91.95	90.95	8388
16	1	490	96.99	96.99	8588
17	1	494	96.95	91.95	6945
18	1	502	98.95	95.95	7697
19	1	505	94.99	96.99	9655
20	1	529	93.99	97.95	11516
21	1	532	91.99	95.99	11952
22	1	533	92.99	97.99	13547
23	1	542	93.99	92.95	9168
24	1	544	90.95	95.95	11942
25	1	547	94.99	93.95	9917
26	1	554	89.95	90.95	10666
27	1	556	96.95	95.95	9717
28	1	560	91.99	97.95	13457
29	1	561	98.99	97.95	10319
30	1	566	93.95	91.99	9731
31	1	566	94.99	94.99	10279
32	1	582	98.99	91.99	7202
33	1	609	89.95	92.99	12103
34	1	612	92.95	92.99	11482
35	1	617	92.95	94.95	11944
36	1	623	94.99	91.99	9188

12.1 Classical Statistics/Bayesian with Diffuse Prior

12.1.1 Initiating the Bayesian Calculations

The first step in a classical regression solution to the decision problem is to calculate the (X^TX) and $(X^TX)^{-1}$ matrices based on Table 12.1, which are shown in Table 12.2. We emphasize that when we use the term classical regression, we are specifically referring to Bayesian analysis with a diffuse prior. We note the variable names in the table.

- Int for intercept
- Ad for advertising
- P for price
- CP for competition price
- S for sales (the dependent variable)

Table 12.2: (X^TX) and $(X^TX)^{-1}$ for Clemen's Problem–Classical Statistics

(X^TX)	Int	Ad	P	CP
Int	36.00	18,296.00	3,400.96	3,411.80
Ad	18,296.00	9,454,524.00	1,728,295.16	1,733,148.44
P	3,400.96	1,728,295.16	321,558.00	322,343.61
CP	3,411.80	1,733,148.44	322,343.61	323,576.31

$(X^TX)^{-1}$	Int	Ad	P	CP
Int	69.95068263	-0.00556735	-0.31914782	-0.38981045
Ad	-0.00556735	0.00000652	0.00000136	0.00002241
P	-0.31914782	0.00000136	0.00381152	-0.00043917
CP	-0.38981045	0.00002241	-0.00043917	0.00443074

$\left[(X^TX)^{-1}\right]_{ii} = z_{ii}$	Int	Ad	P	CP
	69.95068263	0.00000652	0.00381152	0.00443074

These matrices in Table 12.2 are central to the generalization of the classical regression solution computed by Bayes with a diffuse prior, just as they are central to the classical regression solution in Chapter 2. The mathematical developments in previous sections make clear the central importance of (X^TX) and $(X^TX)^{-1}$.

We will need the ii^{th} diagonal element of the inverse matrix$(X^TX)^{-1}$ to calculate the individual univariate Student's t distributions, which we have denoted z_{ii}, in order to calculate marginal distributions over the model coefficients

β. The accumulation of these diagonal elements of $(X^TX)^{-1}$ occurs at the bottom of Table 12.2. To emphasize, the vector at the bottom of Table 12.2 is simply the vector of diagonals of the $(X^TX)^{-1}$ matrix in the middle of Table 12.2.

We next see in Table 12.3 the requisite calculation of X^Ty for the classical statistics mean at the left center of the table. This is no more than the matrix transpose of the X matrix area in the data table in Table 12.1 times the right-hand column designated y in Table 12.1. This calculation is obvious; no equation is referenced. We emphasize that the matrix X^TX is symmetric and, if nonsingular, positive definite with all positive eigenvalues.

Table 12.3: Key Calculations Based on (X^TX) for Clemen's Problem–Classical Statistics

	X^Ty		
Int	345,966.00	n	36
Ad	177,849,135.00	p	4
P	32,561,320.38	$\nu_C = n - p$	32
CP	32,878,377.14		
		$R = \nu_C s_C^2$	6,744,669.36
	$\bar{\beta} = (X^TX)^{-1} X^Ty$		
Int	2199.3423	${s_C}^2 = \frac{R}{\nu_C}$	210,770.92
Ad	15.0466		
P	-503.7640	$\frac{\nu_C}{\nu_C - 2} s_C^2$	224,822.31
CP	499.6713		
		$\frac{2}{\nu_C - 4}\left(s_C^2 \frac{\nu_C}{\nu_C - 2}\right)^2$	3,610,362,280.37
		$\frac{\nu_C s_C^2}{\nu_C - 2}\sqrt{\frac{2}{\nu_C - 4}}$	60,086.29

Once we have the inverse in Table 12.2 and the X^Ty vector in Table 12.3, we are able to calculate the classical regression solution at the lower left of Table 12.3 according to equation (8.6). This is extremely standard in statistical calculations, ranging from Excel to more sophisticated systems. Nothing complicated yet! Clemen[20] summarizes this calculation.

The basic accounting for the classical regression solution appears at the right in Table 12.3. We see in Table 12.3 that

- There are $n = 36$ observations,
- There are $p = 4$ coefficients being estimated.
- The computed degrees of freedom $\nu_C = 32$. See equation (8.2). This is hardly controversial or difficult. This result has been known for years. The degrees of freedom in the classical statistics case (which is identical to the Bayesian with diffuse prior case) is identical.
- The residual sum of squared errors R=6,744,669.36. See equation (8.4).

- The estimate of the σ^2 term, namely s_C^2. See equation (8.5).
- The $\frac{\nu_C s_C^2}{\nu_C - 2}$ term. See equation (8.11).
- The variance of σ^2 in the inverse gamma distribution. See equation (8.12).
- The standard deviation of σ^2 in the inverse gamma distribution.

The variance-covariance matrix of the multivariate Student's t distribution over the four model coefficients appears in Table 12.4. (Keep in mind that this matrix considers only the coefficients and not the error term σ, as characterized in equation (8.15). Because the joint distribution is the product of a β term and a σ term, and because equation (8.17) has zeros off the diagonals, we know that the distributions over β and σ are probabilistically independent. Knowing that, we need not carry all the off-diagonal zero terms around again and again. The coefficient variance-covariance matrix is calculated according to equation (8.15).

Table 12.4: Variance-Covariance Matrix Over Model Coefficients in Clemen's Problem–Classical Statistics

$\frac{\nu_C s_C^2}{\nu_C - 2}(X^T X)^{-1}$	Int	Ad	P	CP
Int	15,726,474.1874	-1,251.6655	-71,751.5498	-87,638.0871
Ad	-1,251.6655	1.4666	0.3054	5.0380
P	-71,751.5498	0.3054	856.9148	-98.7354
CP	-87,638.0871	5.0380	-98.7354	996.1284

This is a bit different from the variance-covariance matrix one would get from the regression add-in in Excel as of 2021. Our calculation, Bayes with diffuse priors, adjusts for small data sizes.[1] When we say that it adjusts for small data sizes, that means that there is a multiplicative adjustment that is larger than 1 based on the ν_C term at the left, and that adjustment emerges when we convert to a Student's t distribution.

We accumulate the variances (the diagonals of the variance-covariance matrix) in Table 12.5 and the standard deviations (their square roots). We also calculate the t-statistic, which is the mean divided by the standard deviation. (The t-statistic presages how tightly our probability distribution over the estimated value is clustered around the mean.) The foregoing calculations appear in the top panel of Table 12.5. The correlation matrix appears in the bottom panel.[2] The correlation matrix allows one to examine the magnitude and direction of probabilistic dependence (mutual relevance) in an intuitively appealing way. We are not interested in a detailed correlation and covariance analysis here but rather

[1]The calculation in Excel as of this writing is problematic in this dimension, as we have discerned. It lacks the adjustment from the normal to the Student'st.

[2]The ij^{th} element of the matrix is $cov(i, j)$ divided by the product of the standard deviations of i and j.

a "What depends on what?" snapshot. Positive correlations mean that a positive change in one of the variables means a positive change in the correlated variable. "Things move together." Negative correlation means that a positive change in one of the variables means a negative change in the other variable. "Things move oppositely." Correlations near 1 or -1 can indicate redundancy (multicollinearity) in the formulation and a reconsideration of the independent variables. The three reasons nicely put forth for calculating and observing a correlation matrix are given by Bock[3], which are reproduced here.

There are three broad reasons for computing a correlation matrix:

1. *To summarize a large amount of data where the goal is to see patterns. In our example above, the observable pattern is that all the variables highly correlate with each other.*

2. *To input into other analyses. For example, people commonly use correlation matrixes as inputs for exploratory factor analysis, confirmatory factor analysis, structural equation models, and linear regression when excluding missing values pairwise.*

3. *As a diagnostic when checking other analyses. For example, with linear regression, a high amount of correlations suggests that the linear regression estimates will be unreliable.*

...Most correlation matrixes use Pearson's Product-Moment Correlation. Our matrix in this section uses the Pearson correlation representation. Some may resort to techniques like ridge regression.[4] We do not address ridge regression herein.

Turning now to Table 12.6, we see all the calculations necessary to calculate the marginal densities over the model coefficients (which, as we have derived, are univariate Student's t densities). We see the requisite formula for the univariate Student's t distribution at the bottom of the panel. The two terms necessary to calculate the z_{ii} factor in the top row are contained in Table 12.2. In particular, we see in the top row of Table 12.2 the z_{ii} factors calculated for the four coefficients. We see the $R = \nu_C s_C^2$ factor at the middle right of Table 12.2. Simple multiplication yields the top row in Table 12.6. Moving down to the middle of the table, we see the two arguments of the two gamma functions, and below that, we see the two gamma function calculations. The term named "First Term" in the table is the first term in the equation for the univariate Student's t

[3] See https://www.displayr.com/what-is-a-correlation-matrix/.

[4] See https://en.wikipedia.org/wiki/Ridge_regression. *Ridge regression is a method of estimating the coefficients of multiple-regression models in scenarios where the independent variables are highly correlated. It has been used in many fields including econometrics, chemistry, and engineering...It is particularly useful to mitigate the problem of multicollinearity in linear regression, which commonly occurs in models with large numbers of parameters. In general, the method provides improved efficiency in parameter estimation problems in exchange for a tolerable amount of bias...*

Table 12.5: Variances and Correlations in Clemen's Problem–Classical Statistics

	Int	Ad	P	CP
Variances	15,726,474.187405	1.466580	856.914807	996.128380
Std. Dev	3,965.661885	1.211025	29.273107	31.561502
t-statistic	0.554597	12.424689	-17.209107	15.831669

Correlation	Int	Ad	P	CP
Int	1.0000	-0.2606	-0.6181	-0.7002
Ad	-0.2606	1.0000	0.0086	0.1318
P	-0.6181	0.0086	1.0000	-0.1069
CP	-0.7002	0.1318	-0.1069	1.0000

density at the bottom. It is the quotient of the two gamma functions divided by the square root of π in the subsequent line. It is the truly "constant" part of the constant term for the univariate t density, being independent of which coefficient one is considering. To compute the "Constant," we must divide this number by the square root of the terms in the first row. That will create the leading term constants in the densities for each of the four coefficients, which are depicted in the second-from-bottom row of numbers. The final row is the reciprocal of the first row, which has been shown to be the "precision matrix" or the constant in the quadratic term in the four Student's t densities. These are all the terms we need to calculate the full marginal probability density functions for each of the four constants.

12.1.2 Alternatives and Distinctions for the Clemen Problem

To address the Clemen[20] problem, we need to review the alternatives and distinctions for which we need to develop probability distributions. There are two alternatives: our own price and the amount we spend on advertising. We must enumerate (akin to a Cartesian product or an outer product) the two decision variables—own price and advertising level.

There is one uncertain variable, i.e., competitor price. This is what is called a distinction in Decision Analysis. For the sake of argument, posit a discrete probability distribution over this distinction: competitor price. For illustrative purposes, we will need to calculate the probability distribution over sales that we will get if the competitor's price is set at each of the prospective discrete levels. Keep in mind that we will not obtain a deterministic estimate of sales given the two decisions and the single distinction. It will be probabilistic. Given the enumerated alternatives plus the full set of distinctions for the uncertain variable,

Table 12.6: Calculation of Marginal Densities Over the Model Coefficients in Clemen's Problem–Classical Statistics

$\nu_C s_C^2 z_{ii}$	Int	Ad	P	CP
	471,794,225.6221	43.9974	25,707.4442	29,883.8514
$\frac{\nu_C+1}{2}$	16.5	$\frac{\nu_C}{2}$	16	
$\Gamma\left(\frac{\nu_C+1}{2}\right)$	5.189998E+12	$\Gamma\left(\frac{\nu_C}{2}\right)$	1.307674E+12	
$\pi^{\frac{1}{2}}$	1.772454			
First Term	2.239199	$\frac{\Gamma\left(\frac{\nu_C+1}{2}\right)}{\pi^{\frac{1}{2}}\Gamma\left(\frac{\nu_C}{2}\right)}$		
	Int	Ad	P	CP
Constant	0.00010309	0.33758188	0.01396572	0.01295312
	Int	Ad	P	CP
Exp. Coeff.	2.119568E-09	2.272861E-02	3.889924E-05	3.346289E-05

$$\{\beta_i\} = \frac{\Gamma\left(\frac{\nu_C+1}{2}\right)}{\pi^{\frac{1}{2}}\Gamma\left(\frac{\nu_C}{2}\right)} \frac{1}{\left(z_{ii}\nu_C s_C^2\right)^{\frac{1}{2}}} \left[1 + \left(\beta_i - \bar{\beta}_i\right) \frac{1}{\nu_C s_C^2 z_{ii}} \left(\beta_i - \bar{\beta}_i\right)\right]^{-\frac{\nu_C+1}{2}}$$

we will be defining a full probability distribution over sales at every set of alternatives and uncertain competitor prices we might posit. This generalizes the approach in Clemen[20] and is, in fact, the typical approach frequently seen in classical regression.

For purposes of solving this problem, we can think of three possible discrete advertising alternatives:

- $585 (strong advertising)
- $505 (moderate advertising))
- $425 (low advertising)

times (in the Cartesian sense) three possible discrete own price alternatives

- $93 (low, aggressive price)
- $95 (moderate price)
- $97 (high price).

This implies nine combinations of alternatives, each combination representing what people have termed a "policy." How might one generalize this to more discrete settings of either? Could this be generalized to continuous settings of the decision variables? Perhaps. If there were, say, 10 discrete settings of each, would we have to deal with a Cartesian product of 100 joint alternatives? Would we have to perform a regression analysis for each of the 100 alternatives individually? (Not necessarily, but the probability distribution over sales works across a discrete range of variables.) We will not pursue such extensions here, concerned as we are with developing the probabilities and supporting Decision Analysis. To do so, we assume that we have to statistically analyze every combination of alternatives, i.e., every "policy."

To represent the set of two alternatives and a single distinction in the Clemen [20] model, we build the δ vector shown in Table 12.7 for each combination of alternatives and distinctions. This δ vector is simply an enumeration of the discrete values the influence diagram will be delivering to the sales node. It is de facto a place to store the values of each particular setting of alternatives and distinctions that comprise the independent variables so that we can "clock" through them as required. This is going to become very important in solving the problem. The influence diagram is going to offer settings for the competitor price distinction and settings for both of the alternatives. Each combination of alternatives and distinctions is going to generate a probability distribution over sales. We will assume that the discrete levels of competitor price are each going to have a probability of occurrence. We shall see that the value of sales will be a weighted sum of three univariate Student's t probability distributions. This is very general and can certainly accommodate expected utility calculations, which is precisely what we are seeking here.[5]

Table 12.7: One Setting of Alternatives and Distinctions for the Clemen Problem

δ

Int	Ad	P	CP
1	505	95	93

q	1

δ^T

Int	1
Ad	505
P	95
CP	93

Clemen[20] himself discusses using three discrete settings of advertising

[5]Because the predictive distributions will be univariate Student's t in form, we will need to simulate expected utility rather than calculate it analytically. While a bit of a pain, this is imminently doable. We will not undertake that here because we do not want to discuss simulation in detail. We could also discretize the Student's t distributions, should we desire, for purposes of solving the decision problem.

(decision), three discrete settings of our price (decision), and three discrete settings of competitor price, each characterized by a probability of occurrence. This would create 27 δ vectors of advertising, own price, and competitor price (one of which is a distinction) that would have to be run through the model to create 9 probability distributions over sales, one for every Cartesian combination of the two decisions with three alternatives each. The reason we have nine probability distributions is that we have nine combinations of alternatives, and for each we have a conditional probability distribution over the competitor's price. (The alternatives enumerate the combinations of alternatives.) This is what would be required to solve Clemen's problem explicitly, but that is not the point quite yet. At present, we want to evaluate a single setting of the two decision variables and a single level of competitor price. The point of doing so is to generate a probability distribution over sales for a given, fixed, invariant advertising and pricing decision. We are not solving the decision problem here but rather the statistical problem for a given set of alternatives and distinctions. For later discussion, let us assume that the probabilities of the three elemental possibilities are

- 0.3 for $93 (low competitor price)
- 0.5 for $95 (moderate competitor price)
- 0.2 for $97 (high competitor price)

and critically important that they are NOT dependent on any alternative or combination of alternatives chosen.[6] Alternatively, in Decision Analysis parlance, none of the alternatives is assumed to have any relevance to the competitor's price. Some might argue this is a "heroic" assumption in fully general real-world problems, but it is certainly the intent of the Clemen's[20] problem as he posed it.

This raises a very important point with regard to Decision Analysis as articulated by Clemen[20], as contrasted with Decision Analysis as articulated by Howard and Abbas[45]. The latter, Howard and Abbas[45] have a far more appealing and robust representation of distinctions and probabilities in decision trees and influence diagrams, recognizing the fundamental origin of uncertainty. The former invariably wants to have the probabilities over distinctions be dependent on **alternatives**, but the latter is very, very careful to do so sparingly if ever and only when justified, noting that such is rarely justified.[7] There are so

[6]This is not necessarily a dramatic oversimplification. We saw that there were probabilistic correlations among the two alternatives and one distinction in the probabilistic model of sales given those variables. We know that the probability distribution over competitor prices will be a function of the nine alternatives times three levels of distinction. Competitors will look at our company's efforts to increase sales and model us before setting their price. We have ignored that phenomenon here. We have not, however, ignored the correlations and covariances that contribute to predicting market share.

[7]Prof. Howard has argued, quite correctly, that alternatives should rarely or never point directly into distinctions in influence diagrams. We know of virtually no decisions, except for perhaps R&D

many distinctions in the world that are simply not affected by the alternatives of the decision-maker. In real-world applications, probabilities are usually not conditional on alternatives. Uncertainties are intrinsic and are not affected by one's alternatives! The roll of the dice does not care what decisions you have made before, during, or after. This is ignored in the Clemen[20] formulation. Throughout his book, virtually all the examples have alternatives pointing to distinctions. (We rarely see that in real-world decision-making or in Howard and Abbas[45].) In spite of that issue, we solve the Clemen[20] problem as he posed it. In this particular application, the influence of alternatives on distinctions may actually be justified. The competitor may set his or her price based on yours. This, of course, begs the question of whether this is in fact a game theory problem—the competitor is not acting randomly but in fact acting strategically. That possibility is ignored in Clemen's[20] formulation and therefore in ours.

12.1.3 Return to the Bayesian Calculations

We now turn to the question of solving for the predictive distribution for each of the nine alternatives. Taking the advertising price of \$505 and the own price of \$95 as an illustration for one of the nine alternatives we have characterized as combinatorial or cartesian in nature, we need to consider the three levels of distinction (competitor price), which has three levels or outcomes. To analyze that specific one of the nine alternatives, we need to define and analyze three δ vectors.

	Int	Ad	P	CP
δ_1	1	505	95	93
δ_2	1	505	95	95
δ_3	1	505	95	97

To illustrate the calculations required, we begin with the first vector from the foregoing list and designate it δ for the forthcoming calculations

	Int	Ad	P	CP
δ	1	505	95	93

Given that vector, we can calculate everything we need to develop the numbers for the Student's t probability density function over the desired prediction. We will need the conditioning variables (precursors), and we will need the algebra

decisions, in which decision arcs point into systematically different distinctions, each fundamentally different. In general, alternatives should point to value nodes. Prof. Howard has said, "You have to really sell me that an alternative should point to a distinction. Alternatives have no effect on uncertainty. The uncertainties are out there, and decisions do not get to change them." We agree, but we solve the Clemen problem notwithstanding his more problematic representation of what Howard and Abbas[45] have called "influence." Influence occurs when there is an arrow from an alternative node into a probability node, and Howard has commented that this almost never happens in the real world.

for the coefficient densities. The requisite calculation of the key parameters begins in Table 12.8. The equation for the univariate density over the prediction variable d is shown at the bottom of Table 12.8. It is the same as equation (8.41) with the single row vector δ, which implies that $q = 1$ in the predictive equation, and with the classical regression precision matrix. The objective of Table 12.8 is to summarize the calculation of the inverse precision matrix of the univariate Student's t predictive distribution, which contains the term

$$[1+\delta(X^TX)^{-1}\delta^T]^{-1} \quad .$$

We note the substitution for the identity matrix (which is now a scalar 1 because we only have a single row in δ) in the foregoing equation.

Table 12.8: Calculation of the Predictive Distribution for Clemen's Problem–Classical Statistics

$\frac{\nu_C+1}{2}$	16.5	$\frac{\nu_C}{2}$	16
$\Gamma\left(\frac{\nu_C+1}{2}\right)$	5.19E+12	$\Gamma\left(\frac{\nu_C}{2}\right)$	1.30767E+12
$\left\{\pi\nu_C s_C^2\left[1+\delta(X^TX)^{-1}\delta^T\right]\right\}^{\frac{1}{2}}$	4703.110386		
Const	0.000843883		
$\delta\bar{\beta}$	8409.719471		
$\nu_C s_C^2\left[1+\delta(X^TX)^{-1}\delta^T\right]$	7,040,775.09	Sq. Rt.	2653.44589
Precision	1.4203E-07	Sq. Rt. Prec.	0.000376868

$$\{d\,|X,y,\delta\} = \frac{\Gamma\left(\frac{\nu_C+1}{2}\right)}{\left\{\pi\nu_C s_C^2\left[1+\delta(X^TX)^{-1}\delta^T\right]\right\}^{\frac{1}{2}}\Gamma\left(\frac{\nu_C}{2}\right)}\left\{1+\frac{(d-\delta\bar{\beta})^2}{\nu_C s_C^2\left[1+\delta(X^TX)^{-1}\delta^T\right]}\right\}^{-\frac{\nu_C+1}{2}}$$

We use the information in Table 12.8 to calculate the parameters for the full univariate predictive distribution in Table 12.9. We see the two gamma functions being calculated, and we see the constant of the univariate predictive distribution being calculated (that is what the reference "Const" in the table means). The constant is calculated as the square root of the product of $\pi\nu_C s_C^2$ times the foregoing matrix. We next see the inner product of δ times the classical regression mean to give the predictive mean, and finally we see the inverse precision (scalar) matrix calculated at the bottom. These are all the parameters that are going to be required to calculate the predictive distribution for any row vector δ of own price, advertising level, and competitor price if we want them to be determined from the classical regression solution, which is what Clemen[20] wanted. The actual parameters of the classical Bayesian posterior, given at the left of equation

(4.52), outside the braces, and (4.53), appear in Table 12.9. The constant represents all the terms outside the braces. The precision term represents all the terms save for the quadratic product inside the braces. The precision and variance are calculated according to the referenced formulas.

Using the parameters in Table 12.8 and Table 12.9, we calculate and display the univariate Student's t distribution over revenue in Figure 12.2. The mean of this distribution is obvious; it is $\delta\bar{\beta} = \$8,409.72$ reported in Table 12.9. The variance of this distribution is shown in Table 12.9, and its square root (the standard deviation) is shown to be \$484.4507. This is actually a fairly narrow probability distribution.

Table 12.9: Summary of Distribution for Clemen's Problem–Classical Statistics

Const	0.000843883		
$\delta\bar{\beta}$	8,409.7195	$\frac{\nu_C+1}{2}$	16.5
Precision	1.420298E-07		
Variance	234,692.5031	Std. Dev.	484.4507

$$\{d\,|X,y,\delta\} = \frac{\Gamma(\frac{\nu_C+1}{2})}{\{\pi\nu_C s_C^2[I+\delta(X^TX)^{-1}\delta^T]\}^{\frac{1}{2}}\Gamma(\frac{\nu_C}{2})}\left\{1+\frac{(d-\delta\bar{\beta})^2}{\nu_C s_C^2[I+\delta(X^TX)^{-1}\delta^T]}\right\}^{-\frac{\nu_C+1}{2}}$$

We can calculate the probability distributions for the three competitor prices \$93, \$95, and \$97 assuming an advertising level of \$505 and an own price of \$95 to obtain the plot in Figure 12.3. Let us belabor the calculation a bit. We have actually created three δ vectors, all three using the same setting of the decision variables (advertising level of \$505 and own price of \$95), but having the third element of the delta be the uncertain variable, the distinction, the competitor price, varying across the range \$93, \$95, and \$97. Three different, discrete values of the distinction, namely \$93, \$95, and \$97, but with the decision (advertising level \$505 and own price \$95) held constant. The foregoing three competitor prices occur with probabilities of 0.3, 0.5, and 0.2, respectively. This creates three probability distributions over sales as shown in Figure 12.3, and they are weighted by the probabilities 0.3, 0.5, and 0.2 from left to right. If we want to develop the overall probability distribution over sales, we multiply the entirety of the first curve by 0.3, the entirety of the second curve by 0.5, and the entirety of the third curve by 0.2. It feels as if we are calculating the expected value of the three curves, but we are in fact calculating the entire continuous probability distribution over sales given the alternative under consideration.[8]

[8] We could have used a three-row D matrix instead of δ three times in succession. However, use of a D matrix would disallow the particular assignments of probability Clemen[20] wanted in his formulation.

Figure 12.2: Plot of Conditional Predictive Distribution in the Clemen Problem–Classical Statistics

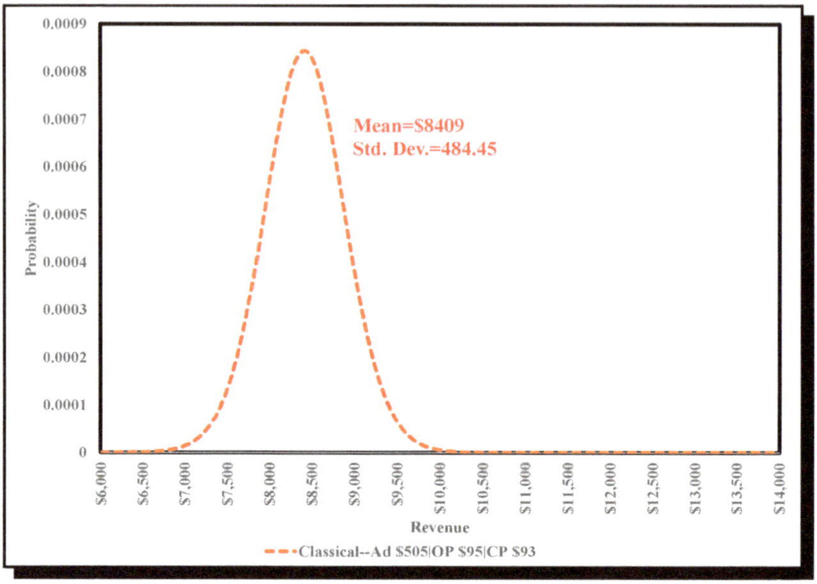

Figure 12.3: Plot of Conditional Probability Distributions Over Sales in the Clemen Problem–Classical Statistics

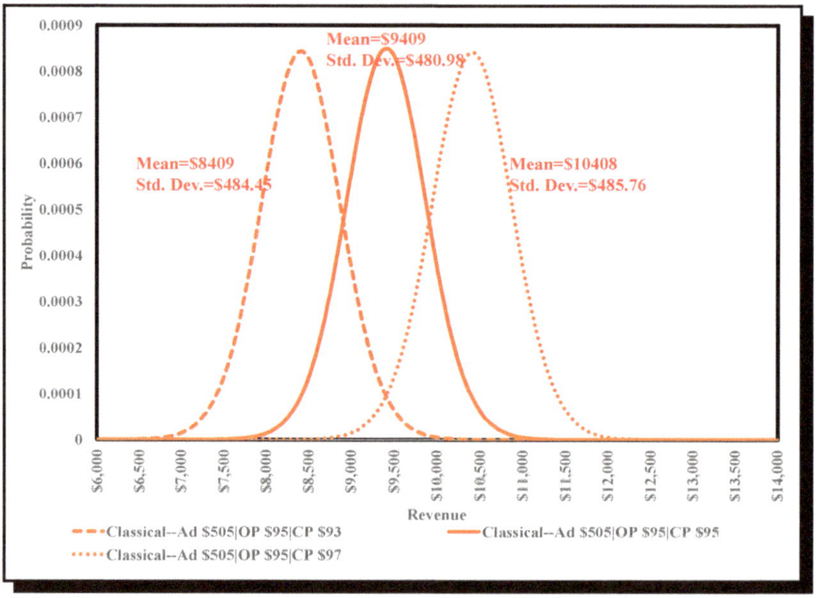

Figure 12.4: Plot of Overall Probability Distribution Over Sales for Clemen's Problem–Classical Statistics

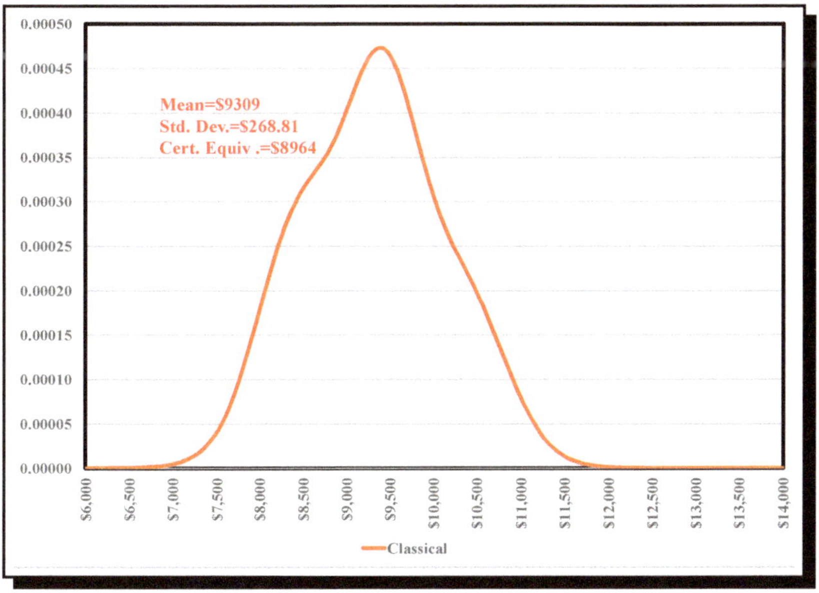

If we use the foregoing probabilities of occurrence of each of the three competitor prices in Figure 12.3, we can calculate the probability distribution over sales over all three competitor prices. The calculation appears in Figure 12.4. This is the classical regression solution: the sales you can expect to get given the probability distribution over competitor prices and given your own pricing and advertising alternatives.

We note in Figure 12.4 the term "certain equivalent." Certain equivalents are a concept from Decision Analysis originally invented by von Neumann[100] and put forth in detail in Von Neumann and Morgenstern[100] and Howard and Abbas[45]. To calculate it, we begin by postulating what is called a "cardinal utility function." For this calculation, we use the well-known exponential cardinal utility function:

$$u(x) = -e^{-\frac{x}{\rho}}$$

in which ρ is the (constant) risk tolerance.[9] For this exercise, we assume $\rho =$ \$1000. This will allow us to calculate a certain equivalent in sales. To calculate certain equivalents, we must first calculate expected utility and thereafter find the single sales level that gives you that level of utility.

Keep this firmly in mind: the decision is how much we advertise and where

[9] See https://en.wikipedia.org/wiki/Exponential_utility.

we set our price. To make this decision, we have to generate probability distributions over sales for the three uncertain levels of that distinction for each of the nine combinations of alternatives: three levels of advertising times three levels of pricing. We would have to make 27 runs of the predictive distribution. We can calculate the expected value and the expected utility over the three uncertain variables to obtain an expected revenue for each of the nine combinations of alternatives. Doing so, we obtain the result in Figure 12.4 for the \$505 advertising and \$95 own price alternative. Keep in mind that we have used classical statistics to build a probabilistic model of sales conditional on the competitor's price, our own price, and our own level of advertising. We have assumed that past data observed in the real world is sufficient to estimate the coefficients of the probability distribution that the model implies. But, as we have seen, that is not the only way. We can impose judgment in the form of a prior and combine the judgment and data in the form of a Bayesian posterior.

12.2 Prior (Small Prefatory Data Set)

Having completed the classical regression solution, we turn to the notion of the prior, which Clemen[20] probably never thought of (because Bayesian statistics was not easily available or widely known, only classical regression–just like BSR in the IQ/Study Hours example in Chapter 11). Suppose we had a small, inaccurate, and cursory data set. Having such a data set is not at all unusual before one begins to collect more definitive and accurate data. Quite often, people enter problems with "lousy," poorly gathered, small data sets that were gathered willy-nilly without any specific purpose in mind.[10] Can you use such objectiveless data to create a prior? Often, the answer is yes. We illustrate how in this section. We will use that small, inaccurate, cursory data set to establish a prior using classical statistics. Not a bad idea! You must use the only data you have, which is generally small in size and somewhat incomplete, to put together a prior. That prior will generally have a high variance and a bit of an inaccurate mean. Assuredly, more observations will update and ultimately override or supersede it where necessary. For our illustration here, the small, prefatory data set we have is that in Table 12.10. Notice there are a mere 10 observations of the four parameters in that data set, which is indeed small for the purpose at hand. It would only have six degrees of freedom, and the "fitting" is likely to be poor. Nonetheless, it may well embody the only prior information Clemen[20] has. That old

[10]More often than we might like to think, real-world databases are gathered without any particular purpose in mind. Just gather some data because someone gave you some money to do it! We are familiar with multiattribute data that is gathered with no particular purpose in mind. We have seen a bit of that in research activities at Stanford. The money to gather such data is appropriated in the "hope" that such data might lead to statistical research on one of the parameters within the data base that can understand variables based on what they may be conditioned on or what their probability distribution might be. Unguided, unmotivated data gathering stands in sharp contrast to activities such as clinical trials, in which data are gathered with a specific purpose in mind. Generally, the latter data is better.

data may be acceptable for putting together a prior, but not great for solving the problem or making decisions. We notice that the data set has the same four independent variables and the same computed variable (sales). It is structurally similar to Clemen's data set, but alas, completely different and ostensibly from a quite different source.

Table 12.10: Tiny Prefatory Data Set in the Clemen Problem

Observation	Constant	Advertising ($1000s)	Price ($)	Competition Price ($)	Sales ($1000s)
	Int	Ad	P	CP	S
1	1	360	92	97	10000
2	1	425	93	97	10600
3	1	460	92	90	7000
4	1	500	97	97	8200
5	1	525	92	96	11400
6	1	550	94	92	9150
7	1	500	97	92	6500
8	1	600	93	95	11500
9	1	560	97	96	9500
10	1	550	94	93	8600

Following the classical statistics model, we must calculate the $(X^T X)$ and $(X^T X)^{-1}$ matrices for the tiny data set as shown in Table 12.11. This is the same thing as we did previously for the full classical regression solution and full data set. (We will show how $(X^T X)$ becomes the precision matrix for our data prior, exactly as one would predict!)

Table 12.11: $(X^T X)$ and $(X^T X)^{-1}$ for the Tiny, Prefatory Data Set for the Clemen Problem

$(X^T X)$	Int	Ad	P	CP
Int	10	5,030	941	945
Ad	5,030	2,576,050	473,785	474,955
P	941	473,785	88,589	88,927
CP	945	474,955	88,927	89,361

$(X^T X)^{-1}$	Int	Ad	P	CP
Int	351.68641108	-0.00161035	-2.18888164	-1.53230308
Ad	-0.00161035	0.00002640	-0.00030945	0.00018468
P	-2.18888164	-0.00030945	0.02814176	-0.00321274
CP	-1.53230308	0.00018468	-0.00321274	0.01843097

Turning to Table 12.12, we see at the left the $X^T y$ matrix and the mean of the classical regression solution $\bar{\beta}$. (We expected to see $\bar{\beta}$ as the mean of the prior, and we will indeed see that.) On the right, we see the residual sum of squared errors, the number of observations (10), and the number of parameters (4). This, of course, means that the degrees of freedom is 6. This is an important insight worth reiterating and justifying. The number of data points is 10. The number of parameters, including the intercept, is 4. The degrees of freedom in classical regression in Chapter 2 is a mere 6. (We expect to see this same degrees of freedom in the prior, and we will!) We see immediately below the estimate of the σ^2 term, the estimate adjusted for the small sample size, the variance of the inverse gamma distribution over σ^2 and the standard deviation thereof. This panel contains a number of the classical statistical results for this prior data set. We have been through these numbers and concepts in the classical statistics section for a larger data set. We should reiterate that the mean value of the coefficients $\bar{\beta}$, the degrees of freedom ν_C, and s_C^2 in this table, together with the variance-covariance matrix in the next table, are sufficient statistics to fully characterize the density we will be using as our prior.

Table 12.12: Key Parameters for the Tiny, Prefatory Data Set for the Clemen Problem

	$X^T y$		
Int	92,450	$R = \nu_C s_C^2$	1,691,244
Ad	46,642,500		
P	8,684,000	n	10
CP	8,762,100		
		p	4
	$\bar{\beta} = \left(X^T X\right)^{-1} X^T y$		
Int	3856.8226	$\nu_C = n - p$	6
Ad	13.2330		
P	-562.9990	${s_C}^2 = \frac{R}{\nu_C}$	281,874.01
CP	547.1975		
		$\frac{\nu_C s_C^2}{\nu_C - 2}$	422,811.02
		$\frac{2}{\nu_C - 4}\left(\frac{\nu_C s_C^2}{\nu_C - 2}\right)^2$	178,769,156,487.00
		$\frac{\nu_C s_C^2}{\nu_C - 2}\sqrt{\frac{2}{\nu_C - 4}}$	422,811.02

The previous two tables are sufficient to calculate the variance-covariance matrix of the prior, which appears in Table 12.13. This is precisely the same calculation we made above for the classical statistics case. Recall from our earlier development that all we need to implement the prior is

- The mean values of the coefficients.
- The variance-covariance matrix of the coefficients.
- The mean of the inverse gamma distribution.

- The variance of the inverse gamma distribution.
- The number p of independent variables.

The final two bullets can be calculated from the parameters ν_C and s_C^2. These two parameters are equivalent to the mean and variance of the inverse gamma and substitute for sufficient statistics.

Table 12.13: Variance-Covariance Matrix for the Tiny, Prefatory Data Set in the Clemen Problem

Variance-Covariance				
	Int	Ad	P	CP
Int	148,696,889.295	-680.875	-925,483.271	-647,874.625
Ad	-680.875	11.160	-130.839	78.086
P	-925,483.271	-130.839	11,898.645	-1,358.383
CP	-647,874.625	78.086	-1,358.383	7,792.816

We saw that the four parameters in Table 12.12 and 12.13 plus p are sufficient to fully specify the prior. We are going to examine them more carefully here. These foregoing four enumerated variables for the prior are going to be apparent as the prior for the Bayesian update, which appears in Table 12.14. The figure accumulates the inputs to the Clemen[20] problem from what we term the "data prior" in Tables 12.12 and 12.13. We see the inputs needed to characterize the prior enumerated in Table 12.14. It is important to note their exact relationship to the areas in Tables 12.12 and 12.13. The equivalences are as follows:

- The prior mean β_0, which will be equated to the mean $\bar{\beta}$ in the middle of Table 12.12, which is shown in Table 12.14.
- The prior degrees of freedom ν_0, which will be equated to the degrees of freedom ν_C in Table 12.12, which is shown in Table 12.14.
- The prior variance term s_0^2, which will be equated to the error term s_C^2 in Table 12.12, which is shown in Table 12.14.
- The prior variance-covariance matrix of the model coefficients, which will be equated to the variance-covariance matrix in Table 12.13, is shown in Table 12.14.

These were calculated directly from the prior database. We see near the top of Table 12.14 the calculation of

- the expected value of the variance term S_E^2. This is the expected value of the embedded inverse gamma.
- the standard deviation of the variance term $\sqrt{V_S^2}$, which is the expected value of the embedded inverse gamma.

Table 12.14: Inputting the Prior from the Tiny, Prefatory Data Set for Use with the Large Data Set in the Clemen Problem

	β_0			
Intercept	3,856.8226			
Ad	13.2330			
P	-562.9990			
CP	547.1975			
	422,811.02	$\frac{\nu_0 s_0^2}{\nu_0-2} = S_E^2$		
	422,811.02	$SD(\sigma^2) = \sqrt{V_{S^2}}$		
	178,769,156,487	V_{S^2}		
	6	ν_0		
	281,874.01	s_0^2		
Variance-Covariance (VCV)				
	Intercept	Ad	P	CP
Intercept	148,696,889.2950	-680.8753	-925,483.2714	-647,874.6253
Ad	-680.8753	11.1604	-130.8390	78.0862
P	-925,483.2714	-130.8390	11,898.6452	-1,358.3834
CP	-647,874.6253	78.0862	-1,358.3834	7,792.8161
	11	$m = \nu_0 + p + 1$		
	1,691,244	$M = \nu_0 s_0^2$		
	422,811	$\frac{\nu_0 s_0^2}{\nu_0-2}$		
	178,769,156,487	$\frac{2}{\nu_0-4}\left(\frac{\nu_0 s_0^2}{\nu_0-2}\right)^2$		

- the variance term V_{S^2}. This is the variance of the embedded inverse gamma.

At the bottom, we see the calculation of the key parameters of the prior:

- $m = \nu_0 + p + 1$
- $M = \nu_0 s_0^2$
- $\frac{\nu_0 s_0^2}{\nu_0-2}$
- $\frac{2}{\nu_0-4}\left(\frac{\nu_0 s_0^2}{\nu_0-2}\right)^2$

that characterize the Student's t distribution. We are going to need the precision matrix so that we can calculate the z_{ii} terms for the univariate densities properly. We go through the process of calculating the precision matrix Q in the second panel of Table 12.15.

Table 12.15: Calculating the Precision Matrix in the Clemen Problem–Prior

$VCV = \frac{\nu_0 s_0^2}{\nu_0 - 2} Q^{-1} \Rightarrow Q = VCV^{-1} \frac{\nu_0 s_0^2}{\nu_0 - 2}$

VCV^{-1}	Intercept	Ad	P	CP
Intercept	0.00002365	0.01189657	0.00222558	0.00223504
Ad	0.01189657	6.09267473	1.12055973	1.12332692
P	0.00222558	1.12055973	0.20952387	0.21032328
CP	0.00223504	1.12332692	0.21032328	0.21134974

$Q = VCV^{-1} \frac{\nu_0 s_0^2}{\nu_0 - 2}$	Intercept	Ad	P	CP
Intercept	10.00	5,030.00	941.00	945.00
Ad	5,030.00	2,576,050.00	473,785.00	474,955.00
P	941.00	473,785.00	88,589.00	88,927.00
CP	945.00	474,955.00	88,927.00	89,361.00

Q^{-1}	Intercept	Ad	P	CP
Intercept	351.68641108	-0.00161035	-2.18888164	-1.53230308
Ad	-0.00161035	0.00002640	-0.00030945	0.00018468
P	-2.18888164	-0.00030945	0.02814176	-0.00321274
CP	-1.53230308	0.00018468	-0.00321274	0.01843097

$\left[Q^{-1}\right]_{ii} = z_{ii}$	Int	Ad	P	CP
	351.68641108	0.00002640	0.02814176	0.01843097

	$Q\beta_0$
Intercept	92,450
Ad	46,642,500
P	8,684,000
CP	8,762,100

The first panel of the table simply inverts the variance-covariance matrix. The reason that it is necessary is apparent in the relevant equation at the top left of Table 12.15. The precision matrix is not a simple inverse of the variance-covariance matrix; it also requires the multiplicative factor evident in the referenced equation. The second panel of the table thereafter calculates the precision matrix. The third panel of the table is the inverse of the precision matrix. It is the diagonals of this pure precision matrix that we will need in order to calculate the distributions over the coefficients under the prior. At the bottom of the table, we see the product of the Q matrix and the prior mean β_0. This will serve as the mean of the prior forecast. We see in Table 12.16 the variances, standard deviations, and t-statistics. We see at the bottom of the table the correlation matrix.

Table 12.16: Variances and Correlations for the Clemen Problem–Prior

	Int	Ad	P	CP
Variances	148,696,889.2950	11.1604	11,898.6452	7,792.8161
Std. Dev	12,194.1334	3.3407	109.0809	88.2769
t-statistic	0.3163	3.9611	-5.1613	6.1986

Correlation	Int	Ad	P	CP
Int	1.0000	-0.0167	-0.6958	-0.6019
Ad	-0.0167	1.0000	-0.3590	0.2648
P	-0.6958	-0.3590	1.0000	-0.1411
CP	-0.6019	0.2648	-0.1411	1.0000

Table 12.17 gives the key parameters of the marginal density over model coefficients. We see in the first row the reciprocals of the precision matrices (which are actually scalars) of the four individual marginal univariate Student's t distributions of the coefficients. Below that, we see the constituents of the constant for the marginal density. We see in the first row the product of the ν_0 from Table 12.14 times the s_0^2 from Table 12.14 times the z_{ii} term from Table 12.15. This is the reciprocal of the precision matrix for each of the model coefficients.

The calculations for this δ, the same value we used for the classical regression analysis previously, appear in Table 12.18. The first panel computes the matrix $Q^{-1}\delta^T$. The second panel computes the matrix $\delta Q^{-1}\delta^T$. The third panel adds 1 to the matrix product, akin to adding an identity matrix in higher dimensions. The determinant and its square root are indicated in the third panel. The final panel shows the inverse.

We are now in a position to calculate all the parameters for the univariate Student's t distribution of each of the predicted variables. That calculation is summarized in Table 12.19. The numbers in this figure are sufficient to calculate and plot the prior predictive densities for each tree scenario going into the (analytically expressed) calculation of revenue. We note in the first row the calculation of the arguments of the gamma function in the constant of the predictive distributions. The second row calculates those inverse gamma distributions themselves. It is noteworthy that the degrees of freedom in the prior are small relative to the data base calculated in the classical regression solution. Of course! The prior database was much smaller. In the third panel, we see the calculation of the denominator of the constant of the univariate Student's t distribution. We see the constant in the next panel. We note in the next panel the product of the δ matrix and the prior mean. Below that, we see the calculation of the denominator of the constant term inside the braces that multiplies the quadratic. The parameters of the predictive distribution going into the univariate Student's t distribution are summarized in Table 12.19. Notice that the last term is the denominator of the univariate Student's t distribution, which appears in the equation at the bottom

Table 12.17: Calculate the Marginal Densities Over Model Coefficients–Prior

	Int	Ad	P	CP
$\nu_0 s_0^2 z_{ii}$	594,787,557.18	44.64	47,594.58	31,171.26
$\frac{\nu_0+1}{2}$	3.5	$\frac{\nu_0}{2}$	3	
$\Gamma\left(\frac{\nu_0+1}{2}\right)$	3.32335097	$\Gamma\left(\frac{\nu_0}{2}\right)$	2	
$\pi^{\frac{1}{2}}$	1.772453851			
First Term	0.93750000	$\frac{\Gamma\left(\frac{\nu_0+1}{2}\right)}{\pi^{\frac{1}{2}}\Gamma\left(\frac{\nu_0}{2}\right)}$		
	Int	Ad	P	CP
Constant	3.84406E-05	0.140314208	0.004297269	0.005309994
	Int	Ad	P	CP
Exp. Coeff.	.68127E-09	0.022400656	2.10108E-05	3.20808E-05

$$\{\beta_i\} = \frac{\Gamma\left(\frac{\nu_0+1}{2}\right)}{\pi^{\frac{1}{2}}\Gamma\left(\frac{\nu_0}{2}\right)} \frac{1}{\left(z_{ii}\nu_0 s_0^2\right)^{\frac{1}{2}}} \left[1 + (\beta_i - \beta_0)\frac{1}{\nu_0 s_0^2 z_{ii}}(\beta_i - \beta_0)\right]^{-\frac{\nu_0+1}{2}}$$

Table 12.18: Calculate the Marginal Densities Over Model Coefficients–Prior

$Q^{-1}\delta^T$			
Int	0.42524044		
Ad	-0.00050274		
P	0.02952780		
CP	-0.03016855		
$\delta Q^{-1}\delta^T$	0.17082237		
			Determinant
$1 + \delta Q^{-1}\delta^T$	1.17082237		1.17082237
			Sqrt. Determ.
			1.08204546
$\left[1 + \delta Q^{-1}\delta^T\right]^{-1}$	0.85410053		

$$\{d|X, y, \delta\} = \frac{\Gamma\left(\frac{\nu_0+1}{2}\right)}{\left[\pi\nu_0 s_0^2(I+\delta Q^{-1}\delta^T)\right]^{\frac{1}{2}}\Gamma\left(\frac{\nu_0}{2}\right)}\left[1 + \frac{(d-\delta\beta_0)^2}{\nu_0 s_0^2(I+\delta Q^{-1}\delta^T)}\right]^{-\frac{\nu_0+1}{2}}$$

center of Table 12.18.

Table 12.19: Calculation of Parameters of the Predictive Distribution–Prior

$\frac{\nu_0+1}{2}$	3.5	$\frac{\nu_0}{2}$	3
$\Gamma\left(\frac{\nu_0+1}{2}\right)$	3.32335097	$\Gamma\left(\frac{\nu_0}{2}\right)$	2
$\left\{\pi\nu_0 s_0^2\left[1+\delta Q^{-1}\delta^T\right]\right\}^{\frac{1}{2}}$	2494.155838		
Constant	0.000666228		
$\delta\beta_0$	7943.97		
$\nu_0 s_0^2\left(1+\delta Q^{-1}\delta^T\right)$	1980146.39	Sq. Rt.	1407.18
Precision	5.050132E-07	Sq. Rt. Prec.	0.000710643

The actual parameters of the univariate Student's t distribution, which we calculate for plotting purposes, appear in Table 12.20. The constant represents everything outside the braces. The precision matrix (actually a scalar here) is inside the braces and includes everything save for the quadratic term.

Table 12.20: Summary of Parameters of the Predictive Distribution–Prior

Constant	0.000666228		
$\delta\beta_0$	7,943.9708	$\frac{\nu_0+1}{2}$	3.5
Precision	5.050132E-07		
Variance	495,036.5970	Std. Dev.	703.5884

$$\{d\,|X,y,\delta\} = \frac{\Gamma\left(\frac{\nu_0+1}{2}\right)}{\left[\pi\nu_0 s_0^2(I+\delta Q^{-1}\delta^T)\right]^{\frac{1}{2}}\Gamma\left(\frac{\nu_0}{2}\right)}\left[1+\frac{(d-\delta B_0)^2}{\nu_0 s_0^2(I+\delta Q^{-1}\delta^T)}\right]^{-\frac{\nu_0+1}{2}}$$

The probabilistically weighted predictive distribution for the combination of decisions represented in the δ vector appears in Figure 12.5. The predictive distribution for the base case and the two other competitor pricing cases appears in Figure 12.6. The probabilistically weighted predictive distribution over all three of the (uncertain) competitor prices appears in Figure 12.7.

Figure 12.5: Plot of Conditional Predictive Distribution–Prior

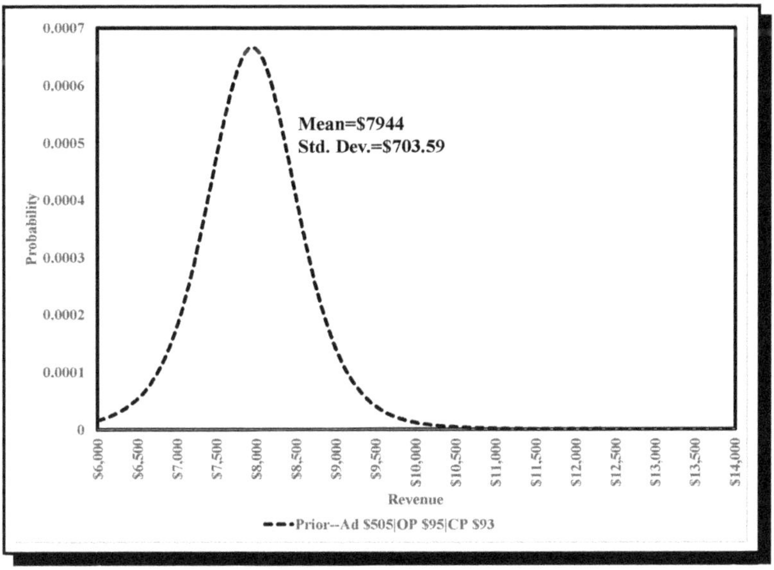

Figure 12.6: Plot of Three Conditional Predictive Distributions–Prior

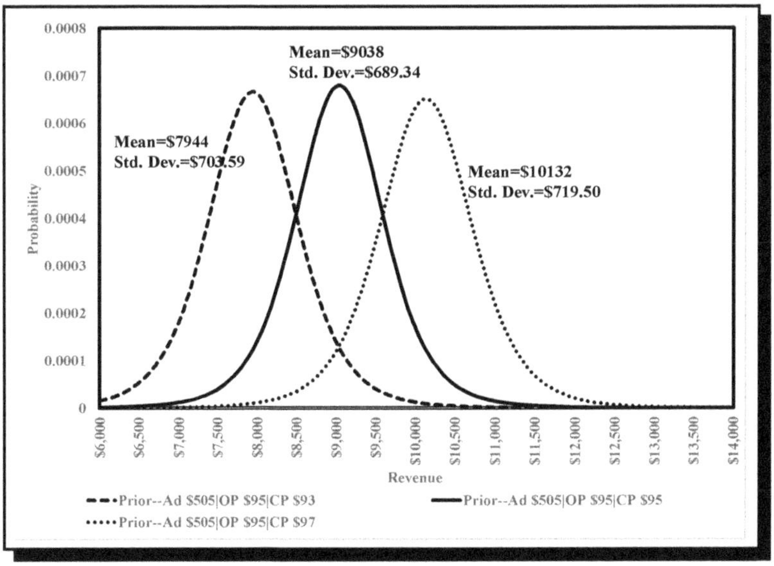

Figure 12.7: Plot of Overall Probability Distribution Over Sales–Prior

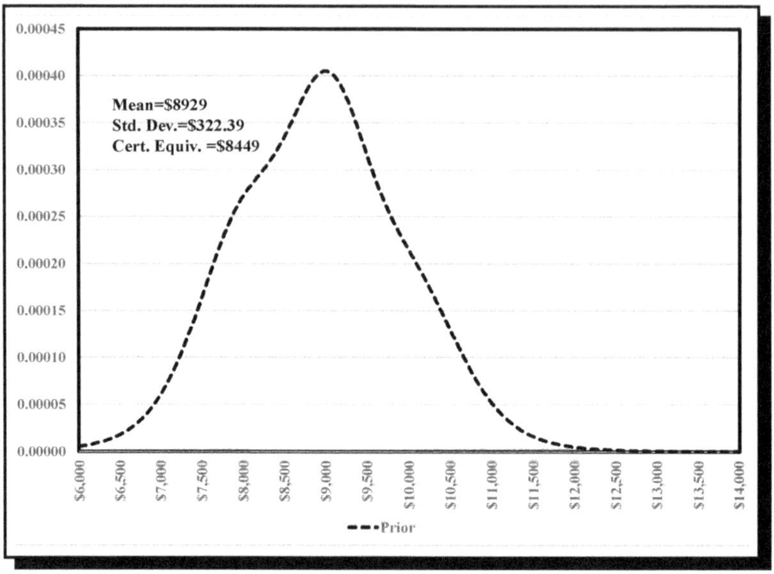

12.3 The Posterior (Using the Small Data Set to Compute the Prior)

At last, we have come to the posterior. With all the work we have done with the classical and prior, articulating the posterior will be a bit simpler. We begin just as before with the precision matrix and the variance-covariance matrix of the posterior. See Table 12.21. The precision matrix, computed according to equation (6.7), appears in the top panel of the figure. This is the so-called matrix weighted average derived carefully herein. The second panel is the inverse of this precision matrix. The diagonals of this inverse precision matrix are going to give us the coefficients of the univariate marginal distributions over the various coefficients. And, precisely as expected, the lower panel extracts those inverse precision diagonals to accumulate the z_{ii} terms that will comprise the precision of the univariate Student's t distributions over the coefficients.

The next figure, Table 12.22, details the important parameters of the multivariate posterior. Beginning at the top, we see the calculation of the A constant and the various critical terms that comprise it.

- Prior degrees of freedom ν_0. See Table 12.14.
- Prior m. See Table 12.14.
- Prior s_0^2. See Table 12.14.
- Prior $M = \nu_0 s_0^2$. See Table 12.14.

Table 12.21: Precision Matrix, Inverse, Mean, and Calculations–Posterior

(X^TX+Q)	Int	Ad	P	CP
Int	46.00	23,326.00	4,341.96	4,356.80
Ad	23,326.00	12,030,574.00	2,202,080.16	2,208,103.44
P	4,341.96	2,202,080.16	410,147.00	411,270.61
CP	4,356.80	2,208,103.44	411,270.61	412,937.31

$(X^TX+Q)^{-1}$	Int	Ad	P	CP
Int	57.2246868841	-0.0038751268	-0.2719857963	-0.3121540906
Ad	-0.0038751268	0.0000050794	-0.0000075275	0.0000212217
P	-0.2719857963	-0.0000075275	0.0032968417	-0.0003736281
CP	-0.3121540906	0.0000212217	-0.0003736281	0.0035545237

$\left[(X^TX+Q)^{-1}\right]_{ii} = z_{ii}$	Int	Ad	P	CP
	57.2246868841	0.0000050794	0.0032968417	0.0035545237

	β^*		$Q\beta_0$	X^Ty	$Q\beta_0 + X^Ty$
Int	1898.203378	Int	92,450.00	345,966.00	438,416.00
Ad	14.5645759	Ad	46,642,500.00	177,849,135.00	224,491,635.00
P	-511.5341817	P	8,684,000.00	32,561,320.38	41,245,320.38
CP	512.4004551	CP	8,762,100.00	32,878,377.14	41,640,477.14

- Classical regression degrees of freedom ν_C. See Table 12.3 .
- Classical regression s_C^2. See Table 12.3 .
- Classical regression $R = \nu_C s_C^2$. See Table 12.3 .
- Posterior degrees of freedom ν. See equation (6.2).
- Posterior s^2. See equation (6.5).
- Posterior $A = \nu s^2$. See equation (6.4).
- Posterior mean of σ^2 term, $\frac{\nu s^2}{\nu - 2}$.
- Posterior variance of σ^2 term, $\frac{2}{\nu-4}\left(\frac{\nu s^2}{\nu-2}\right)^2$.
- $(\bar{\beta} - \beta^*)$, which is the difference between the classical statistics mean and the posterior mean.
- $(\bar{\beta} - \beta^*)^T X^T y$, which is the classical statistics term in the posterior expression for A in equation (8.4).
- $(\beta_0 - \beta^*)$, which is the difference between the prior mean and the posterior mean.
- $(\beta_0 - \beta^*)^T Q\beta_0$, which is the prior term in the posterior expression for A in equation (8.4).

Table 12.23 depicts the variance-covariance matrix for the posterior for the coefficients, calculated according to equation (6.16) and in the conjugate-archetype case equation (4.19). The variances, standard deviations, t-statistics, and correlations appear in Table 12.24.

We now turn to the joint and marginal distributions over the model coefficients, particularly the marginals. Table 12.25 begins with the equation for the

Table 12.22: Precision Matrix, Inverse, Mean, and Calculations–Posterior

		42	$\nu = m + n - p - 1 = \nu_0 + n$		
$R = \nu_C s_C^2$		6,744,669.36			
$M = \nu_0 s_0^2$		1,691,244.07			
	$(\bar{\beta} - \beta^*)$		$(\bar{\beta} - \beta^*)^T X^T y$		
Int	301.138873		24,402,469.7565		
Ad	0.482027				
P	7.770144				
CP	-12.729204				
	$(\beta_0 - \beta^*)$		$(\beta_0 - \beta^*)^T Q \beta_0$		
Int	1958.619245		-23,057,071.4594		
Ad	-1.331535				
P	-51.464812				
CP	34.797036				
$A = \nu s^2$	9,781,311.7238	$s^2 \equiv \frac{A}{\nu}$	232,888.3744	$\frac{\nu s^2}{\nu-2}$	244,532.7931
	3,147,172,994.67	$\frac{2}{\nu-4}\left(s^2 \frac{\nu}{\nu-2}\right)^2$			

marginal density over each of the model coefficients, a formula appearing as (6.31) in the text. The top row of the figure is the inverse of the precision matrix for the marginals. It is the product of the z_{ii} terms accumulated and enumerated in Table 12.21 for the inverse precision matrix times the posterior νs^2 in Table 12.22. We see in Table 12.25 the arguments of the two gamma functions in the next row, and we see the values of the two gamma functions in the next row. We see two rows below the value of the pure constant based on the gamma function quotient and the π term. It is termed the First Term. Below that, we see the constituents of the constants in the various marginal densities. In the final row, we see the precision matrices (actually the constants of the quadratics) of the marginal distributions.

Table 12.23: Variance-Covariance Matrix of the Coefficients–Posterior

$\frac{\nu s^2}{\nu-2}\left(X^T X + Q\right)^{-1}$	Int	Ad	P	CP
Int	13,993,312.517751	-947.595568	-66,509.446443	-76,331.911661
Ad	-947.595568	1.242069	-1.840709	5.189401
P	-66,509.446443	-1.840709	806.185909	-91.364318
CP	-76,331.911661	5.189401	-91.364318	869.197620

The predictive distribution requires the calculations initiated in Table 12.26. We see at the bottom of the figure the equation for the marginal densities over each of the predicted scenarios as represented when we insert the row vector δ as in equation (6.29) in place of the matrix D. We see in the first panel, the

Table 12.24: Variances, Standard Deviations, t-statistics, Correlation–Posterior

	Int	Ad	P	CP
Variances	13,993,312.5178	1.2421	806.1859	869.1976
Std. Dev	3,740.7636	1.1145	28.3934	29.4822
t-statistic	0.5074	13.0685	-18.0159	17.3800

Correlation	Int	Ad	P	CP
Int	1.0000	-0.2273	-0.6262	-0.6921
Ad	-0.2273	1.0000	-0.0582	0.1579
P	-0.6262	-0.0582	1.0000	-0.1091
CP	-0.6921	0.1579	-0.1091	1.0000

Table 12.25: Calculate the Marginal Densities Over Model Coefficients–Posterior

	Int	Ad	P	CP
$\nu s^2 z_{ii}$	559,732,500.71	49.68	32,247.44	34,767.90

$\frac{\nu+1}{2}$	21.5	$\frac{\nu}{2}$	21
$\Gamma\left(\frac{\nu+1}{2}\right)$	1.108280E+19	$\Gamma\left(\frac{\nu}{2}\right)$	2.432902E+18
$\pi^{\frac{1}{2}}$	1.772453851		
First Term	2.57009910	$\frac{\Gamma\left(\frac{\nu+1}{2}\right)}{\pi^{\frac{1}{2}}\Gamma\left(\frac{\nu}{2}\right)}$	

	Int	Ad	P	CP
Constant	0.000108632	0.364625493	0.014312064	0.013783535

	Int	Ad	P	CP
Exp. Coeff.	1.786568E-09	2.012771E-02	3.101022E-05	2.876216E-05

$$\{\beta_i\} = \frac{\Gamma\left(\frac{\nu+1}{2}\right)}{\pi^{\frac{1}{2}}\Gamma\left(\frac{\nu}{2}\right)} \frac{1}{(z_{ii}\nu s^2)^{\frac{1}{2}}} \left[1 + (\beta_i - \beta_i*) \frac{1}{\nu s^2 z_{ii}} (\beta_i - \beta_i*)\right]^{-\frac{\nu+1}{2}}$$

first calculation, the product of the inverse precision times the transpose of δ. The second panel premultiplies this by δ. The third panel adds 1 (the equivalent of adding the identity matrix). The final panel indicates the inverse precision matrix.

Table 12.26: Calculating Parameters of the Predictive Distribution–Posterior

$(X^TX+Q)^{-1}\delta^T$		
Int	0.39876680	
Ad	-0.00005154	
P	0.00266539	
CP	-0.00636109	
$\delta(X^TX+Q)^{-1}\delta^T$	0.03436826	
		Determinant
$1+\delta(X^TX+Q)^{-1}\delta^T$	1.03436826	1.03436826
		Sqrt. Determ.
		1.01703896
$\left[1+\delta(X^TX+Q)^{-1}\delta^T\right]^{-1}$	0.96677367	

$$\{d\,|X,y,\delta\} = \frac{\Gamma(\frac{\nu+1}{2})}{\{\pi\nu s^2[1+\delta(X^TX+Q)^{-1}\delta^T]\}^{\frac{1}{2}}\Gamma(\frac{\nu}{2})}\left\{1+\frac{(d-\delta\beta^*)^2}{\nu s^2[1+\delta(X^TX+Q)^{-1}\delta^T]}\right\}^{-\frac{\nu+1}{2}}$$

The parameters of the predictive distribution going into the univariate Student's t distribution summarized in Table 12.26 allow us to calculate all the parameters for the univariate Student's t distribution for the predicted variable. The calculation is summarized in Table 12.27. We note the two different degrees of freedom in the arguments for the gamma function and the two different gamma function values. We see the calculation of the denominator in the leading constant term following that. The leading constant term is shown immediately below that, with the leading constant being the constant that lies entirely outside the braces in the formula. We note the calculated mean value and, finally, the denominator of the term inside the braces that is associated with the quadratic.

Table 12.28 summarizes the actual parameters of the univariate Student's t distribution for the predictive distribution. We see the calculation of the leading constant, the precision term, which is the multiplier of the quadratic term, and the variance.

The predictive distribution for the base case appears in Figure 12.8. The

Table 12.27: Summary of Parameters of the Predictive Distribution–Posterior

$\frac{\nu+1}{2}$	21.5	$\frac{\nu}{2}$	21
$\Gamma(\frac{\nu+1}{2})$	1.108280E+19	$\Gamma(\frac{\nu}{2})$	2.432902E+18
	5637.818342	$\left\{\pi\nu s^2\left[1+\delta(X^TX+Q)^{-1}\delta^T\right]\right\}^{\frac{1}{2}}$	
Const	0.000808004		
$\delta\beta^*$	8310.809269		
$\nu s^2\left[1+\delta(X^TX+Q)^{-1}\delta^T\right]$	10,117,478.35	Sq. Rt.	3180.798382
Precision	9.883886E-08	Sq. Rt. Prec.	0.000314386

$$\{d\,|X,y,\delta\} = \frac{\Gamma(\frac{\nu+1}{2})}{\{\pi\nu s^2[1+\delta(X^TX+Q)^{-1}\delta^T]\}^{\frac{1}{2}}\Gamma(\frac{\nu}{2})}\left\{1+\frac{(d-\delta\beta^*)^2}{\nu s^2[1+\delta(X^TX+Q)^{-1}\delta^T]}\right\}^{-\frac{\nu+1}{2}}$$

Table 12.28: Summary of Parameters of the Predictive Distribution–Posterior

Constant	0.000808004		
$\delta\beta^*$	8310.8093	$\frac{\nu_0+1}{2}$	21.5
Precision	9.8838857E-08		
Variance	252,936.9588	Std. Dev.	502.9284

predictive distribution for the base case and the two other competitor pricing cases appears in Figure 12.9. The expected predictive distribution over all three of the (uncertain) competitor prices appears in Figure 12.10.

12.4 The Best Decision for the Clemen Problem

After all this, how can we calculate the best decision and its certain equivalent? We see in Table 12.29 the nine expected values we obtain for the nine policies under classical regression in the Classical column. We see the nine in the prior column, and we see the nine in the Bayesian posterior column. We see that the best strategy in all cases is

- Conduct the highest level of advertising, namely \$580.
- Price (aggressively) at the lowest level, namely \$93.

We have not, in this example (as Clemen[20] did not), included product cost or other distinctions. Had we done so, the answer might have been different. However, the point is made. Generalizing to include risk attitudes requires that we calculate expected utility. When we do so, we generate the table in Table 12.30. While straightforward, this requires an additional calculation for each strategy. One disadvantage of the Student's t distribution is that one cannot calculate the expected utility for an exponential utility function in a simple closed form. This is because there is no tractable moment-generating function for a Student's t distribution. However, we can calculate expected utility by discretization (which we have done here) or simulation (which we do not do in this textbook).

Figure 12.8: Plot of Conditional Distribution over Sales–Posterior

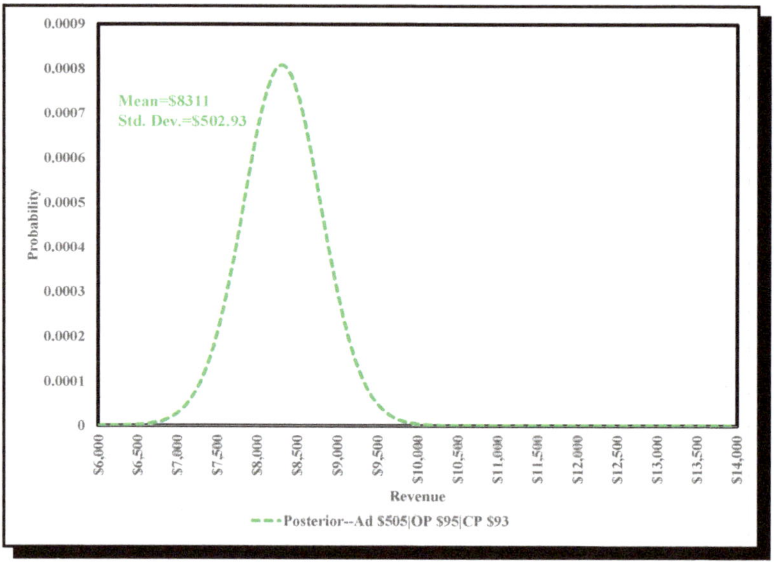

Table 12.29: The Best Decision in the Clemen Problem–Expected Sales

Advertising	Our Price	Classical	Prior	Bayes
$425	$93	$9,113	$8,996	$9,091
$425	$95	$8,105	$7,870	$8,068
$425	$97	$7,098	$6,744	$7,045
$505	$93	$10,317	$10,055	$10,256
$505	$95	$9,309	$8,929	$9,233
$505	$97	$8,302	$7,803	$8,210
$580	$93	$11,445	$11,047	$11,349
$580	$95	$10,438	$9,921	$10,325
$580	$97	$9,430	$8,795	$9,302

Figure 12.9: Plot of Conditional Probability Distributions Over Sales–Posterior

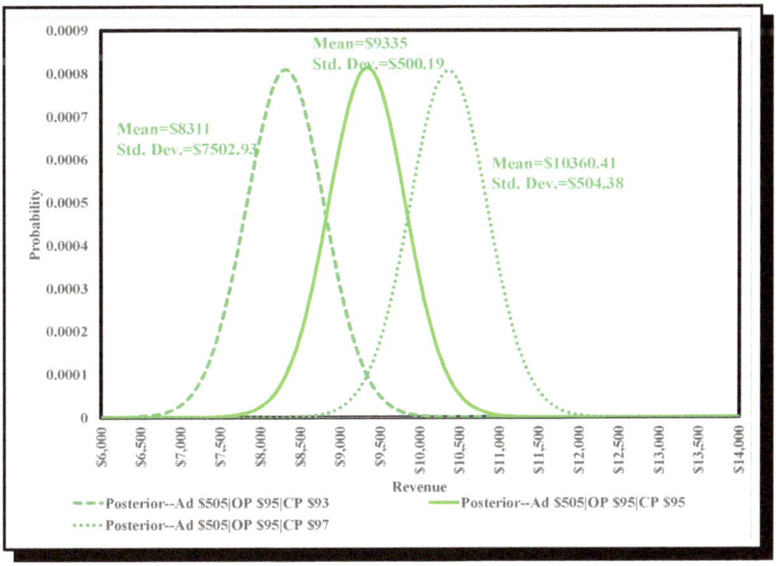

Figure 12.10: Plot of Overall Probability Distribution Over Sales–Posterior

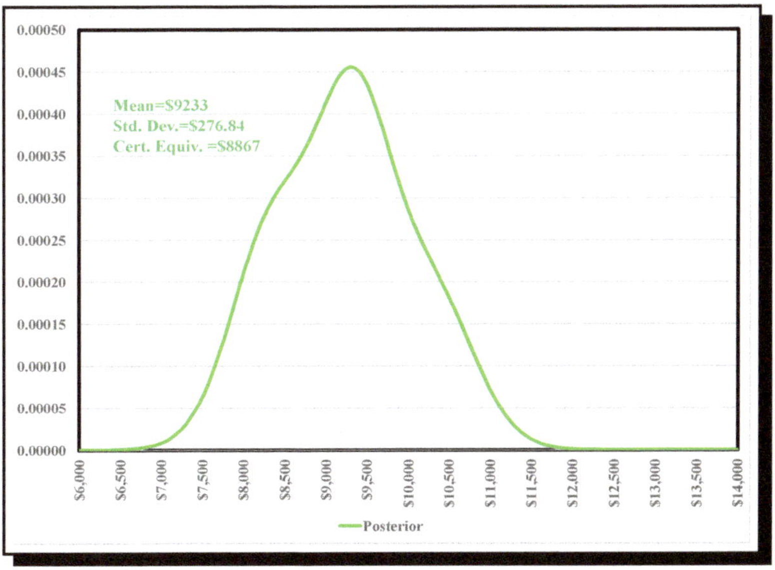

Table 12.30: The Best Decision in the Clemen Problem–Certain Equivalent Sales

Advertising	Our Price	Classical	Prior	Bayes
\$425	\$93	\$8,688	\$8,571	\$8,666
\$425	\$95	\$7,680	\$7,445	\$7,643
\$425	\$97	\$6,673	\$6,319	\$6,620
\$505	\$93	\$9,812	\$9,550	\$9,751
\$505	\$95	\$8,804	\$8,424	\$8,728
\$505	\$97	\$7,797	\$7,298	\$7,705
\$580	\$93	\$10,865	\$10,467	\$10,769
\$580	\$95	\$9,858	\$9,341	\$9,745
\$580	\$97	\$8,850	\$8,215	\$8,722

12.5 Comparative Analysis of the Coefficients

We would like to know how the marginal probability distribution over the individual coefficients β_i changes before and after we gather data. We want to get a feel for how much the data means to our knowledge of each of the coefficients. What is the data really worth in terms of changing our estimate of each of the coefficients? How much does it contribute to our knowledge? Does gathering data lead to a large change in our estimate of each or any of the coefficients, or are the values of the coefficients more dependent on the prior (perhaps meaning that the data is rather worthless)? That is the question we address in this section. This section presents and discusses the marginal probability distribution over each of the model coefficients predicted (probabilistically) under classical statistics, under the prior, and under the posterior. We are going to proceed through each of the three model coefficients plus the intercept.

We focus first on the intercept coefficient in Figure 12.11. The figure does not show (and therefore disproves in general) the most frequent result people seem to expect. People seem to expect that the marginal posterior should always lie between the marginal prior and the marginal classical. They recall the calculation of the posterior mean, which balances the classical solution and the prior solution to create the posterior. People want to think that the mean of the marginal posterior should lie between the mean of the marginal prior and the mean of the marginal classical for every marginal distribution. We often see that—the mean of the marginal posterior lying between the mean of the marginal prior and the mean of the marginal classical—but it is not true in general. Figure 12.11 shows that the posterior mean (green) does not lie between the prior mean (black) and the classical regression mean (amber). This disproves any such conjectured theorem by example.

The advertising coefficient appears in Figure 12.12. The figure does indeed show the most frequent result that the posterior mean (green) should lie between the prior mean (black) and the classical regression mean (amber). As

Figure 12.11: Comparative Analysis of the Intercept Coefficient

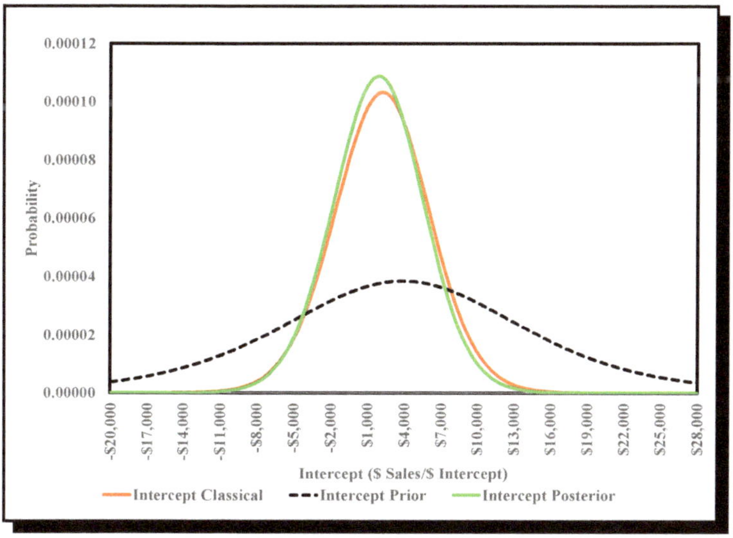

stated above, we generally see (but it is NOT true in general) that the posterior mean lies between the prior mean and the classical regression mean. We see here no violation of that hoped-for general principle, i.e., that the posterior should lie between the prior and the classical regression solution. The lightest gray line (posterior) lies between the black dotted line and the medium gray line (classical).

Figure 12.12: Comparative Analysis of the Advertising Coefficient

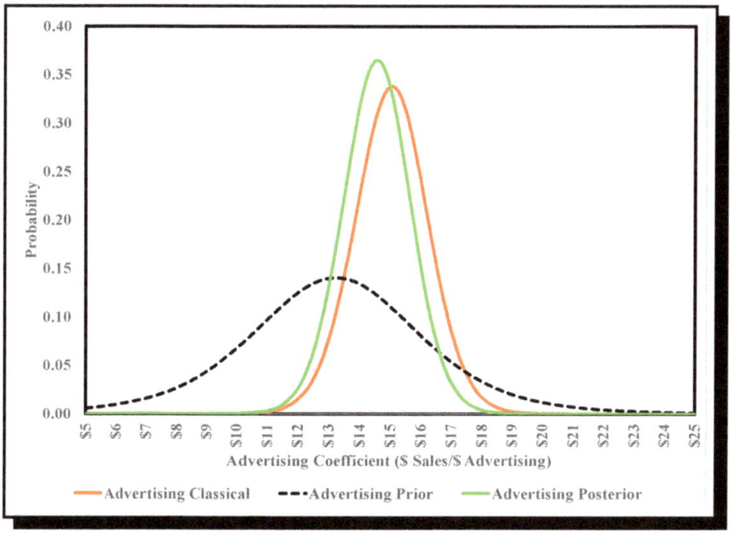

The own price coefficient appears in Figure 12.13. The figure reinforces the frequent result that the posterior mean lies between the prior mean and the classical regression mean:

Figure 12.13: Comparative Analysis of the Own Price Coefficient

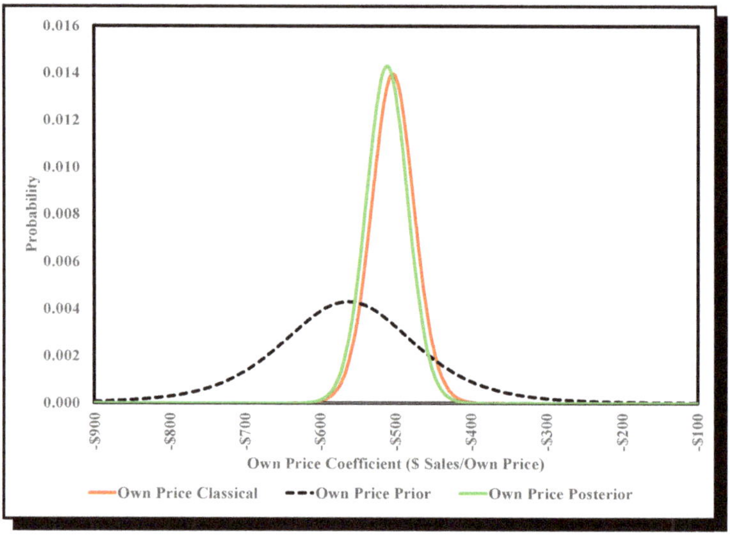

The competitor price coefficient appears in Figure 12.14. The figure again shows the posterior mean lying between the prior mean and the classical regression mean.

Figure 12.14: Comparative Analysis of the Competitor Price Coefficient

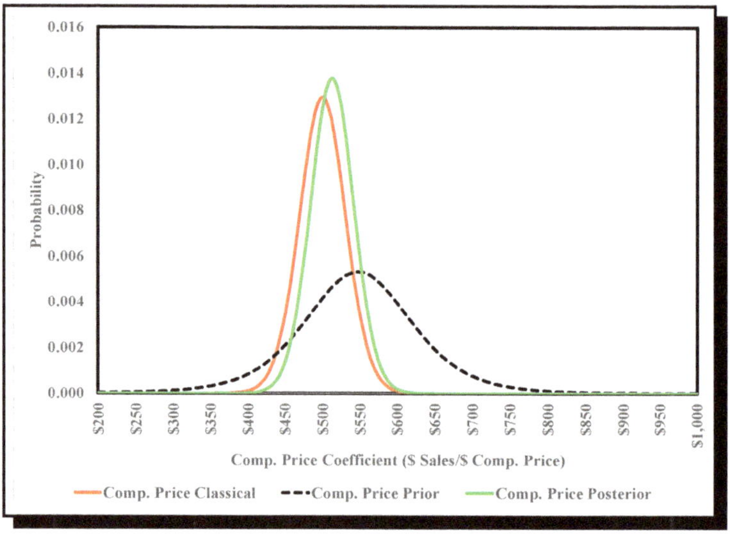

12.6 Comparative Analysis of the Predictions

This section compares the high, medium, and low competitor prices for the own price and the advertising level in Table 12.7. We begin in Figure 12.15 with the competitor price of $93. We have plotted the densities in that case under classical regression, the prior, and the posterior.

Figure 12.15: Predictions for Clemen Case 1–Classical, Prior, and Posterior

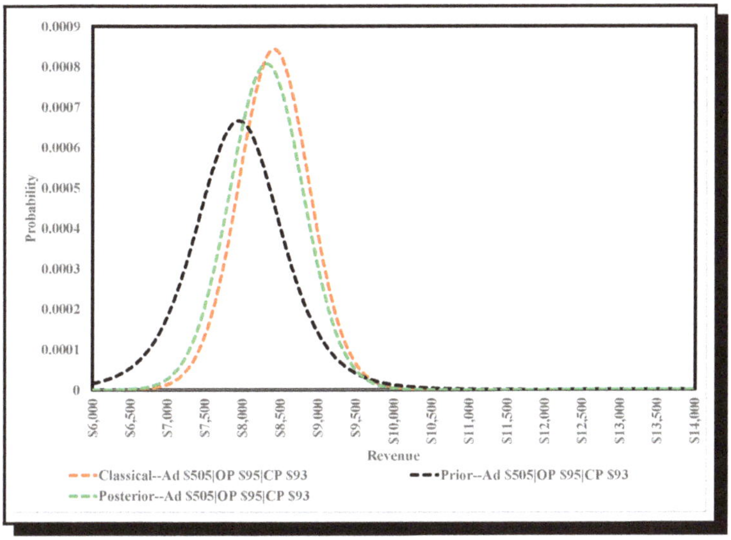

We move next to Figure 12.16, which assumes the same alternatives but the competitor's price of $95. We have plotted the densities in that case under classical regression, the prior, and the posterior:

We conclude with Figure 12.17, which assumes the competitor price of $97. We have plotted the densities in that case under classical regression, the prior, and the posterior:

We see all three competitor prices plotted on a common graph in Figure 12.18. We note in every case that the posterior lies between the prior and the classical regression solution, and it tends to be "tighter" than either. This is actually a profound observation. The Bayesian posterior has "less uncertainty" than either the prior or the classical regression solution, a narrower distribution. We can really see it there. The posterior has a lower variance than either the prior or the classical solution. This is not always true, and we will detail a case in which it is not true, but it is a bit of an edge case.

The case in which the posterior will have more uncertainty than the prior or the classical is the situation in which the prior is very narrow and centered around one mean value and the classical is centered around an entirely different mean. This is a situation we dub "an arrogant prior," one that is too tight and

Figure 12.16: Predictions for Clemen Case 2–Classical, Prior, and Posterior

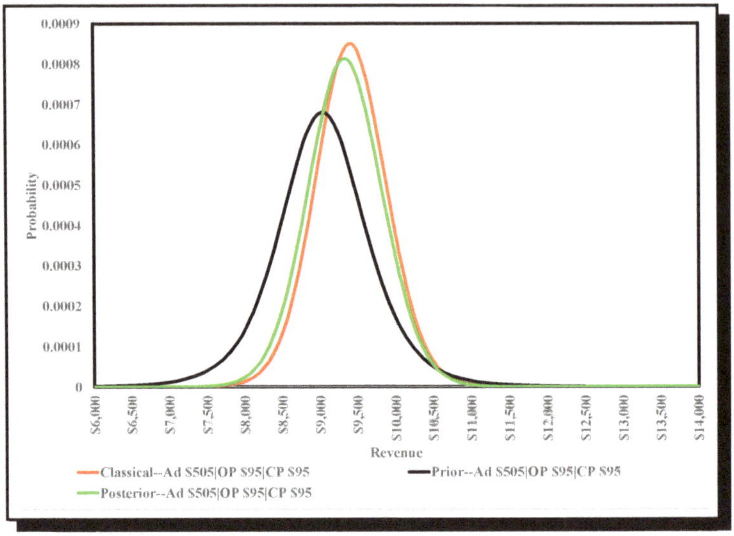

Figure 12.17: Predictions for Clemen Case 3–Classical, Prior, and Posterior

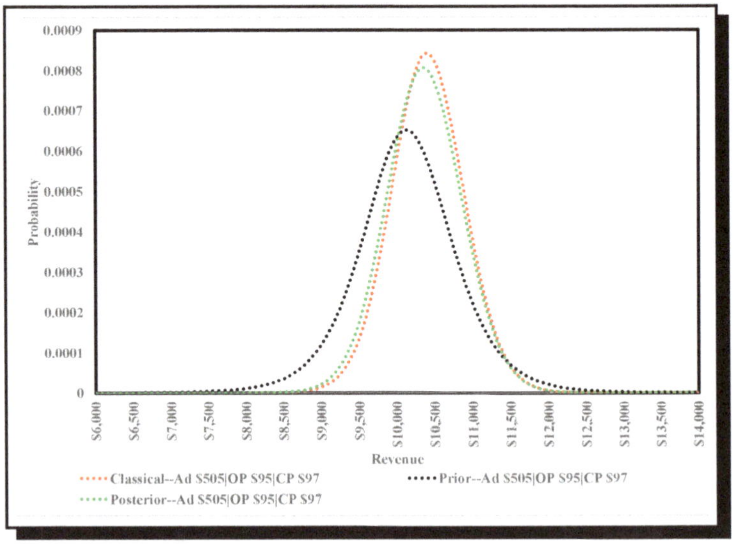

Figure 12.18: Predictions for Prior, Classical, and Posterior

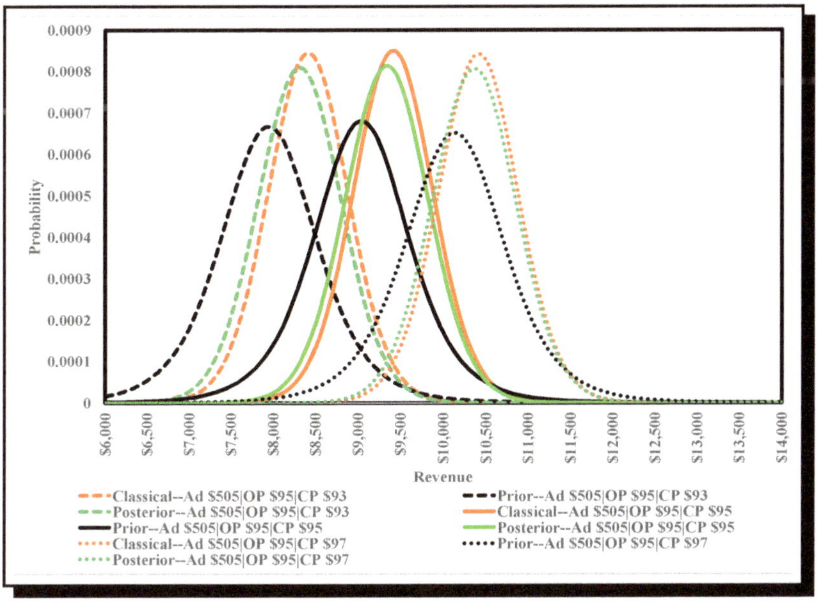

centered around a "contrarian" value. We saw this with KGP in Chapter 11. You fervently believe you are absolutely right about your prior information. You are darn sure of your estimate (as reflected by the low variance). You "don't want to hear it" that you could be wrong, and yet the data gives evidence that you are, in fact, wrong. Kahneman, Slovic, and Tversky[51] have studied the phenomenon and termed it "central tendency," meaning subject matter experts express less uncertainty than they truly believe. They observe that people think they know more than they truly know, express less uncertainty than they actually have, or are embarrassed or reluctant to admit in their minds the true degree of uncertainty that they actually have. Experts think they have to appear "certain." How many times have we seen experts who are far too certain about their expertise? Experts fear that expertise MEANS low variance in one's judgment. An expert is sure, and he or she knows they are sure! What self-respecting expert billing out at $4500/day plus expenses could ever offer a wide probability assessment over something he or she is supposed to be an expert in? Experts have to be sure! That's what they are paid for.

The arrogant prior is not always present, but it frequently is, empirically speaking. In the real world, we would expect the prior to actually be "wider," not narrower, empirically speaking. People don't really know much in advance about data gathering or experimentation. Prof. Howard's[45] party host, Kim, isn't particularly certain about the weather until after a few Howard and Abbas[45] Acme Rain Detector readings. She initially "hip shoots" an estimate based on her mental model of the weather, one that may not be particularly well informed. Who knows? (It is her private, personal, individual probability assessment, and she has no choice but to go with it until she improves it because she has to make a decision now.) We also see quite clearly in the observed case that the Bayesian updating process has left us with a fully indigenously probabilistic prediction. It is not deterministic, as is so often erroneously assumed with regression. It is inherently probabilistic. This is another huge problem with classical regression. It does not sufficiently emphasize the inherently probabilistic nature of decision-making. People default to the regression coefficients and de facto assume there is no uncertainty. Clemen[20] approaches his decision problem that way. His analysis does not consider uncertainty in the regression results and the impact of such uncertainty on the final probability distribution over sales for each alternative and therefore the answer.

12.7 Comparative Analysis of Classical on the Whole Data Set, Prior and Collected

In Chapter 10, we proved analytically that performing classical analysis on the entire data set, previous plus newly collected data, is identical with Bayesian analysis using the previous data (old data) as a prior and updating with the newly collected data. If we compare the sufficient statistics for the classical solution on the entire data base to those calculated previously as the posterior of the Bayesian update, we will have the complete story. In this section, we will verify using a numerical example the result proven in Chapter 10, and that is the point of this section. We proceed through the five sufficient statistics in both cases:

- First, p is trivially the same.
- Second, the precision matrix for the entire, aggregate data set is calculated in Table 12.31. If we compare this term by term with the precision matrix for the Bayesian posterior in the center of Table 12.21, we see that every term in the entire matrix is identical. The two "data matrices" are identical.
- Third, the mean of the Bayesian calculation appears at the bottom left of Table 12.21. The mean calculated using the entire, aggregate data set appears in Table 12.32. Term by term comparison shows them to be identical.
- Fourth, the degrees of freedom for the Bayesian calculation appears at the top of Table 12.22. We see the degrees of freedom calculated for the entire, aggregate data set using classical statistics in Table 12.32. The two degrees of freedom are identical.
- Fifth, the residual sum of squared errors, as represented in the $\frac{\nu s^2}{\nu - 2}$ terms are identical. We see that term at the top of Table 12.22 for the Bayesian case, and we see that term in the center right in Table 12.32. They are identical.

Table 12.31: Precision Matrix and Inverse Calculated from Entire, Aggregate Data Set

$X^T X$	Int	Ad	P	CP
Int	46.00	23,326.00	4,341.96	4,356.80
Ad	23,326.00	12,030,574.00	2,202,080.16	2,208,103.44
P	4,341.96	2,202,080.16	410,147.00	411,270.61
CP	4,356.80	2,208,103.44	411,270.61	412,937.31
$(X^T X)^{-1}$	Int	Ad	P	CP
Int	57.2246868841	-0.0038751268	-0.2719857963	-0.3121540906
Ad	-0.0038751268	0.0000050794	-0.0000075275	0.0000212217
P	-0.2719857963	-0.0000075275	0.0032968417	-0.0003736281
CP	-0.3121540906	0.0000212217	-0.0003736281	0.0035545237
$\left[(X^T X)^{-1}\right]_{ii} = z_{ii}$	Int	Ad	P	CP
	57.2246868841	0.0000050794	0.0032968417	0.0035545237

Table 12.32: Degrees of Freedom and Parameters of the Inverse Gamma Calculated from Entire, Aggregate Data Set

	$X^T y$
Int	438,416.00
Ad	224,491,635.00
P	41,245,320.38
CP	41,640,477.14
	$\bar{\beta} = (X^T X)^{-1} X^T y$
Int	1,898.2034
Ad	14.5646
P	-511.5342
CP	512.4005

n	46
p	4
$\nu_C = n - p$	42
$R = (y - X\bar{\beta})^T (y - X\bar{\beta})$	9,781,311.73
$s_C^2 = \frac{R}{\nu_C}$	232,888.37
$\frac{\nu_C s_C^2}{\nu_C - 2}$	244,532.79
$\frac{2}{\nu_C - 4}\left(\frac{\nu_C s_C^2}{\nu_C - 2}\right)^2$	3,147,172,996.45
$\frac{\nu_C s_C^2}{\nu_C - 2}\sqrt{\frac{2}{\nu_C - 4}}$	56,099.67

The five elements of the sufficient statistic are shown to be identical for both the classical analysis of the entire aggregate data base and the posterior of the Bayesian update using the prior as input and the data to update that prior. This is numerical confirmation of what was proven in Chapter 10.

For completeness, it is useful to compare the variance-covariance matrices of both calculations. The variance-covariance matrix of the Bayesian updating calculation appears in Table 12.24. The classical calculation on the entire aggregate data base yields the variance-covariance matrix in Table 12.33, which is identical.

Table 12.33: Variance-Covariance Matrix Calculated from Entire, Aggregate Data Set

$\left(X^T X\right)^{-1} \frac{\nu_C s_C^2}{\nu_C - 2}$	Int	Ad	P	CP
Int	13,993,312.521710	-947.595568	-66,509.446462	-76,331.911682
Ad	-947.595568	1.242069	-1.840709	5.189401
P	-66,509.446462	-1.840709	806.185910	-91.364318
CP	-76,331.911682	5.189401	-91.364318	869.197621

12.8 Summary

What did we learn from this example? We learned how to build and parameterize a sophisticated probabilistic model of sales and to conduct a Decision Analysis based on that. We did so by starting with a priori estimates of the key parameters, complete with the uncertainty in those estimates, and thereafter gathering data and performing successive Bayesian updates on the prior to create a posterior. The posterior generally had lower uncertainty than either the prior or the data alone (classical regression). We were then able to "predict" sales conditional on the three variables in the influence diagram that comprised the model, namely level of advertising, our price, and competitor price. We were able to project sales for every possible combination of the decision variables—our level of advertising and our price—on a fully probabilistic basis. We were able to find the decision that maximized expected sales and the decision that maximized a certain equivalent of sales, taking account of our attitude about risk as characterized by an exponential utility function.

Chapter 13

ASSET PERFORMANCE PREDICTION USING BAYESIAN PRIOR AND POSTERIOR AS WELL AS CLASSICAL STATISTICS

> ***Without data, you're just another person with an opinion.***
>
> ***W. Edwards Deming***

This section puts forth a highway asset decision problem to illustrate more sophisticated application of Bayesian Linear Regression methods. The "pavement deterioration" problem is a convenient medium to render the Bayesian statistical and prediction discussion concrete (if the reader will excuse the expression). This discussion is going to focus on Bayesian probability, but it is also going to focus on a profoundly important (and expensive) asset highway decision problem, including maintenance and rehabilitation.

Did you ever think about the roads you drive on? In the vernacular, there are zillions of prospective contributory variables to deterioration. Which ones matter? Which contributory variables do you use to signal maintenance? Do you maintain according to certain technical signposts, such as the age of the pavement or cumulative traffic? Do you maintain according to temperature variation signposts such as cumulative degree days? Assuredly, highway departments do this kind of thing, driven mostly by experience (or just plain hip-shooting).

And this section presages how they might predict the performance of pavements and determine their maintenance decisions more scientifically and accurately. In contrast to the previous section, this one will focus solely on the prediction of performance and the precursor variables. We will not focus on decisions but rather on predicting asset deterioration and the principal contributory variables to such deterioration. In short, this chapter is about prediction; the previous chapter is about decision-making and Decision Analysis, drawn as it was from Clemen's[20] book on Decision Analysis.

The pavement management decision is a very real trillion-dollar decision that lay people scarcely notice or think about. The pavement asset management decision is a very real, very prominent, very costly, and very important one. The problem is very simple: assets deteriorate. But what makes them deteriorate? What contributes to faster deterioration? Slower deterioration? Does more frequent or intensive preventive maintenance retard the rate of deterioration? Does it improve the life cycle cost or the annual maintenance cost? What are the key contributing variables? What variables does one consider in order to predict when one will have to maintain or rehabilitate pavement?

The other observable is that pavements are hellishly expensive to route, to build, to maintain, and to rehabilitate. The cost of the interstate or interprovincial highway system is astronomically high. Rights of way, materials, construction costs, maintenance costs, etc. are high. Highway department decisions include initial construction, full replacement or rebuild, rehabilitation (e.g., pavement overlay), and preventive maintenance such as sealing or resurfacing.

Departments (and the citizens they represent) want to know what causes or contributes to the highway deteriorating and degrading. If they knew, they would know how to manage use, conduct preventive maintenance, and schedule overlays or reconstructions. How can they intervene to change or thwart certain variables (along the way, if possible)? What variables should preventive maintenance focus on? Water and drainage? Temperature? Traffic? Earth shifting? After preventive maintenance has run its course, what should be their "rebuild" or rehabilitation action? How does rehabilitation action affect future preventive maintenance?

Highway department realities have been harsh. Building new roads or overlaying old roads is "super cool." It presents engineering and construction challenges that highway departments relish. The citizenry sees palpable progress being made. By contrast, preventive maintenance is far from "super cool"—boring, inconvenient to the public, and a pain in the neck for engineers and drivers—and, alas, does not get citizens excited or get highway people promoted. In practice, highway decision makers have often embraced a sort of "run to failure" policy—let the highway wear out using just sufficient maintenance along the way and then replace or overlay it when it becomes unserviceable. Highway decision makers realize that this just **cannot** be right. Would you drive your Porsche to failure, never maintaining it until it simply stopped running, and then tow it to the dealer?

Superposed over the notion of highway cost is the notion of safety. If roads deteriorate to a certain point, cars and trucks become unstable. They can be thrust in different directions or off the road. Traction can be reduced (e.g., rutted roads that fill with water during winter rains and then freeze). Highway departments are charged with promoting public safety, and the costs and benefits of doing so need to be incorporated into their asset management strategy. This illustration will not focus on the safety issue, but it can be incorporated into the performance measures.

How does one measure the performance of an asset such as a roadway? To measure performance, highway people have over the years devised indexes of performance, i.e., systematic technical measures of pavement serviceability and performance, as engineers frequently do. These indexes have included empirical and scientific measures of highway serviceability and degradation therein. And of course, highway engineers gather a plethora of data on some or all of the performance measures simultaneously with the ongoing variables that may be causal or contributory to pavement serviceability and deterioration. We include a copy from a salient, early data collection effort on the topic for a particular jurisdiction in Canada by Papagiannakis[73]:

...the Newfoundland Department of Works Services and Transportation (DWST) commissioned Pavement Management Systems (PMS) ... to develop a pavement management system for the Trans Canada Highway (TCH). Part of this project involved the development of pavement performance curves on the basis of historic evaluation of data obtained during 1979, 1982, and 1986 surveys. Pavement performance data was summarized in the form of three indices, namely the Riding Comfort Index (RCI), the Surface Distress Index (SDI), and the Structural Adequacy Index (SAI). The objective of the performance modelling study was to develop regression equations expressing RCI, SDI, and SAI as functions of structural, environmental, and traffic variables.

We have deliberately chosen an old application and data set (and old serviceability parameters) so that there is no attribution of responsibilities, actions, or predictions of people still working in the organizations referenced. We are confident the method applies equally well today, but things have probably been learned and applied over the years. The application is nonetheless fresh and illustrative.

The people responsible for building and maintaining the asset had undertaken due diligence to understand and predict pavement performance and to determine efficacious maintenance and rehabilitation strategies. Toward that end, they had defined three measures of performance: RCI, SDI, and SAI. These are quantitative but partially empirical-subjective measures of ride stability, comfort, and implicit safety. The objective is to maintain the system up to a certain minimum level of each index and to do so in a maximally cost-effective manner. The problem is to predict and schedule maintenance and rehabilitation strategies because pavements are inherently proceeding forward on intrinsic deterioration schedules, and the highway department has to intercede at certain points and maintain or rehabilitate. Otherwise, the pavement they are charged with managing ceases to be serviceable.

The contractor for the referenced work, PMS, toured throughout the province to various "pavement sections" along the Trans Canada Highway, accumulating measurements of a vector of potentially contributory (or perhaps causal) variables in pavement asset deterioration. To wit, they measured a vector of independent variables such as temperature, truck traffic, precipitation, age, and so forth. At the same time, they measured the foregoing trio of performance indexes: RCI, SDI, and SAI. They measured three variables and a collection of independent variables prospectively associated with them. They were most interested in understanding the deterioration of RCI. (We will later talk about joint prediction of two or more variables, but for now we will focus on predicting the deterioration of RCI and what the key contributory variables might be.)

From their executive summary: *This report documents the task of fitting regression curves to the Riding Comfort Index (RCI) data obtained on the Trans Canada Highway in Newfoundland in the 1979, 1982, and 1986 pavement evaluation surveys. Problems were experienced, however, with the RCI data obtained in the 1986 survey. They were shown to be significantly higher than the RCI values obtained in the previous survey years (i.e., 1982). This was attributed to the different methodologies used in arriving at the RCIs before and after 1982. This data incompatibility forced the abandonment of the 1986 data in fitting the regression curves. Hence, the variable name RCI indicates the combined 1979 and 1982 data.*

They discovered that their temporal collection of RCI and contributory variable data was definitionally inconsistent over time! Measuring the RCI during the final year employed a different method than measuring the RCI during the earlier years. (Heteroskedasticity anyone?) Consequently, their data set was nowhere near as large as they thought or hoped; they had differences in measurement methods right in the middle of their measurement period!

And they had a "survivor bias" problem because they could not let the TransCanada Highway (or any road) deteriorate very far before they had to rehabilitate and fix it. There simply aren't any "old" pavements out there in the population to measure. Even though they had gathered the numbers, the numbers were intrinsically inconsistent, and the truly consistent data set was smaller than they had realized.

Not knowing in advance what the most important precursor variables were, they gathered many numbers that they could never use. Realistically, they did not have enough data for a classical regression solution! In the final analysis, they had a mere 90 observations that they could use. Is this a typical occurrence? Having to throw away data because of intrinsic measurement differences is a real-world problem. Because of the cost and difficulty of measurement, they did not have a huge abundance of sections and sectional data. They only had two temporal cross sections, and those were the older two using the older methods. In spite of the slimness of their data base, they were bound and determined to conduct classical regression to infer the variables that were most important to RCI. They had to make pavement decisions, and they wanted to use their data to the maximum extent possible. "We want to be data-driven" has become a rather unthinking refrain. Alas, though, they did not have enough data. The Bayesian Linear Regression approach is the only viable solution, as we shall see herein.

13.1 Bayesian Statistical Approach

Implicit in their formulation was that classical regression was the way to approach the problem. Classical regression assumes that everything that is known or knowable resides in the data they have collected. This is the age-old, ubiquitous refrain mentioned previously. A diffuse prior (abject ignorance) combines with a data set to give the answer! This intended approach indeed invites questions.

- When should you use classical statistics? Regression? The answer, as we shall see, is "only as a conceptual check, a corner solution." Use Bayes; if the data are definitive, they will dominate the prior.
- When you use classical regression, what are you REALLY doing?
- Is classical regression inherently probabilistic, inherently Bayesian, or is it just sort of mystical?
- If your senior management asks you to use it or asks why you are not using it, what should you say? What would motivate you to say yes or no if asked to use it?

In thinking about maintenance and rehabilitation strategies, you must approach the questions of performance prediction and asset deterioration prediction. How would you approach the problem? First, of course, you would quantify what you meant by "performance." In business, it is generally profit. In this context, it is the serviceability of the pavement. In this illustration, they used the measure RCI, which to them was easily measured by the highway department and well defined. They wanted to predict RCI and manage the serviceability of the Trans Canada Highway based on it. RCI, or Ride Comfort Index, is, from a technical perspective, a measure of road smoothness and lack of "beating up your car" or "veering your car." To measure it and to conceptualize it, a test vehicle equipped with motion and other monitors is driven over pavement. The vehicle shakes, rocks, weaves, vibrates, pitches, yaws, veers, etc. Physical motion sensors attached to the vehicle and subjective sensors (people) measure ride stability and roughness for each pavement. After extensive application of the test vehicle on a pavement section, the engineers come up with a quantification of the ride comfort index, or RCI, for that section. At the same time they measured the RCI of a given section, they also measured a long list of prospective contributory variables to the RCI.

- Temperature
- Truck traffic
- Auto traffic
- Precipitation
- Structure of the pavement (e.g., thickness of the total road system lying on top of "bare ground"
- Age (time since last overlay)
- Material in the base
- Material beneath the base

For the case study herein, there were 16 prospective (measured) potential contributory variables

- AD: asphalt thickness
- B: base thickness
- SUB: subbase thickness
- SN: structural number
- CBR: California bearing ratio index
- PRED: precipitation (mm/yr)
- AGE: years from last rehab/overlay
- STT: number of semi trucks/day
- SUT: single unit trucks/day
- ACC1: semi trucks since last overlay
- ACC2: single unit trucks since last overlay
- ACC: total number of trucks since last overlay
- AGEI: values of 1,2,3 for various years
- SHI: values of 1,2,3 for structure number ranges
- DDI: values of 1,2,3 for degree day ranges.

They narrowed it down to the three most important contributory variables from a performance prediction and decision making perspective:

- AGE: Time in years since most recent overlay, an asphalt overlay ostensibly creating a "new" road. Intuitively, RCI should drop with AGE (older roads rougher and less serviceable).
- SN: Structure number, a measure of the physical thickness and structural strength of all the road material put down at all previous overlays plus all the sub-base down to original bare ground. It is a civil engineering term for how much mass and intrinsic strength exists in the roadway. Conceptually, how much mass is in the roadway to bear load. RCI should increase with SN (stronger and thicker roads with more load bearing mass better).
- DD: Degree days, a measure of how cold it was. This is a proxy for water seeping into the road and undergoing freeze-thaw cycles. RCI should decrease with DD (cold temperature renders materials brittle and susceptible to freeze-thaw cycles and therefore breaks down roads more rapidly).

Traffic was observed to be the same across all the segments and therefore invariant. One is not able to measure the impact of a variable like traffic unless it differs systematically or serendipitously across the section. Variation does NOT have to be systematic. To think so is one of the major misconceptions of statistical analysis. You need not care why there is variation. You only need to know that there IS variation. You do not need to establish a specific test or control group to ensure that there is variability. You simply analyze the database to examine whether there is variability. If the variability is random, who cares? The fact that there is variability is what matters. This is a single section of the Trans-Canada Highway. If it is the same across all segments, there is no variability and therefore no ability to perform statistical analysis of the type illustrated herein. The engineers knew that truck traffic was a significant precursor to road deterioration, but because there was no variability across these sections, their hands were tied on that variable for this analysis. There was no information related to truck traffic in the data! Measurements of the same traffic across many different road sections did not contain the information they truly wanted. Plus, and this is important, they could not control truck traffic anyway. It is what it is. You know it hurts the pavement, but it is a constant. It will be a hidden variable in the analysis.

There were important variables that did differ across the sections, and those were the contributory variables that had to drive the prediction. They were able to use RCI, AGE, SN, and DD data for 1979 and 1982, with 90 observations. When one thinks of the cost of data gathering and the elapsed time of data gathering, the idea that they only got 90 observations is very typical. "Designing an experiment," especially for an asset that is in continuous operation and cannot be stopped or controlled, is very difficult. We see this in disease management and diagnosis as well. (It was replete during the initial time of COVID.) The analysts inserted an intercept term by inserting a "1" column in the independent variable matrix. They used ln (DD) rather than DD in order to minimize scaling problems. Regression can be sensitive to the scale of numbers, and they wanted to adjust the units of measurement so that the scale of their numbers was more or less comparable. They also chose to use ln (RCI), but for an entirely different reason. They wanted to be sure that they were projecting only nonnegative values of RCI. Negative values of the RCI index were unreasonable and not defined. This is a common technique for imposing nonnegativity constraints on various variables. Their data, their 90 observations, loaded into an Excel spreadsheet, appears in Table 13.1, Table 13.2, and Table 13.3.

Once they loaded dependent and independent variables into Excel, what was left to do? They wouldn't surprise you! They pulled the trigger on classical, linear regression using a simple and direct, commercial classical regression program such as the regression add-in within Excel. (It was actually a FORTRAN program as we recall it! Excel with its regression add-in wasn't around yet!) "No use trying anything else," most people thought. No one can be criticized for doing that. They approached the problem from a classical linear regression perspective, a perspective we have articulated in detail previously. They postulated a linear model relating a dependent variable (RCI) to a trio of independent to predictive variables (Age, SN, DD), and an intercept. They used a program similar to the Excel embedded linear regression tool to discern the linear coefficients on Age, SN, DD, and the intercept. Theirs was not a large data set (to say the least!) We actually developed the Bayesian Linear Regression approach herein in part to figure out what they were truly doing. This section gives a precise answer to that question. The regression equation they postulated was

$$RCI = \beta_1 \times AGE + \beta_2 \times SN + \beta_3 \times \ln(DD) + \beta_4 \times 1 \quad .$$

Why a linear equation? Clearly, it is the easiest possible approach. Probabilistic and predictive calculations to follow are the easiest to implement. The linear model is the 0th and 1st order Taylor Series term for any nonlinear deterioration model. They wanted to get the 0th and 1st order terms to discern what might be contributory variables and help them manage. Perhaps they recognized that CORRELATION is not CAUSALITY. However, these kinds of studies give engineers things to look for to better understand assets and their deterioration.

Of course the first problem they faced was "the data base is too small!" How many zillion times does one hear this in real world problems such as this? Classical regression has always been bothersome, indeed pesky, in terms of size of data base. We remembered our probability and Decision Analysis courses—Bayes Theorem, scrupulous adherence to probability theory and calculus, "evidence" being used to update priors, Prof. Howard's[45] "Acme Rain Detector" spewing forth readings, etc. Classical regression was troubling—it just did not seem to fit into the paradigm of probability in the very direct and straightforward way that Bayes does. In the real world, it cannot possibly be true that "everything that is known or knowable is embedded in the 90 data points, the 90 observations!" We committed to wrestle this to the ground once and for all. We conferred with Dr. Zellner, and all the lights came on with regard to Bayesian rather than classical

Table 13.1: Observed Pavement Data–Part 1

		y	X Matrix			
Obs.	RCI	ln(RCI)	AGE	SN	lnDD	Int.
1	6.8	1.916923	12	7.03	6.115892	1
2	6.7	1.902108	12	7.85	6.131226	1
3	6.6	1.887070	11	7.36	6.049733	1
4	6.3	1.840550	11	5.09	6.063785	1
5	7.0	1.945910	11	5.07	6.063785	1
6	6.2	1.824549	14	4.52	6.040255	1
7	6.0	1.791759	14	3.79	5.950643	1
8	6.7	1.902108	0	5.52	5.950643	1
9	6.6	1.887070	0	5.33	5.950643	1
10	5.9	1.774952	14	3.95	6.232448	1
11	5.5	1.704748	14	4.70	6.291569	1
12	6.6	1.887070	9	6.54	6.291569	1
13	4.5	1.504077	21	4.54	6.380123	1
14	5.4	1.686399	0	4.51	6.396930	1
15	6.7	1.902108	10	3.86	6.396930	1
16	5.6	1.722767	9	3.86	6.563856	1
17	4.7	1.547563	8	5.96	6.563856	1
18	5.9	1.774952	8	5.96	6.551080	1
19	5.8	1.757858	9	6.48	6.538140	1
20	6.7	1.902108	9	6.48	6.538140	1
21	6.9	1.931521	9	1.98	6.538140	1
22	5.8	1.757858	9	6.00	6.538140	1
23	3.5	1.252763	14	4.99	6.538140	1
24	4.2	1.435085	14	4.39	6.588926	1
25	3.2	1.163151	9	4.64	6.588926	1
26	5.7	1.740466	14	4.18	6.588926	1
27	4.3	1.458615	14	3.68	6.620073	1
28	5.1	1.629241	14	4.22	6.721426	1
29	4.9	1.589235	11	3.86	6.721426	1
30	6.3	1.840550	10	6.58	6.709304	1

regression.

Table 13.2: Observed Pavement Data–Part 2

		y	X Matrix			
Obs.	RCI	ln(RCI)	AGE	SN	lnDD	Int.
31	6.9	1.931521	10	5.87	6.701960	1
32	5.5	1.704748	14	4.60	6.701960	1
33	6.1	1.808289	14	3.94	6.684612	1
34	6.8	1.916923	9	5.89	6.654153	1
35	7.4	2.001480	9	5.89	6.652863	1
36	4.4	1.481605	9	4.59	6.395262	1
37	4.1	1.410987	10	6.03	6.395262	1
38	4.6	1.526056	10	3.51	6.395262	1
39	5.9	1.774952	9	6.59	6.395262	1
40	6.5	1.871802	9	6.59	6.396930	1
41	5.3	1.667707	16	2.86	6.408529	1
42	4.5	1.504077	15	3.51	6.309918	1
43	6.1	1.808289	9	5.41	6.212606	1
44	6.4	1.856298	9	6.06	6.212606	1
45	6.0	1.791759	10	4.11	6.214608	1
46	6.0	1.791759	15	7.03	6.115892	1
47	5.8	1.757858	15	7.85	6.131226	1
48	6.4	1.856298	14	7.36	6.049733	1
49	6.0	1.791759	14	5.09	6.063785	1
50	6.2	1.824549	14	5.07	6.063785	1
51	5.2	1.648659	17	4.52	6.040255	1
52	4.5	1.504077	17	3.79	5.950643	1
53	7.1	1.960095	3	5.52	5.950643	1
54	6.6	1.887070	3	5.33	5.950643	1
55	4.6	1.526056	17	3.95	6.232448	1
56	4.3	1.458615	17	4.70	6.291569	1
57	6.9	1.931521	12	6.54	6.291569	1
58	3.0	1.098612	24	4.54	6.380123	1
59	6.4	1.856298	2	4.51	6.396930	1
60	5.5	1.704748	3	3.86	6.396930	1

We simply could not countenance the classical regression mantra that everything that is known or knowable is embedded in data. Everything that could be possibly be known resided in 90 observations?!?!? Taken by an unsupervised, independent, commercial, "money-grubbing" contractor? (Money grubbing is an all-to-typical characterization of contractors by clients!) That just flat lacked credulity. Nothing except for the 90 points contains any knowledge of value? The highway people got unsatisfying results with classical regression using a scant 90 data points. Their formulation was a difficult if not impossible formulation; we will see why when we get to Bayes herein. They learned little about contributory variables and ended up with a less than definitive model. They

Table 13.3: Observed Pavement Data–Part 3

		y	X Matrix			
Obs.	RCI	ln(RCI)	AGE	SN	lnDD	Int.
61	5.1	1.629241	13	3.86	6.563856	1
62	5.2	1.648659	12	5.96	6.563856	1
63	3.5	1.252763	11	5.96	6.551080	1
64	5.2	1.648659	11	6.48	6.538140	1
65	5.2	1.648659	12	6.48	6.538140	1
66	6.0	1.791759	12	1.98	6.538140	1
67	6.5	1.871802	12	6.00	6.538140	1
68	5.6	1.722767	12	4.99	6.538140	1
69	2.5	0.916291	17	4.39	6.588926	1
70	2.7	0.993252	17	4.64	6.588926	1
71	2.4	0.875469	12	4.18	6.588926	1
72	4.4	1.481605	17	3.68	6.620073	1
73	3.6	1.280934	17	4.22	6.721426	1
74	3.6	1.280934	17	3.86	6.721426	1
75	4.3	1.458615	14	6.58	6.709304	1
76	6.0	1.791759	13	5.87	6.701960	1
77	6.3	1.840550	13	4.60	6.701960	1
78	4.4	1.481605	17	3.94	6.684612	1
79	5.1	1.629241	17	5.89	6.654153	1
80	6.4	1.856298	12	5.89	6.652863	1
81	6.7	1.902108	12	4.59	6.395262	1
82	2.3	0.832909	12	6.03	6.395262	1
83	3.1	1.131402	13	3.51	6.395262	1
84	5.6	1.722767	12	6.59	6.395262	1
85	6.7	1.902108	12	6.59	6.396930	1
86	4.5	1.504077	19	2.86	6.408529	1
87	1.3	0.262364	18	3.51	6.309918	1
88	5.1	1.629241	12	5.41	6.212606	1
89	5.5	1.704748	12	6.06	6.212606	1
90	5.7	1.740466	13	4.11	6.214608	1

had a very severe selection bias problem, a sort of survivor bias. To caricature: "You're too sick to be in this disease study! We need to tabulate people with your disease who aren't quite so sick." The really bad events simply are suppressed, i.e., not reported, because the consequences of such events can be adverse. They were just taking the same measurement over and over and over and over–same traffic, same truck traffic, differences in some variables but not others, all on perfectly serviceable roads. "Design of experiment" was badly needed here.

Importantly, there was (and almost always is in asset and safety types of settings) what we term an "adverse selection" problem. Deterioration of assets such as a pavement or a living organism tends to follow an increasingly declining curve such as that in Figure 13.1.[1] Typically, a new asset starts to deteriorate slowly, and then the rate of deterioration begins to accelerate as deterioration feeds on itself. There is a protracted period at the front during which we see slow but slightly increasing deterioration. After a point, deterioration starts to "feed on itself" and the rate of deterioration increases. As the pavement becomes rougher, car and truck tires start to bash on raised and depressed portions of the pavement harder, and that accelerates the rate of deterioration (of the vehicle as well as pavement).

Figure 13.1: Pavement Deterioration Curve (Per Benchmark Inc.)

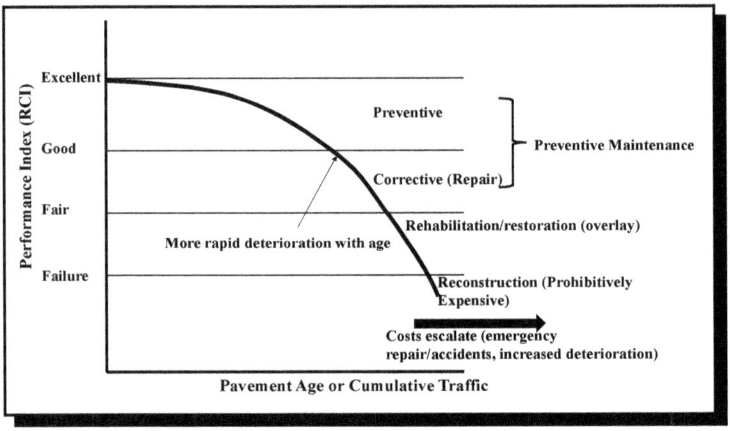

Alas, however, the real world with real operating assets is not so ideal. The highway department has the obligation to maintain the pavement in a safe and serviceable condition so as to foster public safety. As indicated in the figure, all real-world measurements are, of necessity, confined STRICTLY to the early part of the pavement life cycle. Because of public safety concerns, one simply cannot let the pavement deteriorate past a certain highly serviceable state. Otherwise, people die. All real-world measurements are, of necessity, confined strictly to

[1] The figure is reproduced from https://www.benchmark-inc.com/2019/01/17/.

the early part of the life cycle. Pavement management people have no choice but to overlay the pavement before it deteriorates very much because they are neither willing nor chartered to impose a public safety risk. They have to minimize or eliminate safety risks; they cannot kill people simply in order to gather better data! In reality, they are always measuring a nearly new, highly functional pavement.

What if you were always measuring the performance of a brand new car, trying to discern the contributory variables to automobile depreciation? Not likely. Would you learn much about depreciation? Suppose you were studying diseases of old age but were only using 1-year-old infants to gather data. What if you let a few passenger jets run to failure so that you could discern what might happen in the latter part of their life cycle so as to determine a more efficacious maintenance or rehabilitation strategy for onboard jet engines? Highway departments have to fix things the minute they break to protect the public's safety. Doctors have to treat the patient before the clinical data are in. They have no choice; they must use judgment, precisely as they did with COVID. With pavements, they were intrinsically measuring "no variation" scenarios, inescapably so. So even the 90 points they obtained lie at the infantile portion of the pavement life cycle. There is insufficient variability to know much about anything. One would have to have an absolutely monstrous data set to achieve sufficient variability to determine much of anything. The intrinsic value of the data set is not as high as one might think because this "early in the life cycle" bias is so prevalent. We can clearly see in Figure 13.2 the embryonic points in the temporal lifestyle during which all the highway measurements were being taken and the later points in the life cycle in which they were definitely not taken.

Figure 13.2: Measurements Are Taken and Data Are Reported Only for the Extreme Front End of the Asset Life Cycle Curve

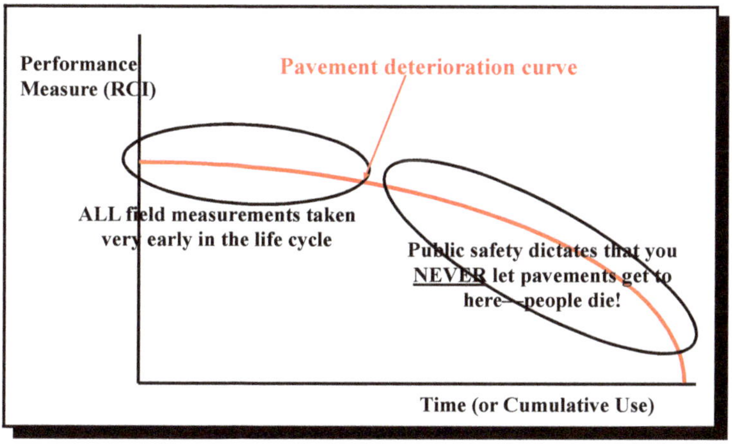

Because of liability, public perception, and other problems, senior highway

managers simply are not incentivized to report asset failure "disasters," in which there is generally much more intrinsic information embedded. People are incentivized to "cover up" disastrous pavement sections lest they expose malfeasance or expose themselves or the organization to legal action or public relations disasters. There is a huge and very real bias against having outliers reported. And yet, there may be asymmetrically high information in those outliers! Elimination of such data represents an obvious adverse selection bias, one motivated by liability and other considerations. The bad cases are omitted, and the data set is "nice." We emphasize this using Figure 13.3. The double bogies and kicking your ball out of the sand trap are omitted from the scorecard! Only the pars and birdies are reported! The author can vouch for how much this can help your golf score, but he can also tell you how unrepresentative of your skills or abilities the score becomes.

Figure 13.3: Adverse Selection of Data Measurements Can Be Colossal

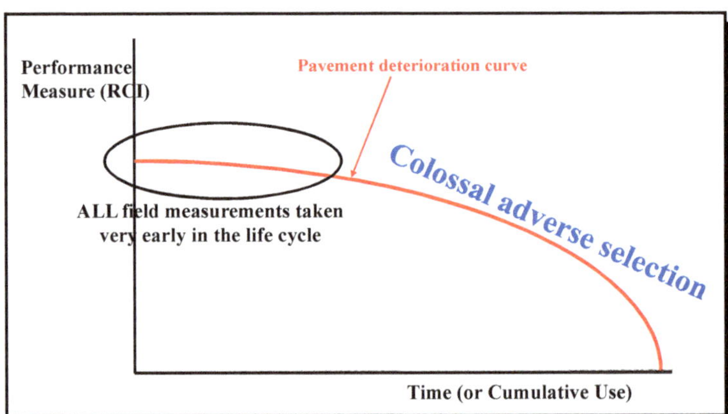

Figure 13.4 emphasizes the prediction problems that may result from this bias. We can see that similar measurements at the front end of the life cycle can lead to dramatically different (and, in fact, unreliable) predictions of asset life. No one wants this when they are trying to manage an asset in the public interest. Quite the contrary, they want something far better than this.

So what do people do when they find there isn't enough data or the data they have is insufficient for the task at hand? There are three enlightened answers to the question:

1. Gather data on zillions of sections—gather a massive cross-sectional data base. This is the standard "gather more cross-sectional data" rallying cry. Statisticians and governments seem to invariably recommend this. "More data will overcome the problem!" In reality, it is not that simple. More of the same data might not add more information. (If the additional data comes from the early part of the life cycle, a lot of numbers may come in,

Figure 13.4: Predictions Can Be Highly Inaccurate When Based Strictly on Observations Early in the Life Cycle

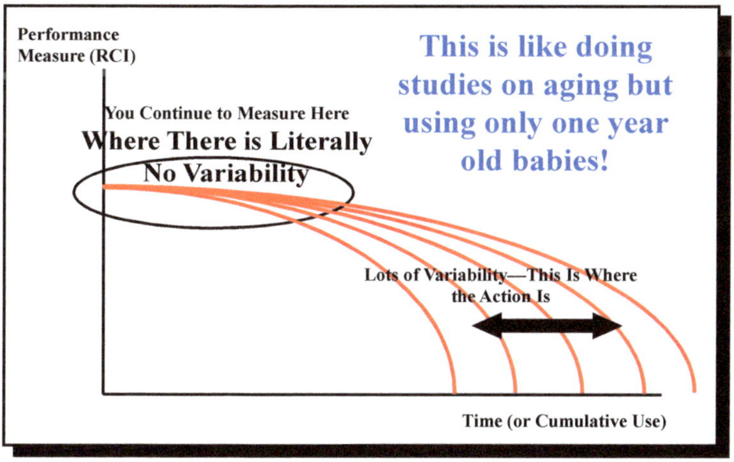

but they are all subject to adverse selection bias.) The call for more data, which is often heeded, needs to be carefully considered. Increased data gathering by an operational system is akin to gathering more and more early life cycle data. They have zero idea regarding the incremental value of new data. They have no idea whether they should seek more readings on the "Acme Rain Detector" in Howard and Abbas[45] context.

2. Test tracks—run test pavements in a test facility (not an in-service highway) all the way through their life cycle. People don't die on test tracks when pavements become dangerously unserviceable. One can systematically vary variables with much less restriction. Test tracks are akin to clinical trials in a controlled environment.

3. Bayesian statistics—start with judgment borne of informed observation by engineers and managers, and thereafter update with field-measured data. You get your information and the information embedded in the data right away.

In the coming three sections, we will examine each of these options in order.

13.1.1 Zillions of Sections—Gather Massive Cross Sectional Data Base

In pursuit of large, comprehensive data, the Strategic Highway Research Program (SHRP) and the Canadian Strategic Highway Research Program (C-SHRP) were formed in the 1980s and 1990s to build and establish

- Something like 600 one-mile-long highway segments were selected and monitored in real time throughout the United States.
- Something like 100 segments in Canada were selected and monitored similarly to the United States.
- Something like 200 prospective contributory variables are monitored at each site.
- Whatever variability there is in the early life cycle, identify it through massively redundant diversity, looking for statistical outliers ("dead people early in the life cycle").

SHRP and C-SHRP were well-intentioned and gathered a boatload of data over a boatload of sections, all, by design, early in the pavement life cycle. In the early days, it fell short of expectations because the data was systematically biased toward the front of the life cycle—severe selection bias. There is no viable way to eliminate the selection bias without a "Tuskegee experiment" type of approach.[2] We will address that below. Randomized clinical trials are analogous to test tracks in the pavement industry.

13.1.2 Test Tracks

A test track puts a set of dual wheels on roller axles, weights them with heavy concrete blocks from above, and runs them repeatedly over test sections of pavement all the way through full, simulated life cycles. In so doing, they try to simulate an external road environment on the test tracks. Test tracks had proved less than satisfactory. It is difficult, if not impossible, to simulate all 200 outdoor variables indoors, and the statistical results have been less than satisfactory. This is akin to "simulated disease," of course, in this situation without a Tuskegee type of procedure replete with all its profound ethical indecency. In this situation, ethical indecency would more likely occur in the former approach above if we systematically ran roads to failure knowing that people would be injured or killed in order to gather "later-in-the-life cycle" road data. (There have been allegations that a similar phenomenon occurred with COVID as well during the early days of the pandemic.) Alas, test track simulations and results have not been satisfactory. It has been difficult, for example, to simulate repeated random freeze-thaw-precipitation cycles on enough test tracks to get a database that actually contains the desired information.

Parenthetically, there have been similar concerns expressed with randomized double-blind placebo control (RDBPC) clinical studies, sometimes shortened to randomized controlled trials (RCT), each with its own form of real-life "test track" in essence. Some have argued that RDBPCs are the "gold standard" of epidemiologic clinical trials. Bayes hints that this may not be the case. Bayes hints that there is no such thing as a non-random, fully controlled background,

[2]See https://en.wikipedia.org/wiki/Tuskegee_Syphilis_Study.

because background is intrinsically random at its core. There is no such thing as complete systematic control over chemicals or environmental substances that enter each of the subjects. There is no measurement or control over the onset of other genetic or environmentally transmitted diseases that were formerly latent. And importantly, there are ethical concerns with any procedure that systematically exposes people to the risk of mortality, disease, or potentially curative treatment, whether voluntary or not, knowing you are intending to impose such a risk from the outset. Is it akin to a firing squad in which five live rounds are given to five shooters and one blank round is given to a sixth shooter, all at random, thereby providing "plausible deniability" to all six shooters? ("My rifle held the blank! It wasn't I who killed him!" Would it be ethical for you to be one of the shooters? (Most would not serve as one of the shooters, and most would strive to avoid a "test track" study that systematically imposes risk or harm on participants, whether voluntary or not.)

13.1.3 Appeal to Bayesian Statistics–Meld Prior Judgment with Data

Let us begin with an apocryphal but very telling anecdote from a different industry altogether (mining and milling). The analogy to highways will be easily drawn. The author grew up in the company town of Weed Heights, Nevada, where his father became the general superintendent (the big boss). His father's company dug the hole at the mine we see in Figure 13.5 and operated the processing plant not visible at the left-hand side of the now closed, now flooded mine in Figure 13.5. That processing plant maintained and operated a fleet of heavy off-road vehicles (KW Darts) and a flotation plant whose function was to concentrate copper sulfide ore. The flotation plant begins with rod mills and ball mills, pictured in Figure 13.6, which grind the incoming ore into a fine powder (1/300 of an inch particle size). The copper sulfide ore in that powder is averse to getting wet, so the copper sulfide preferentially adheres to air bubbles. The non-ore matrix (called the "tailings") preferentially adheres to the water. If one simply blows bubbles into the aqueous, ground-up ore powder, the ore will adhere to the air bubbles and float off the top, letting the wetted non-ore matrix flow out in the liquid stream at the bottom. The non-ore tailings will remain suspended in the water and thereby flow right on through the flotation cell. They call the process "concentration" because the highly dilute ore in the original rock matrix is concentrated in the bubbles, and the non-ore is left to flow out in the slurry. A truck from the heavy truck fleet is pictured in Figure 13.7. Such trucks are much larger and more formidable today!

As Bayesians, let us think about the heavy truck fleet. A driver would, during his trip up and out of the mine, radio the shift officer that his truck was not delivering the power and was not pulling the massive weight up the grade. The obviously "broken" truck would be swapped for a fresh truck at the crusher, and the disabled truck would be driven to the garage for repair. "What's wrong

Figure 13.5: The Former Open Pit Copper Mine at Weed Heights, Nevada Where the Author Grew Up and Worked

Figure 13.6: The Mills and the Flotation Processes at the Weed Heights Mine

Grinding and Milling (Ball Mills) | **Flotation**

Figure 13.7: A Truck from the Off Road Vehicle Fleet Like the Ones at the Weed Heights Mine

with the damn thing? Why isn't it working? What do we need to do?" They had logged every truck that had broken down and what was wrong with it for a decade or more, and hourly during 24 hour, 7 day per week shifts. They had historical data galore—operating hours since the last breakdown, symptoms, performance notes, age of the truck, age of retrofit equipment on the truck, etc. And yet, they didn't know what was wrong with it! They had a complete, continuous history of every symptom every truck in the fleet had exhibited and everything that had been done to repair it. They had a full historical data set for every truck in the fleet, and yet the data could neither predict nor diagnose truck breakdowns.

There was an old KW Dart mechanic who had been in the garage for two decades, whom we refer to as Lloyd. "Pull 'er up in front of the door and let me go around behind 'er. Leave 'er idle and just get the heck away, shut up, and let me listen to 'er idle for a spell. Rev 'er up and then let off when I yell to you." They would pull the truck up in front of the shop and leave it idling in front of Lloyd, listening intently to the rear-mounted exhaust pipes. Within 30 seconds, he would say, "Valve out, left front bank of cylinders." (It was a V-16 turbo-supercharged diesel engine.) People would say, "How do you know that, Lloyd?" "Cancha hear it? Jesus, boy, you ain't listening!" Lloyd was virtually never wrong. They would tear the engine down based on what he was hearing in the exhaust manifold, and the damaged valve was literally ALWAYS where he said it was. Lloyd had 20 years worth of listening to valve sounds in the exhaust manifolds of diesel trucks. "Can't you hear that valve rattling in there? It's so obvious. A baby could hear that!" Nope, nobody could hear it but Lloyd, and he heard it perfectly. The reality was that Lloyd had a "prior," and the shop had a "data base." They updated Lloyd's prior with their evolving data base and moved forward based on the posterior. It was noted that the posterior wasn't varying too far from Lloyd's prior, bringing into question the data they were collecting. They just weren't getting the "audio" data! Audio electronics were not particularly advanced in those days. He was listening to the obvious failures. All of them. Lloyd had listened to virtually every failed truck in the history of the operation. He wasn't just listening to the "business as usual" data set. They did not eliminate certain trucks or certain problems from the data set. Management wasn't disguising bad outcomes. The data gathering by the shop included all means and manners of contributory data. In anticipation of Lloyd retiring or being killed in a barroom fight after work, the company was trying to replace or augment Lloyd with the gathered data. Lloyd had "prior" information that wasn't in any database. If the database had 20 years worth of truck diagnoses in it, it might approach Lloyd. But it didn't, and it wouldn't for 20 years. In the interim, the company needed to update Lloyd with incoming data. The update was infinitesimal.

A crucially important point has been made by Christian Ostberg that is relevant right here. "Over the past decade or so, there has been a tremendous push towards data—big data, small data, and everything in between. This has enabled countless projects and led to some stunning results in many different fields. But perhaps the pendulum has swung too far towards 'data' and too far away from people like Lloyd. Human intuition is being replaced by data (that's how classical statistics works!) Problem-solving has shifted too much towards 'trusting the data' instead of 'trusting the experts.' Bayesian statistics isn't in any way advocating that data is bad (in fact, quite the opposite), but it is arguing that there needs to be more of a balance between the numbers and the intuition, the understanding. That's what Bayes does—it balances both forces in a beautiful way." The author could not have said it better and would add that quite often one thinks one has a big data base when in fact the data base is merely taking the same measurements and saying the same thing over and over and over again. The volume of the database may be big, but the information content is (unknowingly) super small. We will see that in this pavement case. Bayes does not care whether you have big data or small data. It works the same either way. It starts with the expert's prior and updates it with data coming in. The key message—Lloyd had the prior. Lloyd was receiving data (such as frequency patterns) from the exhaust pipes, directly connected to the manifolds, for a decade or more. The real down-time data gathered hadn't even thought of that. (They never even talked to or monitored Lloyd, amazingly.)

Let us turn to the flotation plant as a second example. Without warning, the froth would boil over in three-foot cascades of bubbles, taking not only the copper sulfide ore particles over the top but also a lot of the waste tailings. It was like an explosion of bubbles, interrupting the flotation and separation pattern. There was so much froth that the process would quit working altogether, i.e., stop separating copper sulfide from worthless rock matrix. Milt, the flotation manager, would say, "Cat tipped over in the hoppers and poured his diesel all over the ore in the hopper. Cut the frothing agent and cut the air. Push the diesel-soaked ore off the hopper and get clean ore in there." Or he would say, "The pH of the ore just increased. That damn limestone ore is again coming in from the crusher. Throw in some sulfuric acid to cut the pH and increase the frothing agent to make more bubbles." Or he might say, "The bubbles ain't gold no more. Nothin's floatin'. Check the grade in the hoppers. Some dumbo might have dumped some overburden (waste) into the hoppers by mistake. Cut the grinders, bypass the cells, and flush this entire flotation system, this whole toilet, out. We'll restart in 15 minutes with full froth and full air." The company had a continuous measuring system—a continuous data set of pH, inflow rate, inflow ore grade, concentrate outflow rate, tailings flow rate, tailings assay, concentrate assay, inflow ore chemistry, organic content (e.g., diesel spill), froth and reagent injections, etc. Many of these measurements were taken in real time during the long history of the plant. The company had gathered continuous data on interruptions or changes in operation and recovery. Milt worked one shift out of three, and he looked at the data for shifts other than his. Milt had a prior, and they had the data. Both needed the other in order to accurately predict what various actions might accomplish and which action was best. Bayes. The data never really worked; the prior did.

This happens time and again. An expensive and detailed database, just like the pavement database, is being gathered. Often, it is too small, too restricted, and sometimes unknowingly lacks the critical contributory variables to be of any predictive value on its own. A prior is needed from experienced people, the quintessential experts. (Missing variables.) As the database becomes more definitive, Milt will fade in importance (perhaps never to zero). But in the next ten years, the company will have to meld judgment with data in order to make effective decisions.

13.2 Back to the Highway Problem

What does this have to do with the highway department? Everything! The highway department was relegated to relying on judgment by cognizant, experienced observers, namely experienced field highway engineers who have seen lots and lots of trials that have never been reported in any formal database (and cannot possibly be) and have been systematically shelved, buried, or destroyed by managers or their legal counsel. They have seen all the "disasters," early failures, and liability-laden and substandard construction instances. They know where the governor's son-in-law's construction company cut corners on construction standards. They know what happened and why with regard to instances that were never reported and perhaps never even measured. In some cases, they may have been told to "bury" them to stave off criticism, lawsuits, or worse. But they have assuredly put them into their "visceral computer," and they have learned from them. To wit, many of the things they have seen are not reported (perhaps covered up) by the asset owner. And yet these highway engineers had actually seen and walked on all these "difficult situations," these malfeasant outcomes. The highway engineers muse about them in the lunchroom. They have photographed them on their cell phones, rules or no rules. They have knowledge about them and what might have caused them. These day-to-day engineers have reliable yet incomplete anecdotal intuition and cognition regarding the variables that contribute to pavement deterioration or failure. They have seen the outliers; they don't have the luxury of avoiding deadly situations or notable failures. Their judgment needs to be formalized as a "prior" that satisfies the laws of probability. Bayesian statistics can be used to "update" their prior into a "posterior" using Bayes Theorem (Bayesian statistics). That is so elegant as to be almost incomprehensible. And what do these pavement experts look like? They are often not the most attractive or articulate people. They are day-to-day operating engineers.

Visualize the pavement engineer who has the accurate prior. He is often very empirical and impressionistic in his judgment. He is often narrowly focused and specifically trained. He is often not a senior member of the organization. He is Lloyd. He is Milt. He is Paul (Lloyd's and Milt's counterpart at the highway department, the wizened old 30-year veteran of highway field engineering, whom we now introduce!), the guy who has seen literally everything. He is the person who has seen literally every "disaster" in the pavement deterioration business, either directly or through discussion with someone else. He has seen confidential or private photographs. (Yes, employees take unauthorized private photographs of disasters and keep them on their cell phones.) He has formed judgments, quite often accurate through sheer repetition, as to what has occurred and why. This is the person from whom we want to extract a prior, to be updated by the incoming pavement data. It is the person who should tell us what data to gather.

After learning this during our pavement project, we wanted to go back to fundamental probabilities and Bayes, but we didn't yet know how. Zellner[107] and others were doing the math in the 1970s, borrowing from earlier work by Raiffa and Schlaiffer[81] in 1961 and probably others. The author was aware of the work but didn't quite know how to apply it to the pavement problem. However, when we really delved into Zellner's math (scattered throughout the literature) and related it to classical regression, the lights went on. Zellner[107], whom the author visited and worked with, had the solution in hand. This section and this book lay it out in detail.

One thing that really misled people was the classical statisticians' refrain: *Everything that is known or knowable is resident solely in the gathered data.* Under that paradigm, Lloyd, Milt, and Paul (the regional highway engineer), according to the classical statistician, have absolutely nothing to contribute. Fire 'em, get 'em the hell off the property, and save their inflated salaries! Who needs them? Experience, insight, and judgments add nothing. In classical regression, people were saying there is no judgment allowed in regression; it is purely SCIENTIFIC. The whole line of thinking we have attributed to Dale Jorgenson's vignette is problematic on its face. Who has not witnessed flat-boldface prevarication among econometricians or statisticians? Who among them is not willing to surreptitiously change things (meaning "cheat") before or after the fact? If demand elasticity comes out positive, people are more than willing to surreptitiously change data or change models.[3] To amend that, make it negative. People will alter data, and they will modify coefficients. They will rationalize, as contrasted with reason, that something is awry with the data. That is not what we want. We want the Bayesian approach enumerated herein, which makes judgment absolutely apparent and makes objective data absolutely transparent. There is no substitute for objectivity and transparency.

13.2.1 Classical Regression (Bayes with Diffuse Prior) Solution

We reiterate that we are using the logarithm of the RCI rather than the RCI in these calculations. The reason is that we do not want the value of RCI to ever be nonpositive; such values have no meaning. The (X^TX) and $(X^TX)^{-1}$ matrices, calculated from the X observations in Table 13.1, Table 13.2, and Table 13.3, appear in Table 13.4. We also see in Table 13.4 the calculation of X^Ty and the classical regression mean $\bar{\beta}$. This is what the highway department calculated for their regression coefficients (except that we used the logarithm of RCI and they used RCI itself, as far as we can tell). The latter is not a good idea if one wants to project RCI forward for various pavement sections and does not want any negative values. We shall see later in this chapter that we project only positive values.

We next turn to Table 13.5, which contains the key results

- n, the number of data points
- ν_C, the degrees of freedom
- R, the residual sum of squared errors
- s_C^2, which is related to the variance of the gamma distribution
- mean of inverse gamma, which is $\langle \sigma^2 \rangle = \frac{\nu_C}{\nu_C - 2} {s_C}^2$
- variance of inverse gamma, which is $Var\left(\sigma^2\right) = \frac{2}{\nu_C - 4}\left(\frac{\nu_C}{\nu_C - 2} {s_C}^2\right)^2$.

Notice that the number of data points and the degrees of freedom are exactly what we would expect. The mean and variance of the gamma term are shown in the final two terms in the table.

In Table 13.6, we see the variance-covariance matrix of the model coefficients. Immediately below, we see the correlation matrix. We discussed the

[3]Christian Ostberg has commented that not only do people change the DATA, but they can also change the MODELS to give them a result they like or think is "correct" or "more palatable to the sponsor." We all know that different models can give quite different results, so we can fidget with models and cherry-pick the one we want or the one whose results we like the most.

Table 13.4: (X^TX), $(X^TX)^{-1}$, and $\bar{\beta}$–Classical Regression

X^TX	Age	SN	lnDD	Int.
Age	14520.0000	5351.6500	6900.6073	1076.0000
SN	5351.6500	2459.7874	2918.4856	456.7400
lnDD	6900.6073	2918.4856	3689.8763	575.8929
Int.	1076.0000	456.7400	575.8929	90.0000

$(X^TX)^{-1}$	Age	SN	lnDD	Int.
Age	0.0006493453	0.0004494985	-0.0016940394	0.0007953975
SN	0.0004494985	0.0075364616	0.0049498497	-0.0752938591
lnDD	-0.0016940394	0.0049498497	0.2159311093	-1.3865689606
Int.	0.0007953975	-0.0752938591	-1.3865689606	9.2561014016

	$\bar{\beta} = \left(X^TX\right)^{-1} X^Ty$		X^Ty
Age	-0.02600207	Age	1717.2927
SN	0.04919262	SN	761.4576
lnDD	-0.20030268	lnDD	945.1198
Int.	2.98679482	Int.	147.9486

Table 13.5: Key Results–Classical Regression

n	90
$\nu_C = n - p$	86
$R = \left(y - X\bar{\beta}\right)^T \left(y - X\bar{\beta}\right) = \nu_C s_C^2$	5.839659634
$s_C{}^2 = \frac{R}{\nu_C}$	0.067903019
$\frac{\nu_C}{\nu_C - 2} s_C^2$	0.069519758
$\frac{2}{\nu_C - 4}\left(s_C^2 \frac{\nu_C}{\nu_C - 2}\right)^2$	0.000117878

correlation matrix in Chapter 12 related to the Clemen[20] problem, and that discussion pertains here. Immediately below that, we see the variances, which are the diagonals of the variance-covariance matrix. We see the standard deviations, which are the square roots of the variances, and in the last row we see the t-statistics, which are the means divided by the standard deviations.

Table 13.6: Variance-Covariance and Correlation–Classical Regression

$(X^TX)^{-1}\frac{\nu_C s_C^2}{\nu_C-2}$	Age	SN	lnDD	Int.
Age	0.0000451423	0.0000312490	-0.0001177692	0.0000552958
SN	0.0000312490	0.0005239330	0.0003441123	-0.0052344108
lnDD	-0.0001177692	0.0003441123	0.0150114784	-0.0963939380
Int.	0.0000552958	-0.0052344108	-0.0963939380	0.6434819252

Correlation	Age	SN	lnDD	Int.
Age	1.000000	0.203192	-0.143063	0.010260
SN	0.203192	1.000000	0.122702	-0.285077
lnDD	-0.143063	0.122702	1.000000	-0.980776
Int.	0.010260	-0.285077	-0.980776	1.000000
		Diagonals of Variance-Covariance Matrix		
	0.00004514	0.00052393	0.01501148	0.64348193
			Sqrts (Std. Dev.)	
	0.00671880	0.02288958	0.12252134	0.80217325
			Calculated t-statistics	
	-3.87004409	2.14912691	-1.63483915	3.72337873

In Table 13.7, we turn to the issue of calculating the marginal Student's t distributions over the four β parameters. The requisite equation for the marginal density appears at the bottom of the table. The terms at the top of the table sequentially calculate the terms required by the equation at the bottom of the table.

- the first row $-\frac{\nu_C+1}{2}$ is the exponent of the univariate Student's t distribution
- the second row represents the diagonals of the inverse precision matrix $(X^TX)^{-1}$
- the third row represents the denominator of the precision term that appears in the Student's t distribution quadratic term
- the fourth and fifth rows calculate the key components of the integrating constant
- the final term gives the integrating constant for each of the four marginal distributions in this problem

Table 13.7: Marginal Student's t Distributions over the Four β Parameters–Classical Regression

	Univariate t Exponent		-43.5		$-\frac{\nu_C+1}{2}$

$z_{ii} = i^{th}$ diagonal element of $(X^T X^{-1})$

	Age	SN	lnDD	Intercept
	0.000649	0.007536	0.215931	9.256101
	Age	SN	lnDD	Intercept
$z_{ii}\nu_C s_C^2$	0.003792	0.044010	1.260964	54.052482

$\ln\Gamma\left(\frac{\nu_C+1}{2}\right)$	119.649575	$\ln\Gamma\left(\frac{\nu_C}{2}\right)$	117.771881	π	3.141593

$$\ln c = \ln\Gamma\left(\frac{\nu_C+1}{2}\right) - \ln\Gamma\left(\frac{\nu_C}{2}\right) - \frac{1}{2}\ln\left(z_{ii}\nu_C s_C^2\right) - \frac{1}{2}\ln\pi$$

4.092765	2.866993	1.189390	-0.689650
Univariate Student's t constant			
59.905296	17.584067	3.285076	0. 501752

$$\{\beta_i\} = \frac{\Gamma(\frac{\nu_C+1}{2})}{\pi^{\frac{1}{2}}\Gamma(\frac{\nu_C}{2})}\frac{1}{\left(z_{ii}\nu_C s_C^2\right)^{\frac{1}{2}}}\left[1 + (\beta_i - \bar{\beta}_i)\frac{1}{z_{ii}\nu_C s_C^2}(\beta_i - \bar{\beta}_i)\right]^{-\frac{\nu_C+1}{2}}$$

The precision term and the integrating constant are all that are needed to calculate the full marginal probability distributions over the four coefficients (showing the logs of the gamma functions to keep the numbers in a reasonable range). The equation used is the bottom equation in Table 13.7. The precision term is in the quadratic, and the constant term is out front.

It is interesting and enlightening to observe what comes out of the classical regression calculation in Excel and to compare it to our specific classical regression calculation (Bayes with diffuse prior) formulation. We refer to Table 13.8 to do so. If we consider the top (unshaded) portion of that table, we see our version of the classical regression solution as defined by the Bayesian method—the posterior with a diffuse prior. We see the mean value of the coefficients, the variances of the coefficients, the standard deviations of the coefficients, and the t-statistics of the coefficients arrayed at the top in the unshaded section for our version of the classical regression solution. These are precisely the calculations made herewith.

The gray calculations made immediately below in Table 13.8 are the answers reported out by Excel (2020 vintage regression add-in). We see that the Excel-reported standard error is slightly smaller than our Bayes-reported standard deviation. The difference between the Excel term and our term is the degrees of freedom divided by the degrees of freedom minus 2, which is called the small sample size adjustment or just the sample size adjustment. The classical regression solution in Excel does not adjust for sample size and therefore is not technically correct.

Table 13.8: Classical Bayes versus Excel Calculation

Summary of Data calculations				
$s_C^2\left(X^TX\right)^{-1}\frac{\nu}{\nu-2}$	Means	Variances	Std. Dev.	t stat
Age	-0.02600207	0.00004514	0.00671880	-3.87004409
SN	0.04919262	0.00052393	0.02288958	2.14912691
lnDD	-0.20030268	0.01501148	0.12252134	-1.63483915
Int.	2.98679482	0.64348193	0.80217325	3.72337873

From Excel	Coefficients	Standard Error	t Stat
Age	-0.02600207	0.00665881	-3.90491300
SN	0.04919262	0.02268519	2.16849045
lnDD	-0.20030268	0.12142728	-1.64956897
Int.	2.98679482	0.79501025	3.75692619

$Var\frac{\nu-2}{\nu}$	$\sqrt{Var\frac{\nu-2}{\nu}}$	They are using $s_C^2\left(X^TX\right)^{-1}$
0.00004409	0.00664022	
0.00051175	0.02262186	
0.01466237	0.12108829	
0.62851723	0.79279079	Excel doesn't adjust for small sample size

13.2.2 Developing an Informative Bayesian Prior

Returning to the Bayesian problem context, we need to get in touch with Paul and the other highway engineers right away. It is Paul's past that we want. We want Paul to tell us impressionistically what he thinks about RCI and its relationship to Age, SN, DD, and "everything else," namely the intercept term. Paul has seen and worked on a zillion pavements and pavement sections during his career and has seen all the "highway department failures" and has an experienced idea what caused them. We need to understand his state of information, which we interpret as his prior. We give a rather intuitive way to extract this information from Paul in the next section, but for now consider that we ask him directly for the information in Table 13.9. We start by telling Paul that there are four independent variables (which we state as four variables that may affect RCI deterioration), and we want him to give us his judgment. In particular, we tell Paul that we want his assessment of

- The mean values of the coefficients, which we have designated $\langle\beta_0\rangle$,

- The variance-covariance matrix of the coefficients, which we have designated VCV,
- The mean value of the error term, which we have designated $\langle \sigma^2 \rangle$,
- The variance of the error term, which we have designated V_S^2.

Paul begins by saying he thinks that the logarithm of RCI

- drops by the fraction -0.03 for every year of pavement age, the - sign connoting dropping.
- increases by the fraction 0.02 for every inch of structure in the pavement, the + sign connoting increasing.
- drops by the fraction -0.2 for every point of ln(DD), the - sign connoting dropping.
- increases by the fraction 2.14 for the aggregate of the unidentified variables.

Paul has just given us his prior estimate of the mean values of the model coefficients. We place Paul's prior estimates of the coefficients into the vector β_0 that appears in the upper left of Table 13.9, within which we will build up the prior information. We will attend to the rest of the entries in that table shortly.

We then ask, "Hey, Paul, how sure are you about your estimates of these RCI-eroding variables?" He replies, "Well, I ain't too sure. Let's see, plus or minus 1/3." We might regard him as saying that the standard deviation of each coefficient is 1/3 below and 1/3 above each coefficient. We make that interpretation in the upper right portion of Table 13.9. We calculate 1/3 of the magnitude of the mean values of the coefficients and assume those are our standard deviations. Of course, our variances are the squares of those calculated standard deviations. We insert those variances onto the diagonals of the variance-covariance matrix at the bottom of Table 13.9, designated VCV for short. We ask Paul if he knows any "covariances" among the coefficients, any interdependencies. He says "No!" so we enter zeros on all the off-diagonals of the variance-covariance matrix. He only knows coefficients and standard deviations. So far, so good. We have his variance-covariance matrix at the bottom of Table 13.9, with nonzero variances and zero covariances.

We then ask Paul what the average measurement error is that he anticipates. He guffaws: "What the blazes are you talking about?" We, the analysts, estimate such things as error terms, often arbitrarily. We assume the mean error term $\langle \sigma^2 \rangle = 1$, and we assume the standard deviation of that term is $\sqrt{V_{S^2}} = 0.75$. (We could have assumed a diffuse prior on the variance term if we desired. (We chose not to for this example.) This is not a trivial point. We could have used the diffuse prior σ^{-1} on the error term and thereby created a "hybrid" of diffuse and informative priors, diffuse on the error term but informative on the coefficient.

Table 13.9: The Pavement Engineer's Subjective Prior

	p	4		
			SD	
	β_0		0.333333	
Age	-0.03		0.010000	
SN	0.04		0.013333	
lnDD	-0.20		0.066667	
Int.	2.14		0.713355	
$<\sigma^2>$	1			
$\sqrt{V_{S^2}}$	0.75			
V_{S^2}	0.562500			
ν_0	7.555556			
s_0^2	0.735294			

VCV	Age	SN	lnDD	Int.
Age	0.00010000	0	0	0
SN	0	0.00017778	0	0
lnDD	0	0	0.00444444	0
Int.	0	0	0	0.50887591

This actually has something to recommend it from a conceptual perspective. We are not too certain that 1 is the right answer; we have a lot of uncertainty relative to the mean. When we have completed the assessment, we have completely filled out the table in Table 13.9. We shall see that he has provided us with a complete prior. And it is substantive. Paul has told us what he thinks makes roads deteriorate and how certain he is of the contribution of each of the variables. This is actually a lot of information.

We are now in a position to use the information in Table 13.9 to calculate the parameters of the prior that represent Paul's judgment according to the four equations.

$$\beta_0 =< \beta >$$

$$s_0^2 =< \sigma^2 > \frac{< \sigma^2 >^2 + V_{S^2}}{< \sigma^2 >^2 + 2V_{S^2}}$$

$$Q =< \sigma^2 > VCV^{-1}$$

$$\nu_0 = 2\frac{< \sigma^2 >^2}{V_{S^2}} + 4$$

In particular, we can use the four estimates given to us by Paul together with the four foregoing equations within Table 13.9 to calculate the four parameters that truly and completely characterize the prior

- ν_0
- s_0^2
- β_0
- Q .

We saw how the four questions for Paul put forth in the text yield the four key prior parameters in Table 13.9 that fully characterize the prior. To wit, we see in Table 13.10 how the foregoing four parameters assessed by Paul are used to calculate the following key parameters of the prior:

- m
- M
- mean of error term
- variance of error term.

We see second from the bottom the precision matrix of the prior, and in the bottom the product of the precision matrix and the prior mean of the coefficients, which is needed for later calculations.

Referring to Table 13.11, we see the variances (the diagonals of the variance-covariance matrix), the square roots, which are the standard deviations, and the t-statistics. We see in the middle of the figure the exponent of the univariate Student's t distribution for the prior. We also see the inverse of the precision matrix Q, which is needed for the calculations of the precision coefficients for the univariate Student's t distributions for the prior.

Referring to Table 13.11, we see the variances (the diagonals of the variance-covariance matrix), the square roots, which are the standard deviations, and the t-statistics. We see in the middle of the figure the exponent of the univariate Student's t distribution for the prior. We also see the inverse of the precision matrix Q, which is needed for the calculations of the precision coefficients for the univariate Student's t distributions for the prior.

The full set of parameters for the univariate Student's t distributions for the coefficients of the prior is calculated in Table 13.12. The equation for the univariate Student's t distribution over model coefficients appears at the bottom of the figure. The table calculates all the requisite parameters for that distribution.

13.2.3 Bayesian Posterior

We begin our analysis of the posterior with the calculated precision matrix and the variance-covariance matrix of the posterior. See Table 13.13. The precision

Table 13.10: The Key Parameters of the Prior Including Precision Matrix

$m = \nu_0 + p + 1$	12.555556
$M = \nu_0 s_0^2$	5.555556
$\frac{\nu_0}{\nu_0 - 2} s_0^2$	1.000000
$\frac{2}{\nu_0 - 4}\left(s_0^2 \frac{\nu_0}{\nu_0 - 2}\right)^2$	0.562500

VCV^{-1}	Age	SN	lnDD	Int.
Age	10000.000000	0	0	0
SN	0	5625.000000	0	0
lnDD	0	0	225.000000	0
Int.	0	0	0	1.965116

$Q = VCV^{-1} \frac{\nu_0 s_0{}^2}{\nu_0 - 2}$	Age	SN	lnDD	Int.
Age	10000	0	0	0
SN	0	5625	0	0
lnDD	0	0	225	0
Int.	0	0	0	1.965115624

	$Q\beta_0$
Age	-300.000000
SN	225.000000
lnDD	-45.000000
Int.	4.205477

Table 13.11: Mean, Standard Deviation, and Precision Matrix–Prior

	Age	SN	lnDD	Int.
Diagonals of Variance-Covariance				
	0.000100	0.000178	0.004444	0.508876
Standard Deviations				
	0.010000	0.013333	0.066667	0.713355
Implicit t-statistics in Prior (Mean/Std. Dev.)				
	-3.000000	3.000000	-3.000000	3.000000

	$-\frac{\nu_0 + 1}{2}$
Univariate t Exponent	-4.277778

Q^{-1}	Age	SN	lnDD	Int.
Age	0.000100	0	0	0
SN	0	0.000178	0	0
lnDD	0	0	0.004444	0
Int.	0	0	0	0.508876

Table 13.12: Calculation of Parameters for Distribution Over Coefficients–Prior

$z_{ii} = i^{th}$ diagonal element of inverse Q

$z_{ii} = \left[Q^{-1}\right]_{ii}$

0.000100	0.000178	0.004444	0.508876

Precision

$z_{ii}\nu_0 {s_0}^2$

0.000556	0.000988	0.024691	2.827088

$\ln\Gamma\left(\frac{\nu_0+1}{2}\right)$	$\ln\Gamma\left(\frac{\nu_0}{2}\right)$	π
2.151356	1.519781	3.141593

$$\ln c = \ln\Gamma\left(\frac{\nu_0+1}{2}\right) - \ln\Gamma\left(\frac{\nu_0}{2}\right) - \frac{1}{2}\ln\left(z_{ii}\nu_0 s_0^2\right) - \frac{1}{2}\ln\pi$$

3.806980	3.519298	1.909860	-0.460414

Constant c

45.014310	33.760732	6.752146	0.631022

$$\{\beta_i\} = \frac{\Gamma\left(\frac{\nu_0+1}{2}\right)}{\pi^{\frac{1}{2}}\Gamma\left(\frac{\nu_0}{2}\right)}\frac{1}{\left(z_{ii}\nu_0 s_0^2\right)^{\frac{1}{2}}}\left\{1 + (\beta_i - \beta_{0i})\frac{1}{z_{ii}\nu_0 s_0^2}(\beta_i - \beta_{0i})\right\}^{-\frac{\nu_0+1}{2}}$$

Table 13.13: Precision and Inverse Precision Matrices–Posterior

$X^TX + Q$	Age	SN	lnDD	Intercept
Age	24520.0000	5351.6500	6900.6073	1076.0000
SN	5351.6500	8084.7874	2918.4856	456.7400
lnDD	6900.6073	2918.4856	3914.8763	575.8929
Intercept	1076.0000	456.7400	575.8929	91.9651

$\left(X^TX + Q\right)^{-1}$	Age	SN	lnDD	Intercept
Age	0.000084424	0.000000334	-0.000044550	-0.000710456
SN	0.000000334	0.000172255	-0.000032744	-0.000654367
lnDD	-0.000044550	-0.000032744	0.003270226	-0.019794556
Intercept	-0.000710456	-0.000654367	-0.019794556	0.146391064

$\left[\left(X^TX + Q\right)^{-1}\right]_{ii}$	Age	SN	lnDD	Intercept
	0.0000844243	0.0001722553	0.0032702257	0.1463910639

matrix, computed according to equation (6.7), appears in the top panel of the figure. This is the so-called matrix weighted average derived carefully herein. The second panel is the inverse of this precision matrix. The diagonals of this inverse precision matrix are going to give us the coefficients of the univariate marginal distributions over the various coefficients. And, precisely as expected, the lower panel extracts those inverse precision diagonals to accumulate the z_{ii} terms that will comprise the precision of the univariate Student's t distributions over the coefficients.

The next figure, Table 13.14, details the important parameters of the multivariate posterior. We see at the top the calculation of the A constant and its key terms:

Table 13.14: Key Parameters for Bayesian Posterior

$\beta^* = (X^TX+Q)^{-1}(Q\beta_0 + X^Ty)$	
Age	-0.02821548
SN	0.04135877
lnDD	-0.16366921
Intercept	2.80410362

	$Q\beta_0$	X^Ty	$Q\beta_0 + X^Ty$
Age	-300.000000	1717.292727	1417.292727
SN	225.000000	761.457588	986.457588
lnDD	-45.000000	945.119772	900.119772
Intercept	4.205477	147.948648	152.154125

$\nu = m + n - p - 1$	97.555556

Components of Posterior	
5.555556	$M = \nu_0 s_0^2$
5.839660	$R = \nu_C s_C^2$

	$(\bar{\beta} - \beta^*)$	$(\beta_0 - \beta^*)$
Age	0.00221341	-0.00178452
SN	0.00783384	-0.00135877
lnDD	-0.03663347	-0.03633079
Intercept	0.18269120	-0.66403745

$(\bar{\beta} - \beta^*)^T X^T y$	$(\beta_0 - \beta^*)^T Q\beta_0$
2.172111355	-0.928076377

A	12.639250	$s^2 = \frac{A}{\nu}$	0.129559512	$\frac{\nu s^2}{\nu-2}$	0.132271223
$\frac{2}{\nu-4}\left(\frac{\nu s^2}{\nu-2}\right)^2$	0.000374017	$\sqrt{\frac{2}{\nu-4}}\left(\frac{\nu s^2}{\nu-2}\right)$	0.019339515		

- Prior degrees of freedom ν_0. See Table 13.9.
- Prior m. See Table 13.10.
- Prior s_0^2. See Table 13.10.
- Prior $M = \nu_0 s_0^2$. See Table 13.10.
- Classical regression degrees of freedom ν_C. See Table 13.5.
- Classical regression s_C^2. See Table 13.5.
- Classical regression $R = \nu_C s_C^2$. See Table 13.5.
- Posterior degrees of freedom ν. See equation (6.2) and Table 13.14.
- Posterior s^2. See equation (6.5) and Table 13.14.
- Posterior $A = \nu s^2$. See equation (6.4) and Table 13.14.
- Posterior mean of σ^2 term, $\frac{\nu s^2}{\nu - 2}$.

- Posterior variance of σ^2 term, $\frac{2}{\nu-4}\left(\frac{\nu s^2}{\nu-2}\right)^2$ and Table 13.14.
- $(\bar{\beta}-\beta^*)$, which is the difference between the classical statistics mean and the posterior mean and Table 13.14.
- $(\bar{\beta}-\beta^*)^T X^T y$, which is the classical statistics term in the posterior expression for A in equation (6.4) and Table 13.14.
- $(\beta_0-\beta^*)$, which is the difference between the prior mean and the posterior mean and Table 13.14.
- $(\beta_0-\beta^*)^T Q\beta_0$, which is the prior term in the posterior expression for A in equation (6.4) and Table 13.14.

Table 13.15 depicts the variance-covariance matrix for the posterior, calculated according to equation (6.16). The variances, standard deviations, t-statistics, and correlations appear in Table 13.15.

Table 13.15: Variance-Covariance and Correlation–Posterior

$VCV = (X^T X + Q)^{-1}\frac{\nu s^2}{\nu-2}$	Age	SN	lnDD	Intercept
Age	0.00001117	0.00000004	-0.00000589	-0.00009397
SN	0.00000004	0.00002278	-0.00000433	-0.00008655
lnDD	-0.00000589	-0.00000433	0.00043256	-0.00261825
Intercept	-0.00009397	-0.00008655	-0.00261825	0.01936333

Correlation	Age	SN	lnDD	Intercept
Age	1.0000	0.0028	-0.0848	-0.2021
SN	0.0028	1.0000	-0.0436	-0.1303
lnDD	-0.0848	-0.0436	1.0000	-0.9047
Intercept	-0.2021	-0.1303	-0.9047	1.0000

	Diagonals of Variance-Covariance Matrix			
	0.00001117	0.00002278	0.00043256	0.01936333
	Sqrts (Std. Error/Std. Dev.)			
	0.00334169	0.00477330	0.02079800	0.13915216
	t-statistics: Mean/Std. Dev.			
	-8.44346892	8.66460323	-7.86946931	20.15134740

13.2.4 Bayesian Posterior–Marginal Distributions Over the Model Coefficients

In Table 13.16, we turn to the issue of calculating the marginal Student's t distributions over the four β parameters. The requisite equation for the marginal density appears at the bottom of the table. The terms at the top of the table sequentially calculate the terms required by the equation at the bottom of the table.

- the first row $-\frac{\nu+1}{2}$ is the exponent of the univariate Student's t distribution

- the second row represents the diagonals of the inverse precision matrix
- the third row represents the entire precision row that appears in the Student's t distribution at the bottom
- the fourth and fifth rows calculate the key components of the integrating constant
- the final row gives the integrating constant for each of the four marginal distributions in this problem

Table 13.16: Parameters of Distribution Over Model Coefficients–Posterior

Univariate t exponent		-49.27777778	$-\frac{\nu+1}{2}$
$z_{ii} = i^{th}$ diagonal element of $\left(X^T X + Q\right)^{-1}$			
0.0000844243	0.0001722553	0.0032702257	0.1463910639
Precision $z_{ii}\nu s^2$			
0.00106706	0.002177178	0.041333201	1.850273279
$\ln\Gamma\left(\frac{\nu+1}{2}\right)$		$\ln\Gamma\left(\frac{\nu}{2}\right)$	π
141.7529347		139.8118598	3.141592654
$\ln c = \ln\Gamma\left(\frac{\nu+1}{2}\right) - \ln\Gamma\left(\frac{\nu}{2}\right) - \frac{1}{2}\ln\left(z_{ii}\nu s^2\right) - \frac{1}{2}\ln\pi$			
4.79013385	4.43357278	2.96175448	1.06104320
Constant c			
120.31747228	84.23182153	19.33185929	2.88938363

$$\{\beta_i\} = \frac{\Gamma\left(\frac{\nu+1}{2}\right)}{\pi^{\frac{1}{2}}\Gamma\left(\frac{\nu}{2}\right)} \frac{1}{\left(z_{ii}\nu s^2\right)^{\frac{1}{2}}} \left\{1 + (\beta_i - \beta_i *) \frac{1}{z_{ii}\nu s^2} (\beta_i - \beta_i *)\right\}^{-\frac{\nu+1}{2}}$$

The precision term and the integrating constant are all that are needed to calculate the full marginal probability distributions over the four coefficients.

13.2.5 Bayesian Posterior–Univariate Distributions over Model Coefficients

This section calculates each of the four univariate distributions over the four specified model coefficients. We will do so sequentially, beginning with the coefficient for AGE. The prior, classical, and posterior distributions over the model coefficients appear in Figure 13.8. Notice that the prior is a little bit wider than the classical regression solution, and its mean value is smaller. Paul thinks that roads deteriorate faster than the data suggests. Notice further that the posterior distribution lies between the prior and the classical regression solutions and is slightly tighter (with a higher peak). The higher or taller the distribution, the lower the variance. This has an absolutely classic look to it. The prior and the data both tell us the same thing. The posterior, therefore, tells us "something in the middle," and with a tighter variance. The prior and the data are sending a mutually reinforcing message, and the posterior reflects that.

Figure 13.8: Prior, Classical, and Posterior Distributions over Age Coefficient

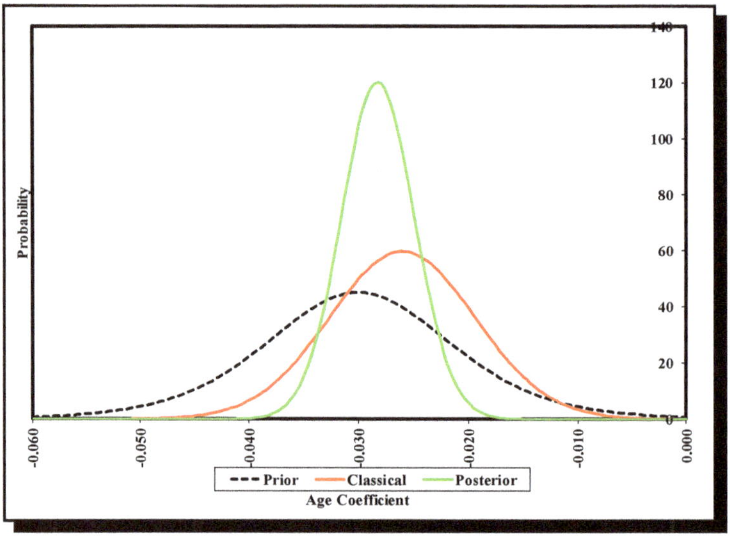

We turn to the coefficient for structure number SN. The prior, classical, and posterior distributions over the model coefficients for structure number SN appear in Figure 13.9. We see here a markedly different pattern. We see the prior as being narrower than the classical regression solution. That means that the data is not particularly definitive with respect to the SN term. The data are "small" with respect to the SN parameter; the data alone are not particularly definitive with regard to the SN coefficient. The prior is more definitive, as we see in the figure. However, the posterior is tighter than either the prior or the classical. This type of Bayesian updating is rather typical. The prior is more definitive than the classical, but together they reinforce and define a rather tighter posterior centered around a mean between the prior and classical means, which are close to being equal.

We turn to the coefficient for the logarithm of degree days, ln (DD). The prior, classical, and posterior distributions over the model coefficients for ln(DD) appear in Figure 13.10. The situation here is markedly different. We see that the mean of the posterior is outside the interval between the prior and the classical solution. Neither the prior nor the data are particularly definitive for the posterior. Notice that the prior is fairly wide, and the classical solution is fairly wide. The Bayesian solution in the middle, which combines priors with data, is not par-

Figure 13.9: Prior, Classical, and Posterior Distributions over Structure Number, SN, Coefficient

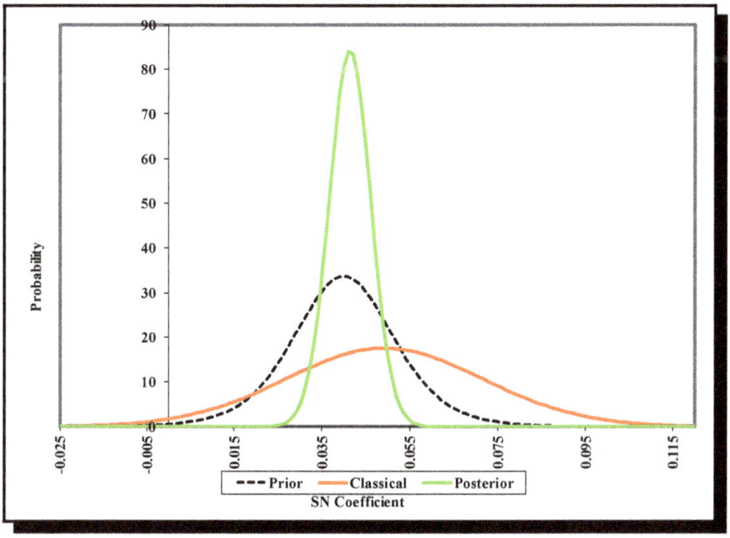

ticularly wide. Neither knows where the answer lies. This case illustrates rather definitively that the posterior mean does not have to lie between the prior mean and the classical mean. It is also a bit startling at first blush how low the variance on the posterior is. There are no easy rules of thumb in Bayesian updating. You have to do the math, and you have to interpret the answer after you do it.

We conclude with the coefficient for the intercept term. The prior, classical, and posterior distributions over the model coefficients for the intercept term appear in Figure 13.11. The wide prior and wide classical combine into a rather narrow posterior.

We conclude by showing the inverse gamma distributions for the prior, posterior, and classical conclusions. We will see that our prior assessment of the error term and its standard deviation are more close to a diffuse prior assumption than one might think at first blush. Inspection of Figure 13.12 shows how wide the prior distribution over the error standard deviation σ truly is. The dotted curve along the horizontal axis represents the prior, and it is indeed wide. That prior is certainly quite "diffuse-like." It represents a lot of ignorance or uncertainty what σ might be. Notice at the left, the classical and posterior densities over σ are much more narrow and quite similar. This is absolutely illustrative and to be expected. When you have a very wide prior or a diffuse prior, the data become much more definitive in establishing a posterior. This is exactly what one would expect, and in fact one sees the shape of the prior is not evident in Figure 13.12, so we rescale and replot it in Figure 13.13.

Figure 13.10: Prior, Classical, and Posterior Distributions over ln(DD) Coefficient

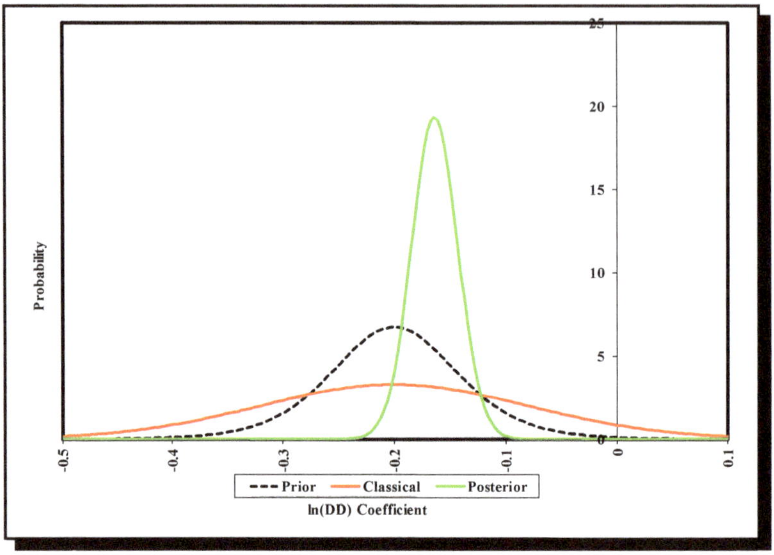

Figure 13.11: Prior, Classical, and Posterior Distributions over Intercept Coefficient

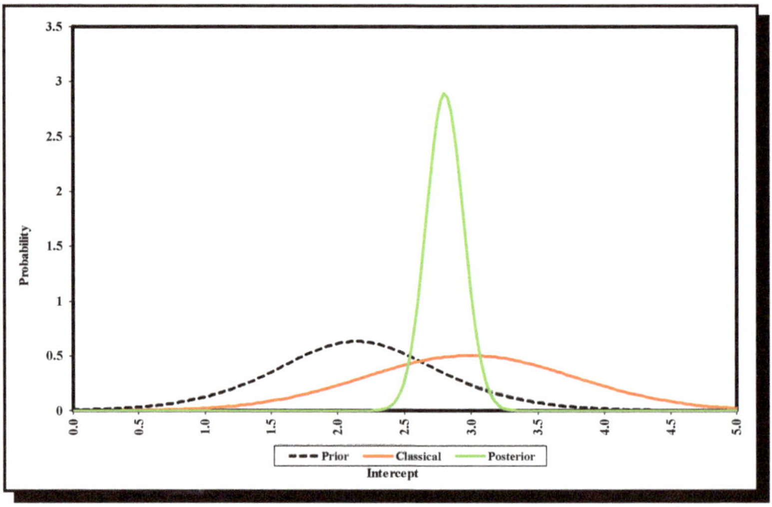

13.3 Pavement Performance Prediction

We are going to analyze the aging pavement here. We are going to consider a pavement that ages 0, 5, 10, 15, 20, 25, and 30 years. This implies $q = 7$ and implies the D matrix and its transpose in Table 13.17. For this illustration, we want the structure number to be 7.03 and the logarithm of degree days to be 6.115892 across the 7 different ages. The seven ages of pavements are arrayed as rows in the D matrix.

Table 13.17: The Seven Aging Pavements to be Predicted

Pavement	Age	SN	lnDD	Intercept
1	0	7.03	6.115892	1
2	5	7.03	6.115892	1
3	10	7.03	6.115892	1
4	15	7.03	6.115892	1
5	20	7.03	6.115892	1
6	25	7.03	6.115892	1
7	30	7.03	6.115892	1

Pavement	1	2	3	4	5	6	7
Age	0	5	10	15	20	25	30
SN	7.03	7.03	7.03	7.03	7.03	7.03	7.03
lnDD	6.115892	6.115892	6.115892	6.115892	6.115892	6.115892	6.115892
Intercept	1	1	1	1	1	1	1

Figure 13.12: Inverse Gamma Distributions with Emphasis on the Prior

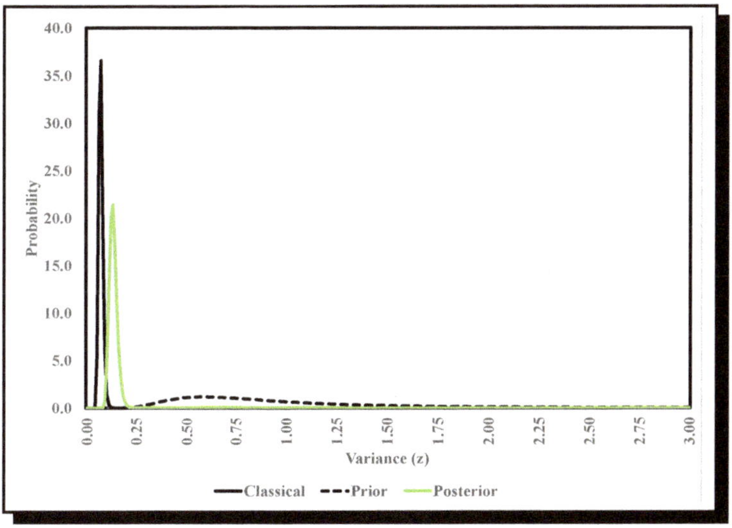

Figure 13.13: Inverse Gamma Distributions–Prior, Posterior, and Classical

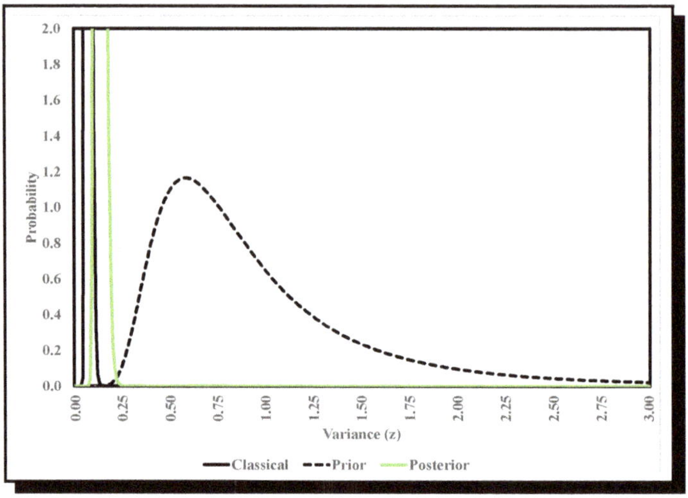

13.3.1 Predictions Based on the Classical Regression Calculations

This section puts forth the classical regression-based prediction of the performance of the seven pavements. In calculating the predictions under the classical regression assumptions (Bayes with diffuse prior), we have to calculate the precision and inverse precision matrices under classical regression. We begin by calculating the requisite matrix products of the D matrix in Table 13.18.

Table 13.18: Components of Precision and Inverse Precision Matrix–Classical Regression

$I + D(X^TX)^{-1}D^T$

	1	2	3	4	5	6	7	
1	1.112068	0.080042	0.048016	0.015990	-0.016036	-0.048062	-0.080088	
2	0.080042	1.064249	0.048457	0.032665	0.016872	0.001080	-0.014712	
3	0.048016	0.048457	1.048898	0.049340	0.049781	0.050222	0.050664	
4	0.015990	0.032665	0.049340	1.066015	0.082690	0.099364	0.116039	
5	-0.016036	0.016872	0.049781	0.082690	1.115598	0.148507	0.181415	
6	-0.048062	0.001080	0.050222	0.099364	0.148507	1.197649	0.246791	Determinant
7	-0.080088	-0.014712	0.050664	0.116039	0.181415	0.246791	1.312167	2.072190

$\left[I + D(X^TX)^{-1}D^T\right]^{-1}$

	1	2	3	4	5	6	7	
1	0.911067	-0.065435	-0.041937	-0.018440	0.005058	0.028556	0.052053	
2	-0.065435	0.947548	-0.039470	-0.026487	-0.013504	-0.000521	0.012462	
3	-0.041937	-0.039470	0.962998	-0.034534	-0.032066	-0.029598	-0.027130	
4	-0.018440	-0.026487	-0.034534	0.957419	-0.050628	-0.058675	-0.066722	
5	0.005058	-0.013504	-0.032066	-0.050628	0.930810	-0.087752	-0.106314	
6	0.028556	-0.000521	-0.029598	-0.058675	-0.087752	0.883172	-0.145905	Determinant
7	0.052053	0.012462	-0.027130	-0.066722	-0.106314	-0.145905	0.814503	0.482581

In Table 13.19, we see the key parameters of the univariate predictive distributions for the seven pavements given the classical regression solution. Table 13.20 enumerates the key parameters of the univariate predictive distributions for the seven pavements given the prior.

Table 13.19: Calculation of Precision and Inverse Precision Matrix–Classical Regression

ν_C	86					
s_C^2	0.067903					
z_{ii}						
1.112068	1.064249	1.048898	1.066015	1.115598	1.197649	1.312167
$D\beta$						
2.107589		Exponent	-43.500000			
1.977579						
1.847569						
1.717558						
1.587548						
1.457538						
1.327527						

Table 13.20: Key Parameters of Univariate Coefficient Predictive Distributions–Classical Regression

$\nu_C s_C^2 z_{ii}$						
6.494096	6.214854	6.125209	6.225162	6.514713	6.993862	7.662609
$\ln\Gamma\left(\frac{\nu_C+1}{2}\right)$	119.649575	$\ln\Gamma\left(\frac{\nu_C}{2}\right)$	117.771881	π	3.141593	
$\ln\left(\nu_C s_C^2 z_{ii}\right)$						
1.870893	1.826942	1.812413	1.828600	1.874063	1.945033	2.036352
		$\ln\Gamma\left(\frac{\nu_C+1}{2}\right)-\ln\Gamma\left(\frac{\nu_C}{2}\right)-\frac{1}{2}\ln\pi-\frac{1}{2}\ln\left(z_{ii}\nu_C s_C^2\right)$				
0.369881	0.391857	0.399122	0.391028	0.368297	0.332812	0.287152
Constant						
1.447563	1.479726	1.490515	1.478501	1.445271	1.394885	1.332627
Exponent						
-43.500000	-43.500000	-43.500000	-43.500000	-43.500000	-43.500000	-43.500000
Precision						
6.494096	6.214854	6.125209	6.225162	6.514713	6.993862	7.662609
Mean						
2.107589	1.977579	1.847569	1.717558	1.587548	1.457538	1.327527

13.3.2 Predictions Based on the Prior

This section puts forth the data for the predictions based on the prior The calculation of the precision and inverse precision matrix appears in Table 13.21. We begin again with the requisite matrix products in Table 13.21. Table 13.22 puts forth the key parameters of the univariate predictive distributions for the seven pavements given the prior. Table 13.23 details the calculation of the precision matrix and the inverse precision matrix for the prior.

Table 13.21: Calculation of Precision and Inverse Precision Matrix–Prior

$I + DQ^{-1}D^T$

1.683902	0.683902	0.683902	0.683902	0.683902	0.683902	0.683902	
0.683902	1.686402	0.688902	0.691402	0.693902	0.696402	0.698902	
0.683902	0.688902	1.693902	0.698902	0.703902	0.708902	0.713902	
0.683902	0.691402	0.698902	1.706402	0.713902	0.721402	0.728902	
0.683902	0.693902	0.703902	0.713902	1.723902	0.733902	0.743902	
0.683902	0.696402	0.708902	0.721402	0.733902	1.746402	0.758902	Determinant
0.683902	0.698902	0.713902	0.728902	0.743902	0.758902	1.773902	6.349929

$\left[I + DQ^{-1}D^T\right]^{-1}$

0.867795	-0.126550	-0.120896	-0.115242	-0.109587	-0.103933	-0.098278	
-0.126550	0.876826	-0.119799	-0.116423	-0.113047	-0.109671	-0.106295	
-0.120896	-0.119799	0.881299	-0.117604	-0.116506	-0.115409	-0.114312	
-0.115242	-0.116423	-0.117604	0.881215	-0.119966	-0.121147	-0.122328	
-0.109587	-0.113047	-0.116506	-0.119966	0.876574	-0.126885	-0.130345	
-0.103933	-0.109671	-0.115409	-0.121147	-0.126885	0.867377	-0.138361	Determinant
-0.098278	-0.106295	-0.114312	-0.122328	-0.130345	-0.138361	0.853622	0.157482

Table 13.22: Calculation of Precision and Inverse Precision Matrix–Prior

ν_0	7.555556					
s_0^2	0.735294					
z_{ii}						
1.683902	1.686402	1.693902	1.706402	1.723902	1.746402	1.773902
$D\beta_0$						
1.198088		Exponent	-4.277778			
1.048088						
0.898088						
0.748088						
0.598088						
0.448088						
0.298088						

13.3.3 Predictions Based on the Posterior Calculations

This section puts forth the posterior prediction of the performance of the seven pavements. The calculation of the precision and inverse precision matrix appears in Table 13.24. In Table 13.25 are the key parameters of the univariate predictive distributions for the seven pavements given the posterior. Table 13.26 enumerates the key parameters of the univariate predictive distributions for the seven pavements given the posterior.

13.3.4 Pavement Performance Prediction

13.3.4.1 Classical Regression

We begin with the classical regression solution, the Bayesian with diffuse prior. The prediction of pavement performance in the classical regression case appears

Table 13.23: Calculation of Precision and Inverse Precision Matrix–Prior

$\nu_0 s_0^2 z_{ii}$						
9.355014	9.368903	9.410569	9.480014	9.577236	9.702236	9.855014
$\ln\Gamma\left(\frac{\nu_0+1}{2}\right)$	2.151356	$\ln\Gamma\left(\frac{\nu_0}{2}\right)$	1.519781	π	3.141593	
$\ln\left(\nu_0 s_0^2 z_{ii}\right)$						
2.235912	2.237396	2.241833	2.249186	2.259389	2.272356	2.287980
$\ln\Gamma\left(\frac{\nu_0+1}{2}\right)-\ln\Gamma\left(\frac{\nu_0}{2}\right)-\frac{1}{2}\ln\pi-\frac{1}{2}\ln\left(\nu_0 s_0^2 z_{ii}\right)$						
-1.058747	-1.059489	-1.061707	-1.065383	-1.070485	-1.076969	-1.084781
Constant						
0.346890	0.346633	0.345865	0.344596	0.342842	0.340626	0.337976
Students t Exponent						
-4.277778	-4.277778	-4.277778	-4.277778	-4.277778	-4.277778	-4.277778
Precision						
9.355014	9.368903	9.410569	9.480014	9.577236	9.702236	9.855014
Mean						
1.198088	1.048088	0.898088	0.748088	0.598088	0.448088	0.298088

Table 13.24: Components of Precision and Inverse Precision Matrix–Posterior

$I+D(X^TX+Q)^{-1}D^T$

	1	2	3	4	5	6	7	
1	1.023085	0.018182	0.013280	0.008377	0.003474	-0.001429	-0.006332	
2	0.018182	1.015390	0.012598	0.009806	0.007013	0.004221	0.001429	
3	0.013280	0.012598	1.011916	0.011235	0.010553	0.009871	0.009190	
4	0.008377	0.009806	0.011235	1.012664	0.014093	0.015522	0.016951	
5	0.003474	0.007013	0.010553	0.014093	1.017632	0.021172	0.024711	
6	-0.001429	0.004221	0.009871	0.015522	0.021172	1.026822	0.032472	Determ
7	-0.006332	0.001429	0.009190	0.016951	0.024711	0.032472	1.040233	1.152581

$\left[I+D(X^TX+Q)^{-1}D^T\right]^{-1}$

	1	2	3	4	5	6	7	
1	0.978022	-0.017275	-0.012571	-0.007868	-0.003164	0.001540	0.006243	
2	-0.017275	0.985448	-0.011830	-0.009107	-0.006385	-0.003662	-0.000940	
3	-0.012571	-0.011830	0.988912	-0.010347	-0.009606	-0.008865	-0.008123	
4	-0.007868	-0.009107	-0.010347	0.988413	-0.012827	-0.014067	-0.015306	
5	-0.003164	-0.006385	-0.009606	-0.012827	0.983952	-0.019269	-0.022490	
6	0.001540	-0.003662	-0.008865	-0.014067	-0.019269	0.975529	-0.029673	Determ
7	0.006243	-0.000940	-0.008123	-0.015306	-0.022490	-0.029673	0.963144	0.867618

Table 13.25: Calculation of Precision and Inverse Precision Matrix–Posterior

ν	97.555556					
s^2	0.129560					
z_{ii}						
1.023085	1.015390	1.011916	1.012664	1.017632	1.026822	1.040233
$D\beta^*$						
2.093873		Exponent	-49.277778			
1.952795						
1.811718						
1.670640						
1.529563						
1.388486						
1.247408						

Table 13.26: Calculation of Precision and Inverse Precision Matrix–Posterior

$\nu s^2 z_{ii}$						
12.931031	12.833771	12.789864	12.799310	12.862109	12.978261	13.147766
$\ln\Gamma\left(\frac{\nu+1}{2}\right)$	141.752935	$\ln\Gamma\left(\frac{\nu}{2}\right)$	139.811860	π	3.141593	
$\ln\left(z_{ii}\nu s^2\right)$						
2.559630	2.552080	2.548653	2.549391	2.554286	2.563276	2.576252
$\ln\Gamma\left(\frac{\nu+1}{2}\right)-\ln\Gamma\left(\frac{\nu}{2}\right)-\frac{1}{2}\ln\pi-\frac{1}{2}\ln\left(z_{ii}\nu s^2\right)$						
0.088895	0.092670	0.094383	0.094014	0.091567	0.087072	0.080584
Constant						
1.092966	1.097099	1.098981	1.098575	1.095890	1.090975	1.083920
Exponent						
-49.277778	-49.277778	-49.277778	-49.277778	-49.277778	-49.277778	-49.277778
Precision						
12.931031	12.833771	12.789864	12.799310	12.862109	12.978261	13.147766
Mean						
2.093873	1.952795	1.811718	1.670640	1.529563	1.388486	1.247408

in Figure 13.14. Notice that the predictions are fairly narrow with low variance. Notice also, because the tops of the predictive distributions trend downward as we move to the left (successively older pavements), that the older pavements have more variance. This is to be expected, but not necessarily intrinsic to Bayesian methods. The older the pavement, the more uncertain we are about its condition. That is extremely intuitively appealing.

13.3.4.2 Prior

We next consider the prediction based on the prior, which represents Paul's judgment as shown in Figure 13.15. Notice that the predictions have a rather wide variance. This would be expected for the prior of an intercept term, which is probably intrinsically wide. Paul is not arrogant; his perspective is wide. Notice that the new (0-year-old) pavement appears at the right (the rightmost curve), and the older pavement shifts leftward. The variances for each successive age are surprisingly constant. This is not always the case. It is a manifestation of the specific prior used here.

13.3.4.3 Posterior

We finally consider the Bayesian with an informative prior. This prediction appears in Figure 13.16. Notice that the predictions are substantially narrower with a lower variance. Paul becomes more certain by updating his judgment with data.

Figure 13.14: Predicted RCI for an Aging Pavement–Classical Regression

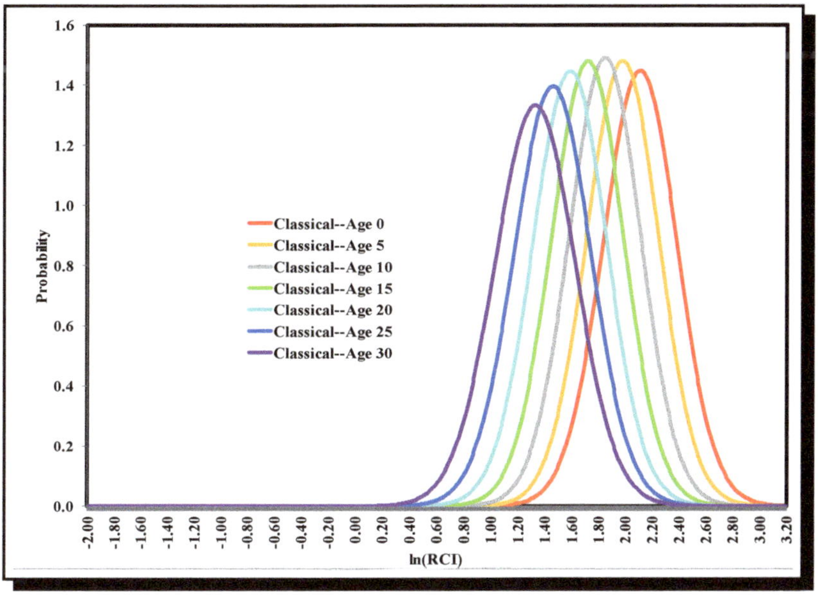

Figure 13.15: Predicted RCI for an Aging Pavement–Prior

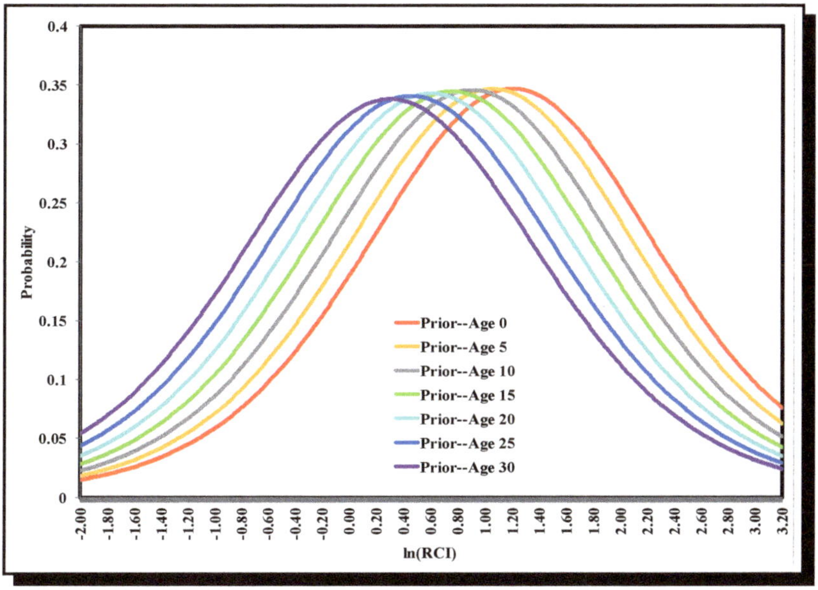

Figure 13.16: Predicted RCI for an Aging Pavement–Posterior

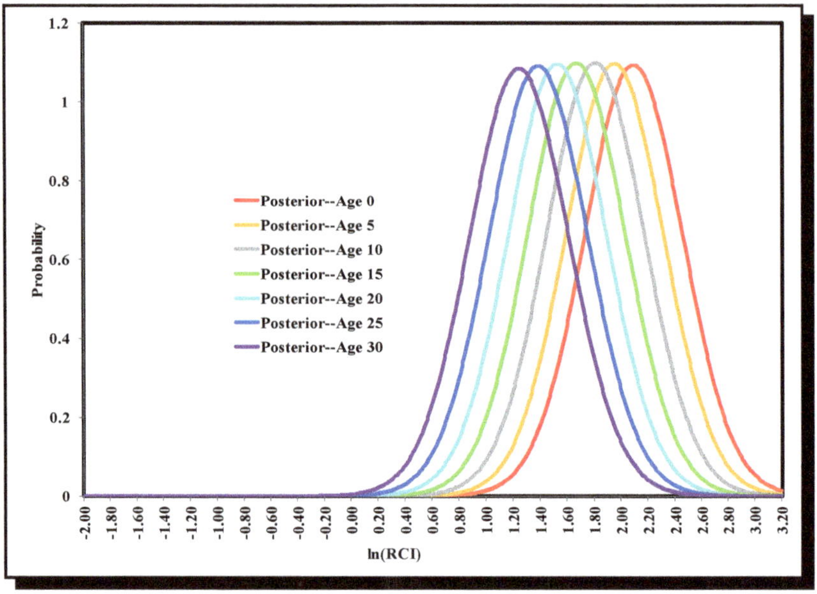

13.3.4.4 A 10 Year Old Pavement–Classical, Prior, Posterior

This section considers the classical regression solution, the prior prediction, and the posterior prediction of the RCI, focusing on a pavement that is ten years old. There is no new analysis in this section, but rather a comparison of predictions for a ten-year-old pavement. See Figure 13.17. We see that the prior is fairly wide. We see that the classical regression solution is centered at the higher point and is much narrower. We see that the prior has the ability to increase the variance relative to the data set (lower and wider density). In a sense, the data is "non-arrogant". The prior contains knowledge, but it is not arrogant in the sense that it has a narrow variance. It widens the probability distribution, and it moves the mean. That is exactly what we expect to see. The prior says one thing, and the wide variance indicates some uncertainty, but the data says something quite different. The posterior "balances" the prior and the data. So nicely intuitive. This provides invaluable data for planning the maintenance and rehabilitation of the pavement.

Figure 13.17: 10 Year Old Pavement–Prior, Classical, Posterior

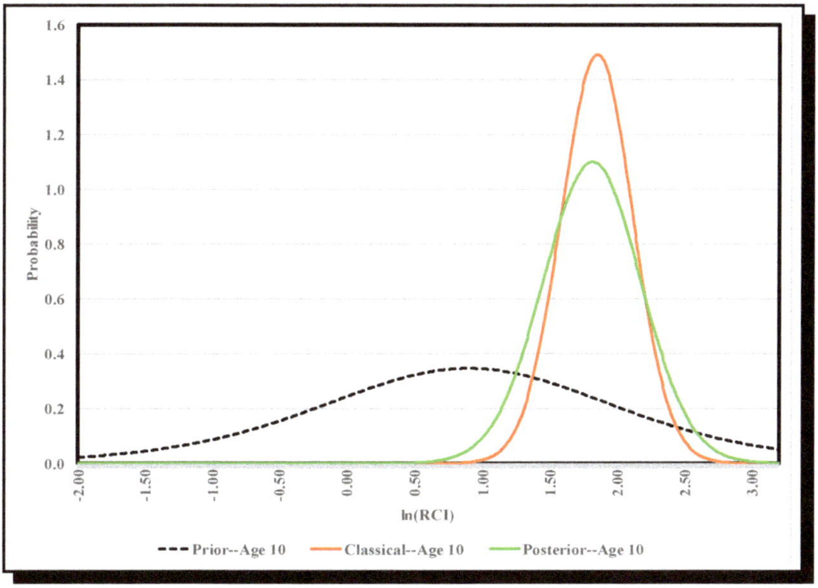

Chapter 14

ENCODING A UNIVARIATE PRIOR USING THE PROBABILITY WHEEL AND THE LOGISTIC DISTRIBUTION

You aren't your past, you are probability of your future.

Oprah Winfrey

Professor Howard[45] has put forth the problem of encoding (i.e., quantifying) a probability distribution over the weight of a desk situated at the front of a classroom. It is most often a standard-size, apparently wooden desk. (One cannot tell by looking whether the desk is solid wood or simply veneered!) The desk is often bolted to the floor in most Stanford lecture rooms. (No! It is not because students or staff would steal them! However, no one can be completely sure about faculty.) We ask the students as they enter the class one day: "How many of you walked in here this morning keenly wondering about the weight of this desk at the front of the class? Show of hands? Who wants to know the weight of this desk? Anyone want to hazard a guess? Anyone want to offer a range of uncertainty? Anyone want to offer a probability distribution? Anyone want to serve as the 'expert' upon whom we will rely to generate a probability distribution over the weight of this desk?" We ask for an "expert" on desk construction and weight. If one does not emerge from the class, we appoint one.

We use a probability wheel to ascertain points, in the opinion of the appointed "expert," on the cumulative distribution function and the complementary cumulative distribution function over the weight of the desk. We ask him or her a series of questions that require a bet either on the wheel, if spun, coming to a stop in the red or on the weight of the desk being higher than some number we specify. The probability wheel is a useful analogy for comparing a visual deal with the probability that the weight of the desk exceeds some specified value (a point on the complementary cumulative distribution) or is lower than some other specified value (a point on the cumulative distribution). (Ponder why the former comparison yields a point on the complementary cumulative density and the latter yields a point on the cumulative density.) See Figure 14.1 for a photo of a probability wheel. The use of the wheel itself to assess the weight of the desk is detailed in Howard and Abbas[45] and Spetzler and von Holstein[94], and we need not repeat it here.

Figure 14.1: A Probability Wheel

The probability wheel questions and answers in Howard and Abbas[45], Spetzler and von Holstein[94] led to the following estimates of points on the cumulative distribution function and the complementary cumulative distribution function for the weight of the desk:

$$
\begin{aligned}
\{weight \geq 150\} &= 0.3\\
\{weight \leq 70\} &= 0.2\\
\{weight \geq 100\} &= 0.5\\
\{weight \leq 220\} &= 0.84\\
\{weight \leq 130\} &= 0.58\\
\{weight \leq 180\} &= 0.79\\
\{weight \leq 250\} &= 0.9\\
\{weight \leq 100\} &= 0.4\\
\{weight \geq 50\} &= 0.95 \quad .
\end{aligned}
$$

These answers can be plotted as a cumulative distribution on a scatter diagram, as shown in Figure 14.2. We have attached numbers to the order in which the assessments were taken from the expert in Howard and Abbas[45] in the foregoing list. That is their only significance—to associate them with the assessments in the foregoing list of assessments.

Can we think of a continuous curve that passes as closely as possible through all these points? Yes. Can we adopt a specific functional form that can represent an approximation, a best fit, for the solid curve? Yes. Can we adopt a formal fitting method that is linear in the coefficients? Yes. Can we discretize the ultimate red curve? Yes. Can we assess joint distributions over precursor variables even though this assessment is univariate? Yes! What is the nature of the probability distribution over the model coefficients we will want to use? We will now show it very carefully.

Figure 14.2: Assessments from the Probability Wheel for the Weight of the Desk

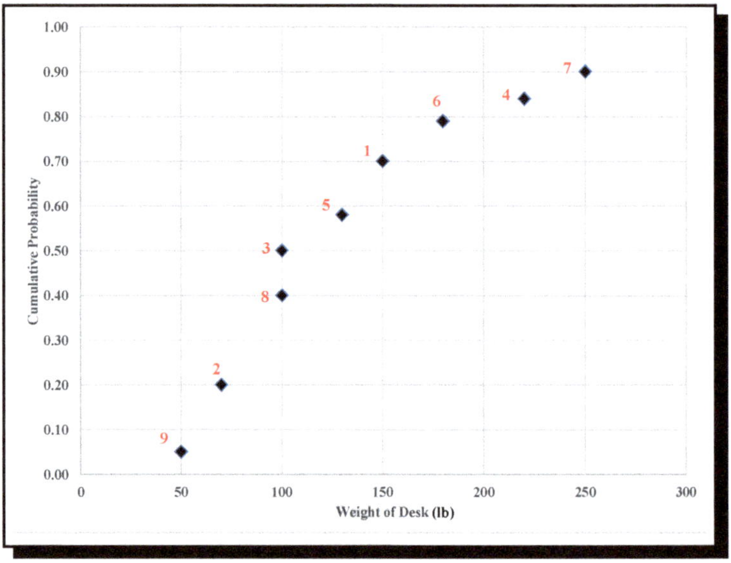

Keelin[53] has made progress on this problem with his contribution to probability assessment. (The author became aware of Keelin's contributions from his presentations to the Stanford Decision Analysis Working Group, or DAWG.) We illustrate his method herein, which is a nice appeal to Bayesian Linear Regression, if not Bayesian Linear Regression. Keelin began a few years ago by examining the logistic probability distribution, or more properly, the cumulative logistic.[1] The functional form of cumulative logistic distribution is

[1] See https://en.wikipedia.org/wiki/Logistic_distribution.

$$p = \frac{1}{1 + e^{-\frac{w-\mu}{s}}} \quad ,$$

in which w is the weight of the desk, p is a cumulative probability, and μ and s are parameters of the logistic distribution. (This function is nonlinear in the constants.) It is crucially important to note that $s > 0$. We will see that μ is actually the mean. Can we fit the cumulative of the logistic distribution to the assessed points, thereby generating a cumulative distribution curve, the solid curve, that is the best fit to the points, and that solid curve has the form of a logistic distribution? The answer is a definitive yes. And owing to Keelin[53], things get even better as we contemplate what that implies. However, as we will see, it does not allow easy use of the linear, conjugate statistical approach developed herein and devolves into a nonlinear regression problem and requires simulation to solve.

It is convenient to develop the functional form of the logistic density and its cumulative effect and calculate its mean and variance. We do so by differentiating the cumulative distribution in the standard way. To facilitate that, we begin by inverting the functional form of the cumulative density immediately above.

$$\begin{aligned}
p &= \frac{1}{1 + e^{-\frac{w-\mu}{s}}} \\
&\Rightarrow 1 + e^{-\frac{w-\mu}{s}} = \frac{1}{p} \\
&\Rightarrow e^{-\frac{w-\mu}{s}} = \frac{1}{p} - 1 \\
&\Rightarrow -\frac{w-\mu}{s} = \ln\left(\frac{1-p}{p}\right) \\
&\Rightarrow w = \mu - s\ln\left(\frac{1-p}{p}\right) \\
&\Rightarrow w = \mu + s\ln\left(\frac{p}{1-p}\right) \quad .
\end{aligned}$$

The notion of a "fit" appears in Figure 14.3. It is the solid curve in this diagram, in the form of a cumulative logistic function, that we strive to identify using Bayesian statistical methods.

This is a cumulative density function, and we want the density function itself, which is related to the derivative of the cumulative. We start with the total differential and then write the derivative of w with respect to p:

$$\begin{aligned}
p &= \left(1 + e^{-\frac{w-\mu}{s}}\right)^{-1} \\
&\Rightarrow dp = -\left(1 + e^{-\frac{w-\mu}{s}}\right)^{-2} e^{-\frac{w-\mu}{s}} \left(-\frac{dw}{s}\right)
\end{aligned}$$

Figure 14.3: A Cumulative Curve Fit to the Weight of the Desk

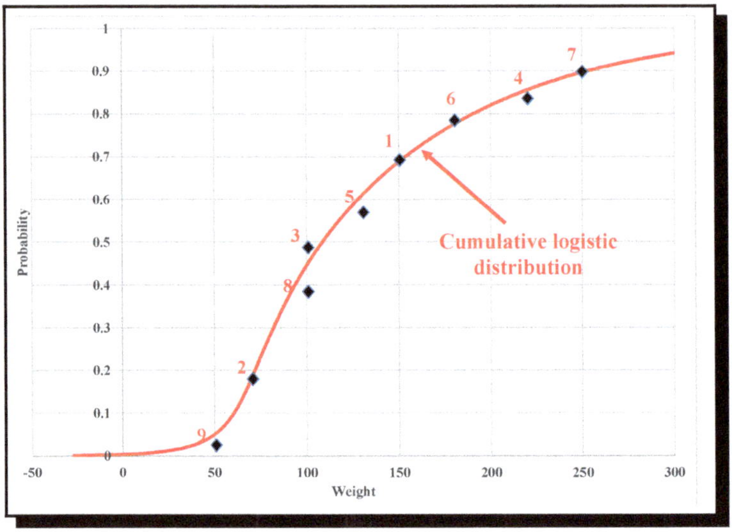

$$\Rightarrow \frac{dp}{dw} = f(w) = \frac{e^{-\frac{w-\mu}{s}}}{s\left(1+e^{-\frac{w-\mu}{s}}\right)^2} \quad .$$

In the equation, $f(w)$ is the logistic probability density function. Using this function, we can define and calculate the mean and variance as follows

$$\text{Mean} = \mu \triangleq \int_{-\infty}^{\infty} wf(w)\,dw \quad .$$

and

$$\text{Variance} = \sigma^2 \triangleq \int_{-\infty}^{\infty} (w-\mu)^2 f(w)\,dw = \frac{s^2\pi^2}{3} \quad .$$

We have used an unbounded form of the logistic density, so the integral must span desk weights from $-\infty$ to ∞. (This assumption occurs frequently with normal distributions.) It is easy to lower the weight to 0, but we will not discuss that here, choosing the simpler mathematics that occurs without a lower or upper bound. It is not particularly easy to perform the foregoing two integrals, but we will simply reference the answers from Wiki[2] because they are well known and not totally germane here. The mean and variance are reported to be

[2]See https://en.wikipedia.org/wiki/Logistic_distribution.

$$\text{Mean} = \mu$$
$$\text{Variance} = \frac{s^2\pi^2}{3} \quad .$$

The variance is the square of the standard deviation, so we can express the s parameter in terms of the standard deviation

$$\sigma^2 = \frac{s^2\pi^2}{3} \Rightarrow s = \sigma\frac{\sqrt{3}}{\pi} \quad .$$

This allows us to substitute and write the cumulative distribution function in terms of mean and standard deviation terms explicitly

$$p = \frac{1}{1 + e^{-\frac{w-\mu}{\sigma}\frac{\pi}{\sqrt{3}}}} \quad .$$

We now want to invert this cumulative distribution in preparation for determining the coefficients from the data

$$\begin{aligned}
p &= \frac{1}{1 + e^{-\frac{w-\mu}{\sigma}\frac{\pi}{\sqrt{3}}}} \\
&\Rightarrow p\left(1 + e^{-\frac{w-\mu}{\sigma}\frac{\pi}{\sqrt{3}}}\right) = 1 \\
&\Rightarrow e^{-\frac{w-\mu}{\sigma}\frac{\pi}{\sqrt{3}}} = \frac{1-p}{p} \\
&\Rightarrow -\frac{w-\mu}{\sigma}\frac{\pi}{\sqrt{3}} = \ln\left(\frac{1-p}{p}\right) \\
&\Rightarrow \frac{w-\mu}{\sigma}\frac{\pi}{\sqrt{3}} = \ln\left(\frac{p}{1-p}\right) \\
&\Rightarrow w = \mu + \sigma\frac{\sqrt{3}}{\pi}\ln\left(\frac{p}{1-p}\right) \quad .
\end{aligned}$$

Keelin[53] points out two key insights, and we append a third.

1. The equation at the bottom shows that w is linear in the two (as yet unknown) parameters μ and σ. Keelin has exploited this to develop his metalog formulation, which we shall discuss after we develop the key relationships for the logistic distribution.

2. We can (seemingly) use Bayesian Linear Regression to "fit" for μ and σ. Wouldn't that ever convenient! Such a fit (or more precisely than a fit, a Bayesian Linear Regression with a diffuse or informative prior) is positioned to give us a continuous curve, the solid curve, that approximates the nine points of the assessment from the probability wheel. This is a

solid, conceptually appealing way for us to obtain a prior to be used for Bayesian Linear Regression. Alas, it isn't quite true, as we shall see.

3. $\sigma > 0$.

Let us summarize. We can assume the solid curve in Figure 14.3 is a logistic cumulative density function, and it appears that we can fit for μ and σ. When we say fit, what we really mean is that we perform a Bayesian statistical analysis to determine the mean values (and, in fact, the entire distribution) of those two coefficients. This is crucially important; the fitting that occurs here is occurring using the cumulative density function, which is linear in the two parameters, μ and σ, and not the density function itself (which is highly nonlinear in the two parameters). The beauty of the cumulative distribution of the logistic density is that there are two (2) constants that fully define the distribution. The distribution is "conditional" on those two constants. The two coefficients are μ (the mean) and σ (the standard deviation), which must be strictly positive. As pointed out by Keelin[53], the logistic cumulative density function is a simple analytical form that is easily invertible and is linear in the unknown coefficients. It is therefore immediately amenable to classical or Bayesian Linear Regression (curve fitting). Keelin[53] has extended this thinking to more complex, more implicit functional forms, as we shall address shortly.

Before proceeding to the fit, we wish to express the equation for the logistic density in terms of the mean μ and standard deviation σ rather than the original terms μ and s. We obtain the properly expressed probability density function by differentiating the cumulative distribution.

$$
\begin{aligned}
p &= \left(1 + e^{-\frac{w-\mu}{\sigma}\frac{\pi}{\sqrt{3}}}\right)^{-1} \\
\Rightarrow dp &= -\left(1 + e^{-\frac{w-\mu}{\sigma}\frac{\pi}{\sqrt{3}}}\right)^{-2} e^{-\frac{w-\mu}{\sigma}\frac{\pi}{\sqrt{3}}}\left(-\frac{dw}{\sigma}\frac{\pi}{\sqrt{3}}\right) \\
\Rightarrow \frac{dp}{dw} &= f(w) = \frac{1}{\sigma}\frac{\pi}{\sqrt{3}}\frac{e^{-\frac{w-\mu}{\sigma}\frac{\pi}{\sqrt{3}}}}{\left(1 + e^{-\frac{w-\mu}{\sigma}\frac{\pi}{\sqrt{3}}}\right)^2} \quad .
\end{aligned}
$$

It is useful for future reference to express this using conditional notation as follows

$$
\{w|\mu,\sigma\} = \frac{1}{\sigma}\frac{\pi}{\sqrt{3}}\frac{e^{-\frac{w-\mu}{\sigma}\frac{\pi}{\sqrt{3}}}}{\left(1 + e^{-\frac{w-\mu}{\sigma}\frac{\pi}{\sqrt{3}}}\right)^2} \quad .
$$

It is important to note that this distribution is not linear in either μ (the mean) or σ (the standard deviation). On the contrary, it is highly nonlinear in those coefficients. As a result of this nonlinearity, there is no prospect of applying Bayesian Linear Regression to the logistic density per se. We examine the inverse of the cumulative distribution, which is the curve we will ultimately want to fit to the points taken from the probability wheel.

$$w = \mu + \sigma \frac{\sqrt{3}}{\pi} \ln\left(\frac{p}{1-p}\right) \quad .$$

If we are going to fit or conduct Bayesian Linear Regression to obtain probability distributions over the two unknowns μ and σ, we are going to need a data set. That is precisely what has come from the estimates of cumulative and complementary cumulative distribution functions using the wheel. This is a quintessential data set. And it is not a particularly big data set. To see why, the answers from the wheel are arranged (in ascending order of the probability p) in Table 14.1.

Table 14.1: The Data from the Probability Wheel

Assessment	p	< w
1	0.05	50
2	0.20	70
3	0.40	100
4	0.50	100
5	0.58	130
6	0.70	150
7	0.79	180
8	0.84	220
9	0.90	250

We see in Table 14.2 the easily derived parameters of the statistical fit that is being done in the regression equation

$$w_i = \mu + \sigma \frac{\sqrt{3}}{\pi} \ln\left(\frac{p_i}{1-p_i}\right) + \varepsilon_i \quad i = 1, ..., n \quad .$$

We define

$$w = \begin{bmatrix} w_1 \\ w_2 \\ \cdot \\ \cdot \\ \cdot \\ w_n \end{bmatrix} ,$$

$$X = \begin{bmatrix} 1 & \frac{\sqrt{3}}{\pi}\ln\left(\frac{p_1}{1-p_1}\right) \\ 1 & \frac{\sqrt{3}}{\pi}\ln\left(\frac{p_2}{1-p_2}\right) \\ \cdot & \cdot \\ \cdot & \cdot \\ \cdot & \cdot \\ 1 & \frac{\sqrt{3}}{\pi}\ln\left(\frac{p_n}{1-p_n}\right) \end{bmatrix} ,$$

$$\beta = \begin{bmatrix} \mu \\ \sigma \end{bmatrix} \quad , and$$

$$\varepsilon = \begin{bmatrix} \varepsilon_1 \\ \varepsilon_2 \\ \cdot \\ \cdot \\ \cdot \\ \varepsilon_n \end{bmatrix} .$$

Table 14.2: Manipulation of Data for Regression

Assessment	Intercept	$\frac{\sqrt{3}}{\pi}\ln\left(\frac{p}{1-p}\right)$	w
1	1	-1.62	50
2	1	-0.76	70
3	1	-0.22	100
4	1	0.00	100
5	1	0.18	130
6	1	0.47	150
7	1	0.73	180
8	1	0.91	220
9	1	1.21	250

$$w = \mu + \sigma\frac{\sqrt{3}}{\pi}\ln\left(\frac{p}{1-p}\right) + \varepsilon$$

Using these definitions, we write in very standard form

$$\varepsilon^T\varepsilon = (w - X\beta)^T (w - X\beta) \quad .$$

This allows us to write the likelihood function in standard form

$$\{y, X|s, \mu, \sigma\} = \frac{1}{s^n(2\pi)^{\frac{n}{2}}} e^{-\frac{1}{2s^2}(w-X\beta)^T(w-X\beta)} \quad .$$

in which we have used the notation s in place of the standard deviation in the usual notation σ, which is necessary because we have used σ as one of the parameters in the logistic density.

Let us manipulate the exponent as we have done previously by completing the square. We begin by defining $z = \beta - b$ and rearranging $\beta = z + b$. Substituting this into the exponent yields the relationship

$$\begin{aligned}(w - X\beta)^T (w - X\beta) &= [w - X(z+b)]^T [w - X(z+b)] \\ &= \left[w^T - (z+b)^T X^T\right] [w - X(z+b)] \\ &= w^T [(w - Xb) - Xz] \\ &- (z+b)^T X^T [(w - Xb) - Xz] \\ &= (w - Xb)^T (w - Xb) \\ &+ 2z^T \left(X^T X b - X^T w\right) \\ &+ z^T X^T X z \quad .\end{aligned}$$

We want to eliminate the term that is linear in z^T

$$\begin{aligned}&\left(X^T X\right) \bar{b} - X^T w = 0 \\ &\Rightarrow \bar{b} = \left(X^T X\right)^{-1} X^T w = \begin{bmatrix} \bar{\mu} \\ \bar{\sigma} \end{bmatrix} \quad ,\end{aligned}$$

so that the exponent reduces to the standard form

$$\left(w - X\bar{b}\right)^T \left(w - X\bar{b}\right) + \left(\beta - \bar{b}\right)^T \left(X^T X\right) \left(\beta - \bar{b}\right) \quad .$$

This means that the likelihood function can be written in the standard form

$$\{y, X | s, \mu, \sigma\} = \frac{1}{s^n (2\pi)^{\frac{n}{2}}} e^{-\frac{1}{2s^2}\left[\left(w - X\bar{b}\right)^T \left(w - X\bar{b}\right) + \left(\beta - \bar{b}\right)^T X^T X \left(\beta - \bar{b}\right)\right]} \quad .$$

If we want to use the square of s, i.e.,

$$z = s^2 \Rightarrow s^n = z^{\frac{n}{2}} \quad ,$$

then we can write the likelihood function

$$\{y, X | z, \mu, \sigma\} = \frac{1}{z^{\frac{n}{2}} (2\pi)^{\frac{n}{2}}} e^{-\frac{1}{z}\left[\frac{\left(w - X\bar{b}\right)^T \left(w - X\bar{b}\right) + \left(\beta - \bar{b}\right)^T X^T X \left(\beta - \bar{b}\right)}{2}\right]} \quad .$$

Now that we have the likelihood function, the problem shifts to the problem of assuming a prior distribution, which we denote $\{z, \mu, \sigma\}$ and multiplying it by the likelihood function to obtain the joint

$$\begin{aligned}&\{y, X|z, \mu, \sigma\}\{z, \mu, \sigma\}\\&= \{y, X, z, \mu, \sigma\}\\&= \{z, \mu, \sigma|y, X\}\{y, X\}\\&= \{z, \mu, \sigma|y, X\}c_0\end{aligned}$$

in which c_0 is a constant of integration. This means that the sought-after conditional posterior distribution is

$$\begin{aligned}&\{z, \mu, \sigma|y, X\}\\&= \frac{1}{c_0}\{z, \mu, \sigma\}\{y, X|z, \mu, \sigma\}\\&= \frac{1}{c_0}\{z, \mu, \sigma\}\frac{1}{z^{\frac{n}{2}}(2\pi)^{\frac{n}{2}}}e^{-\frac{1}{z}\left[\frac{(w-X\bar{b})^T(w-X\bar{b})+(\beta-\bar{b})^T X^T X(\beta-\bar{b})}{2}\right]}\quad .\end{aligned}$$

And now come the storm clouds. If we consider the logistic density function carefully, we see that it is not defined for negative values of σ. This nonnegativity requirement ruins the conjugacy of this problem, necessitating simulation, but we want to create as easy a problem as we can for simulation. In the situation in which there is no requirement on the sign of the σ term, we are easily able, as in Chapter 4, to posit a conjugate prior as we did there and obtain a conjugate posterior. Once the positivity constraint is imposed, adios to the conjugate prior-posterior pairing. That is why we have taken care to reduce the likelihood to a simple form—to facilitate simulation. To be clear, the posterior will not take the form of the prior and therefore will not be conjugate as a result of the positivity requirement for σ. Reiterating, there is no conjugate form of the prior that results in this density, with $\sigma > 0$, that has the same form as the prior.

The Bayesian prior in our archetypical form simply does not require that any one of its coefficients, namely σ, be unconditionally positive. It is the estimated coefficient, not the data, that we want to be unconditionally positive. There is no conjugate form we know of that does that. We see in the literature much ado about "constrained least squares" and that type of philosophy. We are not doing anything like "least squares" with Bayes, and more importantly, we are not doing anything nonlinear that would disrupt our conjugacy assumption.

How can we deal with this easily? We are going to borrow from the results of Appendix E to craft a prior that ensures that the standard deviation variable σ is strictly positive. We are going to first think about a separable prior, isolating the model coefficients from the z term, which is a data error term. In particular, we will want to express the following:

$$\{z, \mu, \sigma\} = \{z\}\{\mu, \sigma\}\quad .$$

This is precisely the same thing we saw in Chapter 4 when we noted that the coefficients and the error term in the prior were probabilistically independent. There was no correlation between the two in previous formulations. Such independence seems, by analogy to the conjugate case, to be a reasonable assumption. Once we assume that the error term and the coefficient terms are probabilistically independent, why wouldn't one assume that the variance of the error term z is governed by an inverse gamma function? To see what this implies, we start with the equation for the inverse gamma function.

$$\{z\} = \frac{B_0^{\alpha_0}}{\Gamma(\alpha_0)}\left(\frac{1}{z}\right)^{\alpha_0+1} e^{-\frac{B_0}{z}} \quad .$$

We bring in the constants α_0 and B_0 to characterize the prior. Substituting into the equation for the joint distribution gives the relationship yields:

$$\begin{aligned}
\{z,\mu,\sigma,y,X\} &= \{\mu,\sigma\}\{z\}\frac{1}{c_0}\frac{1}{z^{\frac{n}{2}}(2\pi)^{\frac{n}{2}}} \\
&\times e^{-\frac{1}{2z}(w-X\bar{b})^T(w-X\bar{b})}e^{-\frac{1}{2z}(\beta-\bar{b})^T(X^TX)(\beta-\bar{b})} \\
&= \{\mu,\sigma\}\frac{B_0{}^{\alpha_0}}{\Gamma(\alpha_0)}\left(\frac{1}{z}\right)^{\alpha_0+1} e^{-\frac{B_0}{z}}\frac{1}{c_0}\frac{1}{z^{\frac{n}{2}}(2\pi)^{\frac{n}{2}}} \\
&\times e^{-\frac{1}{2z}(w-X\bar{b})^T(w-X\bar{b})}e^{-\frac{1}{2z}(\beta-\bar{b})^T(X^TX)(\beta-\bar{b})} \\
&= \{\mu,\sigma\}\frac{B_0{}^{\alpha_0}}{\Gamma(\alpha_0)\,c_0(2\pi)^{\frac{n}{2}}}\left(\frac{1}{z}\right)^{\alpha_0+1} e^{-\frac{B_0}{z}}\frac{1}{z^{\frac{n}{2}}} \\
&\times e^{-\frac{1}{2z}(w-X\bar{b})^T(w-X\bar{b})}e^{-\frac{1}{2z}(\beta-\bar{b})^T(X^TX)(\beta-\bar{b})} \\
&= \{\mu,\sigma\}\,c_1 z^{-\left(\alpha_0+1+\frac{n}{2}\right)}e^{-\frac{1}{z}\left[\frac{(w-X\bar{b})^T(w-X\bar{b})}{2}+B_0\right]} \\
&\times e^{-\frac{1}{2z}(\beta-\bar{b})^T(X^TX)(\beta-\bar{b})}
\end{aligned}$$

in which we have combined all the constant terms into c_1. (We subjectively estimate the parameters of the inverse gamma function embedded in the prior. That is, we assume values for the α_0 and the B_0 term in order to characterize the prior. We also want to emphasize that we have previously easily and directly obtained the mean values of the mean coefficient $\bar{\mu}$ and the standard deviation coefficient $\bar{\sigma}$. The vector $\bar{\beta}$ contains those two variables in its first and second positions. We want to integrate out the z variable, i.e., to solve:

$$\int_0^\infty dz\,\{y,X,z,\mu,\sigma\} =$$

$$\int_0^\infty dz\,\{\mu,\sigma\}\,c_1 z^{-\left(\alpha_0+\frac{n}{2}+1\right)}e^{-\frac{1}{z}\left[\frac{(w-X\bar{b})^T(w-X\bar{b})+(\beta-\bar{b})^T\left(X^TX\right)(\beta-\bar{b})+2B_0}{2}\right]} \quad .$$

Let us make the substitution

$$G = \frac{\left(w - X\bar{b}\right)^T \left(w - X\bar{b}\right) + \left(\beta - \bar{b}\right)^T \left(X^T X\right)\left(\beta - \bar{b}\right) + 2B_0}{2}$$

so that we can write the integral

$$\int_0^\infty dz \{y, X, z, \mu, \sigma\} = \int_0^\infty \{\mu, \sigma\} c_1 z^{-\left(\alpha_0 + \frac{n}{2} + 1\right)} e^{-\frac{1}{z}G} dz \quad .$$

.

Let us substitute

$$h = z^{-1} G$$

which implies

$$z = h^{-1} G \quad .$$

We take the differential of this equation to write

$$dz = -h^{-2} G dh \quad .$$

We raise z to the appropriate power to write

$$z^{-\left(\alpha_0 + \frac{n}{2} + 1\right)} = h^{\alpha_0 + \frac{n}{2} + 1} G^{-\left(\alpha_0 + \frac{n}{2} + 1\right)} \quad .$$

We make all the requisite substitutions (and constant changes) as follows

$$\begin{aligned}
&\int_0^\infty dz \{y, X, z, \mu, \sigma\} = \int_\infty^0 \{\mu, \sigma\} c_1 \left[h^{\alpha_0 + \frac{n}{2} + 1} G^{-\left(\alpha_0 + \frac{n}{2} + 1\right)}\right] e^{-h} \left(-h^{-2} G dh\right) \\
&= c_1 \{\mu, \sigma\} G^{-\alpha_0 - \frac{n}{2}} \int_0^\infty h^{\alpha_0 + \frac{n}{2} - 1} e^{-h} dh \\
&= c_1 \{\mu, \sigma\} G^{-\alpha_0 - \frac{n}{2}} \Gamma\left(\alpha_0 + \frac{n}{2}\right) \\
&= c_2 \{\mu, \sigma\} G^{-\frac{1}{2}(n + 2\alpha_0)} \\
&= c_3 \{\mu, \sigma\} \\
&\times \left[\left(w - X\bar{b}\right)^T \left(w - X\bar{b}\right) + \left(\beta - \bar{b}\right)^T \left(X^T X\right)\left(\beta - \bar{b}\right) + 2B_0\right]^{-\frac{1}{2}(n + 2\alpha_0)} \quad .
\end{aligned}$$

We want to carefully denote the structure of the unknown vector

$$\int_{0}^{\infty} dz \{y, X, z, \mu, \sigma\} = c_3 \{\mu, \sigma\}$$

$$\left\{ \begin{array}{l} \left[w - X\begin{pmatrix} \bar{\mu} \\ \bar{\sigma} \end{pmatrix}\right]^T \left[w - X\begin{pmatrix} \bar{\mu} \\ \bar{\sigma} \end{pmatrix}\right] \\ + \left[\begin{pmatrix} \mu \\ \sigma \end{pmatrix} - \begin{pmatrix} \bar{\mu} \\ \bar{\sigma} \end{pmatrix}\right]^T (X^T X) \left[\begin{pmatrix} \mu \\ \sigma \end{pmatrix} - \begin{pmatrix} \bar{\mu} \\ \bar{\sigma} \end{pmatrix}\right] + 2B_0 \end{array} \right\}^{-\frac{1}{2}(n+2\alpha_0)}$$

Consider the marginal prior distribution over the coefficients $\{\mu, \sigma\}$. We begin with the mean term μ in this joint distribution. We need to specify the prior that specifies our ex ante, prior perception of the μ term. At the end of Appendix E, we learned how to express the prior over the mean term μ using a univariate normal with any mean $M_{\mu\mu}$ we want and any variance $V_{\mu\mu}$ we want. Appendix E shows that we can write the marginal prior over the mean term μ as follows

$$\{\mu\} = \frac{1}{V_{\mu\mu}^{\frac{1}{2}} (2\pi)^{\frac{1}{2}}} e^{-\frac{(\mu - M_{\mu\mu})^2}{2V_{\mu\mu}}} .$$

Appendix E also developed a univariate lognormal distribution (lognormal over the standard deviation term σ in order to ensure that it was nonnegative) as equation (E.1) in Appendix E. Bringing that discussion to the fore, we may define the mean of the lognormal distribution over the standard deviation term σ to be $M_{\sigma\sigma}$, any nonnegative value we want, and the variance of the standard deviation term σ^2 to be $V_{\sigma\sigma}$, any nonnegative value we want. Both are subjectively specified. There is no requirement that they be the same as the term $\bar{\sigma}$ in the likelihood function. They can be ANY level one might want. Under that assumption, we define the quantity from Appendix E

$$\theta_{\sigma\sigma} = \frac{V_{\sigma\sigma} + M_{\sigma\sigma}^2}{M_{\sigma\sigma}^2}$$

and thereafter write the lognormal density over the standard deviation σ as a function of its mean and variance

$$\{\sigma\} = \frac{1}{\sigma\sqrt{2\pi \ln(\theta_{\sigma\sigma})}} e^{-\frac{\left[\ln\left(\sigma \frac{\sqrt{\theta_{\sigma\sigma}}}{M_{\sigma\sigma}}\right)\right]^2}{2 ln(\theta_{\sigma\sigma})}} .$$

We reiterate that the prior estimate of the standard deviation can be anything and it not necessarily consistent with any value in the likelihood function. It is useful to summarize what the two independent components of our prior are:

- The term μ is distributed from $-\infty$ to ∞ and is governed by a univariate normal distribution. Its mean is $M_{\mu\mu}$ and its variance is $V_{\mu\mu}$, any subjective values we assess.

- The term σ is distributed from 0 to ∞ and is governed by a lognormal distribution. This distribution has a zero value for any value of σ that is negative. Every nonnegative real value is allowed. Nonpositive values are zero in the distribution, ensuring that the posterior never allows any negative values. Its mean is $M_{\sigma\sigma}$ and its variance is $V_{\sigma\sigma}$, any subjective values we assess.

The joint distribution over $\{\mu, \sigma\}$ is the product of these two independent distributions:

$$\{\mu, \sigma\} = \frac{1}{V_{\mu\mu}^{\frac{1}{2}} (2\pi)^{\frac{1}{2}}} e^{-\frac{(\mu - M_{\mu\mu})^2}{2V_{\mu\mu}}} \frac{1}{\sigma\sqrt{2\pi \ln(\theta_{\sigma\sigma})}} e^{-\frac{\left[\ln\left(\sigma \frac{\sqrt{\theta_{\sigma\sigma}}}{M_{\sigma\sigma}}\right)\right]^2}{2ln(\theta_{\sigma\sigma})}}$$

$$= c_4 \frac{1}{\sigma} e^{-\frac{(\mu - M_{\mu\mu})^2}{2V_{\mu\mu}}} e^{-\frac{\left[\ln\left(\sigma \frac{\sqrt{\theta_{\sigma\sigma}}}{M_{\sigma\sigma}}\right)\right]^2}{2ln(\theta_{\sigma\sigma})}} .$$

Voila! All 0 or negative values of the standard deviation term σ have been eliminated from the domain of the prior. When we multiply it by the likelihood, those domains are automatically eliminated from the posterior as well. The zero domain in the prior is preserved by multiplication as a zero domain in the posterior. All zeroed out domains in the prior will be zeroed out in the posterior. This ensures that the negative values of the σ term are completely eliminated, conceptually and mathematically correctly, from the likelihood function and from the posterior. This is precisely what we want and need. We cannot allow any nonpositive values of σ to enter the prior or posterior. They must have zero probability of occurrence.

Combining this prior and the foregoing expression of the posterior, we write the posterior in its complete form as follows:

$$\int_0^\infty dz \{y, X, z, \mu, \sigma\} \propto \{\mu, \sigma | y, X\}$$

$$= c_5 \frac{1}{\sigma} e^{-\frac{(\mu - M_{\mu\mu})^2}{2V_{\mu\mu}}}$$

$$\times e^{-\frac{\left[ln\left(\sigma \frac{\sqrt{\theta_{\sigma\sigma}}}{M_{\sigma\sigma}}\right)\right]^2}{2\ln(\theta_{\sigma\sigma})} \left\{ \begin{array}{c} \left[w - X\begin{pmatrix}\bar{\mu}\\ \bar{\sigma}\end{pmatrix}\right]^T \left[w - X\begin{pmatrix}\bar{\mu}\\ \bar{\sigma}\end{pmatrix}\right] + \\ + \left[\begin{pmatrix}\mu\\ \sigma\end{pmatrix} - \begin{pmatrix}\bar{\mu}\\ \bar{\sigma}\end{pmatrix}\right]^T (X^T X) \left[\begin{pmatrix}\mu\\ \sigma\end{pmatrix} - \begin{pmatrix}\bar{\mu}\\ \bar{\sigma}\end{pmatrix}\right] \end{array} \right\}}$$

In developing this distribution, we have not sought to obtain the integrating constant. This distribution is a "shape," entirely suitable for simulation and integration using Monte Carlo importance sampling or some other such technique. **All this trouble, and loss of analytical insight, just because σ has to be positive!**

Conjugate Bayesian analysis may not work if any coefficients are constrained, such as to be positive. That forces us into simulation, which in this book we do not choose to do. We know that if we perform Monte Carlo importance sampling on this "shape" equation for the posterior, we can develop the constant of integration, the means, the variance-covariance matrix among the three parameters, and scatter plots of the three marginals. We see no obvious way to obtain a closed, analytic form for the mean or variance-covariance matrix of this density or any of the marginal distributions.

We reiterate that there is no conjugate relationship between prior and posterior here. The posterior is a non-tractable analytic function of the parameters μ and σ. However, each parameter is confined absolutely to the domain we want, and zero otherwise. In particular, the parameter σ is strictly nonnegative, carried forward from the prior into the posterior. The parameter μ is unbounded. The beauty of this formulation is that it is foolproof with regard to the domains of the uncertain variables considered. The logistic density actually exists for each and every point in the domain of the prior times the likelihood.

14.1 Application to Keelin's Metalog

Bayesian Linear Regression does not work on the logistic distribution without simulation because the conjugacy breaks down when we require $\sigma > 0$, as the previous section shows.

Keelin[53] has made the observation that he can extend the logistic distribution to what he terms a metalog distribution by positing a polynomial expansion representing each of the two unknowns–μ and σ–rather than merely treating them as constants. He recognized that polynomials in $(p - 0.5)$ introduces more constants, yet the cumulative remains linear in the newly added constants. The number 0.5 represents the midpoint of the range of p from 0 to 1 exclusively at both bounds, and higher powers of p become smaller in absolute value. See Figure 14.4 for a representation of his motivation. Keelin was evidently thinking that the mean μ should be a function $\mu(p)$ that could be represented as a Taylor expansion around $p = 0.5$ linear in its coefficients and that the standard deviation σ should be a separate function $\sigma(p)$ that could be represented as a Taylor expansion around $p = 0.5$ linear in its coefficients as well. The value $p = 0.5$ is arbitrarily but intuitively specified, reflecting as it does the midpoint of the interval from 0 to 1.

To render such assumptions tractable, we specifically illustrate what happens when each Taylor expansion contains a constant and linear term (no second or higher powers in either Taylor series. Extension to more terms than two is straightforward later if desired.)

$$w = a_0 + b_0 \frac{\sqrt{3}}{\pi} \ln\left(\frac{p}{1-p}\right) \tag{14.1}$$

Figure 14.4: Keelin's Taylor-like Expansion of the Mean and Standard Deviation Terms

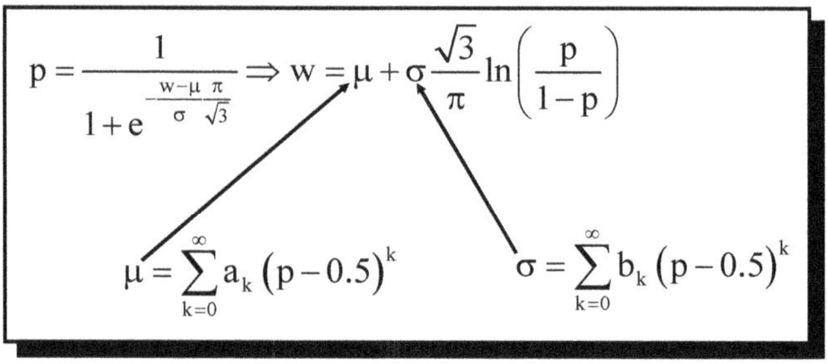

$$+ a_1 (p - 0.5) + b_1 \frac{\sqrt{3}}{\pi} \ln\left(\frac{p}{1-p}\right) (p - 0.5) \quad .$$

Notice that there are now four rather than two terms to be estimated, namely a_0, a_1, b_0, b_1. With the logistic density, we only had two, namely a_0, a_1. Notice further that if both b_0 and b_1 are simultaneously 0 or negative, the standard deviation may become negative. There may be a region in the four dimensional orthant over a_0, a_1, b_0, b_1 that occurs with nonzero probability in which this always occurs, i.e, in a four dimensional, unbounded multivariate Student's t distribution there is a region in which both b_0 and b_1 are simultaneously 0 or negative. We conjecture, and will not pursue the matter further, that the higher dimensional expansion does not work either and that we will need conditions on the prior that will not allow these variables to stray into nonfeasible ranges and therefore will be solvable only by simulation.

Assuming, however, that all is well, the foregoing means the method will consume four degrees of freedom to make the requisite coefficient estimates if we choose classical regression. (If we choose Bayesian Linear Regression, there is no such requirement.) That means we can use simple regression with as many observations (i.e., data points) as we want or have. We continue to ignore any notion of bounding the desk weight above or below for the time being. We can rectify that later. If we write Keelin's new equation in regression format, we have

$$w_i = a_0 + b_0 \frac{\sqrt{3}}{\pi} \ln\left(\frac{p_i}{1-p_i}\right) + a_1 (p_i - 0.5)$$
$$+ b_1 \frac{\sqrt{3}}{\pi} \ln\left(\frac{p_i}{1-p_i}\right) (p_i - 0.5) + \varepsilon_i \quad i = 1, ..., n \quad .$$

This equation tells us exactly how we have to modify the nine data points taken from the wheel. The first step is to calculate the key elements of the coefficients of the foregoing equation, which appear in Table 14.3. The elements of the figure relate very directly to the parameters of the foregoing regression equation.

Table 14.3: Keelin's Polynomial Expansion of the Mean and Standard Deviation Terms

Assessment	p	w	1	$\frac{\sqrt{3}}{\pi}\ln\left(\frac{p}{1-p}\right)$	$p-5$	$\frac{\sqrt{3}}{\pi}\ln\left(\frac{p}{1-p}\right)(p-0.5)$
1	0.05	50	1	-1.62335	-0.45	0.730509
2	0.20	70	1	-0.76430	-0.30	0.229291
3	0.40	100	1	-0.22354	-0.10	0.022354
4	0.50	100	1	0	0	0
5	0.58	130	1	0.17795	0.08	0.014236
6	0.70	150	1	0.46714	0.20	0.093428
7	0.79	180	1	0.73047	0.29	0.211836
8	0.84	220	1	0.91423	0.34	0.310838
9	0.90	250	1	1.21139	0.40	0.484557
			a_0	b_0	a_1	b_1

We have in Figure 14.3 organized the observed data in Table 14.3 so that the data represents the X matrix and the y vector for the regression. Once we have created the X matrix and y vector, we can perform classical regression as summarized in Table 14.4 to obtain the mean values of the model coefficients. We see in the figure the data matrix, the square of the data matrix, and the calculations leading to the mean.

Table 14.4: Key Classical Regression Results Including Mean Values of Coefficients

X^TX	a_0	b_0	b_1	a_1
a_0	9.000000	0.889983	2.097051	0.460000
b_0	0.889983	6.356174	-0.294037	2.097051
b_1	2.097051	-0.294037	0.971940	-0.018987
a_1	0.460000	2.097051	-0.018987	0.708600
			Determ.	0.259207
$(X^TX)^{-1}$	a_0	b_0	b_1	a_1
a_0	0.243967	-0.119788	-0.559082	0.181146
b_0	-0.119788	10.943979	2.939638	-32.231389
b_1	-0.559082	2.939638	2.963147	-8.257311
a_1	0.181146	-32.231389	-8.257311	96.458866
	X^Ty			$\bar{\beta}=(X^TX)^{-1}X^Ty$
a_0	1250.000000		a_0	108.439124
b_0	571.644842		b_0	89.023816
b_1	298.330413		b_1	99.334774
a_1	213.900000		a_1	-29.330193

We will be needing the residual sum of squared errors, the R term, and that requires us to calculate the matrix square of $y - X\bar{\beta}$ along the way, a calculation that is embedded in the calculation of R in Table 14.5. Other key results from the classical regression solution appear in Table 14.5. We see the degrees of freedom, the mean of the error distribution, and the variance of the error distribution in the figure.

Table 14.5: Additional Results from the Classical Regression Solution

4	p
5	$\nu_C = n - p$
300.2338	$R = (y - X\bar{\beta})^T (y - X\bar{\beta})$
60.0468	$s_C^2 = \frac{R}{\nu_C}$
100.0779	$\frac{\nu_C s_C^2}{\nu_C - 2}$
20031.1874	$\frac{2}{\nu_C - 4}\left(\frac{\nu_C s_C^2}{\nu_C - 2}\right)^2$

The variance-covariance matrix over the four parameters appears in Table 14.6 along with the variance, mean, standard deviation, and t-statistic for the four coefficients.

Table 14.6: Variance-Covariance Results from the Classical Regression Solution

$\frac{\nu_C s_C^2}{\nu_C - 2}(X^T X)^{-1}$	a_0	b_0	b_1	a_1
a_0	24.415731	-11.988091	-55.951752	18.128748
b_0	-11.988091	1095.250830	294.192902	-3225.650978
b_1	-55.951752	294.192902	296.545685	-826.374672
a_1	18.128748	-3225.650978	-826.374672	9653.404396
	a_0	b_0	b_1	a_1
Variance	24.415731	1095.250830	296.545685	9653.404396
Std. Dev.	4.941228	33.094574	17.220502	98.251740
Mean	108.439124	89.023816	99.334774	-29.330193
t-statistic	21.945786	2.689982	5.768402	0.298521

We recall, however, the key result. When we ponder correctly, the classical regression solution (and any Bayesian solution not yet presented for this four variable metalog case) gives a four-dimensional multivariate Student's t distribution over the model coefficients. That is, the Bayesian solution, of which classical regression is one special case, yields a probability distribution $\{a_0, b_0, b_1, a_1\}$. There is no escape from this–the probability density function over the four model coefficients is a 4-dimensional, multivariate Student's t probability distribution. The specific parameters of that probability distribution over the four coefficients, if it existed, is derived in Table 14.7.

Table 14.7: The Derived Multivariate Student's t Distribution Over the Model Coefficients

$\frac{X^T X}{\nu_C s_C^2}$	a_0	b_0	b_1	a_1
a_0	0.02997664	0.00296430	0.00698473	0.00153214
b_0	0.00296430	0.02117075	-0.00097936	0.00698473
b_1	0.00698473	-0.00097936	0.00323728	-0.00006324
a_1	0.00153214	0.00698473	-0.00006324	0.00236016

	$\lvert X^T X\rvert^{\frac{1}{2}}$	$\Gamma\left(\frac{\nu_C+p}{2}\right)$	$\left(\pi \nu_C s_C^2\right)^{\frac{p}{2}}$	$\Gamma\left(\frac{\nu_C}{2}\right)$
	0.50912392	11.6317284	889649.529	1.329340388

Constant	5.007404E-06		
		-4.5	exponent

	$\bar{\beta} = \left(X^T X\right)^{-1} X^T y$
a_0	108.439124
b_0	89.023816
b_1	99.334774
a_1	-29.330193

The joint probability distribution over w and the coefficients can be written, via Bayes theorem, as follows.

$$\{w, a_0, b_0, b_1, a_1\} = \{w|a_0, b_0, b_1, a_1\}\{a_0, b_0, b_1, a_1\} \quad .$$

We must multiply the marginal joint distribution over the four model coefficients just developed by the conditional distribution over the weight of the desk given the coefficients to obtain the joint distribution over the weight and the coefficient. (The conditional distribution $\{w|a_0, b_0, b_1, a_1\}$ must be well defined for all positive, zero, and negative values of all the four parameters a_0, b_0, b_1, a_1. This was not the case for the logistic distribution.) Thereafter, we would integrate out the coefficients to obtain the probability distribution over the weight of the desk. That is, the probability distribution over the weight of the desk is

$$\begin{aligned}\{w\} &= \int da_0 \int da_1 \int db_0 \int db_1 \{w, a_0, b_0, b_1, a_1\} \\ &= \int da_0 \int da_1 \int db_0 \int db_1 \{w|a_0, b_0, b_1, a_1\}\{a_0, b_0, b_1, a_1\} \quad .\end{aligned}$$

The conditional distribution $\{w|a_0, b_0, b_1, a_1\}$ is analytically intractable and is developed only as a parametric equation in p. The multivariate Student's t distribution $\{a_0, b_0, b_1, a_1\}$ analytically intractable as well. It is interesting to observe what the conditional distribution of w given a_0, b_0, b_1, a_1 looks like.

Let us examine the conditional probabilities analytically. We recall that the equation for w is given in Figure (14.1). This is a cumulative probability distribution that is parametric in p. We will want to plot this equation along the horizontal axis against the equation $\frac{\partial p}{\partial w}$ up the vertical axis. This will give us a plot of the conditional probability of w given the four coefficients. It is worthwhile to see the shape of this conditional distribution conditional on a setting of the variables a_0, b_0, b_1, a_1. The first derivative of equation (14.1) is the following

$$\frac{dw}{dp} = a_1 + b_0 \frac{\sqrt{3}}{\pi} \frac{1}{p(1-p)} + b_1 \frac{\sqrt{3}}{\pi} \frac{(p-0.5)}{p(1-p)} + b_1 \frac{\sqrt{3}}{\pi} \ln\left(\frac{p}{1-p}\right) \quad .$$

Taking the reciprocal, we write the sought after derivative

$$\frac{dp}{dw} = \frac{p(1-p)}{a_1 p(1-p) + b_0 \frac{\sqrt{3}}{\pi} + b_1 \frac{\sqrt{3}}{\pi}(p-0.5) + b_1 \frac{\sqrt{3}}{\pi} p(1-p) \ln\left(\frac{p}{1-p}\right)} \quad .$$

If we plot the $\frac{\partial p}{\partial w}$ up the vertical axis and w across the horizontal axis, running the parameter p from 0 to 1, we obtain the plot of the conditional distribution in Figure 14.5.

Figure 14.5: Plot of the 4 Term Metalog Using the Mean Values of the Coefficients

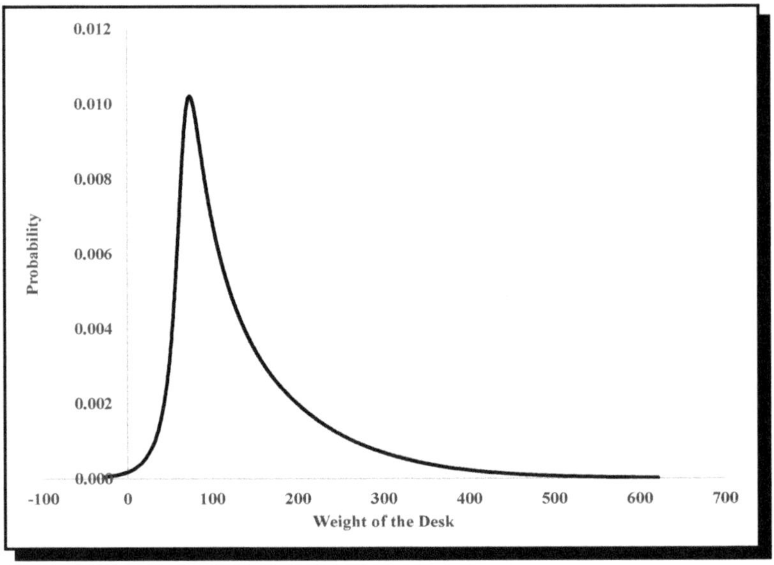

For the Keelin[53] metalog analysis, we have used only classical regression herein. We could just as easily have assessed a prior over the four coefficients, updated that prior using the 9 observations from the probability wheel, and calculated a Bayesian posterior. We are not sure whether that would work, i.e., whether the entire range of the 4-variate Student's t distribution would lead to a non-negative estimate of variance. We fear that there will always be a region where the four coefficients will line up so that one of the central moments, say the variance, is negative and the density is undefined. If we bound the weight of the desk so as to ensure nonnegativity, this problem evaporates. Research remains to be done to see whether additional terms can overcome or sidestep the nonnegativity problem with the standard deviation term we saw in the metalog itself. At a very minimum, regression gives an effective and efficient way to estimate the mean value of the metalog Taylor coefficients, all linearly related to desk weight. There is no proof; only hope.

14.2 Concluding Remarks

Suppose we wanted to bound the distribution say at 35 lb from below. Instead of using the variable w, we would use the variable $ln(w-35)$, all else equal. That would ensure that w never would be below 35, i.e., the range of the probability distribution over weight would never be below 35 lb.

Similarly, if we wanted to bound the distribution say at 500 lb from above, instead of using the variable w, we would use the variable $-ln(500-w)$, all else equal. That would ensure that w never would be above 500, i.e., the range of the probability distribution over weight would never be above 500 lb.

If we wanted a dually bounded distribution, instead of using the variable w, we would use the variable $-ln(\frac{500-w}{w-35})$, all else equal. This will ensure that the range of the probability distribution lies strictly between 35 lb and 500 lb. This curve is monotonic in w, so we can invert the curve to get the pdf over w.

With regard to the logistic density and the metalog (a Taylor-like expansion of terms that comprise the logistic distribution), there is much research to be done, but it could be really productive. It could cement the probability wheel (or its cousin interval estimation) as a way to get really complex and sophisticated distributions encoded and quantified.

In general, the conditional distribution of the weight of the desk given the various parameter settings may not be defined for all parameters a_0, a_1, b_0, b_1 varying between $-\infty$ and ∞. We don't know, but we suspect that this will be true in general, as it was shown to be true for the logistic density, but we are not going to attempt a proof. This would mean that there is no multivariate Student's t distribution that can lead to the requisite joint distribution that must be integrated to give the marginal distribution over the weight of the desk. This does not obviate the usefulness of the logistic density, not at all. The mean values of the coefficients estimated by classical means (Bayes with diffuse prior) or by Bayesian means (with a non-diffuse, non-conjugate prior) are imminently useful in practice. The metalog has been a breakthrough in quantifying distributions assessed by the probability wheel (or by interval estimation), both of which are articulated in detail in Howard and Abbas[45].

In light of the positivity requirement for the standard deviation coefficient in the logistic function, Bayesian inference, linear or not, may be relegated to be numerical, not analytical. With regard to the logistical distribution, we are condemned to simulation rather than closed analytical form in order that the standard deviation term always be strictly positive. With regard to higher order metalog functions, the answer is not clear because we do not know in advance any nonpositivity or nonnegativity constraints on any of the coefficients in a complex Taylor-like expansion form.

Chapter 15

MULTIVARIATE PROBABILISTIC REGRESSION

The probability that the bowman's arrow hits the deer does not lie in the arrow or the deer. It lies in the bowman's mind.

Bart Kosko

Suppose we wanted to develop a joint distribution of the height and weight of people in California prisons. We could start by sampling prisoners, measuring their height and weight, and creating a database of multiple observations. Clearly, the joint distribution of height and weight among California prisoners would be mutually dependent, i.e., mutually relevant. If I told you the height was 5'1" rather than 6'2", you would have very different conditional densities. What if we wanted to predict the distance and hang time that punters could achieve as a function of certain punter attributes (e.g., leg strength and leg flexibility)? What if we wanted to jointly predict total crime rate and violent crime rate in a society with particular socio-economic parameters? We work an example like that in Chapter 19. Making such joint estimates is the topic of this chapter. The chapter generalizes the univariate results developed thus far in this book. This chapter will construct the basic mathematics as an extension (but not a particularly simple or straightforward extension) of the previous univariate mathematics. (Such formal extension, step by step, has been missing from the textbooks as we see it. We aim to correct that omission herein.) As in previous chapters, we will search for analytical forms and avoid simulation to the maximum extent possible.

This and subsequent sections are going to focus on the multivariate problem, generalizing the univariate analysis in the foregoing chapters to the multivariate case. This first chapter, Chapter 15, frames the multivariate estimation problem and develops the multivariate likelihood function that is analogous to and an extension of the univariate likelihood function in equation (2.7).

To initiate the multivariate estimation problem and develop the multivariate likelihood function, consider a case in which we have two attributes being jointly predicted $m = 2$, for instance, the distance and hang time of a punted football. Suppose there are $p = 3$ distinct variables, conditioning variables, that affect the jointly predicted variables. We say that predictions of the distance and hang time will be dependent on $p = 3$ distinct variables such as left leg strength, right leg strength, and right leg flexibility. Assume that $n = 8$ observations of punted footballs have been made and put into a data base. Each observation in the data base includes the $m = 2$ punt variables and the $p = 3$ precursor variables. We can write a multivariate regression equation as follows:

$$\begin{bmatrix} y_{11} & y_{12} \\ y_{21} & y_{22} \\ y_{31} & y_{32} \\ \cdot & \cdot \\ y_{81} & y_{82} \end{bmatrix} = \begin{bmatrix} x_{11} & x_{12} & x_{13} \\ x_{21} & x_{22} & x_{23} \\ x_{31} & x_{32} & x_{33} \\ \cdot & \cdot & \cdot \\ x_{81} & x_{82} & x_{83} \end{bmatrix} \begin{bmatrix} \beta_{11} & \beta_{12} \\ \beta_{21} & \beta_{22} \\ \beta_{31} & \beta_{32} \end{bmatrix} + \begin{bmatrix} \varepsilon_{11} & \varepsilon_{12} \\ \varepsilon_{21} & \varepsilon_{22} \\ \varepsilon_{31} & \varepsilon_{32} \\ \cdot & \cdot \\ \varepsilon_{81} & \varepsilon_{82} \end{bmatrix} .$$

We write this in matrix form

$$Y = XB + E \quad ,$$

in which

$$Y = \begin{bmatrix} y_{11} & y_{12} \\ y_{21} & y_{22} \\ y_{31} & y_{32} \\ \cdot & \cdot \\ y_{81} & y_{82} \end{bmatrix} \quad (n \times m = 8 \times 2) \quad ,$$

$$X = \begin{bmatrix} x_{11} & x_{12} & x_{13} \\ x_{21} & x_{22} & x_{23} \\ x_{31} & x_{32} & x_{33} \\ \cdot & \cdot & \cdot \\ x_{81} & x_{82} & x_{83} \end{bmatrix} \quad (n \times p = 8 \times 3) \quad ,$$

$$B = \begin{bmatrix} \beta_{11} & \beta_{12} \\ \beta_{21} & \beta_{22} \\ \beta_{31} & \beta_{32} \end{bmatrix} \quad (p \times m = 3 \times 2) \quad ,$$

and

$$E = \begin{bmatrix} \varepsilon_{11} & \varepsilon_{12} \\ \varepsilon_{21} & \varepsilon_{22} \\ \varepsilon_{31} & \varepsilon_{32} \\ \cdot & \cdot \\ \varepsilon_{81} & \varepsilon_{82} \end{bmatrix} \qquad (n \times m = 8 \times 2) \quad .$$

We can write this in the form

$$\begin{bmatrix} \varepsilon_{11} & \varepsilon_{12} \\ \cdot & \cdot \\ \cdot & \cdot \\ \cdot & \cdot \\ \varepsilon_{n1} & \varepsilon_{n2} \end{bmatrix} = \begin{bmatrix} y_{11} & y_{12} \\ \cdot & \cdot \\ \cdot & \cdot \\ \cdot & \cdot \\ y_{n1} & y_{n2} \end{bmatrix} - \begin{bmatrix} x_{11} & \cdot\cdot\cdot & x_{1p} \\ \cdot & & \cdot \\ \cdot & & \cdot \\ \cdot & & \cdot \\ x_{n1} & \cdot\cdot\cdot & x_{np} \end{bmatrix} \begin{bmatrix} \beta_{11} & \beta_{12} \\ \cdot & \cdot \\ \cdot & \cdot \\ \cdot & \cdot \\ \beta_{p1} & \beta_{p2} \end{bmatrix}$$

(in which we have used $n = 8$ and $p = 3$, moving implicitly to general indexing.) Equivalently in matrix form, we write

$$E = Y - XB \quad .$$

Keep in mind, the matrix E is 8×2 or $n \times m$ in the case in which $n = 8$ and $m = 2$.

Ponder the JOINT density represented by the ith row. It is going to be a multivariate (actually a bivariate in this special case because $m = 2$) with a nonzero covariance:

$$\{\varepsilon_{i1}, \varepsilon_{i2}\} = \frac{1}{(2\pi)^{\frac{2}{2}} |\Sigma|^{\frac{1}{2}}} e^{-\frac{1}{2}\left[\begin{array}{cc} \varepsilon_{i1} & \varepsilon_{i2} \end{array}\right] \Sigma^{-1} \left[\begin{array}{c} \varepsilon_{i1} \\ \varepsilon_{i2} \end{array}\right]} \quad .$$

We have written the row vector $\varepsilon_i = [\varepsilon_{i1}, \varepsilon_{i2}]$ as being $m = 2$ columns long. The column vector ε_i^T has m rows and a single column. We rewrite this using m in place of 2 in the exponent in the denominator so as to be general.

$$\{\varepsilon_{i1}, ..., \varepsilon_{im}\} = \frac{1}{(2\pi)^{\frac{m}{2}} |\Sigma|^{\frac{1}{2}}} e^{-\frac{1}{2}\left[\begin{array}{c} \varepsilon_{i1} ... \varepsilon_{im} \end{array}\right] \Sigma^{-1} \left[\begin{array}{c} \varepsilon_{i1} \\ \cdot \\ \cdot \\ \cdot \\ \varepsilon_{im} \end{array}\right]} \quad .$$

Keep in mind that the matrix Σ is 2×2 or $m \times m$. The Σ matrix is the matrix of variances and covariances of the error term. In the former univariate case, this error term was a scalar. (Now it is an $m \times m$ matrix, where m is the number of probabilistically dependent joint variables.) The reason it is bivariate in this chapter is because we cannot ignore the covariances among the Y variables, which are m in number. That is what is going to make the multivariate case substantially more difficult—the jointness among the Y variables. The likelihood function should be the product of $n = 8$ rows.

$$LF = \left[\frac{1}{(2\pi)^{\frac{m}{2}}|\Sigma|^{\frac{1}{2}}}\right]^n e^{-\frac{1}{2}\sum_{i=1}^{n}\left[\begin{array}{c}\varepsilon_{i1}...\varepsilon_{im}\end{array}\right]\Sigma^{-1}\left[\begin{array}{c}\varepsilon_{i1}\\ \cdot\\ \cdot\\ \cdot\\ \varepsilon_{im}\end{array}\right]}.$$

We write this generally by expanding the exponent using m in place of 2 and incorporating the n into the exponents appropriately

$$LF = \frac{1}{(2\pi)^{\frac{mn}{2}}|\Sigma|^{\frac{n}{2}}} e^{-\frac{1}{2}\sum_{i=1}^{n}\left[\begin{array}{c}\varepsilon_{i1}...\varepsilon_{im}\end{array}\right]\Sigma^{-1}\left[\begin{array}{c}\varepsilon_{i1}\\ \cdot\\ \cdot\\ \cdot\\ \varepsilon_{im}\end{array}\right]}.$$

We need to work out the exponent for all rows and columns, not just the ith. We temporarily use $A = \Sigma^{-1}$ to simplify the notation. We will later substitute Σ^{-1} for A after the subsequent developments

$$\sum_{i=1}^{8}\left[\begin{array}{cc}\varepsilon_{i1} & \varepsilon_{i2}\end{array}\right] A \left[\begin{array}{c}\varepsilon_{i1}\\ \varepsilon_{i2}\end{array}\right].$$

We need to emphasize that in general the dimension of the A matrix is $m \times m$, where m is the number of joint variables that are being predicted. We calculate this by considering the larger matrix, which is $n \times n$

$$\left[\begin{array}{cc}\varepsilon_{11} & \varepsilon_{12}\\ \cdot & \cdot\\ \cdot & \cdot\\ \cdot & \cdot\\ \varepsilon_{81} & \varepsilon_{82}\end{array}\right] A \left[\begin{array}{ccccc}\varepsilon_{11} & \cdot & \cdot & \cdot & \varepsilon_{81}\\ \varepsilon_{12} & \cdot & \cdot & \cdot & \varepsilon_{82}\end{array}\right].$$

We write the previous exponent in shorthand notation using the specific rows and columns of E, noting that E is 8×2 (or $n \times m$) and A is 2×2 (or $m \times m$). This means that EAE^T is 8×2 (or $n \times m$) $\times$ 2×2 (or $m \times m$) $\times$ 2×8 (or $m \times n$), which results in a matrix that is 8×8 (or $n \times n$). We write the ith row vector of E, denoted ε_i and having length m, to be $[\varepsilon_{i1}, ..., \varepsilon_{im}]$. Using this notation, we write a matrix that is substantially larger than the foregoing exponent but whose trace will be shown to be equal to the foregoing exponent.

$$\begin{bmatrix} \varepsilon_1 \\ \varepsilon_2 \\ \varepsilon_3 \\ . \\ \varepsilon_8 \end{bmatrix} A \begin{bmatrix} \varepsilon_1{}^T & \varepsilon_2{}^T & \varepsilon_3{}^T & . & \varepsilon_8{}^T \end{bmatrix} = EAE^T \quad .$$

We expand this out to expose the rows and columns

$$EAE^T = \begin{bmatrix} \varepsilon_1 A \varepsilon_1{}^T & . \; . \; . & \varepsilon_1 A \varepsilon_8{}^T \\ . & & . \\ . & & . \\ . & & . \\ \varepsilon_8 A \varepsilon_1{}^T & . \; . \; . & \varepsilon_8 A \varepsilon_8{}^T \end{bmatrix} \quad .$$

The sought after exponent in the likelihood function is clearly the trace of this matrix EAE^T. We add the diagonals

$$Trace\left(EAE^T\right) = \sum_{i=1}^{8} \varepsilon_i A \varepsilon_i{}^T \quad .$$

The dimensional analysis of the matrix EAE^T is needed to both understand the trace in the previous equation and to continue forth with the trace from this point forward. The dimensional analysis in the special case here and in the general case is the following:

$$\begin{aligned} EAE^T &= (8 \times 2)\,(2 \times 2)\,(2 \times 8) \\ &= (n \times m)\,(m \times m)\,(m \times n) \\ &= (n \times n) \quad . \end{aligned}$$

We see that the matrix EAE^T is going to end up as an $n \times n$ matrix as in all previous sections and that the trace is the sum of n diagonal elements.

We are going to want to calculate the trace by reordering the matrix product of the matrices if they are conformable. They certainly are because EAE^T certainly exists, and so does E^TEA, owing to the dimensionality of the matrices involved. This is a result of the trace theorem presented previously, which states that $Trace(AB) = Trace(BA)$. We can put the rightmost matrix onto the left because of this conformability, as follows:

$$\begin{aligned} EAE^T &= (8 \times 2)\,(2 \times 2)\,(2 \times 8) \\ \Rightarrow E^TEA &= (2 \times 8)\,(8 \times 2)\,(2 \times 2) = (2 \times 2) \quad . \end{aligned}$$

That means that

$$Trace\left(EAE^T\right) = Trace\left(E^T EA\right) = \sum_{i=1}^{8} \varepsilon_i A \varepsilon_i{}^T \quad .$$

As discussed above, the matrix A is the inverse of the variance-covariance matrix, and its dimension is therefore $m \times m$, where m is the number of dependent variables. Therefore the exponent is

$$Trace\left(E^T E \Sigma^{-1}\right) = \sum_{i=1}^{8} \varepsilon_i \Sigma^{-1} \varepsilon_i{}^T \quad .$$

The matrix $E^T E$ is

$$\begin{aligned} E^T E = \; & (Y - XB)^T \; (Y - XB) \\ & \;(2 \times 8) \qquad (8 \times 2) \qquad = (2 \times 2) \quad . \end{aligned}$$

The matrix whose trace we want is thus

$$\begin{aligned} E^T E \Sigma^{-1} = \; & (Y - XB)^T \; (Y - XB)\, \Sigma^{-1} \\ & \;(2 \times 8) \qquad (8 \times 2) \quad (2 \times 2) = (2 \times 2) \quad . \end{aligned}$$

The likelihood function is thus written

$$\begin{aligned} LF &\triangleq \{X, Y | B, \Sigma\} \\ &= \frac{1}{(2\pi)^8 |\Sigma|^{\frac{8}{2}}} e^{-\frac{1}{2} Trace\left[(Y-XB)^T (Y-XB) \Sigma^{-1}\right]} \\ &= \frac{1}{(2\pi)^{\frac{nm}{2}} |\Sigma|^{\frac{n}{2}}} e^{-\frac{1}{2} Trace\left[(Y-XB)^T (Y-XB) \Sigma^{-1}\right]} \quad , \end{aligned} \tag{15.1}$$

in which we have given care to break out the trace of the larger matrix we have developed carefully and correctly. It is the probability of observing the data X and Y given the model parameters B and Σ. We take care to denote the number of observations n and the number of joint output variables m correctly and to place them correctly into the exponents in the denominator.

Equation (15.1) is a pretty complex-looking likelihood function, but it admits a conjugate prior and posterior. We need to do some more work on this likelihood function to manipulate it into a form that can lead to a closed-form solution. We should emphasize that the Moore-Penrose[1] theory allows us to generalize from the inverses of square matrices to the pseudoinverses of products of matrices that may not be square. We turn to the matrix portion of the exponent in equation(15.1):

[1] See https://en.wikipedia.org/wiki/Moore-Penrose_inverse.

$$\text{Exponent} = -\frac{1}{2}Trace\left[(Y - XB)^T (Y - XB) \Sigma^{-1}\right] \quad .$$

We first focus on the matrix expression within the exponent (excluding for the moment the Σ term, which we will reinsert later) so as to economize on notation.

$$\begin{aligned}\text{Matrix} &= (Y - XB)^T (Y - XB) \\ &= \left(Y^T - B^T X^T\right)(Y - XB) \quad .\end{aligned}$$

We expand this matrix

$$\begin{aligned}&\left(Y^T - B^T X^T\right)(Y - XB) \\ &= Y^T Y - Y^T XB - B^T X^T Y + B^T X^T XB \quad .\end{aligned}$$

We then define a variable

$$z = B - b \quad .$$

so that we can define

$$B = z + b \quad .$$

Substituting into the foregoing equation, we can write

$$\begin{aligned}&\text{Matrix} = Y^T Y - Y^T XB - B^T X^T Y + B^T X^T XB \\ &= Y^T Y - Y^T X (z + b) - (z + b)^T X^T Y \\ &+ (z + b)^T X^T X (z + b) \\ &= Y^T Y - Y^T X z - Y^T X b - z^T X^T Y - b^T X^T Y \\ &+ z^T X^T X z + z^T X^T X b + b^T X^T X z + b^T X^T X b \quad .\end{aligned}$$

We group according to terms that are independent of z, linear in z, and quadratic in z

$$\begin{aligned}&\text{Matrix} \\ &= Y^T Y - Y^T X b - b^T X^T Y + b^T X^T X b \\ &- Y^T X z - z^T X^T Y + z^T X^T X b + b^T X^T X z \\ &+ z^T X^T X z \\ &= Y^T Y - Y^T X b - b^T X^T Y + b^T X^T X b \\ &+ z^T \left(X^T X b - X^T Y\right) + \left(b^T X^T X - Y^T X\right) z \\ &+ z^T X^T X z \quad .\end{aligned}$$

We set the matrix b so that the leftmost of the two terms in the second to last line that are linear in z (specifically the term that is linear in z^T) becomes 0. That is

$$\begin{aligned} &X^T X b - X^T Y = 0 \Rightarrow X^T X b = X^T Y \\ &\Rightarrow b = \bar{B} = \left(X^T X\right)^{-1} X^T Y \\ &\Rightarrow b^T = \bar{B}^T = Y^T X \left(X^T X\right)^{-1} \quad . \end{aligned}$$

Interestingly, if we substitute the expression for b^T into the rightmost of the two terms in the second to last line that are linear in z, specifically the term that is linear in z, it too becomes 0. Therefore, with this value of b, we can eliminate both linear terms and write the matrix:

$$\begin{aligned} \text{Matrix} &= Y^T Y - Y^T X b \\ &- b^T X^T Y + b^T X^T X b \\ &+ z^T X^T X z \\ &= Y^T (Y - Xb) - b^T X^T (Y - Xb) + z^T X^T X z \\ &= \left(Y^T - b^T X^T\right)(Y - Xb) + z^T X^T X z \quad . \end{aligned}$$

Substituting the expression for z, we obtain the final answer for the matrix

$$\text{Matrix} = (Y - Xb)^T (Y - Xb) + (B - b)^T X^T X (B - b) \quad .$$

Substituting the alternative notation for $b = \bar{B}$, we have

$$\text{Matrix} = \left(Y - X\bar{B}\right)^T \left(Y - X\bar{B}\right) + \left(B - \bar{B}\right)^T X^T X \left(B - \bar{B}\right) \quad .$$

Therefore the exponent can be written very much analogously to the univariate case discussed in previous chapters. Substituting this relationship into the exponent term above, that exponent term becomes

$$\begin{aligned} \text{Exponent} = &-\frac{1}{2} Trace \left[\left(Y - X\bar{B}\right)^T \left(Y - X\bar{B}\right) \Sigma^{-1}\right] \\ &-\frac{1}{2} Trace \left[\left(B - \bar{B}\right)^T X^T X \left(B - \bar{B}\right) \Sigma^{-1}\right] \quad , \end{aligned}$$

in which we have exploited the additivity of the trace. By expanding to this two-part form, we are able to solve equations containing each of the two parts separately and then put them together logically at the end. Let us review in succession each of these two constituent terms of the exponent, beginning with the dimensionality of each of the two terms whose trace is being sought. In the first term, recall that the matrix Y is 8×2 (or $n \times m$), X is 8×3 (or $n \times p$), and B is 3×2 (or $p \times m$), meaning that $Y - XB$ is therefore 8×2 (or $n \times m$). Its transpose at the left is 2×8 (or $m \times n$). Thus the first of the two terms is 2×8 (or $m \times n$) by 8×2 (or $n \times m$) by 2×2 (or $m \times m$).

The second of the two exponent terms merits similar dimensional attention. B and $\bar{B}$ are 3×2 (or $p \times m$), their transpose is 2×3 (or $m \times n$), and $X^T X$ is 3×3 (or $p \times p$). The quadratic term (the first three terms at the left of the second exponential term) is therefore 2×2 (or $m \times m$). The final term is also $m \times m$, and so the second exponent is also $m \times m$. The likelihood function, comprised in part of these two exponent terms, becomes

$$\begin{aligned} LF &\triangleq \{X, Y | B, \Sigma\} \\ &= \frac{1}{(2\pi)^{\frac{nm}{2}} |\Sigma|^{\frac{n}{2}}} e^{-\frac{1}{2} Trace\left[(Y - X\bar{B})^T (Y - X\bar{B})\Sigma^{-1}\right]} \\ &\qquad e^{-\frac{1}{2} Trace\left[(B - \bar{B})^T X^T X (B - \bar{B})\Sigma^{-1}\right]} \quad . \end{aligned}$$

By eliminating the linear term in z above, we showed that

$$\bar{B} = \left(X^T X\right)^{-1} X^T Y \quad , \tag{15.2}$$

which allows us to start to build the analogy to the classical regression solution in the univariate case by defining the matrix

$$S = Y - X\bar{B} \tag{15.3}$$

and the matrix product

$$S^T S = (Y - X\bar{B})^T (Y - X\bar{B}) \quad . \tag{15.4}$$

This allows us to write the likelihood function in this multivariate case as follows

$$\begin{aligned} LF &\triangleq \{X, Y | B, \Sigma\} \\ &= \frac{1}{(2\pi)^{\frac{nm}{2}} |\Sigma|^{\frac{n}{2}}} e^{-\frac{1}{2} Trace\left[(S^T S)\Sigma^{-1}\right]} \\ &\quad e^{-\frac{1}{2} Trace\left[(B - \bar{B})^T X^T X (B - \bar{B})\Sigma^{-1}\right]} \quad . \end{aligned} \tag{15.5}$$

We are now going to see that the matrix $S^T S$ is analogous to and is in fact an extension of the scalar variable R in the univariate case. It is akin to the sum of squared errors, but it now considers both variance and covariance in the error matrix. The dimensionality of each of the pertinent matrices is worth restating

- $S^T S \; is \; m \times m$
- $(B - \bar{B})^T \left(X^T X\right) \left(B - \bar{B}\right) \; is \; m \times m$
- $\Sigma^{-1} \; is \; m \times m$
- $X^T X \; is \; p \times p \quad .$

The analogy of equation (15.2) to equation (2.5) in the univariate case (the regression solution), the analogy of equation (15.3) to equation (2.6) in the univariate case (the composition of the residual sum of squared errors), and the analogy of equation (15.4) to equation (2.7) in the univariate case (the likelihood function) are obvious. We note that the expression for $\bar{B}$ in equation (15.2) and the likelihood function in equation (15.3) are extremely similar in form to their counterparts in equations in the univariate case, namely equations (2.5) and (2.7), respectively. Keep in mind that the equations in this section are matrix equations rather than scalar equations. This is what makes this extension difficult.

Chapter 16

IMPLIED CONJUGATE FORM OF PRIOR IN MULTIVARIATE CASE

The future is certain. It is just not known.

Johnny Rich

The multivariate likelihood function in equation (15.5) strongly suggests a potential conjugate functional form for the prior that is precisely parallel in form to that multivariate likelihood function.[1] To begin, we note the key terms in the multivariate likelihood function in equation (15.5):

- There is a term in the denominator of equation (15.5), namely $|\Sigma|^{\frac{n}{2}}$, in which Σ is a measurement error term. We want a parallel term in the prior that has the form $|\Sigma|^{\frac{m_0}{2}}$.[2]
- There is a term in equation (15.5), namely $S^T S$, an $m \times m$ matrix, a residual sum of squared error term. We want a parallel term in the prior that has the form M_0, an $m \times m$ matrix. The dimensionality of $S^T S$, the

[1] This section has not been published in the step by step fashion herein to our knowledge.

[2] The analogy in the univariate case was the scalar σ^2. In the univariate likelihood function in equation (2.12), that scalar σ^2 was seen to be raised to the power $-\frac{n}{2}$. This means σ itself was actually raised to the power $-n$. In the multivariate case in equation (15.5), the dimensionality of Σ, not a scalar, is $m \times m$. Analogous to the univariate case involving σ^2, the determinant of Σ is raised to the same power $-\frac{n}{2}$. Σ represents a variance-covariance term, a "squared" term, quite analogous and in fact very much like σ^2.

same as that of Σ and M_0, and which will become important as we develop the variance-covariance matrices below, is $m \times m$.[3]

- B_0 the mean of the prior distribution, which in the univariate case was designated as a p-vector β_0 in Figure 3.1. The dimensionality of B and B_0, which will become important as we develop mean and variance-covariance matrices below, is $p \times m$.
- $X^T X$ is the precision matrix, which in the univariate case was designated Q in the prior in Figure 3.1. The dimensionality of $X^T X$ is still $p \times p$, just as it was in the univariate case. We continue with the notation Q_0 here.

Specifically, if we conceive a probability distribution that uses

- Σ as in equation (15.5), which is $m \times m$,
- m_0 in place of n in equation (15.5), which is a scalar,
- M_0 in place of $S^T S$ in equation (15.5), which is $m \times m$,
- B_0 in place of $\bar{B}$ in equation (15.5), which is $p \times m$, and
- Q_0 in place of $X^T X$ in equation (15.5), which is $p \times p$,

we will have a prior probability distribution with terms in exactly the same positions as in our likelihood function in equation (15.5). That prior probability distribution will have the following critical parameters (enumerated here for future reference)

$$\text{mean} = B_0 \quad ,$$

$$\text{matrix} = Q_0 \quad , and \tag{16.1}$$

$$\text{residual} = M_0 \quad , \tag{16.2}$$

and will take the form

$$\{B, \Sigma\} = c_0 \frac{1}{|\Sigma|^{\frac{m_0}{2}}} e^{-\frac{1}{2} Trace\left(M_0 \Sigma^{-1}\right)} e^{-\frac{1}{2} Trace\left[(B - B_0)^T Q_0 (B - B_0) \Sigma^{-1}\right]} \quad , \tag{16.3}$$

in which c_0 is an integrating constant to be determined. This is a direct, simple conjugate form of prior, the conjugacy we are about to develop and prove. By that, we mean that it is identical in form, term by term, to the likelihood function in equation (15.5). If we multiply the prior by the likelihood function, we thereby

[3] $S^T S$ is the residual sum of the squared error term, which in the univariate case was designated as the scalar M in the prior in Figure 3.1.

get a posterior distribution that has exactly the same functional form as the prior.[4] That is exactly what conjugacy means. This section will establish a wonderful parallel and extension of the univariate case, one that has not been clearly, step by step, developed in the literature to our knowledge.

16.1 Analysis of the Conjugate Prior–Calculating the Integrating Constant Using Vectorization and Kronecker Products

We are going to proceed through the entire development of the probability distribution using the functional form for the prior in equation (16.3). It is our conjugate archetype function for the multivariate case. The reason we do that is that the posterior is conjugate and will have precisely analogous terms in the same positions. To wit, the prior will have precisely the same functional form as the posterior. This section draws in part from Graham[37]. Recognizing the commonality of form between the prior and posterior will allow us, without much additional work, to apply the results for the prior directly to the posterior. Such is the beauty of conjugacy! To wit, we will dispense with a complete development of a conjugate-archetype but will rather do all the requisite development with the prior.

Because of the increased dimensionality introduced by the jointness in the prior and posterior, we are going to need some expanded mathematics. Toward that goal, we recite the definition and appeal to a vectorization operator. In mathematics, especially in linear algebra and matrix theory, the vectorization of a matrix is a linear transformation that converts the matrix into a column vector. Specifically, the vectorization of an $r \times c$ matrix A, denoted $vec(A)$, is the $rc \times 1$ column vector obtained by stacking the columns of the matrix A on top of one another, moving from left to right from the first column, as shown in Figure 16.1. See the definition of vectorization in Wiki.[5]

There is an important trace theorem to which we will now appeal. See Wiki.[6]

$$(16.4) \qquad Trace\left(A^T B\right) = vec(B)^T vec\left(A\right) = vec(A)^T vec\left(B\right) \quad .$$

Applying this relationship, we can consider in the exponent at the right of equation (16.3) and group the terms thereof as follows

[4]Zellner[107] introduces a generalization of this pure, direct, unadulterated conjugate prior form. We do not want to consider his generalization here, interested as we are in the purity of the pure conjugate form. Zellner[107] is worried about allowing a fully general prior, even if it involves different, nonconjugate functional forms. We are not interested in that. We are interested in a natively conjugate prior and posterior. The prior and posterior will be sufficiently general for the practical problems of which we are aware.

[5]https://en.wikipedia.org/wiki/Vectorization_(mathematics).

[6]https://en.wikipedia.org/wiki/Trace_(linear_algebra).

$$
\begin{aligned}
A^T &= (B - B_0)^T \\
B &= Q_0 \left(B - B_0\right) \Sigma^{-1} \quad ,
\end{aligned}
$$

and thus write the foregoing

$$
\begin{aligned}
&Trace\left[(B - B_0)^T Q_0 \left(B - B_0\right) \Sigma^{-1}\right] \\
&= \left[vec\left(B - B_0\right)\right]^T vec\left[Q_0 \left(B - B_0\right) \Sigma^{-1}\right] \quad .
\end{aligned}
$$

Figure 16.1: Vectorization

$$
vec\left(A\right) = \begin{bmatrix} a_{11} \\ \bullet \\ \bullet \\ a_{r1} \\ a_{12} \\ \bullet \\ \bullet \\ a_{r2} \\ \bullet \\ \bullet \\ \bullet \\ a_{1c} \\ \bullet \\ \bullet \\ a_{rc} \end{bmatrix} = \left[a_{11}, ..., a_{r1}, a_{12}, ..., a_{r2}, ...a_{1c}, ..., a_{rc}\right]^T
$$

We need to reduce the rightmost term by appealing to results from the Kronecker product of two matrices, A and B, defined in Wiki.[7] The Kronecker product is a very useful concept, and it will greatly simplify and streamline the variance-covariance developments in this section and beyond. If A is an $m \times n$ matrix and B is a $p \times q$ matrix, then the Kronecker product $A \otimes B$ (sometimes called the right Kronecker product) is the $mp \times nq$ block matrix:

$$
A \otimes B = \begin{bmatrix} a_{11}B & . & . & . & a_{1n}B \\ . & & & & . \\ . & & & & . \\ . & & & & . \\ a_{m1}B & . & . & . & a_{mn}B \end{bmatrix} \quad .
$$

There is a result in Wiki[8] and Graham[37] p. 25 that

[7] See https://en.wikipedia.org/wiki/Kronecker_product.

[8] https://en.wikipedia.org/wiki/Vectorization_(mathematics).

$$(16.5) \qquad vec\,(AXB) = \left(B^T \otimes A\right) vec\,(X) \quad .$$

If we consider

$$\begin{aligned} A &= Q_0 \\ X &= (B - B_0) \\ B &= \Sigma^{-1} \quad , \end{aligned}$$

and substitute, we can write (using $-T$ to designate inverse transpose)[9]

$$\begin{aligned} &vec\left[Q_0\,(B - B_0)\,\Sigma^{-1}\right] = \left(\Sigma^{-T} \otimes Q_0\right) vec\,(B - B_0) \\ &= (\Sigma^{-1} \otimes Q_0)[vec(B) - vec(B_0)] \quad . \end{aligned}$$

The inverse transpose of a symmetric matrix is simply the inverse, so we eliminate the $-T$ notation. The latter equation uses the fact that the vectorization of a sum of two identically conformed matrices is the sum of the vectorizations of each element of the sum. We vectorize each matrix and add them to get the vectorization of the sum. Therefore, we have the salient result:

$$\begin{aligned} &Trace\left[(B - B_0)^T Q_0\,(B - B_0)\,\Sigma^{-1}\right] \\ &= \left[vec(B)^T - vec(B_0)^T\right]\left(\Sigma^{-1} \otimes Q_0\right)\left[vec\,(B) - vec\,(B_0)\right] \quad . \end{aligned}$$

This allows us to write the prior in equation (16.3) in the following equivalent form

$$(16.6) \qquad \begin{aligned} \{B, \Sigma\} = c_0 \frac{1}{|\Sigma|^{\frac{m_0}{2}}} e^{-\frac{1}{2} Trace\left(M_0 \Sigma^{-1}\right)} \\ e^{-\frac{1}{2}\left[vec(B)^T - vec(B_0)^T\right]\left(\Sigma^{-1} \otimes Q_0\right)\left[vec(B) - vec(B_0)\right]} \quad . \end{aligned}$$

B is $p \times m$; Σ is $m \times m$. The constant c_0 is the same integrating constant as in the density in equation (16.3). Manipulation of the exponent has not changed this integrating constant. Observe the form of the second exponential term, the final line of the equation. It is a row vector times a matrix times a column vector that is the transpose of the row vector. This is precisely the form of the exponent in a usual and customary multivariate normal distribution. By taking note of that, we are in effect recognizing the existence of a multivariate normal distribution within the joint distribution. This will greatly facilitate future reductions via integration from joint densities to marginal densities; we will create and simply integrate out the normal. We will emphasize the normal density that devolves from the second exponential term and note that there is an additional exponential

[9] It is easy to see and prove that $vec(A + B) = vec(A) + vec(B)$, i.e., that vectorization is additive.

term, namely the first term in the exponent, that will emerge upon integration. To reiterate how we got here, we began with the fundamental equation (16.3) and manipulated it to form the product of a multivariate normal distribution times another distribution in (16.6), the latter containing a multivariate normal that can be easily integrated. We therefore think of the final term as a multivariate normal distribution with parameters:

Mean $(\mu) = vec\,(B_0) \qquad 1 \times mp$

Inverse variance-covariance: ${\Sigma_C}^{-1} = \Sigma^{-1} \otimes Q_0 \qquad mp \times mp$

multiplied by some other distribution, whose form we will find now. We want to define (appealing to the notation we formerly used for the classical regression case with the subscript C because we will be returning to that case herein)

$$\Sigma_C^{-1} = \Sigma^{-1} \otimes Q_0 \qquad mp \times mp \quad .$$

This implies that

$$\Sigma_C = \left(\Sigma^{-1} \otimes Q_0\right)^{-1} \quad .$$

There is an inverse theorem for Kronecker products in Graham[37] p. 25 that

$$(A \otimes B)^{-1} = A^{-1} \otimes B^{-1} \quad ,$$

which implies that

$$\Sigma_C = \Sigma \otimes {Q_0}^{-1} \qquad mp \times mp \quad .$$

The Σ_C and Σ_C^{-1} matrices are $mp \times mp$ matrices because they are the Kronecker products of an $m \times m$ and a $p \times p$ matrix. We will see this mp term appearing in the multivariate normal distribution associated with the vectorized quadratic. We also note that the length of the vectorized B term is mp. This is because the vectorization of a $p \times m$ matrix has length mp.

There is a determinant theorem for Kronecker products[10] that is expressed

$$|A_{m\times m} \otimes B_{n\times n}| = |A_{m\times m}|^n |B_{n\times n}|^m \quad .$$

Applying this theorem implies that

$$|\Sigma_C| = |\Sigma|^p \left|{Q_0}^{-1}\right|^m = \frac{|\Sigma|^p}{|Q_0|^m}$$

$$\Rightarrow |\Sigma_C|^{\frac{1}{2}} = \frac{|\Sigma|^{\frac{p}{2}}}{|Q_0|^{\frac{m}{2}}} \quad .$$

[10] See https://en.wikipedia.org/wiki/Kronecker_product.

It is well to remember that Σ_C is a nonsingular $m \times m$ matrix and that Q_0 is a nonsingular $p \times p$ matrix. This helps us see that the determinant rule is applied correctly. We can therefore write the multivariate normal as follows:

$$N\left[vec(B); vec(B_0), \Sigma_C, mp\right]$$
$$= (2\pi)^{-\frac{mp}{2}} \left|\Sigma_C\right|^{-\frac{1}{2}} e^{-\frac{1}{2}\left[vec(B)^T - vec(B_0)^T\right]\Sigma_C{}^{-1}\left[vec(B) - vec(B_0)\right]} \quad .$$

We can substitute the result of the product of determinants theorem above for the matrix Σ_C and reorganize this to write the exponent as a function of the other terms

$$e^{-\frac{1}{2}\left[vec(B)^T - vec(B_0)^T\right]\Sigma_C{}^{-1}\left[vec(B) - vec(B_0)\right]}$$
$$= N\left[vec(B); vec(B_0), \Sigma_C, mp\right] \frac{(2\pi)^{\frac{mp}{2}} \left|\Sigma\right|^{\frac{p}{2}}}{\left|Q_0\right|^{\frac{m}{2}}} \quad .$$

We substitute this expression for the exponential term (the final term in the prior in equation (16.6)) to write

$$\{\Sigma, B\} = c_0 \frac{1}{\left|\Sigma\right|^{\frac{m_0}{2}}} e^{-\frac{1}{2}Trace\left(M_0\Sigma^{-1}\right)} \tag{16.7}$$
$$\times N\left[vec(B); vec(B_0), \Sigma_C, mp\right] \frac{(2\pi)^{\frac{mp}{2}} \left|\Sigma\right|^{\frac{p}{2}}}{\left|Q_0\right|^{\frac{m}{2}}}$$
$$= c_0 \frac{(2\pi)^{\frac{mp}{2}}}{\left|Q_0\right|^{\frac{m}{2}} \left|\Sigma\right|^{\frac{m_0 - p}{2}}} e^{-\frac{1}{2}Trace\left(M_0\Sigma^{-1}\right)}$$
$$N\left[vec(B); vec(B_0), \Sigma_C, mp\right] \quad .$$

We are making progress toward the correct expression for the integrating constant of the joint density over B and Σ because we are now in a position to integrate out the normal distribution at the right and thereby the B dependency and develop the marginal density over Σ.

To move toward the marginal distribution over Σ, we need to integrate the above joint density over the entire domain of B. Because B appears only in the multivariate normal distribution at the right, it will integrate out without any change in any other term in the equation. Integrating this multivariate normal at the right out, we are left with a marginal distribution over Σ, which is simply the leading set of terms in the foregoing joint density without the multivariate normal distribution. We say we can simply "drop" the normal distribution, but in fact we have properly integrated it out to create the marginal. The result appears as if we simply drop the multivariate normal density to obtain the correct expression for the marginal distribution over Σ.

$$\{\Sigma\} = c_0 \frac{(2\pi)^{\frac{mp}{2}}}{|Q_0|^{\frac{m}{2}} |\Sigma|^{\frac{m_0 - p}{2}}} e^{-\frac{1}{2} Trace\left(M_0 \Sigma^{-1}\right)} \quad . \tag{16.8}$$

Let us wrestle with this $m \times m$ distribution. We first turn to Wiki[11] to get the equation for the inverse Wishart distribution:

$$f_X(X; \Psi, \nu, p) = \frac{|\Psi|^{\frac{\nu}{2}}}{2^{\frac{\nu p}{2}} \Gamma_p\left(\frac{\nu}{2}\right)} |X|^{-\frac{\nu + p + 1}{2}} e^{-\frac{1}{2} Trace\left(\Psi X^{-1}\right)} \quad , \tag{16.9}$$

in which

$$\begin{aligned} &X: \quad p \times p, \text{positive definite} \\ &\Psi: \quad p \times p, \text{scale matrix, positive definite} \\ &\Gamma_p(a): \text{multivariate gamma function, defined as follows} \\ &\Gamma_p(a) = \pi^{\frac{p(p-1)}{4}} \prod_{j=1}^{p} \Gamma\left(a + \frac{1-j}{2}\right) \quad . \end{aligned}$$

The multivariate gamma function appears in Wiki.[12] We want to make substitutions to render it into the form of our distribution. The substitutions are

$$\begin{aligned} p &\leftarrow m \\ \Psi &\leftarrow M_0 \\ X &\leftarrow \Sigma \quad . \end{aligned}$$

Under these substitutions, we write the foregoing inverse Wishart distribution from Wiki using the multivariate gamma function as follows

$$f_\Sigma(\Sigma; M_0, \nu, m) = \frac{|M_0|^{\frac{\nu}{2}}}{2^{\frac{\nu m}{2}} \Gamma_m\left(\frac{\nu}{2}\right)} |\Sigma|^{-\frac{\nu + m + 1}{2}} e^{-\frac{1}{2} Trace\left(M_0 \Sigma^{-1}\right)} \quad , \tag{16.10}$$

in which

$$\begin{aligned} &\Sigma: \quad m \times m, \text{positive definite} \\ &M_0: \quad m \times m, \text{scale matrix, positive definite} \quad . \end{aligned}$$

If we equate our inverse Wishart-like distribution in equation (16.8) with the standard inverse Wishart distribution in equation (16.10), we have

$$f_\Sigma(\Sigma; M_0, \nu, m) = \{\Sigma\} \tag{16.11}$$

[11] See https://en.wikipedia.org/wiki/Inverse-Wishart_distribution.

[12] See https://en.wikipedia.org/wiki/Multivariate_gamma_function. We will not derive this function here because it is tedious, mechanical, and well known.

$$= \frac{|M_0|^{\frac{\nu}{2}}}{2^{\frac{\nu m}{2}} \Gamma_m\left(\frac{\nu}{2}\right) |\Sigma|^{\frac{\nu+m+1}{2}}} e^{-\frac{1}{2}Trace\left(M_0 \Sigma^{-1}\right)}$$

$$= c_0 \frac{(2\pi)^{\frac{mp}{2}}}{|Q_0|^{\frac{m}{2}} |\Sigma|^{\frac{m_0-p}{2}}} e^{-\frac{1}{2}Trace\left(M_0 \Sigma^{-1}\right)} \quad .$$

In order to equate the foregoing two equations, we must equate the exponents on the Σ terms in the denominator so that they become identical. This implies that the degrees of freedom must satisfy the following equation

$$m_0 - p = \nu + m + 1$$
$$m_0 = \nu + p + m + 1 \quad .$$

We append the subscript 0 to the ν term to connote that we are dealing with the prior. This can be rearranged to calculate the prior degrees of freedom.

$$\begin{aligned} \nu_0 &= m_0 - p - m - 1 \\ m_0 - p &= \nu_0 + m + 1 \quad . \end{aligned} \tag{16.12}$$

Substituting into equation (16.11) and dividing out common terms and rearranging gives the expression for the integrating constant

$$c_0 = \frac{|M_0|^{\frac{\nu_0}{2}} |Q_0|^{\frac{m}{2}}}{(2\pi)^{\frac{pm}{2}} 2^{\frac{\nu_0 m}{2}} \Gamma_m\left(\frac{\nu_0}{2}\right)} \quad . \tag{16.13}$$

We calculate the full and correct expression for the prior by substituting into equation (16.3):

$$\begin{aligned} \{B, \Sigma\} = & \frac{|M_0|^{\frac{\nu_0}{2}} |Q_0|^{\frac{m}{2}}}{(2\pi)^{\frac{pm}{2}} 2^{\frac{\nu_0 m}{2}} \Gamma_m\left(\frac{\nu_0}{2}\right) |\Sigma|^{\frac{m_0}{2}}} \\ & \times e^{-\frac{1}{2}Trace\left(M_0 \Sigma^{-1}\right)} \\ & \times e^{-\frac{1}{2}Trace\left[(B-B_0)^T Q_0 (B-B_0) \Sigma^{-1}\right]} \quad . \end{aligned} \tag{16.14}$$

B is $p \times m$; Σ is $m \times m$. This distribution integrates to 1 by construction. We can express this in the form that contains an embedded normal distribution by substituting the definition of c_0 developed in equation (16.13) into equation (16.7) and simplifying as follows

$$\begin{aligned} \{B, \Sigma\} = & c_0 \frac{(2\pi)^{\frac{mp}{2}}}{|Q_0|^{\frac{m}{2}} |\Sigma|^{\frac{m_0-p}{2}}} e^{-\frac{1}{2}Trace\left(M_0 \Sigma^{-1}\right)} \\ & \times N\left[vec\left(B\right); vec\left(B_0\right); \Sigma_C; mp\right] \end{aligned}$$

$$
\begin{aligned}
&= \left[\frac{|M_0|^{\frac{\nu_0}{2}}|Q_0|^{\frac{m}{2}}}{(2\pi)^{\frac{mp}{2}}2^{\frac{m\nu_0}{2}}\Gamma_m\left(\frac{\nu_0}{2}\right)}\right]\frac{(2\pi)^{\frac{mp}{2}}}{|Q_0|^{\frac{m}{2}}|\Sigma|^{\frac{m_0-p}{2}}}e^{-\frac{1}{2}Trace\left(M_0\Sigma^{-1}\right)} \\
&\times N\left[vec\left(B\right);vec\left(B_0\right);\Sigma_C;mp\right] \\
&= \frac{|M_0|^{\frac{\nu_0}{2}}}{2^{\frac{m\nu_0}{2}}\Gamma_m\left(\frac{\nu_0}{2}\right)|\Sigma|^{\frac{m_0-p}{2}}}e^{-\frac{1}{2}Trace\left(M_0\Sigma^{-1}\right)} \\
&\times N\left[vec\left(B\right);vec\left(B_0\right);\Sigma_C;mp\right] \quad .
\end{aligned}
$$

B is $p \times m$; Σ is $m \times m$. We rewrite for future reference

$$
\begin{aligned}
\{B,\Sigma\} = &\frac{|M_0|^{\frac{\nu_0}{2}}}{2^{\frac{\nu_0 m}{2}}\Gamma_m\left(\frac{\nu_0}{2}\right)|\Sigma|^{\frac{\nu_0+m+1}{2}}}e^{-\frac{1}{2}Trace\left(M_0\Sigma^{-1}\right)} \\
&N\left[vec\left(B\right);vec\left(B_0\right),\Sigma_C,mp\right] \quad .
\end{aligned}
$$

in which we have substituted $m_0 - p = \nu_0 + m + 1$. B is $p \times m$; Σ is $m \times m$. Let us carefully look at the terms outside the normal. We note that it is term by term equivalent to equation (16.11), meaning it is an inverse Wishart distribution with the following parameters and notation:

$$
\begin{aligned}
&f_\Sigma\left(\Sigma;M_0,\nu_0,m\right) \\
&= W^{-1}\left(\Sigma;M_0,\nu_0,m\right) \quad .
\end{aligned}
$$

That means we can write the former equation in the following classical form

$$
\begin{aligned}
\{B,\Sigma\} &= W^{-1}\left(\Sigma;M_0,\nu_0,m\right) \\
&N\left[vec\left(B\right);vec\left(B_0\right),\left(\Sigma\otimes Q_0^{-1}\right),mp\right] \quad .
\end{aligned}
\tag{16.15}
$$

B is $p \times m$; Σ is $m \times m$.

The joint density is the product of an inverse Wishart distribution in standard form times a vectorized multivariate normal distribution. Inverse Wishart times normal. (See equations (4.12) and (5.17), the first for the conjugate-archetype and the second for the prior, for direct analogies.) Quite like the univariate case, but generalized. The foregoing two equations will drive the rest of our results in this section. Before leaving this section, it is well to integrate B out to create the marginal distribution over Σ, i.e., the inverse Wishart. We note that B appears only in the normal, and therefore the normal integrates out, leaving the marginal distribution over Σ, which is a standard inverse Wishart distribution.

$$
\{\Sigma\} = \frac{|M_0|^{\frac{\nu_0}{2}}}{2^{\frac{\nu_0 m}{2}}\Gamma_m\left(\frac{\nu_0}{2}\right)|\Sigma|^{\frac{m_0-p}{2}}}e^{-\frac{1}{2}Trace\left(M_0\Sigma^{-1}\right)}
\tag{16.16}
$$

$$= \frac{|M_0|^{\frac{\nu_0}{2}}}{2^{\frac{\nu_0 m}{2}}\Gamma_m\left(\frac{\nu_0}{2}\right)|\Sigma|^{\frac{\nu_0+m+1}{2}}} e^{-\frac{1}{2}Trace\left(M_0\Sigma^{-1}\right)} \quad .$$

Σ is $m \times m$. We will be reproducing this exact flow of logic when we get to the projective analysis below.

16.2 Analysis of the Conjugate Prior–Marginal Distribution over the Prior Coefficients

This section undertakes to formulate the marginal probability distribution over the matrix of model coefficients B. As we do so, we will keep in mind the analogy to the univariate case. We begin by defining a constant T_0 for notational convenience:

$$T_0 = M_0 + (B - B_0)^T Q_0 (B - B_0) \quad ,$$

and thereafter rewriting the joint distribution by substituting this expression into equation (16.14) to write

$$\text{(16.17)} \qquad \{B, \Sigma\} = \frac{|M_0|^{\frac{\nu_0}{2}}|Q_0|^{\frac{m}{2}}}{2^{\frac{(\nu_0+p)m}{2}}\pi^{\frac{mp}{2}}\Gamma_m\left(\frac{\nu_0}{2}\right)} \frac{1}{|\Sigma|^{\frac{m_0}{2}}} e^{-\frac{1}{2}Trace\left(T_0\Sigma^{-1}\right)} \quad .$$

We need to interpret this in terms of the inverse Wishart. The inverse Wishart is given (in general form) in Wiki[13] and repeated here for convenient reference as

$$X \sim W^{-1}(X; \Psi, \xi, p) = \frac{|\Psi|^{\frac{\xi}{2}}}{2^{\frac{\xi p}{2}}\Gamma_p\left(\frac{\xi}{2}\right)}|X|^{-\frac{\xi+p+1}{2}} e^{-\frac{1}{2}Trace\left(\Psi X^{-1}\right)} \quad .$$

We have substituted our own variable ξ in place of the degrees of freedom notation ν in the Wiki inverse Wishart. We do so to avoid variable name confounding in forthcoming developments. We need to put the parameters from our distribution into the standard form of inverse Wishart so we can equate terms. For our prior, we are using the specific notation

$$\begin{aligned} \Psi &\leftarrow T_0 \\ p &\leftarrow m \\ X &\leftarrow \Sigma \quad , \end{aligned}$$

which when we put them into the standard form inverse Wishart distribution allows it to be written

[13] https://en.wikipedia.org/wiki/Inverse-Wishart_distribution.

$$\Sigma \sim W^{-1}\left(\Sigma; T_0, \xi, m\right) = \frac{|T_0|^{\frac{\xi}{2}}}{2^{\frac{\xi m}{2}}\Gamma_m\left(\frac{\xi}{2}\right)|\Sigma|^{\frac{\xi+m+1}{2}}} e^{-\frac{1}{2}Trace\left(T_0\Sigma^{-1}\right)} \quad .$$

We rearrange this equation to write the exponential term as follows:

$$e^{-\frac{1}{2}Trace\left(T_0\Sigma^{-1}\right)} = \frac{2^{\frac{\xi m}{2}}\Gamma_m\left(\frac{\xi}{2}\right)|\Sigma|^{\frac{\xi+m+1}{2}}}{|T_0|^{\frac{\xi}{2}}} W^{-1}\left(\Sigma; T_0, \xi, m\right) \quad .$$

We substitute this for the exponential term in equation (16.17) to write

$$\begin{aligned}\{B, \Sigma\} &= \frac{|M_0|^{\frac{m_0-p-m-1}{2}}|Q_0|^{\frac{m}{2}}}{2^{\frac{(m_0-m-1)m}{2}}\pi^{\frac{mp}{2}}\Gamma_m\left(\frac{m_0-p-m-1}{2}\right)|\Sigma|^{\frac{m_0}{2}}} \\ &\times \left[\frac{2^{\frac{\xi m}{2}}\Gamma_m\left(\frac{\xi}{2}\right)|\Sigma|^{\frac{\xi+m+1}{2}}}{|T_0|^{\frac{\xi}{2}}} W^{-1}\left(\Sigma; T_0, \xi, m\right)\right] \\ &= \frac{|M_0|^{\frac{m_0-p-m-1}{2}}|Q_0|^{\frac{m}{2}}2^{\frac{\xi m}{2}}\Gamma_m\left(\frac{\xi}{2}\right)|\Sigma|^{\frac{\xi-m_0+m+1}{2}}}{2^{\frac{(m_0-m-1)m}{2}}\pi^{\frac{mp}{2}}\Gamma_m\left(\frac{m_0-p-m-1}{2}\right)|T_0|^{\frac{\xi}{2}}} W^{-1}\left(\Sigma; T_0, \xi, m\right) \quad .\end{aligned}$$

B is $p \times m$; Σ is $m \times m$. We want to select ξ in order to zero out the exponent on $|\Sigma|$ in this equation so that the only occurrence of Σ is in the inverse Wishart itself. We can do so by selecting ξ such that $\xi - m_0 + m + 1 = 0$. This implies that

$$\xi = m_0 - m - 1 = \nu_0 + p \quad .$$

The latter equality is implied by equation (16.12), which gives the definition of degrees of freedom. This eliminates the exponent on the Σ term, places the only occurrence of Σ into the inverse Wishart, remains consistent with the degrees of freedom definition, and simplifies the previous equation to

$$\{B, \Sigma\} = \frac{|M_0|^{\frac{\nu_0}{2}}|Q_0|^{\frac{m}{2}}\Gamma_m\left(\frac{\nu_0+p}{2}\right)}{\pi^{\frac{mp}{2}}\Gamma_m\left(\frac{\nu_0}{2}\right)|T_0|^{\frac{\nu_0+p}{2}}} W^{-1}\left(\Sigma; T_0, \nu_0+p, m\right) \quad .$$

B is $p \times m$; Σ is $m \times m$. Notice the substitutions of $\nu_0 + p$ for ξ in the inverse Wishart and in the various terms of the equation. We know that the inverse Wishart integrates to 1 over the variable Σ. We integrate out the inverse Wishart over Σ to give the marginal density over the model coefficients B.

$$\{B\} = \frac{\Gamma_m\left(\frac{\nu_0+p}{2}\right)}{\Gamma_m\left(\frac{\nu_0}{2}\right)} \frac{|M_0|^{\frac{\nu_0}{2}}|Q_0|^{\frac{m}{2}}}{\pi^{\frac{mp}{2}}} |T_0|^{-\frac{\nu_0+p}{2}} \quad .$$

B is $p \times m$. Substituting for T_0

$$\{B\} = \frac{\Gamma_m\left(\frac{\nu_0+p}{2}\right)}{\Gamma_m\left(\frac{\nu_0}{2}\right)} \frac{|M_0|^{\frac{\nu_0}{2}} |Q_0|^{\frac{m}{2}}}{\pi^{\frac{mp}{2}}}$$
$$\left|M_0\left[I_{mm} + M_0{}^{-1}(B - B_0)^T Q_0 (B - B_0)\right]\right|^{-\frac{\nu_0+p}{2}} .$$

B is $p \times m$. We can simplify by fully factoring out the M_0 term and melding it into the constants

(16.18)
$$\{B\} = \frac{\Gamma_m\left(\frac{\nu_0+p}{2}\right)}{\Gamma_m\left(\frac{\nu_0}{2}\right)} \frac{|Q_0|^{\frac{m}{2}}}{\pi^{\frac{mp}{2}} |M_0|^{\frac{p}{2}}}$$
$$\left|I_{mm} + M_0{}^{-1}(B - B_0)^T Q_0 (B - B_0)\right|^{-\frac{\nu_0+p}{2}} .$$

B is $p \times m$. This is the complete, unpartitioned distribution. We need to determine exactly what type of distribution this is. We will show that it is in fact a matrix t distribution as defined in equations (5.34) and (5.37) in Kotz and Natarajah[59], which they wrote in general in the form

$$f(X) = \frac{\Gamma_q\left(\frac{m}{2}\right)}{\pi^{\frac{pq}{2}} \Gamma_q\left(\frac{m-p}{2}\right)} |Q|^{\frac{m-p}{2}} |R|^{-\frac{q}{2}} \left|Q + (X-\mu)^T R^{-1} (X-\mu)\right|^{-\frac{m}{2}} .$$

We want to make substitutions that convert this general matrix t distribution into our form. We being by making substitutions that do not confuse the Kotz and Natarajah notation with ours:

$$\begin{aligned} p &\leftarrow i \\ m &\leftarrow j \quad . \end{aligned}$$

This allows us to write

$$f(X) = \frac{\Gamma_q\left(\frac{j}{2}\right)}{\pi^{\frac{iq}{2}} \Gamma_q\left(\frac{j-i}{2}\right)} |Q|^{\frac{j-i}{2}} |R|^{-\frac{q}{2}} \left|Q + (X-\mu)^T R^{-1} (X-\mu)\right|^{-\frac{j}{2}} .$$

We can now replace j with our definition and usage of p:

$$j \leftarrow \nu_0 + p$$

to write

$$f(X) = \frac{\Gamma_q\left(\frac{\nu_0+p}{2}\right)}{\pi^{\frac{iq}{2}} \Gamma_q\left(\frac{\nu_0+p-i}{2}\right)} |Q|^{\frac{\nu_0+p-i}{2}} |R|^{-\frac{q}{2}} \left|Q + (X-\mu)^T R^{-1} (X-\mu)\right|^{-\frac{\nu_0+p}{2}} .$$

We then replace i with our definition and usage of p

$$i \leftarrow p$$

to write

$$f(X) = \frac{\Gamma_q\left(\frac{\nu_0+p}{2}\right)}{\pi^{\frac{pq}{2}}\Gamma_q\left(\frac{\nu_0}{2}\right)} |Q|^{\frac{\nu_0}{2}} |R|^{-\frac{q}{2}} \left|Q + (X-\mu)^T R^{-1}(X-\mu)\right|^{-\frac{\nu_0+p}{2}} .$$

We then substitute our definition and usage of m

$$q \leftarrow m$$

to write

$$f(X) = \frac{\Gamma_m\left(\frac{\nu_0+p}{2}\right)}{\pi^{\frac{pm}{2}}\Gamma_m\left(\frac{\nu_0}{2}\right)} |Q|^{\frac{\nu_0}{2}} |R|^{-\frac{m}{2}} \left|Q + (X-\mu)^T R^{-1}(X-\mu)\right|^{-\frac{\nu_0+p}{2}} .$$

We are now in a position to simplify by factoring out the Q matrix and continuing

$$\begin{aligned}
f(X) &= \frac{\Gamma_m\left(\frac{\nu_0+p}{2}\right)}{\pi^{\frac{pm}{2}}\Gamma_m\left(\frac{\nu_0}{2}\right)} \\
&\times |Q|^{\frac{\nu_0}{2}} |R|^{-\frac{m}{2}} \left|Q + QQ^{-1}(X-\mu)^T R^{-1}(X-\mu)\right|^{-\frac{\nu_0+p}{2}} \\
&= \frac{\Gamma_m\left(\frac{\nu_0+p}{2}\right)}{\pi^{\frac{pm}{2}}\Gamma_m\left(\frac{\nu_0}{2}\right)} \\
&\times |Q|^{\frac{\nu_0}{2}} |R|^{-\frac{m}{2}} \left|Q\left[I_{mm} + Q^{-1}(X-\mu)^T R^{-1}(X-\mu)\right]\right|^{-\frac{\nu_0+p}{2}} \\
&= \frac{\Gamma_m\left(\frac{\nu_0+p}{2}\right)}{\pi^{\frac{pm}{2}}\Gamma_m\left(\frac{\nu_0}{2}\right)} \\
&\times |Q|^{\frac{\nu_0}{2}} |R|^{-\frac{m}{2}} |Q|^{-\frac{\nu_0+p}{2}} \left|I_{mm} + Q^{-1}(X-\mu)^T R^{-1}(X-\mu)\right|^{-\frac{\nu_0+p}{2}} \\
&= \frac{\Gamma_m\left(\frac{\nu_0+p}{2}\right)}{\pi^{\frac{pm}{2}}\Gamma_m\left(\frac{\nu_0}{2}\right)} \\
&\times |Q|^{-\frac{p}{2}} |R|^{-\frac{m}{2}} \left|I_{mm} + Q^{-1}(X-\mu)^T R^{-1}(X-\mu)\right|^{-\frac{\nu_0+p}{2}} .
\end{aligned}$$

We are finally in a position to substitute the two matrices

$$\begin{aligned}
&Q \leftarrow M_0 \\
&R^{-1} \leftarrow Q_0 \Rightarrow R \leftarrow {Q_0}^{-1}
\end{aligned}$$

to show that ours is exactly a matrix t distribution:

$$f(x) = \frac{\Gamma_m\left(\frac{\nu_0+p}{2}\right)|Q_0|^{\frac{m}{2}}}{\pi^{\frac{pm}{2}}\Gamma_m\left(\frac{\nu_0}{2}\right)|M_0|^{\frac{p}{2}}}\left|I_{mm} + M_0{}^{-1}(X-\mu)^T Q_0\,(X-\mu)\right|^{-\frac{\nu_0+p}{2}} \quad .$$

Term by term, we have shown that our distribution in equation (16.18) is a matrix t distribution. For distributions such as this, it is appropriate to apply the Weinstein-Aronszajn identity,[14] which is written, for matrices

$$A : m \times p \quad B : p \times m$$

as follows:

$$|I_{mm} + AB| = |I_{pp} + BA| \quad .$$

(We want to do so to better analyze the prior probability distribution itself, not to change the prior times the likelihood calculation.) If we write

$$\begin{aligned} A &= M_0^{-1}(B - B_0)^T \\ B &= Q_0\,(B - B_0) \quad , \end{aligned}$$

then the determinant in the denominator in equation (16.18) can be substituted and rewritten

$$\begin{aligned} &\left|I_{mm} + M_0{}^{-1}(B-B_0)^T Q_0\,(B-B_0)\right| \\ &= \left|I_{pp} + Q_0\,(B-B_0)\,M_0{}^{-1}(B-B_0)^T\right| \quad . \end{aligned}$$

This means that equation (16.18) can be rewritten in the equivalent form

$$\text{(16.19)} \qquad \{B\} = \frac{\Gamma_m\left(\frac{\nu_0+p}{2}\right)}{\Gamma_m\left(\frac{\nu_0}{2}\right)}\frac{|Q_0|^{\frac{m}{2}}}{\pi^{\frac{mp}{2}}|M_0|^{\frac{p}{2}}} \left|I_{pp} + Q_0\,(B-B_0)\,M_0{}^{-1}(B-B_0)^T\right|^{-\frac{\nu_0+p}{2}} \quad .$$

B is $p \times m$. This is the full, unpartitioned, unintegrated distribution. Equation (16.18) is the prior distribution we want to use to implement Bayes theorem to infer the posterior. However, equation (16.19) is a distribution we may want to use to examine the prior. We have made the transformation to facilitate that. Both of these distributions are in the form of matrix Student's t distributions.[15] Keep in mind that both equation (16.18) and equation (16.19) are joint distributions over $m \times p$ or mp coefficients B_{ij}, $i = 1, ..., p, j = 1, ..., m$. The key message is that the distribution is joint and there are no independencies assumed. Full interdependency. We will address this later. We should also mention, as proven

[14] See https://en.wikipedia.org/wiki/Weinstein-Aronszajn_identity.
[15] See https://en.wikipedia.org/wiki/Matrix_t-distribution.

in the univariate marginal case below, that if we were to calculate the marginal probability of each model coefficient, it would be Student's t in form.

16.2.1 Analysis of the Conjugate Prior–Developing the Marginal m-Variate Joint Distribution

In the univariate case, we developed a univariate marginal distribution for each of the p individual model coefficients. By construction, in this multivariate estimation case, each of the distributions over the pth model coefficient is an m-variate distribution, not a univariate distribution. What we strive to do in this section is write the m-variate (joint) distribution for the pth model coefficient. To reiterate, it was univariate and therefore not joint in the univariate case. Alas, it is multivariate (m-variate) in the m-variate case and therefore characterized by a marginal joint distribution for the pth model coefficient. To begin, we rewrite the joint density for the prior from equation (16.14) for convenience, substituting the expression for $\nu_0 = m_0 - p - m - 1$ as we do so.

$$
\begin{aligned}
\{B, \Sigma\} = & \frac{|M_0|^{\frac{m_0-p-m-1}{2}} |Q_0|^{\frac{m}{2}}}{(2\pi)^{\frac{pm}{2}} 2^{\frac{(m_0-p-m-1)m}{2}} \Gamma_m\left(\frac{m_0-p-m-1}{2}\right) |\Sigma|^{\frac{m_0}{2}}} \\
& \times e^{-\frac{1}{2}Trace\left(M_0\Sigma^{-1}\right)} \\
& \times e^{-\frac{1}{2}Trace\left[(B-B_0)^T Q_0 (B-B_0)\Sigma^{-1}\right]} \quad .
\end{aligned}
$$

B is $p \times m$; Σ is $m \times m$. We consider the exponent in the final term

$$
E_F = (B - B_0)^T Q_0 (B - B_0) \Sigma^{-1} \quad .
$$

The dimensionality of the terms in this exponent are

$$
\begin{aligned}
B, B_0 &= p \times m \\
Q_0 &= p \times p \\
\Sigma &= m \times m \quad ,
\end{aligned}
$$

in which the matrix is

$$
Q_0 = \begin{bmatrix} q_{11} & {q_{p-1}}^T \\ q_{p-1} & Q_1 \end{bmatrix} \quad .
$$

We know from the analysis of the inverse of a partitioned matrix in Appendix A that the ii^{th} element of Q_0^{-1} is

$$
\left[Q_0^{-1}\right]_{11} = \left(q_{11} - {q_{p-1}}^T Q_1^{-1} q_{p-1}\right)^{-1} \quad . \tag{16.20}
$$

Since this is a scalar, the inverse is the reciprocal

$$q_{11} - {q_{p-1}}^T Q_1^{-1} q_{p-1} = \frac{1}{\left[Q_0^{-1}\right]_{11}} \quad . \tag{16.21}$$

We will substitute this into the forthcoming development for notational simplicity. The dimensionality of the terms in this exponent are

$$\begin{aligned} q_{11} : &\quad 1 \times 1 \\ q_{p-1} : &\quad (p-1) \times 1 \\ {q_{p-1}}^T : &\quad 1 \times (p-1) \\ Q_1 : &\quad (p-1) \times (p-1) \quad . \end{aligned}$$

We define the following two matrices

$$\begin{aligned} &v_1 : 1 \times m \\ &V_1 : (p-1) \times m \quad , \end{aligned}$$

so that we can write the matrix

$$B_1 - B_1^0 = \begin{bmatrix} v_1 \\ V_1 \end{bmatrix} \quad .$$

The row vector v represents the first row of $B - B_0$. The matrix V_1 represents the final $p-1$ rows of $B - B_0$. The dimensionality of the key terms is

$$\begin{aligned} v_1 : &\quad 1 \times m \\ V_1 : &\quad (p-1) \times m \quad . \end{aligned}$$

For simplicity, we denote the first element of m to be the variable D, meaning "distance," and the second element of m to be the variable H, meaning "hang time," implicitly adopting the football punting problem of developing a joint distribution over distance and hang time for illustrative purposes. This means that m is 2, the first dependent variable is D, and the second dependent variable is H. We can regard the $m = 2$ columns of a matrix as containing the variables D and H. This mnemonic will help us get and understand the right answer and be able to extrapolate to higher values of m. This means we can write the mean term:

$$\begin{aligned} &B_1 - B_1^0 = \begin{bmatrix} v_1 \\ V_1 \end{bmatrix} \\ &\Rightarrow v_1 = \begin{bmatrix} B_{1D} - B_{1D}^0 & B_{1H} - B_{1H}^0 \end{bmatrix} \quad . \end{aligned} \tag{16.22}$$

v_1 is $1 \times m$. We have enumerated the elements of the first row of $B - B_0$ using the row index 1 and the column indexes in order D and H. The exponent is therefore written

$$E_F = \begin{bmatrix} v_1^T & V_1^T \end{bmatrix} \begin{bmatrix} q_{11} & {q_{p-1}}^T \\ q_{p-1} & Q_1 \end{bmatrix} \begin{bmatrix} v_1 \\ V_1 \end{bmatrix} \Sigma^{-1} \quad ,$$

which expands out to the following form

$$E_F = \left[v_1^T q_{11} v_1 + v_1^T {q_{p-1}}^T V_1 + V_1^T q_{p-1} v_1 + V_1^T Q_1 V_1\right] \Sigma^{-1} \quad .$$

Let us examine the internals of the term in the brackets

$$Z = v_1^T q_{11} v_1 + v_1^T {q_{p-1}}^T V_1 + V_1^T q_{p-1} v_1 + V_1^T Q_1 V_1 \quad .$$

The dimensionalities of the various terms in this equation are as follows

$$\begin{aligned} Z = v_1^T q_{11} v_1 &\qquad (m \times 1)\,(1 \times 1)\,(1 \times m) \\ + v_1^T {q_{p-1}}^T V_1 &\qquad (m \times 1)\,(1 \times p-1)\,(p-1 \times m) \\ + V_1^T q_{p-1} v_1 &\qquad (m \times p-1)\,(p-1 \times 1)\,(1 \times m) \\ + V_1^T Q_1 V_1 &\qquad (m \times p-1)\,(p-1 \times p-1)\,(p-1 \times m) \quad . \end{aligned}$$

Suppose we add and subtract a particular term (which will allow regrouping and simplification)

$$\begin{aligned} Z = {} & v_1^T q_{11} v_1 + v_1^T {q_{p-1}}^T V_1 + V_1^T q_{p-1} v_1 + V_1^T Q_1 V_1 \\ & - v_1^T {q_{p-1}}^T Q_1^{-1} q_{p-1} v_1 + v_1^T {q_{p-1}}^T Q_1^{-1} q_{p-1} v_1 \quad . \end{aligned}$$

We reorganize as follows

$$\begin{aligned} Z &= v_1^T q_{11} v_1 - v_1^T {q_{p-1}}^T Q_1^{-1} q_{p-1} v_1 \\ &\quad + v_1^T {q_{p-1}}^T V_1 + V_1^T q_{p-1} v_1 + V_1^T Q_1 V_1 + v_1^T {q_{p-1}}^T Q_1^{-1} q_{p-1} v_1 \\ &= v_1^T \left(q_{11} - {q_{p-1}}^T Q_1^{-1} q_{p-1}\right) v_1 \\ &\quad + v_1^T {q_{p-1}}^T V_1 + V_1^T q_{p-1} v_1 + V_1^T Q_1 V_1 + v_1^T {q_{p-1}}^T Q_1^{-1} q_{p-1} v_1 \\ &= \frac{1}{\left[Q_0^{-1}\right]_{11}} v_1^T v_1 \\ &\quad + V_1^T Q_1 V_1 + v_1^T {q_{p-1}}^T V_1 + V_1^T q_{p-1} v_1 + v_1^T {q_{p-1}}^T Q_1^{-1} q_{p-1} v_1 \quad . \end{aligned}$$

Define the final four terms, contained in the final line

$$Y = V_1^T Q_1 V_1 + v_1^T {q_{p-1}}^T V_1 + V_1^T q_{p-1} v_1 + v_1^T {q_{p-1}}^T Q_1^{-1} q_{p-1} v_1 \quad .$$

This allows us to write

$$Z = \frac{1}{\left[Q_0^{-1}\right]_{11}} v_1^T v_1 + Y \quad .$$

We want to multiply the matrix Q_1 by its inverse so that we can insert the relationship $I = Q_1 Q_1^{-1}$ in three positions in this matrix to be sure that there is at least one Q_1 term in every one of the terms in the foregoing equation. This insertion will allow regrouping and simplification.

$$\begin{aligned} Y &= V_1^T Q_1 V_1 + v_1^T {q_{p-1}}^T \left(Q_1^{-1} Q_1\right) V_1 \\ &+ V_1^T \left(Q_1 Q_1^{-1}\right) q_{p-1} v_1 + v_1^T {q_{p-1}}^T Q_1^{-1} \left(Q_1 Q_1^{-1}\right) q_{p-1} v_1 \quad . \end{aligned}$$

For notational and factoring convenience, we can temporarily define the quantity

$$\theta = Q_1^{-1} q_{p-1} \Rightarrow \theta^T = {q_{p-1}}^T Q_1^{-1} \quad .$$

Substituting into the previous equation allows us to write it:

$$\begin{aligned} Y &= V_1^T Q_1 V_1 + V_1^T Q_1 \theta v_1 + v_1^T \theta^T Q_1 V_1 + v_1^T \theta^T Q_1 \theta v_1 \\ &= V_1^T Q_1 \left(V_1 + \theta v_1\right) + v_1^T \theta^T Q_1 \left(V_1 + \theta v_1\right) \\ &= \left(V_1^T + v_1^T \theta^T\right) Q_1 \left(V_1 + \theta v_1\right) \\ &= \left(V_1 + \theta v_1\right)^T Q_1 \left(V_1 + \theta v_1\right) \quad . \end{aligned}$$

We substitute to eliminate the temporary variable θ

$$\begin{aligned} Y &= \left(V_1 + \theta v_1\right)^T Q_1 \left(V_1 + \theta v_1\right) \\ &= \left(V_1 + Q_1^{-1} q_{p-1} v_1\right)^T Q_1 \left(V_1 + Q_1^{-1} q_{p-1} v_1\right) \\ &= \left[V_1 - \left(-Q_1^{-1} q_{p-1} v_1\right)\right]^T Q_1 \left[V_1 - \left(-Q_1^{-1} q_{p-1} v_1\right)\right] \quad . \end{aligned}$$

Let us define the conditional means

$$\bar{V}_1 = -Q_1^{-1} q_{p-1} v_1 \quad .$$

So that we can now write the Y term

$$Y = \left(V_1 - \bar{V}_1\right)^T Q_1 \left(V_1 - \bar{V}_1\right) \quad .$$

Referring to the equation above in which Z was defined, we have

$$\begin{aligned} Z &= \frac{1}{\left[Q_0^{-1}\right]_{11}} v_1^T v_1 + Y \\ &= \frac{1}{\left[Q_0^{-1}\right]_{11}} v_1^T v_1 + \left(V_1 - \bar{V}_1\right)^T Q_1 \left(V_1 - \bar{V}_1\right) \quad . \end{aligned}$$

The exponent thus becomes

$$E_F = \frac{1}{\left[Q_0^{-1}\right]_{11}} v_1^T v_1 + \left(V_1 - \bar{V}_1\right)^T Q_1 \left(V_1 - \bar{V}_1\right) \quad .$$

We can now substitute this back into the original distribution

$$\{B, \Sigma\} = \frac{|M_0|^{\frac{m_0-m-p-1}{2}} |Q_0|^{\frac{m}{2}}}{(2\pi)^{\frac{pm}{2}} 2^{\frac{(m_0-m-p-1)m}{2}} \Gamma_m\left(\frac{m_0-m-p-1}{2}\right) |\Sigma|^{\frac{m_0}{2}}}$$
$$\times e^{-\frac{1}{2}Trace\left(M_0\Sigma^{-1}\right)}$$
$$\times e^{-\frac{1}{2}Trace\left[\frac{1}{\left[Q_0^{-1}\right]_{11}} v_1^T v_1 \Sigma^{-1} + \left(V_1-\bar{V}_1\right)^T Q_1 \left(V_1-\bar{V}_1\right)\Sigma^{-1}\right]} \quad .$$

The final term can be broken into two as follows

$$\{B, \Sigma\} = \frac{|M_0|^{\frac{m_0-m-p-1}{2}} |Q_0|^{\frac{m}{2}}}{(2\pi)^{\frac{pm}{2}} 2^{\frac{(m_0-m-p-1)m}{2}} \Gamma_m\left(\frac{m_0-m-p-1}{2}\right) |\Sigma|^{\frac{m_0}{2}}}$$
$$\times e^{-\frac{1}{2}Trace\left(M_0\Sigma^{-1}\right)}$$
$$\times e^{-\frac{1}{2}Trace\left(\frac{1}{\left[Q_0^{-1}\right]_{11}} v_1^T v_1 \Sigma^{-1}\right)}$$
$$\times e^{-\frac{1}{2}Trace\left[\left(V_1-\bar{V}_1\right)^T Q_1 \left(V_1-\bar{V}_1\right)\Sigma^{-1}\right]} \quad .$$

We can eliminate the powers of 2 in the denominator of the first term, which are reciprocals, to simplify

$$\{B, \Sigma\} = \frac{|M_0|^{\frac{m_0-m-p-1}{2}} |Q_0|^{\frac{m}{2}}}{\pi^{\frac{pm}{2}} 2^{\frac{(m_0-m-1)m}{2}} \Gamma_m\left(\frac{m_0-m-p-1}{2}\right) |\Sigma|^{\frac{m_0}{2}}}$$
$$\times e^{-\frac{1}{2}Trace\left(M_0\Sigma^{-1}\right)}$$
$$\times e^{-\frac{1}{2}Trace\left(\frac{1}{\left[Q_0^{-1}\right]_{11}} v_1^T v_1 \Sigma^{-1}\right)}$$
$$\times e^{-\frac{1}{2}Trace\left[\left(V_1-\bar{V}_1\right)^T Q_1 \left(V_1-\bar{V}_1\right)\Sigma^{-1}\right]} \quad .$$

We combine the middle two terms by adding their exponents

$$\{B, \Sigma\} = \frac{|M_0|^{\frac{m_0-m-p-1}{2}} |Q_0|^{\frac{m}{2}}}{\pi^{\frac{pm}{2}} 2^{\frac{(m_0-m-1)m}{2}} \Gamma_m\left(\frac{m_0-m-p-1}{2}\right) |\Sigma|^{\frac{m_0}{2}}}$$
$$\times e^{-\frac{1}{2}Trace\left[\left(\frac{1}{\left[Q_0^{-1}\right]_{11}} v_1^T v_1 + M_0\right)\Sigma^{-1}\right]}$$
$$\times e^{-\frac{1}{2}Trace\left[\left(V_1-\bar{V}_1\right)^T Q_1 \left(V_1-\bar{V}_1\right)\Sigma^{-1}\right]} \quad .$$

In pursuit of the distribution over m coefficients for independent variable 1, we are going to want to integrate the final term out. This is the same process we used in the univariate case—integrate $p-1$ of the p variables out to obtain the marginal over the first variable and Σ. We are going to want to insert a normal distribution for the last term so that we can integrate it out. We do so with the vectorization theorems we have used previously. We begin by defining

$$\begin{aligned} A^T &= \left(V_1 - \bar{V}_1\right)^T \Rightarrow A = \left(V_1 - \bar{V}_1\right) \\ B &= Q_1\left(V_1 - \bar{V}_1\right)\Sigma^{-1} \end{aligned}$$

If we apply the trace-vectorization theorem in equation (16.4), we can write

$$\begin{aligned} &Trace\left[\left(V_1 - \bar{V}_1\right)^T Q_1\left(V_1 - \bar{V}_1\right)\Sigma^{-1}\right] \\ &= vec\left(V_1 - \bar{V}_1\right)^T vec\left[Q_1\left(V_1 - \bar{V}_1\right)\Sigma^{-1}\right] \end{aligned}$$

If we terminate the just-used definitions and now denote

$$\begin{aligned} B &= Q_1 \\ V &= \left(V_1 - \bar{V}_1\right) \\ A^T &= \Sigma^{-1} = A \end{aligned}$$

then we can appeal to the Kronecker-vectorization theorem in equation (16.5) to write

$$vec\left[Q_1\left(V_1 - \bar{V}_1\right)\Sigma^{-1}\right] = \left(\Sigma^{-1} \otimes Q_1\right) vec\left(V_1 - \bar{V}_1\right)$$

This allows us to write the sought after trace as follows

$$\begin{aligned} &Trace\left[\left(V_1 - \bar{V}_1\right)^T Q_1\left(V_1 - \bar{V}_1\right)\Sigma^{-1}\right] \\ &= vec\left(V_1 - \bar{V}_1\right)^T \left(\Sigma^{-1} \otimes Q_1\right) vec\left(V_1 - \bar{V}_1\right) \end{aligned}$$

We are able us to rewrite our joint distribution as follows

$$\begin{aligned} \{B, \Sigma\} &= \frac{|M_0|^{\frac{m_0-m-p-1}{2}} |Q_0|^{\frac{m}{2}}}{\pi^{\frac{pm}{2}} 2^{\frac{(m_0-m-1)m}{2}} \Gamma_m\left(\frac{m_0-m-p-1}{2}\right) |\Sigma|^{\frac{m_0}{2}}} \\ &\times e^{-\frac{1}{2}Trace\left[\left(\frac{1}{\left[Q_0^{-1}\right]_{11}} v_1^T v_1 + M_0\right)\Sigma^{-1}\right]} \\ &\times e^{-\frac{1}{2}vec\left(V_1 - \bar{V}_1\right)^T\left(\Sigma^{-1}\otimes Q_1\right)vec\left(V_1 - \bar{V}_1\right)} \end{aligned}$$

The last term looks like a simple multivariate normal distribution, which motivates us to define the inverse variance-covariance matrix, which is a Kronecker product, as follows

$$\Sigma_C{}^{-1} = \Sigma^{-1} \otimes Q_1 \Rightarrow \Sigma_C = \Sigma \otimes Q_1^{-1} \qquad m(p-1) \times m(p-1) \quad .$$

The determinant theorem for Kronecker products is

$$\left|\Sigma \otimes Q_1^{-1}\right| = |\Sigma|^{p-1} \left|Q_1^{-1}\right|^m = \frac{|\Sigma|^{p-1}}{|Q_1|^m} \quad .$$

The multivariate normal distribution is written[16]

$$N = (2\pi)^{-\frac{k}{2}} |\Sigma_C|^{-\frac{1}{2}} e^{-\frac{1}{2}(x-\mu)^T \Sigma_C{}^{-1}(x-\mu)} \quad .$$

If we use the parameters of this problem $k = m(p-1)$ and the foregoing expression for the variance-covariance matrix, we can write the normal distribution

$$\begin{aligned} N &= (2\pi)^{-\frac{m(p-1)}{2}} \left|\Sigma \otimes Q_1^{-1}\right|^{-\frac{1}{2}} e^{-\frac{1}{2}(x-\mu)^T \left(\Sigma^{-1} \otimes Q_1\right)(x-\mu)} \\ &= (2\pi)^{-\frac{m(p-1)}{2}} \left(\frac{|\Sigma|^{p-1}}{|Q_1|^m}\right)^{-\frac{1}{2}} e^{-\frac{1}{2}(x-\mu)^T \left(\Sigma^{-1} \otimes Q_1\right)(x-\mu)} \\ &= \frac{|Q_1|^{\frac{m}{2}}}{(2\pi)^{\frac{m(p-1)}{2}} |\Sigma|^{\frac{p-1}{2}}} e^{-\frac{1}{2}(x-\mu)^T \left(\Sigma^{-1} \otimes Q_1\right)(x-\mu)} \quad . \end{aligned}$$

We want to rearrange this to express the exponential using the other terms in the equation

$$\begin{aligned} N &= \frac{|Q_1|^{\frac{m}{2}}}{(2\pi)^{\frac{m(p-1)}{2}} |\Sigma|^{\frac{p-1}{2}}} e^{-\frac{1}{2}(x-\mu)^T \left(\Sigma^{-1} \otimes Q_1\right)(x-\mu)} \\ \Rightarrow e^{-\frac{1}{2}(x-\mu)^T \left(\Sigma^{-1} \otimes Q_1\right)(x-\mu)} &= \frac{(2\pi)^{\frac{m(p-1)}{2}} |\Sigma|^{\frac{p-1}{2}}}{|Q_1|^{\frac{m}{2}}} N \quad . \end{aligned}$$

We can substitute this into the distribution and write

$$\begin{aligned} \{B, \Sigma\} &= \frac{|M_0|^{\frac{m_0-m-p-1}{2}} |Q_0|^{\frac{m}{2}}}{\pi^{\frac{pm}{2}} 2^{\frac{(m_0-m-1)m}{2}} \Gamma_m\left(\frac{m_0-m-p-1}{2}\right) |\Sigma|^{\frac{m_0}{2}}} \\ &\times e^{-\frac{1}{2} Trace\left[\left(\frac{1}{\left[Q_0^{-1}\right]_{11}} v_1^T v_1 + M_0\right)\Sigma^{-1}\right]} \\ &\times \frac{(2\pi)^{\frac{m(p-1)}{2}} |\Sigma|^{\frac{p-1}{2}}}{|Q_1|^{\frac{m}{2}}} N \end{aligned}$$

[16] See https://en.wikipedia.org/wiki/Multivariate_normal_distribution.

$$= \frac{|M_0|^{\frac{m_0-m-p-1}{2}}}{\pi^{\frac{m}{2}} 2^{\frac{m(m_0-m-p)}{2}} \Gamma_m\left(\frac{m_0-m-p-1}{2}\right) |\Sigma|^{\frac{m_0-p+1}{2}}} \frac{|Q_0|^{\frac{m}{2}}}{|Q_1|^{\frac{m}{2}}}$$
$$\times e^{-\frac{1}{2}Trace\left[\left(\frac{1}{\left[Q_0^{-1}\right]_{11}} v_1^T v_1 + M_0\right)\Sigma^{-1}\right]_N}$$

We need to simplify the determinants that involve the Q_0 and Q_1 matrices. To do so, we begin with the partitioned matrix determinant theorem in Appendix A in equation (A.1), taking account of the partitioning in equation (16.20). If we substitute the partitioned terms one by one into the general equation (A.1),

$$|A| = \left|A_{11} - A_{12}{A_{22}}^{-1}A_{21}\right| |A_{22}| \quad ,$$

from the specific equation (16.20), we obtain the relationship

$$|Q_0| = \left|q_{11} - {q_{p-1}}^T Q_1^{-1} q_{p-1}\right| |Q_1| \quad .$$

Because the first term on the right is a scalar, the determinant of the term is the term itself. Therefore

$$|Q_0| = \left(q_{11} - {q_{p-1}}^T Q_1^{-1} q_{p-1}\right) |Q_1| \quad .$$

We substitute the relationship in equation (16.21) to write

$$|Q_0| = \frac{1}{\left[Q_0^{-1}\right]_{11}} |Q_1| \quad .$$

We reorganize to write

$$|Q_1| = \left[Q_0^{-1}\right]_{11} |Q_0| \quad .$$

We raise to a power to write the solution that we can use to substitute

$$|Q_1|^{\frac{m}{2}} = \left(\left[Q_0^{-1}\right]_{11}\right)^{\frac{m}{2}} |Q_0|^{\frac{m}{2}} \quad .$$

We substitute this expression to eliminate determinants of Q from the distribution

$$\{B, \Sigma\} = \frac{|M_0|^{\frac{m_0-p-m-1}{2}}}{\pi^{\frac{m}{2}} \left(\left[Q_0^{-1}\right]_{11}\right)^{\frac{m}{2}} 2^{\frac{(m_0-m-p)m}{2}} \Gamma_m\left(\frac{m_0-p-m-1}{2}\right) |\Sigma|^{\frac{m_0-p+1}{2}}}$$
$$\times e^{-\frac{1}{2}Trace\left[\left(\frac{1}{\left[Q_0^{-1}\right]_{11}} v_1^T v_1 + M_0\right)\Sigma^{-1}\right]_N} \quad .$$

We integrate out the conditional normal distribution to get an expression for the joint distribution over B_{1D}, B_{1H} and Σ

$$\{B, \Sigma\} = \frac{|M_0|^{\frac{m_0-p-m-1}{2}}}{\pi^{\frac{m}{2}}\left(\left[Q_0^{-1}\right]_{11}\right)^{\frac{m}{2}} 2^{\frac{(m_0-m-p)m}{2}} \Gamma_m\left(\frac{m_0-p-m-1}{2}\right) |\Sigma|^{\frac{m_0-p+1}{2}}}$$
$$\times e^{-\frac{1}{2}Trace\left[\left(\frac{1}{\left[Q_0^{-1}\right]_{11}} v_1^T v_1 + M_0\right)\Sigma^{-1}\right]} \quad .$$

We need to consider the combined exponent of the final two terms, namely

$$\Psi = M_0 + \frac{1}{\left[Q_0^{-1}\right]_{11}} v_1^T v_1 \quad .$$

We notice in equation (16.22) that v_1 is a row vector and that

(16.23)

$$v_1^T v_1 = \Upsilon =$$
$$\begin{bmatrix} \left(B_{1D} - B_{1D}^0\right)^2 & \left(B_{1H} - B_{1H}^0\right)\left(B_{1D} - B_{1D}^0\right) \\ \left(B_{1H} - B_{1H}^0\right)\left(B_{1D} - B_{1D}^0\right) & \left(B_{1H} - B_{1H}^0\right)^2 \end{bmatrix} \quad .$$

$v_1^T v_1$ is $m \times m$. We can therefore write

$$\begin{aligned} \Psi &= M_0 + \frac{1}{\left[Q_0^{-1}\right]_{11}} v_1^T v_1 \\ &= M_0 + \frac{1}{\left[Q_0^{-1}\right]_{11}} \Upsilon \\ &= M_0 \times \left(I_{mm} + \frac{M_0{}^{-1}}{\left[Q_0^{-1}\right]_{11}} \Upsilon\right) \quad , \end{aligned}$$

and substitute to write the full joint density

$$\{B_{1D}, B_{1H}, \Sigma\} = \frac{|M_0|^{\frac{m_0-p-m-1}{2}}}{\pi^{\frac{m}{2}}\left(\left[Q_0^{-1}\right]_{11}\right)^{\frac{m}{2}} 2^{\frac{(m_0-m-p)m}{2}} \Gamma_m\left(\frac{m_0-p-m-1}{2}\right)}$$
$$\times \frac{1}{|\Sigma|^{\frac{m_0-p+1}{2}}} e^{-\frac{1}{2}Trace\left(\Psi\Sigma^{-1}\right)} \quad .$$

We write the inverse Wishart distribution[17] in the following form

(16.24) $$f(\Sigma) = \frac{|\Psi|^{\frac{\xi}{2}}}{2^{\frac{\xi m}{2}} \Gamma_m\left(\frac{\xi}{2}\right) |\Sigma|^{\frac{\xi+m+1}{2}}} e^{-\frac{1}{2}Trace\left(\Psi\Sigma^{-1}\right)} \quad ,$$

[17] See https://en.wikipedia.org/wiki/Inverse-Wishart_distribution.

in which ξ denotes the degrees of freedom parameter. Our distribution just prior to the previous equation will be an inverse Wishart with the same degrees of freedom if we equate the exponents of the $|\Sigma|$ term in the denominators as follows:

$$\frac{\xi+m+1}{2}=\frac{m_0-p+1}{2} \quad .$$

This occurs when

$$\begin{aligned}
&\xi = m_0 - p - m \\
&\Rightarrow m_0 = \xi + p + m \quad .
\end{aligned}$$

If we substitute for m_0 in the previous density, we can write

$$\begin{aligned}
\{B_{1D}, B_{1H}, \Sigma\} &= \frac{|M_0|^{\frac{\xi-1}{2}}}{\pi^{\frac{m}{2}}\left(\left[Q_0^{-1}\right]_{11}\right)^{\frac{m}{2}} 2^{\frac{\xi m}{2}}\Gamma_m\left(\frac{\xi-1}{2}\right)} \\
&\times \frac{1}{|\Sigma|^{\frac{\xi+m+1}{2}}} e^{-\frac{1}{2}Trace\left(\Psi\Sigma^{-1}\right)} \quad .
\end{aligned}$$

$\{B_{1D}, B_{1H}\}$ is $1 \times m$; Σ is $m \times m$. We now see exactly how to identify the inverse Wishart distribution within our distribution. If we reverse the inverse Wishart distribution in equation (16.24), we can write

$$\frac{1}{|\Sigma|^{\frac{\xi+m+1}{2}}} e^{-\frac{1}{2}Trace\left(T_0\Sigma^{-1}\right)} = f(\Sigma)\,\frac{2^{\frac{\xi m}{2}}\Gamma_m\left(\frac{\xi}{2}\right)}{|\Psi|^{\frac{\xi}{2}}} \quad .$$

We can insert this into the density and simplify to write

$$\begin{aligned}
\{B_{1D}, B_{1H}, \Sigma\} &= \frac{|M_0|^{\frac{m_0-p-m-1}{2}}}{\left(\pi\left[Q_0^{-1}\right]_{11}\right)^{\frac{m}{2}} 2^{\frac{(m_0-m-p)m}{2}}\Gamma_m\left(\frac{m_0-p-m-1}{2}\right)} \\
&\times f(\Sigma)\,\frac{2^{\frac{\xi m}{2}}\Gamma_m\left(\frac{\xi}{2}\right)}{|\Psi|^{\frac{\xi}{2}}} \\
&= \frac{|M_0|^{\frac{m_0-p-m-1}{2}}\Gamma_m\left(\frac{\xi}{2}\right)}{\left(\pi\left[Q_0^{-1}\right]_{11}\right)^{\frac{m}{2}} 2^{\frac{(m_0-m-p-\xi)m}{2}}\Gamma_m\left(\frac{m_0-p-m-1}{2}\right)} |\Psi|^{-\frac{\xi}{2}} f(\Sigma) \quad .
\end{aligned}$$

We obtain the marginal distribution over the two model coefficients B_{1D}, B_{1H} as described above by integrating out the inverse Wishart

$$\{B_{1D}, B_{1H}\} = \frac{|M_0|^{\frac{m_0-p-m-1}{2}}\Gamma_m\left(\frac{\xi}{2}\right)}{\left(\pi\left[Q_0^{-1}\right]_{11}\right)^{\frac{m}{2}} 2^{\frac{(m_0-m-p-\xi)m}{2}}\Gamma_m\left(\frac{m_0-p-m-1}{2}\right)} |\Psi|^{-\frac{\xi}{2}} \quad .$$

$\{B_{1D}, B_{1H}\}$ is $1 \times m$. We now substitute for $\xi = m_0 - p - m$ and simplify

$$\{B_{1D}, B_{1H}\} = \frac{|M_0|^{\frac{m_0-p-m-1}{2}} \Gamma_m\left(\frac{m_0-p-m}{2}\right)}{\left(\pi\left[Q_0^{-1}\right]_{11}\right)^{\frac{m}{2}} \Gamma_m\left(\frac{m_0-p-m-1}{2}\right)} |\Psi|^{-\frac{m_0-p-m}{2}} .$$

We finally substitute for the degrees of freedom $\nu_0 = m_0 - p - m - 1$ to get the marginal distribution

$$\{B_{1D}, B_{1H}\} = \frac{|M_0|^{\frac{\nu_0}{2}} \Gamma_m\left(\frac{\nu_0+1}{2}\right)}{\left(\pi\left[Q_0^{-1}\right]_{11}\right)^{\frac{m}{2}} \Gamma_m\left(\frac{\nu_0}{2}\right)} |\Psi|^{-\frac{\nu_0+1}{2}} .$$

We substitute the definition of the matrix Ψ, using the representation of Υ from equation (16.23) to write

$$\begin{aligned}\{B_{1D}, B_{1H}\} &= \frac{|M_0|^{\frac{\nu_0}{2}} \Gamma_m\left(\frac{\nu_0+1}{2}\right)}{\left(\pi\left[Q_0^{-1}\right]_{11}\right)^{\frac{m}{2}} \Gamma_m\left(\frac{\nu_0}{2}\right)} |M_0|^{-\frac{\nu_0+1}{2}} \\ &\quad \times \left| I_{mm} + \frac{M_0^{-1}}{\left[Q_0^{-1}\right]_{11}} \Upsilon \right|^{-\frac{\nu_0+1}{2}} \\ &= \frac{\Gamma_m\left(\frac{\nu_0+1}{2}\right)}{\left(\pi\left[Q_0^{-1}\right]_{11}\right)^{\frac{m}{2}} |M_0|^{\frac{1}{2}} \Gamma_m\left(\frac{\nu_0}{2}\right)} \\ &\quad \times \left| I_{mm} + \frac{M_0^{-1}}{\left[Q_0^{-1}\right]_{11}} \Upsilon \right|^{-\frac{\nu_0+1}{2}} .\end{aligned}$$

We adjust the M_0 term and write the joint probability distribution for the $m = 2$ variables for each of the p independent variables

$$\begin{aligned}\{B_{1D}, B_{1H}\} &= \frac{\Gamma_m\left(\frac{\nu_0+1}{2}\right)}{\left(\pi\left[Q_0^{-1}\right]_{11} |M_0|^{\frac{1}{m}}\right)^{\frac{m}{2}} \Gamma_m\left(\frac{\nu_0}{2}\right)} \\ &\quad \times \left| I_{mm} + \frac{M_0^{-1}}{\left[Q_0^{-1}\right]_{11}} \Upsilon \right|^{-\frac{\nu_0+1}{2}} .\end{aligned}$$

$\{B_{1D}, B_{1H}\}$ is $1 \times m$. The analogy to equation (5.18) for the univariate case becomes apparent, helping us be confident that this analysis is correct. This is the simplest representation that is analytic—the m-variate joint density over each of the p values of X. It makes sense when one ponders it. The fundamental distribution we used in the formulation of the error term was an m-variate joint distribution for each of the p coefficients in the model.

We need to analyze the determinant in this equation, specifically applying the Weinstein-Aronszajn identity,[18] to the determinant. We make the following to the determinant term

$$\begin{aligned}
&\left| I_{mm} + \frac{M_0^{-1}}{\left[Q_0^{-1}\right]_{11}} \Upsilon \right| \\
&= \left| I_{mm} + \frac{M_0^{-1}}{\left[Q_0^{-1}\right]_{11}} v_1^T v_1 \right| \\
&= \left| I_{mm} + \frac{1}{\left[Q_0^{-1}\right]_{11}} M_0^{-1} v_1^T v_1 \right| \\
&= \left| I_{mm} + \frac{1}{\left[Q_0^{-1}\right]_{11}} \left(M_0^{-1} v_1^T\right) v_1 \right| \quad .
\end{aligned}$$

The matrix is $m \times m$. We notice the dimensions of the two matrices in the final determinant

$$\begin{aligned}
&v_1 = (\beta_2 - \mu_2)^T \text{ is } 1 \times m \\
&M_0^{-1} v_1^T = M_0^{-1}(\beta_2 - \mu_2) \text{ is } m \times 1 \quad .
\end{aligned}$$

This means we can apply the Weinstein-Aronszajn identity to express the determinant differently as follows. The re-expression of the determinant becomes

$$\begin{aligned}
&\left| I_{11} + \frac{1}{\left[Q_0^{-1}\right]_{11}} v_1 \left(M_0^{-1} v_1^T\right) \right| \\
&= \left| 1 + \frac{1}{\left[Q_0^{-1}\right]_{11}} v_1 M_0^{-1} v_1^T \right| \\
&= \left[1 + \frac{1}{\left[Q_0^{-1}\right]_{11}} v_1 M_0^{-1} v_1^T \right] \quad .
\end{aligned}$$

The matrix argument of the determinants now 1×1, a scalar. The argument of the determinant is no longer actually a matrix. It is a determinant with a scalar argument. This dramatically simplifies things, and is quite intuitive as we shall see. It means we can eliminate the determinant in favor of a scalar. We substitute this back into the original equation above to write the marginal multivariate distribution for row 1 in the form of a multivariate Student's t distribution, for which we can formulate the constant

$$\{B_{1D}, B_{1H}\} = \frac{\Gamma_m\left(\frac{\nu_0+1}{2}\right)}{\Gamma_m\left(\frac{\nu_0}{2}\right)\left(\pi\left[Q_0^{-1}\right]_{11}\right)^{\frac{m}{2}} |M_0|^{\frac{1}{2}}}$$

[18]See https://en.wikipedia.org/wiki/Weinstein-Aronszajn_identity.

$$\times \left[1+\frac{1}{\left[Q_0^{-1}\right]_{11}} v_1 M_0^{-1} v_1^T\right]^{-\frac{\nu_0+1}{2}}$$

$$= c_0 \left[1+v_1 \frac{M_0^{-1}}{\left[Q_0^{-1}\right]_{11}} v_1^T\right]^{-\frac{\nu_0+1}{2}} \quad .$$

$\{B_{1D}, B_{1H}\}$ is $1 \times m$. It is insightful to re-develop the constant of integration afresh in order that the constant be intrinsically consistent with the exponent term in the multivariate Student's t distribution. If we define a degrees of freedom parameter $\nu = \nu_0+1-m$, which is consistent with the exponent on the quadratic, we will be able to rewrite the foregoing multivariate density

(16.25)

$$\{B_{1D}, B_{1H}\} = c_0 \left[1+v \frac{M_0^{-1}}{\left[Q_0^{-1}\right]_{11}} v^T\right]^{-\frac{\nu+m}{2}}$$

$$= c_0 \left\{1+\left[B_{1D}-B_{1D}^0, B_{1H}-B_{1H}^0\right] \frac{M_0^{-1}}{\left[Q_0^{-1}\right]_{11}} \begin{bmatrix} B_{1D}-B_{1D}^0 \\ B_{1H}-B_{1H}^0 \end{bmatrix}\right\}^{-\frac{\nu+m}{2}}$$

$$= c_0 \left\{1+\left(B_1-B_1^0\right)^T \frac{M_0^{-1}}{\left[Q_0^{-1}\right]_{11}} \left(B_1-B_1^0\right)\right\}^{-\frac{\nu+m}{2}} \quad .$$

We are going to redevelop the constant of integration with this in mind. This is looking very much like a multivariate Student's t distribution, which is given in Wiki.[19] For clarity. We begin with the m-variate distribution

$$\{B_1\} = \frac{\Gamma\left(\frac{\nu+m}{2}\right)}{\Gamma\left(\frac{\nu}{2}\right) \nu^{\frac{m}{2}} \pi^{\frac{m}{2}} |\Sigma|^{\frac{1}{2}}} \left[1+\frac{1}{\nu}\left(B_1-B_1^0\right)^T \Sigma^{-1}\left(B_1-B_1^0\right)\right]^{-\frac{\nu+m}{2}} \quad .$$

The variance is

$$var = \frac{\nu}{\nu-2} \Sigma \quad .$$

This suggests that we make the substitution

$$\Sigma^{-1} = s^{-2}\Theta$$
$$\Rightarrow \Sigma = s^2 \Theta^{-1}$$
$$\Rightarrow |\Sigma|^{\frac{1}{2}} = s^{\frac{2m}{2}} |\Theta|^{-\frac{1}{2}}$$

to write

[19] https://en.wikipedia.org/wiki/Multivariate_t-distribution

$$(16.26)\quad \{B_1\} = \frac{\Gamma\left(\frac{\nu+m}{2}\right)|\Theta|^{\frac{1}{2}}}{\Gamma\left(\frac{\nu}{2}\right)\left(\pi\nu s^2\right)^{\frac{m}{2}}}\left[1+\left(B_1-B_1^0\right)^T\frac{\Theta}{\nu s^2}\left(B_1-B_1^0\right)\right]^{-\frac{\nu+m}{2}}$$

and

$$(16.27)\qquad VCV = \frac{\nu s^2}{\nu-2}\Theta^{-1} \quad .$$

Keep squarely in mind that the matrix Θ is m by m.

For this m-variate multivariate distribution, the m-variate distribution over the m outputs, the jth output is governed by the jth marginal distribution. (We substitute the proper degrees of later):

$$(16.28)\qquad \{x\} = \frac{\Gamma\left(\frac{\nu+1}{2}\right)}{\left(\pi\nu s^2[\Theta^{-1}]_{jj}\right)^{\frac{1}{2}}\Gamma\left(\frac{\nu}{2}\right)} \left[1+(x-\mu)^T\frac{1}{\nu s^2[\Theta^{-1}]_{jj}}(x-\mu)\right]^{-\frac{\nu+1}{2}} \quad .$$

The vector x is 1 by 1, a scalar. The variance of this univariate distribution is

$$(16.29)\qquad variance = \frac{\nu s^2}{\nu-2}z_{jj} = \frac{\nu s^2}{\nu-2}\left[\Theta^{-1}\right]_{jj} \quad .$$

We will return to this univariate case shortly.

Comparison of equation (16.25) with equation (16.26) suggests the following substitutions in equation (16.25)

$$\begin{aligned}
\frac{M_0^{-1}}{\left[Q_0^{-1}\right]_{11}} &= \frac{\Theta}{\nu s^2}\\
\Rightarrow \Theta &= \frac{\nu s^2}{\left[Q_0^{-1}\right]_{11}}M_0^{-1}\\
\Rightarrow \Theta^{-1} &= \frac{\left[Q_0^{-1}\right]_{11}}{\nu s^2}M_0\\
\Rightarrow |\Theta|^{\frac{1}{2}} &= \frac{\left(\nu s^2\right)^{\frac{m}{2}}}{\left(\left[Q_0^{-1}\right]_{11}\right)^{\frac{m}{2}}}|M_0|^{-\frac{1}{2}}\\
\Rightarrow \left[\Theta^{-1}\right]_{jj} &= \frac{\left[Q_0^{-1}\right]_{11}[M_0]_{jj}}{\nu s^2}
\end{aligned}$$

Making these substitutions, equation (16.26) becomes

$$\{B_1\} = \frac{\Gamma\left(\frac{\nu+m}{2}\right)}{\Gamma\left(\frac{\nu}{2}\right)\left(\pi\left[Q_0^{-1}\right]_{11}\right)^{\frac{m}{2}}|M_0|^{\frac{1}{2}}} \times \left[1+\left(B_1-B_1^0\right)^T \frac{M_0^{-1}}{\left[Q_0^{-1}\right]_{11}}\left(B_1-B_1^0\right)\right]^{-\frac{\nu+m}{2}} .$$

Substituting the degrees of freedom $\nu = \nu_0 + 1 - m$ gives

$$\{B_1\} = \frac{\Gamma\left(\frac{\nu_0+1}{2}\right)}{\Gamma\left(\frac{\nu_0+1-m}{2}\right)\left(\pi\left[Q_0^{-1}\right]_{11}\right)^{\frac{m}{2}}|M_0|^{\frac{1}{2}}} \times \left[1+\left(B_1-B_1^0\right)^T \frac{M_0^{-1}}{\left[Q_0^{-1}\right]_{11}}\left(B_1-B_1^0\right)\right]^{-\frac{\nu_0+1}{2}}$$

Equation (16.27) becomes

$$VCV = \frac{\left[Q_0^{-1}\right]_{11}}{\nu-2} M_0 .$$

Substituting the degrees of freedom $\nu = \nu_0 + 1 - m$ gives

$$VCV = \frac{\left[Q_0^{-1}\right]_{11}}{\nu_0-m-1} M_0$$

Equation (16.28) becomes

$$\{B_{1j}\} = \frac{\Gamma\left(\frac{\nu+1}{2}\right)}{\left(\pi\left[Q_0^{-1}\right]_{11}[M_0]_{jj}\right)^{\frac{1}{2}}\Gamma\left(\frac{\nu}{2}\right)} \times \left[1+\left(B_{1j}-B_{1j}^0\right)^T \frac{1}{\left[Q_0^{-1}\right]_{11}[M_0]_{jj}}\left(B_{1j}-B_{1j}^0\right)\right]^{-\frac{\nu+1}{2}} .$$

Substituting the degrees of freedom $\nu = \nu_0 + 1 - m$ gives

$$\{B_{1j}\} = \frac{\Gamma\left(\frac{\nu_0+2-m}{2}\right)}{\left(\pi\left[Q_0^{-1}\right]_{11}[M_0]_{jj}\right)^{\frac{1}{2}}\Gamma\left(\frac{\nu_0+1-m}{2}\right)} \times \left[1+\left(B_{1j}-B_{1j}^0\right)^T \frac{1}{\left[Q_0^{-1}\right]_{11}[M_0]_{jj}}\left(B_{1j}-B_{1j}^0\right)\right]^{-\frac{\nu_0+2-m}{2}}$$

Equation (16.29) becomes

$$variance = \frac{[Q_0^{-1}]_{11}[M_0]_{jj}}{\nu - 2} \quad .$$

Substituting the degrees of freedom $\nu = \nu_0 + 1 - m$ gives

$$variance = \frac{[Q_0^{-1}]_{11}[M_0]_{jj}}{\nu_0 - m - 1}$$

16.2.2 Analysis of the Conjugate Prior–Mean of the Marginal Distribution Over the Model Coefficients

We must return to the functional form for the prior and calculate the mean or expected value of the model coefficients. Just as with the univariate case, we must make a change of variables from the inverse Wishart to the Wishart and then calculate the mean. To do so, we are going to have to find a transformation of variables and find its Jacobian. We begin by again rewriting the inverse Wishart distribution from Wiki[20]

$$\Sigma \sim W^{-1}(\Sigma; \Psi, \nu, m) = \frac{|\Psi|^{\frac{\nu}{2}}}{2^{\frac{\nu m}{2}}\Gamma_m\left(\frac{\nu}{2}\right)}|\Sigma|^{-\frac{\nu+m+1}{2}} e^{-\frac{1}{2}Trace\left(\Psi\Sigma^{-1}\right)} \quad .$$

We are going to want to do a "change of variables" substitution in this integral that will change it from an inverse Wishart distribution to a Wishart distribution. The reason is that it is easier to calculate expectations and higher-order moments in a Wishart than in an inverse Wishart. Dr. von Rosen[101] figured this out in his insightful paper. This will facilitate our mean and variance-covariance developments. This is the matrix analogy of what we did in the univariate case to change from an inverse gamma distribution to a gamma distribution. To effect the change of variables we desire, we first merely substitute Σ everywhere in the inverse Wishart equation. The substitution we want is based on the observation in the inverse Wishart that the argument of the trace is $\Psi\Sigma^{-1}$. Its inverse is $\Sigma\Psi^{-1}$, implying the substitution we want to use is

$$\begin{aligned}
&\Psi\Sigma^{-1} = H\Psi^{-1} \\
&\Rightarrow \Psi^{-1}\Psi\Sigma^{-1} = \Psi^{-1}H\Psi^{-1} \\
&\Rightarrow \Sigma^{-1} = \Psi^{-1}H\Psi^{-1} \\
&\Rightarrow I = \Sigma\Psi^{-1}H\Psi^{-1} \\
&\Rightarrow \Sigma = \Psi H^{-1}\Psi \quad .
\end{aligned}$$

We want to substitute for Σ in the inverse Wishart equation in order to obtain a Wishart distribution, and we note that we will need the Jacobian J of the transformation from the inverse Wishart distribution to the Wishart distribution, yet-

[20] https://en.wikipedia.org/wiki/Inverse-Wishart_distribution.

to-be-determined, in order to effect the change of variables from the inverse to the non-inverse. We will infer the specific form of that Jacobian from inverse Wishart to Wishart shortly. Following is the direct substitution of the foregoing transformation into the inverse Wishart equation above, with a placeholder for the not-yet-determined Jacobian from the inverse Wishart to the Wishart:

$$f(H) = \frac{|H|^{\frac{\nu+m+1}{2}}}{2^{\frac{\nu m}{2}}\Gamma_m\left(\frac{\nu}{2}\right)|\Psi|^{\frac{\nu}{2}+m+1}} e^{-\frac{1}{2}Trace\left(H\Psi^{-1}\right)} J \quad .$$

We need to manipulate this equation, determining J in the process, to become a true Wishart distribution over H. The easiest way we can do so is to appeal to the published Wishart distribution and infer the true and correct Jacobian from that. The published Wishart distribution is given in Wiki[21] to be

$$X \sim W\left(V, \nu, m\right) = \frac{1}{2^{\frac{\nu m}{2}}\Gamma_m\left(\frac{\nu}{2}\right)} \frac{|X|^{\frac{\nu-m-1}{2}}}{|V|^{\frac{\nu}{2}}} e^{-\frac{1}{2}Trace\left(V^{-1}X\right)} \quad .$$

If we substitute H for X and Ψ for V, we obtain a Wishart using our variables

$$H \sim \left(H; \Psi, \nu, m\right) = \frac{1}{2^{\frac{\nu m}{2}}\Gamma_m\left(\frac{\nu}{2}\right)} \frac{|H|^{\frac{\nu-m-1}{2}}}{|\Psi|^{\frac{\nu}{2}}} e^{-\frac{1}{2}Trace\left(\Psi^{-1}H\right)} \quad .$$

We should point out that in the trace operation on the right, it does not matter in which order the matrix multiplication occurs. We write the two Wishart equations, one superposed over the other, in preparation for equating them to find the Jacobian

$$\begin{aligned} f(H) &= \frac{1}{2^{\frac{\nu m}{2}}\Gamma_m\left(\frac{\nu}{2}\right)} \frac{|H|^{\frac{\nu+m+1}{2}}}{|\Psi|^{\frac{\nu}{2}+m+1}} e^{-\frac{1}{2}Trace\left(H\Psi^{-1}\right)} J \\ H \sim W\left(\Psi, \nu, m\right) &= \frac{1}{2^{\frac{\nu m}{2}}\Gamma_m\left(\frac{\nu}{2}\right)} \frac{|H|^{\frac{\nu-m-1}{2}}}{|\Psi|^{\frac{\nu}{2}}} e^{-\frac{1}{2}Trace\left(\Psi^{-1}H\right)} \quad . \end{aligned}$$

If we equate the two and solve for J, we obtain the equation for the Jacobian of the transformation from the matrix $\Sigma^{-1}\Psi$ to the matrix $H\Psi^{-1}$, taking account of the symmetric nature of the matrices in the distributions:

$$J = \frac{|\Psi|^{m+1}}{|H|^{m+1}} \quad .$$

If we substitute this Jacobian into the first of the foregoing equations, we obtain the second, which is a standard Wishart. We inferred this Jacobian by comparing a published Wishart with our substituted inverse Wishart because it is easier and more direct than deriving the Jacobian directly from matrix algebra. This Jaco-

[21] https://en.wikipedia.org/wiki/Wishart_distribution

bian and the foregoing change of variables are going to prove very convenient for all of our future developments in this section. We can derive this Jacobian directly as well, but this is the easiest way. (Symmetric matrices yield a power of $m+1$ for the Jacobian. General matrices yield the power $2m$ for the Jacobian. Anderson[3] verifies this.)

16.2.3 Analysis of the Conjugate Prior–The Mean Values of the Coefficients

We rewrite the joint density as the product of an inverse Wishart times a matrix normal as derived in equation (16.15) and in which we use the Kronecker product in place of Σ_C

$$\begin{aligned}\{B,\Sigma\} &= \frac{|M_0|^{\frac{\nu_0}{2}}}{2^{\frac{\nu_0 m}{2}}\Gamma_m\left(\frac{\nu_0}{2}\right)|\Sigma|^{\frac{\nu_0+m+1}{2}}}e^{-\frac{1}{2}Trace\left(M_0\Sigma^{-1}\right)} \\ &\times N\left[vec\left(B\right);vec\left(B_0\right),\Sigma\otimes {Q_0}^{-1},mp\right] \quad .\end{aligned}$$

We now calculate the mean value of the coefficients by setting up the integral equation for the mean, which is $vec(B)$ times the immediately foregoing density expressed in vectorized form using the embedded multivariate normal

$$\begin{aligned}\langle vec\left(B\right)\rangle &= \int d\Sigma\int dB\frac{|M_0|^{\frac{\nu_0}{2}}}{2^{\frac{\nu_0 m}{2}}\Gamma_m\left(\frac{\nu_0}{2}\right)|\Sigma|^{\frac{\nu_0+m+1}{2}}}e^{-\frac{1}{2}Trace\left(M_0\Sigma^{-1}\right)} \\ &\times vec\left(B\right)N\left[vec\left(B\right);vec\left(B_0\right),\Sigma\otimes {Q_0}^{-1},mp\right] \quad .\end{aligned}$$

$vec\left(B\right)$ is $1\times mp$. The only occurrence of the variable B is in the final line in the foregoing equation, the line containing the normal term. That means that the final term integrates to the mean of the normal distribution, so therefore

$$\langle vec\left(B\right)\rangle = vec\left(B_0\right)\int d\Sigma\frac{|M_0|^{\frac{\nu_0}{2}}}{2^{\frac{\nu_0 m}{2}}\Gamma_m\left(\frac{\nu_0}{2}\right)|\Sigma|^{\frac{\nu_0+m+1}{2}}}e^{-\frac{1}{2}Trace\left(M_0\Sigma^{-1}\right)} \quad .$$

The expression inside the integral is the standard inverse Wishart, which integrates to 1. Therefore

$$\langle vec\left(B\right)\rangle = vec\left(B_0\right) \quad .$$

This implies the following expression for the mean

$$\langle B\rangle = B_0 \quad . \tag{16.30}$$

16.2.4 Analysis of the Conjugate Prior–Variance-Covariance Matrix of the Marginal Distribution Over the Model Coefficients

We now turn to the variance-covariance matrix over the model coefficients. The variance-covariance matrix is defined as the expectation of an "outer product" of a column vector against itself with respect to the inverse Wishart distribution. The integrand is the inverse Wishart distribution times the outer product; it is the expectation using the inverse Wishart distribution of the outer product:

$$\begin{aligned} VCV = &\int dB \int d\Sigma \frac{|M_0|^{\frac{\nu_0}{2}}}{2^{\frac{\nu_0 m}{2}} \Gamma_m\left(\frac{\nu_0}{2}\right) |\Sigma|^{\frac{\nu_0+m+1}{2}}} e^{-\frac{1}{2}Trace\left(M_0\Sigma^{-1}\right)} \\ &\times \left[vec(B) - vec(B_0)\right]\left[vec(B) - vec(B_0)\right]^T \\ &\times N\left[vec(B); vec(B_0), \Sigma \otimes {Q_0}^{-1}, mp\right] \quad . \end{aligned}$$

VCV is $mp \times mp$. We know that the "outer product" representation of the variance-covariance integrand times the multivariate normal distribution must integrate to the variance-covariance matrix of the normal itself, which is the Kronecker product $\Sigma \otimes Q_0^{-1}$. This occurs because the B terms have been isolated in the normal equation; they occur nowhere else. Therefore, they can be integrated out, and the equation for the variance-covariance matrix immediately simplifies to

$$VCV = \int d\Sigma \frac{|M_0|^{\frac{\nu_0}{2}}}{2^{\frac{\nu_0 m}{2}} \Gamma_m\left(\frac{\nu_0}{2}\right) |\Sigma|^{\frac{\nu_0+m+1}{2}}} e^{-\frac{1}{2}Trace\left(M_0\Sigma^{-1}\right)} \left(\Sigma \otimes {Q_0}^{-1}\right) \quad .$$

The integral equation that remains is the expectation of the Kronecker product with respect to the inverse Wishart, a rather simple problem to pose. To solve this integral equation, we want to make similar substitutions as previously to change variables from the inverse Wishart to the Wishart inside the integrand. Once we do so, we will be solving for the means of terms under the Wishart distribution. The specific substitutions to be made in this variance-covariance calculation, which are very similar to previous substitutions, are

$$\begin{aligned} &M_0\Sigma^{-1} = H{M_0}^{-1} \\ &\Rightarrow {M_0}^{-1}M_0\Sigma^{-1} = {M_0}^{-1}H{M_0}^{-1} \\ &\Rightarrow \Sigma^{-1} = {M_0}^{-1}H{M_0}^{-1} \\ &\Rightarrow \Sigma = M_0H^{-1}M_0 \\ &\Rightarrow I = \Sigma{M_0}^{-1}H{M_0}^{-1} \quad . \end{aligned}$$

If we proceed through the same logic we used between the inverse Wishart and the Wishart, we will show that the Jacobian is precisely analogous, namely

$$J = \frac{|M_0|^{m+1}}{|H|^{m+1}} \quad .$$

Would the Jacobian merely substitute M_0 for Ψ and all the results go through with that substitution? The answer is yes. We first make the substitution for Σ, and we thereafter insert the Jacobian

$$\begin{aligned} VCV &= \int d\Sigma \frac{|M_0|^{\frac{\nu_0}{2}}}{2^{\frac{\nu_0 m}{2}} \Gamma_m\left(\frac{\nu_0}{2}\right) |M_0 H^{-1} M_0|^{\frac{\nu_0+m+1}{2}}} e^{-\frac{1}{2}Trace\left(HM_0{}^{-1}\right)} \\ &\times \left(M_0 H^{-1} M_0 \otimes Q_0{}^{-1}\right) \\ &= \int dH \frac{|H|^{\frac{\nu_0+m+1}{2}}}{2^{\frac{\nu_0 m}{2}} \Gamma_m\left(\frac{\nu_0}{2}\right) |M_0|^{\frac{\nu_0}{2}+m+1}} e^{-\frac{1}{2}Trace\left(HM_0{}^{-1}\right)} J \\ &\times \left(M_0 H^{-1} M_0 \otimes Q_0{}^{-1}\right) \quad . \end{aligned}$$

We apply the Jacobian and simplify

$$\begin{aligned} VCV &= \int dH \frac{|H|^{\frac{\nu_0+m+1}{2}}}{2^{\frac{\nu_0 m}{2}} \Gamma_m\left(\frac{\nu_0}{2}\right) |M_0|^{\frac{\nu_0}{2}+m+1}} e^{-\frac{1}{2}Trace\left(HM_0{}^{-1}\right)} \\ &\times \left(M_0 H^{-1} M_0 \otimes Q_0{}^{-1}\right) \frac{|M_0|^{m+1}}{|H|^{m+1}} \\ &= \int dH \frac{|H|^{\frac{\nu_0-m-1}{2}}}{2^{\frac{\nu_0 m}{2}} \Gamma_m\left(\frac{\nu_0}{2}\right) |M_0|^{\frac{\nu_0}{2}}} e^{-\frac{1}{2}Trace\left(HM_0{}^{-1}\right)} \left(M_0 H^{-1} M_0 \otimes Q_0{}^{-1}\right) \quad . \end{aligned}$$

The distribution inside the integral is the Wishart distribution using the parameters of the prior, written here for clarity

$$W\left(H; M_0, \nu_0, m\right) = \frac{|H|^{\frac{\nu_0-m-1}{2}}}{2^{\frac{\nu_0 m}{2}} \Gamma_m\left(\frac{\nu_0}{2}\right) |M_0|^{\frac{\nu_0}{2}}} e^{-\frac{1}{2}Trace\left(HM_0{}^{-1}\right)} \quad .$$

So the foregoing equation that calculates the variance-covariance matrix VCV is no more than the expectation under the Wishart distribution of the Kronecker product matrix at the right in the previous equation

$$VCV = \int dH \left(M_0 H^{-1} M_0 \otimes Q_0{}^{-1}\right) W\left(H; M_0, \nu_0, m\right) \quad .$$

The Wishart distribution is the following

$$H \sim W\left(H, M_0, \nu_0, m\right) \tag{16.31}$$

$$= \frac{1}{2^{\frac{\nu_0 m}{2}} \Gamma_m\left(\frac{\nu_0}{2}\right)} \frac{|H|^{\frac{\nu_0 - m - 1}{2}}}{|M_0|^{\frac{\nu_0}{2}}} e^{-\frac{1}{2} Trace\left(M_0{}^{-1} H\right)} \quad .$$

It is insightful to consider that this Wishart distribution is a scalar value, and it is therefore a constant with regard to the first term of the Kronecker product (by the theorem that says you can associate the constant with either matrix in the Kronecker product) and therefore with regard to the leading matrix. We can move the constant to the right in the product of three matrices at the left and associate it with the second term because it is a constant. Therefore, we can bring in the Wishart and associate it with the inverse of the matrix H, as indicated in the following equation. It is clear through this mechanism that one can bring the integral Wishart distribution into the Kronecker product as follows:

$$VCV = M_0 \left[\int dH H^{-1} W\left(H; M_0, \nu_0, m\right) \right] M_0 \otimes Q_0{}^{-1} \quad . \tag{16.32}$$

The variance-covariance matrix of the coefficients relies on the expectation of an inverse matrix under a standard Wishart distribution. Von Rosen,[101] in his outstanding paper, generalized this to higher moments, a result we will observe shortly. We have to take a technical detour in order to be able to solve this. It is clear from the structure of equation (16.32) that we need the expectation of H^{-1} with respect to the Wishart distribution.

16.2.4.1 Matrix Derivative of a Wishart Distribution

This section proceeds through the mathematics of establishing the matrix derivative of a Wishart distribution. See Rogers[85] for a discussion of matrix derivatives, a discussion we specialize in our own Appendix C. We use the specific definition of the matrix derivative in Appendix C, equation (C.1), which gives a detailed and intuitive explanation of this particular matrix derivative. We begin using the Kronecker matrix derivative, Type 1, defined in Appendix C. This is the same matrix derivative that was introduced in equation (2.2) in von Rosen[101]. He terms it a rescaled matrix derivative, and Appendix C makes it clear what we mean by the notion of rescaling. Understanding this matrix derivative is going to be necessary background for calculating the mean and variance-covariance matrix of the prior (and later, by analogy, the posterior). Let us begin with the matrix derivative of the Wishart distribution. The Wishart distribution is from equation (16.31)

$$W(H) = c|M_0|^{-\frac{\nu_0}{2}} |H|^{\frac{\nu_0 - m - 1}{2}} e^{-\frac{1}{2} Trace(M_0^{-1} H)} \quad ,$$

in which

$$c = \frac{1}{2^{\frac{\nu_0 m}{2}} \Gamma_m(\frac{\nu_0}{2})} \quad .$$

Taking the matrix derivative of the Wishart distribution, we write

$$\frac{\partial W(H)}{\partial H} = c|M_0|^{-\frac{\nu_0}{2}} \frac{\partial}{\partial H} \left[|H|^{\frac{\nu_0 - m - 1}{2}} e^{-\frac{1}{2} Trace(M_0^{-1} H)} \right] \quad .$$

Inside the brackets, we have a scalar determinant on the left and a scalar exponential on the right. We can apply the product rule for two scalar functions[22], which is quite analogous to the univariate case. Applying the product rule, we have:

$$\frac{\partial W(H)}{\partial H} = c|M_0|^{-\frac{\nu_0}{2}} \left\{ \begin{array}{l} \frac{\partial |H|^{\frac{\nu_0 - m - 1}{2}}}{\partial H} e^{-\frac{1}{2} Trace(M_0^{-1} H)} \\ +|H|^{\frac{\nu_0 - m - 1}{2}} \frac{\partial \left[e^{-\frac{1}{2} Trace(M_0^{-1} H)} \right]}{\partial H} \end{array} \right\} \quad .$$

We must apply the matrix derivative of a determinant raised to a power to simplify the top term in the braces. This derivative appears as Lemma 2.1e in von Rosen[101], rewritten here for convenience

$$\frac{\partial \left(|X|^{\theta} \right)}{\partial X} = \theta |X|^{\theta} X^{-T} \quad .$$

and it is developed in detail in Appendix C as equation (C.7). We substitute this derivative with H in place of X to write:

$$\frac{\partial W(H)}{\partial H} = c|M_0|^{-\frac{\nu_0}{2}} \left\{ \begin{array}{l} \frac{\nu_0 - m - 1}{2} |H|^{\frac{\nu_0 - m - 1}{2}} H^{-1} e^{-\frac{1}{2} Trace\left(M_0{}^{-1} H\right)} \\ +|H|^{\frac{\nu_0 - m - 1}{2}} \frac{\partial}{\partial H} \left[e^{-\frac{1}{2} Trace\left(M_0{}^{-1} H\right)} \right] \end{array} \right\} \quad .$$

The presence of H^{-1} rather than its transpose inside the integral relies on the fact that the matrix H is symmetric. This takes care of the top term on the right. We factor to write

$$\begin{aligned} \frac{\partial W(H)}{\partial H} &= c|M_0|^{-\frac{\nu_0}{2}} |H|^{\frac{\nu_0 - m - 1}{2}} \\ &\times \left\{ \frac{\nu_0 - m - 1}{2} H^{-1} e^{-\frac{1}{2} Trace\left(M_0{}^{-1} H\right)} + \frac{\partial}{\partial H} \left[e^{-\frac{1}{2} Trace\left(M_0{}^{-1} H\right)} \right] \right\} \quad . \end{aligned}$$

We turn to the derivative of the exponential term at the right to begin to develop that partial derivative at using the chain rule for scalar functions.[23] Quite analogous to the univariate case, we write

$$\frac{\partial}{\partial H} \left[e^{-\frac{1}{2} Trace\left(M_0{}^{-1} H\right)} \right] = e^{-\frac{1}{2} Trace\left(M_0{}^{-1} H\right)} \frac{\partial}{\partial H} \left[-\frac{1}{2} Trace\left(M_0{}^{-1} H\right) \right]$$

[22]The product rule is $\frac{\partial a}{\partial H} b + \frac{\partial b}{\partial H} a = \frac{\partial (ab)}{\partial H}$,

[23]The chain rule for scalar functions is $\frac{\partial (g(u))}{\partial H} = \frac{\partial g(u)}{\partial u} \frac{\partial u}{\partial H}$

$$= -\frac{1}{2} e^{-\frac{1}{2} Trace\left(M_0{}^{-1} H\right)} \frac{\partial}{\partial H} \left[Trace\left(M_0{}^{-1} H\right) \right] \quad .$$

We substitute and factor to write

$$\frac{\partial W(H)}{\partial H} = c\frac{1}{2} |M_0|^{-\frac{\nu_0}{2}} |H|^{\frac{\nu_0 - m - 1}{2}} e^{-\frac{1}{2} Trace\left(M_0{}^{-1} H\right)} \times \left\{ (\nu_0 - m - 1) H^{-1} - \frac{\partial Trace\left(M_0{}^{-1} H\right)}{\partial H} \right\} \quad .$$

We apply the derivative of the trace term to the lower term on the right. That derivative is given as Lemma 2.1d (ii) in von Rosen[101], and it is derived as equation (C.6) in Appendix C. Keep in mind that this development uses the matrix derivative definition in equation (2.2) in von Rosen[101], which is the same weighted sum derivative articulated in detail in equation (C.1) in Appendix C. We begin with the derivative in equation (C.6)

$$\frac{\partial Trace\left(A^T H\right)}{\partial H} = \frac{1}{2} \left(A + A^T\right) \quad .$$

We want to substitute

$$A^T = M_0^{-1} \Rightarrow A = M_0^{-1}$$

which implies that

$$\frac{\partial Trace\left(A^T H\right)}{\partial H} = \frac{1}{2} \left(M_0^{-1} + M_0^{-1}\right) = M_0^{-1} \quad .$$

Substituting this expression, we have the following derived expression for the matrix derivative of the Wishart distribution:

$$\frac{\partial W(H)}{\partial H} = c |M_0|^{-\frac{\nu_0}{2}} |H|^{\frac{\nu_0 - m - 1}{2}} e^{-\frac{1}{2} Trace\left(M_0{}^{-1} H\right)} \times \frac{1}{2} \left[(\nu_0 - m - 1) H^{-1} - M_0^{-1} \right] \quad .$$

Finally, if we go back to equation (16.31) and compare it to the four terms in the foregoing equation, we see that those first four terms ARE a Wishart distribution. Therefore, we can write the matrix derivative of the Wishart distribution:

$$\frac{\partial W(H)}{\partial H} = \frac{1}{2} W(H) \left[(\nu_0 - m - 1) H^{-1} - M_0^{-1} \right] \quad . \tag{16.33}$$

This derivative of the Wishart distribution is now going to be used to develop the first and second moments of H^{-1} with respect to the Wishart distribution. This section is included for one simple reason: to calculate the matrix derivative (first type) of the standard Wishart distribution. We will need this in order to calculate the expected values of H^{-1} and $H^{-1} \otimes H^{-1}$ in the immediately forthcoming sections. When we have these two expected values, we will be able to calculate the mean vector and variance-covariance matrix of the model coefficients and of the m error terms.

16.2.4.2 Calculating the First Moment of H^{-1} Using the Fundamental Theorem of the Calculus

We begin by applying the fundamental theorem of the calculus to the derivative of a scalar valued function f with a scalar argument x

$$\int_a^b \frac{df(x)}{dx} dx = f(b) - f(a) \quad .$$

Suppose we wanted to pass b to the limit ∞ and a to the limit $-\infty$. Then, written informally

$$\int_{-\infty}^{\infty} \frac{df(x)}{dx} dx = f(\infty) - f(-\infty) \quad .$$

If f were a probability density function such that f vanishes at ∞ and $-\infty$, then we have

$$\int_{-\infty}^{\infty} \frac{df(x)}{dx} dx = 0 \quad ,$$

because $f(\infty) = f(-\infty) = 0$. It also pertains to the product of a functional f times a scalar function g as long as g is finite and equal in both limits. We can easily extend this to a functional f with a (multivariate) vector or matrix argument whose value vanishes when any of the limits of integration go to ∞ or $-\infty$.[24] This leads us to the equation:

$$\int_{-\infty}^{\infty} \cdots \int_{-\infty}^{\infty} \frac{d}{dH} \left[f(H) \right] dH = 0 \quad ,$$

which we are going to write in shorthand without the limits of the integration or the multiple integration notation

$$\int \frac{d}{dH} \left[f(H) \right] dH = 0 \quad . \tag{16.34}$$

To put this into our specific context, we substitute the derivative of the Wishart distribution as given in equation (16.33) into this equation as the integrand. This is going to give us the ability to find the first moment

[24] It is possible to use Stokes' Theorem to derive the same result, but we will not do so here. See Kollo and von Rosen[57].

$$
\begin{aligned}
0 &= \int \frac{\partial}{\partial H}\left[W(H)\right] dH \\
&= \int \frac{1}{2} W(H)\left[(\nu_0 - m - 1)H^{-1} - M_0^{-1}\right] dH \\
&= \frac{1}{2}\int \left[(\nu_0 - m - 1)H^{-1}W(H) - M_0^{-1}W(H)\right] dH \\
&= \frac{1}{2}(\nu_0 - m - 1)\int \left[H^{-1}W(H)\right] dH - \frac{1}{2}M_0^{-1}\int W(H)dH \\
&= \frac{1}{2}(\nu_0 - m - 1)\int \left[H^{-1}W(H)\right] dH - \frac{1}{2}M_0^{-1} \quad .
\end{aligned}
$$

If we rearrange to express the integral term as a function of all the other terms in the equation, we write

$$
\int \left[H^{-1}W(H)\right] dH = \frac{1}{\nu_0 - m - 1} M_0^{-1} \quad .
$$

Using the result in equation (16.32), we write the expectation of the inverse of H as follows

$$
E(H^{-1}) = \frac{1}{\nu_0 - m - 1} {M_0}^{-1} \quad . \tag{16.35}
$$

Therefore, by substitution into equation (16.32), the variance-covariance matrix over the coefficients is computed to be

$$
\begin{aligned}
VCV &= M_0 \frac{{M_0}^{-1}}{\nu_0 - m - 1}\left(M_0 \otimes {Q_0}^{-1}\right) \\
&= \frac{1}{\nu_0 - m - 1}\left(M_0 \otimes {Q_0}^{-1}\right) \quad .
\end{aligned} \tag{16.36}
$$

VCV is $mp \times mp$. This looks quite analogous to the univariate case, which has the "degrees of freedom minus 1 minus 2" in the denominator ($m = 1$ in that case). This is the variance-covariance matrix you get when you begin with the joint distribution in equation (16.15).

Before validating this against other published results, consider what the Kronecker product implies in the special case of $m = 2$ and $p = 5$ with symmetric matrices M_0 and Q_0

$$
\begin{bmatrix} m_{11} & m_{12} \\ m_{12} & m_{22} \end{bmatrix} \otimes \begin{bmatrix} \zeta_{11} & \zeta_{12} & \zeta_{13} & \zeta_{14} & \zeta_{15} \\ \zeta_{12} & \zeta_{22} & \zeta_{23} & \zeta_{24} & \zeta_{25} \\ \zeta_{13} & \zeta_{23} & \zeta_{33} & \zeta_{34} & \zeta_{35} \\ \zeta_{14} & \zeta_{24} & \zeta_{34} & \zeta_{44} & \zeta_{45} \\ \zeta_{15} & \zeta_{25} & \zeta_{35} & \zeta_{45} & \zeta_{55} \end{bmatrix} \quad ,
$$

in which we have designated the elements of Q_0^{-1} using the notation ζ. If we begin the Kronecker product, we can write

$$\begin{bmatrix} m_{11}\begin{bmatrix} \zeta_{11} & \zeta_{12} & \zeta_{13} & \zeta_{14} & \zeta_{15} \\ \zeta_{12} & \zeta_{22} & \zeta_{23} & \zeta_{24} & \zeta_{25} \\ \zeta_{13} & \zeta_{23} & \zeta_{33} & \zeta_{34} & \zeta_{35} \\ \zeta_{14} & \zeta_{24} & \zeta_{34} & \zeta_{44} & \zeta_{45} \\ \zeta_{15} & \zeta_{25} & \zeta_{35} & \zeta_{45} & \zeta_{55} \end{bmatrix} & m_{12}\begin{bmatrix} \zeta_{11} & \zeta_{12} & \zeta_{13} & \zeta_{14} & \zeta_{15} \\ \zeta_{12} & \zeta_{22} & \zeta_{23} & \zeta_{24} & \zeta_{25} \\ \zeta_{13} & \zeta_{23} & \zeta_{33} & \zeta_{34} & \zeta_{35} \\ \zeta_{14} & \zeta_{24} & \zeta_{34} & \zeta_{44} & \zeta_{45} \\ \zeta_{15} & \zeta_{25} & \zeta_{35} & \zeta_{45} & \zeta_{55} \end{bmatrix} \\ m_{12}\begin{bmatrix} \zeta_{11} & \zeta_{12} & \zeta_{13} & \zeta_{14} & \zeta_{15} \\ \zeta_{12} & \zeta_{22} & \zeta_{23} & \zeta_{24} & \zeta_{25} \\ \zeta_{13} & \zeta_{23} & \zeta_{33} & \zeta_{34} & \zeta_{35} \\ \zeta_{14} & \zeta_{24} & \zeta_{34} & \zeta_{44} & \zeta_{45} \\ \zeta_{15} & \zeta_{25} & \zeta_{35} & \zeta_{45} & \zeta_{55} \end{bmatrix} & m_{22}\begin{bmatrix} \zeta_{11} & \zeta_{12} & \zeta_{13} & \zeta_{14} & \zeta_{15} \\ \zeta_{12} & \zeta_{22} & \zeta_{23} & \zeta_{24} & \zeta_{25} \\ \zeta_{13} & \zeta_{23} & \zeta_{33} & \zeta_{34} & \zeta_{35} \\ \zeta_{14} & \zeta_{24} & \zeta_{34} & \zeta_{44} & \zeta_{45} \\ \zeta_{15} & \zeta_{25} & \zeta_{35} & \zeta_{45} & \zeta_{55} \end{bmatrix} \end{bmatrix}.$$

If we make the full expansion, and we isolate only the diagonals (the variances), those diagonals appear in Figure 16.2.

Figure 16.2: The Diagonal Kronecker Product Matrix

$$\begin{bmatrix} \mathrm{m}_{11}\varsigma_{11} \\ & \mathrm{m}_{11}\varsigma_{22} \\ && \mathrm{m}_{11}\varsigma_{33} \\ &&& \mathrm{m}_{11}\varsigma_{44} \\ &&&& \mathrm{m}_{11}\varsigma_{55} \\ &&&&& \mathrm{m}_{22}\varsigma_{11} \\ &&&&&& \mathrm{m}_{22}\varsigma_{22} \\ &&&&&&& \mathrm{m}_{22}\varsigma_{33} \\ &&&&&&&& \mathrm{m}_{22}\varsigma_{44} \\ &&&&&&&&& \mathrm{m}_{22}\varsigma_{55} \end{bmatrix}$$

There are seven quantities that enter into these ten variances. If we organize these down the row dimension according to the indexes of p and across the column dimension according to the indexes m, we can write the variances

$$\begin{bmatrix} m_{11}\varsigma_{11} & m_{22}\varsigma_{11} \\ m_{11}\varsigma_{22} & m_{22}\varsigma_{22} \\ m_{11}\varsigma_{33} & m_{22}\varsigma_{33} \\ m_{11}\varsigma_{44} & m_{22}\varsigma_{44} \\ m_{11}\varsigma_{55} & m_{22}\varsigma_{55} \end{bmatrix} = \begin{bmatrix} m_{11}\begin{bmatrix} \varsigma_{11} \\ \varsigma_{22} \\ \varsigma_{33} \\ \varsigma_{44} \\ \varsigma_{55} \end{bmatrix} & m_{22}\begin{bmatrix} \varsigma_{11} \\ \varsigma_{22} \\ \varsigma_{33} \\ \varsigma_{44} \\ \varsigma_{55} \end{bmatrix} \end{bmatrix}.$$

We will be referencing these variances in the example in the next section. It is clear that the variances of the $m = 2$ term are a linear multiple of the variances of the $m = 1$ term. The coefficient variances do not "scatter." They relate to each other. We will see this in the example in the next section.

To close this section, the expectation of the inverse of a Wishart (matrix) random variable is given in von Rosen[101] and Cook[22] to be

$$\frac{{M_0}^{-1}}{\nu_0 - m - 1} \quad .$$

This matrix is $m \times m$. This result is verified in Haff[39], Corollary 1. This section has succeeded in equation (16.35) at calculating the expectation of H^{-1} to be the immediately preceding.

We need to interpret exactly what this variance-covariance matrix means. We do so in Chapter 20 using a "units" analysis taken from the example in Chapter 20. It is the covariance of each and every one of the ten elements in the mean matrix in equation (15.2) with regard to every one of the other ten elements in the mean matrix in equation (15.2). These covariances are coefficient by coefficient variances and covariances between every pair of elements of the mean matrix in equation (15.2).

16.2.5 Analysis of the Conjugate Prior–Mean of the Error Terms

This section uses the result in the immediately preceding section, i.e., equation (16.35), to calculate the mean of the error terms specified in the Wishart distribution. We have thus far dealt with the coefficients B. But what about the error term Σ in the inverse Wishart? Such errors are important quantities in their own right, just as the σ terms were in the univariate case, containing variances of error terms and, importantly, covariances among different outputs of the joint distribution. That is, these error terms tell us how much covariance there is among the two or more joint outputs, a rather important point. If these covariances are small, we can infer probabilistic independence. However, if they are large, whether a large positive value or a small negative value, we will see the joint density over outputs being further away from probabilistic independence. This section begins with the inverse Wishart density in equation (16.16). When we do so, we have the following expression for the expected value:

$$\langle \Sigma \rangle = \int d\Sigma \frac{|M_0|^{\frac{\nu_0}{2}}}{2^{\frac{\nu_0 m}{2}} \Gamma_m\left(\frac{\nu_0}{2}\right) |\Sigma|^{\frac{\nu_0+m+1}{2}}} \Sigma e^{-\frac{1}{2} Trace\left(M_0 \Sigma^{-1}\right)} \quad .$$

We want to make the foregoing substitution, rewritten here for easy reference

$$\begin{aligned} & M_0 \Sigma^{-1} = H {M_0}^{-1} \\ & \Rightarrow {M_0}^{-1} M_0 \Sigma^{-1} = {M_0}^{-1} H {M_0}^{-1} \\ & \Rightarrow \Sigma^{-1} = {M_0}^{-1} H {M_0}^{-1} \\ & \Rightarrow \Sigma = M_0 H^{-1} M_0 \\ & \Rightarrow I = \Sigma {M_0}^{-1} H {M_0}^{-1} \quad . \end{aligned}$$

When we make this substitution plus the Jacobian into the integrand, we obtain

$$\langle\Sigma\rangle = \int \left[\frac{|M_0|^{\frac{\nu_0}{2}}}{2^{\frac{\nu_0 m}{2}}\Gamma_m\left(\frac{\nu_0}{2}\right)|M_0 H^{-1} M_0|^{\frac{\nu_0+m+1}{2}}} \left(M_0 H^{-1} M_0\right) e^{-\frac{1}{2}Trace\left(H {M_0}^{-1}\right)} \frac{|M_0|^{m+1}}{|H|^{m+1}} dH \right]$$

$$= \int \left(M_0 H^{-1} M_0\right) \frac{|H|^{\frac{\nu_0-m-1}{2}}}{2^{\frac{\nu_0 m}{2}}\Gamma_m\left(\frac{\nu_0}{2}\right)|M_0|^{\frac{\nu_0}{2}}} e^{-\frac{1}{2}Trace\left(H {M_0}^{-1}\right)} dH \quad .$$

The Wishart distribution is rewritten for convenience again

$$H \sim W\left(H; M_0, \nu_0, m\right) = \frac{1}{2^{\frac{\nu_0 m}{2}}\Gamma_m\left(\frac{\nu_0}{2}\right)} \frac{|H|^{\frac{\nu_0-m-1}{2}}}{|M_0|^{\frac{\nu_0}{2}}} e^{-\frac{1}{2}Trace\left({M_0}^{-1} H\right)} \quad .$$

$\langle\Sigma\rangle$ is no more than the expectation of $M_0 H^{-1} M_0$ under the Wishart distribution

$$\langle\Sigma\rangle = \int \left(M_0 H^{-1} M_0\right) W\left(H; M_0, \nu_0, m\right) dH \quad .$$

We move the integral inside as reasoned previously in this chapter

$$\langle\Sigma\rangle = M_0 \left[\int H^{-1} W\left(H; M_0, \nu_0, m\right) dH\right] M_0 \quad .$$

We see that the term inside the brackets is the expectation of the inverse of H under a Wishart distribution, which we have derived in the immediately preceding section as equation (16.35)

$$\frac{{M_0}^{-1}}{\nu_0 - m - 1} \quad .$$

Therefore

$$\langle\Sigma\rangle = M_0 \frac{{M_0}^{-1}}{\nu_0 - m - 1} M_0 = \frac{1}{\nu_0 - m - 1} M_0 \quad . \tag{16.37}$$

This matrix is $m \times m$. This is an interesting form. It is very similar in form to the variance-covariance matrix over the coefficients, but the second term in the Kronecker product and therefore the Kronecker product as a whole is omitted. That gives us confidence this is correct and correctly interpreted.

16.2.6 Analysis of the Conjugate Prior–Variance-Covariance Matrix of the Error Terms

This section contains the most sophisticated development in the textbook. We want to know the variance-covariance matrix of the inverse Wishart distribution over the error terms. We have already calculated the expectation under the

Wishart distribution of $H^{-1}$and reported it in equation (16.35). We will additionally need the expectation under the Wishart distribution of $H^{-1} \otimes H^{-1}$. This need will emerge from the definition and development of the variance-covariance matrix of the inverse Wishart error term, with which we will begin this section.

16.2.6.1 Basic Characterization of the Variance-Covariance Matrix

The development of the variance-covariance matrix of the inverse Wishart in this section begins with the definition of the variance-covariance matrix of the inverse Wishart distribution. We choose the following definition

$$VCV = \int \left[\begin{array}{l} \left(\Sigma - \frac{1}{\nu_0 - m - 1} M_0\right) \otimes \left(\Sigma - \frac{1}{\nu_0 - m - 1} M_0\right) \\ \times W^{-1}\left(\Sigma; M_0, \nu_0, m\right) \end{array} \right] d\Sigma \quad .$$

Why is this the definition we want to use? We illustrate in detail in Appendix D for the case of $m = 3$. The reason is that von Rosen[101] developed how to calculate the expectation of $\Sigma \otimes \Sigma$ with respect to the inverse Wishart distribution. We will be reproducing that derivation and attempting to carefully clarify it herein. To do so, we begin by expanding out the Kronecker product in the integrand, which is known to be left and right associative[25]

$$\begin{array}{l} \Sigma \otimes \Sigma - \dfrac{1}{\nu_0 - m - 1}\left(\Sigma \otimes M_0\right) \\ \quad - \dfrac{1}{\nu_0 - m - 1}\left(M_0 \otimes \Sigma\right) + \left(\dfrac{1}{\nu_0 - m - 1}\right)^2 \left(M_0 \otimes M_0\right) \quad . \end{array}$$

If we take the expectation with respect to the inverse Wishart distribution, which we write in shorthand using $E(.)$ and knowing that the expectation of Σ is

$$E(\Sigma) = \frac{1}{\nu_0 - m - 1} M_0 \quad .$$

we write each of the four successive terms out with the foregoing expression of $E(\Sigma)$ substituted, keeping in mind that the expectation is with respect to the inverse Wishart distribution

$$\begin{aligned} VCV &= E\left(\Sigma \otimes \Sigma\right) - \frac{1}{\nu_0 - m - 1}\left[E\left(\Sigma\right) \otimes M_0\right] \\ &\quad - \frac{1}{\nu_0 - m - 1}\left[M_0 \otimes E\left(\Sigma\right)\right] + \left(\frac{1}{\nu_0 - m - 1}\right)^2 \left(M_0 \otimes M_0\right) \\ &= E\left(\Sigma \otimes \Sigma\right) - \frac{1}{\nu_0 - m - 1}\left[\frac{1}{\nu_0 - m - 1} M_0 \otimes M_0\right] \end{aligned}$$

[25]See https://en.wikipedia.org/wiki/Kronecker_product.

$$
-\frac{1}{\nu_0-m-1}\left[M_0\otimes\frac{1}{\nu_0-m-1}M_0\right]+\left(\frac{1}{\nu_0-m-1}\right)^2(M_0\otimes M_0)
$$
$$
=E(\Sigma\otimes\Sigma)-\left(\frac{1}{\nu_0-m-1}\right)^2(M_0\otimes M_0) \quad .
$$

Thus the derived expression for the variance-covariance matrix, with the integral for expectation written out in detail, becomes

$$
VCV=-\left(\frac{1}{\nu_0-m-1}\right)^2(M_0\otimes M_0)+\int(\Sigma\otimes\Sigma)\,W^{-1}(\Sigma;M_0,\nu_0,m)\,d\Sigma
$$
$$
=-\left(\frac{1}{\nu_0-m-1}\right)^2(M_0\otimes M_0)
$$
$$
+\int(\Sigma\otimes\Sigma)\frac{|M_0|^{\frac{\nu_0}{2}}}{2^{\frac{\nu_0 m}{2}}\Gamma_m\left(\frac{\nu_0}{2}\right)}|\Sigma|^{-\frac{\nu_0+m+1}{2}}e^{-\frac{1}{2}Trace\left(M_0\Sigma^{-1}\right)}d\Sigma \quad .
$$

Notice how the integral term in this equation is the expectation with respect to the inverse Wishart distribution of the Kronecker product of Σ with itself. If we think about the analogy with the univariate case, we were calculating the expectation of σ^2. Here we are calculating the expectation of $\Sigma\otimes\Sigma$, which looks like an "outer product" or a "square" in principle. That gives us confidence that we are on the right track here. We go about solving this in the same way we solved for the mean in the previous section. We are going to want to transform from the inverse Wishart to the Wishart, so we make the identical substitutions as in the previous section:

$$
\begin{aligned}
&M_0\Sigma^{-1}=HM_0{}^{-1}\\
&M_0{}^{-1}M_0\Sigma^{-1}=M_0{}^{-1}HM_0{}^{-1}\\
&\Sigma^{-1}=M_0{}^{-1}HM_0{}^{-1}\\
&\Sigma=M_0H^{-1}M_0 \quad ,
\end{aligned}
$$

and substitute the Jacobian

$$
J=\frac{|M_0|^{m+1}}{|H|^{m+1}} \quad .
$$

We can substitute, reorganize to see the Wishart, and write

$$
VCV=-\frac{1}{(\nu_0-m-1)^2}(M_0\otimes M_0)
$$

$$+\int\left[\begin{array}{l}\left(M_0H^{-1}M_0\right)\otimes\left(M_0H^{-1}M_0\right)\\ \times\frac{|M_0|^{\frac{\nu_0}{2}}}{2^{\frac{\nu_0 m}{2}}\Gamma_m\left(\frac{\nu_0}{2}\right)}\left|M_0H^{-1}M_0\right|^{-\frac{\nu_0+m+1}{2}}\\ \times e^{-\frac{1}{2}tr\left(HM_0^{-1}\right)}\frac{|M_0|^{m+1}}{|H|^{m+1}}dH\end{array}\right]$$

$$=-\frac{1}{\left(\nu_0-m-1\right)^2}\left(M_0\otimes M_0\right)$$
$$+\int\left[\begin{array}{l}\left(M_0H^{-1}M_0\right)\otimes\left(M_0H^{-1}M_0\right)\\ \times\frac{1}{2^{\frac{\nu_0 m}{2}}\Gamma_m\left(\frac{\nu_0}{2}\right)|M_0|^{\frac{\nu_0}{2}}}|H|^{\frac{\nu_0-m-1}{2}}\times e^{-\frac{1}{2}tr\left(HM_0^{-1}\right)}dH\end{array}\right] \quad .$$

We have to do some simplifying. There is a relationship in Wiki[26] that pertains to Kronecker product, which we write in reversed order

$$(AC)\otimes(BD)=(A\otimes B)(C\otimes D) \quad .$$

We have to apply this twice. First,

$$\begin{aligned}&\left(M_0H^{-1}M_0\right)\otimes\left(M_0H^{-1}M_0\right)\\ &=M_0\left(H^{-1}M_0\right)\otimes M_0\left(H^{-1}M_0\right)\\ &=\left(M_0\otimes M_0\right)\left[\left(H^{-1}M_0\right)\otimes\left(H^{-1}M_0\right)\right] \quad .\end{aligned}$$

We apply the theorem one more time to the Kronecker product on the right

$$\left(H^{-1}M_0\right)\otimes\left(H^{-1}M_0\right)=\left(H^{-1}\otimes H^{-1}\right)\left(M_0\otimes M_0\right) \quad .$$

As a result

$$\begin{aligned}&\left(M_0H^{-1}M_0\right)\otimes\left(M_0H^{-1}M_0\right)\\ &=\left(M_0\otimes M_0\right)\left(H^{-1}\otimes H^{-1}\right)\left(M_0\otimes M_0\right) \quad .\end{aligned}$$

This simplifies the foregoing integral term to

$$VCV=$$
$$=-\frac{1}{\left(\nu_0-m-1\right)^2}\left(M_0\otimes M_0\right)$$
$$+\int\left[\begin{array}{l}\left(M_0\otimes M_0\right)\left(H^{-1}\otimes H^{-1}\right)\left(M_0\otimes M_0\right)\\ \times\frac{1}{2^{\frac{\nu_0 m}{2}}\Gamma_m\left(\frac{\nu_0}{2}\right)|M_0|^{\frac{\nu_0}{2}}}|H|^{\frac{\nu_0-m-1}{2}}\times e^{-\frac{1}{2}tr\left(HM_0^{-1}\right)}dH\end{array}\right]$$
$$=-\frac{1}{\left(\nu_0-m-1\right)^2}\left(M_0\otimes M_0\right)$$

[26] See https://en.wikipedia.org/wiki/Kronecker_product.

$$
\begin{aligned}
&+\int\left[\begin{array}{l}\left(M_0\otimes M_0\right)\left(H^{-1}\otimes H^{-1}\right)\left(M_0\otimes M_0\right)\\ \times\frac{1}{2^{\frac{\nu_0 m}{2}}\Gamma_m\left(\frac{\nu_0}{2}\right)\left|M_0\right|^{\frac{\nu_0}{2}}}\left|H\right|^{\frac{\nu_0-m-1}{2}}\times e^{-\frac{1}{2}tr\left(HM_0^{-1}\right)}dH\end{array}\right]\\
&=-\frac{1}{\left(\nu_0-m-1\right)^2}\left(M_0\otimes M_0\right)\\
&+\left(M_0\otimes M_0\right)\\
&\times\int\left[\begin{array}{l}\left(H^{-1}\otimes H^{-1}\right)\\ \times\frac{1}{2^{\frac{\nu_0 m}{2}}\Gamma_m\left(\frac{\nu_0}{2}\right)\left|M_0\right|^{\frac{\nu_0}{2}}}\left|H\right|^{\frac{\nu_0-m-1}{2}}\times e^{-\frac{1}{2}tr\left(HM_0^{-1}\right)}dH\end{array}\right]\\
&\times\left(M_0\otimes M_0\right)\quad .
\end{aligned}
$$

We have moved the integral term to the inside, which we now need to calculate and define temporarily using the notation S_0, not to be confused with the former S vector

$$
\begin{aligned}
S_0&=\int\left[\begin{array}{l}\left(H^{-1}\otimes H^{-1}\right)\\ \frac{1}{2^{\frac{\nu_0 m}{2}}\Gamma_m\left(\frac{\nu_0}{2}\right)\left|M_0\right|^{\frac{\nu_0}{2}}}\left|H\right|^{\frac{\nu_0-m-1}{2}}e^{-\frac{1}{2}tr\left(HM_0^{-1}\right)}dH\end{array}\right]\\
&=\int\left(H^{-1}\otimes H^{-1}\right)W\left(H\right)dH\quad .
\end{aligned}
\tag{16.38}
$$

This is simply the expectation of the Kronecker product $\left(H^{-1}\otimes H^{-1}\right)$ under the Wishart distribution. This sort of looks like the expectation of the "square," namely $\left(H^{-1}\otimes H^{-1}\right)$, under the Wishart distribution. (It was the expectation of the square in the univariate case.) This is multivariate generalization incarnate. It is going to be the job of this section to calculate this integral, which is neither simple nor straightforward, a result attributed to von Rosen[101], who unfortunately omitted many steps, which are required for intuition, out of his development. We will attempt to sequentially include all the requisite steps here so that this book is complete with regard to methodology and sequence of development. We do not want anyone to use equations that are undeveloped. Von Rosen's proof of the integral of this Kronecker product of inverses was impressive. We intend to be "step-by-step," so we will not omit any significant steps and so that readers may understand and hopefully generalize.

The final expression for the variance-covariance matrix of the error term is therefore, using the S_0 notation

$$
\begin{aligned}
VCV&=-\left(\frac{1}{\nu_0-m-1}\right)^2\left(M_0\otimes M_0\right)\\
&+\left(M_0\otimes M_0\right)S_0\left(M_0\otimes M_0\right)\quad ,
\end{aligned}
\tag{16.39}
$$

in which S_0 is given in equation (16.38). S_0 is the expectation of the Kronecker product of the inverse times itself with respect to the Wishart. This is the second term we referred to that we need to calculate to get the variance-covariance matrix of the Wishart error distribution.

16.2.6.2 Calculating the Expectation of the Kronecker Product of Inverses $\Sigma^{-1} \otimes \Sigma^{-1}$

Calculating the expectation of the Kronecker product of $\Sigma^{-1} \otimes \Sigma^{-1}$ with respect to the inverse Wishart requires the development of three equations and three unknowns. We will develop the equations one by one.

16.2.6.2.1 First of Three Equations. We get the expectation of the Kronecker product of $\Sigma^{-1} \otimes \Sigma^{-1}$ with respect to the inverse Wishart by applying an equation analogous in concept to equation (16.34). Instead of using the integrand $W(H)$ as in equation (16.34), we instead use the integrand $H^{-1}W(H)$. The integral is still zero because the limits at ∞ and $-\infty$ are both still zero. This is intuitive because as any element of H goes to infinity, the value of this function goes to 0. Using this integrand, we want our new integrand to be

$$\frac{\partial}{\partial H}[H^{-1}W(H)] \quad .$$

In particular, we want to calculate the integral

$$\int \frac{\partial}{\partial H}\left[H^{-1}W(H)\right] dH = 0 \quad .$$

The integral is 0 because in the infinite limit $H^{-1}W(H)$ goes to 0. Let us focus on the integrand.

$$\frac{\partial}{\partial H}\left[H^{-1}W(H)\right] \quad .$$

This integrand is the product of a $p \times p$ matrix H times a scalar $W(H)$. We use the matrix times scalar theorem, which is Lemma 2.1a(ii) in von Rosen[101][27] and is derived as equation (C.3) in Appendix C herein to write

$$\frac{\partial}{\partial H}\left[H^{-1}W(H)\right] = \frac{\partial H^{-1}}{\partial H}W(H) + H^{-1} \otimes \frac{\partial W(H)}{\partial H} \quad .$$

We know from Appendix C and from Lemma 2.2(i) in von Rosen[101], which is derived as equation (C.9) in Appendix C herein that

$$\frac{\partial H^{-1}}{\partial H} = -\frac{1}{2}vec\left(H^{-1}\right) vec\left(H^{-1}\right)^T - \frac{1}{2}K_{mm}\left(H^{-1} \otimes H^{-1}\right) \quad ,$$

[27] The derivative of a matrix times a scalar is $\frac{\partial(Zy)}{\partial X} = \frac{\partial Z}{\partial X}y + Z \otimes \frac{\partial y}{\partial X}$. This is a well-known result available in the literature on matrix derivatives, including Kollo and von Rosen[57].

in which the matrix K_{mm} is the square, $m \times m$ "commutation matrix." (The development of this expression in the Appendix is non-trivial and should be mastered by readers who really want to understand the basis of the variance-covariance of the Wishart distribution.) The square commutation matrix transforms the mm vector $vec(A)$ into the mm vector $vec(A^T)$, i.e.,

$$K_{mm} vec(A) = vec(A^T) \quad .$$

This is definitional, as shown in Appendix C. We show the commutation matrix for $m = 3$ in the Appendix. Thus far we have half of the equation simplified

$$\begin{aligned} \frac{\partial}{\partial H}\left[H^{-1}W(H)\right] &= -\frac{1}{2} vec\left(H^{-1}\right) vec(H^{-1})^T W(H) \\ &\quad -\frac{1}{2}K_{mm}\left(H^{-1}\otimes H^{-1}\right)W(H) \\ &\quad + H^{-1}\otimes\frac{\partial W(H)}{\partial H} \quad . \end{aligned} \tag{16.40}$$

We must now substitute the expression for the matrix derivative from equation (16.33) into the Kronecker product at the end of equation (16.40) above and break the two components thereof apart to write

$$H^{-1}\otimes\frac{\partial W(H)}{\partial H} = H^{-1}\otimes\left\{\frac{1}{2}\left[\left(\nu_0 - m - 1\right)H^{-1} - M_0^{-1}\right]W(H)\right\} \quad .$$

We first move the Wishart distribution inside the brackets

$$H^{-1}\otimes\frac{\partial W(H)}{\partial H} = H^{-1}\otimes\left[\frac{1}{2}W(H)\left(\nu_0 - m - 1\right)H^{-1} - \frac{1}{2}W(H)M_0^{-1}\right] \quad .$$

Thereafter, we distribute the Kronecker product as in the theorem in Wiki[28] Thereafter, we pull the Wishart distribution back out so that we can see that the Kronecker products are correctly specified. The first step is

$$\begin{aligned} H^{-1}\otimes\frac{\partial W(H)}{\partial H} &= \frac{1}{2}W(H)\left(\nu_0 - m - 1\right)\left(H^{-1}\otimes H^{-1}\right) \\ &\quad -\frac{1}{2}W(H)(H^{-1}\otimes M_0^{-1}) \quad . \end{aligned}$$

Factoring yields the relationship

$$H^{-1}\otimes\frac{\partial W\left(H\right)}{\partial H} = \frac{1}{2}W\left(H\right)\left[\begin{array}{l}\left(H^{-1}\otimes H^{-1}\right)\left(\nu_0 - m - 1\right) \\ -\left(H^{-1}\otimes M_0^{-1}\right)\end{array}\right] \quad .$$

This is substituted into the last term in equation (16.40) above to yield

[28] https://en.wikipedia.org/wiki/Kronecker_product.

$$\begin{aligned}
&\frac{\partial}{\partial H}\left[H^{-1}W(H)\right] \\
&= -\frac{1}{2}\left[vec\left(H^{-1}\right)vec\left(H^{-1}\right)^{T}W(H)\right] \\
&- \frac{1}{2}K_{mm}\left[\left(H^{-1}\otimes H^{-1}\right)W(H)\right] \\
&+ \frac{1}{2}\left(\nu_{0}-m-1\right)\left[\left(H^{-1}\otimes H^{-1}\right)W(H)\right] \\
&- \frac{1}{2}\left[\left(H^{-1}\otimes M_{0}^{-1}\right)W(H)\right] \quad .
\end{aligned}$$

This is the integrand that we wanted to create when we began this section. We therefore integrate it to obtain

$$\begin{aligned}
&\int \frac{\partial}{\partial H}\left[H^{-1}W\left(H\right)\right]dH = 0 \\
&= -\frac{1}{2}\int\left\{vec\left(H^{-1}\right)\left[vec\left(H^{-1}\right)\right]^{T}W\left(H\right)\right\}dH \\
&- \frac{1}{2}\int K_{mm}\left[\left(H^{-1}\otimes H^{-1}\right)W\left(H\right)\right]dH \\
&+ \frac{1}{2}\left(\nu_{0}-m-1\right)\int\left[\left(H^{-1}\otimes H^{-1}\right)W\left(H\right)\right]dH \\
&- \frac{1}{2}\int\left[\left(H^{-1}\otimes M_{0}{}^{-1}\right)W\left(H\right)\right]dH \quad .
\end{aligned}$$

This simplifies to

$$\begin{aligned}
0 = &-\left[\int vec\left(H^{-1}\right)vec\left(H^{-1}\right)^{T}W(H)dH\right] \\
&- K_{mm}\left[\int\left(H^{-1}\otimes H^{-1}\right)W(H)dH\right] \\
&+ \left(\nu_{0}-m-1\right)\left[\int\left(H^{-1}\otimes H^{-1}\right)W(H)dH\right] \\
&- \left[\int H^{-1}W(H)dH\right]\otimes M_{0}^{-1} \quad .
\end{aligned}$$

Moving the Kronecker product inside the integral in the last term, we obtain

$$\begin{aligned}
0 = &-\int\left\{vec\left(H^{-1}\right)\left[vec\left(H^{-1}\right)\right]^{T}W\left(H\right)\right\}dH \\
&- K_{mm}\int\left[\left(H^{-1}\otimes H^{-1}\right)W\left(H\right)\right]dH \\
&+ \left(\nu_{0}-m-1\right)\int\left[\left(H^{-1}\otimes H^{-1}\right)W\left(H\right)\right]dH
\end{aligned}$$

$$- \int \left\{ H^{-1} \left[W \left(H \right) \right] dH \right\} \otimes {M_0}^{-1} \quad .$$

Substituting the expression for the mean of H^{-1}, which is what the integral in the last term represents, we write

$$\begin{aligned}
0 = &- \int \left\{ vec \left(H^{-1} \right) \left[vec \left(H^{-1} \right) \right]^T W \left(H \right) \right\} dH \\
&- K_{mm} \int \left[\left(H^{-1} \otimes H^{-1} \right) W \left(H \right) \right] dH \\
&+ \left(\nu_0 - m - 1 \right) \int \left[\left(H^{-1} \otimes H^{-1} \right) W \left(H \right) \right] dH \\
&- \left(\frac{1}{\nu_0 - m - 1} {M_0}^{-1} \otimes {M_0}^{-1} \right) \quad ,
\end{aligned}$$

and finally we write

$$\begin{aligned}
&- \int \left\{ vec \left(H^{-1} \right) \left[vec \left(H^{-1} \right) \right]^T W \left(H \right) \right\} dH \\
&- K_{mm} \int \left[\left(H^{-1} \otimes H^{-1} \right) W \left(H \right) \right] dH \\
&+ \left(\nu_0 - m - 1 \right) \int \left[\left(H^{-1} \otimes H^{-1} \right) W \left(H \right) \right] dH \\
&= \frac{1}{\nu_0 - m - 1} \left({M_0}^{-1} \otimes {M_0}^{-1} \right) \quad .
\end{aligned} \tag{16.41}$$

This is equation 3.2 in von Rosen[101], and here we have developed it from first principles, leaving out no steps. To reiterate, this equation is based on the definition of matrix derivative in von Rosen[101] and articulated in Appendix C in some detail.

16.2.6.2.2 Second of Three Equations. Moving onward to the next equation, we start by multiplying the foregoing equation by the $m \times m$ commutation matrix K_{mm} and rearrange, we get a second equation

$$\begin{aligned}
&- K_{mm} \left[\int vec \left(H^{-1} \right) vec \left(H^{-1} \right)^T W(H) dH \right] \\
&- K_{mm} K_{mm} \left[\int \left(H^{-1} \otimes H^{-1} \right) W(H) dH \right] \\
&+ \left(\nu_0 - m - 1 \right) K_{mm} \left[\int \left(H^{-1} \otimes H^{-1} \right) W(H) dH \right] \\
&= \frac{1}{\nu_0 - m - 1} K_{mm} \left(M_0^{-1} \otimes M_0^{-1} \right) \quad .
\end{aligned}$$

We develop an important relationship for the K_{mm} matrix

$$\begin{aligned}
&K_{mm}vec\,(A) = vec\left(A^T\right) \\
&\Rightarrow K_{mm}K_{mm}vec\,(A) = K_{mm}vec\left(A^T\right) \\
&\Rightarrow K_{mm}K_{mm}vec\,(A) = vec\,(A) \\
&\Rightarrow K_{mm}K_{mm} = I \quad .
\end{aligned}$$

Therefore we can write the previous equation, after interchanging of terms on the left, as follows

$$\begin{aligned}
&-K_{mm}\left[\int vec\left(H^{-1}\right) vec(H^{-1})^T W(H)dH\right] \\
&+\left(\nu_0 - m - 1\right) K_{mm}\left[\int \left(H^{-1}\otimes H^{-1}\right) W(H)dH\right] \\
&-\left[\int \left(H^{-1}\otimes H^{-1}\right) W(H)dH\right] \\
&= \frac{1}{\nu_0 - m - 1} K_{mm}\left(M_0^{-1}\otimes M_0^{-1}\right) \quad .
\end{aligned}$$

We now take the commutation matrix in the first term inside the integral to write

$$\begin{aligned}
&-\left[\int \left[K_{mm}vec\left(H^{-1}\right)\right] vec(H^{-1})^T W(H)dH\right] \\
&+\left(\nu_0 - m - 1\right) K_{mm}\left[\int \left(H^{-1}\otimes H^{-1}\right) W(H)dH\right] \\
&-\left[\int \left(H^{-1}\otimes H^{-1}\right) W(H)dH\right] \\
&= \frac{1}{\nu_0 - m - 1} K_{mm}\left(M_0^{-1}\otimes M_0^{-1}\right) \quad .
\end{aligned}$$

There is another important result with regard to commutation matrices, and it is definitional for square commutation matrices

$$K_{mm}vec\,(A) = vec\left(A^T\right) \quad .$$

If we specialize this to our situation here, we can write

$$K_{mm}vec\left(H^{-1}\right) = vec\left[\left(H^{-1}\right)^T\right] = vec\left(H^{-1}\right) \quad ,$$

the latter equation occurring because H^{-1} is symmetric. It is as if the commutation matrix never existed in the first term. If we substitute the forgoing relationship with the commutation matrix into the first term, we can write

$$-\left[\int vec\left(H^{-1}\right) vec(H^{-1})^T W(H)dH\right] \tag{16.42}$$

$$+\left(\nu_0-m-1\right)K_{mm}\left[\int\left(H^{-1}\otimes H^{-1}\right)W(H)dH\right]$$
$$-\left[\int\left(H^{-1}\otimes H^{-1}\right)W(H)dH\right]$$
$$=\frac{1}{\nu_0-m-1}K_{mm}\left(M_0^{-1}\otimes M_0^{-1}\right)\quad .$$

This is the second of three critically important equations in three unknowns needed to determine the variance-covariance matrix of the inverse Wishart.

16.2.6.2.3 Third of Three Equations. We have previously computed two matrix derivative equations, (16.41) and (16.42), using the Wishart distribution, but in this section we have to use a different matrix derivative. We previously used the von Rosen[101] matrix derivative of the first type discussed in Appendix C. In this section, to get the necessary third equation, we need to use a different derivative definition. In particular, we need to start from scratch using the so-called "vec over vec" derivative definition, to be discussed in some detail here and in Appendix C as the third type of matrix derivative. We turn in this section to developing a third equation, which involves a new and different matrix derivative, the "vec over vec" form discussed by Kollo and von Rosen[57] as equation (1.4.4) on p. 125 and later generalized at the top of p. 258. We have derived it as equation (C.15) in Appendix C herein.

Armed with this "vec over vec" notion of matrix derivative, we need to develop the matrix derivative of a matrix times a scalar so that we can derive the matrix derivative of the product of the inverse variable times the Wishart. We calculate the derivative using the "vec over vec" notion, a method that focuses on the vectorization of H rather than the Kronecker product. We see in the equation immediately following equation (1.4.19) in Kollo and von Rosen[57] the derivative of a matrix times a scalar function of a matrix when the derivative is the "vec over vec" definition, an equation we have derived as equation (C.11) in Appendix C for completeness of understanding and rewrite here

$$\frac{\partial\left(Yz\right)}{\partial X}=\frac{\partial Y}{\partial X}z+\frac{\partial z}{\partial X}vec(Y)^T\quad .$$

If we define

$$Y=H^{-1}$$
$$X=H$$
$$z=W\left(H\right)\quad ,$$

then we can write

$$\frac{\partial\left[H^{-1}W\left(H\right)\right]}{\partial H}=\frac{\partial H^{-1}}{\partial H}W\left(H\right)+\frac{\partial W\left(H\right)}{\partial H}vec\left(H^{-1}\right)^T\quad . \tag{16.43}$$

First we have to deal with the derivative of the Wishart distribution on the right. We begin by writing the full Wishart distribution for convenience

$$W(H) = c|M_0|^{-\frac{\nu_0}{2}}|H|^{\frac{\nu_0-m-1}{2}}e^{-\frac{1}{2}Trace(M_0^{-1}H)} \quad ,$$

in which

$$c = \frac{1}{2^{\frac{\nu_0 m}{2}}\Gamma_m(\frac{\nu_0}{2})} \quad .$$

We need to take the matrix derivative of the Wishart distribution using the new rules established for the "vec over vec" matrix derivative of a matrix times a scalar. We do so using the chain rule in equation (16.43) as follows

$$\text{(16.44)} \qquad \frac{\partial W(H)}{\partial H} = c|M_0|^{-\frac{\nu_0}{2}}\left[\begin{array}{l}\frac{\partial|H|^{\frac{\nu_0-m-1}{2}}}{\partial H}e^{-\frac{1}{2}Trace\left(M_0{}^{-1}H\right)} \\ +|H|^{\frac{\nu_0-m-1}{2}}\frac{\partial e^{-\frac{1}{2}Trace\left(M_0{}^{-1}H\right)}}{\partial H}\end{array}\right]$$

$$= c|M_0|^{-\frac{\nu_0}{2}}e^{-\frac{1}{2}Trace\left(M_0{}^{-1}H\right)}$$

$$\left[\begin{array}{l}\frac{\partial|H|^{\frac{\nu_0-m-1}{2}}}{\partial H} \\ -\frac{1}{2}|H|^{\frac{\nu_0-m-1}{2}}\frac{\partial Trace\left(M_0{}^{-1}H\right)}{\partial H}\end{array}\right] \quad .$$

Note that this is the identical application of the chain rule as we saw previously using the previous matrix derivative. We observe in equation 1.4.31 in Kollo and von Rosen[57] and developed in detail in Appendix C and reported as equation (C.14) that

$$\frac{\partial|X|^r}{\partial X} = r|X|^r vec\left(X^{-1}\right)^T \quad ,$$

for a scalar constant r and a symmetric matrix X. Substituting the proper parameters, we have

$$\text{(16.45)} \qquad \frac{\partial|H|^{\frac{\nu_0-m-1}{2}}}{\partial H} = \frac{\nu_0-m-1}{2}|H|^{\frac{\nu_0-m-1}{2}}vec\left(H^{-1}\right)^T \quad .$$

Equation 1.4.28 in Kollo and von Rosen[57] and developed in detail in Appendix C and reported as equation (C.13) is

$$\frac{\partial Trace\left(A^T X\right)}{\partial X} = vec\left(A\right) \quad ,$$

for symmetric matrices A and X. We substitute the following definitions

$$\begin{aligned}A^T &= M_0{}^{-1} \Rightarrow A = M_0{}^{-1} \\ X &= H \quad ,\end{aligned}$$

and write

$$\frac{\partial Trace\left({M_0}^{-1}H\right)}{\partial H} = vec\left({M_0}^{-1}\right) \quad . \tag{16.46}$$

We put equation (16.45) and equation (16.46) into equation (16.44) and simplify

$$\begin{aligned}
\frac{\partial W(H)}{\partial H} &= c|M_0|^{-\frac{\nu_0}{2}} e^{-\frac{1}{2}Trace\left({M_0}^{-1}H\right)} \\
&\times \left[\begin{array}{l} \frac{\partial |H|^{\frac{\nu_0-m-1}{2}}}{\partial H} \\ -\frac{1}{2}|H|^{\frac{\nu_0-m-1}{2}} \frac{\partial Trace\left({M_0}^{-1}H\right)}{\partial H} \end{array}\right] \\
&= c|M_0|^{-\frac{\nu_0}{2}} e^{-\frac{1}{2}Trace\left({M_0}^{-1}H\right)} \\
&\times \left\{\begin{array}{l} \left[\frac{\nu_0-m-1}{2}|H|^{\frac{\nu_0-m-1}{2}} vec\left(H^{-1}\right)\right] \\ -\frac{1}{2}|H|^{\frac{\nu_0-m-1}{2}} vec\left({M_0}^{-1}\right) \end{array}\right\} \\
&= \frac{1}{2}c|M_0|^{-\frac{\nu_0}{2}} |H|^{\frac{\nu_0-m-1}{2}} e^{-\frac{1}{2}Trace\left({M_0}^{-1}H\right)} \\
&\quad \times \left[(\nu_0 - m - 1)\, vec\left(H^{-1}\right) - vec\left({M_0}^{-1}\right)\right] \\
&= \frac{1}{2}W(H)\left[(\nu_0 - m - 1)\, vec\left(H^{-1}\right) - vec\left({M_0}^{-1}\right)\right] \quad .
\end{aligned} \tag{16.47}$$

The second equation relies on the symmetry of H^{-1} inside the $vec(.)$ operator. We substitute this into equation (16.44) to write

$$\begin{aligned}
\frac{\partial\left[H^{-1}W(H)\right]}{\partial H} &= \frac{\partial H^{-1}}{\partial H} W(H) + \frac{\partial W(H)}{\partial H} vec\left(H^{-1}\right)^T \\
&= \frac{\partial H^{-1}}{\partial H} W(H) \\
&+ \frac{1}{2}W(H)\left[\begin{array}{l} (\nu_0 - m - 1)\, vec\left(H^{-1}\right) \\ -vec\left(M_0^{-1}\right) \end{array}\right] vec\left(H^{-1}\right)^T \\
&= W(H)\left\{\begin{array}{l} \frac{\partial H^{-1}}{\partial H} \\ +\frac{1}{2}\left[\begin{array}{l} (\nu_0 - m - 1)\, vec\left(H^{-1}\right) \\ -vec\left(M_0^{-1}\right) \end{array}\right] vec\left(H^{-1}\right)^T \end{array}\right\} \quad .
\end{aligned} \tag{16.48}$$

Equation (1.4.21) in Kollo and von Rosen[57], which we have derived in detail in Appendix C and written as equation (C.12), tells us that for this specific "vec over vec" derivative, we have the following relationship (for a symmetric matrix H)

$$\frac{\partial H^{-1}}{\partial H} = -\frac{\partial H}{\partial H}\left(H^{-1} \otimes H^{-1}\right) \quad . \tag{16.49}$$

We have proven in Appendix C as equation (C.16), and Kollo and von Rosen[57] have noted the result in general as equation (2.4.30) on p. 258, for this "vec over vec" derivative, the following

$$\frac{\partial H}{\partial H} = \frac{1}{2}\left(I + K_{mm}\right) \quad . \tag{16.50}$$

We need to reemphasize that the derivative of a matrix with itself is not the identity matrix. It is more complicated than that, as we have shown in Appendix C. We substitute equations (16.49) and (16.50) into equation (16.48) and simplify to write

$$\begin{aligned}\frac{\partial\left[H^{-1}W(H)\right]}{\partial H} = &-\frac{1}{2}\left(I + K_{mm}\right)\left(H^{-1} \otimes H^{-1}\right)W(H)\\ &+ \frac{\partial W(H)}{\partial H} vec\left(H^{-1}\right)^{T} \quad .\end{aligned}$$

We substitute this expression for the derivative in equation (16.47) above into the immediately preceding equation to write

$$\begin{aligned}&\frac{\partial}{\partial H}\left[H^{-1}W(H)\right]\\ &= -\frac{1}{2}\left(I + K_{mm}\right)\left(H^{-1} \otimes H^{-1}\right)W(H)\\ &+ \frac{1}{2}W(H)\left[\left(\nu_0 - m - 1\right)vec\left(H^{-1}\right) - vec\left(M_0^{-1}\right)\right]vec\left(H^{-1}\right)^{T} \quad .\end{aligned}$$

Expanding and reordering all the terms yields the equation

$$\begin{aligned}\frac{\partial}{\partial H}\left[H^{-1}W(H)\right] = &\frac{1}{2}\left(\nu_0 - m - 1\right)vec\left(H^{-1}\right)vec\left(H^{-1}\right)^{T}W(H)\\ &- \frac{1}{2}K_{mm}\left(H^{-1} \otimes H^{-1}\right)W(H)\\ &- \frac{1}{2}\left(H^{-1} \otimes H^{-1}\right)W(H)\\ &- \frac{1}{2}vec\left(M_0^{-1}\right)vec\left(H^{-1}\right)^{T}W(H) \quad .\end{aligned}$$

If we integrate both sides with respect to H, using the knowledge that

$$\int_{-\infty}^{\infty} \frac{\partial}{\partial H}\left[H^{-1}W(H)\right]dH = 0 \quad ,$$

we obtain the sought after third and final equation

$$0 = \frac{1}{2}\left(\nu_0 - m - 1\right)\left[\int vec\left(H^{-1}\right)vec\left(H^{-1}\right)^{T}W(H)dH\right]$$

$$
\begin{aligned}
&-\frac{1}{2}\left[\int K_{mm}\left(H^{-1}\otimes H^{-1}\right)W(H)dH\right] \\
&-\frac{1}{2}\left[\int \left(H^{-1}\otimes H^{-1}\right)W(H)dH\right] \\
&-\frac{1}{2}vec\left(M_0^{-1}\right)\left[\int vec\left(H^{-1}\right)^T W(H)dH\right] \quad .
\end{aligned}
$$

The term on the right is amenable to simplification because it is merely the expectation of the inverse of H, which we have already calculated

$$
\begin{aligned}
0 &= \frac{1}{2}\left(\nu_0 - m - 1\right)\left[\int vec\left(H^{-1}\right)vec\left(H^{-1}\right)^T W(H)dH\right] \\
&-\frac{1}{2}\left[\int K_{mm}\left(H^{-1}\otimes H^{-1}\right)W(H)dH\right] \\
&-\frac{1}{2}\left[\int \left(H^{-1}\otimes H^{-1}\right)W(H)dH\right] \\
&= \frac{1}{2}vec\left(M_0^{-1}\right)\left[\frac{1}{\left(\nu_0 - m - 1\right)}vec\left(M_0^{-1}\right)^T\right] \quad .
\end{aligned}
$$

Eliminating the 1/2 factors and rearranging gives the final form of the third equation

$$
\begin{aligned}
&\left(\nu_0 - m - 1\right)\left[\int vec\left(H^{-1}\right)vec\left(H^{-1}\right)^T W(H)dH\right] \\
&-\left[\int K_{mm}\left(H^{-1}\otimes H^{-1}\right)W(H)dH\right] \\
&-\left[\int \left(H^{-1}\otimes H^{-1}\right)W(H)dH\right] \\
&= \frac{1}{\nu_0 - m - 1}vec\left(M_0^{-1}\right)vec\left(M_0^{-1}\right)^T \quad .
\end{aligned}
\tag{16.51}
$$

16.2.6.2.4 Solving the Three Equations and Three Unknowns We now have everything we need to determine the second moment of the inverse Wishart. We will want to write equations (16.41), (16.42), and (16.51) as a system of equations in matrix form. To do so, we will want to define

$$
\begin{aligned}
T_1 &= \left[\int vec\left(H^{-1}\right)vec\left(H^{-1}\right)^T W(H)dH\right] \\
T_2 &= K_{mm}\left[\int \left(H^{-1}\otimes H^{-1}\right)W(H)dH\right] \\
T_3 &= S_0 = \left[\int \left(H^{-1}\otimes H^{-1}\right)W(H)dH\right] \quad ,
\end{aligned}
$$

and

$$M_1 = \left(M_0^{-1} \otimes M_0^{-1}\right)$$
$$M_2 = K_{mm}\left(M_0^{-1} \otimes M_0^{-1}\right)$$
$$M_3 = vec\left(M_0^{-1}\right) vec\left(M_0^{-1}\right)^T \quad .$$

The equivalence of T_3 and S_0 is obvious by noticing the final expression for T_3 immediately above and the definition of S_0 in equation (16.38). We will use this equivalency later, but for now we will maintain the T_3 notation and write

$$\begin{aligned} &\begin{bmatrix} -1 & -1 & \nu_0 - m - 1 \\ -1 & \nu_0 - m - 1 & -1 \\ \nu_0 - m - 1 & -1 & -1 \end{bmatrix} \begin{bmatrix} T_1 \\ T_2 \\ T_3 \end{bmatrix} \\ &= \frac{1}{\nu_0 - m - 1} \begin{bmatrix} M_1 \\ M_2 \\ M_3 \end{bmatrix} \quad . \end{aligned} \tag{16.52}$$

This is written in expanded form

$$\begin{aligned} &\begin{bmatrix} -1 & -1 & \nu_0 - m - 1 \\ -1 & \nu_0 - m - 1 & -1 \\ \nu_0 - m - 1 & -1 & -1 \end{bmatrix} \\ &\times \begin{bmatrix} \left[\int vec\left(H^{-1}\right) vec\left(H^{-1}\right)^T W(H) dH\right] \\ K_{mm}\left[\int \left(H^{-1} \otimes H^{-1}\right) W(H) dH\right] \\ \left[\int \left(H^{-1} \otimes H^{-1}\right) W(H) dH\right] \end{bmatrix} \\ &= \frac{1}{\nu_0 - m - 1} \begin{bmatrix} M_0^{-1} \otimes M_0^{-1} \\ K_{mm}\left(M_0^{-1} \otimes M_0^{-1}\right) \\ vec\left(M_0^{-1}\right) vec\left(M_0^{-1}\right)^T \end{bmatrix} \quad . \end{aligned}$$

Consider the matrix

$$\begin{bmatrix} -a & -a & b \\ -a & b & -a \\ b & -a & -a \end{bmatrix} \quad .$$

We can easily establish the inverse by considering the matrix product

$$\begin{aligned} &\begin{bmatrix} -a & -a & b \\ -a & b & -a \\ b & -a & -a \end{bmatrix} \begin{bmatrix} a & a & b-a \\ a & b-a & a \\ b-a & a & a \end{bmatrix} \\ &= (b - 2a)(b + a) \begin{bmatrix} 1 & 0 & 0 \\ 0 & 1 & 0 \\ 0 & 0 & 1 \end{bmatrix} \quad . \end{aligned}$$

Therefore we know the inverse

$$\begin{bmatrix} -a & -a & b \\ -a & b & -a \\ b & -a & -a \end{bmatrix}^{-1} = \frac{1}{(b-2a)(b+a)} \begin{bmatrix} a & a & b-a \\ a & b-a & a \\ b-a & a & a \end{bmatrix} .$$

If we set $a = 1$ and $b = \nu_0 - m - 1$, we write the previous equation

$$\begin{bmatrix} -1 & -1 & \nu_0 - m - 1 \\ -1 & \nu_0 - m - 1 & -1 \\ \nu_0 - m - 1 & -1 & -1 \end{bmatrix}^{-1}$$
$$= \frac{1}{(\nu_0 - m - 3)(\nu_0 - m)} \begin{bmatrix} 1 & 1 & \nu_0 - m - 2 \\ 1 & \nu_0 - m - 2 & 1 \\ \nu_0 - m - 2 & 1 & 1 \end{bmatrix} .$$

We can left multiply equation (16.52) by this inverse matrix and write

$$\begin{bmatrix} T_1 \\ T_2 \\ T_3 \end{bmatrix} = \frac{1}{(\nu_0 - m)(\nu_0 - m - 1)(\nu_0 - m - 3)}$$
$$\times \begin{bmatrix} 1 & 1 & \nu_0 - m - 2 \\ 1 & \nu_0 - m - 2 & 1 \\ \nu_0 - m - 2 & 1 & 1 \end{bmatrix} \begin{bmatrix} M_1 \\ M_2 \\ M_3 \end{bmatrix} .$$

We can write the three equations this implies as follows

$$\begin{aligned} T_1 &= \frac{1}{(\nu_0 - m)(\nu_0 - m - 1)(\nu_0 - m - 3)} \\ &\quad [M_1 + M_2 + (\nu_0 - m - 2) M_3] \\ T_2 &= \frac{1}{(\nu_0 - m)(\nu_0 - m - 1)(\nu_0 - m - 3)} \\ &\quad [M_1 + (\nu_0 - m - 2) M_2 + M_3] \\ T_3 &= \frac{1}{(\nu_0 - m)(\nu_0 - m - 1)(\nu_0 - m - 3)} \\ &\quad [(\nu_0 - m - 2) M_1 + M_2 + M_3] \quad . \end{aligned}$$

The third term is the sought after value of $T_3 = S_0$:

$$\begin{aligned} T_3 = S_0 &= \left[\int \left(H^{-1} \otimes H^{-1} \right) W(H) dH \right] \\ &= \frac{(\nu_0 - m - 2)}{(\nu_0 - m)(\nu_0 - m - 1)(\nu_0 - m - 3)} \left(M_0^{-1} \otimes M_0^{-1} \right) \\ &+ \frac{1}{(\nu_0 - m)(\nu_0 - m - 1)(\nu_0 - m - 3)} K_{mm} \left(M_0^{-1} \otimes M_0^{-1} \right) \end{aligned}$$

$$+\frac{1}{(\nu_0-m)(\nu_0-m-1)(\nu_0-m-3)} vec\left(M_0^{-1}\right) vec\left(M_0^{-1}\right)^T \quad .$$

This is precisely the answer given in Wiki[29], presumably merely copied from the literature somewhere. We have chosen to develop it step by step rather than copying it from some other reference. We can apply the Kronecker product rule

$$\left({M_0}^{-1} \otimes {M_0}^{-1}\right) = (M_0 \otimes M_0)^{-1} \quad .$$

Returning to the definition of the variance-covariance matrix in equation (16.39) and substituting the derived expression for S_0, we can write

$$\begin{aligned} VCV =& -\frac{1}{(\nu_0-m-1)^2}(M_0 \otimes M_0) \\ &+ (M_0 \otimes M_0) \\ &\times \int \left(H^{-1} \otimes H^{-1}\right) W(H) dH \\ &\times (M_0 \otimes M_0) \quad . \end{aligned}$$

Substituting the foregoing result

$$\begin{aligned} VCV =& -\frac{1}{(\nu_0-m-1)^2}(M_0 \otimes M_0) \\ &+ (M_0 \otimes M_0) \\ &\times \left[\begin{array}{l} \frac{(\nu_0-m-2)}{(\nu_0-m)(\nu_0-m-1)(\nu_0-m-3)}(M_0 \otimes M_0)^{-1} \\ +\frac{1}{(\nu_0-m)(\nu_0-m-1)(\nu_0-m-3)} K_{mm} (M_0 \otimes M_0)^{-1} \\ +\frac{1}{(\nu_0-m)(\nu_0-m-1)(\nu_0-m-3)} \left[vec\left({M_0}^{-1}\right) vec\left({M_0}^{-1}\right)^T\right] \end{array}\right] \\ &\times (M_0 \otimes M_0) \quad , \end{aligned}$$

which simplifies to

$$\begin{aligned} VCV =& -\frac{1}{(\nu_0-m-1)^2}(M_0 \otimes M_0) \\ &+ \frac{(\nu_0-m-2)}{(\nu_0-m)(\nu_0-m-1)(\nu_0-m-3)}(M_0 \otimes M_0) \\ &+ \frac{1}{(\nu_0-m)(\nu_0-m-1)(\nu_0-m-3)}(M_0 \otimes M_0) K_{mm} \\ &+ \frac{1}{(\nu_0-m)(\nu_0-m-1)(\nu_0-m-3)} \\ &\times (M_0 \otimes M_0)\left[vec\left({M_0}^{-1}\right) vec\left({M_0}^{-1}\right)^T\right](M_0 \otimes M_0) \quad . \end{aligned} \tag{16.53}$$

[29] See https://en.wikipedia.org/wiki/Kronecker_product.

This matrix is $mm \times mm$. We have not seen this specific result before, but it is certainly a direct consequence of the second moment of the inverse Wishart developed by von Rosen[101]. In the next section, in which we work with a numerical example, the structure of these equations will be far more apparent. That numerical example is going to be helpful in understanding the mathematics here.

16.2.7 Analysis of the Prior–The Predictive Distribution

As with the univariate case, the reason we have gone to all this trouble is to "predict something." We need to develop a generalized representation of prediction such as we did in Chapter 4. Prediction is going to be materially harder in this multivariate case. Suppose we wanted to define q vectors of length p, the elements of each one of the q vectors comprised of p independent variables, and we wanted to predict their joint probability. To see how, define the following matrices

- D is a $q \times p$ matrix, and it represents the specific independent variable settings, i.e., the conditioning variables. There are q distinct vectors of independent variable settings, and each is of length p
- B is a $p \times m$ matrix, and it represents the pm coefficients being inferred
- W is a $q \times m$ matrix, and it represents the q variables to be projected, each of length m
- V is a $q \times m$ matrix, and it represents the random terms related to the error terms

Analogous to the univariate case, we want to predict the rows of the matrix W according to the formula

$$\begin{aligned} &W = DB + V \\ &(q \times m) = (q \times p)(p \times m) + (q \times m) \quad . \end{aligned}$$

Bayes theorem tells us that

$$\begin{aligned} \{W, D, B, \Sigma\} &= \{W, D|B, \Sigma\}\{B, \Sigma\} \\ &= \{B, \Sigma|W, D\}\{W, D\} \\ &= \{B, \Sigma|W, D\}\, c_1 \quad . \end{aligned}$$

We rearrange this into standard Bayes theorem form in which the answer is a conditional probability distribution

$$\{B, \Sigma|W, D\} = const\, \{W, D|B, \Sigma\}\{B, \Sigma\} \quad .$$

The first term on the right hand side is a likelihood function whose form is easy to develop. The second term is the density over model coefficients that was developed in equations (16.15).

We begin by developing the functional form for the likelihood function for the predictive distribution. The marginal distribution over the underlying variables will be easy. Returning to equation (15.5), which we rewrite here so that it is contiguous to our development and easy to refer to

$$LF = \{X, Y|B, \Sigma\} = \frac{1}{(2\pi)^{\frac{nm}{2}} |\Sigma|^{\frac{n}{2}}} e^{-\frac{1}{2}Trace\left[(Y-XB)^T(Y-XB)\Sigma^{-1}\right]} \quad .$$

Suppose we made the variable substitutions

$$\begin{aligned} n &\leftarrow q \\ Y &\leftarrow W \quad (q \times m) \\ X &\leftarrow D \quad (q \times p) \\ B &\leftarrow B \quad (p \times m) \\ \Sigma &\leftarrow \Sigma \quad (m \times m) \quad . \end{aligned}$$

This would allow us to write the likelihood function as follows

$$LF = \{W, D|B, \Sigma\} = \frac{1}{(2\pi)^{\frac{qm}{2}} |\Sigma|^{\frac{q}{2}}} \times e^{-\frac{1}{2}Trace\left[(W-DB)^T(W-DB)\Sigma^{-1}\right]} \quad . \tag{16.54}$$

Every row of D (and there are q such rows) is a p-vector whose conditional pdf we want to consider. W is the actual $q \times m$ variable we want to predict. This is precisely the likelihood function we need to consider here. We rewrite the second term of the Bayesian posterior above, which is the joint prior density over the coefficients expressed in equation (16.14) and implicitly (16.15):

$$\{B, \Sigma\} = \frac{1}{|\Sigma|^{\frac{m_0}{2}}} \frac{|M_0|^{\frac{\nu_0}{2}} |Q_0|^{\frac{m}{2}}}{(2\pi)^{\frac{pm}{2}} 2^{\frac{\nu_0 m}{2}} \Gamma_m\left(\frac{\nu_0}{2}\right)} \times e^{-\frac{1}{2}Trace\left(M_0\Sigma^{-1}\right)} e^{-\frac{1}{2}Trace\left[(B-B_0)^T Q_0 (B-B_0)\Sigma^{-1}\right]} \quad .$$

B is $p \times m$; Σ is $m \times m$. The product of the likelihood function times the prior is therefore

$$\{W, D|B, \Sigma\}\{B, \Sigma\} = \{W, D, B, \Sigma\}$$
$$= \frac{1}{(2\pi)^{\frac{qm}{2}} |\Sigma|^{\frac{q}{2}}} e^{-\frac{1}{2}Trace\left[(W-DB)^T(W-DB)\Sigma^{-1}\right]}$$

$$
\begin{aligned}
&\times \frac{1}{|\Sigma|^{\frac{m_0}{2}}} \frac{|M_0|^{\frac{\nu_0}{2}} |Q_0|^{\frac{m}{2}}}{(2\pi)^{\frac{pm}{2}} 2^{\frac{\nu_0 m}{2}} \Gamma_m\left(\frac{\nu_0}{2}\right)} \\
&\times e^{-\frac{1}{2}Trace\left[M_0\Sigma^{-1}\right]} e^{-\frac{1}{2}Trace\left[(B-B_0)^T Q_0 (B-B_0)\Sigma^{-1}\right]} \\
&= \frac{|M_0|^{\frac{\nu_0}{2}} |Q_0|^{\frac{m}{2}}}{2^{\frac{m(m_0-m-1+q)}{2}} \pi^{\frac{(p+q)m}{2}} \Gamma_m\left(\frac{\nu_0}{2}\right)} \\
&\times \frac{1}{|\Sigma|^{\frac{m_0+q}{2}}} e^{-\frac{1}{2}Trace\left[(W-DB)^T (W-DB)\Sigma^{-1}\right]} \\
&\times e^{-\frac{1}{2}Trace\left[M_0\Sigma^{-1}\right]} e^{-\frac{1}{2}Trace\left[(B-B_0)^T Q_0 (B-B_0)\Sigma^{-1}\right]} \quad .
\end{aligned}
$$

W is $q \times m$; D is $q \times p$. If we define

$$
c_1 = \frac{|M_0|^{\frac{\nu_0}{2}} |Q_0|^{\frac{m}{2}}}{2^{\frac{m(m_0-m-1+q)}{2}} \pi^{\frac{(p+q)m}{2}} \Gamma_m\left(\frac{\nu_0}{2}\right)} \quad , \tag{16.55}
$$

then we can write the product of the prior times the likelihood in the following form

$$
\begin{aligned}
&\{W, D|B, \Sigma\}\{B, \Sigma\} = \{W, D, B, \Sigma\} \\
&= c_1 |\Sigma|^{-\frac{q+m_0}{2}} e^{-\frac{1}{2}Trace\left(M_0\Sigma^{-1}\right)} \\
&\times e^{-\frac{1}{2}Trace(W-DB)^T (W-DB)\Sigma^{-1}} e^{-\frac{1}{2}Trace\left[(B-B_0)^T Q_0 (B-B_0)\Sigma^{-1}\right]} \quad .
\end{aligned}
$$

W is $q \times m$; D is $q \times p$. We will come back and reinsert the constant later. Temporarily ignoring the matrix terms that comprise the exponent (ignoring the multiplicative term $\frac{1}{2}$ and Σ and summing the final three matrices inside the trace operator for notational simplicity) allows us to write

$$
\begin{aligned}
T_0 &= M_0 \\
&+ (W - DB)^T (W - DB) \\
&+ (B - B_0)^T Q_0 (B - B_0) \quad .
\end{aligned}
$$

The dimensionality of the T_0 matrix is $m \times m$. We rewrite the transpose terms at the left in preparation for correct expansion

$$
\begin{aligned}
T_0 &= M_0 \\
&+ \left(W^T - B^T D^T\right) (W - DB) \\
&+ \left(B^T - {B_0}^T\right) Q_0 (B - B_0) \quad .
\end{aligned}
$$

We perform the first expansion into individual terms

$$\begin{aligned}T_0 = M_0 \\ &+ W^TW - W^TDB - B^TD^TW + B^TD^TDB \\ &+ B^TQ_0B - B^TQ_0B_0 - {B_0}^TQ_0B + {B_0}^TQ_0B_0 \quad .\end{aligned}$$

We reorganize around terms that are constant in B, linear in B, and quadratic in B

$$\begin{aligned}T_0 &= M_0 + {B_0}^TQ_0B_0 + W^TW \\ &- \left(W^TD + {B_0}^TQ_0\right)B \\ &- B^T\left(D^TW + Q_0B_0\right) \\ &+ B^T\left(D^TD + Q_0\right)B \quad .\end{aligned}$$

In preparation for completing the square, we add and subtract an as yet unknown $p \times m$ matrix G from B

$$B = B - G + G \quad ,$$

and express

$$B - G = \Lambda \Rightarrow B = G + \Lambda \quad ,$$

which allows us to write

$$\begin{aligned}B &= \Lambda + G \\ B^T &= \Lambda^T + G^T \quad .\end{aligned}$$

We substitute for B in the foregoing quadratic equation

$$\begin{aligned}T_0 &= M_0 + {B_0}^TQ_0B_0 + W^TW \\ &- \left(W^TD + {B_0}^TQ_0\right)(\Lambda + G) \\ &- \left(\Lambda^T + G^T\right)\left(D^TW + Q_0B_0\right) \\ &+ \left(\Lambda^T + G^T\right)\left(D^TD + Q_0\right)(\Lambda + G) \quad .\end{aligned}$$

We now expand

$$\begin{aligned}T_0 &= M_0 + {B_0}^TQ_0B_0 + W^TW \\ &- \left(W^TD + {B_0}^TQ_0\right)\Lambda - \left(W^TD + {B_0}^TQ_0\right)G \\ &- \Lambda^T\left(D^TW + Q_0B_0\right) - G^T\left(D^TW + Q_0B_0\right) \\ &+ \Lambda^T\left(D^TD + Q_0\right)\Lambda + \Lambda^T\left(D^TD + Q_0\right)G \\ &+ G^T\left(D^TD + Q_0\right)\Lambda + G^T\left(D^TD + Q_0\right)G \quad .\end{aligned}$$

Reorganizing and regrouping terms that are constant in Λ, linear in Λ, and quadratic in Λ, we have

$$\begin{aligned} T_0 &= M_0 + {B_0}^T Q_0 B_0 + W^T W + G^T \left(D^T D + Q_0\right) G \\ &- \left(W^T D + {B_0}^T Q_0\right) G - G^T \left(D^T W + Q_0 B_0\right) \\ &- \left(W^T D + {B_0}^T Q_0\right) \Lambda + G^T \left(D^T D + Q_0\right) \Lambda \\ &- \Lambda^T \left(D^T W + Q_0 B_0\right) + \Lambda^T \left(D^T D + Q_0\right) G \\ &+ \Lambda^T \left(D^T D + Q_0\right) \Lambda \quad . \end{aligned}$$

Regrouping terms that are linear in Λ so we can solve for G that completes the square, we write

$$\begin{aligned} T_0 &= M_0 + {B_0}^T Q_0 B_0 + W^T W + G^T \left(D^T D + Q_0\right) G \\ &- \left(W^T D + {B_0}^T Q_0\right) G - G^T \left(D^T W + Q_0 B_0\right) \\ &+ \left[G^T \left(D^T D + Q_0\right) - \left(W^T D + {B_0}^T Q_0\right)\right] \Lambda \\ &+ \Lambda^T \left[\left(D^T D + Q_0\right) G - \left(D^T W + Q_0 B_0\right)\right] \\ &+ \Lambda^T \left(D^T D + Q_0\right) \Lambda \quad . \end{aligned}$$

Define the $p \times p$ matrix Ω that appears repeatedly in the previous equation

$$\Omega = D^T D + Q_0 \quad . \tag{16.56}$$

It is clear that this matrix Ω is $p \times p$ and is symmetric because each of the two constituents matrices is symmetric. Substitute the definition of Ω into the previous equation

$$\begin{aligned} T_0 &= M_0 + {B_0}^T Q_0 B_0 + W^T W + G^T \Omega G \\ &- \left(W^T D + {B_0}^T Q_0\right) G - G^T \left(D^T W + Q_0 B_0\right) \\ &+ \left[G^T \Omega - \left(W^T D + {B_0}^T Q_0\right)\right] \Lambda \\ &+ \Lambda^T \left[\Omega G - \left(D^T W + Q_0 B_0\right)\right] \\ &+ \Lambda^T \Omega \Lambda \quad . \end{aligned}$$

It is now easy to see how to solve for G to eliminate the second to last term that is linear in Λ^T. We set the term in brackets associated with the Λ^T term in the second to last line of the foregoing equation to 0. This allows us to solve for G and thereby complete the square

$$\begin{aligned} G &= \Omega^{-1} \left(D^T W + Q_0 B_0\right) \\ \Rightarrow G^T &= \left(W^T D + {B_0}^T Q_0\right) \Omega^{-1} \quad . \end{aligned}$$

We substitute this solution back into the second and third to last equations, recognizing that the term that is linear in Λ^T is zero, and the term that is linear in Λ is also zero

$$\begin{aligned}T_0 &= M_0 + {B_0}^T Q_0 B_0 + W^T W \\ &+ \left(W^T D + {B_0}^T Q_0\right) \Omega^{-1} \Omega \Omega^{-1} \left(D^T W + Q_0 B_0\right) \\ &- \left(W^T D + {B_0}^T Q_0\right) \Omega^{-1} \left(D^T W + Q_0 B_0\right) \\ &- \left(W^T D + {B_0}^T Q_0\right) \Omega^{-1} \left(D^T W + Q_0 B_0\right) \\ &+ \Lambda^T \Omega \Lambda \\ &= M_0 + {B_0}^T Q_0 B_0 + W^T W \\ &- \left(W^T D + {B_0}^T Q_0\right) \Omega^{-1} \left(D^T W + Q_0 B_0\right) \\ &+ \Lambda^T \Omega \Lambda \quad .\end{aligned}$$

We substitute the definition of Λ back into to write the final expression for the exponent

$$\begin{aligned}T_0 &= M_0 + {B_0}^T Q_0 B_0 + W^T W \\ &- \left(D^T W + Q_0 B_0\right)^T \Omega^{-1} \left(D^T W + Q_0 B_0\right) \\ &+ (B - G)^T \Omega (B - G) \quad .\end{aligned}$$

This allows us to write the density

$$\begin{aligned}\{W, D, B, \Sigma\} &= c_1 |\Sigma|^{-\frac{m_0+q}{2}} \\ &e^{-\frac{1}{2} Trace\left\{\left[M_0 + W^T W + {B_0}^T Q_0 B_0 - \left(W^T D + {B_0}^T Q_0\right) \Omega^{-1} \left(D^T W + Q_0 B_0\right)\right] \Sigma^{-1}\right\}} \\ &e^{-\frac{1}{2} Trace\left[(B-G)^T \Omega (B-G) \Sigma^{-1}\right]} \quad .\end{aligned}$$

We first focus on the latter exponent, the one that contains the quadratic in B. Consider the matrices

$$\begin{aligned}&(B - G)^T \\ &\Omega (B - G) \Sigma^{-1} \quad .\end{aligned}$$

Applying the Kronecker-vectorization theorems mentioned previously, we can write

$$\begin{aligned}&Trace\left[(B - G)^T \Omega (B - G) \Sigma^{-1}\right] \\ &= \left[vec (B - G)\right]^T vec \left[\Omega (B - G) \Sigma^{-1}\right] \quad ,\end{aligned}$$

and using the product of matrices theorem to write the following

$$\left[vec (B - G)\right]^T \left(\Sigma^{-1} \otimes \Omega\right) vec (B - G) \quad .$$

In order to keep the density current with the already-completed substitution, the density can be rewritten by altering the quadratic term to reflect the foregoing development (which will facilitate subsequent manipulations of the normal)

$$\{W, D, B, \Sigma\} = c_1 |\Sigma|^{-\frac{m_0+q}{2}}$$
$$e^{-\frac{1}{2}Trace\left\{\left[M_0+W^TW+B_0{}^TQ_0B_0-\left(W^TD+B_0{}^TQ_0\right)\Omega^{-1}\left(D^TW+Q_0B_0\right)\right]\Sigma^{-1}\right\}}$$
$$e^{-\frac{1}{2}[vec(B-G)]^T\left(\Sigma^{-1}\otimes\Omega\right)vec(B-G)} \quad .$$

As before, we have an eye toward writing the last exponential in this equation as a multivariate normal distribution. If we begin with the definition of the multivariate normal distribution in Wiki,[30] we see that the inverse variance in the Wiki normal Σ^{-1} corresponds to the term $\Sigma^{-1} \otimes \Omega$ in our equation, i.e.,

$$\Sigma^{-1} \to \Sigma^{-1} \otimes \Omega \quad .$$

This is the inverse of the variance-covariance matrix (by observation of the last exponent). The inverse of this expression, which is the variance-covariance of a multivariate normal with this as the inverse variance-covariance matrix would be

$$\Sigma = \Sigma \otimes \Omega^{-1} \quad .$$

We know that we will have to have the determinant of this term in the denominator of a multivariate normal, that determinant being

$$\left|\Sigma \otimes \Omega^{-1}\right| = \frac{|\Sigma|^p}{|\Omega|^m} \Rightarrow \left|\Sigma \otimes \Omega^{-1}\right|^{-1} = \frac{|\Omega|^m}{|\Sigma|^p} \quad .$$

The standard normal distribution from Wiki can be written

$$N(x; \mu, \sigma) = \frac{1}{(2\pi)^{\frac{p}{2}}} |\Sigma|^{-\frac{1}{2}} e^{-\frac{1}{2}(x-\mu)^T\Sigma^{-1}(x-\mu)} \quad .$$

We can make the substitution

$$N(x; \mu, \sigma) = \frac{1}{(2\pi)^{\frac{mp}{2}}} \frac{|\Omega|^{\frac{m}{2}}}{|\Sigma|^{\frac{p}{2}}} e^{-\frac{1}{2}(x-\mu)^T\left(\Sigma^{-1}\otimes\Omega\right)(x-\mu)} \quad .$$

We want to reverse the foregoing equation and express the exponential term in terms of the normal distribution times a constant

$$\frac{N(x; \mu, \sigma)(2\pi)^{\frac{mp}{2}}|\Sigma|^{\frac{p}{2}}}{|\Omega|^{\frac{m}{2}}} = e^{-\frac{1}{2}(x-\mu)^T\left(\Sigma^{-1}\otimes\Omega\right)(x-\mu)} \quad ,$$

which allows us to substitute for the final term, the normal term, to write

[30]https://en.wikipedia.org/wiki/Multivariate_normal_distribution.

$$\{W, D, B, \Sigma\} = c_1 |\Sigma|^{-\frac{m_0+q}{2}}$$
$$e^{-\frac{1}{2}Trace\left\{\left[M_0+W^TW+B_0{}^TQ_0B_0-\left(W^TD+B_0{}^TQ_0\right)\Omega^{-1}\left(D^TW+Q_0B_0\right)\right]\Sigma^{-1}\right\}}$$
$$\frac{N\left(B\right)\left(2\pi\right)^{\frac{mp}{2}}|\Sigma|^{\frac{p}{2}}}{|\Omega|^{\frac{m}{2}}} \quad .$$

This in turn allows us to write the density

$$\{W, D, B, \Sigma\} = c_1 \frac{(2\pi)^{\frac{mp}{2}}}{|\Omega|^{\frac{m}{2}}} |\Sigma|^{-\frac{m_0+q-p}{2}} N\left(B\right)$$
$$e^{-\frac{1}{2}Trace\left\{\left[M_0+W^TW+B_0{}^TQ_0B_0-\left(W^TD+B_0{}^TQ_0\right)\Omega^{-1}\left(D^TW+Q_0B_0\right)\right]\Sigma^{-1}\right\}} \quad .$$

W is $q \times m$; D is $q \times p$. Because $N(B)$ is the only term that contains B (remembering that B_0 is not the same as B; B_0 is a constant), it is easy to see how we integrate out the density over B to write the marginal distribution over Σ

(16.57)
$$\{W, D, \Sigma\} = c_1 \frac{(2\pi)^{\frac{mp}{2}}}{|\Omega|^{\frac{m}{2}}} |\Sigma|^{-\frac{m_0+q-p}{2}}$$
$$e^{-\frac{1}{2}Trace\left\{\left[M_0+W^TW+B_0{}^TQ_0B_0-\left(W^TD+B_0{}^TQ_0\right)\Omega^{-1}\left(D^TW+Q_0B_0\right)\right]\Sigma^{-1}\right\}} \quad .$$

W is $q \times m$; D is $q \times p$; Σ is $m \times m$. If we inspect the exponent, we see the following term resident therein

$$\begin{aligned} T_1 &= M_0 + B_0{}^T Q_0 B_0 + W^T W \\ &\quad - \left(W^T D + B_0{}^T Q_0\right) \Omega^{-1} \left(D^T W + Q_0 B_0\right) \quad . \end{aligned}$$

In preparation for completing the square, we add and subtract an as yet unknown $q \times m$ matrix N from W

$$W = W - N + N \quad ,$$

and express

$$W - N = \Psi \Rightarrow W = N + \Psi \quad ,$$

which allows us to write

$$\begin{aligned} W &= \Psi + N \\ W^T &= \Psi^T + N^T \quad . \end{aligned}$$

We substitute for W in the foregoing quadratic equation, modifying the $m \times m$

exponent to the form

$$\begin{aligned}
T_1 = M_0 &+ \left(\Psi^T + N^T\right)(\Psi + N) + {B_0}^T Q_0 B_0 \\
&- \left[\left(\Psi^T + N^T\right) D + {B_0}^T Q_0\right] \Omega^{-1} \left[D^T (\Psi + N) + Q_0 B_0\right] \quad .
\end{aligned}$$

We expand all the terms

$$\begin{aligned}
T_1 = M_0 &+ {B_0}^T Q_0 B_0 + \Psi^T \Psi + \Psi^T N + N^T \Psi + N^T N \\
&- \Psi^T D \Omega^{-1} D^T \Psi - \Psi^T D \Omega^{-1} D^T N \\
&- N^T D \Omega^{-1} D^T \Psi - N^T D \Omega^{-1} D^T N \\
&- \Psi^T D \Omega^{-1} Q_0 B_0 - N^T D \Omega^{-1} Q_0 B_0 \\
&- {B_0}^T Q_0 \Omega^{-1} D^T \Psi - {B_0}^T Q_0 \Omega^{-1} D^T N \\
&- {B_0}^T Q_0 \Omega^{-1} Q_0 B_0 \quad .
\end{aligned}$$

We combine terms into a quadratic in Ψ, a linear term in Ψ, and a constant

$$\begin{aligned}
T_1 = M_0 &+ {B_0}^T Q_0 B_0 + N^T \left(I_{qq} - D \Omega^{-1} D^T\right) N \\
&- N^T D \Omega^{-1} Q_0 B_0 - {B_0}^T Q_0 \Omega^{-1} D^T N - {B_0}^T Q_0 \Omega^{-1} Q_0 B_0 \\
&+ \left[N^T \left(I_{qq} - D \Omega^{-1} D^T\right) - {B_0}^T Q_0 \Omega^{-1} D^T\right] \Psi \\
&+ \Psi^T \left[\left(I_{qq} - D \Omega^{-1} D^T\right) N - D \Omega^{-1} Q_0 B_0\right] \\
&+ \Psi^T \left(I_{qq} - D \Omega^{-1} D^T\right) \Psi \quad .
\end{aligned}$$

The I_{qq} matrix can be verified by checking the dimension of the second matrix that is subtracted from it. We want to eliminate the term that is linear in Ψ third from last by finding the N that eliminates it

$$\begin{aligned}
&N^T \left(I_{qq} - D \Omega^{-1} D^T\right) - {B_0}^T Q_0 \Omega^{-1} D^T = 0 \\
&\Rightarrow \left(I_{qq} - D \Omega^{-1} D^T\right) N - D \Omega^{-1} Q_0 B_0 = 0 \\
&\Rightarrow N = \left(I_{qq} - D \Omega^{-1} D^T\right)^{-1} D \Omega^{-1} Q_0 B_0 \\
&\Rightarrow N^T = {B_0}^T Q_0 \Omega^{-1} D^T \left(I_{qq} - D \Omega^{-1} D^T\right)^{-1} \quad .
\end{aligned}$$

If we substitute for N and N^T in the former equation, we eliminate the terms that are linear in Ψ

$$\begin{aligned}
T_1 = M_0 &+ {B_0}^T Q_0 B_0 + N^T \left(I_{qq} - D \Omega^{-1} D^T\right) N \\
&- N^T D \Omega^{-1} Q_0 B_0 - {B_0}^T Q_0 \Omega^{-1} D^T N - {B_0}^T Q_0 \Omega^{-1} Q_0 B_0 \\
&+ \Psi^T \left(I_{qq} - D \Omega^{-1} D^T\right) \Psi \quad .
\end{aligned} \tag{16.58}$$

We need a simpler expression for N, which we get by considering an expression for the matrix embedded in the expression for N and its inverse. In particular, we postulate (without proof quite yet, but it is forthcoming) the following inverse relationship

$$\left[I_{qq} - D\left(D^T D + Q_0\right)^{-1} D^T\right]^{-1} = \left(I_{qq} + D{Q_0}^{-1} D^T\right) \quad . \tag{16.59}$$

This matrix is $q \times q$. We can verify this by multiplying and manipulating

$$\begin{aligned}
&\left[I_{qq} - D\left(D^T D + Q_0\right)^{-1} D^T\right] \left(I_{qq} + D{Q_0}^{-1} D^T\right) \\
&= I_{qq} + D{Q_0}^{-1} D^T - D\left(D^T D + Q_0\right)^{-1} D^T \left(I_{qq} + D{Q_0}^{-1} D^T\right) \\
&= I_{qq} + D\left[\left(D^T D + Q_0\right)^{-1} \left(D^T D + Q_0\right)\right] {Q_0}^{-1} D^T \\
&- D\left(D^T D + Q_0\right)^{-1} D^T \left(I_{qq} + D{Q_0}^{-1} D^T\right) \\
&= I_{qq} + D\left(D^T D + Q_0\right)^{-1} \left[\begin{array}{l} \left(D^T D + Q_0\right) {Q_0}^{-1} D^T \\ -D^T \left(I_{qq} + D{Q_0}^{-1} D^T\right) \end{array}\right] \\
&= I_{qq} + D\left(D^T D + Q_0\right)^{-1} \left[\begin{array}{l} D^T D{Q_0}^{-1} D^T + Q_0{Q_0}^{-1} D^T \\ -D^T - D^T D{Q_0}^{-1} D^T \end{array}\right] \\
&= I_{qq} + D\left(D^T D + Q_0\right)^{-1} \left[\begin{array}{l} D^T D{Q_0}^{-1} D^T + D^T \\ -D^T - D^T D{Q_0}^{-1} D^T \end{array}\right] \\
&= I_{qq} + D\left(D^T D + Q_0\right)^{-1} 0 \\
&= I_{qq} \quad .
\end{aligned}$$

Using the inverse identity in equation (16.59) as well as the definition of Ω in equation (16.56), we can develop a far simpler (and more intuitive) expression for N

$$\begin{aligned}
N &= \left(I_{qq} - D\Omega^{-1} D^T\right)^{-1} D\Omega^{-1} Q_0 B_0 \\
&= \left(I_{qq} + D{Q_0}^{-1} D^T\right) D\Omega^{-1} Q_0 B_0 \\
&= D\Omega^{-1} Q_0 B_0 + D{Q_0}^{-1} (D^T D) \Omega^{-1} Q_0 B_0 \\
&= D\Omega^{-1} Q_0 B_0 + D{Q_0}^{-1} \left[\Omega - Q_0\right] \Omega^{-1} Q_0 B_0 \\
&= D\Omega^{-1} Q_0 B_0 + D{Q_0}^{-1} \left[\Omega\Omega^{-1} Q_0 B_0 - Q_0 \Omega^{-1} Q_0 B_0\right] \\
&= D\Omega^{-1} Q_0 B_0 + D{Q_0}^{-1} \Omega\Omega^{-1} Q_0 B_0 - D{Q_0}^{-1} Q_0 \Omega^{-1} Q_0 B_0 \\
&= D\Omega^{-1} Q_0 B_0 + D{Q_0}^{-1} Q_0 B_0 - D\Omega^{-1} Q_0 B_0 \\
&= DB_0 \quad .
\end{aligned}$$

We rewrite this simple expression for N for future reference

$$N = DB_0 \Rightarrow N^T = {B_0}^T D^T \quad . \tag{16.60}$$

Keep in mind that even this simple expression for N eliminates all terms that are linear in Ψ. When we substitute it in the forthcoming expressions, we continue to eliminate all terms linear in Ψ. We substitute this simple expression for N from equation (16.60) into equation (16.58) and reduce the exponent materially

$$\begin{aligned}T_1 = M_0 + {B_0}^T Q_0 B_0 + N^T \left(I_{qq} - D\Omega^{-1}D^T\right) N \\ - N^T D\Omega^{-1}Q_0 B_0 - {B_0}^T Q_0 \Omega^{-1} D^T N - {B_0}^T Q_0 \Omega^{-1} Q_0 B_0 \\ + \Psi^T \left(I_{qq} - D\Omega^{-1}D^T\right) \Psi \quad .\end{aligned}$$

The process of reducing this to final form is arduous

$$\begin{aligned}T_1 = M_0 &+ {B_0}^T Q_0 B_0 + \left({B_0}^T D^T\right) \left(I_{qq} - D\Omega^{-1}D^T\right) \left(D B_0\right) \\ &- \left({B_0}^T D^T\right) D\Omega^{-1} Q_0 B_0 - {B_0}^T Q_0 \Omega^{-1} D^T \left(D B_0\right) - {B_0}^T Q_0 \Omega^{-1} Q_0 B_0 \\ &+ \Psi^T \left(I_{qq} - D\Omega^{-1}D^T\right) \Psi \\ =&M_0 + {B_0}^T Q_0 B_0 + {B_0}^T D^T D B_0 - {B_0}^T D^T D \Omega^{-1} D^T D B_0 \\ &- {B_0}^T D^T D \Omega^{-1} Q_0 B_0 - {B_0}^T Q_0 \Omega^{-1} D^T D B_0 - {B_0}^T Q_0 \Omega^{-1} Q_0 B_0 \\ &+ \Psi^T \left(I_{qq} - D\Omega^{-1}D^T\right) \Psi \\ =&M_0 + {B_0}^T \left(Q_0 + D^T D\right) B_0 - {B_0}^T D^T D \Omega^{-1} D^T D B_0 \\ &- {B_0}^T D^T D \Omega^{-1} Q_0 B_0 - {B_0}^T Q_0 \Omega^{-1} \left(D^T D + Q_0\right) B_0 \\ &+ \Psi^T \left(I_{qq} - D\Omega^{-1}D^T\right) \Psi \\ =&M_0 + {B_0}^T \Omega B_0 - {B_0}^T D^T D \Omega^{-1} D^T D B_0 - {B_0}^T D^T D \Omega^{-1} Q_0 B_0 \\ &- {B_0}^T Q_0 B_0 \\ &+ \Psi^T \left(I_{qq} - D\Omega^{-1}D^T\right) \Psi \\ =&M_0 + {B_0}^T \Omega B_0 - {B_0}^T Q_0 B_0 \\ &- {B_0}^T D^T D \Omega^{-1} \left(D^T D + Q_0\right) B_0 \\ &+ \Psi^T \left(I_{qq} - D\Omega^{-1}D^T\right) \Psi \\ =&M_0 + {B_0}^T \Omega B_0 - {B_0}^T Q_0 B_0 \\ &- {B_0}^T D^T D \Omega^{-1} \Omega B_0 \\ &+ \Psi^T \left(I_{qq} - D\Omega^{-1}D^T\right) \Psi \\ =&M_0 + {B_0}^T \Omega B_0 - {B_0}^T Q_0 B_0 - {B_0}^T D^T D B_0 \\ &+ \Psi^T \left(I_{qq} - D\Omega^{-1}D^T\right) \Psi \\ =&M_0 + {B_0}^T \Omega B_0 - {B_0}^T \left(Q_0 + D^T D\right) B_0 \\ &+ \Psi^T \left(I_{qq} - D\Omega^{-1}D^T\right) \Psi \\ =&M_0 + {B_0}^T \Omega B_0 - {B_0}^T \left(Q_0 + D^T D\right) B_0 \\ &+ \Psi^T \left(I_{qq} - D\Omega^{-1}D^T\right) \Psi\end{aligned}$$

$$=M_0 + \Psi^T \left(I_{qq} - D\Omega^{-1}D^T\right)\Psi \quad .$$

If we substitute for $\Psi = W - N = W - DB_0$, then we can write the final answer

$$T_1 = M_0 + (W - DB_0)^T \left(I_{qq} - D\Omega^{-1}D^T\right)(W - DB_0) \quad .$$

We take the inverse of both sides of equation (16.59) and substitute to write the final equation

$$T_1 = M_0 + (W - DB_0)^T \left(I_{qq} + D{Q_0}^{-1}D^T\right)^{-1}(W - DB_0) \quad . \tag{16.61}$$

This is quite simple. Our density in equation (16.57) has thus reduced to the following form (substituting the expanded expression for Ω from equation (16.56))

$$\{W, D, \Sigma\} = c_1 \frac{(2\pi)^{\frac{mp}{2}}}{\left|D^T D + Q_0\right|^{\frac{m}{2}}} |\Sigma|^{-\frac{m_0+q-p}{2}} e^{-\frac{1}{2}Trace\left(T_1\Sigma^{-1}\right)} \quad . \tag{16.62}$$

W is $q \times m$; D is $q \times p$; Σ is $m \times m$. The next order of business is to eliminate the rather complex determinant in the denominator. We do that by appealing to the matrix determinant lemma[31]

$$\left|A + UV^T\right| = \left|I + V^T A^{-1} U\right| |A| \quad .$$

If we substitute the parameters of our problem into the lemma

$$\begin{aligned} U &= D^T \\ V &= D^T \\ A &= Q_0 \quad , \end{aligned}$$

we can write

$$\left|Q_0 + D^T D\right| = \left|I_{qq} + DQ_0^{-1}D^T\right| |Q_0| \quad .$$

We substitute this into equation (16.62) to write

$$\{W, D, \Sigma\} = c_1 \frac{(2\pi)^{\frac{mp}{2}} |Q_0|^{-\frac{m}{2}}}{\left|I_{qq} + DQ_0^{-1}D^T\right|^{\frac{m}{2}}} |\Sigma|^{-\frac{m_0+q-p}{2}} e^{-\frac{1}{2}Trace\left(T_1\Sigma^{-1}\right)} \quad .$$

W is $q \times m$; D is $q \times p$; Σ is $m \times m$. We then substitute the definition of c_1 from equation (16.55) and simplify to write

[31] See https://en.wikipedia.org/wiki/Matrix_determinant_lemma.

$$\{W, D, \Sigma\} = \frac{|M_0|^{\frac{\nu_0}{2}}}{2^{\frac{m(m_0-m-p-1+q)}{2}}\pi^{\frac{qm}{2}}\Gamma_m\left(\frac{\nu_0}{2}\right)} \times \frac{|\Sigma|^{-\frac{m_0-p+q}{2}}}{\left|I_{qq}+DQ_0^{-1}D^T\right|^{\frac{m}{2}}} e^{-\frac{1}{2}Trace\left(T_1\Sigma^{-1}\right)} \quad .$$

We then appeal to the definition of the prior degrees of freedom, which we rewrite for convenience

$$\nu_0 = m_0 - p - m - 1 \Rightarrow m_0 - p = \nu_0 + m + 1 \quad .$$

Substituting this equation into the previous equation allows us to write

$$\{W, D, \Sigma\} = \frac{|M_0|^{\frac{\nu_0}{2}}}{2^{\frac{m(\nu_0+q)}{2}}\pi^{\frac{qm}{2}}\Gamma_m\left(\frac{\nu_0}{2}\right)} \frac{|\Sigma|^{-\frac{\nu_0+m+q+1}{2}}}{\left|I_{qq}+DQ_0^{-1}D^T\right|^{\frac{m}{2}}} \times e^{-\frac{1}{2}Trace\left(T_1\Sigma^{-1}\right)} \quad .$$

We break the exponent into its original constituents using equation (16.61) to write

$$\{W, \Sigma\} = \frac{|M_0|^{\frac{\nu_0}{2}}}{2^{\frac{m(\nu_0+q)}{2}}\pi^{\frac{qm}{2}}\Gamma_m\left(\frac{\nu_0}{2}\right)} \frac{|\Sigma|^{-\frac{\nu_0+m+q+1}{2}}}{\left|I_{qq}+DQ_0^{-1}D^T\right|^{\frac{m}{2}}} \times e^{-\frac{1}{2}Trace\left(M_0\Sigma^{-1}\right)} \times e^{-\frac{1}{2}Trace\left[(W-DB_0)^T\left(I_{qq}+DQ_0^{-1}D^T\right)^{-1}(W-DB_0)\Sigma^{-1}\right]} \quad . \tag{16.63}$$

W is $q \times m$; Σ is $m \times m$. We have dropped the D in the distribution because it is understood. We want to create a normal distribution in place of the last term in this equation. We start with the vectorization and trace theorem

$$Trace\left(A^T B\right) = vec(A)^T vec(B) \quad .$$

We define

$$A^T = (W - DB_0)^T$$
$$B = \left(I_{qq} + DQ_0^{-1}D^T\right)^{-1}(W - DB_0)\,\Sigma^{-1} \quad .$$

and write the trace in the exponent of the final exponential term as follows

$$Trace\,(.) = vec(W - DB_0)^T vec\left[\left(I_{qq} + DQ_0^{-1}D^T\right)^{-1}(W - DB_0)\,\Sigma^{-1}\right] \quad .$$

We appeal to the other theorem, which stipulates that

$$vec\,(AXB) = \left(B^T \otimes A\right) vec\,(X) \quad .$$

If we define

$$\begin{aligned} A &= \left(I_{qq} + DQ_0^{-1}D^T\right)^{-1} \\ B &= \Sigma^{-1} \\ X &= (W - DB_0) \quad , \end{aligned}$$

then the right hand vector in the trace term becomes

$$vec\,(AXB) = \left[\Sigma^{-1} \otimes \left(I_{qq} + DQ_0^{-1}D^T\right)^{-1}\right] vec\,(W - DB_0)$$

and the trace term itself becomes

$$Trace\,(.) = vec(W - DB_0)^T \left[\Sigma^{-1} \otimes \left(I_{qq} + DQ_0^{-1}D^T\right)^{-1}\right] vec\,(W - DB_0) \quad .$$

For notational convenience, we write

$${\Sigma_W}^{-1} = \Sigma^{-1} \otimes \left(I_{qq} + DQ_0^{-1}D^T\right)^{-1} \Rightarrow \Sigma_W = \Sigma \otimes \left(I_{qq} + DQ_0^{-1}D^T\right) \quad .$$

We note that the left term is $m \times m$ and the right term is $q \times q$. We apply the Kronecker determinant theorem to write

$$|\Sigma_W| = |\Sigma|^q \left|I_{qq} + DQ_0^{-1}D^T\right|^m \Rightarrow |\Sigma_W|^{\frac{1}{2}} = |\Sigma|^{\frac{q}{2}} \left|I_{qq} + DQ_0^{-1}D^T\right|^{\frac{m}{2}} \quad .$$

We are now seeing evidence of the emerging multivariate normal distribution of dimension mq, namely

$$\begin{aligned} &N\,(W; mq, DB_0, \Sigma_W) = \frac{1}{(2\pi)^{\frac{mq}{2}} |\Sigma_W|^{\frac{1}{2}}} e^{-\frac{1}{2}vec(W-DB_0)^T {\Sigma_W}^{-1} vec(W-DB_0)} \\ &\Rightarrow e^{-\frac{1}{2}vec(W-DB_0)^T {\Sigma_W}^{-1} vec(W-DB_0)} \\ &= \times N\,(W; mq, DB_0, \Sigma_W)\,(2\pi)^{\frac{mq}{2}} |\Sigma_W|^{\frac{1}{2}} \quad . \end{aligned}$$

Substituting this into the density in equation (16.63) yields the result

$$\begin{aligned} \{W, \Sigma\} &= \frac{|M_0|^{\frac{\nu_0}{2}}}{2^{\frac{m(\nu_0+q)}{2}} \pi^{\frac{qm}{2}} \Gamma_m\left(\frac{\nu_0}{2}\right)} \frac{|\Sigma|^{-\frac{\nu_0+m+q+1}{2}}}{\left|I_{qq} + DQ_0^{-1}D^T\right|^{\frac{m}{2}}} \\ &\quad \times e^{-\frac{1}{2}Trace\left(M_0\Sigma^{-1}\right)} N\,(W; mq, DB_0, \Sigma_W)\,(2\pi)^{\frac{qm}{2}} |\Sigma_W|^{\frac{1}{2}} \\ &= \frac{|M_0|^{\frac{\nu_0}{2}} |\Sigma_W|^{\frac{1}{2}}}{2^{\frac{m\nu_0}{2}} \Gamma_m\left(\frac{\nu_0}{2}\right)} \frac{|\Sigma|^{-\frac{\nu_0+m+q+1}{2}}}{\left|I_{qq} + DQ_0^{-1}D^T\right|^{\frac{m}{2}}} \\ &\quad \times e^{-\frac{1}{2}Trace\left(M_0\Sigma^{-1}\right)} N\,(W; mq, DB_0, \Sigma_W) \quad . \end{aligned}$$

We substitute the foregoing expression for $|\Sigma_W|^{\frac{1}{2}}$ and simplify to write

$$\begin{aligned}\{W,\Sigma\} = {} & \frac{|M_0|^{\frac{\nu_0}{2}}}{2^{\frac{m\nu_0}{2}}\Gamma_m\left(\frac{\nu_0}{2}\right)}|\Sigma|^{-\frac{\nu_0+m+1}{2}} \\ & \times e^{-\frac{1}{2}Trace\left(M_0\Sigma^{-1}\right)}N\left(W;mq,DB_0,\Sigma_W\right) \quad .\end{aligned}$$

We integrate out the coefficients in the normal at the far right hand side to write

$$\{\Sigma\} = \frac{|M_0|^{\frac{\nu_0}{2}}}{2^{\frac{m\nu_0}{2}}\Gamma_m\left(\frac{\nu_0}{2}\right)}|\Sigma|^{-\frac{\nu_0+m+1}{2}}e^{-\frac{1}{2}Trace\left(M_0\Sigma^{-1}\right)} \quad .$$

This matrix is $m \times m$. From Wiki,[32] the inverse Wishart is

$$W^{-1}\left(X;\Psi,\nu,p\right) = \frac{|\Psi|^{\frac{\nu}{2}}}{2^{\frac{\nu p}{2}}\Gamma_p\left(\frac{\nu}{2}\right)}|X|^{-\frac{\nu+p+1}{2}}e^{-\frac{1}{2}Trace\left(\Psi X^{-1}\right)} \quad .$$

If we make the following substitution,

$$\begin{aligned} X &\leftarrow \Sigma \\ \Psi &\leftarrow M_0 \\ p &\leftarrow m \quad , \end{aligned}$$

we can write

$$W^{-1}\left(X;\Psi,\nu_0,m\right) = \frac{|M_0|^{\frac{\nu_0}{2}}}{2^{\frac{\nu_0 m}{2}}\Gamma_m\left(\frac{\nu_0}{2}\right)}|\Sigma|^{-\frac{\nu_0+m+1}{2}}e^{-\frac{1}{2}Trace\left(M_0\Sigma^{-1}\right)} \quad .$$

If we compare our distribution withe the Wishart distribution with the same parameters, we see our distribution is precisely Wishart if we equate the exponents on the Σ terms in the two equations

$$-\frac{m_0-p}{2} = -\frac{\nu+m+1}{2} \Rightarrow \nu = m_0 - p - m - 1 \triangleq \nu_0 \quad .$$

We have thus shown the result that

$$\{W,\Sigma\} = W^{-1}\left(\Sigma;M_0,\nu_0,m\right)N\left(W;DB_0,\Sigma_W,mq\right) \tag{16.64}$$

in which

$$\nu_0 = m_0 - p - m - 1$$

and

[32] See https://en.wikipedia.org/wiki/Inverse-Wishart_distribution.

$$\Sigma_W = \Sigma \otimes \left(I_{qq} + DQ_0^{-1}D^T\right)$$

using the following definitions

(16.65)
$$W^{-1}\left(\Sigma; M_0, \nu_0, m\right)$$
$$= \frac{|M_0|^{\frac{\nu_0}{2}}}{2^{\frac{\nu_0 m}{2}}\Gamma_m\left(\frac{\nu_0}{2}\right)}|\Sigma|^{-\frac{\nu_0+m+1}{2}}e^{-\frac{1}{2}Trace\left(M_0\Sigma^{-1}\right)}$$

and

$$N\left(W; mq, DB_0, \Sigma_W\right)$$
$$= \frac{1}{(2\pi)^{\frac{mq}{2}}|\Sigma_W|^{\frac{1}{2}}}e^{-\frac{1}{2}vec(W-DB_0)^T\Sigma_W{}^{-1}vec(W-DB_0)} \quad .$$

Notice that the inverse Wishart does not contain the variable q. This presages that the underlying inverse Wishart is common to all predictions. We notice by substitution in equation (16.36) that the variance-covariance matrix is

$$VCV = \frac{1}{\nu_0 - m - 1}\left[M_0 \otimes \left(I_{qq} + DQ_0^{-1}D^T\right)\right] \quad .$$

Trivially, the mean is

(16.66)
$$\langle W\rangle = DB_0 \quad .$$

16.2.7.1 Using $\{\beta, \Sigma\}$ to Develop $\{W\}$ by Substitution

We could begin with the joint density in equation (16.63) and proceed through the full, grassroots development of the marginals. It is easier to use an analogy developed for the joint density over the coefficients. In this section, we take an analogy/substitution approach to develop the salient results related to the density $\{W\}$, beginning with the expression for the joint density $\{B, \Sigma\}$ developed in equations (16.14) and (16.15). Instead of deriving $\{W\}$ from scratch, we are going to use the previously analyzed distribution $\{B, \Sigma\}$ in equations (16.14) and (16.15), substitute expressions into it term by term so that it represents the predictive distribution in equation (16.63), and then we are going to make those same substitutions in the results for the original distribution equation (16.14). Beginning with equation (16.14):

$$\{B, \Sigma\} = \frac{|M_0|^{\frac{\nu_0}{2}}|Q_0|^{\frac{m}{2}}}{(2\pi)^{\frac{pm}{2}}2^{\frac{\nu_0 m}{2}}\Gamma_m\left(\frac{\nu_0}{2}\right)|\Sigma|^{\frac{m_0}{2}}}$$
$$\times\, e^{-\frac{1}{2}Trace\left(M_0\Sigma^{-1}\right)}$$
$$\times\, e^{-\frac{1}{2}Trace\left[(B-B_0)^T Q_0 (B-B_0)\Sigma^{-1}\right]} \quad .$$

If we make the following substitutions into this to convert it exactly to the distribution in equation (16.63)

$$
\begin{aligned}
&p \leftarrow q \\
&B_0 \leftarrow DB_0 \\
&B \leftarrow W \\
&Q_0 \leftarrow \left(I_{qq} + D{Q_0}^{-1}D^T\right)^{-1} \Rightarrow {Q_0}^{-1} \leftarrow I_{qq} + D{Q_0}^{-1}D^T \\
&M_0 \leftarrow M_0 \\
&\nu_0 \leftarrow \nu_0 \quad ,
\end{aligned}
$$

we obtain the equation for the joint density

$$
\begin{aligned}
\{W, \Sigma\} = &\frac{|M_0|^{\frac{\nu_0}{2}}}{(2\pi)^{\frac{qm}{2}} 2^{\frac{\nu_0 m}{2}} \Gamma_m\left(\frac{\nu_0}{2}\right) \left|I_{qq} + D{Q_0}^{-1}D^T\right|^{\frac{m}{2}} |\Sigma|^{\frac{m_0}{2}}} \\
&\times e^{-\frac{1}{2}Trace\left(M_0\Sigma^{-1}\right)} \\
&\times e^{-\frac{1}{2}Trace\left[(W-DB_0)^T\left(I_{qq}+D{Q_0}^{-1}D^T\right)^{-1}(W-DB_0)\Sigma^{-1}\right]} \quad .
\end{aligned}
$$

If we begin with the marginal distribution over coefficients in equations (16.14) and (16.15) and make the same foregoing substitutions into the prior, we will obtain all the key results for the marginal predictive distribution. Making such substitution, we write the marginal distribution over the coefficients as follows

$$
\begin{aligned}
\{W\} = &\frac{\Gamma_m\left(\frac{\nu_0+q}{2}\right)}{\Gamma_m\left(\frac{\nu_0}{2}\right) \pi^{\frac{qm}{2}} \left[I_{qq} + DQ_0^{-1}D^T\right]^{\frac{m}{2}} |M_0|^{\frac{q}{2}}} \\
&\times \left| \begin{array}{l} I_{mm} + M_0^{-1}(W - DB_0)^T \\ \times\left(I_{qq} + DQ_0^{-1}D^T\right)^{-1}(W - DB_0) \end{array} \right|^{-\frac{\nu_0+q}{2}} \quad .
\end{aligned}
\tag{16.67}
$$

W is $q \times m$. The dimensionality of the determinant is $m \times m$. We are going to transform equation (16.67) using the Weinstein-Aronszajn identity[33] to use a $q \times q$ determinant rather than the embedded $m \times m$ determinant:

$$
\begin{aligned}
\{W\} = &\frac{\Gamma_m\left(\frac{\nu_0+q}{2}\right)}{\Gamma_m\left(\frac{\nu_0}{2}\right) \pi^{\frac{qm}{2}} \left[I_{qq} + DQ_0^{-1}D^T\right]^{\frac{m}{2}} |M_0|^{\frac{q}{2}}} \\
&\times \left| \begin{array}{l} I_{qq} + \left(I_{qq} + DQ_0^{-1}D^T\right)^{-1}(W - DB_0) \\ \times M_0^{-1}(W - DB_0)^T \end{array} \right|^{-\frac{\nu_0+q}{2}} \quad .
\end{aligned}
\tag{16.68}
$$

These are classic matrix Student's t distributions, and the second of the two, equation (16.68), is going to become critically important. Before proceeding,

[33] See https://en.wikipedia.org/wiki/Weinstein-Aronszajn_identity.

we review the dimensionality of the key terms

- B_0 is $p \times m$
- D is $q \times p$
- DB_0 is $q \times m$
- W is $q \times m$.

16.2.7.2 Special Case–Single Row Prediction

We will want to construct the special case in which D is a single row and is designated by the single row vector δ. We not only want to project for a particular row vector δ but also we will want to project for a unit row vector ε_k to get the probability distribution over a single coefficient. In preparation for the development in this section, we need to specialize the inverse gamma distribution to the single row case.

We begin with the joint density in equation (16.63). We substitute $q = 1$ and $D = \delta$ to write

$$\begin{aligned}\{\omega, \Sigma\} = & \frac{|M_0|^{\frac{\nu_0}{2}}}{2^{\frac{m(\nu_0+1)}{2}} \pi^{\frac{m}{2}} \Gamma_m\left(\frac{\nu_0}{2}\right)} \frac{|\Sigma|^{-\frac{\nu_0+m+2}{2}}}{\left(1+\delta Q_0^{-1}\delta^T\right)^{\frac{m}{2}}} \\ & \times e^{-\frac{1}{2}Trace\left(M_0\Sigma^{-1}\right)} \\ & \times e^{-\frac{1}{2}Trace\left[\frac{1}{1+\delta Q_0^{-1}\delta^T}(\omega-\delta B_0)^T(\omega-\delta B_0)\Sigma^{-1}\right]} \quad .\end{aligned}$$

ω is $1 \times m$; Σ is $m \times m$. We turn now to equation (16.64), which we specialize to $q = 1$ and $D = \delta$ as follows

$$\{\omega, \Sigma\} = W^{-1}\left(\Sigma; M_0, \nu_0, m\right) N\left(\omega; \delta B_0, \Sigma_\omega, m\right) \quad .$$

We need to examine the normal embedded here, which is

$$N\left(\omega; \delta B_0, \Sigma_\omega, m\right) = \frac{1}{(2\pi)^{\frac{m}{2}} |\Sigma_\omega|^{\frac{1}{2}}} e^{-\frac{1}{2}vec(\omega-\delta B_0)^T \Sigma_\omega{}^{-1} vec(\omega-\delta B_0)} \quad .$$

We set $q = 1$ and $D = \delta$ and substitute into

$$\Sigma_W = \Sigma \otimes \left(I_{qq} + DQ_0^{-1}D^T\right)$$

to obtain

$$\Sigma_\omega = \Sigma \otimes \left(1 + \delta Q_0^{-1}\delta^T\right)$$

$$= \left(1 + \delta Q_0^{-1}\delta^T\right)\Sigma$$

and

$$|\Sigma_\omega| = \left(1 + \delta Q_0^{-1}\delta^T\right)^m |\Sigma| \quad .$$

This implies that

$$\Sigma_\omega{}^{-1} = \frac{1}{1 + \delta Q_0^{-1}\delta^T}\Sigma^{-1} \quad .$$

This matrix is $m \times m$. Substituting into the normal distribution

$$N\left(\omega; m, DB_0, \Sigma_\omega\right) = \frac{1}{(2\pi)^{\frac{m}{2}} |\Sigma_\omega|^{\frac{1}{2}}} e^{-\frac{1}{2} vec(\omega - DB_0)^T \Sigma_\omega{}^{-1} vec(\omega - DB_0)} \quad ,$$

we can rewrite for clarity

$$\begin{aligned} & N\left[\omega; m, \delta B_0, \Sigma\left(1 + \delta Q_0^{-1}\delta^T\right)\right] \\ & = \frac{1}{\left[2\pi\left(1 + \delta Q_0^{-1}\delta^T\right)\right]^{\frac{m}{2}} |\Sigma|^{\frac{1}{2}}} \\ & \times e^{-\frac{1}{2} vec(\omega - \delta B_0)^T \frac{\Sigma^{-1}}{1 + \delta Q_0^{-1}\delta^T} vec(\omega - \delta B_0)} \quad . \end{aligned} \tag{16.69}$$

This is a m-variate normal distribution with

$$\begin{aligned} & \text{mean} = \delta B_0 = \mu_0 \\ & \text{variance-covariance} = \Sigma\left(1 + \delta Q_0^{-1}\delta^T\right) \quad . \end{aligned}$$

This distribution is used in equation (16.64) when we are considering a single row prediction matrix, which becomes equation (16.65) times equation (16.69).

Let us now turn to equation (16.67) and see how it specializes when $q = 1$ and $D = \delta$. After such substitutions, we have

$$\begin{aligned} \{\omega\} = & \frac{\Gamma_m\left(\frac{\nu_0+1}{2}\right)}{\Gamma_m\left(\frac{\nu_0}{2}\right)\pi^{\frac{m}{2}}\left(1 + \delta Q_0^{-1}\delta^T\right)^{\frac{m}{2}} |M_0|^{\frac{1}{2}}} \\ & \times \left| \begin{array}{l} I_{mm} + \frac{1}{1 + \delta Q_0^{-1}\delta^T} M_0^{-1} \\ \quad \times (\omega - \delta B_0)^T (\omega - \delta B_0) \end{array} \right|^{-\frac{\nu_0+1}{2}} \quad . \end{aligned} \tag{16.70}$$

We insert δ into equation (16.68), which has applied the Weinstein-Aronszajn identity,[34] to equation (16.70), to write

[34] See https://en.wikipedia.org/wiki/Weinstein-Aronszajn_identity.

$$(16.71)\qquad \{\omega\} = \frac{\Gamma_m\left(\frac{\nu_0+1}{2}\right)}{\Gamma_m\left(\frac{\nu_0}{2}\right)\pi^{\frac{m}{2}}\left[1+\delta Q_0^{-1}\delta^T\right]^{\frac{m}{2}}|M_0|^{\frac{1}{2}}}$$
$$\times\left|1+(\omega-\delta B_0)\frac{M_0^{-1}}{1+\delta Q_0^{-1}\delta^T}(\omega-\delta B_0)^T\right|^{-\frac{\nu_0+1}{2}}$$
$$= c_1\left[1+(\omega-\delta B_0)\frac{M_0^{-1}}{1+\delta Q_0^{-1}\delta^T}(\omega-\delta B_0)^T\right]^{-\frac{\nu_0+1}{2}} \quad .$$

By virtue of Weinstein-Aronszajn, we have moved to a situation in which the argument of the determinant is a scalar and therefore the equation represents a standard multivariate Student's t distribution. The analogous structure between this equation and equation (16.25) is obvious. We are going to be able to use that former equation, make some substitutions, and get to the bottom line for the projective distribution.

The expression in equation (16.67) remains a matrix Student's t distribution with dimension m, but the expression in equation (16.71) has simplified to an m-variate Student's t distribution. The mean remains

$$\text{mean} = \delta B_0 \quad .$$

This is $1 \times m$.

We are going to re-develop and simplify this shortly. In particular we are going to turn this into a standard multivariate Student's t distribution and identify the correct coefficient c_0. We are going to match the terms of a multivariate Student's t distribution term for term. The multivariate Student's t distribution from Wiki[35] is

$$\{x\} = \frac{\Gamma\left(\frac{\nu+p}{2}\right)}{\Gamma\left(\frac{\nu}{2}\right)\nu^{\frac{p}{2}}\pi^{\frac{p}{2}}|\Sigma|^{\frac{1}{2}}}\left[1+\frac{1}{\nu}(x-\mu)^T\Sigma^{-1}(x-\mu)\right]^{-\frac{\nu+p}{2}}$$

with variance-covariance matrix

$$VCV = \Sigma\frac{\nu}{\nu-2} \quad .$$

If we make the following variable substitutions

$$\begin{aligned} x &\leftarrow \omega \\ \mu &\leftarrow \delta B_0 \\ \nu &\leftarrow \xi \\ p &\leftarrow m \end{aligned}$$

[35] See https://en.wikipedia.org/wiki/Multivariate_t-distribution.

$$\Sigma^{-1} = \leftarrow \frac{M_0^{-1}\xi}{(1+\delta Q_0^{-1}\delta^T)}$$
$$\Rightarrow \Sigma = M_0\left(\frac{1+\delta Q_0^{-1}\delta^T}{\xi}\right)$$
$$\Rightarrow |\Sigma|^{\frac{1}{2}} = |M_0|^{\frac{1}{2}}\left(1+\delta Q_0^{-1}\delta^T\right)^{\frac{m}{2}}\xi^{-\frac{m}{2}}$$

we can write

$$\{\omega\} = \frac{\Gamma\left(\frac{\xi+m}{2}\right)}{\Gamma\left(\frac{\xi}{2}\right)\xi^{\frac{m}{2}}\pi^{\frac{m}{2}}|M_0|^{\frac{1}{2}}\left(1+\delta Q_0^{-1}\delta^T\right)^{\frac{m}{2}}\xi^{-\frac{m}{2}}}$$
$$\times\left[1+\frac{1}{\xi}(\omega-\delta B_0)^T\frac{M_0^{-1}\xi}{(1+\delta Q_0^{-1}\delta^T)}(\omega-\delta B_0)\right]^{-\frac{\xi+m}{2}}$$

and the variance-covariance matrix

$$VCV = M_0\left(\frac{1+\delta Q_0^{-1}\delta^T}{\xi}\right)\frac{\xi}{\xi-2} = M_0\frac{1+\delta Q_0^{-1}\delta^T}{\xi-2} \quad .$$

We can substitute for ξ into the exponent in equation (16.71) as follows

$$\frac{\nu_0+1}{2} = \frac{\xi+m}{2}$$
$$\Rightarrow \nu_0+1 = \xi+m$$
$$\Rightarrow \nu_0 = \xi+m-1$$
$$\Rightarrow \xi = \nu_0-m+1$$

to write

(16.72)
$$\{\omega|\delta\} = \frac{\Gamma\left(\frac{\nu_0+1}{2}\right)}{\Gamma\left(\frac{\nu_0+1-m}{2}\right)\left[\pi\left(1+\delta Q_0^{-1}\delta^T\right)\right]^{\frac{m}{2}}|M_0|^{\frac{1}{2}}}$$
$$\times\left[1+(\omega-\delta B_0)^T\frac{M_0^{-1}}{(1+\delta Q_0^{-1}\delta^T)}(\omega-\delta B_0)\right]^{-\frac{\nu_0+1}{2}} \quad .$$

and

(16.73)
$$VCV = M_0\frac{1+\delta Q_0^{-1}\delta^T}{\nu_0-m-1} \quad .$$

and

$$\text{mean} = \delta B_0$$

Recall that ω is a single row vector with m columns and δ is a single row vector with p columns. Equation (16.71) has now become a "regular" multivariate Student's t distribution, and it is a distribution over the $1 \times m$ matrix ω. It is an m-variate distribution. It is important to summarize the dimensionality of the key terms

- B_0 is $p \times m$
- δ is $1 \times p$
- δB_0 is $1 \times m$
- ω is $1 \times m$.

To solve for the marginal, consider the p-variate distribution in equation (4.15) and its relationship to the univariate marginal in equation (4.52). We want to adapt this relationship to the m variate equation developed previously, so suppose we make the following substitutions

$$\begin{aligned} \beta &\leftarrow x \\ p &\leftarrow m \\ \nu &\leftarrow \xi \\ H &\leftarrow H_{mm} \end{aligned}$$

in equations (4.15) and (4.52) to write

$$\{x\} = \frac{|H|^{\frac{1}{2}}\Gamma\left(\frac{\xi+m}{2}\right)}{(\pi\xi s^2)^{\frac{m}{2}}\Gamma\left(\frac{\xi}{2}\right)}\left[1 + (x-\mu)^T\left(\frac{H}{\xi s^2}\right)(x-\mu)\right]^{-\frac{\xi+m}{2}}$$

and

$$\{x_i\} = \frac{\Gamma\left(\frac{\xi+1}{2}\right)}{\Gamma\left(\frac{\xi}{2}\right)(\pi[H^{-1}]_{ii}\xi s^2)^{\frac{1}{2}}}\left[1 + \frac{1}{\xi s^2[H^{-1}]_{ii}}(x_i-\mu_i)^2\right]^{-\frac{\xi+1}{2}} .$$

We need to set the exponents of the two m-variates to be equal, i.e., $\xi = \nu_0 + 1 - m$ so that

$$\{x\} = \frac{|H|^{\frac{1}{2}}\Gamma\left(\frac{\nu_0+1}{2}\right)}{(\pi\xi s^2)^{\frac{m}{2}}\Gamma\left(\frac{\nu_0+1-m}{2}\right)}\left[1 + (x-\mu)^T\left(\frac{H}{\xi s^2}\right)(x-\mu)\right]^{-\frac{\nu_0+1}{2}}$$

and

$$\{x_i\} = \frac{\Gamma\left(\frac{\nu_0+2-m}{2}\right)}{\Gamma\left(\frac{\nu_0+1-m}{2}\right)\left(\pi[H^{-1}]_{ii}\xi s^2\right)^{\frac{1}{2}}}\left[1+\frac{1}{\xi s^2[H^{-1}]_{ii}}(x_i-\mu_i)^2\right]^{-\frac{\nu_0+2-m}{2}} .$$

We are now ready to make the substitutions

$$\begin{aligned}
&H \leftarrow M_0^{-1} \\
&\xi s^2 \leftarrow \left(1+\delta Q_0^{-1}\delta^T\right) \\
&\mu \leftarrow [\delta B_0] \\
&x \leftarrow \omega \quad .
\end{aligned}$$

When we make them and simplify, we can write the two equations in the new forms

$$\begin{aligned}
\text{(16.74)} \qquad \{\omega\} = &\frac{\Gamma\left(\frac{\nu_0+1}{2}\right)}{\Gamma\left(\frac{\nu_0+1-m}{2}\right)\left[\pi\left(1+\delta Q_0^{-1}\delta^T\right)\right]^{\frac{m}{2}}|M_0|^{\frac{1}{2}}} \\
&\times\left[1+(\omega-\delta B_0)^T\frac{M_0^{-1}}{\left(1+\delta Q_0^{-1}\delta^T\right)}(\omega-\delta B_0)\right]^{-\frac{\nu_0+1}{2}}
\end{aligned}$$

and

$$\begin{aligned}
\text{(16.75)} \quad \{\omega_i\} = &\frac{\Gamma\left(\frac{\nu_0+2-m}{2}\right)}{\Gamma\left(\frac{\nu_0+1-m}{2}\right)\left\{\pi[M_0]_{ii}\left(1+\delta Q_0^{-1}\delta^T\right)\right\}^{\frac{1}{2}}} \\
&\times\left[1+\frac{1}{\left(1+\delta Q_0^{-1}\delta^T\right)[M_0]_{ii}}\left(\omega_i-[\delta B_0]_i\right)^2\right]^{-\frac{\nu_0+2-m}{2}} \\
&= c_0\left[1+\frac{1}{\left(1+\delta Q_0^{-1}\delta^T\right)[M_0]_{ii}}\left(\omega_i-[\delta B_0]_i\right)^2\right]^{-\frac{\nu_0+2-m}{2}} .
\end{aligned}$$

The mean of this distribution is

$$\text{mean} = [\delta B_0]_{ii} \quad .$$

We need to develop the variance. We need to return to the fundamental structure of the univariate Student's t distribution in equation (4.44) and the variance calculated immediately thereafter in Chapter 4.

$$\frac{1}{k-4}\frac{P}{Q} \quad .$$

We need to find the term k that equates the exponent of our univariate Student's t distribution with the standard form

$$-\frac{\nu_0 + 2 - m}{2} = -\frac{k-1}{2} \quad .$$

This implies the result

$$k = \nu_0 + 3 - m$$
$$\Rightarrow k - 4 = \nu_0 - m - 1 \quad .$$

We see further that the precision coefficient in the univariate Student's t distribution is

$$\frac{Q}{P} = \frac{1}{\left(1 + \delta Q_0^{-1} \delta^T\right) [M_0]_{ii}} \quad .$$

Its reciprocal is

$$\frac{P}{Q} = \left(1 + \delta Q_0^{-1} \delta^T\right) [M_0]_{ii} \quad .$$

Meaning that the variance is

$$Var = \frac{P}{Q} \frac{1}{k-4} = \frac{\left(1 + \delta Q_0^{-1} \delta^T\right) [M_0]_{ii}}{\nu_0 - m - 1} \quad . \tag{16.76}$$

We need to reconcile the exponent in our univariate Student's t distribution with the standard form. We do so by positing a quantity η as follows

$$-\frac{\nu_0 + 2 - m}{2} = -\frac{\eta + 1}{2}$$

which implies that

$$\eta = \nu_0 + 1 - m \quad .$$

With this definition of η, we can write our univariate Student's t distribution in a fashion in which the constant of integration becomes apparent

$$\{\omega_i\} = \frac{\Gamma\left(\frac{\eta+1}{2}\right)}{\Gamma\left(\frac{\eta}{2}\right) \left[\pi \left(1 + \delta Q_0^{-1} \delta^T\right) [M_0]_{ii}\right]^{\frac{1}{2}}}$$
$$\times \left[1 + \frac{1}{\left(1 + \delta Q_0^{-1} \delta^T\right) [M_0]_{ii}} \left(\omega_i - [\delta B_0]_i\right)^2\right]^{-\frac{\eta+1}{2}} \quad .$$

We then substitute for η to write our final univariate Student's t distribution as follows

$$\{\omega_i\} = \frac{\Gamma\left(\frac{\nu_0-m+2}{2}\right)}{\Gamma\left(\frac{\nu_0-m+1}{2}\right)\left[\pi\left(1+\delta Q_0^{-1}\delta^T\right)[M_0]_{ii}\right]^{\frac{1}{2}}} \tag{16.77}$$
$$\times\left[1+\frac{1}{\left(1+\delta Q_0^{-1}\delta^T\right)[M_0]_{ii}}\left(\omega_i-[\delta B_0]_i\right)^2\right]^{-\frac{\nu_0-m+2}{2}} .$$

The variance of this univariate Student's t distribution appears in equation (16.76).

16.2.7.3 Predictions Using a Single Row in the General Case

Assume for the sake of illustration that $p = 5$ independent variables are present and thus that

$$\delta = [\delta_1, \delta_2, \delta_3, \delta_4, \delta_5] \Rightarrow \delta^T = \begin{bmatrix} \delta_1 \\ \delta_2 \\ \delta_3 \\ \delta_4 \\ \delta_5 \end{bmatrix} .$$

and that

$$B_0 = \begin{bmatrix} B_{11} & B_{12} \\ B_{21} & B_{22} \\ B_{31} & B_{32} \\ B_{41} & B_{42} \\ B_{51} & B_{52} \end{bmatrix} .$$

In this situation, we can write

$$\delta B_0 = [\delta_1, \delta_2, \delta_3, \delta_4, \delta_5] \begin{bmatrix} B_{11} & B_{12} \\ B_{21} & B_{22} \\ B_{31} & B_{32} \\ B_{41} & B_{42} \\ B_{51} & B_{52} \end{bmatrix} = \left[\sum_{i=1}^{5} \delta_i B_{i1} \quad \sum_{i=1}^{5} \delta_i B_{i2} \quad . \right] .$$

We continue

$$Q_0^{-1} = \begin{bmatrix} \zeta_{11} & \zeta_{12} & \zeta_{13} & \zeta_{14} & \zeta_{15} \\ \zeta_{21} & \zeta_{22} & \zeta_{23} & \zeta_{24} & \zeta_{25} \\ \zeta_{31} & \zeta_{32} & \zeta_{33} & \zeta_{34} & \zeta_{35} \\ \zeta_{41} & \zeta_{42} & \zeta_{43} & \zeta_{44} & \zeta_{45} \\ \zeta_{51} & \zeta_{52} & \zeta_{53} & \zeta_{54} & \zeta_{55} \end{bmatrix} .$$

We can write

$$\delta Q_0^{-1} = [\delta_1, \delta_2, \delta_3, \delta_4, \delta_5] \begin{bmatrix} \zeta_{11} & \zeta_{12} & \zeta_{13} & \zeta_{14} & \zeta_{15} \\ \zeta_{21} & \zeta_{22} & \zeta_{23} & \zeta_{24} & \zeta_{25} \\ \zeta_{31} & \zeta_{32} & \zeta_{33} & \zeta_{34} & \zeta_{35} \\ \zeta_{41} & \zeta_{42} & \zeta_{43} & \zeta_{44} & \zeta_{45} \\ \zeta_{51} & \zeta_{52} & \zeta_{53} & \zeta_{54} & \zeta_{55} \end{bmatrix}$$
$$= \begin{bmatrix} \sum_{i=1}^{5} \delta_i\zeta_{i1} & \sum_{i=1}^{5} \delta_i\zeta_{i2} & \sum_{i=1}^{5} \delta_i\zeta_{i3} & \sum_{i=1}^{5} \delta_i\zeta_{i4} & \sum_{i=1}^{5} \delta_i\zeta_{i5} \end{bmatrix} .$$

This means that

$$\begin{bmatrix} \sum_{i=1}^{5} \delta_i\zeta_{i1} & \sum_{i=1}^{5} \delta_i\zeta_{i2} & \sum_{i=1}^{5} \delta_i\zeta_{i3} & \sum_{i=1}^{5} \delta_i\zeta_{i4} & \sum_{i=1}^{5} \delta_i\zeta_{i5} \end{bmatrix} \begin{bmatrix} \delta_1 \\ \delta_2 \\ \delta_3 \\ \delta_4 \\ \delta_4 \end{bmatrix}$$
$$= \sum_{i=1}^{5}\sum_{j=1}^{5} \delta_i\delta_j\zeta_{ij} \quad .$$

This is the general case.

16.2.7.4 Predictions Using a Single Row Containing a Single Nonzero Element

We now ask whether we can "project" a single coefficient and obtain thereby the correct marginal distribution over an individual model coefficient. We would like to investigate one of the p coefficients, recognizing that it is characterized by an m-variate distribution. We choose the third arbitrarily in order to illustrate. In the specific case in which

$$\delta = e_3 = [0, 0, 1, 0, 0] \Rightarrow \delta^T = \begin{bmatrix} 0 \\ 0 \\ 1 \\ 0 \\ 0 \end{bmatrix}$$

we can write

$$\delta B_0 = [0, 0, 1, 0, 0] \begin{bmatrix} B_{11} & B_{12} \\ B_{21} & B_{22} \\ B_{31} & B_{32} \\ B_{41} & B_{42} \\ B_{51} & B_{52} \end{bmatrix} = \begin{bmatrix} B_{31} & B_{32} \end{bmatrix}$$

and

$$\delta Q_0^{-1} = [0,0,1,0,0]\begin{bmatrix} \zeta_{11} & \zeta_{12} & \zeta_{13} & \zeta_{14} & \zeta_{15} \\ \zeta_{21} & \zeta_{22} & \zeta_{23} & \zeta_{24} & \zeta_{25} \\ \zeta_{31} & \zeta_{32} & \zeta_{33} & \zeta_{34} & \zeta_{35} \\ \zeta_{41} & \zeta_{42} & \zeta_{43} & \zeta_{44} & \zeta_{45} \\ \zeta_{51} & \zeta_{52} & \zeta_{53} & \zeta_{54} & \zeta_{55} \end{bmatrix}$$
$$= \begin{bmatrix} \zeta_{31} & \zeta_{32} & \zeta_{33} & \zeta_{34} & \zeta_{35} \end{bmatrix}$$

which implies that

$$\begin{bmatrix} \zeta_{31} & \zeta_{32} & \zeta_{33} & \zeta_{34} & \zeta_{35} \end{bmatrix}\begin{bmatrix} 0 \\ 0 \\ 1 \\ 0 \\ 0 \end{bmatrix} = \zeta_{33} = \left[Q_0^{-1}\right]_{33} \quad .$$

This means that the m-variate prediction of the third coefficient that we observed in equation (16.72) simplifies to

$$\{\omega\} = \frac{\Gamma\left(\frac{\nu_0+1}{2}\right)}{\Gamma\left(\frac{\nu_0+1-m}{2}\right)\left[\pi\left(1+\zeta_{33}\right)\right]^{\frac{m}{2}}|M_0|^{\frac{1}{2}}}$$
$$\times \left[1+(\omega-e_3B_0)^T\frac{M_0^{-1}}{(1+\zeta_{33})}(\omega-e_3B_0)\right]^{-\frac{\nu_0+1}{2}} \quad .$$

and the variance we saw in equation (16.73) simplifies to

$$VCV = M_0\frac{(1+\zeta_{33})}{\nu_0-m-1} \quad .$$

16.2.7.5 Single Line Forecasting to Examine Coefficients

When we use e_3, we are looking for the simple change of variables problem to find a probability distribution over coefficients, which means we eliminate the 1+ term introduced during prediction to include the data error term in the prediction and emphasize that we are looking for the coefficient β_3, an m-vector, for coefficient 3 itself. We also eliminate δ in favor of e_3 and ζ_{33} in favor of to write

$$\{\beta_3\} = \frac{\Gamma\left(\frac{\nu_0+1}{2}\right)}{\Gamma\left(\frac{\nu_0+1-m}{2}\right)\left(\pi\left[Q_0^{-1}\right]_{33}\right)^{\frac{m}{2}}|M_0|^{\frac{1}{2}}}$$
$$\times \left[1+(\beta_3-e_3B_0)^T\frac{M_0^{-1}}{\left[Q_0^{-1}\right]_{33}}(\beta_3-e_3B_0)\right]^{-\frac{\nu_0+1}{2}}$$

with variance-covariance matrix

$$VCV = M_0 \frac{\left[Q_0^{-1}\right]_{33}}{\nu_0 - m - 1} \quad .$$

We can write the ith marginal in equation (16.77) by examining the relationship between $\{\omega\}$ in equation (16.74) and $\{\omega_i\}$ in equation (16.77) to write

$$\{\beta_{3i}\} = \frac{\Gamma\left(\frac{\nu_0 - m + 2}{2}\right)}{\Gamma\left(\frac{\nu_0 - m + 1}{2}\right)\left\{\pi\left[Q_0^{-1}\right]_{33}[M_0]_{ii}\right\}^{\frac{1}{2}}} \times \left[1 + \frac{1}{\left[Q_0^{-1}\right]_{33}[M_0]_{ii}}\left(\beta_{3i} - [e_3 B_0]_i\right)^2\right]^{-\frac{\nu_0 - m + 2}{2}} \quad .$$

This is precisely and exactly in the form of equation (4.44), the generic univariate Student's t distribution. This and the previous section have developed the needed distributions. The mean of this distribution is

$$\langle \beta_{3i} \rangle = [e_3 B_0]_i \quad . \tag{16.78}$$

The variance of this distribution is taken from equation (16.79) to be

$$Var = \frac{[M_0]_{ii}\left[Q_0^{-1}\right]_{33}}{\nu_0 - m - 1} \quad . \tag{16.79}$$

16.3 Enter the Programmers–Summary of Prior Equations to Be Programmed or Simulated

This section summarizes the results developed in Chapter 16 relevant to the prior distribution.

16.3.1 Coefficients–Prior

16.3.1.1 Full Joint Distribution–Prior

The equation for the multivariate prior $\{B, \Sigma\}$ were developed in equation (16.14), which is reproduced here

$$\begin{aligned} \{B, \Sigma\} = {} & \frac{|M_0|^{\frac{\nu_0}{2}}|Q_0|^{\frac{m}{2}}}{(2\pi)^{\frac{pm}{2}} 2^{\frac{\nu_0 m}{2}} \Gamma_m\left(\frac{\nu_0}{2}\right)|\Sigma|^{\frac{m_0}{2}}} \\ & \times e^{-\frac{1}{2} Trace\left(M_0 \Sigma^{-1}\right)} \\ & \times e^{-\frac{1}{2} Trace\left[(B - B_0)^T Q_0 (B - B_0) \Sigma^{-1}\right]} \quad . \end{aligned} \tag{16.80}$$

Keep in mind, B is a $p \times m$ matrix and Σ is an $m \times m$ matrix.

16.3.1.2 Full Joint Distribution in Product Form–Prior

This joint distribution can be written as the product of an inverse Wishart and a vectorized normal according to equation (16.15), which is reproduced here

$$\begin{aligned}\{B,\Sigma\} &= W^{-1}\left(\Sigma; M_0, \nu_0, m\right) \\ &\quad N\left[vec\left(B\right); vec\left(B_0\right), \left(\Sigma \otimes Q_0^{-1}\right), mp\right] \quad .\end{aligned} \tag{16.81}$$

Keep in mind, B is a $p \times m$ matrix and Σ is an $m \times m$ matrix.

16.3.1.3 Inverse Wishart–Prior

The equation for the inverse Wishart appears in equation (16.16), which is reproduced here

$$\begin{aligned}\{\Sigma\} &= IW(\Sigma, M_0, \nu_0, m) \\ &= \frac{|M_0|^{\frac{\nu_0}{2}}}{2^{\frac{\nu_0 m}{2}}\Gamma_m\left(\frac{\nu_0}{2}\right)|\Sigma|^{\frac{\nu_0+m+1}{2}}} e^{-\frac{1}{2}Trace\left(M_0\Sigma^{-1}\right)} \quad .\end{aligned} \tag{16.82}$$

Keep in mind, Σ is an $m \times m$ matrix.

16.3.1.4 Marginal Distribution Over Coefficients in mm Form–Prior

The marginal distribution over the coefficients with the determinant term based on I_{mm} appears in equation (16.18) and is reproduced here:

$$\begin{aligned}\{B\} &= \frac{\Gamma_m\left(\frac{\nu_0+p}{2}\right)}{\Gamma_m\left(\frac{\nu_0}{2}\right)} \frac{|Q_0|^{\frac{m}{2}}}{\pi^{\frac{mp}{2}}|M_0|^{\frac{p}{2}}} \\ &\quad \left|I_{mm} + {M_0}^{-1}(B-B_0)^T Q_0\left(B-B_0\right)\right|^{-\frac{\nu_0+p}{2}} \quad .\end{aligned} \tag{16.83}$$

Keep in mind, B is a $p \times m$ matrix.

16.3.1.5 Marginal Distribution Over Coefficients in pp Form–Prior

The marginal distribution over the coefficients with the determinant term based on I_{pp} over the coefficients only appears in equation (16.19) and is reproduced here:

$$\begin{aligned}\{B\} &= \frac{\Gamma_m\left(\frac{\nu_0+p}{2}\right)}{\Gamma_m\left(\frac{\nu_0}{2}\right)} \frac{|Q_0|^{\frac{m}{2}}}{\pi^{\frac{mp}{2}}|M_0|^{\frac{p}{2}}} \\ &\quad \left|I_{pp} + Q_0\left(B-B_0\right){M_0}^{-1}(B-B_0)^T\right|^{-\frac{\nu_0+p}{2}} \quad .\end{aligned} \tag{16.84}$$

Keep in mind, B is a $p \times m$ matrix.

16.3.2 Mean Coefficients–Prior

The mean values of the coefficients appear in equation (16.30), which is rewritten here:

$$\langle B \rangle = B_0 \quad . \tag{16.85}$$

Keep in mind, B is a $p \times m$ matrix.

16.3.2.1 Variance-Covariance Matrix of Coefficients–Prior

The variance-covariance matrix of the coefficients appears in equation (16.36) and is rewritten here:

$$\begin{aligned} VCV &= M_0 \frac{{M_0}^{-1}}{\nu_0 - m - 1} \left(M_0 \otimes {Q_0}^{-1} \right) \\ &= \frac{1}{\nu_0 - m - 1} \left(M_0 \otimes {Q_0}^{-1} \right) \quad . \end{aligned} \tag{16.86}$$

Keep in mind, M_0 is an $m \times m$ matrix and Q_0 is a $p \times p$ matrix so that VCV is an $mp \times mp$ matrix.

16.3.2.2 Mean Values of the Error Coefficients–Prior

The mean value of the Σ coefficients appear in equation (16.37) and are reproduced here.

$$\langle \Sigma \rangle = M_0 \frac{{M_0}^{-1}}{\nu_0 - m - 1} M_0 = \frac{1}{\nu_0 - m - 1} M_0 \quad . \tag{16.87}$$

Keep in mind, Σ is an $m \times m$ matrix.

16.3.2.3 Variance-Covariance Matrix of the Error Coefficients–Prior

The variance-covariance matrix of the Σ term appears in equation (16.53).

$$\begin{aligned} VCV = &-\frac{1}{(\nu_0 - m - 1)^2} \left(M_0 \otimes M_0 \right) \\ &+ \frac{(\nu_0 - m - 2)}{(\nu_0 - m)(\nu_0 - m - 1)(\nu_0 - m - 3)} \left(M_0 \otimes M_0 \right) \\ &+ \frac{1}{(\nu_0 - m)(\nu_0 - m - 1)(\nu_0 - m - 3)} \left(M_0 \otimes M_0 \right) K_{mm} \\ &+ \frac{1}{(\nu_0 - m)(\nu_0 - m - 1)(\nu_0 - m - 3)} \\ &\times \left(M_0 \otimes M_0 \right) \left[vec\left({M_0}^{-1} \right) vec\left({M_0}^{-1} \right)^T \right] \left(M_0 \otimes M_0 \right) \quad . \end{aligned} \tag{16.88}$$

Keep in mind, M_0 is an $m \times m$ matrix so that VCV is a $mm \times mm$ matrix.

16.3.3 Predictive Distributions–Prior

16.3.3.1 Joint Predictive Distribution–Prior

The complete multivariate joint projective distribution $\{W, \Sigma\}$ is given in equation(16.63) and is reproduced here

$$(16.89)\quad \{W,\Sigma\} = \frac{|M_0|^{\frac{\nu_0}{2}}}{2^{\frac{\nu_0 m}{2}}(2\pi)^{\frac{qm}{2}}\Gamma_m\left(\frac{\nu_0}{2}\right)\left|I_{qq}+DQ_0^{-1}D^T\right|^{\frac{m}{2}}}|\Sigma|^{\frac{-m_0+p-q}{2}} \times e^{-\frac{1}{2}Trace\left(M_0\Sigma^{-1}\right)} \times e^{-\frac{1}{2}Trace\left[(W-DB_0)^T\left(I_{qq}+DQ_0^{-1}D^T\right)^{-1}(W-DB_0)\Sigma^{-1}\right]} .$$

W is $q \times p$; Σ is $m \times m$.

16.3.3.2 Joint Predictive Distribution in Product Form–Prior

The multivariate joint projective distribution $\{W, \Sigma\}$ is shown in equation (16.64) to be an inverse Wishart times a vectorized normal, meaning

$$(16.90)\quad \{W,\Sigma\} = W^{-1}\left(\Sigma; M_0, \nu_0, m\right) \times N\left(W; DB_0, \Sigma\otimes\left(I_{qq}+DQ_0^{-1}D^T\right), mq\right) .$$

W is $q \times p$; Σ is $m \times m$.

16.3.3.3 Marginal Distribution Over q-Variate Prediction in mm Form–Prior

The marginal distribution over the predictive values with the determinant term based on I_{mm} appears in equation (16.67) and is reproduced here:

$$(16.91)\quad \{W\} = \frac{\Gamma_m\left(\frac{\nu_0+q}{2}\right)}{\Gamma_m\left(\frac{\nu_0}{2}\right)\pi^{\frac{qm}{2}}\left|I_{qq}+DQ_0^{-1}D^T\right|^{\frac{m}{2}}|M_0|^{\frac{q}{2}}} \left|\begin{array}{l} I_{mm}+M_0^{-1}(W-DB_0)^T \\ \quad\times\left(I_{qq}+DQ_0^{-1}D^T\right)^{-1}(W-DB_0)\end{array}\right|^{-\frac{\nu_0+q}{2}} .$$

W is $q \times p$.

16.3.3.4 Marginal Distribution Over q-Variate Prediction in qq Form–Prior

The marginal distribution over the predictive values with the determinant term based on I_{qq} appears in equation (16.68) and is reproduced here:

$$\{W\} = \frac{\Gamma_m\left(\frac{\nu_0+q}{2}\right)}{\Gamma_m\left(\frac{\nu_0}{2}\right)\pi^{\frac{qm}{2}}\left|I_{qq}+DQ_0^{-1}D^T\right|^{\frac{m}{2}}|M_0|^{\frac{q}{2}}} \left|\begin{array}{l} I_{qq}+\left(I_{qq}+DQ_0^{-1}D^T\right)^{-1}(W-DB_0) \\ \times M_0^{-1}(W-DB_0)^T \end{array}\right|^{-\frac{\nu_0+q}{2}}. \tag{16.92}$$

16.3.4 Single Row Full Distribution Predictive Distribution–Prior

16.3.4.1 Single Row Predictive Distribution in mm Form–Prior

We turn to the row vector ($q = 1$) and write the equation for the prediction of the m-variate row vector ω having the single row prediction vector δ using I_{mm} in the determinant term from equation (16.91) as follows

$$\{\omega|\delta\} = \frac{\Gamma_m\left(\frac{\nu_0+1}{2}\right)}{\Gamma_m\left(\frac{\nu_0}{2}\right)\pi^{\frac{m}{2}}\left|1+\delta Q_0^{-1}\delta^T\right|^{\frac{m}{2}}|M_0|^{\frac{1}{2}}} \left|\begin{array}{l} I_{mm}+\frac{1}{1+\delta Q_0^{-1}\delta^T}M_0^{-1} \\ (\omega-\delta B_0)^T(\omega-\delta B_0) \end{array}\right|^{-\frac{\nu_0+1}{2}}. \tag{16.93}$$

ω is $1 \times m$.

16.3.4.2 Single Row Predictive Distribution Over the m-Variate ω Vector in $qq = 11$ Form–Prior

We write the equation for the prediction of the m-variate row vector ω having the single row prediction vector δ using $I_{qq} = I_{11} = 1$ in the determinant term from equation (16.71) as follows

$$\begin{aligned} \{\omega|\delta\} = & \frac{\Gamma_m\left(\frac{\nu_0+1}{2}\right)}{\Gamma_m\left(\frac{\nu_0}{2}\right)\pi^{\frac{m}{2}}\left|1+\delta Q_0^{-1}\delta^T\right|^{\frac{m}{2}}|M_0|^{\frac{1}{2}}} \\ & \times\left[1+(\omega-\delta B_0)\frac{M_0^{-1}}{1+\delta Q_0^{-1}\delta^T}(\omega-\delta B_0)^T\right]^{-\frac{\nu_0+1}{2}} \\ = & c_0\left[1+(\omega-\delta B_0)\frac{M_0^{-1}}{1+\delta Q_0^{-1}\delta^T}(\omega-\delta B_0)^T\right]^{-\frac{\nu_0+1}{2}}. \end{aligned}$$

We want to define a variable ξ that will look like a degrees of freedom parameter such that

$$\begin{aligned}
&\xi + m = \nu_0 + 1 \\
&\Rightarrow \xi = \nu_0 - m + 1 \\
&\Rightarrow \nu_0 = \xi + m - 1
\end{aligned}$$

.

When we make this substitution, we obtain the expression

$$\{\omega|\delta\} = c_0 \left[1 + (\omega - \delta B_0) \frac{M_0^{-1}}{1 + \delta Q_0^{-1} \delta^T} (\omega - \delta B_0)^T \right]^{-\frac{\xi+m}{2}} .$$

This is in the form of a standard multivariate Student's t distribution. We emphasize that when $q = 1$, the determinant term has become a one dimensional vector, i.e., a scalar. That renders this a multivariate Student's t distribution instead of a more complex matrix t distribution, exactly of the form in equation (4.16) whose variance-covariance matrix is given in equation (4.17). Let us consider the substitution of m for n and ξ for ν in equation (4.16) to get started:

$$T_m(r; \mu, \Omega, \xi) = \frac{\Gamma\left(\frac{\xi+m}{2}\right) |\Omega|^{-\frac{1}{2}}}{\Gamma\left(\frac{\xi}{2}\right) (\pi\xi)^{\frac{m}{2}}} \left[1 + (r - \mu)^T \frac{\Omega^{-1}}{\xi} (r - \mu) \right]^{-\frac{\xi+m}{2}}$$

with mean value

$$VCV = \frac{\xi}{\xi - 2} \Omega .$$

After this, we want to make the following substitutions into the multivariate Student's t distribution:

$$\begin{aligned}
\frac{\Omega^{-1}}{\xi} &\leftarrow \frac{M_0^{-1}}{\left(1 + \delta Q_0^{-1} \delta^T\right)} \\
\Omega &\leftarrow \frac{\left(1 + \delta Q_0^{-1} \delta^T\right)}{\xi} M_0 \\
|\Omega|^{\frac{1}{2}} &\leftarrow \left(1 + \delta Q_0^{-1} \delta^T\right)^{\frac{m}{2}} \xi^{-\frac{m}{2}} |M_0|^{\frac{1}{2}} .
\end{aligned}$$

When we do so, we obtain

$$\begin{aligned}
\{\omega|\delta\} = & \frac{\Gamma\left(\frac{\xi+m}{2}\right)}{\Gamma\left(\frac{\xi}{2}\right) (\pi\xi)^{\frac{m}{2}} \left(1 + \delta Q_0^{-1} \delta^T\right)^{\frac{m}{2}} \xi^{-\frac{m}{2}} |M_0|^{\frac{1}{2}}} \\
& \times \left[1 + (\omega - \delta B_0)^T \frac{M_0^{-1}}{\left(1 + \delta Q_0^{-1} \delta^T\right)} (\omega - \delta B_0) \right]^{-\frac{\xi+m}{2}} .
\end{aligned}$$

When we substitute for ξ, we obtain the marginal distribution over ω

$$\{\omega|\delta\} = \frac{\Gamma\left(\frac{\nu_0+1}{2}\right)}{\Gamma\left(\frac{\nu_0+1-m}{2}\right)\left[\pi\left(1+\delta Q_0^{-1}\delta^T\right)\right]^{\frac{m}{2}}|M_0|^{\frac{1}{2}}} \times\left[1+(\omega-\delta B_0)^T\frac{M_0^{-1}}{\left(1+\delta Q_0^{-1}\delta^T\right)}(\omega-\delta B_0)\right]^{-\frac{\nu_0+1}{2}} \quad . \tag{16.94}$$

16.3.4.3 Mean of the Single Row Predictive Distribution Over the m-Variate ω Vector in qq Form–Prior

The mean value of the single row predictive distribution is the row vector with m columns

$$\langle\omega\rangle = \delta B_0 \quad . \tag{16.95}$$

16.3.4.4 Variance-Covariance Matrix of the Single Row Predictive Distribution Over the m-Variate ω Vector in qq Form–Prior

It is obvious from the definition of the VCV matrix of the multivariate Student's t distribution immediately above that the variance-covariance matrix has become

$$VCV = \frac{1+\delta Q_0^{-1}\delta^T}{\nu_0+1-m}M_0 \quad . \tag{16.96}$$

16.3.4.5 Marginal Distribution Over the ith Element of the m-Variate ω Vector in qq Form–Prior

The marginal predictive distribution for the ith output from among the m outputs is given in equation (16.77) to be

$$\{\omega_i\} = \frac{\Gamma\left(\frac{\nu_0-m+2}{2}\right)}{\Gamma\left(\frac{\nu_0-m+1}{2}\right)\left[\pi\left(1+\delta Q_0^{-1}\delta^T\right)[M_0]_{ii}\right]^{\frac{1}{2}}} \times\left[1+\frac{1}{\left(1+\delta Q_0^{-1}\delta^T\right)[M_0]_{ii}}\left(\omega_i-[\delta B_0]_i\right)^2\right]^{-\frac{\nu_0-m+2}{2}} \quad . \tag{16.97}$$

16.3.4.6 Mean of the Marginal Distribution Over the ith Element of the m-Variate ω Vector–Prior

The mean value of the single row predictive distribution for the ith output from among the m outputs is obvious from equation (16.75)

$$\langle\omega_i\rangle = [\delta B_0]_i \quad . \tag{16.98}$$

16.3.4.7 Variance of the ith Element of the Marginal Distribution Over the m-Variate ω Vector–Prior

It is obvious from the arguments of the multivariate Student's t distribution that the variance of the ith output from among the m outputs is that given in equation (16.76)

(16.99)
$$\frac{\left(1+\delta Q_0^{-1}\delta^T\right)[M_0]_{ii}}{\nu_0-m-1} \quad .$$

16.3.5 Marginal Coefficient Distribution Inferred from Marginal Predictive Distribution–Prior

Finally, if we consider coefficient 3 and the ith of the m error variates, the marginal density over the coefficient β_{3i} is

(16.100)
$$\begin{aligned}\{\beta_{3i}\} = &\frac{\Gamma\left(\frac{\nu_0-m+2}{2}\right)}{\Gamma\left(\frac{\nu_0-m+1}{2}\right)}\frac{1}{\left(\pi\left[Q_0^{-1}\right]_{33}[M_0]_{ii}\right)^{\frac{1}{2}}}\\ &\times\left[1+\frac{1}{\left[Q_0^{-1}\right]_{33}[M_0]_{ii}}\left(\beta_{3i}-[B_0]_{3i}\right)^2\right]^{-\frac{\nu_0-m+2}{2}} \quad .\end{aligned}$$

16.3.6 Mean of Marginal Coefficient Distribution Inferred from Marginal Predictive Distribution–Prior

The mean of this distribution is taken from equation (16.78) to be

(16.101)
$$\langle\beta_{3i}\rangle = [B_0]_{3i} \quad .$$

16.3.7 Variance of Marginal Coefficient Distribution Inferred from Marginal Predictive Distribution–Prior

The variance of this distribution is taken from equation (16.76), which is reproduced here

(16.102)
$$Var = \frac{[M_0]_{ii}\left[Q_0^{-1}\right]_{33}}{\nu_0-m-1} \quad .$$

Chapter 17

CALCULATION AND INTERPRETATION OF MULTIVARIATE POSTERIOR

> ***The laws of probability, so true in general, so fallacious in particular.***
>
> ***Edward Gibbon***

Now that we have finished all the requisite calculations for the prior, we turn to calculating and reporting the posterior, which is the exact same functional form as the prior by virtue of the conjugacy. The posterior is the product of the prior and the likelihood function. The prior appears in equations (16.14) and (16.15), and the likelihood appears in equation (15.5). The product, in which we are careful to carry along the determinants of the Σ matrix with the appropriate power (but eliminating both constants of integration), gives the posterior.

$$\begin{aligned}\text{Posterior} &= c_0 \frac{1}{|\Sigma|^{\frac{n}{2}}} e^{-\frac{1}{2}Trace\left[\left(S^T S\right)\Sigma^{-1}\right]} e^{-\frac{1}{2}Trace\left[\left(B-\bar{B}\right)^T X^T X\left(B-\bar{B}\right)\Sigma^{-1}\right]} \\ &\times \frac{1}{|\Sigma|^{\frac{m_0}{2}}} e^{-\frac{1}{2}Trace\left(M_0\Sigma^{-1}\right)} e^{-\frac{1}{2}Trace\left[(B-B_0)^T Q_0 (B-B_0)\Sigma^{-1}\right]} \quad ,\end{aligned}$$

in which c_0 is a constant of integration to be determined. We regroup the exponential terms

$$\text{Posterior} = c_0 \frac{1}{|\Sigma|^{\frac{n+m_0}{2}}} e^{-\frac{1}{2}Trace\left[\left(S^T S\right)\Sigma^{-1}\right]} e^{-\frac{1}{2}Trace\left(M_0\Sigma^{-1}\right)}$$

$$\times\, e^{-\frac{1}{2}Trace\left[\left(B-\bar{B}\right)^{T}X^{T}X\left(B-\bar{B}\right)\Sigma^{-1}\right]}e^{-\frac{1}{2}Trace\left[(B-B_0)^{T}Q_0(B-B_0)\Sigma^{-1}\right]} \quad .$$

Notice the exponent on the Σ term. In the prior, it was m_0. In the posterior, it is now m_0+n. Recall that the prior degrees of freedom were $\nu_0 = m_0 - p - m - 1$ from equation (16.12), i.e., we simply add $-p-m-1$ to the exponent of the Σ term in the prior to get the prior degrees of freedom. By analogy, the exponent of the Σ term in the posterior is m_0+n. If we simply add $-p-m-1$ to the exponent of the Σ term in the posterior, we obtain the posterior degrees of freedom, which is

$$\begin{aligned} \nu &= \nu_0 + n \\ &= m_0 + n - p - m - 1 \quad . \end{aligned} \tag{17.1}$$

Knowing that matrix traces are additive (and omitting the $Trace(.)$ and the $-\frac{1}{2}$ multiplier temporarily so that we can focus on the matrices), we want to complete the square on the exponent in the second line, which is written

$$\text{Exponent} = \left(B-\bar{B}\right)^{T}X^{T}X\left(B-\bar{B}\right) + (B-B_0)^{T}Q_0\left(B-B_0\right) \quad .$$

We expand both terms

$$\begin{aligned} \text{Exponent} &= B^{T}Q_0B - B^{T}Q_0B_0 - {B_0}^{T}Q_0B + {B_0}^{T}Q_0B_0 \\ &+ B^{T}\left(X^{T}X\right)B - B^{T}\left(X^{T}X\right)\bar{B} - \bar{B}^{T}\left(X^{T}X\right)B + \bar{B}^{T}\left(X^{T}X\right)\bar{B} \quad . \end{aligned}$$

We reorganize this into terms that do not depend on B, terms that are linear in B, and terms that are quadratic in B

$$\begin{aligned} \text{Exponent} = {} & {B_0}^{T}Q_0B_0 + \bar{B}^{T}\left(X^{T}X\right)\bar{B} \\ & - \left[{B_0}^{T}Q_0 + \bar{B}^{T}\left(X^{T}X\right)\right]B \\ & - B^{T}\left[Q_0B_0 + \left(X^{T}X\right)\bar{B}\right] \\ & + B^{T}\left(X^{T}X + Q_0\right)B \quad . \end{aligned}$$

We want to complete the square by defining

$$\begin{aligned} B &= (B - B^*) + B^* \\ z &= B - B^* \\ B &= z + B^* \quad . \end{aligned}$$

We substitute this into the foregoing equation

$$\begin{aligned} \text{Exponent} = {} & {B_0}^{T}Q_0B_0 + \bar{B}^{T}\left(X^{T}X\right)\bar{B} \\ & - \left[{B_0}^{T}Q_0 + \bar{B}^{T}\left(X^{T}X\right)\right](z + B^*) \\ & - (z + B^*)^{T}\left[Q_0B_0 + \left(X^{T}X\right)\bar{B}\right] \end{aligned}$$

$$+ (z + B^*)^T \left(X^T X + Q_0\right)(z + B^*) \quad .$$

We expand out all the terms

$$\begin{aligned}
\text{Exponent} = {} & {B_0}^T Q_0 B_0 + \bar{B}^T \left(X^T X\right) \bar{B} \\
& - {B_0}^T Q_0 z - {B_0}^T Q_0 B^* - \bar{B}^T \left(X^T X\right) z - \bar{B}^T \left(X^T X\right) B^* \\
& - z^T Q_0 B_0 - {B^*}^T Q_0 B_0 - z^T \left(X^T X\right) \bar{B} - {B^*}^T \left(X^T X\right) \bar{B} \\
& + z^T \left(X^T X\right) z + {B^*}^T \left(X^T X\right) z + z^T Q_0 z + {B^*}^T Q_0 z \\
& + z^T \left(X^T X\right) B^* + {B^*}^T \left(X^T X\right) B^* + z^T Q_0 B^* + {B^*}^T Q_0 B^* \quad .
\end{aligned}$$

Reorganize into terms that do not depend on z, that are linear in z, and that are quadratic in z

$$\begin{aligned}
\text{Exponent} = {} & {B_0}^T Q_0 B_0 + \bar{B}^T \left(X^T X\right) \bar{B} - {B_0}^T Q_0 B^* - \bar{B}^T \left(X^T X\right) B^* \\
& - {B^*}^T Q_0 B_0 - {B^*}^T \left(X^T X\right) \bar{B} + {B^*}^T \left(X^T X + Q_0\right) B^* \\
& - {B_0}^T Q_0 z - \bar{B}^T \left(X^T X\right) z + {B^*}^T \left(X^T X + Q_0\right) z \\
& - z^T Q_0 B_0 - z^T \left(X^T X\right) \bar{B} + z^T \left(X^T X + Q_0\right) B^* \\
& + z^T \left(X^T X + Q_0\right) z \quad .
\end{aligned}$$

Group and reorganize the terms for clarity

$$\begin{aligned}
\text{Exponent} = {} & {B_0}^T Q_0 B_0 + \bar{B}^T \left(X^T X\right) \bar{B} - {B_0}^T Q_0 B^* - \bar{B}^T \left(X^T X\right) B^* \\
& - {B^*}^T Q_0 B_0 - {B^*}^T \left(X^T X\right) \bar{B} + {B^*}^T \left(X^T X + Q_0\right) B^* \\
& + \left[-{B_0}^T Q_0 - \bar{B}^T \left(X^T X\right) + {B^*}^T \left(X^T X + Q_0\right)\right] z \\
& + z^T \left[-Q_0 B_0 - \left(X^T X\right) \bar{B} + \left(X^T X + Q_0\right) B^*\right] \\
& + z^T \left(X^T X + Q_0\right) z \quad .
\end{aligned}$$

We want to set B^* so that the term that is linear in z^T, the second to last term in the foregoing equation, is 0

$$\begin{aligned}
& - Q_0 B_0 - \left(X^T X\right) \bar{B} + \left(X^T X + Q_0\right) B^* = 0 \\
& \Rightarrow \left(X^T X + Q_0\right) B^* = Q_0 B_0 + \left(X^T X\right) \bar{B} \\
& \Rightarrow \left(X^T X + Q_0\right) B^* = Q_0 B_0 + X^T Y \\
& \Rightarrow B^* = \left(X^T X + Q_0\right)^{-1} \left(Q_0 B_0 + X^T Y\right) \\
& \Rightarrow {B^*}^T = \left({B_0}^T {Q_0}^T + Y^T X\right) \left(X^T X + Q_0\right)^{-1} \quad .
\end{aligned}$$

Writing this key result as a separate equation

$$B^* = \left(Q_0 + X^T X\right)^{-1} \left(Q_0 B_0 + X^T Y\right) \quad . \tag{17.2}$$

B^* is $p \times m$. If we substitute for B^* in the expression for the exponent, the third and second to last terms leave the equation and we are left with

$$\begin{aligned}\text{Exponent} &= {B_0}^T Q_0 B_0 + \bar{B}^T \left(X^T X\right) \bar{B} - {B_0}^T Q_0 B^* - \bar{B}^T \left(X^T X\right) B^* \\ &- {B^*}^T Q_0 B_0 - {B^*}^T \left(X^T X\right) \bar{B} + {B^*}^T \left(X^T X\right) B^* + {B^*}^T Q_0 B^* \\ &+ z^T \left(X^T X + Q_0\right) z \quad .\end{aligned}$$

We expand all the terms save for the last

$$\begin{aligned}\text{Exponent} &= {B_0}^T Q_0 B_0 - {B_0}^T Q_0 B^* - {B^*}^T Q_0 B_0 + {B^*}^T Q_0 B^* \\ &+ \bar{B}^T \left(X^T X\right) \bar{B} - {B^*}^T \left(X^T X\right) \bar{B} \\ &+ {B^*}^T \left(X^T X\right) B^* - \bar{B}^T \left(X^T X\right) B^* \\ &+ z^T \left(X^T X + {'Q_0}\right) z \quad .\end{aligned}$$

Reorganize the first eight terms (which are based on the prior matrix and the data matrix)

$$\begin{aligned}\text{Exponent} &= {B_0}^T Q_0 B_0 - {B_0}^T Q_0 B^* - {B^*}^T Q_0 B_0 + {B^*}^T Q_0 B^* \\ &+ \bar{B}^T \left(X^T X\right) \bar{B} - {B^*}^T \left(X^T X\right) \bar{B} \\ &+ {B^*}^T \left(X^T X\right) B^* - \bar{B}^T \left(X^T X\right) B^* \\ &+ z^T \left(X^T X + Q_0\right) z \quad .\end{aligned}$$

The first eight terms other than the quadratic in z can be reorganized into a convenient and insightful form

$$\begin{aligned}&{B^*}^T Q_0 B^* - {B^*}^T Q_0 B_0 - {B_0}^T Q_0 B^* + {B_0}^T Q_0 B_0 \\ &+ {B^*}^T \left(X^T X\right) B^* - {B^*}^T \left(X^T X\right) \bar{B} \\ &- \bar{B}^T \left(X^T X\right) B^* + \bar{B}^T \left(X^T X\right) \bar{B} \quad .\end{aligned}$$

We can then begin to simplify

$$\begin{aligned}&{B^*}^T Q_0 B^* - {B^*}^T Q_0 B_0 - {B_0}^T Q_0 B^* + {B_0}^T Q_0 B_0 \\ &+ {B^*}^T \left(X^T X\right) B^* - {B^*}^T \left(X^T X\right) \bar{B} \\ &- \bar{B}^T \left(X^T X\right) B^* + \bar{B}^T \left(X^T X\right) \bar{B} \\ &= {B^*}^T \left(Q_0 + X^T X\right) B^* \\ &- {B^*}^T \left[Q_0 B_0 + \left(X^T X\right) \bar{B}\right] \\ &- \left[{B_0}^T Q_0 + \bar{B}^T \left(X^T X\right)\right] B^* \\ &+ {B_0}^T Q_0 B_0 + \bar{B}^T \left(X^T X\right) \bar{B} \quad .\end{aligned}$$

We substitute the classical regression solution $\bar{B} = (X^TX)^{-1}X^TY$ into the middle two terms

$$\begin{aligned}
&= {B^*}^T \left(Q_0 + X^TX\right) B^* \\
&- {B^*}^T \left(Q_0B_0 + X^TY\right) \\
&- \left({B_0}^TQ_0 + Y^TX\right) B^* \\
&+ {B_0}^TQ_0B_0 + \bar{B}^T \left(X^TX\right) \bar{B} \quad .
\end{aligned}$$

We substitute the expression for B^* into the first term

$$\begin{aligned}
&= \left({B_0}^T{Q_0}^T + Y^TX\right) \left(X^TX + Q_0\right)^{-1} \left(Q_0 + X^TX\right) B^* \\
&- {B^*}^T \left(Q_0B_0 + X^TY\right) \\
&- \left({B_0}^TQ_0 + Y^TX\right) B^* \\
&+ {B_0}^TQ_0B_0 + \bar{B}^T \left(X^TX\right) \bar{B} \quad .
\end{aligned}$$

The remaining two terms, after noting the first and third terms are the same, is

$$\begin{aligned}
&- {B^*}^T \left(Q_0B_0 + X^TY\right) \\
&+ {B_0}^TQ_0B_0 + \bar{B}^T \left(X^TX\right) \bar{B} \quad .
\end{aligned}$$

Expanding the terms and substituting the aforementioned equation for $\bar{B}$

$$\begin{aligned}
&- {B^*}^TQ_0B_0 - {B^*}^TX^TY \\
&+ {B_0}^TQ_0B_0 + \bar{B}^TX^TY \quad .
\end{aligned}$$

This can be rewritten

$$\left(B_0 - B^*\right)^TQ_0B_0 + \left(\bar{B} - B^*\right)^TX^TY \quad .$$

This means that the exponent as a whole reduces to

$$\begin{aligned}
\text{Exponent} &= \left(B_0 - B^*\right)^TQ_0B_0 + \left(\bar{B} - B^*\right)^TX^TY \\
&+ \left(B - B^*\right)^T \left(X^TX + Q_0\right) \left(B - B^*\right) \quad .
\end{aligned}$$

The posterior is thus

$$\text{Posterior} = c_0 \frac{1}{|\Sigma|^{\frac{m_0+n}{2}}} e^{-\frac{1}{2}Trace\left[\left(S^TS+M_0\right)\Sigma^{-1}\right]}$$

$$e^{-\frac{1}{2}Trace\left\{\left[(B-B^*)^T\left(Q_0+X^TX\right)(B-B^*)+(B_0-B^*)^TQ_0B_0+\left(\bar{B}-B^*\right)^TX^TY\right]\Sigma^{-1}\right\}} \quad .$$

We are going to denote the $p \times p$ matrix

(17.3) $$Z = X^T X + Q_0 \quad ,$$

and the $m \times m$ matrix

(17.4) $$A = S^T S + M_0 + (B_0 - B^*)^T Q_0 B_0 + \left(\bar{B} - B^*\right)^T X^T Y \quad .$$

This allows us to write the posterior probability distribution

$$\{B, \Sigma\} = \frac{c_0}{|\Sigma|^{\frac{m_0+n}{2}}} e^{-\frac{1}{2} Trace\left(A\Sigma^{-1}\right)} e^{-\frac{1}{2} Trace\left[(B-B^*)^T Z (B-B^*)\Sigma^{-1}\right]} \quad .$$

If we begin with equation (16.80) for the prior and make the following substitutions,

(17.5) $$\begin{aligned} m_0 &\leftarrow m_0 + n \\ &\Rightarrow \nu \leftarrow \nu_0 + n \\ M_0 &\leftarrow A \\ Q_0 &\leftarrow Z \\ B_0 &\leftarrow B^* \end{aligned}$$

we get the foregoing posterior joint distribution equation with the correct integrating constant embedded

(17.6) $$\{B, \Sigma\} = \frac{\dfrac{|A|^{\frac{\nu}{2}} |Z|^{\frac{m}{2}}}{(2\pi)^{\frac{pm}{2}} 2^{\frac{\nu m}{2}} \Gamma_m\left(\frac{\nu}{2}\right)}}{|\Sigma|^{\frac{m_0+n}{2}}} \times e^{-\frac{1}{2} Trace\left(A\Sigma^{-1}\right)} e^{-\frac{1}{2} Trace\left[(B-B^*)^T Z (B-B^*)\Sigma^{-1}\right]} \quad .$$

B^* is $p \times m$; Σ is $m \times m$. This means that the integrating constant for the posterior is given by making the substitutions in equation (17.5) into equation (16.13) that specified the integrating constant for the prior

(17.7) $$c_0 = \frac{|A|^{\frac{\nu}{2}} |Z|^{\frac{m}{2}}}{(2\pi)^{\frac{pm}{2}} 2^{\frac{\nu m}{2}} \Gamma_m\left(\frac{\nu}{2}\right)} \quad .$$

We thus have a direct and general generalization of the prior in equation (16.14) to the Bayesian posterior in equation (17.6). This will allow us to develop all the results for the Bayesian posterior we need to by simply making the substitutions from equation (17.5) into the various terms from the prior.

17.1 Enter the Programmers–Summary of Equations to Be Programmed or Simulated–Posterior

This section summarizes the results developed in Chapter 16 and 17 relevant to the Posterior distribution.

17.1.1 Coefficients–Posterior

17.1.1.1 Full Joint Distribution–Posterior

The equation for the multivariate prior $\{B, \Sigma\}$ was developed in equation (16.80). When we make the substitutions from equation (17.5), we obtain the equation for the posterior $\{B, \Sigma\}$, which fully characterizes the posterior

$$\begin{aligned} \{B, \Sigma\} &= \frac{|A|^{\frac{\nu}{2}} |Z|^{\frac{m}{2}}}{(2\pi)^{\frac{pm}{2}} 2^{\frac{\nu m}{2}} \Gamma_m\left(\frac{\nu}{2}\right) |\Sigma|^{\frac{m_0+n}{2}}} \\ &\times e^{-\frac{1}{2} Trace\left(A\Sigma^{-1}\right)} \\ &\times e^{-\frac{1}{2} Trace\left[(B-B^*)^T Z (B-B^*) \Sigma^{-1}\right]} \quad . \end{aligned} \tag{17.8}$$

B is $p \times m$; Σ is $m \times m$.

17.1.1.2 Full Joint Distribution in Product Form–Posterior

This joint distribution for the prior can be written as the product of an inverse Wishart and a vectorized normal according to equation (16.81). When we make the foregoing substitutions in equation (17.5), we obtain for the posterior

$$\begin{aligned} \{B, \Sigma\} &= W^{-1}\left(\Sigma; A, \nu, m\right) \\ &\quad N\left[vec\left(B\right); vec\left(B^*\right), \left(\Sigma \otimes Z^{-1}\right), mp\right] \quad . \end{aligned} \tag{17.9}$$

B is $p \times m$; Σ is $m \times m$.

17.1.1.3 Inverse Wishart–Posterior

The equation for the inverse Wishart appears for the prior appears in equation (16.82). When we make the foregoing substitutions in equation (17.5), we obtain for the posterior

$$\begin{aligned} \{\Sigma\} &= IW(\Sigma, A, \nu, m) \\ &= \frac{|A|^{\frac{\nu}{2}}}{2^{\frac{\nu m}{2}} \Gamma_m\left(\frac{\nu}{2}\right) |\Sigma|^{\frac{\nu+m+1}{2}}} e^{-\frac{1}{2} Trace\left(A\Sigma^{-1}\right)} \quad . \end{aligned} \tag{17.10}$$

Σ is $m \times m$.

17.1.1.4 Marginal Distribution Over Coefficients in mm Form–Posterior

The marginal distribution over the coefficients for the prior with the determinant term based on I_{mm} appears in equation (16.83). When we make the foregoing substitutions in equation (17.5), we obtain for the posterior

$$\{B\} = \frac{\Gamma_m\left(\frac{\nu+p}{2}\right)}{\Gamma_m\left(\frac{\nu}{2}\right)} \frac{|Z|^{\frac{m}{2}}}{\pi^{\frac{mp}{2}}|A|^{\frac{p}{2}}} \left|I_{mm} + A^{-1}(B-B^*)^T Z\,(B-B^*)\right|^{-\frac{\nu+p}{2}} \quad . \tag{17.11}$$

B is $p \times m$.

17.1.1.5 Marginal Distribution Over Coefficients in pp Form–Posterior

The marginal distribution over the coefficients with the determinant term based on I_{pp} over the coefficients only for the prior appears in equation (16.84). When we make the foregoing substitutions in equation (17.5), we obtain for the posterior

$$\{B\} = \frac{\Gamma_m\left(\frac{\nu+p}{2}\right)}{\Gamma_m\left(\frac{\nu}{2}\right)} \frac{|Z|^{\frac{m}{2}}}{\pi^{\frac{mp}{2}}|A|^{\frac{p}{2}}} \left|I_{pp} + Z\,(B-B^*)\,A^{-1}(B-B^*)^T\right|^{-\frac{\nu+p}{2}} \quad . \tag{17.12}$$

B is $p \times m$.

17.1.2 Mean Coefficients–Posterior

The mean values of the coefficients for the prior appear in equation (16.85). When we make the foregoing substitutions in equation (17.5), we obtain for the posterior

$$\langle B \rangle = B^* \quad .$$

B is $p \times m$.

17.1.2.1 Variance-Covariance Matrix of Coefficients–Posterior

The variance-covariance matrix of the coefficients for the prior appears in equation (16.86). When we make the foregoing substitutions in equation (17.5), we obtain for the posterior

$$VCV = A\frac{A^{-1}}{\nu-m-1}\left(A \otimes Z^{-1}\right) = \frac{1}{\nu-m-1}\left(A \otimes Z^{-1}\right) \quad . \tag{17.13}$$

VCV is $mp \times mp$.

17.1.2.2 Mean Values of the Error Coefficients–Posterior

The mean value of the Σ coefficients for the prior appear in equation (16.87). When we make the foregoing substitutions in equation (17.5), we obtain for the posterior

$$\langle \Sigma \rangle = A \frac{A^{-1}}{\nu - m - 1} A = \frac{1}{\nu - m - 1} A \quad . \tag{17.14}$$

Σ is $m \times m$.

17.1.2.3 Variance-Covariance Matrix of the Error Coefficients–Posterior

The variance-covariance matrix of the Σ term for the prior appears in equation (16.88). When we make the foregoing substitutions in equation (17.5), we obtain for the posterior

$$\begin{aligned} VCV = & -\frac{1}{(\nu - m - 1)^2} (A \otimes A) \\ & + \frac{(\nu - m - 2)}{(\nu - m)(\nu - m - 1)(\nu - m - 3)} (A \otimes A) \\ & + \frac{1}{(\nu - m)(\nu - m - 1)(\nu - m - 3)} (A \otimes A) K_{mm} \\ & + \frac{1}{(\nu - m)(\nu - m - 1)(\nu - m - 3)} \\ & \times (A \otimes A) \left[vec\left(A^{-1}\right) vec\left(A^{-1}\right)^T \right] (A \otimes A) \quad . \end{aligned} \tag{17.15}$$

VCV is $mp \times mp$.

17.1.3 Predictive Distributions–Posterior

17.1.3.1 Joint Predictive Distribution–Posterior

The complete multivariate joint projective distribution $\{W, \Sigma\}$ is given in equation (16.63) for the prior mean. When we make the foregoing substitutions in equation (16.63), we obtain for the posterior joint density

$$\begin{aligned} \{W, \Sigma\} = & \frac{|A|^{\frac{\nu}{2}}}{2^{\frac{\nu m}{2}} (2\pi)^{\frac{qm}{2}} \Gamma_m\left(\frac{\nu}{2}\right) \left|I_{qq} + DZ^{-1}D^T\right|^{\frac{m}{2}}} |\Sigma|^{\frac{-\nu+p-q}{2}} \\ & \times e^{-\frac{1}{2} Trace\left(A\Sigma^{-1}\right)} \\ & \times e^{-\frac{1}{2} Trace\left[(W - DB^*)^T \left(I_{qq} + DZ^{-1}D^T\right)^{-1} (W - DB^*)\Sigma^{-1}\right]} \quad . \end{aligned} \tag{17.16}$$

W is $q \times m$; Σ is $m \times m$. If we make the substitutions into equation (16.66), we can write for the mean

$$\langle W \rangle = DB^* \quad .$$

W is $q \times m$. and for the prior variance-covariance matrix

$$VCV = \frac{1}{\nu - m - 1} \left\{ A \otimes \left[I_{qq} + D\left(X^T X + Q_0\right)^{-1} D^T \right] \right\} \quad .$$

VCV is $mq \times mq$.

17.1.3.2 Joint Predictive Distribution in Product Form–Posterior

The multivariate joint projective distribution $\{W, \Sigma\}$ for the prior is shown in equation (16.90), an inverse Wishart times a vectorized normal. When we make the foregoing substitutions in equation (17.5), we obtain for the posterior

$$\begin{aligned} \{W, \Sigma\} &= W^{-1}\left(\Sigma; A, \nu, m\right) \\ &\times N\left(W; DB^*, \Sigma \otimes \left(I_{qq} + DZ^{-1}D^T\right), mq\right) \quad . \end{aligned} \tag{17.17}$$

17.1.3.3 Marginal Distribution Over q-Variate Prediction in mm Form–Posterior

The marginal distribution over the predictive values for the prior with the determinant term based on I_{mm} appears in equation (16.91). When we make the foregoing substitutions in equation (17.16), we obtain for the posterior

$$\begin{aligned} \{W\} &= \frac{\Gamma_m\left(\frac{\nu+q}{2}\right)}{\Gamma_m\left(\frac{\nu}{2}\right) \pi^{\frac{qm}{2}} \left|I_{qq} + DZ^{-1}D^T\right|^{\frac{m}{2}} |A|^{\frac{q}{2}}} \\ &\times \left| \begin{array}{l} I_{mm} + A^{-1}(W - DB^*)^T \\ \times \left(I_{qq} + DZ^{-1}D^T\right)^{-1} (W - DB^*) \end{array} \right|^{-\frac{\nu+q}{2}} \quad . \end{aligned} \tag{17.18}$$

W is $q \times m$.

17.1.3.4 Marginal Distribution Over q-Variate Prediction in qq Form–Posterior

The marginal distribution over the predictive values for the prior with the determinant term based on I_{qq} appears in equation (16.92). When we make the foregoing substitutions in equation (17.5), we obtain for the posterior

$$\begin{aligned} \{W\} &= \frac{\Gamma_m\left(\frac{\nu+q}{2}\right)}{\Gamma_m\left(\frac{\nu}{2}\right) \pi^{\frac{qm}{2}} \left|I_{qq} + DZ^{-1}D^T\right|^{\frac{m}{2}} |A|^{\frac{q}{2}}} \\ &\times \left| \begin{array}{l} I_{qq} + \left(I_{qq} + DZ^{-1}D^T\right)^{-1} (W - DB^*) \\ \times A^{-1}(W - DB^*)^T \end{array} \right|^{-\frac{\nu+q}{2}} \quad . \end{aligned} \tag{17.19}$$

W is $mq \times mq$.

17.1.4 Single Row Full Distribution Predictive Distribution–Posterior

17.1.4.1 Single Row Predictive Distribution in mm Form–Posterior

The complete multivariate joint projective distribution $\{\omega, \Sigma\}$ is given by substituting form equation (17.5) into equation (17.16)

$$\{\omega, \Sigma\} = \frac{|A|^{\frac{\nu}{2}}}{2^{\frac{\nu m}{2}} (2\pi)^{\frac{m}{2}} \Gamma_m\left(\frac{\nu}{2}\right) |1 + \delta Z^{-1}\delta^T|^{\frac{m}{2}}} |\Sigma|^{\frac{-\nu+p-1}{2}}$$
$$\times e^{-\frac{1}{2}Trace\left(A\Sigma^{-1}\right)}$$
$$\times e^{-\frac{1}{2}Trace\left[(\omega-\delta B^*)^T\left(1+\delta Z^{-1}\delta^T\right)^{-1}(\omega-\delta B^*)\Sigma^{-1}\right]} \quad .$$

ω is $1 \times m$; Σ is $m \times m$. That is, ω is an m-vector.

17.1.4.2 Single Row Predictive Distribution in Product Form–Posterior

The product form of the joint density is

$$\{\omega, \Sigma\} = W^{-1}\left(\Sigma; A, \nu, m\right)$$
$$\times N\left(\omega; \delta B^*, \Sigma \otimes \left(1 + \delta Z^{-1}\delta^T\right), m\right) \quad .$$

Keep in mind, ω is an m-vector.

17.1.4.3 Single Row Predictive Distribution in mm Form–Posterior

We turn to the row vector ($q = 1$) and write the equation for the prediction of the m-variate row vector ω having the single row prediction vector δ using I_{mm} in the determinant term from equation (16.93) for the prior. When we make the foregoing substitutions in equation (17.5), we obtain for the posterior

$$\{\omega\} = \frac{\Gamma_m\left(\frac{\nu+1}{2}\right)}{\Gamma_m\left(\frac{\nu}{2}\right) \pi^{\frac{m}{2}} |1 + \delta Z^{-1}\delta^T|^{\frac{m}{2}} |A|^{\frac{1}{2}}} \times \left| \begin{array}{l} I_{mm} + \frac{1}{1+\delta Z^{-1}\delta^T} A^{-1} \\ \times (\omega - \delta B^*)^T (\omega - \delta B^*) \end{array} \right|^{-\frac{\nu+1}{2}} \quad . \tag{17.20}$$

Keep in mind, ω is an m-vector.

17.1.4.4 Single Row Predictive Distribution in qq=11 Form–Posterior

We write the equation for the prediction of the m-variate row vector ω having the single row prediction vector δ using $I_{qq} = I_{11} = 1$ in the determinant term in equation (16.94). When we make the foregoing substitutions in equation (17.5), we obtain for the posterior

$$\{\omega\} = \frac{\Gamma_m\left(\frac{\nu+1}{2}\right)}{\Gamma_m\left(\frac{\nu}{2}\right)\pi^{\frac{m}{2}}\left|1+\delta Z^{-1}\delta^T\right|^{\frac{m}{2}}|A|^{\frac{1}{2}}} \times \left[1+(\omega-\delta B^*)\frac{A^{-1}}{1+\delta Z^{-1}\delta^T}(\omega-\delta B^*)^T\right]^{-\frac{\nu+1}{2}} \quad . \tag{17.21}$$

Keep in mind, ω is an m-vector.

17.1.4.5 Single Row Predictive Distribution Mean–Posterior

The mean value of the single row predictive distribution for the posterior is

$$\langle\beta\rangle = \delta B^* \quad . \tag{17.22}$$

Keep in mind, δB^* is a $1 \times m$ matrix.

17.1.4.6 Single Row Predictive Distribution Variance-Covariance Matrix–Posterior

It is obvious from the arguments of the multivariate Student's t distribution for the posterior that the variance-covariance matrix has become

$$VCV = \frac{1+\delta Z^{-1}\delta^T}{\nu+1-m}A \quad . \tag{17.23}$$

A is $m \times m$. The marginal predictive distribution for the ith output from among the m outputs is given in equation (16.77) to be

$$\{\omega_i\} = \frac{\Gamma\left(\frac{\nu-m+2}{2}\right)}{\Gamma\left(\frac{\nu-m+1}{2}\right)\left[\pi\left(1+\delta Z^{-1}\delta^T\right)[A]_{ii}\right]^{\frac{1}{2}}} \times \left[1+\frac{1}{\left(1+\delta Z^{-1}\delta^T\right)[A]_{ii}}\left(\omega_i-[\delta B^*]_i\right)^2\right]^{-\frac{\nu-m+2}{2}} \quad . \tag{17.24}$$

The mean value of the single row predictive distribution is obvious from equation (16.75)

$$\langle\omega_i\rangle = [\delta B^*]_i \quad . \tag{17.25}$$

It is obvious from the arguments of the multivariate Student's t distribution that the variance-covariance matrix is that given in equation (16.76)

$$\frac{\left(1+\delta Z^{-1}\delta^T\right)[A]_{ii}}{\nu-m-1} \quad . \tag{17.26}$$

17.1.5 Marginal Coefficient Distribution Inferred from Marginal Predictive Distribution–Posterior

Finally, if we consider coefficient 3 and the ith of the m error variates, the marginal density over the coefficient β_{3i} is equation (16.78) for the prior. When we make the foregoing substitutions in equation (17.5), we obtain for the posterior

$$\{\beta_{3i}\} = \frac{\Gamma\left(\frac{\nu-m+2}{2}\right)}{\Gamma\left(\frac{\nu-m+1}{2}\right)} \frac{1}{(\pi[Z^{-1}]_{33}[A]_{ii})^{\frac{1}{2}}} \times \left[1 + \frac{1}{[Z^{-1}]_{33}[A]_{ii}}(\beta_{3i} - [B^*]_{3i})^2\right]^{-\frac{\nu-m+2}{2}} \quad . \tag{17.27}$$

The mean of this distribution is taken from equation (16.78) to be

$$\langle\beta_{3i}\rangle = [B^*]_{3i} \quad . \tag{17.28}$$

The variance of this distribution is taken from equation (16.76), which is reproduced here

$$Var = \frac{[A]_{ii}\left[Z^{-1}\right]_{33}}{\nu - m - 1} \quad . \tag{17.29}$$

This is also easily seen to be a direct generalization of the posterior in equation (3.12) with Σ in place of σ^2. The matrix determinant and the first exponent are going to be related to an inverse Wishart distribution, which we shall shortly characterize very carefully and give a lot of attention to. The inverse Wishart distribution continues to be conceptually very satisfying because it is a multivariate generalization of the gamma distribution that we saw in the univariate case. All the terms are parallel. We will exploit this parallelism in the coming sections.

17.2 Analysis of the Bayesian Posterior with a Diffuse Prior (Termed Classical Regression)

The unidimensional diffuse prior was given as

$$\{\sigma\} = c_0\frac{1}{\sigma} = c_0\frac{1}{(\sigma^2)^{\frac{1}{2}}} = c_0\left(\sigma^2\right)^{-\frac{1}{2}} \quad .$$

This looks (by analogy, ignoring the constant) very much like

$$\Sigma^{-\frac{1}{2}} \quad .$$

Jeffreys[50], Sun and Berger[97], Gelman[33], and Zellner[107] (who has a sign error in his diffuse prior recommendation) and others have extrapolated this model to assert that the multivariate diffuse prior should be

$$\{\Sigma\} = |\Sigma|^{-\frac{m+1}{2}} \quad .$$

This becomes evident when we take the Wiki Inverse Wishart distribution in equation (16.9), substitute m for p, and thereafter aggregate the constant to write

$$\{X|\Psi, \nu, m\} = c_3 |X|^{-\frac{\nu+m+1}{2}} e^{-\frac{1}{2} Trace\left(\Psi X^{-1}\right)} \quad .$$

If we let $\Psi = \nu I$, we obtain

$$\{X|\nu I, \nu, m\} = c_3 |X|^{-\frac{\nu+m+1}{2}} e^{-\frac{1}{2} Trace\left(\nu I X^{-1}\right)} \quad .$$

Passing ν to the limit of 0 yields the foregoing expression for the multivariate diffuse prior

$$\{X|0 \times I, 0, m\} = c_3 |X|^{-\frac{m+1}{2}} \quad .$$

To summarize, the diffuse prior case can be achieved as a special case of the informative prior discussed at length previously by setting

$$\begin{aligned} M_0 &\leftarrow 0 \\ Q_0 &\leftarrow 0 \\ m_0 &\leftarrow m+1 \\ \nu &\leftarrow \nu_C \\ A &\leftarrow S^T S \\ Z &\leftarrow X^T X \\ B^* &\leftarrow \bar{B} \quad . \end{aligned} \tag{17.30}$$

17.2.1 Mean Value–Diffuse Prior

In equation (17.2), we make the substitutions from (17.30) to write for the diffuse prior case

$$B^* = \bar{B} = (X^T X)^{-1} X^T Y \quad . \tag{17.31}$$

B^* is $p \times m$. Note that the mean value is identical with the classical regression solution.

17.2.2 Precision Matrix–Diffuse Prior

Making the foregoing substitutions from (17.30) into equation (17.3), we write for the diffuse prior case

$$Z = X^T X \quad . \tag{17.32}$$

Z is $p \times m$. This precision matrix is identical with the classical regression solution.

17.2.3 A Constant–Diffuse Prior

In equation (17.4), we make this substitution from (17.30) to find that for the diffuse prior

$$A = S^T S \quad . \tag{17.33}$$

This constant is identical with the classical regression solution.

17.2.4 Degrees of Freedom–Diffuse Prior

Substituting from equation (17.30) in equation (17.1), we know that the degrees of freedom is

$$\nu_C = n - p \quad . \tag{17.34}$$

17.2.5 Integrating Constant–Diffuse Prior

$$c_0 = \frac{\left|S^T S\right|^{\frac{\nu_C}{2}} \left|X^T X\right|^{\frac{m}{2}}}{(2\pi)^{\frac{pm}{2}} 2^{\frac{\nu_C m}{2}} \Gamma_m\left(\frac{\nu_C}{2}\right)} \quad . \tag{17.35}$$

17.3 Enter the Programmers–Summary of Diffuse Equations to Be Programmed or Simulated

This section summarizes the results developed in Chapter 16 relevant to the posterior distribution.

17.3.1 Coefficients–Diffuse

17.3.1.1 Full Joint Distribution–Diffuse

The equation for the multivariate posterior $\{B, \Sigma\}$ were developed in equation (16.80), which is reproduced here

$$\begin{aligned} \{B, \Sigma\} = {} & \frac{\left|S^T S\right|^{\frac{\nu_C}{2}} \left|X^T X\right|^{\frac{m}{2}}}{(2\pi)^{\frac{pm}{2}} 2^{\frac{\nu_C m}{2}} \Gamma_m\left(\frac{\nu_C}{2}\right) \left|\Sigma\right|^{\frac{m+1+n}{2}}} \\ & \times e^{-\frac{1}{2} Trace\left[(S^T S)\Sigma^{-1}\right]} \\ & \times e^{-\frac{1}{2} Trace\left[(B-\bar{B})^T (X^T X)(B-\bar{B})\Sigma^{-1}\right]} \quad . \end{aligned} \tag{17.36}$$

B is $p \times m$; Σ is $m \times m$. The coefficient of Σ emanates from the fact that the matrix Σ is raised to the $m_0 + n$ power in the diffuse prior, and $m_0 = m + 1$.

17.3.1.2 Full Joint Distribution in Product Form–Diffuse

This joint distribution can be written as the product of an inverse Wishart and a vectorized normal according to equation (16.81), which is reproduced here

$$\{B, \Sigma\} = W^{-1}\left[\Sigma; \left(S^T S\right), \nu_C, m\right] \\ N\left[vec\left(B\right); vec\left(\bar{B}\right), \Sigma \otimes \left(X^T X\right)^{-1}, mp\right] \quad . \tag{17.37}$$

B is $p \times m$; Σ is $m \times m$.

17.3.1.3 Inverse Wishart–Diffuse

The equation for the inverse Wishart appears in equation (16.82), which is reproduced here

$$\begin{aligned}\{\Sigma\} &= W^{-1}[\Sigma, \left(S^T S\right), \nu_C, m] \\ &= \frac{\left|S^T S\right|^{\frac{\nu_C}{2}}}{2^{\frac{\nu_C m}{2}} \Gamma_m\left(\frac{\nu_C}{2}\right) \left|\Sigma\right|^{\frac{\nu_C + m + 1}{2}}} e^{-\frac{1}{2} Trace\left[\left(S^T S\right)\Sigma^{-1}\right]} \quad .\end{aligned} \tag{17.38}$$

17.3.1.4 Marginal Distribution Over Coefficients in mm Form–Diffuse

The marginal distribution over the coefficients with the determinant term based on I_{mm} appears in equation (16.83) and is reproduced here:

$$\{B\} = \frac{\Gamma_m\left(\frac{\nu_C + p}{2}\right)}{\Gamma_m\left(\frac{\nu_C}{2}\right)} \frac{\left|X^T X\right|^{\frac{m}{2}}}{\pi^{\frac{mp}{2}} \left|S^T S\right|^{\frac{p}{2}}} \\ \left|I_{mm} + \left(S^T S\right)^{-1} \left(B - \bar{B}\right)^T \left(X^T X\right) \left(B - \bar{B}\right)\right|^{-\frac{\nu_C + p}{2}} \quad . \tag{17.39}$$

17.3.1.5 Marginal Distribution Over Coefficients in pp Form–Diffuse

The marginal distribution over the coefficients with the determinant term based on I_{pp} over the coefficients only appears in equation (16.19) and is reproduced here:

$$\{B\} = \frac{\Gamma_m\left(\frac{\nu_C + p}{2}\right)}{\Gamma_m\left(\frac{\nu_C}{2}\right)} \frac{\left|X^T X\right|^{\frac{m}{2}}}{\pi^{\frac{mp}{2}} \left|S^T S\right|^{\frac{p}{2}}} \\ \left|I_{pp} + \left(X^T X\right) \left(B - \bar{B}\right) \left(S^T S\right)^{-1} \left(B - \bar{B}\right)^T\right|^{-\frac{\nu_C + p}{2}} \quad . \tag{17.40}$$

17.3.2 Mean Coefficients–Diffuse

The mean values of the coefficients appear in equation (16.30), which is rewritten here:

$$\langle B \rangle = \bar{B} \quad .$$

17.3.2.1 Variance-Covariance Matrix of Coefficients–Diffuse

The variance-covariance matrix of the coefficients appears in equation (16.36) and is rewritten here:

$$\begin{aligned} VCV &= \left(S^T S\right) \frac{\left(S^T S\right)^{-1}}{\nu_C - m - 1} \left[\left(S^T S\right) \otimes \left(X^T X\right)^{-1}\right] \\ &= \frac{1}{\nu_C - m - 1} \left[\left(S^T S\right) \otimes \left(X^T X\right)^{-1}\right] \quad . \end{aligned} \tag{17.41}$$

17.3.2.2 Mean Values of the Error Coefficients–Diffuse

The mean value of the Σ coefficients appear in equation (16.37) and are reproduced here.

$$\langle \Sigma \rangle = \left(S^T S\right) \frac{\left(S^T S\right)^{-1}}{\nu_C - m - 1} \left(S^T S\right) = \frac{1}{\nu_C - m - 1} \left(S^T S\right) \quad . \tag{17.42}$$

17.3.2.3 Variance-Covariance Matrix of the Error Coefficients–Diffuse

The variance-covariance matrix of the Σ term appears in equation (16.53).

$$\begin{aligned} VCV = &- \frac{1}{\left(\nu_C - m - 1\right)^2} \left[\left(S^T S\right) \otimes \left(S^T S\right)\right] \\ &+ \frac{\left(\nu_C - m - 2\right)}{\left(\nu_C - m\right)\left(\nu_C - m - 1\right)\left(\nu_C - m - 3\right)} \left[\left(S^T S\right) \otimes \left(S^T S\right)\right] \\ &+ \frac{1}{\left(\nu_C - m\right)\left(\nu_C - m - 1\right)\left(\nu_C - m - 3\right)} \\ &\times \left[\left(S^T S\right) \otimes \left(S^T S\right)\right] K_{mm} \\ &+ \frac{1}{\left(\nu_C - m\right)\left(\nu_C - m - 1\right)\left(\nu_C - m - 3\right)} \\ &\times \left[\left(S^T S\right) \otimes \left(S^T S\right)\right] \left\{ vec\left[\left(S^T S\right)^{-1}\right] vec\left[\left(S^T S\right)^{-1}\right]^T \right\} \\ &\times \left[\left(S^T S\right) \otimes \left(S^T S\right)\right] \quad . \end{aligned} \tag{17.43}$$

17.3.3 Predictive Distributions–Diffuse

17.3.3.1 Joint Predictive Distribution–Diffuse

The complete multivariate joint projective distribution $\{W, \Sigma\}$ is given in equation(16.63) and is reproduced here

$$
(17.44) \qquad \{W, \Sigma\} = \frac{\left|S^T S\right|^{\frac{\nu_C}{2}} |\Sigma|^{\frac{-m_0+n+p-q}{2}}}{2^{\frac{\nu_C m}{2}} (2\pi)^{\frac{qm}{2}} \Gamma_m\left(\frac{\nu_C}{2}\right) \left|I_{qq} + D\left(X^T X\right)^{-1} D^T\right|^{\frac{m}{2}}}
\times e^{-\frac{1}{2} Trace\left[\left(S^T S\right)\Sigma^{-1}\right]}
\times e^{-\frac{1}{2} Trace\left\{\left(W - D\bar{B}\right)^T \left[I_{qq} + D\left(X^T X\right)^{-1} D^T\right]^{-1} \left(W - D\bar{B}\right)\Sigma^{-1}\right\}} \quad .
$$

17.3.3.2 Joint Predictive Distribution in Product Form–Diffuse

The multivariate joint projective distribution $\{W, \Sigma\}$ is shown in equation (16.64) to be an inverse Wishart times a vectorized normal, meaning

$$
(17.45) \qquad \{W, \Sigma\} = W^{-1}\left[\Sigma; \left(S^T S\right), \nu_C, m\right]
\times N\left\{W; D\bar{B}, \Sigma \otimes \left[I_{qq} + D\left(X^T X\right)^{-1} D^T\right], mq\right\} \quad .
$$

If we make the substitutions into equation (16.66), we can write for the mean

$$
\langle W \rangle = D\bar{B} \quad .
$$

and for the classical variance-covariance matrix

$$
VCV = \frac{1}{\nu_C - m - 1} \left\{S^T S \otimes \left[I_{qq} + D\left(X^T X\right)^{-1} D^T\right]\right\} \quad .
$$

17.3.3.3 Marginal Distribution Over q-Variate Prediction in mm Form–Diffuse

The marginal distribution over the predictive values with the determinant term based on I_{mm} appears in equation (16.67) and is reproduced here:

$$
(17.46) \qquad \{W\} = \frac{\Gamma_m\left(\frac{\nu_C + q}{2}\right)}{\Gamma_m\left(\frac{\nu_C}{2}\right) \pi^{\frac{qm}{2}} \left|I_{qq} + D(X^T X)^{-1} D^T\right|^{\frac{m}{2}} \left|S^T S\right|^{\frac{q}{2}}}
\times \left| \begin{array}{l} I_{mm} + \left(S^T S\right)^{-1} \left(W - D\bar{B}\right)^T \\ \times \left[I_{qq} + D(X^T X)^{-1} D^T\right]^{-1} \left(W - D\bar{B}\right) \end{array} \right|^{-\frac{\nu_C + q}{2}} \quad .
$$

17.3.3.4 Marginal Distribution Over q-Variate Prediction in qq Form–Diffuse

The marginal distribution over the predictive values with the determinant term based on I_{qq} appears in equation (16.68) and is reproduced here:

$$\{W\} = \frac{\Gamma_m\left(\frac{\nu_C+q}{2}\right)}{\Gamma_m\left(\frac{\nu_C}{2}\right)\pi^{\frac{qm}{2}}\left|I_{qq}+D(X^TX)^{-1}D^T\right|^{\frac{m}{2}}|S^TS|^{\frac{q}{2}}} \times \left|\begin{array}{l} I_{qq}+\left[I_{qq}+D(X^TX)^{-1}D^T\right]^{-1}(W-D\bar{B}) \\ \times(S^TS)^{-1}(W-D\bar{B})^T \end{array}\right|^{-\frac{\nu_C+q}{2}} . \tag{17.47}$$

17.3.4 Single Row Full Distribution Predictive Distribution–Diffuse

17.3.4.1 Single Row Predictive Distribution in mm Form–Diffuse

We turn to the row vector ($q = 1$) and write the equation for the prediction of the m-variate row vector ω having the single row prediction vector δ using I_{mm} in the determinant term from equation (17.46) as follows

$$\{\omega\} = \frac{\Gamma_m\left(\frac{\nu_C+1}{2}\right)}{\Gamma_m\left(\frac{\nu_C}{2}\right)\pi^{\frac{m}{2}}\left|1+\delta(X^TX)^{-1}\delta^T\right|^{\frac{m}{2}}|S^TS|^{\frac{1}{2}}} \times \left|\begin{array}{l} I_{mm}+\frac{1}{1+\delta(X^TX)^{-1}\delta^T}(S^TS)^{-1} \\ \times(\omega-\delta\bar{B})^T(\omega-\delta\bar{B}) \end{array}\right|^{-\frac{\nu_C+1}{2}} . \tag{17.48}$$

17.3.4.2 Single Row Predictive Distribution in qq=11 Form–Diffuse

We write the equation for the prediction of the m-variate row vector ω having the single row prediction vector δ using $I_{qq} = I_{11} = 1$ in the determinant term from equation (16.94) as follows

$$\{\omega|\delta\} = \frac{\Gamma\left(\frac{\nu_C+1}{2}\right)}{\Gamma\left(\frac{\nu_C-m+1}{2}\right)\left\{\pi\left[1+\delta(X^TX)^{-1}\delta^T\right]\right\}^{\frac{m}{2}}|S^TS|^{\frac{1}{2}}} \times \left[1+(\omega-\delta\bar{B})^T\frac{(S^TS)^{-1}}{[1+\delta(X^TX)^{-1}\delta^T]}(\omega-\delta\bar{B})\right]^{-\frac{\nu_C+1}{2}} . \tag{17.49}$$

The mean value of the single row predictive distribution is the row vector with m columns

$$\langle\omega\rangle = \delta\bar{B} . \tag{17.50}$$

It is obvious from the arguments of the multivariate Student's t distribution that the variance-covariance matrix has become

$$VCV = \frac{1+\delta\left(X^TX\right)^{-1}\delta^T}{\nu_C+1-m}\left(S^TS\right) \quad . \tag{17.51}$$

The marginal predictive distribution for the ith output from among the m outputs is given in equation (16.77) to be

$$\begin{aligned} \{\omega_i\} = {} & \frac{\Gamma\left(\frac{\nu_C-m+2}{2}\right)}{\Gamma\left(\frac{\nu_C-m+1}{2}\right)\left\{\pi\left[1+\delta\left(X^TX\right)^{-1}\delta^T\right]\left[\left(S^TS\right)\right]_{ii}\right\}^{\frac{1}{2}}} \\ & \times\left\{1+\frac{\left(\omega_i-\left[\delta\bar{B}\right]_i\right)^2}{\left[1+\delta\left(X^TX\right)^{-1}\delta^T\right]\left[\left(S^TS\right)\right]_{ii}}\right\}^{-\frac{\nu_C-m+2}{2}} \quad . \end{aligned} \tag{17.52}$$

The mean value of the single row predictive distribution is obvious from equation (16.75)

$$\langle\omega_i\rangle = \left[\delta\bar{B}\right]_i \quad . \tag{17.53}$$

It is obvious from the arguments of the multivariate Student's t distribution that the variance-covariance matrix is that given in equation (16.76)

$$Var = \frac{\left[1+\delta\left(X^TX\right)^{-1}\delta^T\right]\left[\left(S^TS\right)\right]_{ii}}{\nu_C-m-1} \quad . \tag{17.54}$$

17.3.5 Marginal Coefficient Distribution Inferred from Marginal Predictive Distribution–Diffuse

Finally, if we consider coefficient 3 and the ith of the m error variates, the marginal density over the coefficient β_{3i} is equation (16.97)

$$\begin{aligned} \{\beta_{3i}\} = {} & \frac{\Gamma\left(\frac{\nu_C-m+2}{2}\right)}{\Gamma\left(\frac{\nu_C-m+1}{2}\right)}\frac{1}{\left(\pi\left[\left(X^TX\right)^{-1}\right]_{33}\left[\left(S^TS\right)\right]_{ii}\right)^{\frac{1}{2}}} \\ & \times\left[1+\frac{1}{\left[\left(X^TX\right)^{-1}\right]_{33}\left[\left(S^TS\right)\right]_{ii}}\left(\beta_{3i}-\left[\bar{B}\right]_{3i}\right)^2\right]^{-\frac{\nu_C-m+2}{2}} \quad . \end{aligned} \tag{17.55}$$

The mean of this distribution is taken from equation (16.78) to be

$$\langle\beta_{3i}\rangle = \left[\bar{B}\right]_{3i} \quad . \tag{17.56}$$

The variance of this distribution is taken from equation (16.76), which is reproduced here

$$Var = \frac{\left[S^T S\right]_{ii} \left[\left(X^T X\right)^{-1}\right]_{33}}{\nu_C - m - 1} \quad . \tag{17.57}$$

Chapter 18

SUMMARY OF KEY EQUATIONS AND PARAMETERS FOR THE MULTIVARIATE CASE

Probability is the most important concept in modern science, especially as nobody has the slightest notion what it means.

Bertrand Russell, 1929

This has been quite a prodigious set of three sections. We have developed the multivariate likelihood function and followed it closely with the multivariate conjugate prior and posterior. We have strived for mathematical purity here. We have developed a multivariate diffuse prior and used it to craft a classical regression solution in the multivariate case. We have developed the predictive cases, and we have generalized all of the results from the univariate case. Our objective has been to put this into a logical, stepwise manuscript so that people can have a deep understanding of these results. We have not seen them put together like this before. One of the reasons we have given so much attention to this multivariate case will become clear in the next chapter. It is needed for the simultaneous regression equation case, which cannot be analyzed without it.

18.1 Mean Values of the Coefficients

Prior: Equation (16.85)
Posterior: Equation (17.2)
Classical: Equation (17.31)

18.2 The Q Matrix

Prior: Equation (16.1)
Posterior: Equation (17.3)
Classical: Equation (17.32)

18.3 The Residual Matrix

Prior: Equation (16.2)
Posterior: Equation (17.4)
Classical: Equation (17.33)

18.4 The Degrees of Freedom

Prior: Equation (16.12)
Posterior: Equation (17.1)
Classical: Equation (17.34)

18.5 The Integrating Constant

Prior: Equation (16.13)
Posterior: Equation (17.7)
Classical: Equation (17.35)

18.6 Full Probability Distribution with Two Traces in the Exponent

Prior: Equation (16.80)
Posterior: Equation (17.8)
Classical: Equation (17.36)

18.7 Full Probability Distribution in Product Form

Prior: Equation (16.81)
Posterior: Equation (17.9)
Classical: Equation (17.37)

18.8 Inverse Wishart Distribution

Prior: Equation (16.82)
Posterior: Equation (17.10)
Classical: Equation (17.38)

18.9 Marginal Distribution over Coefficients in mm Form

Prior: Equation (16.83)
Posterior: Equation (17.11)
Classical: Equation (17.39)

18.10 Marginal Distribution over Coefficients in pp Form

Prior: Equation (16.84)
Posterior: Equation (17.12)
Classical: Equation (17.40)

18.11 Mean Values of Coefficients

Prior: Equation (16.85)
Posterior: Equation (17.2)
Classical: Equation (17.31)

18.12 Variance-Covariance of Coefficients

Prior: Equation (16.86)
Posterior: Equation (17.13)
Classical: Equation (17.41)

18.13 Mean Values Error Term

Prior: Equation (16.87)
Posterior: Equation (17.14)
Classical: Equation (17.42)

18.14 Variance-Covariance of Error Term

Prior: Equation (16.88)
Posterior: Equation (17.15)
Classical: Equation (17.43)

18.15 Joint Predictive Distribution W and Σ

Prior: Equation (16.89)
Posterior: Equation (17.16)
Classical: Equation (17.44)

18.16 Joint Predictive Distribution W and Σ in Product Form

Prior: Equation (16.90)
Posterior: Equation (17.17)
Classical: Equation (17.45)

18.17 Joint Predictive Distribution of W in mm Form

Prior: Equation (16.91)
Posterior: Equation (17.18)
Classical: Equation (17.46)

18.18 Joint Predictive Distribution of W in qq Form

Prior: Equation (16.92)
Posterior: Equation (17.19)
Classical: Equation (17.47)

18.19 Single Row Predictive Distribution of ω in mm Form

Prior: Equation (16.93)
Posterior: Equation (17.20)
Classical: Equation (17.48)

18.20 Single Row Predictive Distribution of ω in qq Form

Prior: Equation (16.94)
Posterior: Equation (17.21)
Classical: Equation (17.49)

18.21 Mean of Single Row Predictive Distribution

Prior: Equation (16.95)
Posterior: Equation (17.22)
Classical: Equation (17.50)

18.22 Variance-Covariance of Single Row Predictive Distribution

Prior: Equation (16.96)
Posterior: Equation (17.23)
Classical: Equation (17.51)

18.23 Marginal Predictive Distribution for the ith Output

Prior: Equation (16.97)
Posterior: Equation (17.24)
Classical: Equation (17.52)

18.24 Mean of the ith Output

Prior: Equation (16.98)
Posterior: Equation (17.25)
Classical: Equation (17.53)

18.25 Variance of the ith Output

Prior: Equation (16.99)
Posterior: Equation (17.26)
Classical: Equation (17.54)

18.26 Marginal Coefficient $3i$ Distribution Inferred from Marginal Predictive Distribution

Prior: Equation (16.100)
Posterior: Equation (17.27)
Classical: Equation (17.55)

18.27 Mean of Marginal Coefficient $3i$ Inferred from Marginal Predictive Distribution

Prior: Equation (16.101)
Posterior: Equation (17.28)
Classical: Equation (17.56)

18.28 Variance of Marginal Coefficient $3i$ Inferred from Marginal Predictive Distribution

Prior: Equation (16.102)
Posterior: Equation (17.29)
Classical: Equation (17.57)

Chapter 19

BIVARIATE EXAMPLE–CRIME IN AMERICAN CITIES

> ***Probability is the very guide of life.***
>
> ***Marcus Tullius Cicero***

We are intrigued by an example from Life in America's Small Cities by G. Scott Thomas, Prometheus Books, 1990. The book presents the following crime statistics for small cities

- TC= total overall reported crime rate expressed per one million residents. This observation represents the number of total reported crimes per million people
- VC=reported violent crime rate per 100,000 residents
- PF = Annual police funding in $ per resident
- HS = Percentage of people 25 years or older with 4 years of high school
- DROP = Percentage of 16 to 19 year olds not in high school and not high school graduates.
- INCOL = Percentage of 18 to 24 year olds in college
- GRCOL = Percent of people 25 years of age or more with at least four years of college

We want to consider total overall reported crime rate TC and the reported violent crime rate VC to be predicted as a function of PF, HS, DROP, INCOL, and GRCOL. There are two variables to be predicted, TC and VC, and we cannot assume they are probabilistically independent. Until proven otherwise, they are joint, mutually relevant, mutually dependent. We will see in this application that in fact they are highly probabilistically dependent, i.e., joint.

The data for 50 observations of these seven variables across 50 sample cities appears in Table 19.1. The X matrix, the matrix of independent variables, appears in Table 19.2. The Y matrix, the matrix of dependent variables, over which there is a joint distribution, appears in Table 19.3. We elect to use the logarithms of the Y matrix, not the values themselves, so that we can be sure that the Y values themselves never become negative. Negative total crime and negative violent crime have no meaning. The lnY values appear in Table 19.4.

19.1 Classical Regression (Bayes with Diffuse Prior)

19.1.1 Model Coefficients–Classical

The key parameters of the classical regression analysis, including the degrees of freedom, of this problem appear in Table 19.5. We calculate the matrix $X^T X$ based on the X matrix that appears in Table 19.2 and depict it in Table 19.6. We note the five independent variables and the intercept in the calculation. We calculate the inverse of the $X^T X$ matrix in Table 19.7.

We see that the matrix $(X^T X)$ and its inverse $(X^T X)^{-1}$ are symmetric. We know they are positive definite and therefore have all positive eigenvalues and an orthonormal diagonalization matrix as discussed in Chapter 2. In order to calculate the classical regression mean values of the coefficients, we need to calculate the $X^T Y$ matrix, which is given in Table 19.8. We then left multiply this by the $(X^T X)^{-1}$ matrix in Table 19.7 to calculate the mean values of the twelve model coefficients, which appear in Table 19.9. We have termed this the $\bar{B}$ matrix in Table 19.9. Because we are going to need the vectorization of the $\bar{B}$ matrix in subsequent developments, we put it forth in Table 19.10.

These coefficients, without the intercept terms (which are much larger in magnitude), are plotted in Figure 19.1. They show the absolute and relative values of the mean coefficient values, as well as the signs. The signs might seem to be counterintuitive, and we will discuss this at some length later in this chapter.

We next must calculate the mean squared error matrix $S^T S$ and its inverse for future calculations. Those calculations appear in Table 19.11.

In order to calculate the variance-covariance matrix of the twelve model coefficients, we must first calculate the Kronecker product matrix $(S^T S) \otimes (X^T X)^{-1}$ and display it in Table 19.12.

The variance-covariance matrix of the coefficients is derived by multiplying the Kronecker matrix in Table 19.12 by the multiplicative constant in equation (17.41) to obtain the variance-covariance matrix among the model coefficients in Table 19.13.

Table 19.1: The (Dependent) Crime Observations and (Independent) Precursor Variable Observations–the Full Data Set

			1	2	3	4	5
Observation	TC	VC	PF	GRHS	DROP	INCOL	GRCOL
1	478	184	40	74	11	31	20
2	494	213	32	72	11	43	18
3	643	347	57	70	18	16	16
4	341	565	31	71	11	25	19
5	773	327	67	72	9	29	24
6	603	260	25	68	8	32	15
7	484	325	34	68	12	24	14
8	546	102	33	62	13	28	11
9	424	38	36	69	7	25	12
10	548	226	31	66	9	58	15
11	506	137	35	60	13	21	9
12	819	369	30	81	4	77	36
13	541	109	44	66	9	37	12
14	491	809	32	67	11	37	16
15	514	29	30	65	12	35	11
16	371	245	16	64	10	42	14
17	457	118	29	64	12	21	10
18	437	148	36	62	7	81	27
19	570	387	30	59	15	31	16
20	432	98	23	56	15	50	15
21	619	608	33	46	22	24	8
22	357	218	35	54	14	27	13
23	623	254	38	54	20	22	11
24	547	697	44	45	26	18	8
25	792	827	28	57	12	23	11
26	799	693	35	57	9	60	18
27	439	448	31	61	19	14	12
28	867	942	39	52	17	31	10
29	912	1017	27	44	21	24	9
30	462	216	36	43	18	23	8
31	859	673	38	48	19	22	10
32	805	989	46	57	14	25	12
33	652	630	29	47	19	25	9
34	776	404	32	50	19	21	9
35	919	692	39	48	16	32	11
36	732	1517	44	49	13	31	14
37	657	879	33	72	13	13	22
38	1419	631	43	59	14	21	13
39	989	1375	22	49	9	46	13
40	821	1139	30	54	13	27	12
41	1740	3545	86	62	22	18	15
42	815	706	30	47	17	39	11
43	760	451	32	45	34	15	10
44	936	433	43	48	26	23	12
45	863	601	20	69	23	7	12
46	783	1024	55	42	23	23	11
47	715	457	44	49	18	30	12
48	1504	1441	37	57	15	35	13
49	1324	1022	82	72	22	15	16
50	940	1244	66	67	26	18	16

Table 19.2: The X Matrix Portion of the Data

	1	2	3	4	5	6
	Intercept	PF	HS	DROP	INCOL	GRCOL
1	1	40	74	11	31	20
2	1	32	72	11	43	18
3	1	57	70	18	16	16
4	1	31	71	11	25	19
5	1	67	72	9	29	24
6	1	25	68	8	32	15
7	1	34	68	12	24	14
8	1	33	62	13	28	11
9	1	36	69	7	25	12
10	1	31	66	9	58	15
11	1	35	60	13	21	9
12	1	30	81	4	77	36
13	1	44	66	9	37	12
14	1	32	67	11	37	16
15	1	30	65	12	35	11
16	1	16	64	10	42	14
17	1	29	64	12	21	10
18	1	36	62	7	81	27
19	1	30	59	15	31	16
20	1	23	56	15	50	15
21	1	33	46	22	24	8
22	1	35	54	14	27	13
23	1	38	54	20	22	11
24	1	44	45	26	18	8
25	1	28	57	12	23	11
26	1	35	57	9	60	18
27	1	31	61	19	14	12
28	1	39	52	17	31	10
29	1	27	44	21	24	9
30	1	36	43	18	23	8
31	1	38	48	19	22	10
32	1	46	57	14	25	12
33	1	29	47	19	25	9
34	1	32	50	19	21	9
35	1	39	48	16	32	11
36	1	44	49	13	31	14
37	1	33	72	13	13	22
38	1	43	59	14	21	13
39	1	22	49	9	46	13
40	1	30	54	13	27	12
41	1	86	62	22	18	15
42	1	30	47	17	39	11
43	1	32	45	34	15	10
44	1	43	48	26	23	12
45	1	20	69	23	7	12
46	1	55	42	23	23	11
47	1	44	49	18	30	12
48	1	37	57	15	35	13
49	1	82	72	22	15	16
50	1	66	67	26	18	16

Table 19.3: The Y Matrix Portion of the Data

	TC	VC
1	478	184
2	494	213
3	643	347
4	341	565
5	773	327
6	603	260
7	484	325
8	546	102
9	424	38
10	548	226
11	506	137
12	819	369
13	541	109
14	491	809
15	514	29
16	371	245
17	457	118
18	437	148
19	570	387
20	432	98
21	619	608
22	357	218
23	623	254
24	547	697
25	792	827
26	799	693
27	439	448
28	867	942
29	912	1017
30	462	216
31	859	673
32	805	989
33	652	630
34	776	404
35	919	692
36	732	1517
37	657	879
38	1419	631
39	989	1375
40	821	1139
41	1740	3545
42	815	706
43	760	451
44	936	433
45	863	601
46	783	1024
47	715	457
48	1504	1441
49	1324	1022
50	940	1244

Table 19.4: The ln(Y) Matrix Portion of the Data

ln(y)	TC	VC
1	6.169610732	5.214935758
2	6.202535517	5.361292166
3	6.466144724	5.84932478
4	5.831882477	6.336825731
5	6.650279049	5.789960171
6	6.401917197	5.560681631
7	6.182084907	5.783825182
8	6.302618976	4.624972813
9	6.049733455	3.63758616
10	6.306275287	5.420534999
11	6.226536669	4.919980926
12	6.708084084	5.910796644
13	6.293419279	4.691347882
14	6.196444128	6.695798917
15	6.242223265	3.36729583
16	5.916202063	5.501258211
17	6.124683391	4.770684624
18	6.079933195	4.997212274
19	6.345636361	5.958424693
20	6.068425588	4.584967479
21	6.428105273	6.410174882
22	5.877735782	5.384495063
23	6.434546519	5.537334267
24	6.304448802	6.546785411
25	6.674561392	6.717804695
26	6.683360946	6.541029999
27	6.084499413	6.104793232
28	6.765038977	6.848005275
29	6.81563999	6.924612396
30	6.135564891	5.375278408
31	6.755768922	6.51174533
32	6.690842277	6.896694332
33	6.480044562	6.445719819
34	6.65415252	6.001414878
35	6.823286122	6.539585956
36	6.595780514	7.324489979
37	6.487684018	6.778784898
38	7.257707677	6.447305863
39	6.896694332	7.22620901
40	6.710523109	7.037905963
41	7.461640392	8.173293439
42	6.703188113	6.559615237
43	6.633318433	6.11146734
44	6.841615476	6.070737728
45	6.760414691	6.398594935
46	6.663132696	6.931471806
47	6.572282543	6.124683391
48	7.315883505	7.273092596
49	7.188412736	6.929516771
50	6.845879875	7.126087273

Table 19.5: Key Parameters of the Regression–Classical

	Key parameters
n	50
p	6
m	2
$\nu_C = n - p$	44
$\nu_C - m$	42
$\nu_C - m - 1$	41
$\nu_C - m - 2$	40
$\nu_C - m - 3$	39

Table 19.6: The $X^T X$ Matrix–Classical

$X^T X$	Intercept	PF	GRHS	DROP	INCOL	GRCOL
Intercept	50	1888	2940	770	1495	691
PF	1888	80650	111826	30346	53670	26527
GRHS	2940	111826	177738	43696	89224	42346
DROP	770	30346	43696	13636	20284	9859
INCOL	1495	53670	89224	20284	55435	22874
GRCOL	691	26527	42346	9859	22874	10853

Table 19.7: The $(X^T X)^{-1}$ Matrix–Classical

$(X^T X)^{-1}$	Intercept	PF	GRHS	DROP	INCOL	GRCOL
Intercept	3.47990004	-0.00157501	-0.04212569	-0.05909975	-0.02329195	0.04943044
PF	-0.00157501	0.00014774	-0.00001327	-0.00012976	0.00004526	-0.00018657
GRHS	-0.04212569	-0.00001327	0.00066781	0.00058026	0.00026323	-0.00097303
DROP	-0.05909975	-0.00012976	0.00058026	0.00157632	0.00038584	-0.00042920
INCOL	-0.02329195	0.00004526	0.00026323	0.00038584	0.00030433	-0.00064663
GRCOL	0.04943044	-0.00018657	-0.00097303	-0.00042920	-0.00064663	0.00295027

Table 19.8: The $X^T Y$ Matrix–Classical

XTY	Total Crime	Violent Crime
Intercept	325.3064248	302.276437
PF	12405.32556	11633.8945
GRHS	19091.53095	17630.2535
DROP	5048.776463	4764.569594
INCOL	9679.583643	8878.829651
GRCOL	4491.263974	4168.932385

Table 19.9: The $\bar{B}$ Matrix–Classical

B bar	Total Crime	Violent Crime
Intercept	6.418661	8.563630
PF	0.012107	0.014571
GRHS	-0.011361	-0.068889
DROP	0.008351	0.002984
INCOL	-0.000460	-0.028537
GRCOL	0.013275	0.129498

Table 19.10: The $vec(\bar{B})$ Vector–Classical

	vec(B bar)
Intercept, TC	6.418661
PF, TC	0.012107
HS, TC	-0.011361
DROP, TC	0.008351
INCOL, TC	-0.000460
GRCOL, TC	0.013275
Intercept, TV	8.563630
PF, TV	0.014571
HS, TV	-0.068889
DROP, TV	0.002984
INCOL, TV	-0.028537
GRCOL, TV	0.129498

Table 19.11: The S^TS and $(S^TS)^{-1}$ Matrices–Classical

S^TS	Total Crime	Violent Crime
Total Crime	4.515634662	6.786307974
Violent Crime	6.786307974	27.74553005
	Determinant	79.234701
	Sqrt Determinant	8.901388
$(S^TS)^{-1}$	TC	TV
TC	0.35016892	-0.08564818
TV	-0.08564818	0.05699062

Figure 19.1: Model Coefficients–Classical

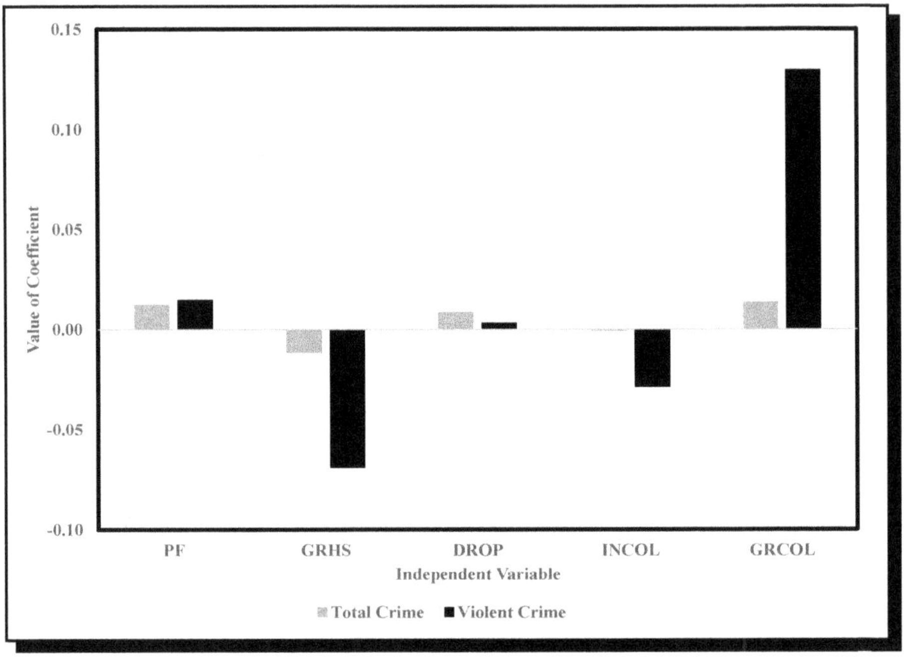

Table 19.12: The $(S^T S) \otimes (X^T X)^{-1}$ Matrix–Classical

$(S^T S) \otimes (X^T X)^{-1}$	Intercept, TC	PF, TC	HS, TC	DROP, TC	INCOL, TC	GRCOL, TC
Intercept, TC	1.571396E+01	-7.112188E-03	-1.902242E-01	-2.668729E-01	-1.051780E-01	2.232098E-01
PF, TC	-7.112188E-03	6.671211E-04	-5.991323E-05	-5.859313E-04	2.043909E-04	-8.425003E-04
HS, TC	-1.902242E-01	-5.991323E-05	3.015597E-03	2.620230E-03	1.188657E-03	-4.393843E-03
DROP, TC	-2.668729E-01	-5.859313E-04	2.620230E-03	7.118082E-03	1.742292E-03	-1.938123E-03
INCOL, TC	-1.051780E-01	2.043909E-04	1.188657E-03	1.742292E-03	1.374235E-03	-2.919956E-03
GRCOL, TC	2.232098E-01	-8.425003E-04	-4.393843E-03	-1.938123E-03	-2.919956E-03	1.332234E-02
Intercept, TV	2.361567E+01	-1.068853E-02	-2.858779E-01	-4.010691E-01	-1.580664E-01	3.354502E-01
PF, TV	-1.068853E-02	1.002581E-03	-9.004042E-05	-8.805651E-04	3.071683E-04	-1.266149E-03
HS, TV	-2.858779E-01	-9.004042E-05	4.531981E-03	3.937805E-03	1.786369E-03	-6.603274E-03
DROP, TV	-4.010691E-01	-8.805651E-04	3.937805E-03	1.069739E-02	2.618398E-03	-2.912702E-03
INCOL, TV	-1.580664E-01	3.071683E-04	1.786369E-03	2.618398E-03	2.065265E-03	-4.388248E-03
GRCOL, TV	3.354502E-01	-1.266149E-03	-6.603274E-03	-2.912702E-03	-4.388248E-03	2.002144E-02

$(S^T S) \otimes (X^T X)^{-1}$	Intercept, TV	PF, TV	HS, TV	DROP, TV	INCOL, TV	GRCOL, TV
Intercept, TC	2.361567E+01	-1.068853E-02	-2.858779E-01	-4.010691E-01	-1.580664E-01	3.354502E-01
PF, TC	-1.068853E-02	1.002581E-03	-9.004042E-05	-8.805651E-04	3.071683E-04	-1.266149E-03
HS, TC	-2.858779E-01	-9.004042E-05	4.531981E-03	3.937805E-03	1.786369E-03	-6.603274E-03
DROP, TC	-4.010691E-01	-8.805651E-04	3.937805E-03	1.069739E-02	2.618398E-03	-2.912702E-03
INCOL, TC	-1.580664E-01	3.071683E-04	1.786369E-03	2.618398E-03	2.065265E-03	-4.388248E-03
GRCOL, TC	3.354502E-01	-1.266149E-03	-6.603274E-03	-2.912702E-03	-4.388248E-03	2.002144E-02
Intercept, TV	9.655167E+01	-4.369960E-02	-1.168800E+00	-1.639754E+00	-6.462476E-01	1.371474E+00
PF, TV	-4.369960E-02	4.099009E-03	-3.681264E-04	-3.600153E-03	1.255844E-03	-5.176596E-03
HS, TV	-1.168800E+00	-3.681264E-04	1.852881E-02	1.609955E-02	7.303495E-03	-2.699720E-02
DROP, TV	-1.639754E+00	-3.600153E-03	1.609955E-02	4.373581E-02	1.070521E-02	-1.190846E-02
INCOL, TV	-6.462476E-01	1.255844E-03	7.303495E-03	1.070521E-02	8.443750E-03	-1.794116E-02
GRCOL, TV	1.371474E+00	-5.176596E-03	-2.699720E-02	-1.190846E-02	-1.794116E-02	8.185682E-02

Table 19.13: The Variance-Covariance Matrix–Classical Regression

VCV Matrix Full	Intercept, TC	PF, TC	HS, TC	DROP, TC	INCOL, TC	GRCOL, TC
Intercept, TC	3.832672E-01	-1.734680E-04	-4.639615E-03	-6.509094E-03	-2.565316E-03	5.444142E-03
PF, TC	-1.734680E-04	1.627125E-05	-1.461298E-06	-1.429101E-05	4.985144E-06	-2.054879E-05
HS, TC	-4.639615E-03	-1.461298E-06	7.355114E-05	6.390805E-05	2.899163E-05	-1.071669E-04
DROP, TC	-6.509094E-03	-1.429101E-05	6.390805E-05	1.736117E-04	4.249492E-05	-4.727129E-05
INCOL, TC	-2.565316E-03	4.985144E-06	2.899163E-05	4.249492E-05	3.351794E-05	-7.121845E-05
GRCOL, TC	5.444142E-03	-2.054879E-05	-1.071669E-04	-4.727129E-05	-7.121845E-05	3.249352E-04
Intercept, TV	5.759920E-01	-2.606959E-04	-6.972632E-03	-9.782173E-03	-3.855277E-03	8.181712E-03
PF, TV	-2.606959E-04	2.445319E-05	-2.196108E-06	-2.147720E-05	7.491909E-06	-3.088168E-05
HS, TV	-6.972632E-03	-2.196108E-06	1.105361E-04	9.604401E-05	4.356999E-05	-1.610555E-04
DROP, TV	-9.782173E-03	-2.147720E-05	9.604401E-05	2.609119E-04	6.386336E-05	-7.104152E-05
INCOL, TV	-3.855277E-03	7.491909E-06	4.356999E-05	6.386336E-05	5.037233E-05	-1.070304E-04
GRCOL, TV	8.181712E-03	-3.088168E-05	-1.610555E-04	-7.104152E-05	-1.070304E-04	4.883279E-04

VCV Matrix Full	Intercept, TV	PF, TV	HS, TV	DROP, TV	INCOL, TV	GRCOL, TV
Intercept, TC	5.759920E-01	-2.606959E-04	-6.972632E-03	-9.782173E-03	-3.855277E-03	8.181712E-03
PF, TC	-2.606959E-04	2.445319E-05	-2.196108E-06	-2.147720E-05	7.491909E-06	-3.088168E-05
HS, TC	-6.972632E-03	-2.196108E-06	1.105361E-04	9.604401E-05	4.356999E-05	-1.610555E-04
DROP, TC	-9.782173E-03	-2.147720E-05	9.604401E-05	2.609119E-04	6.386336E-05	-7.104152E-05
INCOL, TC	-3.855277E-03	7.491909E-06	4.356999E-05	6.386336E-05	5.037233E-05	-1.070304E-04
GRCOL, TC	8.181712E-03	-3.088168E-05	-1.610555E-04	-7.104152E-05	-1.070304E-04	4.883279E-04
Intercept, TV	2.354919E+00	-1.065844E-03	-2.850731E-02	-3.999400E-02	-1.576214E-02	3.345058E-02
PF, TV	-1.065844E-03	9.997584E-05	-8.978693E-06	-8.780861E-05	3.063035E-05	-1.262584E-04
HS, TV	-2.850731E-02	-8.978693E-06	4.519222E-04	3.926718E-04	1.781340E-04	-6.584684E-04
DROP, TV	-3.999400E-02	-8.780861E-05	3.926718E-04	1.066727E-03	2.611026E-04	-2.904502E-04
INCOL, TV	-1.576214E-02	3.063035E-05	1.781340E-04	2.611026E-04	2.059451E-04	-4.375894E-04
GRCOL, TV	3.345058E-02	-1.262584E-04	-6.584684E-04	-2.904502E-04	-4.375894E-04	1.996508E-03

The diagonal of the variance-covariance matrix contains the variances. They are tabulated in Table 19.14. We notice that the t-statistics of the various coefficients range from comfortably high (the intercepts) to uncomfortably low (fraction of people in college with regard to total crime, and dropouts with regard to violent crime.) We would like to see these t-statistics become high, indicating high value of mean/standard deviation, meaning a low standard deviation compared to the mean.

Table 19.14: Classical Key Results

	Intercept, TC	PF, TC	HS, TC	DROP, TC	INCOL, TC	GRCOL, TC
Variance	0.38326725	1.62712E-05	7.35511E-05	0.000173612	3.35179E-05	0.0003249
Standard Deviation	0.61908582	0.00403376	0.00857620	0.01317618	0.00578947	0.01802596
Means	6.41866104	0.01210695	-0.01136068	0.00835127	-0.00046005	0.01327518
t statistic	10.36796653	3.00140406	-1.32467606	0.63381533	-0.07946321	0.73644784

	Intercept, TV	PF, TV	HS, TV	DROP, TV	INCOL, TV	GRCOL, TV
Variance	2.35491881	0.00009998	0.00045192	0.00106673	0.00020595	0.00199651
Standard Deviation	1.53457447	0.00999879	0.02125846	0.03266079	0.01435079	0.04468230
Means	8.56363044	0.01457146	-0.06888913	0.00298441	-0.02853705	0.12949772
t statistic	5.58045934	1.45732216	-3.24055113	0.09137586	-1.98853510	2.89818854

We calculate the correlation matrix based on the variance-covariance matrix in Table 19.13. It appears in Table 19.15. We notice quite high correlation of every independent variable with the intercept. Discussions and analysis are left to the reader (e.g., alternative models, better data, more observed data).

We now turn to the mean of the Wishart distribution in the classical regression case. The Wishart error distribution is the hardest part of multivariate

Table 19.15: The Correlation Matrix–Classical

Correlation	Intercept, TC	PF, TC	HS, TC	DROP, TC	INCOL, TC	GRCOL, TC
Intercept, TC	1	-0.069463727	-0.873849054	-0.797958235	-0.715733557	0.487843106
PF, TC	-0.069463727	1	-0.042240961	-0.268882723	0.213465987	-0.2826034
HS, TC	-0.873849054	-0.042240961	1	0.565550362	0.583901123	-0.693214476
DROP, TC	-0.797958235	-0.268882723	0.565550362	1	0.55706878	-0.199025867
INCOL, TC	-0.715733557	0.213465987	0.583901123	0.55706878	1	-0.682425912
GRCOL, TC	0.487843106	-0.2826034	-0.693214476	-0.199025867	-0.682425912	1
Intercept, TV	0.606286158	-0.042114896	-0.529802585	-0.483791032	-0.433939348	0.295772522
PF, TV	-0.042114896	0.606286158	-0.02561011	-0.163019873	0.129421473	-0.171338529
HS, TV	-0.529802585	-0.02561011	0.606286158	0.342885356	0.354011168	-0.420286341
DROP, TV	-0.483791032	-0.163019873	0.342885356	0.606286158	0.33774309	-0.120666628
INCOL, TV	-0.433939348	0.129421473	0.354011168	0.33774309	0.606286158	-0.413745384
GRCOL, TV	0.295772522	-0.171338529	-0.420286341	-0.120666628	-0.413745384	0.606286158

	Intercept, TV	PF, TV	HS, TV	DROP, TV	INCOL, TV	GRCOL, TV
Intercept, TC	0.606286158	-0.042114896	-0.529802585	-0.483791032	-0.433939348	0.295772522
PF, TC	-0.042114896	0.606286158	-0.02561011	-0.163019873	0.129421473	-0.171338529
HS, TC	-0.529802585	-0.02561011	0.606286158	0.342885356	0.354011168	-0.420286341
DROP, TC	-0.483791032	-0.163019873	0.342885356	0.606286158	0.33774309	-0.120666628
INCOL, TC	-0.433939348	0.129421473	0.354011168	0.33774309	0.606286158	-0.413745384
GRCOL, TC	0.295772522	-0.171338529	-0.420286341	-0.120666628	-0.413745384	0.606286158
Intercept, TV	1	-0.069463727	-0.873849054	-0.797958235	-0.715733557	0.487843106
PF, TV	-0.069463727	1	-0.042240961	-0.268882723	0.213465987	-0.2826034
HS, TV	-0.873849054	-0.042240961	1	0.565550362	0.583901123	-0.693214476
DROP, TV	-0.797958235	-0.268882723	0.565550362	1	0.55706878	-0.199025867
INCOL, TV	-0.715733557	0.213465987	0.583901123	0.55706878	1	-0.682425912
GRCOL, TV	0.487843106	-0.2826034	-0.693214476	-0.199025867	-0.682425912	1

Bayesian Linear Regression to interpret. It is calculated in Table 19.16. There is obviously a high degree of covariance in this example, which we shall see manifest itself in the correlation and variance-covariance matrix. Of course there is! One of our dependent variables (total crime) contains the other (violent crime) as a subset.

Table 19.16: The Mean of the Wishart Over Errors–Classical

$\langle\Sigma\rangle$	TC	TV
TC	0.110137	0.165520
TV	0.165520	0.676720

As we expand our analysis to compete the variance-covariance matrix of the Wishart error terms, we have to carefully consider the dimensional structure of the variance-covariance matrix. The Kronecker product is, in fact, the outer product of $S^T S$ with itself, as indicated in Table 19.17. The shaded areas are the variances of the two dependent variables and the jointness between them. The unshaded variables are more difficult to interpret, and we leave that as an exercise for the reader.

We will be needing the Kronecker product of $(S^T S)$ with itself in order to apply the formula in equation (17.43) for the variance-covariance matrix of the Wishart. We calculate it in Table 19.18 so that it can be substituted into

Table 19.17: The Dimensionality of the Variance-Covariance Over the Wishart Errors

(TC,TC)(TC,TC)	(TC,TC)(TC,VC)	(TC,VC)(TC,TC)	(TC,VC)(TC,VC)
(TC,TC)(VC,TC)	(TC,TC)(VC,VC)	(TC,VC)(VC,TC)	(TC,VC)(VC,VC)
(VC,TC)(TC,TC)	(VC,TC)(TC,VC)	(VC,VC)(TC,TC)	(VC,VC)(TC,VC)
(VC,TC)(VC,TC)	(VC,TC)(VC,VC)	(VC,VC)(VC,TC)	(VC,VC)(VC,VC)

equation (17.43). We can see from the column and row headings the exact same dimensionality as an outer product. That will be so helpful as we search inside the final variance-covariance matrix for the variances and covariances we truly want. There are three shaded numbers, which are important to us. The first is the one at the upper left.

We notice in equation (17.43) that the variance-covariance matrix is the sum of four matrix terms. They are the four individual terms on the right. We refer to them and first, second, third, and fourth matrix terms henceforth.

Table 19.18: The Kronecker Product $(S^TS) \otimes (S^TS)$–Classical

$(S^TS) \otimes (S^TS)$	(TC,TC)	(TC,VC)	(VC,TC)	(VC,VC)
(TC,TC)	20.39095640	30.64448751	30.64448751	46.05397591
(TC,VC)	30.64448751	125.28867721	46.05397591	188.28971183
(VC,TC)	30.64448751	46.05397591	125.28867721	188.28971183
(VC,VC)	46.05397591	188.28971183	188.28971183	769.81443790

We calculate the first matrix term by multiplying the Kronecker product matrix in Table 19.18 by the constant $-\frac{1}{(\nu_C-m-1)^2}$ to write the first matrix term in Table 19.19.

Table 19.19: First Term in Variance-Covariance Matrix–Classical

First	(TC,TC)	(TC,VC)	(VC,TC)	(VC,VC)
(TC,TC)	-0.01213025	-0.01822992	-0.01822992	-0.02739677
(TC,VC)	-0.01822992	-0.07453223	-0.02739677	-0.11201054
(VC,TC)	-0.01822992	-0.02739677	-0.07453223	-0.11201054
(VC,VC)	-0.02739677	-0.11201054	-0.11201054	-0.45795029

We get the second term by multiplying the Kronecker product in equation 19.18 by the constant $\frac{(\nu_C-m-2)}{(\nu_C-m)(\nu_C-m-1)(\nu_C-m-3)}$ to write Table 19.20.

We will be needing the Kronecker product of (S^TS) with itself multiplied by the commutation matrix K_{mm}, and we calculate it in Table 19.21. We multiply the matrix in Table 19.22 by the constant

Table 19.20: Second Term in Variance-Covariance Matrix–Classical

Second	(TC,TC)	(TC,VC)	(VC,TC)	(VC,VC)
(TC,TC)	0.01214506	0.01825217	0.01825217	0.02743022
(TC,VC)	0.01825217	0.07462323	0.02743022	0.11214730
(VC,TC)	0.01825217	0.02743022	0.07462323	0.11214730
(VC,VC)	0.02743022	0.11214730	0.11214730	0.45850945

Table 19.21: The $(S^TS) \otimes (S^TS) \times K_{mm}$ Matrix–Classical

Third	(TC,TC)	(TC,VC)	(VC,TC)	(VC,VC)
(TC,TC)	0.00030363	0.00045630	0.00045630	0.00068576
(TC,VC)	0.00045630	0.00068576	0.00186558	0.00280368
(VC,TC)	0.00045630	0.00186558	0.00068576	0.00280368
(VC,VC)	0.00068576	0.00280368	0.00280368	0.01146274

$$\frac{1}{(\nu_C - m)(\nu_C - m - 1)(\nu_C - m - 3)}$$

to write

Table 19.22: Third Term in Variance-Covariance Matrix–Classical

Third	(TC,TC)	(TC,VC)	(VC,TC)	(VC,VC)
(TC,TC)	0.00030363	0.00045630	0.00045630	0.00068576
(TC,VC)	0.00045630	0.00068576	0.00186558	0.00280368
(VC,TC)	0.00045630	0.00186558	0.00068576	0.00280368
(VC,VC)	0.00068576	0.00280368	0.00280368	0.01146274

To get the fourth term in the variance-covariance matrix, we begin with the matrix $(S^TS)^{-1}$ in Table 19.11 and we calculate the vectorization of that matrix in Table 19.23.

Table 19.23: The $vec[(S^TS)^{-1}]$ Vector–Classical

$vec[(S^TS)^{-1}]$
0.35016892
-0.08564818
-0.08564818
0.05699062

We have to calculate the outer product $vec[(S^TS)^{-1}]vec[(S^TS)^{-1}]^T$, which we do in Table 19.24. We then left multiply the matrix in Table 19.24 by the Kronecker product matrix in Table 19.18 to write the interim matrix in Table 19.25. We then have to right multiply the matrix in Table 19.25 by the Kronecker product matrix in Table 19.18 to obtain the triple matrix product in Table 19.26. We then have to multiply this triple matrix product in Table 19.26 by the constant term $\frac{1}{(\nu_C - m)(\nu_C - m - 1)(\nu_C - m - 3)}$ to write the fourth term in Table 19.27.

Table 19.24: The $vec[(S^T S)^{-1}]vec[(S^T S)^{-1}]^T$ Matrix–Classical

	(TC,TC)	(TC,VC)	(VC,TC)	(VC,VC)
(TC,TC)	0.12261827	-0.02999133	-0.02999133	0.01995634
(TC,VC)	-0.02999133	0.00733561	0.00733561	-0.00488114
(VC,TC)	-0.02999133	0.00733561	0.00733561	-0.00488114
(VC,VC)	0.01995634	-0.00488114	-0.00488114	0.00324793

Table 19.25: The $[(S^T S) \otimes (S^T S)]vec[(S^T S)^{-1}]vec[(S^T S)^{-1}]^T$ Matrix–Classical

	(TC,TC)	(TC,VC)	(VC,TC)	(VC,VC)
(TC,TC)	1.58123493	-0.38675589	-0.38675589	0.25734881
(TC,VC)	2.37635416	-0.58123493	-0.58123493	0.38675589
(VC,TC)	2.37635416	-0.58123493	-0.58123493	0.38675589
(VC,VC)	9.71562239	-2.37635416	-2.37635416	1.58123493

Table 19.26: The $[(S^T S) \otimes (S^T S)]vec[(S^T S)^{-1}]vec[(S^T S)^{-1}]^T[(S^T S) \otimes (S^T S)]$ Matrix–Classical

	(TC,TC)	(TC,VC)	(VC,TC)	(VC,VC)
(TC,TC)	20.39095640	30.64448751	30.64448751	125.28867721
(TC,VC)	30.64448751	46.05397591	46.05397591	188.28971183
(VC,TC)	30.64448751	46.05397591	46.05397591	188.28971183
(VC,VC)	125.28867721	188.28971183	188.28971183	769.81443790

We then sum up the first, second, third, and fourth terms to calculate the variance-covariance matrix of the Wishart distribution. Notice that the matrix is symmetric and organized precisely the way Table 19.17 is organized, and is shown in Table 19.28.

19.1.2 Predictive Analysis–Classical

We now turn to the predictive distribution parameters. We begin by postulating an "old" δ vector, the one we use as a base case, as shown in Table 19.29.

Table 19.27: The Fourth Matrix–Classical

Fourth	(TC,TC)	(TC,VC)	(VC,TC)	(VC,VC)
(TC,TC)	0.00030363	0.00045630	0.00045630	0.00186558
(TC,VC)	0.00045630	0.00068576	0.00068576	0.00280368
(VC,TC)	0.00045630	0.00068576	0.00068576	0.00280368
(VC,VC)	0.00186558	0.00280368	0.00280368	0.01146274

Table 19.28: The Variance-Covariance Matrix, First + Second + Third + Fourth–Classical

VCV	(TC,TC)	(TC,VC)	(VC,TC)	(VC,VC)
(TC,TC)	0.00062206	0.00093487	0.00093487	0.00258479
(TC,VC)	0.00093487	0.00146252	0.00258479	0.00574413
(VC,TC)	0.00093487	0.00258479	0.00146252	0.00574413
(VC,VC)	0.00258479	0.00574413	0.00574413	0.02348463

Table 19.29: δ

Intercept	PF	HS	DROP	INCOL	GRCOL
1	30	60	15	30	15

We calculate the mean predicted value in the classical regression case, namely $\delta\bar{B}$, as shown in Table 19.30.

Table 19.30: Calculating the Mean of the Prediction $\delta\bar{B}$–Classical

δ					
Intercept	PF	HS	DROP	INCOL	GRCOL
1	30	60	15	30	15

$\delta\bar{B}$	
Total Crime	Violent Crime
6.410823918	5.998547093

We then turn to all the terms of the bivariate distribution over total crime and violent crime. They appear in Table 19.31, in which

$$constant = \frac{\Gamma\left(\frac{\nu_C+1}{2}\right)}{\Gamma\left(\frac{\nu_C-m+1}{2}\right)\left[\pi(1+\delta(X^TX)^{-1}\delta^T)\right]^{m/2}\left|S^TS\right|^{1/2}}$$

and

Table 19.31: Terms in the Multivariate Probability Distribution–Classical

$\delta(X^TX)^{-1}$					
Intercept	PF	HS	DROP	INCOL	GRCOL
0.06130990	-0.00132608	-0.00044962	0.00060481	-0.00092229	0.00386852

$\delta(X^TX)^{-1}\delta^T$	$1+\delta(X^TX)^{-1}\delta^T$		
0.033981535	1.033981535		
$\pi(1+\delta(X^TX)^{-1}\delta^T)$	$[\pi(1+\delta(X^TX)^{-1}\delta^T)]^{m/2}$		
3.248348796	3.248348796		
$\lvert S^TS \rvert$	$\lvert S^TS \rvert^{1/2}$		
79.23470130	8.90		
$\frac{\nu_C+1}{2}$	$\frac{\nu_C-m+1}{2}$		
22.5	21.5		
$\Gamma\left(\frac{\nu_C+1}{2}\right)$	$\Gamma\left(\frac{\nu_C-m+1}{2}\right)$		
2.382802E+20	1.108280E+19		
constant	matrix		
0.74356355	0.33866071	-0.08283337	exponent
	-0.08283337	0.05511764	-22.5

$$matrix = \frac{(S^TS)^{-1}}{1+\delta(X^TX)^{-1}\delta^T} \quad .$$

We now turn to the posterior predictive distribution by rewriting the old δ vector and $\delta\bar{B}$ for the classical solution in Table 19.30. The parameters of the bivariate distribution over TC and VC are calculated in Table 19.31. The distribution has the form

$$\{TC, VC\} = const\times$$

$$\left\{1+\begin{bmatrix} TC-\langle TC\rangle \\ VC-\langle VC\rangle \end{bmatrix}^T matrix \begin{bmatrix} TC-\langle TC\rangle \\ VC-\langle VC\rangle \end{bmatrix}\right\}^{exponent} \quad . \tag{19.1}$$

Before proceeding, we calculate the variances and covariances and the correlation of the Wishart error terms, which appear in Table 19.32.

It is difficult to plot the bivariate distribution over total crime and violent crime. We will soon calculate the two marginal distributions. If we divide the joint distribution by the marginal distribution, we get the conditional distributions. Because of the high degree of correlation, we expect a lot of conditionality, i.e., relevance in the language of Howard[45]. In order to do so, we are going to need to calculate the key parameters for both marginal distributions, one for total crime and one for violent crime. If we thereafter divide the bivariate density by the marginal density, Bayes theorem tells us that we get the conditional density (for this bivariate form assumed in this example). The key parameters for the univariate posterior for total crime appears in Table 19.33.

We then turn to the marginal distribution over violent crime, which appears

Table 19.32: Variance-Covariance and Correlation of Wishart Error Terms–Classical

$\frac{1+\delta(X^TX)^{-1}\delta^T}{(\nu_C+1-m)}(S^TS)$	TC	VC
TC	0.10858332	0.16318412
VC	0.16318412	0.66717130
Correlation	TC	VC
TC	1	0.606286158
VC	0.606286158	1
Diagonals	TC	VC
Variance	0.10858332	0.66717130
Std. Deviation	0.32951984	0.81680554
Means	6.41082392	5.99854709
t Statistic	19.45504706	7.343910844

Table 19.33: Key Parameters of the Univariate Posterior for Total Crime–Classical

Total Crime	6.410823918	
$\frac{\nu_C-m+2}{2}$	$\frac{\nu_C-m+1}{2}$	
22	21.5	
$\Gamma\left(\frac{\nu_C-m+2}{2}\right)$	$\Gamma\left(\frac{\nu_C-m+1}{2}\right)$	
5.10909422E+19	1.10827981E+19	
$1+\delta(X^TX)^{-1}\delta^T$	$[S^TS]_{ii}$	$\left[1+\delta(X^TX)^{-1}\delta^T\right][S^TS]_{ii}$
1.03398154	4.51563466	4.669082861
$\left\{\left[1+\delta(X^TX)^{-1}\delta^T\right][S^TS]_{ii}\right\}^{1/2}$	π	$\pi^{1/2}$
2.160806067	3.14159265	1.772453851
$\left\{\pi\left[1+\delta(X^TX)^{-1}\delta^T\right][S^TS]_{ii}\right\}^{1/2}$	$\frac{\Gamma\left(\frac{\nu_C-m+2}{2}\right)}{\Gamma\left(\frac{\nu_C-m+1}{2}\right)\{\pi[1+\delta(X^TX)^{-1}\delta^T][S^TS]_{ii}\}^{1/2}}$	$\frac{1}{[1+\delta(X^TX)^{-1}\delta^T][S^TS]_{ii}}$
3.829929035	1.203660	0.214174824
$\frac{[1+\delta(X^TX)^{-1}\delta^T][S^TS]_{ii}}{\nu_C-m-1}$	Std. Dev	$-\frac{\nu_C-m+2}{2}$
0.113880	0.337461	-22

in Table 19.34

Table 19.34: Key Parameters of the Univariate Posterior for Violent Crime–Classical

Violent Crime	5.998547093				
$\frac{\nu_C-m+2}{2}$	22	$\frac{\nu_C-m+1}{2}$	21.5		
$\Gamma\left(\frac{\nu_C-m+2}{2}\right)$	5.10909422E+19	$\Gamma\left(\frac{\nu_C-m+1}{2}\right)$	1.10827981E+19		
$1+\delta(X^TX)^{-1}\delta^T$	1.03398154	$[S^TS]_{ii}$	27.74553005	$\left[1+\delta(X^TX)^{-1}\delta^T\right][S^TS]_{ii}$	28.68836577
$\left[1+\delta(X^TX)^{-1}\delta^T\right][S^TS]_{ii}^{1/2}$	5.356152142	π	3.14159265	$\pi^{1/2}$	1.772453851
$\left\{\pi\left[1+\delta(X^TX)^{-1}\delta^T\right][S^TS]_{ii}\right\}^{1/2}$	9.49353249	$\frac{\Gamma\left(\frac{\nu_C-m+2}{2}\right)}{\Gamma\left(\frac{\nu_C-m+1}{2}\right)\{\pi[1+\delta(X^TX)^{-1}\delta^T][S^TS]_{ii}\}^{1/2}}$	0.485587	$\frac{1}{[1+\delta(X^TX)^{-1}\delta^T][S^TS]_{ii}}$	0.034857336
$\frac{[1+\delta(X^TX)^{-1}\delta^T][S^TS]_{ii}}{\nu_C-m-1}$	0.699716	Std. Dev	0.836490	$-\frac{\nu_C-m+2}{2}$	-22

The structure of the marginal distribution is obvious from the entries in Table 19.34, the constant being the middle and the matrix the rightmost numbers in the second to last line and the exponent being the final entry at the lower right. Following are the plots of the marginal distribution for TC (Figure 19.2) and the marginal distribution for VC (Figure 19.3) in the classical case.

Figure 19.2: Total Crime Marginal Distribution–Classical

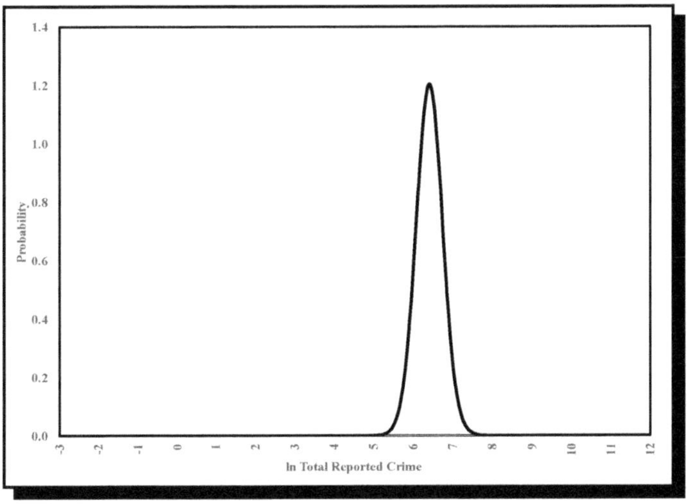

We conclude by writing the variance-covariance, correlation, and t-statistics for the classical case in Table 19.35, the prior in Table 19.36, and the posterior in Table19.37. These three tables illustrate some important covariance concepts.

Figure 19.3: Violent Crime Marginal Distribution–Classical

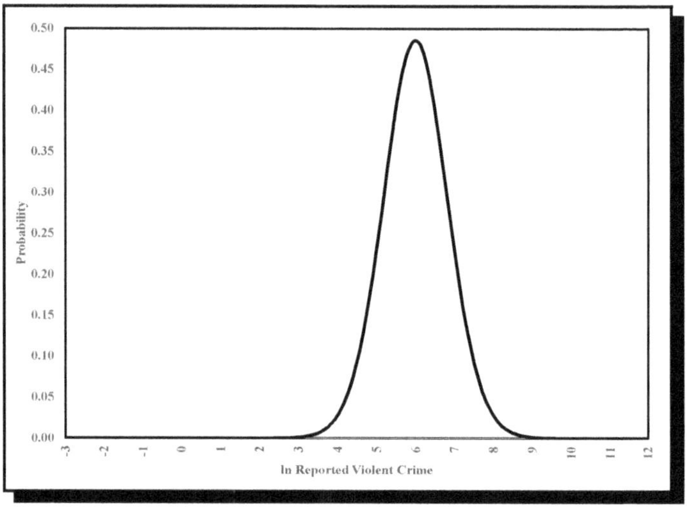

Notice that the t-statistics are highest for the classical, lowest for the prior, and in the middle for the posterior. This is intuitively satisfying, emphasizing that the final answer gives a bit of weight to the prior, recognizing it as legitimate information.

Table 19.35: The Variance-covariance Matrix of TC and VC–Classical

VCV	TC	VC
TC	0.10858332	0.16318412
VC	0.16318412	0.66717130
Correlation	TC	VC
TC	1	0.606286158
VC	0.606286158	1
Diagonals	TC	VC
Variance	0.10858332	0.66717130
Std. Deviation	0.32951984	0.81680554
Means	6.41082392	5.99854709
t Statistic	19.45504706	7.343910844

We also notice that the correlation is 0.606 in the classical regression case, 0 in the prior case (by construction), and the very much "between them" in the posterior case, having a value of 0.324 in the posterior case. Indeed the prior assumption of zero covariance is not at all inconsequential, and it is not absent from the posterior. These correlations are crucial in any predictions you do.

Table 19.36: The Variance-covariance Matrix of TC and VC–Prior

VCV	TC	VC
TC	1.55633333	0.00000000
VC	0.00000000	5.18777778
Correlation	TC	VC
TC	1	0
VC	0	1
Diagonals	TC	VC
Variance	1.55633333	5.18777778
Std. Deviation	1.24753089	2.27766937
Means	5.70000000	5.55000000
t Statistic	4.569025123	2.436701333

Table 19.37: The Variance-covariance Matrix of TC and VC–Posterior

VCV	TC	VC
TC	0.19450861	0.13087910
VC	0.13087910	0.83939152
Correlation	TC	VC
TC	1	0.323905705
VC	0.323905705	1
Diagonals	TC	VC
Variance	0.19450861	0.83939152
Std. Deviation	0.44103130	0.91618312
Means	6.41438834	5.93961391
t Statistic	14.54406599	6.482998596

19.2 Assembling and Using an Informative Prior

19.2.1 Coefficients–Prior

We are going to establish the simplest possible prior, extending the simple reasoning from Chapter 5 above. This section conducts a similar analysis for the prior as we did in the univariate cases. We begin by making estimates of the mean values of the prior model coefficients. We present our subjective estimates of the mean values in Table 19.38. This is what we "think" the coefficients will be before we gather any data.

Table 19.38: Estimates of Mean Values of Model Coefficients–Prior

B_0	Total Crime	Violent Crime
Intercept	6	9
PF	0.01	0.02
GRHS	-0.01	-0.08
DROP	0.01	0.01
INCOL	-0.01	-0.03
GRCOL	0.01	0.1

We will be needing terms in the prior related to the degrees of freedom, which we shall assume subjectively. These estimates will be required in the subsequent calculations of means and variances of coefficients and Wishart error terms. The degrees-of-freedom related terms we have estimated for the prior are given in Table 19.39.

Table 19.39: Degrees of Freedom Related Parameters–Prior

	Prior Parameters
m_0	16
p	6
m	2
ν_0	7
$\nu_0 - m$	5
$\nu_0 - m - 1$	4
$\nu_0 - m - 2$	3
$\nu_0 - m - 3$	2

To establish the probabilistic aspects of the prior, we begin with the $m \times m$ matrix M_0, which we assume herein is a diagonal matrix with two yet-to-be-determined diagonal values

$$M_0 = \begin{bmatrix} m_{TC,TC} & 0 \\ 0 & m_{VC,VC} \end{bmatrix} .$$

We sometimes use the index 1 to represent TC and the index 2 to represent VC. No confusion should result. We assume that the $p \times p$ matrix Q_0 is diagonal as well with six yet to be determined terms

$$Q_0 = \begin{bmatrix} q_1 & 0 & 0 & 0 & 0 & 0 \\ 0 & q_2 & 0 & 0 & 0 & 0 \\ 0 & 0 & q_3 & 0 & 0 & 0 \\ 0 & 0 & 0 & q_4 & 0 & 0 \\ 0 & 0 & 0 & 0 & q_5 & 0 \\ 0 & 0 & 0 & 0 & 0 & q_6 \end{bmatrix}.$$

The inverse of this $p \times p$ matrix Q_0 is

$$Q_0{}^{-1} = \begin{bmatrix} \frac{1}{q_1} & 0 & 0 & 0 & 0 & 0 \\ 0 & \frac{1}{q_2} & 0 & 0 & 0 & 0 \\ 0 & 0 & \frac{1}{q_3} & 0 & 0 & 0 \\ 0 & 0 & 0 & \frac{1}{q_4} & 0 & 0 \\ 0 & 0 & 0 & 0 & \frac{1}{q_5} & 0 \\ 0 & 0 & 0 & 0 & 0 & \frac{1}{q_6} \end{bmatrix}.$$

If we appeal to equation (16.36), we will want to write the following Kronecker product, which will be related to the variance-covariance matrix of the model coefficients, namely $M_0 \otimes Q_0^{-1}$

$$M_0 \otimes Q_0^{-1} =$$

$$\begin{bmatrix} \frac{m_{11}}{q_1} & & & & & & & & & & & \\ & \frac{m_{11}}{q_2} & & & & & & & & & & \\ & & \frac{m_{11}}{q_3} & & & & & & & & & \\ & & & \frac{m_{11}}{q_4} & & & & & & & & \\ & & & & \frac{m_{11}}{q_5} & & & & & & & \\ & & & & & \frac{m_{11}}{q_6} & & & & & & \\ & & & & & & \frac{m_{22}}{q_1} & & & & & \\ & & & & & & & \frac{m_{22}}{q_2} & & & & \\ & & & & & & & & \frac{m_{22}}{q_3} & & & \\ & & & & & & & & & \frac{m_{22}}{q_4} & & \\ & & & & & & & & & & \frac{m_{22}}{q_5} & \\ & & & & & & & & & & & \frac{m_{22}}{q_6} \end{bmatrix}.$$

We can write the variances in this Kronecker product as the diagonals in the previous matrix times the constant $\frac{1}{\nu_0 - m - 1}$ to form the variance-covariance matrix

$$VCV = \frac{1}{\nu_0 - m - 1}(M_0 \otimes Q_0^{-1}) = \frac{1}{\nu_0 - m - 1} \times$$

$$\begin{bmatrix} \frac{m_{11}}{q_1} & & & & & & & & & & & \\ & \frac{m_{11}}{q_2} & & & & & & & & & & \\ & & \frac{m_{11}}{q_3} & & & & & & & & & \\ & & & \frac{m_{11}}{q_4} & & & & & & & & \\ & & & & \frac{m_{11}}{q_5} & & & & & & & \\ & & & & & \frac{m_{11}}{q_6} & & & & & & \\ & & & & & & \frac{m_{22}}{q_1} & & & & & \\ & & & & & & & \frac{m_{22}}{q_2} & & & & \\ & & & & & & & & \frac{m_{22}}{q_3} & & & \\ & & & & & & & & & \frac{m_{22}}{q_4} & & \\ & & & & & & & & & & \frac{m_{22}}{q_5} & \\ & & & & & & & & & & & \frac{m_{22}}{q_6} \end{bmatrix}.$$

We put this into the form in which the original mean matrix was written

$$\text{variances} = \frac{1}{\nu_0 - m - 1} \begin{bmatrix} \frac{m_{11}}{q_1} & \frac{m_{22}}{q_1} \\ \frac{m_{11}}{q_2} & \frac{m_{22}}{q_2} \\ \frac{m_{11}}{q_3} & \frac{m_{22}}{q_3} \\ \frac{m_{11}}{q_4} & \frac{m_{22}}{q_4} \\ \frac{m_{11}}{q_5} & \frac{m_{22}}{q_5} \\ \frac{m_{11}}{q_6} & \frac{m_{22}}{q_6} \end{bmatrix}$$

$$= \begin{bmatrix} \frac{\langle \sigma_{DD} \rangle}{q_1} & \frac{\langle \sigma_{HH} \rangle}{q_1} \\ \frac{\langle \sigma_{DD} \rangle}{q_2} & \frac{\langle \sigma_{HH} \rangle}{q_2} \\ \frac{\langle \sigma_{DD} \rangle}{q_3} & \frac{\langle \sigma_{HH} \rangle}{q_3} \\ \frac{\langle \sigma_{DD} \rangle}{q_4} & \frac{\langle \sigma_{HH} \rangle}{q_4} \\ \frac{\langle \sigma_{DD} \rangle}{q_5} & \frac{\langle \sigma_{HH} \rangle}{q_5} \\ \frac{\langle \sigma_{DD} \rangle}{q_6} & \frac{\langle \sigma_{HH} \rangle}{q_6} \end{bmatrix}.$$

The standard deviations are the square roots of the various terms

$$\text{std. dev.} = \begin{bmatrix} \frac{\sqrt{\langle \sigma_{DD} \rangle}}{\sqrt{q_1}} & \frac{\sqrt{\langle \sigma_{HH} \rangle}}{\sqrt{q_1}} \\ \frac{\sqrt{\langle \sigma_{DD} \rangle}}{\sqrt{q_2}} & \frac{\sqrt{\langle \sigma_{HH} \rangle}}{\sqrt{q_2}} \\ \frac{\sqrt{\langle \sigma_{DD} \rangle}}{\sqrt{q_3}} & \frac{\sqrt{\langle \sigma_{HH} \rangle}}{\sqrt{q_3}} \\ \frac{\sqrt{\langle \sigma_{DD} \rangle}}{\sqrt{q_4}} & \frac{\sqrt{\langle \sigma_{HH} \rangle}}{\sqrt{q_4}} \\ \frac{\sqrt{\langle \sigma_{DD} \rangle}}{\sqrt{q_5}} & \frac{\sqrt{\langle \sigma_{HH} \rangle}}{\sqrt{q_5}} \\ \frac{\sqrt{\langle \sigma_{DD} \rangle}}{\sqrt{q_6}} & \frac{\sqrt{\langle \sigma_{HH} \rangle}}{\sqrt{q_6}} \end{bmatrix}.$$

If we divide the mean values of the coefficients μ_{ii} by the standard deviations of the coefficients, we compute the t-statistic of the coefficients

$$\text{t-statistic} = \begin{bmatrix} \frac{\mu_{11}}{\sqrt{\langle\sigma_{DD}\rangle}}\sqrt{q_1} & \frac{\mu_{12}}{\sqrt{\langle\sigma_{HH}\rangle}}\sqrt{q_1} \\ \frac{\mu_{21}}{\sqrt{\langle\sigma_{DD}\rangle}}\sqrt{q_2} & \frac{\mu_{22}}{\sqrt{\langle\sigma_{HH}\rangle}}\sqrt{q_2} \\ \frac{\mu_{31}}{\sqrt{\langle\sigma_{DD}\rangle}}\sqrt{q_3} & \frac{\mu_{32}}{\sqrt{\langle\sigma_{HH}\rangle}}\sqrt{q_3} \\ \frac{\mu_{41}}{\sqrt{\langle\sigma_{DD}\rangle}}\sqrt{q_4} & \frac{\mu_{42}}{\sqrt{\langle\sigma_{HH}\rangle}}\sqrt{q_4} \\ \frac{\mu_{51}}{\sqrt{\langle\sigma_{DD}\rangle}}\sqrt{q_5} & \frac{\mu_{52}}{\sqrt{\langle\sigma_{HH}\rangle}}\sqrt{q_5} \\ \frac{\mu_{61}}{\sqrt{\langle\sigma_{DD}\rangle}}\sqrt{q_6} & \frac{\mu_{62}}{\sqrt{\langle\sigma_{HH}\rangle}}\sqrt{q_6} \end{bmatrix} .$$

Using these equations, we can begin to assume specific values of the parameters in order to encode our prior. We start by subjectively estimating the expected value of the Wishart error distribution, the diagonals of the matrix

$$\begin{aligned} \langle\sigma_{TC,TC}\rangle &= 1.2 \\ \langle\sigma_{VC,VC}\rangle &= 4 \quad . \end{aligned}$$

We use these estimates to compute the diagonal M_0 matrix using the following logic

$$\begin{aligned} \langle\Sigma\rangle &= \begin{bmatrix} \langle\sigma_{TC,TC}\rangle & 0 \\ 0 & \langle\sigma_{VC,VC}\rangle \end{bmatrix} = \begin{bmatrix} 1.2 & 0 \\ 0 & 4 \end{bmatrix} \\ &= \frac{1}{\nu_0 - m - 1} \begin{bmatrix} m_{TC,TC} & 0 \\ 0 & m_{VC,VC} \end{bmatrix} \\ \Rightarrow M_0 &= \begin{bmatrix} m_{TC,TC} & 0 \\ 0 & m_{VC,VC} \end{bmatrix} = (\nu_0 - m - 1) \begin{bmatrix} \langle\sigma_{TC,TC}\rangle & 0 \\ 0 & \langle\sigma_{VC,VC}\rangle \end{bmatrix} \\ &= 4 \begin{bmatrix} 1.2 & 0 \\ 0 & 4 \end{bmatrix} = \begin{bmatrix} 4.8 & 0 \\ 0 & 16 \end{bmatrix} \quad . \end{aligned}$$

We summarize the M_0 and M_0^{-1} matrix in Table 19.40.

Table 19.40: M_0 and M_0^{-1}–Prior

M_0	TC	VC
TC	4.8	0
VC	0	16
Determinant	76.8	
M_0^{-1}	TC	VC
TC	0.208333	0
VC	0	0.062500
Determinant	0.013021	

We can use these equations to efficiently estimate the key matrices to be used for the prior. We begin by calculating the quotients of the estimated mean values in equation 19.38, denoted with the notation μ, divided by the square roots of the sigmas, denoted by the notation σ, to form the matrix in Table 19.41.

Table 19.41: Quotients of Means and Sigmas–Prior

$\frac{\mu_{c,c}}{\sigma_r}$	Total Crime	Violent Crime
Intercept	5.477226	4.500000
PF	0.009129	0.010000
GRHS	-0.009129	-0.040000
DROP	0.009129	0.005000
INCOL	-0.009129	-0.015000
GRCOL	0.009129	0.050000

We need to take the absolute values of these ratios in preparation for calculating a "not to exceed" t statistic, a guide in which the paucity if information we already have. The absolute values appear in Table 19.42.

Table 19.42: Absolute Values of Quotients of Means and Sigmas–Prior

Abs	Total Crime	Violent Crime
Intercept	5.477225575	4.5
PF	0.009128709	0.01
GRHS	0.009128709	0.04
DROP	0.009128709	0.005
INCOL	0.009128709	0.015
GRCOL	0.009128709	0.05

We want all the t-statistics of all the error terms, no matter what their signs, to be less than some target value. We specify the target t statistics in the left hand column of Table 19.43, we calculate the square roots of the q values that give us that t statistic. The calculated t values appear in the right hand column.

Table 19.43: Calculation of q Values Based on Target t Values–Prior

target t	$\sqrt{q}$	q
12	2.19089	4.80000
3	300	90000
3	75	5625
3	328.6335345	108000
3	200.00	40000.00
3	60	3600

Using the rightmost column, we can populate the Q_0 matrix as in Table 19.44.

Table 19.44: Q_0–Prior

Q_0	Intercept	PF	GRHS	DROP	INCOL	GRCOL
Intercept	4.8	0	0	0	0	0
PF	0	90000	0	0	0	0
GRHS	0	0	5625	0	0	0
DROP	0	0	0	108000	0	0
INCOL	0	0	0	0	40000	0
GRCOL	0	0	0	0	0	3600

We will be needing the inverse Q_0^{-1}, and we display it in Table 19.45

Table 19.45: Q_0^{-1}–Prior

Q_0^{-1}	Intercept	PF	GRHS	DROP	INCOL	GRCOL
Intercept	0.208333	0	0	0	0	0
PF	0	1.111111E-05	0	0	0	0
GRHS	0	0	1.777778E-04	0	0	0
DROP	0	0	0	9.259259E-06	0	0
INCOL	0	0	0	0	2.500000E-05	0
GRCOL	0	0	0	0	0	2.777778E-04

In order to calculate the variances of the coefficients assumed in the prior, we calculate the ratios $\frac{\mu_{c,c}}{q_r}$, which are multiplicatively related to the variance-covariance matrix of the error term. See Table 19.46.

Table 19.46: Calculation of the Ratios $\frac{\mu_{c,c}}{q_r}$–Prior

$\frac{\mu_{c,c}}{q_r}$	Total Crime	Violent Crime
Intercept	1.00000	3.33333
PF	5.33333E-05	1.77778E-04
GRHS	8.53333E-04	2.84444E-03
DROP	4.44444E-05	1.48148E-04
INCOL	1.20000E-04	4.00000E-04
GRCOL	1.33333E-03	4.44444E-03

We next multiply by the constant $\nu_0 - m - 1$ to obtain the variance-covariance matrix for the error terms. We display it in Table 19.47. We take square roots to calculate the standard deviation in Table 19.48. We then calculate the t statistics in Table 19.49. We reiterate that, other than the intercept, the t-statistics are capped at 3 or less. This is what we wanted in our estimation of the prior. The expected values of the Wishart in Table 19.50.

Table 19.47: Variance-Covariance Matrix–Prior

VCV	Total Crime	Violent Crime
Intercept	0.25000000	0.83333333
PF	1.33333E-05	4.44444E-05
GRHS	2.13333E-04	7.11111E-04
DROP	1.11111E-05	3.70370E-05
INCOL	3.00000E-05	1.00000E-04
GRCOL	3.33333E-04	1.11111E-03

Table 19.48: Standard Deviation Matrix–Prior

Std. Dev	Total Crime	Violent Crime
Intercept	0.5	0.912870929
PF	0.003651	0.006667
GRHS	0.014606	0.026667
DROP	0.003333	0.006086
INCOL	0.005477	0.010000
GRCOL	0.018257	0.033333

Table 19.49: t Statistics–Prior

t statistic	Total Crime	Violent Crime
Intercept	12	9.859006035
PF	2.738613	3
GRHS	-0.68465	-3
DROP	3	1.643167673
INCOL	-1.82574	-3
GRCOL	0.547723	3

Table 19.50: $\langle\Sigma\rangle$–Prior

$\langle\Sigma\rangle$	TC	TV
TC	1.2	0
TV	0	4

We next turn to the calculation of the quantity Q_0B_0 that will be needed for the posterior. It appears in Table 19.51.

Table 19.51: Q_0B_0–Prior

Q_0B_0	Total Crime	Violent Crime
Intercept	28.8	43.2
PF	900	1800
GRHS	-56.25	-450
DROP	1080	1080
INCOL	-400	-1200
GRCOL	36	360

We now turn to the calculation of the variance-covariance matrix for the coefficients in the prior. The first step is to calculate the Kronecker product $M_0 \otimes Q_0^{-1}$, which is summarized in Table 19.52.

Table 19.52: $M_0 \otimes Q_0^{-1}$–Prior

$M_0 \otimes Q_0^{-1}$	Intercept, TC	PF, TC	HS, TC	DROP, TC	INCOL, TC	GRCOL, TC
Intercept, TC	1	0	0	0	0	0
PF, TC	0	5.33333E-05	0	0	0	0
HS, TC	0	0	0.000853333	0	0	0
DROP, TC	0	0	0	4.44444E-05	0	0
INCOL, TC	0	0	0	0	0.00012	0
GRCOL, TC	0	0	0	0	0	0.001333333
Intercept, TV	0	0	0	0	0	0
PF, TV	0	0	0	0	0	0
HS, TV	0	0	0	0	0	0
DROP, TV	0	0	0	0	0	0
INCOL, TV	0	0	0	0	0	0
GRCOL, TV	0	0	0	0	0	0

M0 Kron Q0-1	Intercept, TV	PF, TV	HS, TV	DROP, TV	INCOL, TV	GRCOL, TV
Intercept, TC	0	0	0	0	0	0
PF, TC	0	0	0	0	0	0
HS, TC	0	0	0	0	0	0
DROP, TC	0	0	0	0	0	0
INCOL, TC	0	0	0	0	0	0
GRCOL, TC	0	0	0	0	0	0
Intercept, TV	3.333333333	0	0	0	0	0
PF, TV	0	0.000177778	0	0	0	0
HS, TV	0	0	0.002844444	0	0	0
DROP, TV	0	0	0	0.000148148	0	0
INCOL, TV	0	0	0	0	0.0004	0
GRCOL, TV	0	0	0	0	0	0.004444444

We multiply the matrix in Table 19.52 by the constant $\frac{1}{\nu_0 - m - 1}$ to obtain the variance-covariance matrix in Table 19.53.

We then extract the diagonals to write all the variances. We then calculate the standard deviations and the t statistics for the twelve coefficients under the prior in Table 19.54.

Table 19.53: VCV–Prior

VCV	Intercept, TC	PF, TC	HS, TC	DROP, TC	INCOL, TC	GRCOL, TC
Intercept, TC	4	0	0	0	0	0
PF, TC	0	0.000213333	0	0	0	0
HS, TC	0	0	0.003413333	0	0	0
DROP, TC	0	0	0	0.000177778	0	0
INCOL, TC	0	0	0	0	0.00048	0
GRCOL, TC	0	0	0	0	0	0.005333333
Intercept, TV	0	0	0	0	0	0
PF, TV	0	0	0	0	0	0
HS, TV	0	0	0	0	0	0
DROP, TV	0	0	0	0	0	0
INCOL, TV	0	0	0	0	0	0
GRCOL, TV	0	0	0	0	0	0

VCV	Intercept, TV	PF, TV	HS, TV	DROP, TV	INCOL, TV	GRCOL, TV
Intercept, TC	0	0	0	0	0	0
PF, TC	0	0	0	0	0	0
HS, TC	0	0	0	0	0	0
DROP, TC	0	0	0	0	0	0
INCOL, TC	0	0	0	0	0	0
GRCOL, TC	0	0	0	0	0	0
Intercept, TV	13.33333333	0	0	0	0	0
PF, TV	0	0.000711111	0	0	0	0
HS, TV	0	0	0.011377778	0	0	0
DROP, TV	0	0	0	0.000592593	0	0
INCOL, TV	0	0	0	0	0.0016	0
GRCOL, TV	0	0	0	0	0	0.017777778

Table 19.54: Summary of VCV–Prior

	Intercept, TC	PF, TC	HS, TC	DROP, TC	INCOL, TC	GRCOL, TC
Var	4	0.000213333	0.003413333	0.000177778	0.00048	0.005333333
Std. Dev	2	0.014605935	0.058423739	0.013333333	0.021908902	0.073029674
B0	6	0.01	-0.01	0.01	-0.01	0.01
t statistic	3	0.684653197	-0.171163299	0.75	-0.456435465	0.136930639

	Intercept, TV	PF, TV	HS, TV	DROP, TV	INCOL, TV	GRCOL, TV
Var	13.33333333	0.000711111	0.011377778	0.000592593	0.0016	0.017777778
Std. Dev	3.651483717	0.026666667	0.106666667	0.024343225	0.04	0.133333333
B0	9	0.02	-0.08	0.01	-0.03	0.1
t statistic	2.464751509	0.75	-0.75	0.410791918	-0.75	0.75

With the exception of the intercept terms, all t-statistics are low, as we would expect. We are uncertain, and we express that uncertainty in our prior. We are not arrogant about knowing the answer. We now begin the calculation of the variance-covariance matrix for the Wishart error term. We need as a fundamental element of that calculation the Kronecker product. To begin that, we return to the matrices M_0 and M_0^{-1} in Table 19.40. The Kronecker product of $M_0 \otimes M_0$, ubiquitous in the terms of the variance-covariance matrix of the Wishart error terms in equation (16.53) appears in Table 19.55.

Table 19.55: $M_0 \otimes M_0$–Prior

$M_0 \otimes M_0$	(TC,TC)	(TC,VC)	(VC,TC)	(VC,VC)
(TC,TC)	23.04	0	0	0
(TC,VC)	0	76.8	0	0
(VC,TC)	0	0	76.8	0
(VC,VC)	0	0	0	256
$(M_0 \otimes M_0)Kmm$	(TC,TC)	(TC,VC)	(VC,TC)	(VC,VC)
(TC,TC)	23.04	0	0	0
(TC,VC)	0	0	76.8	0
(VC,TC)	0	76.8	0	0
(VC,VC)	0	0	0	256

With the information in the previous two matrices, it is easy to calculate the first, second, and third matrices in the expression in equation (16.53) and array them as in Table 19.56.

To obtain the fourth and final term, we will be needing the vectorization of the matrix M_0, which is calculated in Table 19.57.

We need the outer product of the vectorized M_0 with itself, which is depicted in Table 19.58.

We left multiply this matrix by the Kronecker product to calculate the top panel in Table 19.59. We right multiply the result by the Kronecker product to calculate the bottom panel in Table 19.59.

We thereafter multiply by the multiplier in equation (16.53) to calculate the fourth term, which appears in Table 19.60 in the top panel. We sum the first, second, third, and fourth term to create the variance-covariance term in the middle panel. We compute the t-statistics in the bottom pane.

19.2.2 Predictive Distributions–Prior

We now turn to the predictive distributions for the prior. We repeat the δ vector from the foregoing classical analysis, and we multiply by the mean vector δB_0 to obtain Table 19.61.

Table 19.56: First, Second, and Third Matrices in VCV Matrix–Prior

First	(TC,TC)	(TC,VC)	(VC,TC)	(VC,VC)
(TC,TC)	-1.44	0	0	0
(TC,VC)	0	-4.8	0	0
(VC,TC)	0	0	-4.8	0
(VC,VC)	0	0	0	-16
Second	(TC,TC)	(TC,VC)	(VC,TC)	(VC,VC)
(TC,TC)	1.728	0	0	0
(TC,VC)	0	5.76	0	0
(VC,TC)	0	0	5.76	0
(VC,VC)	0	0	0	19.2
Third	(TC,TC)	(TC,VC)	(VC,TC)	(VC,VC)
(TC,TC)	0.576	0	0	0
(TC,VC)	0	0	1.92	0
(VC,TC)	0	1.92	0	0
(VC,VC)	0	0	0	6.4

Table 19.57: $[vec(M_0^{-1})]$–Prior

$[vec(M_0^{-1})]$
0.20833333
0.00000000
0.00000000
0.06250000

Table 19.58: $[vec(M_0^{-1})][vec(M_0^{-1})]^T$–Prior

$[vec(M_0^{-1})][vec(M_0^{-1})]^T$	(TC,TC)	(TC,VC)	(VC,TC)	(VC,VC)
(TC,TC)	0.04340278	0.00000000	0.00000000	0.01302083
(TC,VC)	0.00000000	0.00000000	0.00000000	0.00000000
(VC,TC)	0.00000000	0.00000000	0.00000000	0.00000000
(VC,VC)	0.01302083	0.00000000	0.00000000	0.00390625

Table 19.59: $(M_0 \otimes M_0)[vec(M_0^{-1})][vec(M_0^{-1})]^T(M_0 \otimes M_0)$–Prior

$(M_0 \otimes M_0)$ $[vec(M_0^{-1})][vec(M_0^{-1})]^T$	(TC,TC)	(TC,VC)	(VC,TC)	(VC,VC)
(TC,TC)	1	0	0	0.3
(TC,VC)	0	0	0	0
(VC,TC)	0	0	0	0
(VC,VC)	3.333333333	0	0	1

$(M_0 \otimes M_0)$ $[vec(M_0^{-1})][vec(M_0^{-1})]^T$ $(M_0 \otimes M_0)$	(TC,TC)	(TC,VC)	(VC,TC)	(VC,VC)
(TC,TC)	23.04	0	0	76.8
(TC,VC)	0	0	0	0
(VC,TC)	0	0	0	0
(VC,VC)	76.8	0	0	256

Table 19.60: Fourth Term, Sum for VCV, and Key Statistics–Prior

Fourth	(TC,TC)	(TC,VC)	(VC,TC)	(VC,VC)
(TC,TC)	0.576	0	0	1.92
(TC,VC)	0	0	0	0
(VC,TC)	0	0	0	0
(VC,VC)	1.92	0	0	6.4

VCV	(TC,TC)	(TC,VC)	(VC,TC)	(VC,VC)
(TC,TC)	1.44	0	0	1.92
(TC,VC)	0	0.96	1.92	0
(VC,TC)	0	1.92	0.96	0
(VC,VC)	1.92	0	0	16

	TT,TT	TT,VV	VV,VV
Variance	1.44	1.92	16
Std. Dev.	1.2	1.385640646	4
Mean	1.2	0	4
t statistic	1	0	1

Table 19.61: δ and δB_0–Prior

δ					
Intercept	PF	HS	DROP	INCOL	GRCOL
1	30	60	15	30	15

δB_0	
Total Crime	Violent Crime
5.7	5.55

We are now in a position to calculate all of the coefficients and constants in the bivariate distribution over model coefficients in the prior. We make those calculations in Table 19.62.

Table 19.62: Calculation of Parameters of Multivariate Prior

δQ_0^{-1}					
Intercept	PF	HS	DROP	INCOL	GRCOL
0.20833333	0.00033333	0.01066667	0.00013889	0.00075000	0.00416667

$\delta Q_0^{-1}\delta^T$	$1+\delta Q_0^{-1}\delta^T$		
0.945416667	1.945416667		
$\pi[1+\delta Q_0^{-1}\delta^T]$	$\{\pi[1+\delta Q_0^{-1}\delta^T]\}^{m/2}$		
6.111706708	6.111706708		
$\|M_0\|$	$\|M_0\|^{m/2}$		
76.80000000	8.76		
$\frac{\nu_0+1}{2}$	$\frac{\nu_0-m+1}{2}$		
4	3		
$\Gamma\left(\frac{\nu_0+1}{2}\right)$	$\Gamma\left(\frac{\nu_0-m+1}{2}\right)$		
6.000000E+00	2.000000E+00		
constant	matrix		
0.05601162	0.10708931	0.00000000	exponent
	0.00000000	0.03212679	-4

We apply the same procedure as we did in the classical to calculate the final bivariate distribution. We can comprise the entire bivariate distribution using the foregoing parameters and substituting them into equation (19.1) given in the previous section. We are, after all this, able to examine the variance-covariance matrix and related parameters for the prior as in Table 19.63. The table gives all the terms, the entirety, of the multivariate Student's t distribution over the two outputs TC and VC.

We now turn to the two univariate marginal distributions over total crime TC and over violent crime VC. First, in Table 19.64, we calculate the marginal distribution over total crime TC.

Table 19.63: VCV and Related Parameters–Prior

VCV	TC	VC
TC	7.33404805	0.00000000
VC	0.00000000	24.44682683
Correlation	TC	VC
TC	1	0
VC	0	1
Diagonals	TC	VC
Variance	7.33404805	24.44682683
Std. Deviation	2.70814476	4.94437325
Means	5.70000000	5.55000000
t Statistic	2.104761932	1.122488073

Table 19.64: Univariate Distribution Over Total Crime–Prior

Total Crime 5.7		
$\frac{\nu_0-m+2}{2}$	$\frac{\nu_0-m+1}{2}$	
3.5	3	
$\Gamma\left(\frac{\nu_0-m+2}{2}\right)$	$\Gamma\left(\frac{\nu_0-m+1}{2}\right)$	
3.32335097E+00	2.00000000E+00	
$1+\delta Q_0^{-1}\delta^T$	$[M_0]_{ii}$	$(1+\delta Q_0^{-1}\delta)^T\,[M_0]_{ii}$
1.94541667	4.80000000	9.338
$\{[1+\delta Q_0^{-1}\delta)]^T\,[M_0]_{ii}\}^{1/2}$	π	$\pi^{1/2}$
3.05581413	3.14159265	1.772453851
$\{\pi(1+\delta Q_0^{-1}\delta)^T\,[M_0]_{ii}\}^{1/2}$	$\frac{\Gamma\left(\frac{\nu_0-m+2}{2}\right)}{\Gamma\left(\frac{\nu_0-m+1}{2}\right)\{\pi(1+\delta Q_0^{-1}\delta^T)[M_0]_{ii}\}^{1/2}}$	$\frac{1}{(1+\delta Q_0^{-1}\delta^T)[M_0]_{ii}}$
5.416289523	0.306792	0.107089312
$\frac{(1+\delta Q_0^{-1}\delta^T)[M_0]_{ii}}{\nu_0-m-1}$	sqrt variance	exponent
2.334500	1.527907	-3.5

Second, in Table 19.65, we calculate the marginal distribution over violent crime VC. We now plot the marginal distribution on total crime in Table 19.4 and the marginal distribution on violent crime in 19.5.

Table 19.65: Univariate Distribution Over Total Crime–Prior–Prior

Violent Crime 5.55		
$\frac{\nu_0-m+2}{2}$	$\frac{\nu_0-m+1}{2}$	
3.5	3	
$\Gamma\left(\frac{\nu_0-m+2}{2}\right)$	$\Gamma\left(\frac{\nu_0-m+1}{2}\right)$	
3.32335097E+00	2.00000000E+00	
$1+\delta Q_0^{-1}\delta^T$	$[M_0]_{ii}$	$(1+\delta Q_0^{-1}\delta)^T\,[M_0]_{ii}$
1.94541667	16.00000000	31.1266667
$\left\{[1+\delta Q_0^{-1}\delta)]^T\,[M_0]_{ii}\right\}^{1/2}$	π	$\pi^{1/2}$
5.579127769	3.14159265	1.77245385
$\left\{\pi(1+\delta Q_0^{-1}\delta)^T\,[M_0]_{ii}\right\}^{1/2}$	$\frac{\Gamma\left(\frac{\nu_0-m+2}{2}\right)}{\Gamma\left(\frac{\nu_0-m+1}{2}\right)\left\{\pi(1+\delta Q_0^{-1}\delta^T)[M_0]_{ii}\right\}^{1/2}}$	$\frac{1}{(1+\delta Q_0^{-1}\delta^T)[M_0]_{ii}}$
9.888746499	0.168037	0.03212679
$\frac{(1+\delta Q_0^{-1}\delta^T)[M_0]_{ii}}{\nu_0-m-1}$	sqrt variance	exponent
4.446667	2.108712	-3.5

Figure 19.4: Total Crime Marginal Distribution–Prior

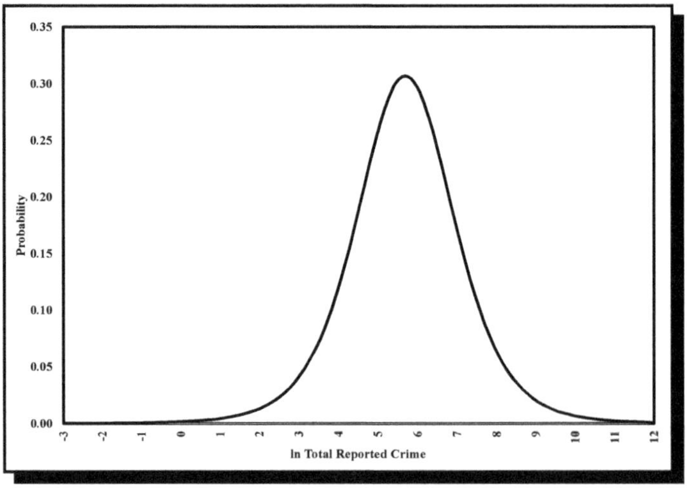

Figure 19.5: Violent Crime Marginal Distribution–Prior

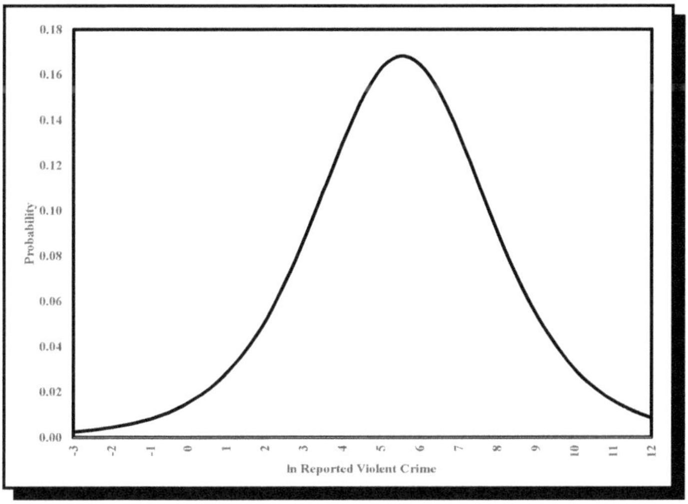

19.3 Posterior

19.3.1 Coefficients–Posterior

We begin this section by creating the $X^TX + Q_0$ matrix, summing up the matrix in Table 19.6 with the matrix in Table 19.44 to obtain the matrix in Table 19.66.

Table 19.66: The $X^TX + Q_0$ Matrix–Posterior

$X^TX + Q_0$	Intercept	PF	GRHS	DROP	INCOL	GRCOL
Intercept	54.8	1888	2940	770	1495	691
PF	1888	170650	111826	30346	53670	26527
GRHS	2940	111826	183363	43696	89224	42346
DROP	770	30346	43696	121636	20284	9859
INCOL	1495	53670	89224	20284	95435	22874
GRCOL	691	26527	42346	9859	22874	14453

We then invert this matrix to create the matrix $(X^TX + Q_0)^{-1}$ matrix, which appears in Table 19.67. We then pull the matrix X^TY from Table 19.8 and the matrix Q_0B_0 from Table 19.51 and sum them to create the matrix $X^TY + Q_0B_0$ in Table 19.68. We next calculate the mean values of the coefficients using the famous formula in equation (17.2) to create Table 19.69.

We need to calculate the $B_0 - B^*$ and $\bar{B} - B^*$ matrices, which we do in Table 19.70. We calculate the A constant in Table 19.71.

The key degrees of freedom parameters of the posterior appear in Table 19.72.

Table 19.67: The $(X^TX + Q_0)^{-1}$ Matrix–Posterior

$(X^TX + Q_0)^{-1}$	Intercept	PF	GRHS	DROP	INCOL	GRCOL
Intercept	1.345151E-01	-1.211932E-04	-1.927223E-03	-8.181920E-05	-1.586826E-04	-2.552052E-04
PF	-1.211932E-04	9.894013E-06	-3.886063E-06	-2.530923E-07	3.459002E-07	-1.354156E-06
GRHS	-1.927223E-03	-3.886063E-06	5.105225E-05	-9.434902E-07	-5.237301E-06	-4.137285E-05
DROP	-8.181920E-05	-2.530923E-07	-9.434902E-07	9.053743E-06	2.426349E-07	5.807103E-07
INCOL	-1.586826E-04	3.459002E-07	-5.237301E-06	2.426349E-07	1.983355E-05	-9.258419E-06
GRCOL	-2.552052E-04	-1.354156E-06	-4.137285E-05	5.807103E-07	-9.258419E-06	2.193520E-04

Table 19.68: The $X^TY + Q_0B_0$ Matrix–Posterior

$X^TY + Q_0B_0$	Total Crime	Violent Crime
Intercept	354.1064248	345.476437
PF	13305.32556	13433.8945
GRHS	19035.28095	17180.2535
DROP	6128.776463	5844.569594
INCOL	9279.583643	7678.829651
GRCOL	4527.263974	4528.932385

Table 19.69: The B^* Matrix–Posterior

B^*	Total Crime	Violent Crime
Intercept	6.20556684	8.88101334
PF	0.01028350	0.01932618
GRHS	-0.00404179	-0.07403008
DROP	0.01006916	0.00953236
INCOL	-0.00766292	-0.02836671
GRCOL	0.01477825	0.10857572

Table 19.70: The $B_0 - B^*$ and $\bar{B} - B^*$ Matrix–Posterior

$B_0 - B^*$	Total Crime	Violent Crime	$\bar{B} - B^*$	Total Crime	Violent Crime
Intercept	-0.205566845	0.118986663	Intercept	0.213094	-0.317383
PF	-0.000283502	0.000673816	PF	0.001823	-0.004755
GRHS	-0.005958213	-0.005969922	GRHS	-0.007319	0.005141
DROP	-6.91623E-05	0.000467644	DROP	-0.001718	-0.006548
INCOL	-0.002337083	-0.001633288	INCOL	0.007203	-0.000170
GRCOL	-0.00477825	-0.008575718	GRCOL	-0.001503	0.020922

Table 19.71: Calculation of the A Matrix–Posterior

	Total Crime	Violent Crime
Total Crime	-5.152206299	-5.699960698
Violent Crime	5.218703166	8.417299497

	Total Crime	Violent Crime
Total Crime	6.509313538	6.095025631
Violent Crime	-4.823638233	-6.105181476

$S^T S$	Total Crime	Violent Crime
Total Crime	4.515634662	6.786307974
Violent Crime	6.786307974	27.74553005

M_0	Total Crime	Violent Crime
Total Crime	4.8	0
Violent Crime	0	16

A	Total Crime	Violent Crime
Total Crime	10.6727419	7.181372906
Violent Crime	7.181372906	46.05764807

Determinant	sqrt Determinant
439.9892737	20.97592128

Table 19.72: Key Parameters Related to Degrees of Freedom–Posterior

$\nu = \nu_0 + n$	57
p	6
m	2
$\nu - m$	55
$\nu - m - 1$	54
$\nu - m - 2$	53
$\nu - m - 3$	52

We turn to the statistics of the posterior Wishart error distribution. The expected values appear in Table 19.73.

Table 19.73: Mean Wishart Error Coefficients–Posterior

$\langle \Sigma \rangle$	TC	TV
TC	0.083623	0.125672
TV	0.125672	0.513806

$\lvert\Sigma\rvert$	$\lvert\Sigma\rvert^{1/2}$
0.027172394	0.164840511

The Kronecker product of A with itself is given in Table 19.74.

Table 19.74: $(A \otimes A)$–Posterior

$(A \otimes A)$	(TC,TC)	(TC,VC)	(VC,TC)	(VC,VC)
(TC,TC)	113.90741969	76.64493952	76.64493952	51.57211682
(TC,VC)	76.64493952	491.56139047	51.57211682	330.75714601
(VC,TC)	76.64493952	51.57211682	491.56139047	330.75714601
(VC,VC)	51.57211682	330.75714601	330.75714601	2121.30694615

The Kronecker product of A with itself times the commutation matrix Kmm is given in Table 19.75.

Table 19.75: $(A \otimes A)Kmm$–Posterior

$(A \otimes A)Kmm$	(TC,TC)	(TC,VC)	(VC,TC)	(VC,VC)
(TC,TC)	113.90741969	76.64493952	76.64493952	51.57211682
(TC,VC)	76.64493952	51.57211682	491.56139047	330.75714601
(VC,TC)	76.64493952	491.56139047	51.57211682	330.75714601
(VC,VC)	51.57211682	330.75714601	330.75714601	2121.30694615

The first three matrix terms in the variance-covariance matrix of the posterior appear in Table 19.76.

Table 19.76: First Three Terms in the Variance-Covariance Sum–Posterior

First	(TC,TC)	(TC,VC)	(VC,TC)	(VC,VC)
(TC,TC)	-0.06776170	-0.04559485	-0.04559485	-0.03067943
(TC,VC)	-0.04559485	-0.29242201	-0.03067943	-0.19676213
(VC,TC)	-0.04559485	-0.03067943	-0.29242201	-0.19676213
(VC,VC)	-0.03067943	-0.19676213	-0.19676213	-1.26193156
Second	(TC,TC)	(TC,VC)	(VC,TC)	(VC,VC)
(TC,TC)	0.06784444	0.04565052	0.04565052	0.03071689
(TC,VC)	0.04565052	0.29277905	0.03071689	0.19700238
(VC,TC)	0.04565052	0.03071689	0.29277905	0.19700238
(VC,VC)	0.03071689	0.19700238	0.19700238	1.26347238
Third	(TC,TC)	(TC,VC)	(VC,TC)	(VC,VC)
(TC,TC)	0.00169611	0.00114126	0.00114126	0.00076792
(TC,VC)	0.00114126	0.00076792	0.00731948	0.00492506
(VC,TC)	0.00114126	0.00731948	0.00076792	0.00492506
(VC,VC)	0.00076792	0.00492506	0.00492506	0.03158681

The vectorization of A^{-1} appears in Table 19.77.

Table 19.77: $vec(A^{-1})$–Posterior

$vec(A^{-1})$
0.10467902
-0.01632170
-0.01632170
0.02425682

The outer product of $vec(A^{-1})$ with its transpose is given in Table 19.78. The fourth matrix term appears in Table 19.79. The variance-covariance matrix of the posterior Wishart distribution, the sum of the four matrices, is given in

Table 19.80. The variance and t statistics for the Wishart error distribution appear in Table 19.81.

Table 19.78: $vec[A^{-1}]vec[A^{-1}]^T$–Posterior

$vec[A^{-1}]vec[A^{-1}]^T$	(TC,TC)	(TC,VC)	(VC,TC)	(VC,VC)
(TC,TC)	0.01095770	-0.00170854	-0.00170854	0.00253918
(TC,VC)	-0.00170854	0.00026640	0.00026640	-0.00039591
(VC,TC)	-0.00170854	0.00026640	0.00026640	-0.00039591
(VC,VC)	0.00253918	-0.00039591	-0.00039591	0.00058839

Table 19.79: Fourth Matrix Term–Posterior

Fourth	(TC,TC)	(TC,VC)	(VC,TC)	(VC,VC)
(TC,TC)	0.00169611	0.00114126	0.00114126	0.00731948
(TC,VC)	0.00114126	0.00076792	0.00076792	0.00492506
(VC,TC)	0.00114126	0.00076792	0.00076792	0.00492506
(VC,VC)	0.00731948	0.00492506	0.00492506	0.03158681

Table 19.80: Variance-Covariance Matrix–Posterior

VCV	(TC,TC)	(TC,VC)	(VC,TC)	(VC,VC)
(TC,TC)	0.00347496	0.00233820	0.00233820	0.00812486
(TC,VC)	0.00233820	0.00189289	0.00812486	0.01009037
(VC,TC)	0.00233820	0.00812486	0.00189289	0.01009037
(VC,VC)	0.00812486	0.01009037	0.01009037	0.06471444

Table 19.81: Variance and t-statistics Terms–Posterior

	TT,TT	TT,VV	VV,VV
Variance	0.00347496	0.00812486	0.06471444
Std. Dev.	0.05894878	0.09013799	0.25439033
Mean	0.083623	0.125672	0.513806
t statistic	1.418568	1.394222	2.019755

19.3.2 Prediction–Posterior

We now turn to the posterior predictive distribution by rewriting the old δ vector and δB^* for the posterior in Table 19.82. The parameters of the bivariate distribution over TC and VC are calculated in Table 19.83. The distribution has the form

$$\{TC, VC\} = const \times \left\{1 + \begin{bmatrix} TC - \langle TC \rangle \\ VC - \langle VC \rangle \end{bmatrix}^T matrix \begin{bmatrix} TC - \langle TC \rangle \\ VC - \langle VC \rangle \end{bmatrix}\right\}^{exponent} \tag{19.2}$$

Table 19.82: δ and δB^*–Posterior

δ						
Intercept		PF	HS	DROP	INCOL	GRCOL
1		30	60	15	30	15
δB^*						
Total Crime	Violent Crime					
6.41438834	5.939613909					

Table 19.83: Calculation of Parameters of Multivariate Posterior–Posterior

$\delta(X^TX+Q_0)^{-1}$					
Intercept	PF	HS	DROP	INCOL	GRCOL
0.00543006	-0.00007127	0.00022747	0.00000577	-0.00000277	0.00024304
$\delta(X^TX+Q_0)^{-1}\delta^T$		$[1+\delta(X^TX+Q_0)^{-1}\delta^T]$			
0.020588913		1.020588913			
$\pi[1+\delta(X^TX+Q_0)^{-1}\delta^T]$		$\{\pi[1+\delta(X^TX+Q_0)^{-1}\delta^T]\}^{m/2}$			
3.206274631		3.206274631			
$\|A\|$		$\|A\|^{1/2}$			
10.67274190		3.27			
$\frac{\nu+1}{2}$		$\frac{\nu-m+1}{2}$			
29		28			
$\Gamma\left(\frac{\nu+1}{2}\right)$		$\Gamma\left(\frac{\nu-m+1}{2}\right)$			
3.048883E+29		1.088887E+28			
constant		matrix			
2.67312526		0.20171263	0.13572647		exponent
		0.13572647	0.87048009		-29

We substitute the parameters at the bottom of Table 19.83 into equation (19.2) to create the complete and correct m-variate posterior probability distribution over TC and VC. The means of the posterior are obvious. The variances and t-statistics of the Wishart are given in Table 19.84. We calculate the univariate distribution over total crime in Table 19.85. We calculate the univariate distribution over total crime in Table 19.86.

We now plot the marginal distribution on total crime in Figure 19.6 and the marginal distribution on violent crime in Figure 19.7.

Table 19.84: Variance-Covariance Matrix and t Statistic–Posterior

VCV	TC	VC
TC	7.33404805	0.00000000
VC	0.00000000	24.44682683
Correlation	TC	VC
TC	1	0
VC	0	1
Diagonals	TC	VC
Variance	7.33404805	24.44682683
Std. Deviation	2.70814476	4.94437325
Means	5.70000000	5.55000000
t Statistic	2.104761932	1.122488073

Table 19.85: Univariate Distribution Over Total Crime–Posterior

Total Crime 6.41438834		
$\frac{\nu-m+2}{2}$	$\frac{\nu-m+1}{2}$	
28.5	28	
$\Gamma\left(\frac{\nu-m+2}{2}\right)$	$\Gamma\left(\frac{\nu-m+1}{2}\right)$	
5.73618428E+28	1.08888695E+28	
$1+\delta(X^TX+Q_0)^{-1}\delta^T$	$[A]_{ii}$	$[1+\delta(X^TX+Q_0)^{-1}\delta]^T\,[A]_{ii}$
1.02058891	10.67274190	10.89248205
$\left\{[1+\delta(X^TX+Q_0)^{-1}\delta)]^T\,[A]_{ii}\right\}^{1/2}$	π	$\pi^{1/2}$
3.300376048	3.14159265	1.772453851
$\left\{\pi(1+\delta(X^TX+Q_0)^{-1}\delta)^T\,[A]_{ii}\right\}^{1/2}$	$\frac{\Gamma\left(\frac{\nu-m+2}{2}\right)}{\Gamma\left(\frac{\nu-m+1}{2}\right)\left\{\pi(1+\delta(X^TX+Q_0)^{-1}\delta^T)[A]_{ii}\right\}^{1/2}}$	$\frac{1}{(1+\delta(X^TX+Q_0)^{-1}\delta^T)[A]_{ii}}$
5.849764235	0.900538	0.09180644
Variance	Std. Dev	exponent
0.201713	0.449124	-28.5

Table 19.86: Univariate Distribution Over Violent Crime–Posterior

Violent Crime 5.939613909		
$\frac{\nu-m+2}{2}$	$\frac{\nu-m+1}{2}$	
28.5	28	
$\Gamma\left(\frac{\nu-m+2}{2}\right)$	$\Gamma\left(\frac{\nu-m+1}{2}\right)$	
5.73618428E+28	1.08888695E+28	
$1+\delta(X^TX+Q_0)^{-1}\delta^T$	$[A]_{ii}$	$(1+\delta(X^TX+Q_0)^{-1}\delta)^T\,[A]_{ii}$
1.02058891	46.05764807	47.00592
$\left\{[1+\delta(X^TX+Q_0)^{-1}\delta)]^T\,[A]_{ii}\right\}^{1/2}$	π	$\pi^{1/2}$
6.85608671	3.14159265	1.772454
$\left\{\pi(1+\delta(X^TX+Q_0)^{-1}\delta)^T\,[A]_{ii}\right\}^{1/2}$	$\frac{\Gamma\left(\frac{\nu-m+2}{2}\right)}{\Gamma\left(\frac{\nu-m+1}{2}\right)\left\{\pi(1+\delta(X^TX+Q_0)^{-1}\delta^T)[A]_{ii}\right\}^{1/2}}$	$\frac{1}{(1+\delta(X^TX+Q_0)^{-1}\delta^T)[A]_{ii}}$
12.15209729	0.433500	0.021274
$\frac{(1+\delta(X^TX+Q_0)^{-1}\delta^T)[A]_{ii}}{\nu-m-1}$	sqrt variance	exponent
0.870480	0.932995	-28.5

Figure 19.6: Total Crime Marginal Distribution–Posterior

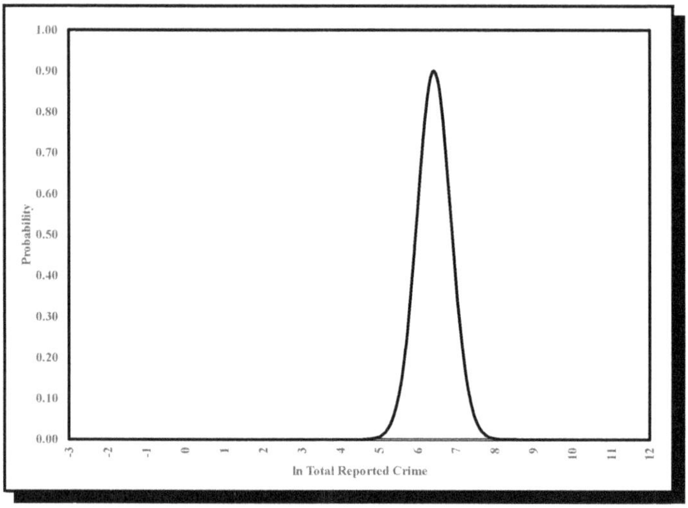

Figure 19.7: Violent Crime Marginal Distribution–Posterior

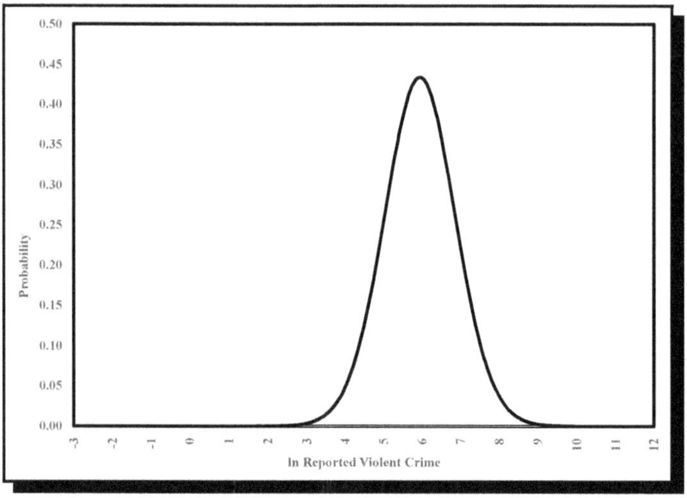

19.4 Summary

We plot the results for the prior, classical, and posterior marginal predicted distribution for TC in Figure 19.8 and for VC in Figure 19.9. These are absolutely archetypical Bayesian distributions.

Figure 19.8: Total Crime Marginal Distribution–All Three

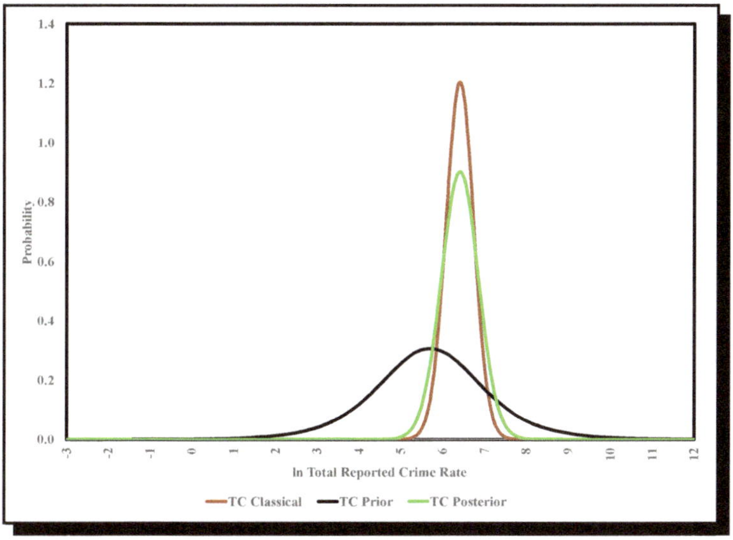

Figure 19.9: Violent Crime Marginal Distribution–All Three

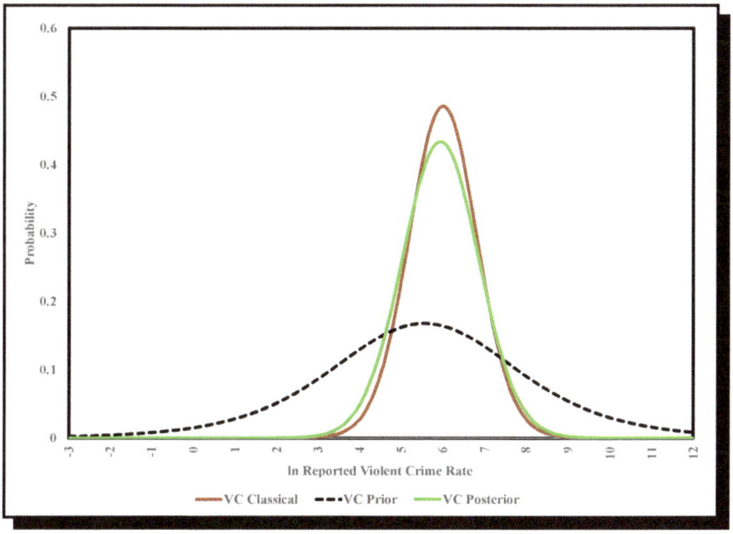

19.5 Example Analyses

We are in a position now to perform a number of analyses. We put forth just a few here, with no intent to be exhaustive.

19.5.1 Increase in Annual Per Resident Police Funding

Instead of having the base case level of police funding, suppose police funding is increased by 50 percent, moving from 30 to 45. The difference between the base case and 50 percent higher police funding appears in Table 19.87. We notice that both total reported crime and reported violent crime increase as police spending increases. Huh? More money, more crime? We see in the first three rows that reported total crime is higher with higher police funding. How could that be? The answer is simple, more crimes that were unreported with lower police funding are now actually reported with higher police funding. Part of police functioning is crime reporting. More cops doing more detective work and paperwork.

In lines 4-6, particularly the first and the last, we see that the standard deviations are largely unchanged in the classical and the posterior. That means in fact in lines 7-9 that the t statistic increases. This means that we have a tighter estimate of the increase in reporting of total crime and violent crime when we increase police funding. Uncertainty is reduced the more police funding that occurs. That makes total sense. The more police funding, the more certain we become about the crimes actually committed.

19.5.2 Fraction of People 25 years of Age or Older with Four Years of High School Goes from 60 to 75

Instead of the base case fraction, suppose that the fraction of people 25 years or age or over who have completed four years of high school rises from 60 to 75 percent. That means that the non high school graduate percent-of-population drops from 40 to 25 percent. Table 19.88 gives us the answer. We see in the classical regression and the posterior that the total crime rate drops significantly, but the violent crime rate drops even more. By having a higher percentage of high school graduate equivalent people in the city, we see lower total crime and much lower violent crime. Thankfully we have all the correlations inherently modeled to ensure that this is correct. Having people who have gotten through high school is apparently a good thing for reducing crime and especially violent crime in the city. Keep in mind, this is not the "dropouts;" some of them go back and get a diploma or GED. This is the people who never finish high school and never go back to get a high school diploma.

Table 19.87: Change in TC and VC if Police Funding Increases by 50 Percent

$\delta_0\bar{B}$		$\delta\bar{B}$		Multiple	
TC	VC	TC	VC	TC	VC
6.410824	5.998547	6.592428	6.217119	1.028328	1.036437
$\delta_0 B_0$		δB_0		Multiple	
TC	VC	TC	VC	TC	VC
5.7	5.55	5.85	5.85	1.026316	1.054054
$\delta_0 B^*$		δB^*		Multiple	
TC	VC	TC	VC	TC	VC
6.414388	5.939614	6.568641	6.229507	1.024048	1.048807
$StdDev_0\bar{B}$		$Std.Dev.\bar{B}$		Multiple	
TC	VC	TC	VC	TC	VC
0.337461	0.83649	0.336392	0.83384	0.996832	0.996832
$StdDev_0 B_0$		$StdDev B_0$		Multiple	
TC	VC	TC	VC	TC	VC
1.527907	2.108712	1.532808	2.115476	1.003208	1.003208
$StdDev_0 B^*$		$StdDev B^*$		Multiple	
TC	VC	TC	VC	TC	VC
0.449124	0.932995	0.449144	0.933035	1.000043	1.000043
$t - statistic_0\bar{B}$		$t - statistic\bar{B}$		Multiple	
TC	VC	TC	VC	TC	VC
18.99722	7.171089	19.59746	7.456009	1.031596	1.039732
$t - statistic_0 B_0$		$t - statistic B_0$		Multiple	
TC	VC	TC	VC	TC	VC
3.730593	2.631938	3.816525	2.765335	1.023034	1.050684
$t - statistic_0 B^*$		$t - statistic B^*$		Multiple	
TC	VC	TC	VC	TC	VC
14.28199	6.366178	14.62481	6.676602	1.024004	1.048761

19.5.3 Cut High School Dropout Rate in Half

The notion of high school dropouts committing crimes is prevalent. However, let us note in Table 19.89 that the effect of high current non-attendance rate is surprisingly small. It doesn't appear that a leading term in total crime or violent crime is the number of current dropouts. In the classical and posterior in the first three rows, the drop in crime rate is a couple of percent, not a large effect. Crime rate is apparently a function of longer term, more systematic, more institutional educational failure.

Table 19.88: Change in TC and VC if Number of People 25 Years of Age or Older Have Four or More Years of High School Goes from 60 to 75

$\delta_0 \bar{B}$		$\delta \bar{B}$		Multiple	
TC	VC	TC	VC	TC	VC
6.410824	5.998547	6.240414	4.96521	0.973418	0.827735
$\delta_0 B_0$		δB_0		Multiple	
TC	VC	TC	VC	TC	VC
5.7	5.55	5.55	4.35	0.973684	0.783784
$\delta_0 B^*$		$\delta_0 B^*$		Multiple	
TC	VC	TC	VC	TC	VC
6.414388	5.939614	6.353762	4.829163	0.990548	0.813043
$StdDev_0 \bar{B}$		$StdDev \bar{B}$		Multiple	
TC	VC	TC	VC	TC	VC
0.337461	0.83649	0.359087	0.890096	1.064084	1.064084
$StdDev_0 B_0$		$StdDev B_0$		Multiple	
TC	VC	TC	VC	TC	VC
1.527907	2.108712	1.66328	2.295544	1.0886	1.0886
$StdDev_0 B^*$		$StdDev B^*$		Multiple	
TC	VC	TC	VC	TC	VC
0.449124	0.932995	0.453135	0.941328	1.008931	1.008931
$t - statistic_0 \bar{B}$		$t - statistic \bar{B}$		Multiple	
TC	VC	TC	VC	TC	VC
18.99722	7.171089	17.37856	5.578287	0.914795	0.777886
$t - statistic_0 B_0$		$t - statistic B_0$		Multiple	
TC	VC	TC	VC	TC	VC
3.730593	2.631938	3.336781	1.894975	0.894437	0.719992
$t - statistic_0 B^*$		$t - statistic B^*$		Multiple	
TC	VC	TC	VC	TC	VC
14.28199	6.366178	14.02177	5.130162	0.98178	0.805846

Table 19.89: Cut High School Dropout Rate from 15 to 7.5

$\delta_0 \bar{B}$		$\delta \bar{B}$		Multiple	
TC	VC	TC	VC	TC	VC
6.410824	5.998547	6.348189	5.976164	0.99023	0.996269
$\delta_0 B_0$		δB_0		Multiple	
TC	VC	TC	VC	TC	VC
5.7	5.55	5.625	5.475	0.986842	0.986486
$\delta_0 B^*$		$\delta_0 B^*$		Multiple	
TC	VC	TC	VC	TC	VC
6.414388	5.939614	6.33887	5.868121	0.988227	0.987963
$StdDev_0 \bar{B}$		$Std.Dev.\bar{B}$		Multiple	
TC	VC	TC	VC	TC	VC
0.337461	0.83649	0.350209	0.86809	1.037776	1.037776
$StdDev_0 B_0$		$StdDev B_0$		Multiple	
TC	VC	TC	VC	TC	VC
1.527907	2.108712	1.527293	2.107865	0.999598	0.999598
$StdDev_0 B^*$		$StdDev B^*$		Multiple	
TC	VC	TC	VC	TC	VC
0.449124	0.932995	0.449217	0.933188	1.000207	1.000207
$t-statistic_0 \bar{B}$		$t-statistic \bar{B}$		Multiple	
TC	VC	TC	VC	TC	VC
18.99722	7.171089	18.12685	6.884268	0.954184	0.960003
$t-statistic_0 B_0$		$t-statistic B_0$		Multiple	
TC	VC	TC	VC	TC	VC
3.730593	2.631938	3.682986	2.597415	0.987239	0.986883
$t-statistic_0 B^*$		$t-statistic B^*$		Multiple	
TC	VC	TC	VC	TC	VC
14.28199	6.366178	14.11092	6.288249	0.988022	0.987759

19.6 Percent of 18-24 Year Olds in College Goes Up By 50 Percent from 30 to 45

What if the kids in the 18-24 year old range have gone off to college and are away from the city? What does that do to the crime rates? The educationally oriented kids are away from the city. Does that affect the crime rate? Table 19.90 gives us the answer. We notice in lines 1 and 3 that the mean total crime drops only slightly, but violent crime drops more significantly. College kids, absent from the town, are no longer victims or perpetrators of crime. The next section will suggest that they may be victims rather than perpetrators.

Table 19.90: Percent of 18-24 Year Olds in College Goes Up By 50 Percent from 30 to 45

$\delta_0\bar{B}$		$\delta\bar{B}$		Multiple	
TC	VC	TC	VC	TC	VC
6.410824	5.998547	6.403923	5.570491	0.998924	0.92864
$\delta_0 B_0$		δB_0		Multiple	
TC	VC	TC	VC	TC	VC
5.7	5.55	5.55	5.1	0.973684	0.918919
$\delta_0 B^*$		$\delta_0 B^*$		Multiple	
TC	VC	TC	VC	TC	VC
6.414388	5.939614	6.299445	5.514113	0.98208	0.928362
$StdDev_0\bar{B}$		$Std.Dev.\bar{B}$		Multiple	
TC	VC	TC	VC	TC	VC
0.337461	0.83649	0.344056	0.852836	1.019541	1.019541
$StdDev_0 B_0$		$StdDevB_0$		Multiple	
TC	VC	TC	VC	TC	VC
1.527907	2.108712	1.538912	2.1239	1.007203	1.007203
$StdDev_0 B^*$		$StdDevB^*$		Multiple	
TC	VC	TC	VC	TC	VC
0.449124	0.932995	0.450087	0.934995	1.002143	1.002143
$t-statistic_0\bar{B}$		$t-statistic\bar{B}$		Multiple	
TC	VC	TC	VC	TC	VC
18.99722	7.171089	18.61305	6.531723	0.979778	0.910841
$t-statistic_0 B_0$		$t-statisticB_0$		Multiple	
TC	VC	TC	VC	TC	VC
3.730593	2.631938	3.606444	2.401243	0.966721	0.912348
$t-statistic_0 B^*$		$t-statisticB^*$		Multiple	
TC	VC	TC	VC	TC	VC
14.28199	6.366178	13.99606	5.89748	0.97998	0.926377

19.6.1 Percentage of College Graduates Living in the City Rises from 15 to 25

What if the town has a higher percentage of college graduates, moving up from 15 percent to 25 percent of the population. Table 19.91 tells the story. Alarmingly, the violent crime rate predicted in the data increases dramatically. What might this mean? It might mean that college graduates, affluent people, are much more frequent targets of criminals. Criminals target college graduates because they have the money and property to steal. Or it means the higher income college graduates are decadent and kill, assault, and rape each other and people who they view as beneath them in stature. Or it means that more college grads means better crime reporting.

It is interesting that the effect is higher for violent crime. Looking at the classical and posterior solutions, the mean value of total crime is only up a bit, but the mean value of violent crime is up substantially. We note in addition that uncertainty in the form of standard deviation is increased. The t-statistic is actually lower for TC, meaning that we are more uncertain about the impact on TC. However, the t-statistic is higher for VC, meaning that we are more certain that violent crime will be higher with more college graduates residing in the city. In summary, the t-statistic is up for VC, and down for TC. VC is more certain, and TC is less certain by that measure.

Table 19.91: Change in TC and VC if the Number of College Graduates in the Town Increases from 15 to 25

$\delta_0 \bar{B}$		$\delta \bar{B}$		Multiple	
TC	VC	TC	VC	TC	VC
6.410824	5.998547	6.543576	7.293524	1.020707	1.215882
Second					
$\delta_0 B_0$		δB_0		Multiple	
TC	VC	TC	VC	TC	VC
5.7	5.55	5.8	6.55	1.017544	1.18018
$\delta_0 B^*$		$\delta_0 B^*$		Multiple	
TC	VC	TC	VC	TC	VC
6.414388	5.939614	6.562171	7.025371	1.023039	1.182799
$StdDev_0 \bar{B}$		$Std.Dev.\bar{B}$		Multiple	
TC	VC	TC	VC	TC	VC
0.337461	0.83649	0.393567	0.975564	1.166258	1.166258
$StdDev_0 B_0$		$StdDev B_0$		Multiple	
TC	VC	TC	VC	TC	VC
1.527907	2.108712	1.570934	2.168095	1.028161	1.028161
$StdDev_0 B^*$		$StdDev B^*$		Multiple	
TC	VC	TC	VC	TC	VC
0.449124	0.932995	0.454982	0.945164	1.013043	1.013043
$t - statistic_0 \bar{B}$		$t - statistic\bar{B}$		Multiple	
TC	VC	TC	VC	TC	VC
18.99722	7.171089	16.62633	7.476213	0.875198	1.042549
$t - statistic_0 B_0$		$t - statistic B_0$		Multiple	
TC	VC	TC	VC	TC	VC
3.730593	2.631938	3.692071	3.021086	0.989674	1.147856
$t - statistic_0 B^*$		$t - statistic B^*$		Multiple	
TC	VC	TC	VC	TC	VC
14.28199	6.366178	14.42292	7.432966	1.009868	1.167571

Chapter 20

BAYESIAN ANALYSIS OF SIMULTANEOUS EQUATION REGRESSION MODELS

> ***It was our use of probability theory as logic that has enabled us to do so easily what was impossible for those who thought of probability as a physical phenomenon associated with "randomness." Quite the opposite; we have thought of probability distributions as carriers of information.***
>
> ***Edwin Thompson Jaynes***

One reason we spent so much effort in Chapters 15 through 17 was to develop the background for simultaneous equation regression models. Simultaneous equation models are important in practice, and there may be a bit of mathematical development yet to be done to handle such models in a closed-form Bayesian fashion (if at all possible). For now, we are relegated to simulation for some of the regression equation structures. To illustrate and motivate the simultaneous model, we begin with a fully identified, simple supply-demand model, which is what we are ultimately interested in here.[1]

[1]The first specific model put forth here will have the property that the model coefficients are fully "identifiable." That means there will be a one-to-one correspondence between the specific model coefficients and the posterior probability distribution being estimated. Because of its intrinsic "identi-

20.1 A Simple Supply-Demand Construct

We conceive of a linear direct demand function of the following form

$$\begin{aligned} & q = \alpha_2 p + \alpha_1 + \alpha_3 DI + \varepsilon_d \quad \alpha_1 > 0 \quad \alpha_2 < 0 \quad \alpha_3 > 0 \\ & \Rightarrow q - \alpha_2 p - \alpha_1 - \alpha_3 DI = \varepsilon_d \quad \alpha_1 > 0 \quad \alpha_2 < 0 \quad \alpha_3 > 0 \quad . \end{aligned}$$

in which the inequality signs on the constants are what we would expect, not what we would require or mandate (constrain).[2] The specific variables are defined as follows:

- q = quantity consumed
- p = price
- DI = disposable income.

We turn to a linear supply curve, which we express as follows

$$\begin{aligned} & q = \beta_2 p + \beta_1 + \beta_3 FP + \varepsilon_s \quad \beta_1 < 0 \quad \beta_2 > 0 \quad \beta_3 < 0 \\ & \Rightarrow q - \beta_2 p - \beta_1 - \beta_3 FP = \varepsilon_s \quad \beta_1 < 0 \quad \beta_2 > 0 \quad \beta_3 < 0 \quad , \end{aligned}$$

in which the inequality signs on the constants are what we expect, not require. We define

- q = quantity produced
- p = price
- FP = factor price of production.

When we combine the demand and supply models, we write the following econometric system of equations

$$\begin{aligned} q_i - \alpha_2 p_i - \alpha_1 - \alpha_3 DI_i &= \varepsilon_{di} \quad i = 1, .., n \\ q_i - \beta_2 p_i - \beta_1 - \beta_3 FP_i &= \varepsilon_{si} \quad i = 1, .., n \quad . \end{aligned}$$

It is a system such as this that we want to analyze both classically and from a Bayesian perspective. The Bayesian perspective is not going to be easy, straightforward, or direct here, but is it ever important! We will take the time to develop

fiability," everything will just work out peachy in this example. The notion of identification, which is subtle, will be made clear as we proceed through the developments.

[2]The demand function is linear in the coefficients. The specific functional form can be log-linear if we simply substitute the logarithms of the observed terms. If we use a log linear form by taking the logarithms of the observed variables, we obtain the demand elasticity in standard economic terms, and by construction, it will be constant across the entire range of prices. We avoid that generality here for simplicity of exposition.

it in detail. This section will appeal to the results in Chapters 15 through 17, the multivariate output case, which is a prerequisite here.

20.2 A "Fully Identifiable" Model of the Foregoing Supply-Demand Type

We will be defining the concept of identification and identifiability as we go along, when the issue will be come obvious. Beginning with the foregoing demand equation

$$q = \alpha_1 + \alpha_2 p + \alpha_3 DI + \varepsilon_d \quad .$$

d stands for "demand."[3] This is a standard linear demand curve with a dependency on own prices and disposable income DI, which is very common in econometrics. (People want to know the price elasticity of demand and the income elasticity of demand in various markets, and this is how we seek to obtain them.) If we consider the ith observation or i^{th} data point, we write this equation:

$$q_i = \alpha_1 + \alpha_2 p_i + \alpha_3 DI_i + \varepsilon_{di} \quad i = 1, ..., n \quad .$$

Turning to the supply equation, we write

$$q = \beta_1 + \beta_2 p + \beta_3 FP + \varepsilon_s \quad .$$

In this supply equation, FP represents the price of a factor of production, p represents price, and q represents the quantity supplied. The subscript s stands for "supply." This is a standard linear supply curve. If we consider the ith observation or i^{th} data point, we write this equation

$$q_i = \beta_1 + \beta_2 p_i + \beta_3 FP_i + \varepsilon_{si} \quad i = 1, ..., n \quad .$$

We now write down the two equations simultaneously

$$\begin{aligned} q_i &= \alpha_1 + \alpha_2 p_i + \alpha_3 DI_i + \varepsilon_{di} \quad i = 1, ..., n \\ q_i &= \beta_1 + \beta_2 p_i + \beta_3 FP_i + \varepsilon_{si} \quad i = 1, ..., n \quad . \end{aligned}$$

These two equations are going to have what has been termed the "exactly-identifiable" property. More to come on that later. We reorganize this equation with the endogenous variables on the left

$$\begin{aligned} q_i - \alpha_2 p_i - \alpha_1 - \alpha_3 DI_i &= \varepsilon_{di} \quad i = 1, ..., n \\ q_i - \beta_2 p_i - \beta_1 - \beta_3 FP_i &= \varepsilon_{si} \quad i = 1, ..., n \quad . \end{aligned}$$

[3] We could use the logarithms of these terms as well, but we will stay with the actual terms.

We write this in matrix form

$$\begin{bmatrix} 1 & -\alpha_2 \\ 1 & -\beta_2 \end{bmatrix} \begin{bmatrix} q_i \\ p_i \end{bmatrix} - \begin{bmatrix} \alpha_1 & \alpha_3 & 0 \\ \beta_1 & 0 & \beta_3 \end{bmatrix} \begin{bmatrix} 1 \\ DI_i \\ FP_i \end{bmatrix} = \begin{bmatrix} \varepsilon_{di} \\ \varepsilon_{si} \end{bmatrix} .$$

Let us write the transpose of this equation in preparation for establishing the proper matrices correctly

$$\begin{bmatrix} q_i & p_i \end{bmatrix} \begin{bmatrix} 1 & 1 \\ -\alpha_2 & -\beta_2 \end{bmatrix} - \begin{bmatrix} 1 & DI_i & FP_i \end{bmatrix} \begin{bmatrix} \alpha_1 & \beta_1 \\ \alpha_3 & 0 \\ 0 & \beta_3 \end{bmatrix} = \begin{bmatrix} \varepsilon_{di} & \varepsilon_{si} \end{bmatrix} .$$

Let us organize the observations $i = 1, ..., n$ so that we can replicate the uncertainty structure in Chapters 15 through 17:

$$\begin{bmatrix} q_1 & p_1 \\ \cdot & \cdot \\ \cdot & \cdot \\ \cdot & \cdot \\ q_n & p_n \end{bmatrix} \begin{bmatrix} 1 & 1 \\ -\alpha_2 & -\beta_2 \end{bmatrix} - \begin{bmatrix} 1 & DI_1 & FP_1 \\ \cdot & \cdot & \cdot \\ \cdot & \cdot & \cdot \\ \cdot & \cdot & \cdot \\ 1 & DI_n & FP_n \end{bmatrix} \begin{bmatrix} \alpha_1 & \beta_1 \\ \alpha_3 & 0 \\ 0 & \beta_3 \end{bmatrix}$$

$$= \begin{bmatrix} \varepsilon_{d1} & \varepsilon_{s1} \\ \cdot & \cdot \\ \cdot & \cdot \\ \cdot & \cdot \\ \varepsilon_{dn} & \varepsilon_{sn} \end{bmatrix} .$$

Let us make the following matrix definitions

$$Y = \begin{bmatrix} q_1 & p_1 \\ q_2 & p_2 \\ \cdot & \cdot \\ \cdot & \cdot \\ \cdot & \cdot \\ q_n & p_n \end{bmatrix} ,$$

$$\Gamma = \begin{bmatrix} 1 & 1 \\ -\alpha_2 & -\beta_2 \end{bmatrix} ,$$

$$X = \begin{bmatrix} 1 & DI_1 & FP_1 \\ 1 & DI_2 & FP_2 \\ \cdot & \cdot & \cdot \\ \cdot & \cdot & \cdot \\ \cdot & \cdot & \cdot \\ 1 & DI_n & FP_n \end{bmatrix} ,$$

$$B = \begin{bmatrix} \alpha_1 & \beta_1 \\ \alpha_3 & 0 \\ 0 & \beta_3 \end{bmatrix} \quad ,$$

and

$$E = \begin{bmatrix} \varepsilon_{d1} & \varepsilon_{s1} \\ \varepsilon_{d2} & \varepsilon_{s2} \\ \cdot & \cdot \\ \cdot & \cdot \\ \cdot & \cdot \\ \varepsilon_{dn} & \varepsilon_{sn} \end{bmatrix} \quad .$$

Under these definitions, this equation becomes

$$\begin{aligned} Y\Gamma - XB &= E \\ \begin{bmatrix} Y & X \end{bmatrix} \begin{bmatrix} \Gamma \\ -B \end{bmatrix} &= E \quad . \end{aligned}$$

The matrices Y and X are filled with observation data. The matrices Γ and B are the unknown constants. We can write

$$\begin{aligned} & Y\Gamma - XB = E \\ & \Rightarrow Y - XB\Gamma^{-1} = E\Gamma^{-1} \\ & \Rightarrow Y = XB\Gamma^{-1} + E\Gamma^{-1} \quad . \end{aligned}$$

If we define

$$\begin{aligned} \Pi &= B\Gamma^{-1} \\ V &= E\Gamma^{-1} \quad . \end{aligned}$$

then we can write what is called the "reduced form equation"

$$\begin{aligned} Y &= X\Pi + V \\ Y - X\Pi &= V \quad , \end{aligned}$$

in which we can estimate the variable Π using the multivariate techniques from Chapters 15 through 17. This equation is what allows us to directly apply the results from Chapter 15 through 17 for the multivariate case and the reduced form analysis of the simultaneous case. If we do so, we need to return to the matrix E and write the probability distribution that is represented by E. We know from Chapters 15 through 17 that the probability distribution over the ith row is

$$N(0,\Sigma) = \frac{1}{(2\pi)^{\frac{m}{2}}} |\Sigma|^{-\frac{1}{2}} e^{-\frac{1}{2}\left[\begin{array}{cc} \varepsilon_{di} & \varepsilon_{si} \end{array}\right] \Sigma^{-1} \left[\begin{array}{c} \varepsilon_{di} \\ \varepsilon_{si} \end{array}\right]} \quad .$$

If we begin with the relationship

$$Y = XB\Gamma^{-1} + E\Gamma^{-1}$$

and wish to switch variables from E to Y, we note that the Jacobian is

$$\begin{aligned}
&\left[\begin{array}{cc} y_d & y_s \end{array}\right] = \left[\begin{array}{cc} \varepsilon_d & \varepsilon_s \end{array}\right] \Gamma^{-1} \\
&\Rightarrow \left[\begin{array}{cc} dy_d & dy_s \end{array}\right] = \left[\begin{array}{cc} d\varepsilon_d & d\varepsilon_s \end{array}\right] \Gamma^{-1} \\
&\Rightarrow \left[\begin{array}{cc} d\varepsilon_d & d\varepsilon_s \end{array}\right] = \left[\begin{array}{cc} dy_d & dy_s \end{array}\right] \Gamma \\
&\Rightarrow \text{Jacobian} = |\Gamma| \quad .
\end{aligned}$$

Therefore the probability distribution over v_i is calculated to be

$$\begin{aligned}
f(v_i) &= \left|\frac{\partial \varepsilon_i}{\partial v_i}\right| f(\varepsilon_i) \\
&= |\Gamma| f(\varepsilon_i) \\
&= |\Gamma| \frac{1}{(2\pi)^{\frac{m}{2}}} |\Sigma|^{-\frac{1}{2}} e^{-\frac{1}{2}\varepsilon_i{}^T \Sigma^{-1} \varepsilon_i} \quad .
\end{aligned}$$

Multiplying for each of the observations, we have the likelihood function

$$\begin{aligned}
LF &= \prod_{i=1}^{n} |\Gamma| \frac{1}{(2\pi)^{\frac{m}{2}}} |\Sigma|^{-\frac{1}{2}} e^{-\frac{1}{2}\varepsilon_i{}^T \Sigma^{-1} \varepsilon_i} \\
&= |\Gamma|^n \frac{1}{(2\pi)^{\frac{mn}{2}}} |\Sigma|^{-\frac{n}{2}} \prod_{i=1}^{n} e^{-\frac{1}{2}\varepsilon_i{}^T \Sigma^{-1} \varepsilon_i} \\
&= |\Gamma|^n \frac{1}{(2\pi)^{\frac{mn}{2}}} |\Sigma|^{-\frac{n}{2}} e^{-\frac{1}{2}\sum_{i=1}^{n} \varepsilon_i{}^T \Sigma^{-1} \varepsilon_i} \\
&= |\Gamma|^n \frac{1}{(2\pi)^{\frac{mn}{2}}} |\Sigma|^{-\frac{n}{2}} e^{-\frac{1}{2} Trace\left(E\Sigma^{-1}E^T\right)} \quad .
\end{aligned}$$

We know that

$$\begin{aligned}
&V = E\Gamma \\
&\Rightarrow E = V\Gamma \Rightarrow E^T = \Gamma^T V^T \quad ,
\end{aligned}$$

and therefore that

$$LF = |\Gamma|^n \frac{1}{(2\pi)^{\frac{mn}{2}}} |\Sigma|^{-\frac{n}{2}} e^{-\frac{1}{2}Trace\left(V\Gamma\Sigma^{-1}\Gamma^T V^T\right)} \quad .$$

We define

$$\begin{aligned}
&\Omega^{-1} = \Gamma\Sigma^{-1}\Gamma^T \\
&\Rightarrow \Omega = \left(\Gamma^T\right)^{-1}\Sigma\Gamma^{-1} \\
&\Rightarrow \Sigma^{-1} = \Gamma^{-1}\Omega^{-1}\left(\Gamma^T\right)^{-1} \\
&\Rightarrow \Sigma = \Gamma^T\Omega\Gamma \quad ,
\end{aligned}$$

which allows the foregoing equation to be reduced to a useful form

$$\begin{aligned}
LF &= |\Gamma|^n \frac{1}{(2\pi)^{\frac{mn}{2}}} |\Sigma|^{-\frac{n}{2}} e^{-\frac{1}{2}Trace\left(V\Omega^{-1}V^T\right)} \\
&= |\Gamma|^n \frac{1}{(2\pi)^{\frac{mn}{2}}} |\Sigma|^{-\frac{n}{2}} e^{-\frac{1}{2}Trace\left[(Y-X\Pi)\Omega^{-1}(Y-X\Pi)^T\right]} \\
&= |\Gamma|^n \frac{1}{(2\pi)^{\frac{mn}{2}}} |\Sigma|^{-\frac{n}{2}} e^{-\frac{1}{2}Trace\left[(Y-X\Pi)^T(Y-X\Pi)\Omega^{-1}\right]} \quad .
\end{aligned}$$

The shift from the second to last to the last equation comes from the reordering of conformable matrices under the trace operator. Let us examine some terms in the constant

$$\begin{aligned}
|\Gamma|^n|\Sigma|^{-\frac{n}{2}} &= |\Gamma|^{\frac{n}{2}}|\Sigma|^{-\frac{n}{2}}\left|\Gamma^T\right|^{\frac{n}{2}} \\
&= \left|\Gamma\Sigma^{-1}\Gamma^T\right|^{\frac{n}{2}} \\
&= \left|\Gamma^{-T}\Sigma\Gamma^{-1}\right|^{-\frac{n}{2}} \\
&= |\Omega|^{-\frac{n}{2}} \quad .
\end{aligned}$$

This puts the likelihood function into the exact same form as equation (16.54) in Chapter 16, meaning all the subsequent results in Chapters 16 and 17 will hold directly by analogy:

$$LF = \frac{1}{(2\pi)^{\frac{mn}{2}}} |\Omega|^{-\frac{n}{2}} e^{-\frac{1}{2}Trace\left[(Y-X\Pi)^T(Y-X\Pi)\Omega^{-1}\right]} \quad .$$

To use these earlier results, we need the substitutions

$$\begin{aligned}
&n \leftarrow n \\
&\Pi \leftarrow B \\
&\Omega \leftarrow \Sigma \quad .
\end{aligned}$$

The analogy to equation (15.2) in this situation is

$$\bar{\Pi} = (X^T X)^{-1}(X^T Y) \quad .$$

The analogy to equation (16.54) in this situation is

$$\bar{S} = Y - X\bar{\Pi} \quad .$$

The likelihood function, analogous to equation (16.54), is therefore

$$\begin{aligned} LF &= \{X, Y|\Pi, \Omega\} \\ &= \frac{1}{(2\pi)^{\frac{mn}{2}} |\Omega|^{\frac{n}{2}}} e^{-\frac{1}{2}Trace\left[\left(\bar{S}^T \bar{S}\right)\Omega^{-1}\right]} \\ &\times e^{-\frac{1}{2}Trace\left[\left(\Pi-\bar{\Pi}\right)^T \left(X^T X\right)\left(\Pi-\bar{\Pi}\right)\Omega^{-1}\right]} \quad . \end{aligned} \tag{20.1}$$

20.3 The Identification Problem

Let us return to the foregoing supply curve-demand curve example and rewrite the salient coefficients matrices

$$\Gamma = \begin{bmatrix} 1 & 1 \\ -\alpha_2 & -\beta_2 \end{bmatrix} \quad ,$$

and

$$B = \begin{bmatrix} \alpha_1 & \beta_1 \\ \alpha_3 & 0 \\ 0 & \beta_3 \end{bmatrix} \quad .$$

Using these coefficient matrices, we can constitute the Π matrix from the basic coefficients in the two simultaneous equations

$$\begin{aligned} \Pi = B\Gamma^{-1} &= \begin{bmatrix} \alpha_1 & \beta_1 \\ \alpha_2 & 0 \\ 0 & \beta_3 \end{bmatrix} \begin{bmatrix} -\beta_2 & -1 \\ \alpha_2 & 1 \end{bmatrix} \frac{1}{\alpha_2 - \beta_2} \\ &= \begin{bmatrix} \frac{\beta_1\alpha_2 - \beta_2\alpha_1}{\alpha_2 - \beta_2} & \frac{\beta_1 - \alpha_1}{\alpha_2 - \beta_2} \\ \frac{-\alpha_3\beta_2}{\alpha_2 - \beta_2} & \frac{-\alpha_3}{\alpha_2 - \beta_2} \\ \frac{\alpha_2\beta_3}{\alpha_2 - \beta_2} & \frac{\beta_3}{\alpha_2 - \beta_2} \end{bmatrix} = \begin{bmatrix} \frac{\beta_1\alpha_2 - \beta_2\alpha_1}{\Delta} & \frac{\beta_1 - \alpha_1}{\Delta} \\ \frac{-\alpha_3\beta_2}{\Delta} & \frac{-\alpha_3}{\Delta} \\ \frac{\alpha_2\beta_3}{\Delta} & \frac{\beta_3}{\Delta} \end{bmatrix} \quad , \end{aligned}$$

in which the term $\Delta = \alpha_2 - \beta_2$ is the determinant of the Γ matrix. We equate term by term to the solution matrix Π of the reduced form problem

$$B\Gamma^{-1} = \begin{bmatrix} \pi_{d1} & \pi_{s1} \\ \pi_{d2} & \pi_{s2} \\ \pi_{d3} & \pi_{s3} \end{bmatrix} \quad .$$

Following are the six equations implied by this

$$\pi_{d1} = \frac{\beta_1\alpha_2 - \beta_2\alpha_1}{\alpha_2 - \beta_2}$$
$$\pi_{d2} = \frac{-\alpha_3\beta_2}{\alpha_2 - \beta_2}$$
$$\pi_{d3} = \frac{\alpha_2\beta_3}{\alpha_2 - \beta_2}$$
$$\pi_{s1} = \frac{\beta_1 - \alpha_1}{\alpha_2 - \beta_2}$$
$$\pi_{s2} = \frac{-\alpha_3}{\alpha_2 - \beta_2}$$
$$\pi_{s3} = \frac{\beta_3}{\alpha_2 - \beta_2} \quad .$$

Do these six equations provide a one-to-one, invertible relationship between the model coefficients $\alpha_1, \alpha_2, \alpha_3, \beta_1, \beta_2, \beta_3$ and the specific elements of Π? Can we think of these six equations as characterizing a transformation of the original model variables $\alpha_1, \alpha_2, \alpha_3, \beta_1, \beta_2, \beta_3$ to the new model variables Π? The existence of such one-to-one relationships would mean that the model is exactly identified. The coefficients uniquely determine the six elements of Π, and the six elements of Π determine the coefficients.

We are going to modify the likelihood function in equation (20.1) to express it in terms of the original variables $\alpha_1, \alpha_2, \alpha_3, \beta_1, \beta_2, \beta_3$. We begin by rewriting the definition of Ω

$$\Omega^{-1} = \Gamma\Sigma^{-1}\Gamma^T \Rightarrow \Omega = \left(\Gamma^T\right)^{-1}\Sigma\Gamma^{-1} \quad .$$

We know from theorems in matrix algebra that the inverse of the transpose of a matrix is the transpose of the inverse of that matrix, which are summarized in the following line

$$A^T(A^{-1})^T = [(A^{-1})A]^T = I^T = I = A^T(A^T)^{-1} \quad .$$

Therefore

$$\Omega^{-1} = \Gamma\Sigma^{-1}\Gamma^T \Rightarrow \Omega = \left(\Gamma^{-1}\right)^T\Sigma\Gamma^{-1} \quad .$$

In preparation for considering the determinants, we know that the determinant of a square matrix is the same as the determinant of the transpose of that square matrix.[4] That is, $|\Gamma| = \left|\Gamma^T\right|$. Therefore, when we turn to the determinants, we can write

[4]See https://en.wikipedia.org/wiki/Transpose.

$$\Omega^{-1} = \Gamma\Sigma^{-1}\Gamma^{T}$$
$$\Omega = \left(\Gamma^{-1}\right)^{T}\Sigma\Gamma^{-1}$$
$$|\Omega| = \frac{|\Sigma|}{|\Gamma|^{2}} \Rightarrow |\Sigma| = |\Omega|\,|\Gamma|^{2} \quad .$$

We can make substitutions from these relationships into the likelihood function to write

$$\begin{aligned}
LF &= \{X, Y|\Pi, \Omega\} \\
&= \frac{1}{(2\pi)^{\frac{mn}{2}}|\Omega|^{\frac{n}{2}}} e^{-\frac{1}{2}Trace\left[\left(S^{T}S\right)\Omega^{-1}\right]} e^{-\frac{1}{2}Trace\left[(\Pi-\bar{\Pi})^{T}\left(X^{T}X\right)(\Pi-\bar{\Pi})\Omega^{-1}\right]} \\
&= \frac{|\Gamma|^{n}}{(2\pi)^{\frac{mn}{2}}|\Sigma|^{\frac{n}{2}}} e^{-\frac{1}{2}Trace\left[\left(S^{T}S\right)\Omega^{-1}\right]} e^{-\frac{1}{2}Trace\left[(\Pi-\bar{\Pi})^{T}\left(X^{T}X\right)(\Pi-\bar{\Pi})\Omega^{-1}\right]} \\
&= \frac{|\Gamma|^{n}}{(2\pi)^{\frac{mn}{2}}|\Sigma|^{\frac{n}{2}}} e^{-\frac{1}{2}Trace\left[\left(S^{T}S\right)\left(\Gamma\Sigma^{-1}\Gamma^{T}\right)\right]} \\
&\times e^{-\frac{1}{2}Trace\left[(\Pi-\bar{\Pi})^{T}\left(X^{T}X\right)(\Pi-\bar{\Pi})\left(\Gamma\Sigma^{-1}\Gamma^{T}\right)\right]} \quad .
\end{aligned}$$

We want to make the same manipulations in the final exponential in the foregoing equation, which we wish to drive toward a normal distribution, as we did in Chapter 16. In particular, if we make the same vectorization substitution and Kronecker product substitution, we can write the final exponent in the foregoing equation in the form

$$\begin{aligned}
&Trace\left[\left(\Pi-\bar{\Pi}\right)^{T}\left(X^{T}X\right)\left(\Pi-\bar{\Pi}\right)\left(\Gamma\Sigma^{-1}\Gamma^{T}\right)\right] \\
&= \left[vec\left(\Pi-\bar{\Pi}\right)\right]^{T} vec\left[\left(X^{T}X\right)\left(\Pi-\bar{\Pi}\right)\left(\Gamma\Sigma^{-1}\Gamma^{T}\right)\right] \\
&= \left[vec\left(\Pi-\bar{\Pi}\right)\right]^{T}\left[\left(\Gamma\Sigma^{-1}\Gamma^{T}\right)\otimes\left(X^{T}X\right)\right]\left[vec\left(\Pi-\bar{\Pi}\right)\right] \\
&= \left[vec\left(\Pi-\bar{\Pi}\right)\right]^{T}\left[\Omega^{-1}\otimes\left(X^{T}X\right)\right]\left[vec\left(\Pi-\bar{\Pi}\right)\right] \quad .
\end{aligned}$$

We substitute this back into the likelihood function and rearrange

$$\begin{aligned}
LF = \frac{|\Gamma|^{n}}{(2\pi)^{\frac{mn}{2}}|\Sigma|^{\frac{n}{2}}} & e^{-\frac{1}{2}Trace\left[\left(S^{T}S\right)\Omega^{-1}\right]} \\
&\times e^{-\frac{1}{2}\left[vec(\Pi-\bar{\Pi})\right]^{T}\left[\Omega^{-1}\otimes\left(X^{T}X\right)\right]\left[vec(\Pi-\bar{\Pi})\right]} \quad .
\end{aligned}$$

To simplify future substitutions and definitions, we define

$$\Psi^{-1} = \Omega^{-1}\otimes\left(X^{T}X\right)$$
$$\Rightarrow \Psi = \Omega\otimes\left(X^{T}X\right)^{-1}$$

$$\Rightarrow \left|\Psi^{-1}\right| = \left|\Omega^{-1}\right|^{p}\left|X^{T}X\right|^{m}$$
$$\Rightarrow |\Psi| = \frac{|\Omega|^{p}}{\left|X^{T}X\right|^{m}}$$
$$\Rightarrow |\Psi|^{\frac{1}{2}} = \frac{|\Omega|^{\frac{p}{2}}}{\left|X^{T}X\right|^{\frac{m}{2}}}$$
$$\Rightarrow |\Psi|^{-\frac{1}{2}} = \frac{\left|X^{T}X\right|^{\frac{m}{2}}}{|\Omega|^{\frac{p}{2}}} = \frac{\left|X^{T}X\right|^{\frac{m}{2}}}{\left(\frac{|\Sigma|^{\frac{p}{2}}}{|\Gamma|^{p}}\right)} = \frac{\left|X^{T}X\right|^{\frac{m}{2}}}{|\Sigma|^{\frac{p}{2}}}(\alpha_2 - \beta_2)^{p} \quad .$$

This allows us to write the likelihood function in simpler form

$$LF = \frac{|\Gamma|^{n}}{(2\pi)^{\frac{mn}{2}}|\Sigma|^{\frac{n}{2}}} e^{-\frac{1}{2}Trace\left[\left(S^{T}S\right)\Omega^{-1}\right]} \times e^{-\frac{1}{2}\left[vec\left(\Pi-\bar{\Pi}\right)\right]^{T}\Psi^{-1}\left[vec\left(\Pi-\bar{\Pi}\right)\right]} \quad .$$

This prompts us to make a substitution so that the last term turns into a multivariate normal, precisely as before

$$\begin{aligned} & N\left[vec\left(\Pi\right); vec\left(\bar{\Pi}\right), \Psi, mp\right] \\ & = (2\pi)^{-\frac{mp}{2}}|\Psi|^{-\frac{1}{2}} e^{-\frac{1}{2}\left[vec(\Pi)-vec\left(\bar{\Pi}\right)\right]^{T}\Psi^{-1}\left[vec(\Pi)-vec\left(\bar{\Pi}\right)\right]} \\ & \Rightarrow e^{-\frac{1}{2}\left[vec(\Pi)-vec\left(\bar{\Pi}\right)\right]^{T}\Psi^{-1}\left[vec(\Pi)-vec\left(\bar{\Pi}\right)\right]} \\ & = N\left[vec\left(\Pi\right); vec\left(\bar{\Pi}\right), \Psi, mp\right]|\Psi|^{\frac{1}{2}}(2\pi)^{\frac{mp}{2}} \quad . \end{aligned} \tag{20.2}$$

Making the substitution and simplifying, we have

$$\begin{aligned} LF &= \frac{|\Gamma|^{n}}{(2\pi)^{\frac{mn}{2}}|\Sigma|^{\frac{n}{2}}} e^{-\frac{1}{2}Trace\left[\left(S^{T}S\right)\Omega^{-1}\right]} \\ &\quad \times N\left[vec\left(\Pi\right); vec\left(\bar{\Pi}\right), \Psi, mp\right]|\Psi|^{\frac{1}{2}}(2\pi)^{\frac{mp}{2}} \\ &= \frac{|\Gamma|^{n}|\Psi|^{\frac{1}{2}}}{(2\pi)^{\frac{m(n-p)}{2}}|\Sigma|^{\frac{n}{2}}} e^{-\frac{1}{2}Trace\left[\left(S^{T}S\right)\Omega^{-1}\right]} \\ &\quad \times N\left[vec\left(\Pi\right); vec\left(\bar{\Pi}\right), \Psi, mp\right] \quad . \end{aligned}$$

We next need to substitute for two determinants. First we substitute for $|\Psi|$

$$\begin{aligned} LF = & \frac{|\Gamma|^{n}|\Psi|^{\frac{1}{2}}}{(2\pi)^{\frac{m(n-p)}{2}}|\Sigma|^{\frac{n}{2}}} e^{-\frac{1}{2}Trace\left[\left(S^{T}S\right)\Omega^{-1}\right]} \\ & \times N\left[vec\left(\Pi\right); vec\left(\bar{\Pi}\right), \Psi, mp\right] \end{aligned}$$

$$= \frac{|\Gamma|^n \left(\frac{|\Omega|^{\frac{p}{2}}}{|X^T X|^{\frac{m}{2}}} \right)}{(2\pi)^{\frac{m(n-p)}{2}} |\Sigma|^{\frac{n}{2}}} e^{-\frac{1}{2}Trace\left[(S^T S)\Omega^{-1}\right]}$$
$$\times N\left[vec(\Pi); vec(\bar{\Pi}), \Psi, mp\right]$$
$$= \frac{|\Gamma|^n |\Omega|^{\frac{p}{2}}}{(2\pi)^{\frac{m(n-p)}{2}} |X^T X|^{\frac{m}{2}} |\Sigma|^{\frac{n}{2}}} e^{-\frac{1}{2}Trace\left[(S^T S)\Omega^{-1}\right]}$$
$$\times N\left[vec(\Pi); vec(\bar{\Pi}), \Psi, mp\right] \quad .$$

Next we substitute for $|\Sigma|$ and simplify

$$LF = \frac{|\Gamma|^n |\Omega|^{\frac{p}{2}}}{(2\pi)^{\frac{m(n-p)}{2}} |X^T X|^{\frac{m}{2}} \left(|\Omega| \, |\Gamma|^2\right)^{\frac{n}{2}}} e^{-\frac{1}{2}Trace\left[(S^T S)\Omega^{-1}\right]}$$
$$\times N\left[vec(\Pi); vec(\bar{\Pi}), \Psi, mp\right]$$
$$= \frac{1}{(2\pi)^{\frac{m(n-p)}{2}} |X^T X|^{\frac{m}{2}}} |\Omega|^{-\frac{n-p}{2}} e^{-\frac{1}{2}Trace\left[(S^T S)\Omega^{-1}\right]}$$
$$\times N\left[vec(\Pi); vec(\bar{\Pi}), \Psi, mp\right] \quad .$$

It is certainly looking like the latter portion of the first term might indicate an inverse Wishart being embedded here. The inverse Wishart is

$$IW(X; \Psi, \nu, m) = \frac{|\Psi|^{\frac{\nu}{2}}}{2^{\frac{\nu m}{2}} \Gamma_m\left(\frac{\nu}{2}\right)} |X|^{-\frac{\nu+m+1}{2}} e^{-\frac{1}{2}Trace\left(\Psi X^{-1}\right)} \quad .$$

Insert the parameters of this problem and rearrange to facilitate a substitution

$$IW\left(\Omega; S^T S, \nu, m\right) = \frac{\left|S^T S\right|^{\frac{\nu}{2}}}{2^{\frac{\nu m}{2}} \Gamma_m\left(\frac{\nu}{2}\right)} |\Omega|^{-\frac{\nu+m+1}{2}} e^{-\frac{1}{2}Trace\left[(S^T S)\Omega^{-1}\right]}$$
$$\Rightarrow |\Omega|^{-\frac{\nu+m+1}{2}} e^{-\frac{1}{2}Trace\left[(S^T S)\Omega^{-1}\right]} = \frac{IW\left(\Omega; S^T S, \nu, m\right) 2^{\frac{\nu m}{2}} \Gamma_m\left(\frac{\nu}{2}\right)}{|S^T S|^{\frac{\nu}{2}}} \quad .$$

We make the substitution for the degrees of freedom

$$\nu = n - p - m - 1$$

and substitute in the previous equation to write it in the form

$$|\Omega|^{-\frac{n-p}{2}} e^{-\frac{1}{2}Trace\left[(S^T S)\Omega^{-1}\right]} = \frac{IW\left(\Omega; S^T S, \nu, m\right) 2^{\frac{\nu m}{2}} \Gamma_m\left(\frac{\nu}{2}\right)}{|S^T S|^{\frac{\nu}{2}}} \quad .$$

We are now prepared to substitute this into our likelihood function to write

$$
\begin{aligned}
LF &= \frac{1}{(2\pi)^{\frac{m(n-p)}{2}} |X^T X|^{\frac{m}{2}}} |\Omega|^{-\frac{n-p}{2}} e^{-\frac{1}{2} Trace\left[(S^T S)\Omega^{-1}\right]} \\
&\times N\left[vec(\Pi); vec(\bar{\Pi}), \Psi, mp\right] \\
&= \frac{1}{(2\pi)^{\frac{m(n-p)}{2}} |X^T X|^{\frac{m}{2}}} \frac{IW\left(\Omega; S^T S, \nu, m\right) 2^{\frac{\nu m}{2}} \Gamma_m\left(\frac{\nu}{2}\right)}{|S^T S|^{\frac{\nu}{2}}} \\
&\times N\left[vec(\Pi); vec(\bar{\Pi}), \Psi, mp\right] \\
&= \frac{2^{\frac{(m-1)m}{2}} \Gamma_m\left(\frac{\nu}{2}\right)}{\pi^{\frac{m(n-p)}{2}} |X^T X|^{\frac{m}{2}} |S^T S|^{\frac{\nu}{2}}} \\
&\times IW\left(\Omega; S^T S, \nu, m\right) N\left[vec(\Pi); vec(\bar{\Pi}), \Psi, mp\right] \quad .
\end{aligned}
$$

Rewriting for convenient future reference

(20.3)
$$
\begin{aligned}
LF &= \frac{2^{\frac{(m-1)m}{2}} \Gamma_m\left(\frac{\nu}{2}\right)}{\pi^{\frac{m(n-p)}{2}} |X^T X|^{\frac{m}{2}} |S^T S|^{\frac{\nu}{2}}} \\
&\times IW\left(\Omega; S^T S, \nu, m\right) N\left[vec(\Pi); vec(\bar{\Pi}), \Psi, mp\right] \quad .
\end{aligned}
$$

We now take a close look at the inverse Wishart to see where all the variables are embedded:

(20.4)
$$
\begin{aligned}
IW\left(\Omega; S^T S, \nu, m\right) &= \frac{|S^T S|^{\frac{\nu}{2}}}{2^{\frac{\nu m}{2}} \Gamma_m\left(\frac{\nu}{2}\right)} |\Omega|^{-\frac{\nu+m+1}{2}} e^{-\frac{1}{2} Trace\left[(S^T S)\Gamma\Sigma^{-1}\Gamma^T\right]} \\
&= \frac{|S^T S|^{\frac{\nu}{2}}}{2^{\frac{\nu m}{2}} \Gamma_m\left(\frac{\nu}{2}\right)} \left(\frac{|\Sigma|^{-\frac{\nu+m+1}{2}}}{|\Gamma|^{-(\nu+m+1)}}\right) e^{-\frac{1}{2} Trace\left[(S^T S)\Gamma\Sigma^{-1}\Gamma^T\right]} \\
&= \frac{|S^T S|^{\frac{\nu}{2}} (\alpha_2 - \beta_2)^{n-p} |\Sigma|^{-\frac{n-p}{2}}}{2^{\frac{\nu m}{2}} \Gamma_m\left(\frac{\nu}{2}\right)} \\
&\times e^{-\frac{1}{2} Trace\left\{(S^T S)\begin{bmatrix} 1 & 1 \\ -\alpha_2 & -\beta_2 \end{bmatrix}\Sigma^{-1}\begin{bmatrix} 1 & -\alpha_2 \\ 1 & -\beta_2 \end{bmatrix}\right\}} \quad .
\end{aligned}
$$

We see three of the probabilistic variables in this expression, namely Σ, α_2, and β_2.

We turn now to the embedded normal distribution which will be slightly more complex:

$$
\begin{aligned}
&N\left[vec(\Pi); vec(\bar{\Pi}), \Psi, mp\right] \\
&= (2\pi)^{-\frac{mp}{2}} |\Psi|^{-\frac{1}{2}} e^{-\frac{1}{2}\left[vec(\Pi) - vec(\bar{\Pi})\right]^T \Psi^{-1}\left[vec(\Pi) - vec(\bar{\Pi})\right]} \quad .
\end{aligned}
$$

The exponent of the normal can be written

$$\exp = \left[vec\left(\Pi\right) - vec\left(\bar{\Pi}\right)\right]^T \Psi^{-1} \left[vec\left(\Pi\right) - vec\left(\bar{\Pi}\right)\right] \quad .$$

We now need to work on the Π terms, which are defined

$$\Pi = \frac{1}{\alpha_2 - \beta_2} \begin{bmatrix} \beta_1\alpha_2 - \beta_2\alpha_1 & \beta_1 - \alpha_1 \\ -\alpha_3\beta_2 & -\alpha_3 \\ \alpha_2\beta_3 & \beta_3 \end{bmatrix} \quad .$$

If we take the vectorization, we can write

$$vec\left(\Pi\right) = \frac{1}{\alpha_2 - \beta_2} \begin{bmatrix} \beta_1\alpha_2 - \beta_2\alpha_1 \\ -\alpha_3\beta_2 \\ \alpha_2\beta_3 \\ \beta_1 - \alpha_1 \\ -\alpha_3 \\ \beta_3 \end{bmatrix} \quad .$$

The solution to the matrix equations that give the mean values yields the mean values as follows

$$vec\left(\bar{\Pi}\right) = \begin{bmatrix} \bar{\pi}_{d1} \\ \bar{\pi}_{d2} \\ \bar{\pi}_{d3} \\ \bar{\pi}_{s1} \\ \bar{\pi}_{s2} \\ \bar{\pi}_{s3} \end{bmatrix} \quad .$$

Therefore the premultiplier and postmultiplier multipliers are

$$\begin{aligned} & vec\left(\Pi\right) - vec\left(\bar{\Pi}\right) \\ &= \frac{1}{\alpha_2 - \beta_2} \begin{bmatrix} \beta_1\alpha_2 - \beta_2\alpha_1 \\ -\alpha_3\beta_2 \\ \alpha_2\beta_3 \\ \beta_1 - \alpha_1 \\ -\alpha_3 \\ \beta_3 \end{bmatrix} - \begin{bmatrix} \bar{\pi}_{d1} \\ \bar{\pi}_{d2} \\ \bar{\pi}_{d3} \\ \bar{\pi}_{s1} \\ \bar{\pi}_{s2} \\ \bar{\pi}_{s3} \end{bmatrix} \\ &= \begin{bmatrix} \frac{\beta_1\alpha_2 - \beta_2\alpha_1}{\alpha_2 - \beta_2} - \bar{\pi}_{d1} \\ \frac{-\alpha_3\beta_2}{\alpha_2 - \beta_2} - \bar{\pi}_{d2} \\ \frac{\alpha_2\beta_3}{\alpha_2 - \beta_2} - \bar{\pi}_{d3} \\ \frac{\beta_1 - \alpha_1}{\alpha_2 - \beta_2} - \bar{\pi}_{s1} \\ \frac{-\alpha_3}{\alpha_2 - \beta_2} - \bar{\pi}_{s2} \\ \frac{\beta_3}{\alpha_2 - \beta_2} - \bar{\pi}_{s3} \end{bmatrix} \quad . \end{aligned}$$

The exponent becomes

$$exp = \begin{bmatrix} \frac{\beta_1\alpha_2-\beta_2\alpha_1}{\alpha_2-\beta_2} - \bar{\pi}_{d1} \\ \frac{-\alpha_3\beta_2}{\alpha_2-\beta_2} - \bar{\pi}_{d2} \\ \frac{\alpha_2\beta_3}{\alpha_2-\beta_2} - \bar{\pi}_{d3} \\ \frac{\beta_1-\alpha_1}{\alpha_2-\beta_2} - \bar{\pi}_{s1} \\ \frac{-\alpha_3}{\alpha_2-\beta_2} - \bar{\pi}_{s2} \\ \frac{\beta_3}{\alpha_2-\beta_2} - \bar{\pi}_{s3} \end{bmatrix}^T \times \Psi^{-1} \times \begin{bmatrix} \frac{\beta_1\alpha_2-\beta_2\alpha_1}{\alpha_2-\beta_2} - \bar{\pi}_{d1} \\ \frac{-\alpha_3\beta_2}{\alpha_2-\beta_2} - \bar{\pi}_{d2} \\ \frac{\alpha_2\beta_3}{\alpha_2-\beta_2} - \bar{\pi}_{d3} \\ \frac{\beta_1-\alpha_1}{\alpha_2-\beta_2} - \bar{\pi}_{s1} \\ \frac{-\alpha_3}{\alpha_2-\beta_2} - \bar{\pi}_{s2} \\ \frac{\beta_3}{\alpha_2-\beta_2} - \bar{\pi}_{s3} \end{bmatrix} .$$

Substituting the inverse precision matrix

$$exp = \begin{bmatrix} \frac{\beta_1\alpha_2-\beta_2\alpha_1}{\alpha_2-\beta_2} - \bar{\pi}_{d1} \\ \frac{-\alpha_3\beta_2}{\alpha_2-\beta_2} - \bar{\pi}_{d2} \\ \frac{\alpha_2\beta_3}{\alpha_2-\beta_2} - \bar{\pi}_{d3} \\ \frac{\beta_1-\alpha_1}{\alpha_2-\beta_2} - \bar{\pi}_{s1} \\ \frac{-\alpha_3}{\alpha_2-\beta_2} - \bar{\pi}_{s2} \\ \frac{\beta_3}{\alpha_2-\beta_2} - \bar{\pi}_{s3} \end{bmatrix}^T \times \left[\Omega^{-1} \otimes (X^T X)\right] \times \begin{bmatrix} \frac{\beta_1\alpha_2-\beta_2\alpha_1}{\alpha_2-\beta_2} - \bar{\pi}_{d1} \\ \frac{-\alpha_3\beta_2}{\alpha_2-\beta_2} - \bar{\pi}_{d2} \\ \frac{\alpha_2\beta_3}{\alpha_2-\beta_2} - \bar{\pi}_{d3} \\ \frac{\beta_1-\alpha_1}{\alpha_2-\beta_2} - \bar{\pi}_{s1} \\ \frac{-\alpha_3}{\alpha_2-\beta_2} - \bar{\pi}_{s2} \\ \frac{\beta_3}{\alpha_2-\beta_2} - \bar{\pi}_{s3} \end{bmatrix} .$$

We need to work on the matrix Ψ as follows

$$\begin{aligned} \Psi^{-1} &= \Omega^{-1} \otimes \left(X^T X\right) \\ &= \left(\Gamma\Sigma^{-1}\Gamma^T\right) \otimes \left(X^T X\right) \\ &= \left(\begin{bmatrix} 1 & 1 \\ -\alpha_2 & -\beta_2 \end{bmatrix} \Sigma^{-1} \begin{bmatrix} 1 & -\alpha_2 \\ 1 & -\beta_2 \end{bmatrix}\right) \otimes \left(X^T X\right) \quad . \end{aligned}$$

We can substitute to see the final form of the exponent

$$exp = \begin{bmatrix} \frac{\beta_1\alpha_2-\beta_2\alpha_1}{\alpha_2-\beta_2} - \bar{\pi}_{d1} \\ \frac{-\alpha_3\beta_2}{\alpha_2-\beta_2} - \bar{\pi}_{d2} \\ \frac{\alpha_2\beta_3}{\alpha_2-\beta_2} - \bar{\pi}_{d3} \\ \frac{\beta_1-\alpha_1}{\alpha_2-\beta_2} - \bar{\pi}_{s1} \\ \frac{-\alpha_3}{\alpha_2-\beta_2} - \bar{\pi}_{s2} \\ \frac{\beta_3}{\alpha_2-\beta_2} - \bar{\pi}_{s3} \end{bmatrix}^T$$

$$\times \left\{ \begin{bmatrix} 1 & 1 \\ -\alpha_2 & -\beta_2 \end{bmatrix} \Sigma^{-1} \begin{bmatrix} 1 & -\alpha_2 \\ 1 & -\beta_2 \end{bmatrix} \otimes \left(X^T X\right) \right\}$$

$$\times \begin{bmatrix} \frac{\beta_1\alpha_2-\beta_2\alpha_1}{\alpha_2-\beta_2} - \bar{\pi}_{d1} \\ \frac{-\alpha_3\beta_2}{\alpha_2-\beta_2} - \bar{\pi}_{d2} \\ \frac{\alpha_2\beta_3}{\alpha_2-\beta_2} - \bar{\pi}_{d3} \\ \frac{\beta_1-\alpha_1}{\alpha_2-\beta_2} - \bar{\pi}_{s1} \\ \frac{-\alpha_3}{\alpha_2-\beta_2} - \bar{\pi}_{s2} \\ \frac{\beta_3}{\alpha_2-\beta_2} - \bar{\pi}_{s3} \end{bmatrix}.$$

We can certainly see all the disparate occurrences of the variables in the prior scattered throughout this exponent. We of course need the rest of the constant in the normal

$$\frac{\left|X^T X\right|^{\frac{m}{2}}}{(2\pi)^{\frac{pm}{2}} |\Sigma|^{\frac{p}{2}}} (\alpha_2 - \beta_2)^p \quad .$$

We append this constant to write the normal distribution beginning with equation (20.2)

$$N\left[vec\left(\Pi\right); vec\left(\bar{\Pi}\right), \Psi, mp\right]$$
$$= (2\pi)^{-\frac{mp}{2}} \frac{\left|X^T X\right|^{\frac{m}{2}} (\alpha_2 - \beta_2)^p}{|\Sigma|^{\frac{p}{2}}}$$

(20.5)

$$\times \exp \left\{ \begin{array}{l} -\frac{1}{2} \begin{bmatrix} \frac{\beta_1\alpha_2-\beta_2\alpha_1}{\alpha_2-\beta_2} - \bar{\pi}_{d1} \\ \frac{-\alpha_3\beta_2}{\alpha_2-\beta_2} - \bar{\pi}_{d2} \\ \frac{\alpha_2\beta_3}{\alpha_2-\beta_2} - \bar{\pi}_{d3} \\ \frac{\beta_1-\alpha_1}{\alpha_2-\beta_2} - \bar{\pi}_{s1} \\ \frac{-\alpha_3}{\alpha_2-\beta_2} - \bar{\pi}_{s2} \\ \frac{\beta_3}{\alpha_2-\beta_2} - \bar{\pi}_{s3} \end{bmatrix}^T \\ \times \left\{ \left(\begin{bmatrix} 1 & 1 \\ -\alpha_2 & -\beta_2 \end{bmatrix} \Sigma^{-1} \begin{bmatrix} 1 & -\alpha_2 \\ 1 & -\beta_2 \end{bmatrix} \right) \otimes (X^T X) \right\} \\ \times \begin{bmatrix} \frac{\beta_1\alpha_2-\beta_2\alpha_1}{\alpha_2-\beta_2} - \bar{\pi}_{d1} \\ \frac{-\alpha_3\beta_2}{\alpha_2-\beta_2} - \bar{\pi}_{d2} \\ \frac{\alpha_2\beta_3}{\alpha_2-\beta_2} - \bar{\pi}_{d3} \\ \frac{\beta_1-\alpha_1}{\alpha_2-\beta_2} - \bar{\pi}_{s1} \\ \frac{-\alpha_3}{\alpha_2-\beta_2} - \bar{\pi}_{s2} \\ \frac{\beta_3}{\alpha_2-\beta_2} - \bar{\pi}_{s3} \end{bmatrix} \end{array} \right\} .$$

We know that the likelihood function in equation (20.3) is the product of

- The constant evident in equation (20.3)
- Times the IW in equation (20.4)
- Times the Normal in equation (20.5).

We have just developed a detailed expression for the likelihood function that contains all the uncertain variables that comprise the prior

- $\alpha_1, \alpha_2, \alpha_3, \beta_1, \beta_2, \beta_3$.
- Σ .

We might, therefore, write this likelihood function, which is completely general for the fully identifiable case, in shorthand form

$$LF(\alpha_1, \alpha_2, \alpha_3, \beta_1, \beta_2, \beta_3, \Sigma) \quad .$$

This means that if we have a prior, any prior, that has the proper and correct domain for the model variables, we can multiply it by this likelihood function and simulate the complete answer.

$$\{\alpha_1, \alpha_2, \alpha_3, \beta_1, \beta_2, \beta_3, \Sigma\} \times LF(\alpha_1, \alpha_2, \alpha_3, \beta_1, \beta_2, \beta_3, \Sigma) \quad .$$

There is obviously no closed form for this. One simply cannot form any analytic marginals, nor is the posterior in the form of the prior. We leave simulation with this rather complicated functional form as an exercise to the reader. If we want a complete and general answer for any prior for the variables α_1, α_2, α_3, β_1, β_2, β_3, and Σ. We have a completely general posterior probability distribution here. Alas, it is not conjugate to the prior, is not integrable, and would have to be updated from scratch with a full simulation with every single new piece of data that comes in.

20.4 Return to the Reduced Form Equation

In this section, we return to the reduced form equations and see what can be done strictly with those equations, which are so simple and easy to apply directly as compared with the difficult simulation in the general case.

20.4.1 Begin with a Prior

In light of the structure of the likelihood function in equation (20.1), and all the analysis in Chapter 15 through 17, it is clear the form the prior is required to take if we are to maintain the conjugacy between prior and posterior in the reduced form equation. It has to take an analogous form as in equation (17.36), namely

$$\begin{aligned}\{\Omega, \Pi\} = & \frac{|M_0|^{\frac{\nu_0}{2}} |Q_0|^{\frac{m}{2}}}{(2\pi)^{\frac{pm}{2}} 2^{\frac{\nu_0 m}{2}} \Gamma_m\left(\frac{\nu_0}{2}\right) |\Omega|^{\frac{m_0}{2}}} \\ & \times e^{-\frac{1}{2} Trace\left(M_0 \Omega^{-1}\right)} \\ & \times e^{-\frac{1}{2} Trace\left[(\Pi - \Pi_0)^T Q_0 (\Pi - \Pi_0) \Omega^{-1}\right]} \quad .\end{aligned}$$

This can be specialized to the marginal distribution over the reduced form term Π as follows

$$\begin{aligned}\{\Pi | X, Y\} = & \frac{\Gamma_m\left(\frac{\nu_0 + p}{2}\right) |Q_0|^{\frac{m}{2}}}{\Gamma_m\left(\frac{\nu_0}{2}\right) \pi^{\frac{mp}{2}} |M_0|^{\frac{p}{2}}} \\ & \times \left| I_{mm} + M_0^{-1} (\Pi - \Pi_0)^T Q_0 \left(\Pi - \Pi_0\right) \right|^{-\frac{\nu_0 + p}{2}} \quad .\end{aligned}$$

The mean value of the coefficients is

$$B_0 \quad .$$

The variance-covariance matrix of the coefficients is

$$VCV = \frac{1}{\nu_0 - m - 1} M_0 \otimes {Q_0}^{-1} \quad .$$

20.4.2 A General Posterior for the Reduced Form Equation

Similarly, the posterior has to take an analogous form as in equation (17.9)

$$\{\Pi, \Omega | X, Y\} = \frac{|A|^{\frac{\nu}{2}} \left|X^T X + Q_0\right|^{\frac{m}{2}}}{(2\pi)^{\frac{pm}{2}} 2^{\frac{\nu m}{2}} \Gamma\left(\frac{\nu}{2}\right) |\Omega|^{\frac{m_0+n}{2}}} \times e^{-\frac{1}{2}Trace\left(A\Omega^{-1}\right)} \times e^{-\frac{1}{2}Trace\left[(\Pi-\Pi^*)^T (X^T X + Q_0)(\Pi-\Pi^*)\Omega^{-1}\right]} \quad ,$$

in which

$$A = \bar{S}^T \bar{S} + M_0 + (\Pi_0 - \Pi^*)^T Q_0 B_0 + (\bar{\Pi} - \Pi^*)^T X^T Y \quad ,$$

and

$$\Pi^* = (X^T X + Q_0)^{-1} (Q_0 B_0 + X^T Y) \quad .$$

This can be specialized to the marginal distribution over the reduced form term Π as follows

$$\{\Pi | X, Y\} = \frac{\Gamma_m\left(\frac{\nu+p}{2}\right) \left|X^T X + Q_0\right|^{\frac{m}{2}}}{\Gamma_m\left(\frac{\nu}{2}\right) \pi^{\frac{mp}{2}} |A|^{\frac{p}{2}}} \times \left|I_{mm} + A^{-1}(\Pi - \Pi^*)^T \left(X^T X + Q_0\right)(\Pi - \Pi^*)\right|^{-\frac{\nu+p}{2}} \quad .$$

The mean value of the coefficients is

$$B^* \quad .$$

The variance-covariance matrix of the coefficients is

$$VCV = \frac{1}{\nu - m - 1} A \otimes \left(X^T X + Q_0\right)^{-1} \quad .$$

We can, as Zellner[107] proposed, specify a prior and a posterior specifically for this reduced-form equation. In particular, Zellner has done so by assuming a diffuse prior and thereafter conducting classical regression analysis on the foregoing reduced form problem. The problem, quite succinctly, is that the variables in the reduced form problem are NOT the variables whose prior and posterior are being sought. This raises the "identification" problem addressed herein. We have given a general solution above, and we develop a few approximate approaches below.

20.4.3 A Posterior for the Reduced Form Equation Assuming a Diffuse Prior–Classical Regression

We can apply the general posterior with the informative prior in the previous section to the case of a diffuse prior. If we wish to use a diffuse prior, as Zellner[107] did, we could use a multidimensional diffuse prior.

$$|\Omega|^{-\frac{m+1}{2}} \quad .$$

This is what we do, de facto, if we choose classical regression at this stage. Given the complex relationship among the true prior variables, namely α_1, α_2, α_3, β_1, β_2, β_3, and Σ, there is no real ex ante justification for classical regression that involves a diffuse prior. (Even if the priors over all these variables were diffuse, the prior over the reduced form coefficients may not be.) If we use the foregoing prior, we can use the informative prior and make the following specializations to get the diffuse prior case:

$$\begin{aligned}
\Pi^* &= \bar{\Pi} = \left(X^T X\right)^{-1} X^T Y \\
A &= \bar{S}^T \bar{S} \\
m_0 &= m + 1 \\
\nu &= \nu_C = n - p \Rightarrow n = \nu_C + p \quad .
\end{aligned}$$

We can write the posterior joint density for this diffuse prior case as follows:

$$\begin{aligned}
\{\Pi, \Omega | X, Y\} = {} & \frac{\left|\bar{S}^T \bar{S}\right|^{\frac{\nu_C}{2}} \left|X^T X\right|^{\frac{m}{2}}}{(2\pi)^{\frac{pm}{2}} 2^{\frac{\nu_C m}{2}} \Gamma\left(\frac{\nu_C}{2}\right) |\Omega|^{\frac{\nu_C + p + m + 1}{2}}} e^{-\frac{1}{2}\left[\left(\bar{S}^T \bar{S}\right)\Omega^{-1}\right]} \\
& \times e^{-\frac{1}{2} Trace\left[\left(\Pi - \bar{\Pi}\right)^T \left(X^T X\right)\left(\Pi - \bar{\Pi}\right)\Omega^{-1}\right]} \quad .
\end{aligned}$$

This can be specialized to the marginal distribution over the reduced form term Π as follows

$$\begin{aligned}
\{\Pi | X, Y\} = {} & \frac{\Gamma_m\left(\frac{\nu_C + p}{2}\right) \left|X^T X\right|^{\frac{m}{2}}}{\Gamma_m\left(\frac{\nu_C}{2}\right) \pi^{\frac{mp}{2}} \left|\bar{S}^T \bar{S}\right|^{\frac{p}{2}}} \\
& \times \left| I_{mm} + \left(\bar{S}^T \bar{S}\right)^{-1} \left(\Pi - \bar{\Pi}\right)^T \left(X^T X\right) \left(\Pi - \bar{\Pi}\right) \right|^{-\frac{\nu_C + p}{2}} \quad .
\end{aligned}$$

The mean value of the coefficients is

$$\bar{B} \quad .$$

The variance-covariance matrix of the coefficients is

$$VCV = \frac{1}{\nu_C - m - 1} \left(S^T S\right) \otimes \left(X^T X\right)^{-1} \quad .$$

20.5 Transformation of Variables from Original Coefficients to Π

It is important to consider whether there is an inverse for the foregoing transformation, which will be needed to allow us to make a transformation from the original coefficients $\alpha_1, \alpha_2, \alpha_3, \beta_1, \beta_2, \beta_3$ to the reduced coefficients Π. Following are all six equalities developed above in which we substituted Δ for the determinant of the Γ matrix and moved it out of the denominator and into the numerator on the other side after we equated the solution matrix Π with these six coefficients: $\alpha_1, \alpha_2, \alpha_3, \beta_1, \beta_2, \beta_3$

$$\beta_1 \alpha_2 - \beta_2 \alpha_1 = \pi_{d1} \Delta \quad ,$$

$$\alpha_3 \beta_2 = -\pi_{d2} \Delta \quad ,$$

$$\alpha_2 \beta_3 = \pi_{d3} \Delta \quad ,$$

$$\beta_1 - \alpha_1 = \pi_{s1} \Delta \quad ,$$

(20.6)
$$\alpha_3 = -\pi_{s2} \Delta \quad \text{, and}$$

(20.7)
$$\beta_3 = \pi_{s3} \Delta \quad .$$

We substitute from equations (20.6) and (20.7) into the first four equations to write them in the following forms

$$\begin{aligned}
&\beta_1 \alpha_2 - \beta_2 \alpha_1 = \pi_{d1} \Delta \\
&\beta_2 = \frac{\pi_{d2}}{\pi_{s2}} \\
&\alpha_2 = \frac{\pi_{d3}}{\pi_{s3}} \\
&\beta_1 - \alpha_1 = \pi_{s1} \Delta \\
&\beta_3 = \pi_{s2} \Delta \\
&\alpha_3 = \pi_{s3} \Delta \quad .
\end{aligned}$$

The second and third equations give us the final expression for the determinant

$$\Delta = \alpha_2 - \beta_2 = \frac{\pi_{d3}}{\pi_{s3}} - \frac{\pi_{d2}}{\pi_{s2}} = \frac{\pi_{d3}\pi_{s2} - \pi_{d2}\pi_{s3}}{\pi_{s2}\pi_{s3}} \quad .$$

We substitute this definition of the determinant as well as the β_2 and β_2 constants in all equations previously to write

$$\begin{aligned}
&\beta_1\frac{\pi_{d3}}{\pi_{s3}} - \frac{\pi_{d2}}{\pi_{s2}}\alpha_1 = \pi_{d1}\frac{\pi_{d3}\pi_{s2} - \pi_{d2}\pi_{s3}}{\pi_{s2}\pi_{s3}} \\
&\beta_2 = \frac{\pi_{d2}}{\pi_{s2}} \\
&\alpha_2 = \frac{\pi_{d3}}{\pi_{s3}} \\
&\beta_1 - \alpha_1 = \pi_{s1}\frac{\pi_{d3}\pi_{s2} - \pi_{d2}\pi_{s3}}{\pi_{s2}\pi_{s3}} \\
&\alpha_3 = -\frac{\pi_{d3}\pi_{s2} - \pi_{d2}\pi_{s3}}{\pi_{s3}} \\
&\beta_3 = \frac{\pi_{d3}\pi_{s2} - \pi_{d2}\pi_{s3}}{\pi_{s2}} \quad .
\end{aligned}$$

We now rearrange the fourth equation so that we can substitute it into the first equation for β_1

$$\beta_1 = \alpha_1 + \pi_{s1}\frac{\pi_{d3}\pi_{s2} - \pi_{d2}\pi_{s3}}{\pi_{s2}\pi_{s3}} \quad .$$

We can substitute for β_1 and thereafter solve for α_1

$$\begin{aligned}
&\left(\alpha_1 + \pi_{s1}\frac{\pi_{d3}\pi_{s2} - \pi_{d2}\pi_{s3}}{\pi_{s2}\pi_{s3}}\right)\frac{\pi_{d3}}{\pi_{s3}} - \alpha_1\frac{\pi_{d2}}{\pi_{s2}} = \pi_{d1}\frac{\pi_{d3}\pi_{s2} - \pi_{d2}\pi_{s3}}{\pi_{s2}\pi_{s3}} \\
&\Rightarrow \alpha_1\frac{\pi_{d3}}{\pi_{s3}} - \alpha_1\frac{\pi_{d2}}{\pi_{s2}} = \pi_{d1}\frac{\pi_{d3}\pi_{s2} - \pi_{d2}\pi_{s3}}{\pi_{s2}\pi_{s3}} \\
&\qquad\qquad - \pi_{s1}\frac{\pi_{d3}}{\pi_{s3}}\frac{\pi_{d3}\pi_{s2} - \pi_{d2}\pi_{s3}}{\pi_{s2}\pi_{s3}} \\
&\Rightarrow \alpha_1\left(\frac{\pi_{d3}}{\pi_{s3}} - \frac{\pi_{d2}}{\pi_{s2}}\right) = \pi_{d1}\left(\frac{\pi_{d3}\pi_{s2} - \pi_{d2}\pi_{s3}}{\pi_{s2}\pi_{s3}}\right) \\
&\qquad\qquad - \pi_{s1}\frac{\pi_{d3}}{\pi_{s3}}\left(\frac{\pi_{d3}\pi_{s2} - \pi_{d2}\pi_{s3}}{\pi_{s2}\pi_{s3}}\right) \\
&\Rightarrow \alpha_1\left(\frac{\pi_{d3}\pi_{s2} - \pi_{d2}\pi_{s3}}{\pi_{s2}\pi_{s3}}\right) = \pi_{d1}\left(\frac{\pi_{d3}\pi_{s2} - \pi_{d2}\pi_{s3}}{\pi_{s2}\pi_{s3}}\right) \\
&\qquad\qquad - \pi_{s1}\frac{\pi_{d3}}{\pi_{s3}}\left(\frac{\pi_{d3}\pi_{s2} - \pi_{d2}\pi_{s3}}{\pi_{s2}\pi_{s3}}\right) \\
&\Rightarrow \alpha_1 = \pi_{d1} - \pi_{s1}\frac{\pi_{d3}}{\pi_{s3}} = \frac{\pi_{d1}\pi_{s3} - \pi_{d3}\pi_{s1}}{\pi_{s3}} \quad .
\end{aligned}$$

We substitute this into the fourth equation to solve for β_1

$$\beta_1 - \frac{\pi_{d1}\pi_{s3} - \pi_{d3}\pi_{s1}}{\pi_{s3}} = \pi_{s1}\frac{\pi_{d3}\pi_{s2} - \pi_{d2}\pi_{s3}}{\pi_{s2}\pi_{s3}}$$

$$
\begin{aligned}
&\Rightarrow \beta_1 = \pi_{s1}\frac{\pi_{d3}\pi_{s2} - \pi_{d2}\pi_{s3}}{\pi_{s2}\pi_{s3}} + \frac{\pi_{d1}\pi_{s3} - \pi_{d3}\pi_{s1}}{\pi_{s3}} \\
&\Rightarrow \beta_1 = \frac{\pi_{d3}\pi_{s1}\pi_{s2} - \pi_{d2}\pi_{s1}\pi_{s3}}{\pi_{s2}\pi_{s3}} + \frac{\pi_{d1}\pi_{s2}\pi_{s3} - \pi_{d3}\pi_{s1}\pi_{s2}}{\pi_{s2}\pi_{s3}} \\
&\Rightarrow \beta_1 = \frac{-\pi_{d2}\pi_{s1}\pi_{s3}}{\pi_{s2}\pi_{s3}} + \frac{\pi_{d1}\pi_{s2}\pi_{s3}}{\pi_{s2}\pi_{s3}} = \pi_{s3}\left(\frac{\pi_{d1}\pi_{s2}}{\pi_{s2}\pi_{s3}} - \frac{\pi_{d2}\pi_{s1}}{\pi_{s2}\pi_{s3}}\right) \\
&= \frac{\pi_{d1}\pi_{s2} - \pi_{d2}\pi_{s1}}{\pi_{s2}} \quad .
\end{aligned}
$$

We have thus solved for the six coefficients $\alpha_1, \alpha_2, \alpha_3, \beta_1, \beta_2, \beta_3$ we want as follows

$$
\begin{aligned}
\alpha_2 &= \frac{\pi_{d3}}{\pi_{s3}} \\
\alpha_1 &= \frac{\pi_{d1}\pi_{s3} - \pi_{d3}\pi_{s1}}{\pi_{s3}} = \pi_{d1} - \pi_{s1}\frac{\pi_{d3}}{\pi_{s3}} = \pi_{d1} - \pi_{s1}\alpha_2 \\
\alpha_3 &= \frac{\pi_{d2}\pi_{s3} - \pi_{d3}\pi_{s2}}{\pi_{s3}} = \pi_{d2} - \pi_{s2}\frac{\pi_{d3}}{\pi_{s3}} = \pi_{d2} - \pi_{s2}\alpha_2 \\
\beta_2 &= \frac{\pi_{d2}}{\pi_{s2}} \\
\beta_1 &= \frac{\pi_{d1}\pi_{s2} - \pi_{d2}\pi_{s1}}{\pi_{s2}} = \pi_{d1} - \frac{\pi_{d2}\pi_{s1}}{\pi_{s2}} = \pi_{d1} - \pi_{s1}\beta_2 \\
\beta_3 &= \frac{\pi_{d3}\pi_{s2} - \pi_{d2}\pi_{s3}}{\pi_{s2}} = \pi_{d3} - \pi_{s3}\beta_2 \quad .
\end{aligned}
$$

The fact that we can solve for the six coefficients individually is a statement that the six coefficients are "identifiable," or more accurately, "exactly identifiable." This is not at all serendipitous (nor generally true); the fact that there are six "slots" in the Π matrix and there are simultaneously six coefficients to be estimated is totally a function of the specific equation structures put forth in the original regression equations. Different equation structures have no guarantee of leading to this result. This is the "identification" condition we want for Bayesian Linear Regression analysis.

20.6 Transformation of Variables from Original Coefficients to Reduced Form Variables Π

We can develop the Jacobian matrix for the transformation from the coefficients $\alpha_1, \alpha_2, \alpha_3, \beta_1, \beta_2, \beta_3$ to the Π variables

$$\begin{bmatrix} & \partial\pi_{d1} & \partial\pi_{d2} & \partial\pi_{d3} & \partial\pi_{s1} & \partial\pi_{s2} & \partial\pi_{s3} \\ \partial\alpha_1 & 1 & 0 & -\frac{\pi_{s1}}{\pi_{s3}} & -\frac{\pi_{d3}}{\pi_{s3}} & 0 & \frac{\pi_{d3}\pi_{s1}}{\pi_{s3}^2} \\ \partial\alpha_2 & 0 & 0 & \frac{1}{\pi_{s3}} & 0 & 0 & -\frac{\pi_{d3}}{\pi_{s3}^2} \\ \partial\alpha_3 & 0 & 1 & -\frac{\pi_{s2}}{\pi_{s3}} & 0 & -\frac{\pi_{d3}}{\pi_{s3}} & \frac{\pi_{d3}\pi_{s2}}{\pi_{s3}^2} \\ \partial\beta_1 & 1 & -\frac{\pi_{s1}}{\pi_{s2}} & 0 & -\frac{\pi_{d2}}{\pi_{s2}} & \frac{\pi_{d2}\pi_{s1}}{\pi_{s2}^2} & 0 \\ \partial\beta_2 & 0 & \frac{1}{\pi_{s2}} & 0 & 0 & -\frac{\pi_{d2}}{\pi_{s2}^2} & 0 \\ \partial\beta_3 & 0 & -\frac{\pi_{s3}}{\pi_{s2}} & 1 & 0 & \frac{\pi_{d2}\pi_{s3}}{\pi_{s2}^2} & -\frac{\pi_{d2}}{\pi_{s2}} \end{bmatrix} .$$

The determinant of this Jacobian that would allow us to move from the distribution over the coefficients $\alpha_1, \alpha_2, \alpha_3, \beta_1, \beta_2, \beta_3$ to the distribution over Π is calculated, with great effort, to take a surprisingly simple form

$$\text{determinant} = \frac{\left(\pi_{d2}\pi_{s3} - \pi_{d3}\pi_{s2}\right)^3}{\pi_{s2}{}^4\pi_{s3}{}^4} .$$

If we began with a distribution over the original model coefficients

$$f_C(\alpha_1, \alpha_2, \alpha_3, \beta_1, \beta_2, \beta_3) \quad ,$$

we could perform the following change of variables calculation to obtain the distribution over the Π variables

$$\begin{aligned} & f_\Pi\left(\pi_{d1}, \pi_{d2}, \pi_{d3}, \pi_{s1}, \pi_{s2}, \pi_{s3}\right) \\ & = f_C\left(\begin{array}{c} \frac{\pi_{d1}\pi_{s3}-\pi_{d3}\pi_{s1}}{\pi_{s3}}, \frac{\pi_{d3}}{\pi_{s3}}, \frac{\pi_{d2}\pi_{s3}-\pi_{d3}\pi_{s2}}{\pi_{s3}}, \\ \frac{\pi_{d1}\pi_{s2}-\pi_{d2}\pi_{s1}}{\pi_{s2}}, \frac{\pi_{d2}}{\pi_{s2}}, \frac{\pi_{d3}\pi_{s2}-\pi_{d2}\pi_{s3}}{\pi_{s2}} \end{array}\right) \\ & \times \frac{1}{\pi_{s2}\pi_{s3}}\left(\frac{\pi_{d2}\pi_{s3} - \pi_{d3}\pi_{s2}}{\pi_{s2}\pi_{s3}}\right)^3 . \end{aligned}$$

As an example, if we were to postulate a 6-variable lognormal prior distribution over the six actual, original coefficients $\alpha_1, \alpha_2, \alpha_3, \beta_1, \beta_2, \beta_3$, substitute the foregoing π variables into that lognormal prior, and multiply by this Jacobian determinant, we would have an explicit analytical expression for the probability distribution over the Π vector. This would be an explicit analytical form for the prior. It is clear that this will NOT be a multivariate normal or lognormal distribution. It is not a named distribution at all, and its mathematics are not analytically tractable.

There is no way that this transformation to the Π variables will lead to the prior in the form we want to see it, namely the conjugate prior as expressed in equation (16.81). The foregoing transformed probability distribution multiplied by an exponential Jacobian is simply not going to look like the desired conjugate prior in equation (16.81). One would have no choice but to simulate the actual distribution over the Π terms and use means and variances to "fill up" the data elements in a prior in the form of the equation (16.81) as an approximation. Even with a fully unique relationship from the original model variables to the reduced model variables, you still cannot get a conjugate form. You cannot even get an analytical form.

This procedure is not the same as the general transformations with which we began this chapter. That development was fully general and did not rely on interim transformations from $\alpha_1, \alpha_2, \alpha_3, \beta_1, \beta_2, \beta_3$ to Π.

20.7 Transformation of Variables from Π to Original Coefficients

If we wanted to start with a distribution over the π variables and transform to a distribution over the coefficients $\alpha_1, \alpha_2, \alpha_3, \beta_1, \beta_2, \beta_3$, the Jacobian matrix whose determinant we would seek is

$$\begin{bmatrix} & \partial\alpha_1 & \partial\alpha_2 & \partial\alpha_3 & \partial\beta_1 & \partial\beta_2 & \partial\beta_3 \\ \partial\pi_{d1} & -\beta_2\Delta & \beta_2(\alpha_1-\beta_1) & 0 & \alpha_2\Delta & -\alpha_2(\alpha_1-\beta_1) & 0 \\ \partial\pi_{d2} & 0 & \alpha_3\beta_2 & -\beta_2\Delta & 0 & -\alpha_2\alpha_3 & 0 \\ \partial\pi_{d3} & 0 & -\beta_3\beta_2 & 0 & 0 & \alpha_2\beta_3 & \alpha_2\Delta \\ \partial\pi_{s1} & -\Delta & -(\beta_1-\alpha_1) & 0 & \Delta & \beta_1-\alpha_1 & 0 \\ \partial\pi_{s2} & 0 & \alpha_3 & -\Delta & 0 & -\alpha_3 & 0 \\ \partial\pi_{s3} & 0 & -\beta_3 & 0 & 0 & \beta_3 & \Delta \end{bmatrix} \frac{1}{\Delta^2} \quad .$$

With a great deal of algebra, we can show that the determinant of this matrix is

$$\text{Determinant} = \alpha_3\beta_3(\alpha_2-\beta_2)^5 \quad .$$

This is what we would have to multiply the distribution over Π by to get the distribution over the coefficients. In particular, if we began with a distribution

$$f_\Pi(\pi_{d1}, \pi_{d2}, \pi_{d3}, \pi_{s1}, \pi_{s2}, \pi_{s3})$$

we would transform to a distribution over the model coefficients as follows

$$\begin{aligned} & f_C(\alpha_1, \alpha_2, \alpha_3, \beta_1, \beta_2, \beta_3) \\ & = f_\Pi\left(\frac{\alpha_2\beta_1-\alpha_1\beta_2}{\alpha_2-\beta_2}, \frac{-\alpha_3\beta_2}{\alpha_2-\beta_2}, \frac{\alpha_2\beta_3}{\alpha_2-\beta_2}, \frac{\beta_1-\alpha_1}{\alpha_2-\beta_2}, \frac{-\alpha_3}{\alpha_2-\beta_2}, \frac{\beta_3}{\alpha_2-\beta_2}\right) \\ & \times \alpha_3\beta_3(\alpha_2-\beta_2)^5 \quad . \end{aligned}$$

This would be a closed form, equational representation of the distribution over the coefficients. For any representation of the distribution f_Π one might postulate, this would be a complex functional form, but it could be simulated.

20.8 Bayesian Analysis of the Exactly Identified Case

We immediately begin to sense the problem with Bayesian analysis for the exactly identified case. Assuredly, the reduced-form equation is easily and directly amenable to Bayesian analysis. One can formulate a prior $\{\Pi, \Omega\}$ expressed in terms of this reduced form equation and use the likelihood function to calculate a posterior $\{\Pi, \Omega | X, Y\}$ expressed in terms of this reduced form equation. This is very easy and is a special case of the analysis in Chapter 16. The calculations are precisely those in Chapters 15 through 17.

We saw in Chapters 15 through 17, however, that everything would work out just fine if we could specify the prior in terms of the Π coefficients and use the likelihood and conjugacy properties to get the posterior in terms of the Π coefficients, i.e., use the reduced form expression for the prior, the likelihood function, and the posterior. In other words, using the reduced form variables Π, we could conduct standard, linear, conjugate Bayesian updating as in Chapters 15 through 17. We could simulate to get the equivalent prior expressed in terms of the original coefficients, and we could simulate to get the posterior expressed in terms of the original coefficients. As long as we are fully and exactly identifiable, this type of calculation will always work. This is the approach Zellner[107] used, and he used only the diffuse prior in the reduced form. Alas, it is not fully general. A diffuse prior in the reduced form is NOT a diffuse prior in the original form.

However, can we make it a bit more general, a "halfway house" to full generality, by performing two change of variables substitutions, one for the prior and one for the posterior? Not quite, as we will now show. We can find coherent functional forms for the prior and posterior distributions over the original model coefficients, but they will not be analytically tractable. They will have to be solved by simulation, as we shall now illustrate.

20.8.1 Halfway House–Change of Variables from Reduced Form to Original Coefficients

What if we were to change the variables to the original coefficients according to the fully identified equations developed previously? That is, we "back calculate" the implicit prior in terms of the actual coefficients themselves rather than the reduced form variables using a nonlinear change of variables calculation. We examine that briefly here. If we begin with the prior immediately above and integrate out the Ω term,[5] We obtain the marginal distribution over the Π variables as follows:

[5] We will not carry the Ω term here. It will bog down the calculation. Because the joint probability distribution is a product of independent distributions over the Omega and the model coefficients, nothing will be sacrificed by this simplification.

$$\{\Pi\} = \frac{\Gamma_m\left(\frac{\nu_0+p}{2}\right)}{\Gamma_m\left(\frac{\nu_0}{2}\right)} \frac{|Q_0|^{\frac{m}{2}}}{\pi^{\frac{pm}{2}}|M_0|^{\frac{p}{2}}}$$
$$\left|I_{mm} + M_0{}^{-1}(\Pi - \Pi_0)^T Q_0\left(\Pi - \Pi_0\right)\right|^{-\frac{\nu_0+p}{2}} .$$

We want to change variables to the vector of original coefficients. We do so by first writing the equation in expanded form

$$\{\Pi\} = \frac{\Gamma_m\left(\frac{\nu_0+p}{2}\right)}{\Gamma_m\left(\frac{\nu_0}{2}\right)} \frac{|Q_0|^{\frac{m}{2}}}{\pi^{\frac{pm}{2}}|M_0|^{\frac{p}{2}}}$$
$$\left|\begin{array}{l} I_{mm} \\ +M_0{}^{-1}\left(\begin{bmatrix} \pi_{d1} & \pi_{s1} \\ \pi_{d2} & \pi_{s2} \\ \pi_{d3} & \pi_{s3} \end{bmatrix} - \begin{bmatrix} \mu_{d1} & \mu_{s1} \\ \mu_{d2} & \mu_{s2} \\ \mu_{d3} & \mu_{s3} \end{bmatrix}\right)^T \\ \times Q_0\left(\begin{bmatrix} \pi_{d1} & \pi_{s1} \\ \pi_{d2} & \pi_{s2} \\ \pi_{d3} & \pi_{s3} \end{bmatrix} - \begin{bmatrix} \mu_{d1} & \mu_{s1} \\ \mu_{d2} & \mu_{s2} \\ \mu_{d3} & \mu_{s3} \end{bmatrix}\right) \end{array}\right|^{-\frac{\nu_0+p}{2}} ,$$

in which the μ terms are the respective terms of $(X^TX)^{-1}X^TY$. We then substitute the expressions for the Π elements in terms of the original model coefficients, and at the end of the substitution, multiply by the determinant of the Jacobian to write the derived distribution over the original model coefficient

$$\{\alpha_1, \alpha_2, \alpha_3, \beta_1, \beta_2, \beta_3\} = \frac{\Gamma_m\left(\frac{\nu_0+p}{2}\right)}{\Gamma_m\left(\frac{\nu_0}{2}\right)} \frac{|Q_0|^{\frac{m}{2}}}{\pi^{\frac{pm}{2}}|M_0|^{\frac{p}{2}}}$$
$$\times\left|\begin{array}{l} I_{mm} \\ +M_0{}^{-1}\left(\begin{bmatrix} \frac{\alpha_2\beta_1-\alpha_1\beta_2}{\alpha_2-\beta_2} & \frac{\beta_1-\alpha_1}{\alpha_2-\beta_2} \\ \frac{-\alpha_3\beta_2}{\alpha_2-\beta_2} & \frac{-\alpha_3}{\alpha_2-\beta_2} \\ \frac{\alpha_2\beta_3}{\alpha_2-\beta_2} & \frac{\beta_3}{\alpha_2-\beta_2} \end{bmatrix} - \begin{bmatrix} \mu_{d1} & \mu_{s1} \\ \mu_{d2} & \mu_{s2} \\ \mu_{d3} & \mu_{s3} \end{bmatrix}\right)^T \\ \times Q_0\left(\begin{bmatrix} \frac{\alpha_2\beta_1-\alpha_1\beta_2}{\alpha_2-\beta_2} & \frac{\beta_1-\alpha_1}{\alpha_2-\beta_2} \\ \frac{-\alpha_3\beta_2}{\alpha_2-\beta_2} & \frac{-\alpha_3}{\alpha_2-\beta_2} \\ \frac{\alpha_2\beta_3}{\alpha_2-\beta_2} & \frac{\beta_3}{\alpha_2-\beta_2} \end{bmatrix} - \begin{bmatrix} \mu_{d1} & \mu_{s1} \\ \mu_{d2} & \mu_{s2} \\ \mu_{d3} & \mu_{s3} \end{bmatrix}\right) \end{array}\right|^{-\frac{\nu_0+p}{2}}$$
$$\times\ \alpha_3\beta_3(\alpha_2 - \beta_2)^{-5} \quad .$$

This is a somewhat complex representation of the implied probability distribution over the original model coefficients, but we can certainly simulate it.

We can make exactly the same calculations for the posterior, which results in the following functional form for the posterior over the original model coefficients: This is the sought-after answer, but it is not analytically tractable. We can use it to simulate, however

$$\{\alpha_1, \alpha_2, \alpha_3, \beta_1, \beta_2, \beta_3\} = \frac{\Gamma_m\left(\frac{\nu+p}{2}\right)}{\Gamma_m\left(\frac{\nu}{2}\right)} \frac{\left|Q_0 + X^T X\right|^{\frac{m}{2}}}{\pi^{\frac{pm}{2}} |A|^{\frac{p}{2}}}$$

$$\times \left| \begin{matrix} I_{mm} \\ +A^{-1}\left(\begin{bmatrix} \frac{\alpha_2\beta_1-\alpha_1\beta_2}{\alpha_2-\beta_2} & \frac{\beta_1-\alpha_1}{\alpha_2-\beta_2} \\ \frac{-\alpha_3\beta_2}{\alpha_2-\beta_2} & \frac{-\alpha_3}{\alpha_2-\beta_2} \\ \frac{\alpha_2\beta_3}{\alpha_2-\beta_2} & \frac{\beta_3}{\alpha_2-\beta_2} \end{bmatrix} - \begin{bmatrix} \mu_{d1}{}^* & \mu_{s1}{}^* \\ \mu_{d2}{}^* & \mu_{s2}{}^* \\ \mu_{d3}{}^* & \mu_{s3}{}^* \end{bmatrix}\right)^T \\ \times \left(Q_0 + X^T X\right) \left(\begin{bmatrix} \frac{\alpha_2\beta_1-\alpha_1\beta_2}{\alpha_2-\beta_2} & \frac{\beta_1-\alpha_1}{\alpha_2-\beta_2} \\ \frac{-\alpha_3\beta_2}{\alpha_2-\beta_2} & \frac{-\alpha_3}{\alpha_2-\beta_2} \\ \frac{\alpha_2\beta_3}{\alpha_2-\beta_2} & \frac{\beta_3}{\alpha_2-\beta_2} \end{bmatrix} - \begin{bmatrix} \mu_{d1}{}^* & \mu_{s1}{}^* \\ \mu_{d2}{}^* & \mu_{s2}{}^* \\ \mu_{d3}{}^* & \mu_{s3}{}^* \end{bmatrix}\right) \end{matrix} \right|^{-\frac{\nu+p}{2}}$$

$$\times \alpha_3\beta_3(\alpha_2 - \beta_2)^5 \quad .$$

20.9 A Second Example–An Under-Identifiable (and Therefore Not Identifiable) Case

Suppose we make a seemingly small change in the previous model. We eliminate the FP variable in the former supply equation, and we assume in its place that the supply equation is also a function of disposable income DI. In such a situation, the regression equations would become

$$\begin{aligned} q_i - \alpha_2 p_i - \alpha_1 - \alpha_3 DI_i &= \varepsilon_{di} \quad i = 1, ..., n \\ q_i - \beta_2 p_i - \beta_1 - \beta_3 DI_i &= \varepsilon_{si} \quad i = 1, ..., n \quad . \end{aligned}$$

We would write them in matrix form

$$\begin{bmatrix} 1 & -\alpha_2 \\ 1 & -\beta_2 \end{bmatrix} \begin{bmatrix} q_i \\ p_i \end{bmatrix} - \begin{bmatrix} \alpha_1 & \alpha_3 \\ \beta_1 & \beta_3 \end{bmatrix} \begin{bmatrix} 1 \\ DI_i \end{bmatrix} = \begin{bmatrix} \varepsilon_{di} \\ \varepsilon_{si} \end{bmatrix} \qquad i = 1, ..., n \quad .$$

We take the transpose to write

$$\begin{bmatrix} q_i & p_i \end{bmatrix} \begin{bmatrix} 1 & 1 \\ -\alpha_2 & -\beta_2 \end{bmatrix} - \begin{bmatrix} 1 & DI_i \end{bmatrix} \begin{bmatrix} \alpha_1 & \beta_1 \\ \alpha_3 & \beta_3 \end{bmatrix} = \begin{bmatrix} \varepsilon_{di} & \varepsilon_{si} \end{bmatrix} \quad .$$

This allows us to add and account for all the observations (data)

$$\begin{bmatrix} q_1 & p_1 \\ \cdot & \cdot \\ \cdot & \cdot \\ \cdot & \cdot \\ q_n & p_n \end{bmatrix} \begin{bmatrix} 1 & 1 \\ -\alpha_2 & -\beta_2 \end{bmatrix} - \begin{bmatrix} 1 & DI_1 \\ \cdot & \cdot \\ \cdot & \cdot \\ \cdot & \cdot \\ 1 & DI_n \end{bmatrix} \begin{bmatrix} \alpha_1 & \beta_1 \\ \alpha_3 & \beta_3 \end{bmatrix}$$

$$= \begin{bmatrix} \varepsilon_{d1} & \varepsilon_{s1} \\ \cdot & \cdot \\ \cdot & \cdot \\ \cdot & \cdot \\ \varepsilon_{dn} & \varepsilon_{sn} \end{bmatrix} .$$

We calculate the inverse

$$\begin{bmatrix} 1 & 1 \\ -\alpha_2 & -\beta_2 \end{bmatrix}^{-1} = \begin{bmatrix} -\beta_2 & -1 \\ \alpha_2 & 1 \end{bmatrix} \frac{1}{\alpha_2 - \beta_2} = \begin{bmatrix} -\frac{\beta_2}{\alpha_2-\beta_2} & -\frac{1}{\alpha_2-\beta_2} \\ \frac{\alpha_2}{\alpha_2-\beta_2} & \frac{1}{\alpha_2-\beta_2} \end{bmatrix} .$$

Multiplying through on the right by this inverse

$$\begin{bmatrix} q_1 & p_1 \\ \\ \\ q_n & p_n \end{bmatrix} - \begin{bmatrix} 1 & DI_1 \\ \\ \\ 1 & DI_n \end{bmatrix} \begin{bmatrix} \alpha_1 & \beta_1 \\ \alpha_3 & \beta_3 \end{bmatrix} \begin{bmatrix} 1 & 1 \\ -\alpha_2 & -\beta_2 \end{bmatrix}^{-1}$$

$$= \begin{bmatrix} \varepsilon_{d1} & \varepsilon_{s1} \\ \\ \\ \varepsilon_{dn} & \varepsilon_{sn} \end{bmatrix} \begin{bmatrix} 1 & 1 \\ -\alpha_2 & -\beta_2 \end{bmatrix}^{-1} .$$

We need to expand the matrix product in the middle, which is where we shall see the Π matrix

$$\begin{bmatrix} q_1 & p_1 \\ \cdot & \cdot \\ \cdot & \cdot \\ \cdot & \cdot \\ q_n & p_n \end{bmatrix} - \begin{bmatrix} 1 & DI_1 \\ \cdot & \cdot \\ \cdot & \cdot \\ \cdot & \cdot \\ 1 & DI_n \end{bmatrix} \begin{bmatrix} \frac{\alpha_2\beta_1-\alpha_1\beta_2}{\alpha_2-\beta_2} & \frac{-\alpha_1+\beta_1}{\alpha_2-\beta_2} \\ \frac{\alpha_2\beta_3-\alpha_3\beta_2}{\alpha_2-\beta_2} & \frac{-\alpha_3+\beta_3}{\alpha_2-\beta_2} \end{bmatrix}$$

$$= \begin{bmatrix} \varepsilon_{d1} & \varepsilon_{s1} \\ \cdot & \cdot \\ \cdot & \cdot \\ \cdot & \cdot \\ \varepsilon_{dn} & \varepsilon_{sn} \end{bmatrix} \begin{bmatrix} 1 & 1 \\ -\alpha_2 & -\beta_2 \end{bmatrix}^{-1} .$$

The matrix in the middle is of course the solution for the Π matrix

$$\begin{bmatrix} \frac{\alpha_2\beta_1-\alpha_1\beta_2}{\alpha_2-\beta_2} & \frac{-\alpha_1+\beta_1}{\alpha_2-\beta_2} \\ \frac{\alpha_2\beta_3-\alpha_3\beta_2}{\alpha_2-\beta_2} & \frac{-\alpha_3+\beta_3}{\alpha_2-\beta_2} \end{bmatrix} .$$

And now we get to the "identification" problem. If we examine this Π matrix, we see that it is comprised of six model constants inside a 2×2 matrix that has only four cells. There are going to be four equalities with the four elements of the Π matrix, which we have calculated as before as

$$\bar{\Pi} = (X^T X) X^T Y$$

in the regression, but there are six numbers we need. We need six numbers, but we only get four from the regression. Two of the constants are absent and are therefore non-identifiable. We cannot solve these regression equations and identify all six of the constants. The reason this occurs, from a geometric perspective, dates back to the original example. Recall that we started with a 2×3 matrix B, which is the ultimate size of the Π matrix.

Let us begin to make the substitutions to solve this problem and see how far we get

$$\begin{aligned}
\frac{\alpha_2\beta_1 - \alpha_1\beta_2}{\Delta} &= x_{d1} \Rightarrow \alpha_2\beta_1 - \alpha_1\beta_2 = \Delta x_{d1} \\
\frac{\alpha_2\beta_3 - \alpha_3\beta_2}{\Delta} &= x_{d2} \Rightarrow \alpha_2\beta_3 - \alpha_3\beta_2 = \Delta x_{d2} \\
\frac{-\alpha_1 + \beta_1}{\Delta} &= x_{s1} \Rightarrow \beta_1 = \Delta x_{s1} + \alpha_1 \\
\frac{-\alpha_3 + \beta_3}{\Delta} &= x_{s2} \Rightarrow \beta_3 = \Delta x_{s2} + \alpha_3 \quad .
\end{aligned}$$

Substituting the third and fourth equations for β_1 and β_3 turns the first two equations into the following

$$\begin{aligned}
\alpha_2 (\Delta x_{s1} + \alpha_1) - \alpha_1\beta_2 = \Delta x_{d1} \Rightarrow \alpha_1 = x_{d1} - \alpha_2 x_{s1} \\
\alpha_2 (\Delta x_{s2} + \alpha_3) - \alpha_3\beta_2 = \Delta x_{d2} \Rightarrow \alpha_3 = x_{d2} - \alpha_2 x_{s2} \quad .
\end{aligned}$$

Substituting these expressions for α_1 and α_3 into the third and fourth equations above, we write

$$\begin{aligned}
\beta_1 = \Delta x_{s1} + \alpha_1 \Rightarrow \beta_1 = (\alpha_2 - \beta_2) x_{s1} + (x_{d1} - \alpha_2 x_{s1}) \Rightarrow \beta_1 = x_{d1} - \beta_2 x_{s1} \\
\beta_3 = \Delta x_{s2} + \alpha_3 \Rightarrow \beta_3 = (\alpha_2 - \beta_2) x_{s2} + (x_{d2} - \alpha_2 x_{s2}) \Rightarrow \beta_3 = x_{d2} - \beta_2 x_{s2} \quad .
\end{aligned}$$

Accumulating the solution thus far, we can write

$$\begin{aligned}
\alpha_1 &= x_{d1} - \alpha_2 x_{s1} \\
\alpha_3 &= x_{d2} - \alpha_2 x_{s2} \\
\beta_1 &= x_{d1} - \beta_2 x_{s1} \\
\beta_3 &= x_{d2} - \beta_2 x_{s2} \quad .
\end{aligned}$$

The solution cannot quantify the terms α_2 and β_2. They are not determined at all. We say they are not "identified" by the reduced-form equation. If we substitute these expressions into the original four equations, we will find that all four hold. And yet, two of the six parameters remain unidentified. This is a problem with the structure of the original supply and demand equations.

If we compare this to the first example, we see that the Π matrix in the first problem was 3×2, which meant that it could contain all six requisite coefficients in the regression equations. However, the Π matrix in the second problem was only 2×2, which meant that it could not possibly contain a solution for six unknowns. The maximum would be four. In order for a solution to be exactly "identified," it must have precisely the same number of cells or positions in the Π matrix as there are unknown equations in the original regression formulation.

20.10 Under-Identification, Exact-Identification, and Over-Identification of Systems of Regression Equations

We have seen a perfectly identified system, and we have seen an under-identified system. We need to develop some theory here as to how and whether a particular system of equations is under-identified, exactly-identified, or over-identified (to be discussed in the next section). There are two conditions that have been developed to examine and ensure where a system is under-, exactly-, or over-identified: (1.) the order condition and (2.) the rank condition. The order condition counts equations and variables within each of the m equations as follows: In our example, let us characterize the key parameters

- The number of equations is m, which is the number of endogenous variables in the model. We saw that there were $m = 2$ simultaneously determined variables in the first model we considered in this section.
- $k = m + p$ = total number of endogenous plus exogenous variables in the model counting the constant as one such variable. It is the length of the column dimension of Γ plus the column dimension of B in the model development. In the first model in this section, $p = 3$ because there is a constant and two actual variables. The number k includes all the endogenous variables plus all the exogenous variables. In practice k is bigger than m.
- n_i = number of endogenous plus exogenous variables that are included (not excluded) in the ith equation.

A necessary condition for the equation to be identified and identifiable is the so-called order condition

$$k - n_i \geq m - 1 \quad .$$

To see how the order condition works, consider our first problem in this section (the exactly identified problem), in which $m = 2$, $p = 3$, and therefore $k = 5$. If we look at each of the two equations, we see that $n_i = 4$ in both equations comprises the two endogenous variables and 2 of a possible 3 exogenous variables in each equation. The identification equation is written with these parameters

$$5 - 4 \geq 2 - 1$$

and therefore the identification equation holds for both equations. Because there is exact equality here, we say that the matrix is exactly identified. If there is strict inequality, it is under-identified. If there is a strict less than, we say the equation is overidentified. We strive for systems of equations that are exactly identified. By contrast, in our second problem, $m = 2$, $p = 2$ and therefore $k = 4$. If we look at each of the two equations, we see that $n_i = 4$ in both equations, including the constant. The identification equation is written with these parameters.

$$4 - 4 \geq 2 - 1 \quad .$$

We see that the rank condition fails for both equations, meaning both equations are under-identified. This means the estimation of parameters is not possible.

The internet paper eGyanKosh[28] is quite elucidating, summarizing the order condition nicely. The paper eGyanKosh[28] also nicely motivates and illustrates the rank condition for identification. We shall base our analysis of the rank condition in part on the illustration herein. The rank condition gives a necessary and sufficient condition for identification for each equation in the simultaneous system. The rank condition must be performed one equation at a time, and we want to formulate a simultaneous system in which we will put it forth here as well because it is so definitive. We use both rank and order conditions to determine identifiability. Without identifiability, we are simply not getting a viable solution to the problem at hand.

The paper eGyanKosh[28] puts forth the following definition of the rank condition, which we think is quite a good and clear definition, so we will quote it: *The rank condition states that in a system of G equations, any particular equation is identified if and only if it is possible to construct at least one non-zero determinant of order (G-1) from the coefficients of the variables excluded from that particular equation but contained in the other equations of the model.*[6]

To illustrate the application of the rank condition test for the first example, consider the matrix of the coefficients of the structural model in the first example in Table 20.1. We want to test the identifiability of the first equation. We do so by

[6]In eGyanKosh[28], this condition is called the rank condition because it refers to the rank of the matrix of parameters of excluded variables. The rank of a matrix is the order of the largest non-zero determinant that can be formed from the matrix. In our case, the relevant matrix is the submatrix of coefficients of the absent variables. Hence the rank condition may also be stated as follows: A sufficient condition for identification of a relationship is that the rank of the matrix of parameters of all the excluded variables (endogenous and predetermined) from that equation be equal to (G - 1).

first redacting all the columns of the first equation that have a nonzero coefficient within them. That is, we want to redact all the variables that are contained in the first equation. When we do so, we are left only with those variables that are not contained in the first equation, i.e., the domain of the model is strictly other than the first equation. When we do so, we obtain the table in Table 20.2. As a second step, we now remove the entirety of the first equation as indicated in Table 20.3. The solution is now at hand. If the determinant of any cofactor of the remaining matrix is 0, the system is non-identifiable. If the determinant of any cofactor of the remaining matrix is nonzero, the system is identifiable. We notice that the sole remaining entry, $-\beta_3$ is nonzero, so the first equation is fully identifiable. The rank of the remaining cofactor matrix is 1. The rank condition tells us that the first equation is fully identifiable.

Table 20.1: Coefficients of the Structural Model–First Example

	Variables				
	q	p	$const$	DI	W
Equation 1	1	$-\alpha_2$	$-\alpha_1$	$-\alpha_3$	0
Equation 2	1	$-\beta_2$	$-\beta_1$	0	$-\beta_3$

We want to test the identifiability of the first equation. We do so by first redacting all the columns of the first equation that have a nonzero coefficient within them. That is, we want to redact all the variables that are contained in the first equation. When we do so, we are left only with those variables that are not contained in the first equation, i.e., the domain of the model strictly other than the first equation. When we do so, we obtain the table in Table 20.2. As a second step, we now remove the entirety of the first equation as indicated in Table 20.3. The solution is now at hand. If the determinant of any cofactor of the remaining matrix is 0, the system is non-identifiable. If the determinant of any cofactor of the remaining matrix is nonzero, the system is identifiable. We notice that the sole remaining entry $-\beta_3$ is nonzero, so the first equation is fully identifiable. The rank of the remaining cofactor matrix is 1. The rank condition tells us that the first equation is fully identifiable.

Table 20.2: Redact Nonzero Columns of the First Equation–First Example

	Variables				
	q	p	$const$	DI	W
Equation 1	*	*	*	*	0
Equation 2	*	*	*	*	$-\beta_3$

Let us turn to the second equation. We address it by first redacting all the columns of the second equation that have a nonzero coefficient within them. That is, we want to redact all the variables that are contained in the second equation. When we do so, we are left only with those variables that are not contained in the second equation, i.e., the domain of the model strictly other than the second

Table 20.3: Redact Zero Rows of the First Equation–First Example

	Variables				
	q	p	$const$	DI	W
Equation 1	$*$	$*$	$*$	$*$	$*$
Equation 2	$*$	$*$	$*$	$*$	$-\beta_3$

equation. When we do so, we obtain the table in Table 20.4. As a second step, we now remove the entirety of the second equation as indicated in Table 20.5. The solution is now at hand. If the determinant of any cofactor of the remaining matrix is 0, the system is non-identifiable. If the determinant of any cofactor of the remaining matrix is nonzero, the system is identifiable. We notice that the sole remaining entry $-\alpha_3$ is nonzero, so the second equation is fully identifiable. The rank of the remaining cofactor matrix is 1. The rank condition tells us that the second equation is fully identifiable. Therefore, both equations are fully identifiable.

Table 20.4: Redact Nonzero Columns of the Second Equation–First Example

	Variables				
	q	p	$const$	DI	W
Equation 1	$*$	$*$	$*$	$-\alpha_3$	$*$
Equation 2	$*$	$*$	$*$	0	$*$

Table 20.5: Coefficients of the Structural Model–First Example

	Variables				
	q	p	$const$	DI	W
Equation 1	$*$	$*$	$*$	$-\alpha_3$	$*$
Equation 2	$*$	$*$	$*$	$*$	$*$

We therefore know that we can perform the regression on the reduced-form version of the model and simulate the implied distributions over the true and correct model parameters. The eGyanKosh[28] document cites Gujrati[38], Kmenta[55], Johnson and Dinaro[48], Pindyck and Rubinfield[74], and Woodridge[104] as references on the problem of identification in simultaneous equation models.

If we return to the second example, we will see that the matrix of coefficients of the structural model has no zeroes in either equation. Therefore, when one removes the columns that correspond to the nonzero element in the first equation, there is nothing left! That means the determinant of the remaining matrix is zero by default. The problem fails the rank test. All determinants of all cofactors are zero because there are no cofactors. This is a classic situation where the coefficients are "under-identified." There is not a unique mapping from the reduced-form solution to the model coefficients. There are "too many" model coefficients.

20.11 A Third Example–One That Is Over-Identified

Suppose we have a supply and demand situation for truffles (a consumer product). The postulated demand curve is

$$q = \alpha_1 + \alpha_2 p + \alpha_3 PS + \alpha_4 DI + \varepsilon_d$$

in which

- q = quantity of truffles supplied and demanded
- p = price of truffles
- PS = price of a substitute for truffles (i.e., a cross-elastic effect)
- DI = disposable income.

The postulated supply curve is

$$q = \beta_1 + \beta_2 p + \beta_3 FP + \varepsilon_s \quad ,$$

in which

- FP = price of the factors of production of truffles.

Thus the demand and supply equations are, for each observation

$$\begin{aligned} &q_i - \alpha_2 p_i - \alpha_1 - \alpha_3 PS_i - \alpha_4 DI_i = \varepsilon_{di} \\ &q_i - \beta_2 p_i - \beta_1 - \beta_3 PF_i = \varepsilon_{si} \quad . \end{aligned}$$

We write this in matrix form

$$\begin{bmatrix} 1 & -\alpha_2 \\ 1 & -\beta_2 \end{bmatrix} \begin{bmatrix} q_i \\ p_i \end{bmatrix} - \begin{bmatrix} \alpha_1 & \alpha_3 & 0 & \alpha_4 \\ \beta_1 & 0 & \beta_3 & 0 \end{bmatrix} \begin{bmatrix} 1 \\ PS_i \\ PF_i \\ DI_i \end{bmatrix} = \begin{bmatrix} \varepsilon_{di} \\ \varepsilon_{si} \end{bmatrix} \quad .$$

We take the transpose of both sides of the equation

$$\begin{aligned} &\begin{bmatrix} q_i & p_i \end{bmatrix} \begin{bmatrix} 1 & 1 \\ -\alpha_2 & -\beta_2 \end{bmatrix} \\ &- \begin{bmatrix} 1 & PS_i & PF_i & DI_i \end{bmatrix} \begin{bmatrix} \alpha_1 & \beta_1 \\ \alpha_3 & 0 \\ 0 & \beta_3 \\ \alpha_4 & 0 \end{bmatrix} \end{aligned}$$

$$= \begin{bmatrix} \varepsilon_{di} & \varepsilon_{si} \end{bmatrix} \quad .$$

Let us add all the observations (assuming there are six) as additional rows

$$\begin{bmatrix} q_1 & p_1 \\ q_2 & p_2 \\ . & . \\ . & . \\ . & . \\ q_n & p_n \end{bmatrix} \begin{bmatrix} 1 & 1 \\ -\alpha_2 & -\beta_2 \end{bmatrix} - \begin{bmatrix} 1 & PS_1 & PF_1 & DI_1 \\ 1 & PS_2 & PF_2 & DI_2 \\ . & & & . \\ . & & & . \\ . & & & . \\ 1 & PS_n & PF_n & DI_n \end{bmatrix} \begin{bmatrix} \alpha_1 & \beta_1 \\ \alpha_3 & 0 \\ 0 & \beta_3 \\ \alpha_4 & 0 \end{bmatrix} = \begin{bmatrix} \varepsilon_{d1} & \varepsilon_{s1} \\ \varepsilon_{d2} & \varepsilon_{s2} \\ . & . \\ . & . \\ . & . \\ \varepsilon_{dn} & \varepsilon_{sn} \end{bmatrix} \quad .$$

We right multiply by the inverse of the matrix in the first term

$$\begin{bmatrix} q_1 & p_1 \\ q_2 & p_2 \\ . & . \\ . & . \\ . & . \\ q_n & p_n \end{bmatrix} - \begin{bmatrix} 1 & PS_1 & PF_1 & DI_1 \\ 1 & PS_2 & PF_2 & DI_2 \\ . & & & . \\ . & & & . \\ . & & & . \\ 1 & PS_n & PF_n & DI_n \end{bmatrix} \begin{bmatrix} \alpha_1 & \beta_1 \\ \alpha_3 & 0 \\ 0 & \beta_3 \\ \alpha_4 & 0 \end{bmatrix} \begin{bmatrix} 1 & 1 \\ -\alpha_2 & -\beta_2 \end{bmatrix}^{-1}$$

$$= \begin{bmatrix} \varepsilon_{d1} & \varepsilon_{s1} \\ \varepsilon_{d2} & \varepsilon_{s2} \\ . & . \\ . & . \\ . & . \\ \varepsilon_{dn} & \varepsilon_{sn} \end{bmatrix} \begin{bmatrix} 1 & 1 \\ -\alpha_2 & -\beta_2 \end{bmatrix}^{-1} \quad .$$

The inverse matrix is

$$\begin{bmatrix} 1 & 1 \\ -\alpha_2 & -\beta_2 \end{bmatrix}^{-1} = \begin{bmatrix} -\beta_2 & -1 \\ \alpha_2 & 1 \end{bmatrix} \frac{1}{\alpha_2 - \beta_2} = \begin{bmatrix} \frac{-\beta_2}{\alpha_2-\beta_2} & \frac{-1}{\alpha_2-\beta_2} \\ \frac{\alpha_2}{\alpha_2-\beta_2} & \frac{1}{\alpha_2-\beta_2} \end{bmatrix} \quad .$$

The equation therefore becomes

$$\begin{bmatrix} q_1 & p_1 \\ q_2 & p_2 \\ & \\ & \\ q_n & p_n \end{bmatrix} - \begin{bmatrix} 1 & PS_1 & PF_1 & DI_1 \\ 1 & PS_2 & PF_2 & DI_2 \\ & & & \\ & & & \\ 1 & PS_n & PF_n & DI_n \end{bmatrix} \begin{bmatrix} \alpha_1 & \beta_1 \\ \alpha_3 & 0 \\ 0 & \beta_3 \\ \alpha_4 & 0 \end{bmatrix} \begin{bmatrix} \frac{-\beta_2}{\alpha_2-\beta_2} & \frac{-1}{\alpha_2-\beta_2} \\ \frac{\alpha_2}{\alpha_2-\beta_2} & \frac{1}{\alpha_2-\beta_2} \end{bmatrix}$$

$$= \begin{bmatrix} \varepsilon_{d1} & \varepsilon_{s1} \\ \varepsilon_{d2} & \varepsilon_{s2} \\ & \\ & \\ \varepsilon_{dn} & \varepsilon_{sn} \end{bmatrix} \begin{bmatrix} \frac{-\beta_2}{\alpha_2-\beta_2} & \frac{-1}{\alpha_2-\beta_2} \\ \frac{\alpha_2}{\alpha_2-\beta_2} & \frac{1}{\alpha_2-\beta_2} \end{bmatrix} .$$

We expand the matrix to write the equation in reduced form

$$\begin{bmatrix} q_1 & p_1 \\ q_2 & p_2 \\ & \\ & \\ q_n & p_n \end{bmatrix} - \begin{bmatrix} 1 & PS_1 & PF_1 & DI_1 \\ 1 & PS_2 & PF_2 & DI_2 \\ & & & \\ & & & \\ 1 & PS_n & PF_n & DI_n \end{bmatrix} \begin{bmatrix} \frac{\beta_1\alpha_2-\alpha_1\beta_2}{\alpha_2-\beta_2} & \frac{\beta_1-\alpha_1}{\alpha_2-\beta_2} \\ \frac{-\alpha_3\beta_2}{\alpha_2-\beta_2} & \frac{-\alpha_3}{\alpha_2-\beta_2} \\ \frac{\beta_3\alpha_2}{\alpha_2-\beta_2} & \frac{\beta_3}{\alpha_2-\beta_2} \\ \frac{-\alpha_4\beta_2}{\alpha_2-\beta_2} & \frac{-\alpha_4}{\alpha_2-\beta_2} \end{bmatrix}$$

$$= \begin{bmatrix} \varepsilon_{d1} & \varepsilon_{s1} \\ \varepsilon_{d2} & \varepsilon_{s2} \\ & \\ & \\ \varepsilon_{dn} & \varepsilon_{sn} \end{bmatrix} \begin{bmatrix} \frac{-\beta_2}{\alpha_2-\beta_2} & \frac{-1}{\alpha_2-\beta_2} \\ \frac{\alpha_2}{\alpha_2-\beta_2} & \frac{1}{\alpha_2-\beta_2} \end{bmatrix} .$$

We substitute the matrix elements with the solution to the reduced value equation

$$\begin{bmatrix} \frac{\beta_1\alpha_2-\alpha_1\beta_2}{\alpha_2-\beta_2} & \frac{\beta_1-\alpha_1}{\alpha_2-\beta_2} \\ \frac{-\alpha_3\beta_2}{\alpha_2-\beta_2} & \frac{-\alpha_3}{\alpha_2-\beta_2} \\ \frac{\beta_3\alpha_2}{\alpha_2-\beta_2} & \frac{\beta_3}{\alpha_2-\beta_2} \\ \frac{-\alpha_4\beta_2}{\alpha_2-\beta_2} & \frac{-\alpha_4}{\alpha_2-\beta_2} \end{bmatrix} = \begin{bmatrix} \pi_{d1} & \pi_{s1} \\ \pi_{d2} & \pi_{s2} \\ \pi_{d3} & \pi_{s3} \\ \pi_{d4} & \pi_{s4} \end{bmatrix} .$$

Storm clouds are hovering here. Notice that there are eight (8) parameters of the solution to the reduced form equation, namely π_{d1}, π_{d2}, π_{d3}, π_{d4}, π_{s1}, π_{s2}, π_{s3}, and π_{s4}. However, there are only seven (7) model coefficients, namely

α_1, α_2, α_3, α_4, β_1, β_2, and β_3. This is again "Troubles in River City." We have eight equations for seven unknowns—too many equations relative to the number of unknowns. To explore what this means, we write these equalities in matrix form (after substituting for Δ), and as the determinant of the Γ matrix, we have

$$\begin{bmatrix} \frac{\beta_1\alpha_2-\alpha_1\beta_2}{\Delta} = \pi_{d1} & \frac{\beta_1-\alpha_1}{\Delta} = \pi_{s1} \\ \frac{-\alpha_3\beta_2}{\Delta} = \pi_{d2} & \frac{-\alpha_3}{\Delta} = \pi_{s2} \\ \frac{\beta_3\alpha_2}{\Delta} = \pi_{d3} & \frac{\beta_3}{\Delta} = \pi_{s3} \\ \frac{-\alpha_4\beta_2}{\Delta} = \pi_{d4} & \frac{-\alpha_4}{\Delta} = \pi_{s4} \end{bmatrix} \Rightarrow \begin{bmatrix} \beta_1\alpha_2 - \alpha_1\beta_2 = \Delta\pi_{d1} & \beta_1 - \alpha_1 = \Delta\pi_{s1} \\ -\alpha_3\beta_2 = \Delta\pi_{d2} & -\alpha_3 = \Delta\pi_{s2} \\ \beta_3\alpha_2 = \Delta\pi_{d3} & \beta_3 = \Delta\pi_{s3} \\ -\alpha_4\beta_2 = \Delta\pi_{d4} & -\alpha_4 = \Delta\pi_{s4} \end{bmatrix}$$

We expand the expressions in this matrix equation

$$\begin{bmatrix} \beta_1\alpha_2 - \alpha_1\beta_2 = (\alpha_2 - \beta_2)\,\pi_{d1} & \beta_1 - \alpha_1 = (\alpha_2 - \beta_2)\,\pi_{s1} \\ \alpha_3\beta_2 = -(\alpha_2 - \beta_2)\,\pi_{d2} & \alpha_3 = -(\alpha_2 - \beta_2)\,\pi_{s2} \\ \beta_3\alpha_2 = (\alpha_2 - \beta_2)\,\pi_{d3} & \beta_3 = (\alpha_2 - \beta_2)\,\pi_{s3} \\ \alpha_4\beta_2 = -(\alpha_2 - \beta_2)\,\pi_{d4} & \alpha_4 = -(\alpha_2 - \beta_2)\,\pi_{s4} \end{bmatrix}.$$

Substituting the three lower right cells into the three lower left cells, we can solve for the three lower left elements

$$\begin{bmatrix} \beta_1\alpha_2 - \alpha_1\beta_2 = (\alpha_2 - \beta_2)\,\pi_{d1} & \beta_1 - \alpha_1 = (\alpha_2 - \beta_2)\,\pi_{s1} \\ \beta_2 = \frac{\pi_{d2}}{\pi_{s2}} & \alpha_3 = -(\alpha_2 - \beta_2)\,\pi_{s2} \\ \alpha_2 = \frac{\pi_{d3}}{\pi_{s3}} & \beta_3 = (\alpha_2 - \beta_2)\,\pi_{s3} \\ \beta_2 = \frac{\pi_{d4}}{\pi_{s4}} & \alpha_4 = -(\alpha_2 - \beta_2)\,\pi_{s4} \end{bmatrix}.$$

Notice that there are two different solutions in the lower left column. That is going to be a problem, but let us continue to solve. Substitute α_2 in the lower left column into every position it occurs throughout the matrix, including into the definition of Δ to obtain

$$\Delta = \alpha_2 - \beta_2 = \frac{\pi_{d3}}{\pi_{s3}} - \beta_2$$

and

$$\begin{bmatrix} \beta_1\frac{\pi_{d3}}{\pi_{s3}} - \alpha_1\beta_2 = \left(\frac{\pi_{d3}}{\pi_{s3}} - \beta_2\right)\pi_{d1} & \beta_1 - \alpha_1 = \left(\frac{\pi_{d3}}{\pi_{s3}} - \beta_2\right)\pi_{s1} \\ \beta_2 = \frac{\pi_{d2}}{\pi_{s2}} & \alpha_3 = \left(\beta_2 - \frac{\pi_{d3}}{\pi_{s3}}\right)\pi_{s2} \\ \alpha_2 = \frac{\pi_{d3}}{\pi_{s3}} & \beta_3 = \left(\frac{\pi_{d3}}{\pi_{s3}} - \beta_2\right)\pi_{s3} \\ \beta_2 = \frac{\pi_{d4}}{\pi_{s4}} & \alpha_4 = \left(\beta_2 - \frac{\pi_{d3}}{\pi_{s3}}\right)\pi_{s4} \end{bmatrix}.$$

We isolate β_1 in the upper right cell, substitute it into the upper left cell, and solve the upper left cell for α_1

$$\begin{bmatrix} \alpha_1 = \frac{\pi_{d1}\pi_{s3}-\pi_{d3}\pi_{s1}}{\pi_{s3}} \Leftarrow & \beta_1 = \left[\alpha_1 + \left(\frac{\pi_{d3}}{\pi_{s3}} - \beta_2\right)\pi_{s1}\right] \\ \beta_2 = \frac{\pi_{d2}}{\pi_{s2}} & \alpha_3 = \left(\beta_2 - \frac{\pi_{d3}}{\pi_{s3}}\right)\pi_{s2} \\ \alpha_2 = \frac{\pi_{d3}}{\pi_{s3}} & \beta_3 = \left(\frac{\pi_{d3}}{\pi_{s3}} - \beta_2\right)\pi_{s3} \\ \beta_2 = \frac{\pi_{d4}}{\pi_{s4}} & \alpha_4 = \left(\beta_2 - \frac{\pi_{d3}}{\pi_{s3}}\right)\pi_{s4} \end{bmatrix}.$$

We substitute the solution for α_1 from the upper left cell into the upper right cell and solve for β_1 in the upper right cell

$$\begin{bmatrix} \alpha_1 = \frac{\pi_{d1}\pi_{s3}-\pi_{d3}\pi_{s1}}{\pi_{s3}} \Rightarrow & \beta_1 = \pi_{d1} - \beta_2\pi_{s1} \\ \beta_2 = \frac{\pi_{d2}}{\pi_{s2}} & \alpha_3 = \left(\beta_2 - \frac{\pi_{d3}}{\pi_{s3}}\right)\pi_{s2} \\ \alpha_2 = \frac{\pi_{d3}}{\pi_{s3}} & \beta_3 = \left(\frac{\pi_{d3}}{\pi_{s3}} - \beta_2\right)\pi_{s3} \\ \beta_2 = \frac{\pi_{d4}}{\pi_{s4}} & \alpha_4 = \left(\beta_2 - \frac{\pi_{d3}}{\pi_{s3}}\right)\pi_{s4} \end{bmatrix}.$$

The solution is now at hand

$$\begin{aligned} \alpha_1 &= \frac{\pi_{d1}\pi_{s3} - \pi_{d3}\pi_{s1}}{\pi_{s3}} \quad \beta_1 = \pi_{d1} - \beta_2\pi_{s1} \\ \alpha_2 &= \frac{\pi_{d3}}{\pi_{s3}} \quad \beta_2 = \frac{\pi_{d2}}{\pi_{s2}} \quad \beta_2 = \frac{\pi_{d4}}{\pi_{s4}} \\ \alpha_3 &= \left(\beta_2 - \frac{\pi_{d3}}{\pi_{s3}}\right)\pi_{s2} \quad \beta_3 = \left(\frac{\pi_{d3}}{\pi_{s3}} - \beta_2\right)\pi_{s3} \\ \alpha_4 &= \left(\beta_2 - \frac{\pi_{d3}}{\pi_{s3}}\right)\pi_{s4} \quad . \end{aligned}$$

If we substitute every element of this solution into the original problem, we see the following results occur

$$\begin{bmatrix} 0 = 0 & 0 = 0 \\ \pi_{s2}\beta_2 = \pi_{d2} & 0 = 0 \\ 0 = 0 & 0 = 0 \\ \pi_{s4}\beta_2 = \pi_{d4} & 0 = 0 \end{bmatrix}.$$

Every equation holds perfectly, save for the two indicated in the matrix. Notice that the value of β_2 is different between the two equations. This means that β_2 is "overidentified." There are more than one, different value for the variable β_2. It is fully identified; in fact, it is overidentified. We can see why from the rank condition, in which we know that

- $m = 2$, number of equations
- $p = 4$, number of additional variables
- $k = 6$, total number of variables

- $n_1 = 5$, total number of variables in the demand equation
- $n_2 = 4$, total number of variables in the supply equation.

These rank terms imply exactly what we have seen

- for the first equation, we have that $k - n_1 = 1 = m - 1 = 1$, exactly identified.
- for the second equation, we have that $k - n_2 = 2 > m - 1 = 1$, overidentified by one.

The term α_2 corresponds to the first equation, which is exactly identified. Notice that its value is uniquely defined in the reduced-form solution. This is the "elasticity of demand," and it is identified by the reduced form equation. However, the term β_2 corresponds to the second equation. We have seen that it has two values; it is not uniquely identified. This is the implication in the rank test of the left-hand side being strictly larger, larger by 1, than the right-hand side. For Bayesian analysis, we want the reduced system of equations to have the property that every equation is exactly identified.

20.12 Generalization of the Reduced Form Model

This section generalizes. Let us write the simultaneous regression equations

$$\gamma_{11}y_{t1} + \gamma_{21}y_{t2} + ... + \gamma_{m1}y_{tm} + \beta_{11}x_{t1} + \beta_{21}x_{t2} + ... + \beta_{p1}x_{tp} = \varepsilon_{t1}$$
$$\gamma_{12}y_{t1} + \gamma_{22}y_{t2} + ... + \gamma_{m2}y_{tm} + \beta_{12}x_{t1} + \beta_{22}x_{t2} + ... + \beta_{p2}x_{tp} = \varepsilon_{t2}$$

.

.

.

$$\gamma_{1m}y_{t1} + \gamma_{2m}y_{t2} + ... + \gamma_{mm}y_{tm} + \beta_{1m}x_{t1} + \beta_{2m}x_{t2} + ... + \beta_{pm}x_{tp} = \varepsilon_{tm} \quad .$$

We can express these equations in matrix format

$$\begin{bmatrix} \gamma_{11} & \gamma_{21} & \cdot & \cdot & \cdot & \gamma_{m1} \\ \gamma_{12} & \gamma_{22} & & & & \gamma_{m2} \\ \cdot & & & & & \cdot \\ \cdot & & & & & \cdot \\ \cdot & & & & & \cdot \\ \gamma_{1m} & \gamma_{2m} & \cdot & \cdot & \cdot & y_{mm} \end{bmatrix} \begin{bmatrix} y_{t1} \\ y_{t2} \\ \cdot \\ \cdot \\ \cdot \\ y_{tm} \end{bmatrix}$$

$$+\begin{bmatrix} \beta_{11} & \beta_{21} & \cdots & \beta_{p1} \\ \beta_{12} & \beta_{22} & & \beta_{p2} \\ \vdots & & & \vdots \\ \beta_{1m} & \beta_{2m} & \cdots & \beta_{pm} \end{bmatrix} \begin{bmatrix} x_{t1} \\ x_{t2} \\ \vdots \\ x_{tp} \end{bmatrix} = \begin{bmatrix} \varepsilon_{t1} \\ \varepsilon_{t2} \\ \vdots \\ \varepsilon_{tm} \end{bmatrix} .$$

It is convenient to take the transpose, which facilitates adding in the observations

$$\begin{bmatrix} y_{t1} & y_{t2} & \cdots & y_{tm} \end{bmatrix} \begin{bmatrix} \gamma_{11} & \gamma_{12} & \cdots & \gamma_{1m} \\ \gamma_{21} & \gamma_{22} & & \gamma_{2m} \\ \vdots & & & \vdots \\ \gamma_{m1} & \gamma_{m2} & \cdots & y_{mm} \end{bmatrix}$$

$$+\begin{bmatrix} x_{t1} & x_{t2} & \cdots & x_{tp} \end{bmatrix} \begin{bmatrix} \beta_{11} & \beta_{12} & \cdots & \beta_{1m} \\ \beta_{21} & \beta_{22} & & \beta_{2m} \\ \vdots & & & \vdots \\ \beta_{p1} & \beta_{p2} & \cdots & \beta_{pm} \end{bmatrix}$$

$$= \begin{bmatrix} \varepsilon_{t1} & \varepsilon_{t2} & \cdots & \varepsilon_{tm} \end{bmatrix} .$$

It is very easy to see how we augment the two row vectors on the left and the row vector at the bottom to accommodate all the observations

$$\begin{bmatrix} y_{11} & y_{12} & \cdots & y_{1m} \\ y_{21} & y_{22} & & y_{2m} \\ \vdots & & & \vdots \\ y_{n1} & y_{n2} & \cdots & y_{nm} \end{bmatrix} \begin{bmatrix} \gamma_{11} & \gamma_{12} & \cdots & \gamma_{1m} \\ \gamma_{21} & \gamma_{22} & & \gamma_{2m} \\ \vdots & & & \vdots \\ \gamma_{m1} & \gamma_{m2} & \cdots & y_{mm} \end{bmatrix}$$

$$+\begin{bmatrix} x_{11} & x_{12} & \cdots & x_{1p} \\ x_{21} & x_{22} & & x_{2p} \\ \vdots & & & \vdots \\ x_{n1} & x_{n2} & \cdots & x_{np} \end{bmatrix} \begin{bmatrix} \beta_{11} & \beta_{12} & \cdots & \beta_{1m} \\ \beta_{21} & \beta_{22} & & \beta_{2m} \\ \vdots & & & \vdots \\ \beta_{p1} & \beta_{p2} & \cdots & \beta_{pm} \end{bmatrix}$$

$$= \begin{bmatrix} \varepsilon_{11} & \varepsilon_{12} & \cdots & \varepsilon_{1m} \\ \varepsilon_{21} & \varepsilon_{22} & & \varepsilon_{2m} \\ \vdots & & & \vdots \\ \varepsilon_{n1} & \varepsilon_{n2} & \cdots & \varepsilon_{nm} \end{bmatrix} .$$

If we define the following matrices from this formulation

$$Y = \begin{bmatrix} y_{11} & y_{12} & \cdot & \cdot & \cdot & y_{1m} \\ y_{21} & y_{22} & & & & y_{2m} \\ \cdot & & & & & \cdot \\ \cdot & & & & & \cdot \\ \cdot & & & & & \cdot \\ y_{n1} & y_{n2} & \cdot & \cdot & \cdot & y_{nm} \end{bmatrix} \quad n \times m \quad ,$$

$$X = \begin{bmatrix} x_{11} & x_{12} & \cdot & \cdot & \cdot & x_{1p} \\ x_{21} & x_{22} & & & & x_{2p} \\ \cdot & & & & & \cdot \\ \cdot & & & & & \cdot \\ \cdot & & & & & \cdot \\ x_{n1} & x_{n2} & \cdot & \cdot & \cdot & x_{n} \end{bmatrix} \quad n \times p \quad ,$$

$$\Gamma = \begin{bmatrix} \gamma_{11} & \gamma_{12} & \cdot & \cdot & \cdot & \gamma_{1m} \\ \gamma_{21} & \gamma_{22} & & & & \gamma_{2m} \\ \cdot & & & & & \cdot \\ \cdot & & & & & \cdot \\ \cdot & & & & & \cdot \\ \gamma_{m1} & \gamma_{m2} & \cdot & \cdot & \cdot & y_{mm} \end{bmatrix} \quad m \times m \quad ,$$

$$B = \begin{bmatrix} \beta_{11} & \beta_{12} & \cdot & \cdot & \cdot & \beta_{1m} \\ \beta_{21} & \beta_{22} & & & & \beta_{2m} \\ \cdot & & & & & \cdot \\ \cdot & & & & & \cdot \\ \cdot & & & & & \cdot \\ \beta_{p1} & \beta_{p2} & \cdot & \cdot & \cdot & \beta_{pm} \end{bmatrix} \quad p \times m \quad , and$$

$$E = \begin{bmatrix} \varepsilon_{11} & \varepsilon_{12} & \cdot & \cdot & \cdot & \varepsilon_{1m} \\ \varepsilon_{21} & \varepsilon_{22} & & & & \varepsilon_{2m} \\ \cdot & & & & & \cdot \\ \cdot & & & & & \cdot \\ \cdot & & & & & \cdot \\ \varepsilon_{n1} & \varepsilon_{n2} & \cdot & \cdot & \cdot & \varepsilon_{nm} \end{bmatrix} \quad n \times m \quad ,$$

then we can write the equation in the standard form

$$Y\Gamma + XB = E \quad ,$$

which can be expressed by concatenating the matrices on the left

$$\begin{bmatrix} Y & X \end{bmatrix} \begin{bmatrix} \Gamma \\ B \end{bmatrix} = E \quad .$$

The previous analysis continues to hold here. As before, the matrices Y and

X are filled with observation data. The matrices Γ and B are the unknown constants. We can write as before for this general case

$$
\begin{aligned}
&Y\Gamma + XB = E \\
&\Rightarrow Y + XB\Gamma^{-1} = E\Gamma^{-1} \\
&\Rightarrow Y = -XB\Gamma^{-1} + E\Gamma^{-1} \quad .
\end{aligned}
$$

Again defining

$$
\begin{aligned}
\Pi &= -B\Gamma^{-1} \\
V &= E\Gamma^{-1} \quad .
\end{aligned}
$$

we can write the following reduced form equation

$$
\begin{aligned}
Y &= X\Pi + V \\
Y - X\Pi &= V \quad .
\end{aligned}
$$

We can directly apply the results from Chapter 15 through 17 for the multivariate case to this, the reduced form analysis of the general simultaneous case. We should add in closing this discussion that there will be no substitute for making sure the model is identifiable. In the absence of that, what one learns from the model becomes more limited.

Chapter 21

A SIMULTANEOUS SUPPLY-DEMAND ESTIMATION CASE STUDY

Fear was stronger than the calculation of probabilities.

George Eliot (pen name of Mary Ann Evans)

This chapter puts forth and solves a problem from a Bayesian perspective, nicely illustrating the methods in Chapters 16 and 17 and the simultaneous system developed in Chapter 20. The problem has been seen in the literature before (see Yobero[106] for a similar formulation), and we have posed and solved the problem using Bayesian methods. We pose an exactly-identified problem (and show some of the nuances with it) and then show what goes wrong with an over-identified problem. The purpose is to demonstrate what to do with these types of problems. We are not trying to be comprehensive here but rather illustrative. There is much work to be done with Bayesian methods in a simultaneous regression equation environment. The particular supply-demand pair model is the same as in Chapters 15 through 17 in the exactly identified initial section

$$q_i - \alpha_2 p_i - \alpha_1 - \alpha_3 DI_i = \varepsilon_{di} \quad i = 1, .., n$$
$$q_i - \beta_2 p_i - \beta_1 - \beta_3 FP_i = \varepsilon_{si} \quad i = 1, .., n \quad .$$

These equations are manipulated precisely as they were in the previous chapter to the same form

$$\begin{bmatrix} q_1 & p_1 \\ \cdot & \cdot \\ \cdot & \cdot \\ \cdot & \cdot \\ q_n & p_n \end{bmatrix} \begin{bmatrix} 1 & 1 \\ -\alpha_2 & -\beta_2 \end{bmatrix} - \begin{bmatrix} 1 & DI_1 & FP_1 \\ \cdot & \cdot & \cdot \\ \cdot & \cdot & \cdot \\ \cdot & \cdot & \cdot \\ 1 & DI_n & FP_n \end{bmatrix} \begin{bmatrix} \alpha_1 & \beta_1 \\ \alpha_3 & 0 \\ 0 & \beta_3 \end{bmatrix}$$
$$= \begin{bmatrix} \varepsilon_{d1} & \varepsilon_{s1} \\ \cdot & \cdot \\ \cdot & \cdot \\ \cdot & \cdot \\ \varepsilon_{dn} & \varepsilon_{sn} \end{bmatrix} .$$

The solution to the mean value implies the six equations developed in Chapter 16

$$\begin{aligned} \pi_{d1} &= \frac{\beta_1\alpha_2 - \beta_2\alpha_1}{\alpha_2 - \beta_2} \\ \pi_{d2} &= \frac{-\alpha_3\beta_2}{\alpha_2 - \beta_2} \\ \pi_{d3} &= \frac{\alpha_2\beta_3}{\alpha_2 - \beta_2} \\ \pi_{s1} &= \frac{\beta_1 - \alpha_1}{\alpha_2 - \beta_2} \\ \pi_{s2} &= \frac{-\alpha_3}{\alpha_2 - \beta_2} \\ \pi_{s3} &= \frac{\beta_3}{\alpha_2 - \beta_2} \end{aligned} .$$

If we write the determinant $\Delta = \alpha_2 - \beta_2$, the matrix equation becomes

$$\begin{bmatrix} \pi_{d1}\Delta = \alpha_2\beta_1 - \alpha_1\beta_2 & \pi_{s1}\Delta = \beta_1 - \alpha_1 \\ -\pi_{d2}\Delta = \alpha_3\beta_2 & -\pi_{s2}\Delta = \alpha_3 \\ \pi_{d3}\Delta = \alpha_2\beta_3 & \pi_{s3}\Delta = \beta_3 \end{bmatrix} .$$

We begin to solve this by substituting the bottom two terms in the right column into the bottom two equations in the left column

$$\begin{bmatrix} \pi_{d1}\Delta = \alpha_2\beta_1 - \alpha_1\beta_2 & \pi_{s1}\Delta = \beta_1 - \alpha_1 \\ \beta_2 = \frac{\pi_{d2}}{\pi_{s2}} & -\pi_{s2}\Delta = \alpha_3 \\ \alpha_2 = \frac{\pi_{d3}}{\pi_{s3}} & \pi_{s3}\Delta = \beta_3 \end{bmatrix} .$$

This implies that the determinant can be written

$$\Delta = \alpha_2 - \beta_2 = \frac{\pi_{d3}}{\pi_{s3}} - \frac{\pi_{d2}}{\pi_{s2}} = \frac{\pi_{d3}\pi_{s2} - \pi_{d2}\pi_{s3}}{\pi_{s3}\pi_{s2}} .$$

If we substitute this determinant as well as α_2 and β_2 everywhere they occur, we obtain

$$\begin{bmatrix} \pi_{d1}\frac{\pi_{d3}\pi_{s2}-\pi_{d2}\pi_{s3}}{\pi_{s3}\pi_{s2}} = \frac{\pi_{d3}}{\pi_{s3}}\beta_1 - \alpha_1\frac{\pi_{d2}}{\pi_{s2}} & \frac{\pi_{d3}\pi_{s1}\pi_{s2}-\pi_{d2}\pi_{s1}\pi_{s3}}{\pi_{s3}\pi_{s2}} = \beta_1 - \alpha_1 \\ \beta_2 = \frac{\pi_{d2}}{\pi_{s2}} & \alpha_3 = \frac{\pi_{d2}\pi_{s3}-\pi_{d3}\pi_{s2}}{\pi_{s3}} \\ \alpha_2 = \frac{\pi_{d3}}{\pi_{s3}} & \beta_3 = \frac{\pi_{d3}\pi_{s2}-\pi_{d2}\pi_{s3}}{\pi_{s2}} \end{bmatrix} .$$

We solve the equation at the upper right for β_1,

$$\beta_1 = \alpha_1 + \pi_{s1}\frac{\pi_{d3}\pi_{s2} - \pi_{d2}\pi_{s3}}{\pi_{s3}\pi_{s2}}$$

and then we substitute into the equation at the upper left to solve for α_1

$$\begin{aligned} \pi_{d1}\frac{\pi_{d3}\pi_{s2} - \pi_{d2}\pi_{s3}}{\pi_{s3}\pi_{s2}} &= \frac{\pi_{d3}}{\pi_{s3}}\left(\alpha_1 + \pi_{s1}\frac{\pi_{d3}\pi_{s2} - \pi_{d2}\pi_{s3}}{\pi_{s3}\pi_{s2}}\right) - \alpha_1\frac{\pi_{d2}}{\pi_{s2}} \\ \Rightarrow \frac{\pi_{d3}\pi_{s2} - \pi_{d2}\pi_{s3}}{\pi_{s3}\pi_{s2}}\left(\pi_{d1} - \frac{\pi_{d3}\pi_{s1}}{\pi_{s3}}\right) &= \alpha_1\frac{\pi_{d3}\pi_{s2} - \pi_{d2}\pi_{s3}}{\pi_{s3}\pi_{s2}} \\ \Rightarrow \alpha_1 = \frac{\pi_{d1}\pi_{s3} - \pi_{d3}\pi_{s1}}{\pi_{s3}} & \quad . \end{aligned}$$

We substitute this expression into the upper right equation and solve for β_1

$$\begin{aligned} \frac{\pi_{d1}\pi_{s3}\pi_{s2} - \pi_{d3}\pi_{s1}\pi_{s2}}{\pi_{s3}\pi_{s2}} + \frac{\pi_{s1}\pi_{d3}\pi_{s2} - \pi_{s1}\pi_{d2}\pi_{s3}}{\pi_{s3}\pi_{s2}} &= \beta_1 \\ \Rightarrow \frac{\pi_{d1}\pi_{s2}\pi_{s3} - \pi_{d3}\pi_{s1}\pi_{s2} + \pi_{d3}\pi_{s1}\pi_{s2} - \pi_{d2}\pi_{s1}\pi_{s3}}{\pi_{s3}\pi_{s2}} &= \beta_1 \\ \Rightarrow \beta_1 = \frac{\pi_{d1}\pi_{s2} - \pi_{d2}\pi_{s1}}{\pi_{s2}} & \quad . \end{aligned}$$

The answer is at hand

$$\begin{aligned} \alpha_1 &= \frac{\pi_{d1}\pi_{s3} - \pi_{d3}\pi_{s1}}{\pi_{s3}} \\ \alpha_2 &= \frac{\pi_{d3}}{\pi_{s3}} \\ \alpha_3 &= \frac{\pi_{d2}\pi_{s3} - \pi_{d3}\pi_{s2}}{\pi_{s3}} \\ \beta_1 &= \frac{\pi_{d1}\pi_{s2} - \pi_{d2}\pi_{s1}}{\pi_{s2}} \\ \beta_2 &= \frac{\pi_{d2}}{\pi_{s2}} \\ \beta_3 &= \frac{\pi_{d3}\pi_{s2} - \pi_{d2}\pi_{s3}}{\pi_{s2}} \quad . \end{aligned}$$

If we get the π values from the reduced form equation, we can get the full probability distribution and the marginal probability distributions over the coefficients via these equations. The inverse relationship, given in the matrix equation above, is

$$\begin{aligned}
\pi_{d1} &= \frac{\alpha_1\beta_2 - \alpha_2\beta_1}{\alpha_2 - \beta_2} & \pi_{s1} &= \frac{\alpha_1 - \beta_1}{\alpha_2 - \beta_2} \\
\pi_{d2} &= \frac{\alpha_3\beta_2}{\alpha_2 - \beta_2} & \pi_{s2} &= \frac{\alpha_3}{\alpha_2 - \beta_2} \\
\pi_{d3} &= -\frac{\alpha_2\beta_3}{\alpha_2 - \beta_2} & \pi_{s3} &= -\frac{\beta_3}{\alpha_2 - \beta_2} \quad .
\end{aligned}$$

21.1 The Observations (The Data)

The data set for this problem is given in Hill et.al.[42] and appears as Table 21.1. The shaded portion at the left is the matrix we have previously labeled Y, and the unshaded portion at the right is the matrix we have previously labeled X, including a column of 1's representing the constant.

The interim calculations that lead to the mean value term Π appear in Table 21.2. If we begin with the mean values Π and convert to their equivalent coefficient values, we have the result in Table 21.3. We cannot think of these as the mean values of the original parameters; they are the nonlinear transformation of the mean values of the reduced form problem. The true mean values of the original coefficients would have to be computed using simulation.

It is interesting and instructive to plot the supply and demand curves and see where they might cross. If we take the coefficient estimates from Table 21.3 and assume a representative value of 4 for the disposable income DI and 22 for the price of the factor PF, we end up with a simple linear supply curve and a simple linear demand curve. Those curves and their crossing points appear in Figure 21.1, which plots data from Table 21.3. Wow! This is like magic! You have identified a supply curve and a demand curve, and you have identified the crossing point. This is precisely what you set out to do. When we consider the probability distribution over the posterior, we have actually identified them probabilistically.

We could start with a prior over the raw model coefficients and perform the following change of variables to characterize the prior in terms of the reduced form variables: Π, hoping (beyond hope) that they might be able to be used to approximate the specific form for the prior that we have developed herein. This will NEVER happen analytically, but it might be possible to approximate means and variances-covariances. This is an open research topic. (The next step would be to multiply this transformed prior by the likelihood in the reduced form problem. This would lead to the posterior over the reduced form problem variables that comprise Π.)

$$f_{\Pi}\left(\pi_{d1}, \pi_{d2}, \pi_{d3}, \pi_{s1}, \pi_{s2}, \pi_{s3}\right)$$

Table 21.1: The Truffle Data Set

Quantity(q)	Price (p)	Constant	Disp. Inc.(DI)	Price Factor (PF)
19.89	29.64	1	2.103	10.52
13.04	40.23	1	2.043	19.67
19.61	34.71	1	1.870	13.74
17.13	41.43	1	1.525	17.95
22.55	53.37	1	2.709	13.71
6.37	38.52	1	2.489	24.95
15.02	54.33	1	2.294	24.17
10.22	40.56	1	2.196	23.61
23.64	67.35	1	3.885	19.52
16.12	49.65	1	3.169	20.03
24.55	58.17	1	2.623	15.38
18.92	66.87	1	3.007	22.98
11.94	49.95	1	3.367	25.76
18.93	64.95	1	3.290	25.17
12.60	52.68	1	3.746	25.82
20.49	61.20	1	3.518	19.31
22.94	80.55	1	4.381	26.02
21.08	89.94	1	4.121	29.65
16.68	70.77	1	3.820	27.45
17.61	57.33	1	4.398	18.00
16.62	46.23	1	3.764	18.87
20.99	77.43	1	4.524	24.58
24.53	83.01	1	4.815	25.25
19.67	70.71	1	3.670	24.24
23.29	66.75	1	4.392	22.63
16.64	76.80	1	4.603	27.35
20.81	83.70	1	4.632	27.80
14.95	81.00	1	4.894	30.34
26.27	88.44	1	5.125	24.12
20.65	105.45	1	4.836	34.01

Table 21.2: Calculation of the Mean Values $\bar{\Pi}$ in the Reduced Form Equation

$X^T X$

30.000	105.809	682.600
105.809	404.600	2508.456
682.600	2508.456	16355.176

$(X^T X)^{-1}$

0.690463	-0.038784	-0.022869
-0.038784	0.052510	-0.006435
-0.022869	-0.006435	0.002003

$X^T Y$		$\bar{\Pi}$	
553.750	1881.720	18.754080	-4.254224
2010.376	7117.444	3.718534	11.639623
12489.518	44929.008	-0.589403	1.139423

Table 21.3: Calculation of the Fully Identified α and β Values

	Coefficients	
	alpha	beta
1	16.553444	20.113186
2	-0.517283	0.319472
3	9.739508	-0.953417

$$= f_C \left(\begin{array}{c} \frac{\pi_{d1}\pi_{s3} - \pi_{d3}\pi_{s1}}{\pi_{s3}}, \frac{\pi_{d3}}{\pi_{s3}}, \frac{\pi_{d2}\pi_{s3} - \pi_{d3}\pi_{s2}}{\pi_{s3}}, \\ \frac{\pi_{d1}\pi_{s2} - \pi_{d2}\pi_{s1}}{\pi_{s2}}, \frac{\pi_{d2}}{\pi_{s2}}, \frac{\pi_{d3}\pi_{s2} - \pi_{d2}\pi_{s3}}{\pi_{s2}} \end{array} \right)$$
$$\times \frac{1}{\pi_{s2}\pi_{s3}} \left(\frac{\pi_{d2}\pi_{s3} - \pi_{d3}\pi_{s2}}{\pi_{s2}\pi_{s3}} \right)^3 \quad .$$

Figure 21.1: Example Demand Curve and Supply Curve Inferred from Regression

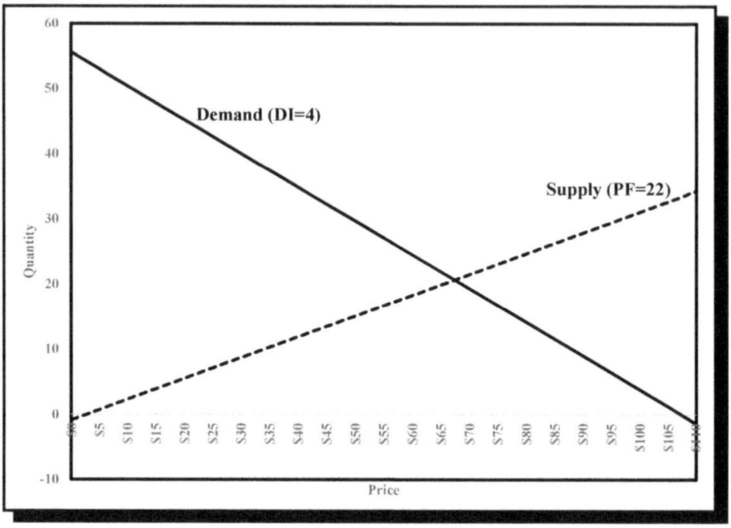

21.2 A Bayesian Approach

If we wanted to be truly and completely Bayesian, we would start with a prior distribution over the six original model coefficients

$$f(\alpha_1, \alpha_2, \alpha_3, \beta_1, \beta_2, \beta_3) \quad .$$

We could contemplate a change of variables problem, converting the density f(.) of the original coefficients to the six parameters of the reduced problem

$$g(\pi_{d1}, \pi_{d2}, \pi_{d3}, \pi_{s1}, \pi_{s2}, \pi_{s3}) \quad .$$

We would hope that such a prior for the reduced form problem would be exactly in the form of the prior in Chapters 15 through 17. Alas, that would not necessarily be the case (unless we had invoked the inverse relationship from the π variables to the coefficient beforehand and required that the prior coefficients be characterized by that distribution).

We would then use the distribution g as the prior in the reduced form problem, and we multiply it by the likelihood function to calculate the posterior distribution expressed in terms of the π parameters.

$$h(\pi_{d1}, \pi_{d2}, \pi_{d3}, \pi_{s1}, \pi_{s2}, \pi_{s3}) \quad .$$

Once we have this distribution, we then have to execute a change of variables transformation from the pi variables to the original coefficients $\alpha_1, \alpha_2, \alpha_3, \beta_1, \beta_2,$

and β_3. This involves the reverse direction of transformation. The resulting distribution over the original coefficients is the desired output. We are going to dispense with this calculation, which would require several steps of simulation. If we desire a closed form, we are relegated to posing the prior and receiving the posterior in the context of the reduced form equation.

Chapter 22

CONCLUSION

Ignorance gives one a large range of probabilities.

George Eliot (pen name of Mary Ann Evans)

We have developed classical linear regression from a fundamental probabilistic perspective, both in its original incarnation and in its more modern Bayesian incarnation. The framing of classical regression from a Bayesian perspective (with a diffuse prior) is both elucidating and revealing about what you are truly assuming in reality. ("Oh, so THAT's what I was doing!") The notion that classical regression is almost exactly consistent with the Bayesian model (conditional probabilities, Bayes theorem, conjugacy between prior and posterior) should be extremely satisfying for everyone. The result is not new or unique to this book. However, it has not, to our knowledge, been put in one place and developed from first principles. The literature seems scattered on this topic, even though the results in this book are known, and the purpose of this book is to put it all in one place in one tidy, focused, integrated form. Bayes offers a high degree of intellectual power and method relative to classical statistics, and putting classical regression into a Bayesian context, we believe, reinforces it.

Any idea that one has absolutely no ex ante information of any kind that logically preexists any data gathering exercise in any problem does not make intellectual sense and discards valuable, reliable, valid information or judgment. Bayes allows us to meld prior judgement with data transparently in a systematic, logical, auditable probabilistic fashion. Therein lies the intellectual and practical power of Bayes. The knowledge brought to the problem is explicit and in the form of a prior.

As shown throughout this book, the objective of Bayesian Linear Regression is to develop a probability distribution over some critical distinction in a statistical problem or decision problem with some uncertainty that has to be considered. As such, Bayesian Linear Regression does not have a lick of optimization in it. The only optimization a decision maker wants to perform is maximizing expected utility, which John von Neumann[100] taught us all about some years ago.

This book has focused on a special case, the case in which the prediction is linear in the coefficients being estimated. There have been a number of applications of Bayesian Linear Regression in which the predictor is not linear in the coefficients. There are no closed forms and no easy interpretation in nonlinear cases. One has to resort to simulation. This book is not about simulation. It is about analytical forms and broad implications of Bayesian updating using data. This work leaves nonlinear Bayes and simulations to other works.

With the equations and results in this book, one can easily program and implement sophisticated computer programs to accomplish the calculations. This will be a boon for big data problems and small data problems alike.

Any idea that one has absolutely no ex ante information of any kind that logically preexists any data gathering exercise in any problem does not make intellectual sense and discards valuable, reliable, and valid information or judgment. Bayes allows us to meld prior judgment with data transparently in a systematic, logical, and auditable probabilistic fashion. Therein lies the intellectual and practical power of Bayes. The knowledge brought to the problem is explicit and in the form of a prior.

As shown throughout this book, the objective of Bayesian Linear Regression is to develop a probability distribution over some critical distinction in a statistical problem or decision problem with some uncertainty that has to be considered. As such, Bayesian Linear Regression does not have a lick of optimization in it. The only optimization a decision-maker wants to perform is maximizing expected utility, which John von Neumann taught us all about some years ago.

This book has focused on a special case: the case in which the prediction is linear in the coefficients being estimated. There have been a number of applications of Bayesian Linear Regression in which the predictor is not linear in the coefficients. There are no closed forms and no easy interpretation in nonlinear cases. One has to resort to simulation. This book is not about simulation. It is about analytical forms. This work leaves nonlinear Bayes to other works.

With the equations and results in this book, one can easily program and implement sophisticated computer programs to accomplish the calculations. This will be a boon for big data problems and small data problems alike.

Appendix A

APPENDIX A—PARTITIONED SUBDIVISIONS AND OTHER MATRIX ALGEBRAIC RESULTS

Those who ignore Statistics
are condemned to reinvent it.

Bradley Efron

A.1 Inverse of a Partitioned Matrix

We want a partitioned representation of the matrix A and of its inverse Z so that we can develop (unconditional) marginal distributions over individual model parameters. We write the fundamental relationship for the inverse of a matrix, writing the matrix and its inverse in partitioned form

$$\begin{bmatrix} Z_{11} & Z_{12} \\ Z_{21} & Z_{22} \end{bmatrix} \begin{bmatrix} A_{11} & A_{12} \\ A_{21} & A_{22} \end{bmatrix} = \begin{bmatrix} I & 0 \\ 0 & I \end{bmatrix} \quad .$$

We write the four equations implied by the four partitions of this matrix

$$\begin{aligned} Z_{11}A_{11} + Z_{12}A_{21} &= I \\ Z_{11}A_{12} + Z_{12}A_{22} &= 0 \\ Z_{21}A_{11} + Z_{22}A_{21} &= I \\ Z_{21}A_{12} + Z_{22}A_{22} &= 0 \quad . \end{aligned}$$

Let us right multiply the second equation by the inverse of A_{22} and manipulate

$$\begin{aligned} &Z_{11}A_{12}{A_{22}}^{-1} + Z_{12}A_{22}{A_{22}}^{-1} = 0 \\ &= Z_{11}A_{12}{A_{22}}^{-1} + Z_{12} \quad , \end{aligned}$$

so

$$Z_{12} = -Z_{11}A_{12}{A_{22}}^{-1} \quad .$$

Substituting this into the first equation gives the relationship

$$\begin{aligned} &Z_{11}A_{11} - Z_{11}A_{12}{A_{22}}^{-1}A_{21} = I \\ &Z_{11}\left(A_{11} - A_{12}{A_{22}}^{-1}A_{21}\right) = I \quad . \end{aligned}$$

Thus we know that the upper right portion of the inverse matrix is

$$Z_{11} = \left(A_{11} - A_{12}{A_{22}}^{-1}A_{21}\right)^{-1} \quad .$$

Substitution back into the second equation yields the result

$$\left(A_{11} - A_{12}{A_{22}}^{-1}A_{21}\right)^{-1}A_{12} + Z_{12}A_{22} = 0 \quad ,$$

so that

$$Z_{12} = -\left(A_{11} - A_{12}{A_{22}}^{-1}A_{21}\right)^{-1}A_{12}{A_{22}}^{-1} \quad .$$

If we reverse the indexes, we obtain the analogous relationships for the lower partition of the matrix

$$\begin{aligned} Z_{22} &= \left(A_{22} - A_{21}{A_{11}}^{-1}A_{12}\right)^{-1} \\ Z_{21} &= -\left(A_{22} - A_{21}{A_{11}}^{-1}A_{12}\right)^{-1}A_{21}{A_{11}}^{-1} \quad . \end{aligned}$$

Thus the partitioned inverse matrix can be written

$$\begin{aligned} &\begin{bmatrix} Z_{11} & Z_{12} \\ Z_{21} & Z_{22} \end{bmatrix} \\ &= \begin{bmatrix} \left(A_{11} - A_{12}{A_{22}}^{-1}A_{21}\right)^{-1} & -\left(A_{11} - A_{12}{A_{22}}^{-1}A_{21}\right)^{-1}A_{12}{A_{22}}^{-1} \\ -\left(A_{22} - A_{21}{A_{11}}^{-1}A_{12}\right)^{-1}A_{21}{A_{11}}^{-1} & \left(A_{22} - A_{21}{A_{11}}^{-1}A_{12}\right)^{-1} \end{bmatrix} \end{aligned}$$

We are going to need this when we compute a marginal distribution over a subset of the p model coefficients and a conditional distribution over the complementary subset of the p model coefficients.

A.2 Determinant of a Partitioned Matrix

We will also be needing the determinant of the same partitioned matrix

$$\begin{bmatrix} A_{11} & A_{12} \\ A_{21} & A_{22} \end{bmatrix} \quad .$$

Consider the following matrix product

$$\begin{bmatrix} I & A_{12}{A_{22}}^{-1} \\ 0 & I \end{bmatrix} \begin{bmatrix} A_{11} - A_{12}{A_{22}}^{-1}A_{21} & 0 \\ A_{21} & A_{22} \end{bmatrix} \quad .$$

We can show that this product is A as follows

$$\begin{aligned} &\begin{bmatrix} I & A_{12}{A_{22}}^{-1} \\ 0 & I \end{bmatrix} \begin{bmatrix} A_{11} - A_{12}{A_{22}}^{-1}A_{21} & 0 \\ A_{21} & A_{22} \end{bmatrix} \\ &= \begin{bmatrix} A_{11} - A_{12}{A_{22}}^{-1}A_{21} + A_{12}{A_{22}}^{-1}A_{21} & A_{12}{A_{22}}^{-1}A_{22} \\ A_{21} & A_{22} \end{bmatrix} \\ &= A \quad . \end{aligned}$$

The determinant of a block triangular matrix is the product of the determinants of the blocks (by principal cofactor analysis). Taking the determinant of both sides of the original matrix product, we have

$$\begin{aligned} &\det(I)\det(I)\det(A_{11} - A_{12}{A_{22}}^{-1}A_{21})\det(A_{22}) \\ &= \det(A) \quad . \end{aligned} \tag{A.1}$$

Thus the determinant of A is therefore

$$\det\left(A\right) = \det\left(A_{11} - A_{12}{A_{22}}^{-1}A_{21}\right)\det\left(A_{22}\right) \quad .$$

Appendix B

APPENDIX B–MULTIVARIATE NORMAL FUNDAMENTALS

The dangers of life are infinite, and among them is safety.

Goethe

B.1 Integrating Constant in Univariate Normal

Let's prove what the integral of $N(0,1)$, namely

$$N(0,1) = c_0 e^{-\frac{1}{2}x^2}$$

is. We want this to integrate to 1. We know that the square of this integral must integrate to 1, so

$$\begin{aligned} 1 &= \left[\int_{-\infty}^{\infty} c_0 e^{-\frac{1}{2}x^2} dx\right]\left[\int_{-\infty}^{\infty} c_0 e^{-\frac{1}{2}y^2} dy\right] \\ &= {c_0}^2 \int_{-\infty}^{\infty} dx \int_{-\infty}^{\infty} dy e^{-\frac{1}{2}(x^2+y^2)} \quad . \end{aligned}$$

Do a polar coordinate substitution

$$\begin{aligned} x = r\cos\theta &\Rightarrow dx = dr\cos\theta - r\sin\theta d\theta \\ y = r\sin\theta &\Rightarrow dy = dr\sin\theta + r\cos\theta d\theta \quad . \end{aligned}$$

The Jacobian and its determinant are

$$J = \begin{bmatrix} \cos\theta & -r\sin\theta \\ \sin\theta & r\cos\theta \end{bmatrix}$$
$$|J| = r \quad .$$

Thus the change of variables integral is

$$1 = {c_0}^2 \int_{-\infty}^{\infty} dx \int_{-\infty}^{\infty} dy e^{-\frac{1}{2}(x^2+y^2)}$$
$$= {c_0}^2 \int_0^{\infty}\int_0^{2\pi} e^{-\frac{1}{2}r^2} r dr d\theta \quad .$$

Let's make the substitution

$$x = \frac{1}{2}r^2$$
$$dx = r dr \quad .$$

Under this substitution, we can write

$$1 = {c_0}^2 \int_0^{2\pi}\int_0^{\infty} e^{-x} dx d\theta$$
$$= {c_0}^2 \int_0^{2\pi} -e^{-x} \Big|_0^{\infty} d\theta$$
$$= {c_0}^2 \int_0^{2\pi} d\theta = {c_0}^2 2\pi \quad .$$

We now equate the left and right hand sides and solve for the constant

$$1 = {c_0}^2 2\pi$$
$$\Rightarrow {c_0}^2 = \frac{1}{2\pi}$$
$$\Rightarrow c_0 = \frac{1}{\sqrt{2\pi}} \quad .$$

B.2 Multivariate Normal from Fundamentals

We are going to develop the multivariate normal distribution from fundamentals. Consider a distribution with the following functional form

$$f(x) = c_0 e^{-\frac{1}{2}(x-\mu)^T A(x-\mu)} \quad ,$$

in which x and μ are a p-vectors and A is a $p \times p$ symmetric, positive definite matrix. We know from the discussion in Section 2.5.9 that the symmetric, positive definite matrix A can be orthonormally diagonalized, meaning that

$$A = E\Lambda E^T$$

in which the matrix E is orthonormal (every column of E is orthogonal to every other column of E, the length of every column of E is 1, and the inverse and transpose of E are identical). We highlight that

$$\begin{aligned} E^{-1} &= E^T \\ |E| &= 1 \quad . \end{aligned}$$

We are going to prove a couple of things about positive definite matrices and then figure out the normalizing constant from first principles. We apply the eigenvalue equation

$$A\xi_i = \lambda_i \xi_i, \quad i = 1, ..., p \quad .$$

We also know that eigenvectors are unique up to a multiplicative constant. This is easy to show because any multiple of the eigenvector satisfies the foregoing equation. We know there are p eigenvectors and p eigenvalues, so we make a matrix E comprised of the p eigenvectors as columns

$$E = [\xi_1 \xi_2 ... \xi_p] \quad .$$

We can write all p of the eigenvalue equations in matrix form

$$\begin{aligned} AE &= A\,[\xi_1 \xi_2 ... \xi_p] \\ &= [\xi_1 \xi_2 ... \xi_p] \begin{bmatrix} \lambda_1 & & & \\ & \lambda_2 & & \\ & & \cdots & \\ & & & \lambda_p \end{bmatrix} \\ &= E\Lambda \quad . \end{aligned}$$

We can right multiply by the inverse of E to write

$$A = E\Lambda E^{-1} \quad .$$

Because the matrix A is symmetric, we know that it is its own transpose. Thus

$$\begin{aligned} A &= E\Lambda\left(E^{-1}\right) \\ A^T &= \left(E^{-1}\right)^T \Lambda^T E^T \quad . \end{aligned}$$

We know that A is symmetric so that

$$A = E\Lambda\left(E^{-1}\right) = \left(E^{-1}\right)^T \Lambda E^T \quad .$$

If we equate the two left multiplying matrices and separately the two right multiplying matrices, we can write

$$\begin{aligned} E &= \left(E^{-1}\right)^T \\ E^{-1} &= E^T \quad . \end{aligned}$$

This means that the diagonalization can be written

$$A = E\Lambda E^{-1} = E\Lambda E^T \quad .$$

The fact that the inverse is the transpose means that

$$E^T E = E E^T = I \quad .$$

This means that the eigenvectors are mutually orthonormal. Like eigenvectors have a length of 1, and unlike eigenvectors inner product to 0. It also means that the determinant of E is going to be 1. This results from the inverse equation

$$\left|E^T\right| |E| = |I| = |E| |E| = |E|^2 \quad .$$

This is going to be critical in getting our normalizing constant. Let us integrate the original postulated density

$$\begin{aligned} &\int_{-\infty}^{\infty} dx_1 ... \int_{-\infty}^{\infty} dx_p f(x) = 1 \\ &= c_0 \int_{-\infty}^{\infty} dx_1 ... \int_{-\infty}^{\infty} dx_p e^{-\frac{1}{2}(x-\mu)^T A (x-\mu)} \quad . \end{aligned}$$

This means that

$$\frac{1}{c_0} = \int_{-\infty}^{\infty} dx_1 ... \int_{-\infty}^{\infty} dx_p e^{-\frac{1}{2}(x-\mu)^T A (x-\mu)} \quad .$$

We now want to substitute the fundamental eigenvalue eigenvector relationship with the diagonal matrix and the transpose properly considered for the matrix A,

$$\begin{aligned}\frac{1}{c_0} &= \int_{-\infty}^{\infty} dx_1 ... \int_{-\infty}^{\infty} dx_p e^{-\frac{1}{2}(x-\mu)^T A(x-\mu)} \\ &= \int_{-\infty}^{\infty} dx_1 ... \int_{-\infty}^{\infty} dx_p e^{-\frac{1}{2}(x-\mu)^T E\Lambda E^T(x-\mu)} \\ &= \int_{-\infty}^{\infty} dx_1 ... \int_{-\infty}^{\infty} dx_p e^{-\frac{1}{2}[(x-\mu)^T E]\Lambda[E^T(x-\mu)]} \quad .\end{aligned}$$

We can then make a convenient change of variables substitution and use the Jacobian to ensure accuracy. The substitution will be

$$\begin{aligned}&y = E^T(x-\mu) = E^T x - E^T \mu \\ &\Rightarrow y = (x-\mu)^T E = x^T E - \mu^T E \quad .\end{aligned}$$

The Jacobian relationship will be

$$dy = |E|\, dx = dx$$

because the determinant of the E matrix is 1. That is really convenient because we don't have to introduce any determinant of the Jacobian terms into the substituted integral, which becomes

$$\begin{aligned}\frac{1}{c_0} &= \int_{-\infty}^{\infty} dy_1 ... \int_{-\infty}^{\infty} dy_p e^{-\frac{1}{2}y^T \Lambda y} \\ &= \int_{-\infty}^{\infty} dy_1 ... \int_{-\infty}^{\infty} dy_p e^{-\frac{1}{2}\left[\lambda_1 {y_1}^2 + ... + \lambda_p {y_p}^2\right]} \\ &= \int_{-\infty}^{\infty} dy_1 ... \int_{-\infty}^{\infty} dy_p e^{-\frac{1}{2}\left[\frac{{y_1}^2}{\left(\frac{1}{\lambda_1}\right)} + ... + \frac{{y_p}^2}{\left(\frac{1}{\lambda_p}\right)}\right]} \\ &= \left[\int_{-\infty}^{\infty} dy_1 e^{-\frac{1}{2}\frac{{y_1}^2}{\left(\frac{1}{\lambda_1}\right)}}\right] ... \left[\int_{-\infty}^{\infty} dy_p e^{-\frac{1}{2}\frac{{y_p}^2}{\left(\frac{1}{\lambda_p}\right)}}\right] \quad .\end{aligned}$$

The second step results from the fact that the matrix Λ is diagonal. Consider the ith of the foregoing integrals

$$\int_{-\infty}^{\infty} dy_i e^{-\frac{1}{2}\frac{y_i^2}{\left(\frac{1}{\lambda_i}\right)}} \quad .$$

Prepare to make the following substitution

$$\begin{aligned} w_i^2 &= \frac{y_i^2}{\left(\frac{1}{\lambda_i}\right)} \\ \Rightarrow y_i &= \frac{1}{\sqrt{\lambda_i}} w_i \\ \Rightarrow dy_i &= \frac{1}{\sqrt{\lambda_i}} dw_i \quad . \end{aligned}$$

Making the substitution implies

$$\int_{-\infty}^{\infty} dy_i e^{-\frac{1}{2}\frac{y_i^2}{\left(\frac{1}{\lambda_i}\right)}} = \frac{1}{\sqrt{\lambda_i}} \int_{-\infty}^{\infty} dw_i e^{-\frac{1}{2}w_i^2} = \left(\frac{2\pi}{\lambda_i}\right)^{\frac{1}{2}} \quad .$$

If we substitute this into the foregoing equation for the integrating constant, we write

$$\begin{aligned} \frac{1}{c_0} &= \left(\frac{2\pi}{\lambda_1}\right)^{\frac{1}{2}} \times \ldots \times \left(\frac{2\pi}{\lambda_p}\right)^{\frac{1}{2}} = \frac{1}{\sqrt{\lambda_1 \ldots \lambda_p}} (2\pi)^{\frac{p}{2}} \\ \Rightarrow c_0 &= (2\pi)^{-\frac{p}{2}} \sqrt{\lambda_1 \ldots \lambda_p} \quad . \end{aligned}$$

But what is the product of the eigenvalues? It is the determinant of the diagonal matrix of eigenvalues. Because the determinant of the matrix E is 1, it is therefore true that

$$\begin{aligned} A &= E \Lambda E^T \\ \Rightarrow |A| &= |E| \, |\Lambda| \, \left|E^T\right| = 1 \, |\Lambda| \, 1 = |\Lambda| \quad . \end{aligned}$$

This means that

$$\begin{aligned} c_0 &= (2\pi)^{-\frac{p}{2}} (\lambda_1 \ldots \lambda_p)^{\frac{1}{2}} \\ &= (2\pi)^{-\frac{p}{2}} |\Lambda|^{\frac{1}{2}} \\ &= (2\pi)^{-\frac{p}{2}} |A|^{\frac{1}{2}} \quad . \end{aligned}$$

The multivariate normal with the normalizing constant is therefore

$$f(x) = (2\pi)^{-\frac{p}{2}} |A|^{\frac{1}{2}} e^{-\frac{1}{2}(x-\mu)^T A (x-\mu)} \quad .$$

The standard representation of the normal distribution is to insert the inverse of the variance-covariance matrix into the exponent, i.e.,

$$A = \Sigma^{-1} \Rightarrow |A| = \frac{1}{|\Sigma|} \quad .$$

This leads to, and proves, the virtually always seen form

$$f(x) = \frac{1}{(2\pi)^{\frac{p}{2}} |\Sigma|^{\frac{1}{2}}} e^{-\frac{1}{2}(x-\mu)^T \Sigma^{-1}(x-\mu)} \quad .$$

Let's prove what the variance-covariance matrix is.

$$\begin{aligned}
\Sigma &\equiv \int_{-\infty}^{\infty} dx_1 ... \int_{-\infty}^{\infty} dx_p \, (x-\mu)\,(x-\mu)^T f(x) \\
&= \int_{-\infty}^{\infty} dx_1 ... \int_{-\infty}^{\infty} dx_p \, (x-\mu)\,(x-\mu)^T (2\pi)^{-\frac{p}{2}} |A|^{\frac{1}{2}} e^{-\frac{1}{2}(x-\mu)^T A (x-\mu)} \quad .
\end{aligned}$$

Make the same eigenvector substitution we made before, namely

$$\begin{aligned}
& y = E^T (x-\mu) \\
& \Rightarrow y = E^T x - E^T \mu \\
& \Rightarrow y^T = (x-\mu)^T E \\
& \Rightarrow Ey = EE^T (x-\mu) = (x-\mu) \\
& \Rightarrow (x-\mu)^T = y^T E^T \quad .
\end{aligned}$$

The Jacobian relationship will be

$$dy = |E| \, dx = dx \quad .$$

We get the relationship

$$\begin{aligned}
\Sigma &= \int_{-\infty}^{\infty} dx_1 ... \int_{-\infty}^{\infty} dx_p \, (x-\mu)\,(x-\mu)^T (2\pi)^{-\frac{p}{2}} |A|^{\frac{1}{2}} e^{-\frac{1}{2}(x-\mu)^T E \Lambda E^T (x-\mu)} \\
&= \int_{-\infty}^{\infty} dy_1 ... \int_{-\infty}^{\infty} dy_p E y y^T E^T (2\pi)^{-\frac{p}{2}} |A|^{\frac{1}{2}} e^{-\frac{1}{2} y^T \Lambda y} \\
&= E \left\{ \int_{-\infty}^{\infty} dy_1 ... \int_{-\infty}^{\infty} dy_p y y^T (2\pi)^{-\frac{p}{2}} |A|^{\frac{1}{2}} e^{-\frac{1}{2} y^T \Lambda y} \right\} E^T \quad .
\end{aligned}$$

We are now going to make one more substitution

$$\begin{aligned}
& y^T \Lambda y = y^T \Lambda^{\frac{1}{2}} \Lambda^{\frac{1}{2}} y \\
& w = \Lambda^{\frac{1}{2}} y \Rightarrow \Lambda^{-\frac{1}{2}} w = y \\
& \Rightarrow w^T \Lambda^{-\frac{1}{2}} = y^T \\
& dw = \left| \Lambda^{\frac{1}{2}} \right| dy \Rightarrow \left| \Lambda^{-\frac{1}{2}} \right| dw = dy \quad .
\end{aligned}$$

We write

$$\begin{aligned}
\Sigma &= E \left\{ \int_{-\infty}^{\infty} dw_1 ... \int_{-\infty}^{\infty} dw_p \left| \Lambda^{-\frac{1}{2}} \right| \Lambda^{-\frac{1}{2}} w w^T \Lambda^{-\frac{1}{2}} (2\pi)^{-\frac{p}{2}} |A|^{\frac{1}{2}} e^{-\frac{1}{2} w^T w} \right\} E^T \\
&= \left| \Lambda^{-\frac{1}{2}} \right| |A|^{\frac{1}{2}} E \Lambda^{-\frac{1}{2}} \left\{ \int_{-\infty}^{\infty} dw_1 ... \int_{-\infty}^{\infty} dw_p w w^T (2\pi)^{-\frac{p}{2}} e^{-\frac{1}{2} w^T w} \right\} \Lambda^{-\frac{1}{2}} E^T \\
&= E \Lambda^{-\frac{1}{2}} \left\{ \int_{-\infty}^{\infty} dw_1 ... \int_{-\infty}^{\infty} dw_p w w^T (2\pi)^{-\frac{p}{2}} e^{-\frac{1}{2} w^T w} \right\} \Lambda^{-\frac{1}{2}} E^T \\
&= E \Lambda^{-\frac{1}{2}} I \Lambda^{-\frac{1}{2}} E^T = E \Lambda^{-1} E^T \quad .
\end{aligned}$$

Now we have it. Let's take the inverse of both sides

$$\begin{aligned}
\Sigma &= E \Lambda^{-1} E^T \\
\Sigma^{-1} &= E \Lambda E^T \quad .
\end{aligned}$$

This was precisely the equation for the matrix A. Thus

$$\begin{aligned}
\Sigma^{-1} &= A \\
\Sigma &= A^{-1} \quad .
\end{aligned}$$

Appendix C

APPENDIX C–MATRIX DERIVATIVES OF SEVERAL TYPES

> ***Probability fractions arise from our knowledge and from our ignorance.***
>
> ***Ian Hacking***

This section contains a brief derivation of important matrix derivatives. A matrix derivative is a convenient way to organize the derivatives of functions that occur in one matrix by variables that occur in another matrix. It is very convenient, and it provides the basis for our derivations in Chapters 15 through 17 of the multivariate Bayesian inference.

C.1 Commutation Matrix

The commutation matrix will be needed throughout this development. We develop and illustrate it here. It will emerge in the forthcoming developments. The square commutation matrix K_{mm} is defined by the relationship between the vectorization of a matrix A and the vectorization of the transpose of the matrix A, namely

$$K_{mm} vec\left(A\right) = vec\left(A^T\right) \quad .$$

The commutation matrix drives the vectorization of a matrix into the vectorization of the transpose of that matrix. For any square matrix with any particular structure, it can be verified that the commutation matrix has the following entries (illustrated for $m = 3$). Performing the foregoing multiplication symbolically for K_{mm} and A will verify the relationship

$$K_{mm} = \begin{bmatrix} 1 & 0 & 0 & 0 & 0 & 0 & 0 & 0 & 0 \\ 0 & 0 & 0 & 1 & 0 & 0 & 0 & 0 & 0 \\ 0 & 0 & 0 & 0 & 0 & 0 & 1 & 0 & 0 \\ 0 & 1 & 0 & 0 & 0 & 0 & 0 & 0 & 0 \\ 0 & 0 & 0 & 0 & 1 & 0 & 0 & 0 & 0 \\ 0 & 0 & 0 & 0 & 0 & 0 & 0 & 1 & 0 \\ 0 & 0 & 1 & 0 & 0 & 0 & 0 & 0 & 0 \\ 0 & 0 & 0 & 0 & 0 & 1 & 0 & 0 & 0 \\ 0 & 0 & 0 & 0 & 0 & 0 & 0 & 0 & 1 \end{bmatrix} .$$

We see the commutation matrix entering all our future developments. It is instructive to left multiply and right multiply a matrix by the commutation matrix to watch how rows and columns are shifted. It is well known that

$$K_{mm}K_{mm} = I_{mm} \quad ,$$

and therefore that the commutation matrix K_{mm} is its own inverse.

C.2 Development of Matrix Derivatives–Kronecker Type 1 Derivative

In this section, we develop the derivative of a matrix with respect to a matrix. In particular, we want to think of a matrix derivative as a collection of partial matrix derivatives of each of the elements of that matrix. We will expand this out in detail, and we will see why it is a potentially important way to array matrices of partial derivatives in a comprehensive way. To begin, ponder the following type of derivative structure (which uses the Kronecker product):

$$\frac{\partial Y}{\partial X} = Y \otimes \frac{\partial}{\partial X} \quad ,$$

in which we interpret the differentiation matrix to be the following

$$\frac{\partial}{\partial X} = \begin{bmatrix} \frac{\partial}{\partial x_{11}} & \cdots & \frac{\partial}{\partial x_{1m}} \\ \vdots & & \vdots \\ \frac{\partial}{\partial x_{m1}} & \cdots & \frac{\partial}{\partial x_{mm}} \end{bmatrix} .$$

If we consider the partition of a matrix indexed tu, meaning the tth row and u^{th} column of the y matrix (numerator), we can write that partition as follows

$$\left[\frac{\partial Y}{\partial X}\right]_{tu} = \begin{bmatrix} \frac{\partial y_{tu}}{\partial x_{11}} & \cdots & \frac{\partial y_{tu}}{\partial x_{1m}} \\ \vdots & & \vdots \\ \frac{\partial y_{tu}}{\partial x_{m1}} & \cdots & \frac{\partial y_{tu}}{\partial x_{mm}} \end{bmatrix} \quad t = 1, ..., m; \; u = 1, ..., m; \quad .$$

We are going to create mosaic matrixes for different tu pairs to create the full matrix derivative. When we do so, we obtain the following equation for the matrix derivative of the first type as characterized in Kollo and von Rosen[57] p. 124

$$\frac{\partial Y}{\partial X} = Y \otimes \frac{\partial}{\partial X} = \sum_{i=1}^{m}\sum_{j=1}^{m}\sum_{k=1}^{m}\sum_{\ell=1}^{m} \frac{\partial y_{ij}}{\partial x_{k\ell}}(r_i {s_j}^T) \otimes (f_k {g_\ell}^T) \quad .$$

in which r_i is a unit vector of length m with a 1 in the ith position, s_j is a unit vector of length m with a 1 in the jth position, f_k is a unit vector of length m with a 1 in the kth position, and g_ℓ is a unit vector of length m with a 1 in the $\ell^t h$ position. We see that this summation enumerates each and every partial derivative once (and only once) and places it into a matrix in a uniquely defined position. Therein lies the power (and concept) of the matrix derivative.

We now structure a very small illustrative example in the 2×2 case, the key terms of which are

$$X = \begin{bmatrix} x_{11} & x_{12} \\ x_{21} & x_{22} \end{bmatrix} \quad \frac{\partial}{\partial X} = \begin{bmatrix} \frac{\partial}{\partial x_{11}} & \frac{\partial}{\partial x_{12}} \\ \frac{\partial}{\partial x_{21}} & \frac{\partial}{\partial x_{22}} \end{bmatrix} \quad Y = \begin{bmatrix} y_{11} & y_{12} \\ y_{21} & y_{22} \end{bmatrix} \quad .$$

When we substitute these terms into the foregoing definition, we obtain the equation

$$\frac{\partial Y}{\partial X} = Y \otimes \frac{\partial}{\partial X} = \begin{bmatrix} y_{11}\begin{bmatrix} \frac{\partial}{\partial x_{11}} & \frac{\partial}{\partial x_{12}} \\ \frac{\partial}{\partial x_{21}} & \frac{\partial}{\partial x_{22}} \end{bmatrix} & y_{12}\begin{bmatrix} \frac{\partial}{\partial x_{11}} & \frac{\partial}{\partial x_{12}} \\ \frac{\partial}{\partial x_{21}} & \frac{\partial}{\partial x_{22}} \end{bmatrix} \\ y_{21}\begin{bmatrix} \frac{\partial}{\partial x_{11}} & \frac{\partial}{\partial x_{12}} \\ \frac{\partial}{\partial x_{21}} & \frac{\partial}{\partial x_{22}} \end{bmatrix} & y_{22}\begin{bmatrix} \frac{\partial}{\partial x_{11}} & \frac{\partial}{\partial x_{12}} \\ \frac{\partial}{\partial x_{21}} & \frac{\partial}{\partial x_{22}} \end{bmatrix} \end{bmatrix}$$

$$= \begin{bmatrix} \frac{\partial y_{11}}{\partial x_{11}} & \frac{\partial y_{11}}{\partial x_{12}} & \frac{\partial y_{12}}{\partial x_{11}} & \frac{\partial y_{12}}{\partial x_{12}} \\ \frac{\partial y_{11}}{\partial x_{21}} & \frac{\partial y_{11}}{\partial x_{22}} & \frac{\partial y_{12}}{\partial x_{21}} & \frac{\partial y_{12}}{\partial x_{22}} \\ \frac{\partial y_{21}}{\partial x_{11}} & \frac{\partial y_{21}}{\partial x_{12}} & \frac{\partial y_{22}}{\partial x_{11}} & \frac{\partial y_{22}}{\partial x_{12}} \\ \frac{\partial y_{21}}{\partial x_{21}} & \frac{\partial y_{21}}{\partial x_{22}} & \frac{\partial y_{22}}{\partial x_{21}} & \frac{\partial y_{22}}{\partial x_{22}} \end{bmatrix} \quad .$$

This is the first method of matrix differentiation. Notice how the tu blocks are laid out as the left-hand multiplier in a Kronecker product would lay them out. That is why we get this form of matrix derivative. This is one way to enumerate the partial derivative with respect to every matrix element. This generalizes to any level of m, not just $m = 2$.

Each type of matrix derivative is a subtle and sometimes ambiguous concept. Each requires crystal-clear definition and clarification. We begin with this Kronecker Type 1 definition of matrix derivatives, which appears as equation (1.4.2) in Kollo and von Rosen[57]. We shall see its basis clearly when we write the matrices. This is the first of two Kronecker relationships from Wiki.[1]

$$
\begin{aligned}
(AC) \otimes (BD) &= (A \otimes B)\,(C \otimes D) \\
(A \otimes B)^T &= A^T \otimes B^T \quad .
\end{aligned}
$$

If we apply the first relationship to the unit vectors, we obtain

$$
\begin{aligned}
(AC) \otimes (BD) &= (A \otimes B)\,(C \otimes D) \\
\left(d_i {e_j}^T\right) \otimes \left(f_k {g_\ell}^T\right) &= (d_i \otimes f_k)\left({e_j}^T \otimes {g_\ell}^T\right) \\
&= (d_i \otimes f_k)\,(e_j \otimes g_\ell)^T \quad .
\end{aligned}
$$

This allows us to write equivalently

$$
\frac{\partial Y}{\partial X} = \sum_{i=1}^{m} \sum_{j=1}^{m} \sum_{k=1}^{m} \sum_{\ell=1}^{m} \frac{\partial y_{ij}}{\partial x_{k\ell}} (d_i \otimes f_k)\,(e_j \otimes g_\ell)^T \quad .
$$

The first of these two equations (and the second as well) lead, in the case of $m = 3$, for a nonsymmetric matrix of derivatives, to the following form of the matrix derivative

$$
\frac{\partial Y}{\partial X} =
$$

$$
\begin{bmatrix}
\frac{\partial y_{11}}{\partial x_{11}} & \frac{\partial y_{11}}{\partial x_{12}} & \frac{\partial y_{11}}{\partial x_{13}} & \frac{\partial y_{12}}{\partial x_{11}} & \frac{\partial y_{12}}{\partial x_{12}} & \frac{\partial y_{12}}{\partial x_{13}} & \frac{\partial y_{13}}{\partial x_{11}} & \frac{\partial y_{13}}{\partial x_{12}} & \frac{\partial y_{13}}{\partial x_{13}} \\
\frac{\partial y_{11}}{\partial x_{21}} & \frac{\partial y_{11}}{\partial x_{22}} & \frac{\partial y_{11}}{\partial x_{23}} & \frac{\partial y_{12}}{\partial x_{21}} & \frac{\partial y_{12}}{\partial x_{22}} & \frac{\partial y_{12}}{\partial x_{23}} & \frac{\partial y_{13}}{\partial x_{21}} & \frac{\partial y_{13}}{\partial x_{22}} & \frac{\partial y_{13}}{\partial x_{23}} \\
\frac{\partial y_{11}}{\partial x_{31}} & \frac{\partial y_{11}}{\partial x_{32}} & \frac{\partial y_{11}}{\partial x_{33}} & \frac{\partial y_{12}}{\partial x_{31}} & \frac{\partial y_{12}}{\partial x_{32}} & \frac{\partial y_{12}}{\partial x_{33}} & \frac{\partial y_{13}}{\partial x_{31}} & \frac{\partial y_{13}}{\partial x_{32}} & \frac{\partial y_{13}}{\partial x_{33}} \\
\frac{\partial y_{21}}{\partial x_{11}} & \frac{\partial y_{21}}{\partial x_{12}} & \frac{\partial y_{21}}{\partial x_{13}} & \frac{\partial y_{22}}{\partial x_{11}} & \frac{\partial y_{22}}{\partial x_{12}} & \frac{\partial y_{22}}{\partial x_{13}} & \frac{\partial y_{23}}{\partial x_{11}} & \frac{\partial y_{23}}{\partial x_{12}} & \frac{\partial y_{23}}{\partial x_{13}} \\
\frac{\partial y_{21}}{\partial x_{21}} & \frac{\partial y_{21}}{\partial x_{22}} & \frac{\partial y_{21}}{\partial x_{23}} & \frac{\partial y_{22}}{\partial x_{21}} & \frac{\partial y_{22}}{\partial x_{22}} & \frac{\partial y_{22}}{\partial x_{23}} & \frac{\partial y_{23}}{\partial x_{21}} & \frac{\partial y_{23}}{\partial x_{22}} & \frac{\partial y_{23}}{\partial x_{23}} \\
\frac{\partial y_{21}}{\partial x_{31}} & \frac{\partial y_{21}}{\partial x_{32}} & \frac{\partial y_{21}}{\partial x_{33}} & \frac{\partial y_{22}}{\partial x_{31}} & \frac{\partial y_{22}}{\partial x_{32}} & \frac{\partial y_{22}}{\partial x_{33}} & \frac{\partial y_{23}}{\partial x_{31}} & \frac{\partial y_{23}}{\partial x_{32}} & \frac{\partial y_{23}}{\partial x_{33}} \\
\frac{\partial y_{31}}{\partial x_{11}} & \frac{\partial y_{31}}{\partial x_{12}} & \frac{\partial y_{31}}{\partial x_{13}} & \frac{\partial y_{32}}{\partial x_{11}} & \frac{\partial y_{32}}{\partial x_{12}} & \frac{\partial y_{32}}{\partial x_{13}} & \frac{\partial y_{33}}{\partial x_{11}} & \frac{\partial y_{33}}{\partial x_{12}} & \frac{\partial y_{33}}{\partial x_{13}} \\
\frac{\partial y_{31}}{\partial x_{21}} & \frac{\partial y_{31}}{\partial x_{22}} & \frac{\partial y_{31}}{\partial x_{23}} & \frac{\partial y_{32}}{\partial x_{21}} & \frac{\partial y_{32}}{\partial x_{22}} & \frac{\partial y_{32}}{\partial x_{23}} & \frac{\partial y_{33}}{\partial x_{21}} & \frac{\partial y_{33}}{\partial x_{22}} & \frac{\partial y_{33}}{\partial x_{23}} \\
\frac{\partial y_{31}}{\partial x_{31}} & \frac{\partial y_{31}}{\partial x_{32}} & \frac{\partial y_{31}}{\partial x_{33}} & \frac{\partial y_{32}}{\partial x_{31}} & \frac{\partial y_{32}}{\partial x_{32}} & \frac{\partial y_{32}}{\partial x_{33}} & \frac{\partial y_{33}}{\partial x_{31}} & \frac{\partial y_{33}}{\partial x_{32}} & \frac{\partial y_{33}}{\partial x_{33}}
\end{bmatrix} .
$$

We see the tu mosaicing evident in this matrix. If we modify this to represent a symmetric matrix X and Y

$$
\frac{\partial Y}{\partial X}_{Symmetric} =
$$

[1]See https://en.wikipedia.org/wiki/Kronecker_product.

$$\begin{bmatrix}
\frac{\partial y_{11}}{\partial x_{11}} & \frac{\partial y_{11}}{\partial x_{12}} & \frac{\partial y_{11}}{\partial x_{13}} & \frac{\partial y_{12}}{\partial x_{11}} & \frac{\partial y_{12}}{\partial x_{12}} & \frac{\partial y_{12}}{\partial x_{13}} & \frac{\partial y_{13}}{\partial x_{11}} & \frac{\partial y_{13}}{\partial x_{12}} & \frac{\partial y_{13}}{\partial x_{13}} \\
\frac{\partial y_{11}}{\partial x_{12}} & \frac{\partial y_{11}}{\partial x_{22}} & \frac{\partial y_{11}}{\partial x_{23}} & \frac{\partial y_{12}}{\partial x_{12}} & \frac{\partial y_{12}}{\partial x_{22}} & \frac{\partial y_{12}}{\partial x_{23}} & \frac{\partial y_{13}}{\partial x_{12}} & \frac{\partial y_{13}}{\partial x_{22}} & \frac{\partial y_{13}}{\partial x_{23}} \\
\frac{\partial y_{11}}{\partial x_{13}} & \frac{\partial y_{11}}{\partial x_{23}} & \frac{\partial y_{11}}{\partial x_{33}} & \frac{\partial y_{12}}{\partial x_{13}} & \frac{\partial y_{12}}{\partial x_{23}} & \frac{\partial y_{12}}{\partial x_{33}} & \frac{\partial y_{13}}{\partial x_{13}} & \frac{\partial y_{13}}{\partial x_{23}} & \frac{\partial y_{13}}{\partial x_{33}} \\
\frac{\partial y_{12}}{\partial x_{11}} & \frac{\partial y_{12}}{\partial x_{12}} & \frac{\partial y_{12}}{\partial x_{13}} & \frac{\partial y_{22}}{\partial x_{11}} & \frac{\partial y_{22}}{\partial x_{12}} & \frac{\partial y_{22}}{\partial x_{13}} & \frac{\partial y_{23}}{\partial x_{11}} & \frac{\partial y_{23}}{\partial x_{12}} & \frac{\partial y_{23}}{\partial x_{13}} \\
\frac{\partial y_{12}}{\partial x_{12}} & \frac{\partial y_{12}}{\partial x_{22}} & \frac{\partial y_{12}}{\partial x_{23}} & \frac{\partial y_{22}}{\partial x_{12}} & \frac{\partial y_{22}}{\partial x_{22}} & \frac{\partial y_{22}}{\partial x_{23}} & \frac{\partial y_{23}}{\partial x_{12}} & \frac{\partial y_{23}}{\partial x_{22}} & \frac{\partial y_{23}}{\partial x_{23}} \\
\frac{\partial y_{12}}{\partial x_{13}} & \frac{\partial y_{12}}{\partial x_{23}} & \frac{\partial y_{12}}{\partial x_{33}} & \frac{\partial y_{22}}{\partial x_{13}} & \frac{\partial y_{22}}{\partial x_{23}} & \frac{\partial y_{22}}{\partial x_{33}} & \frac{\partial y_{23}}{\partial x_{13}} & \frac{\partial y_{23}}{\partial x_{23}} & \frac{\partial y_{23}}{\partial x_{33}} \\
\frac{\partial y_{13}}{\partial x_{11}} & \frac{\partial y_{13}}{\partial x_{12}} & \frac{\partial y_{13}}{\partial x_{13}} & \frac{\partial y_{23}}{\partial x_{11}} & \frac{\partial y_{23}}{\partial x_{12}} & \frac{\partial y_{23}}{\partial x_{13}} & \frac{\partial y_{33}}{\partial x_{11}} & \frac{\partial y_{33}}{\partial x_{12}} & \frac{\partial y_{33}}{\partial x_{13}} \\
\frac{\partial y_{13}}{\partial x_{12}} & \frac{\partial y_{13}}{\partial x_{22}} & \frac{\partial y_{13}}{\partial x_{23}} & \frac{\partial y_{23}}{\partial x_{12}} & \frac{\partial y_{23}}{\partial x_{22}} & \frac{\partial y_{23}}{\partial x_{23}} & \frac{\partial y_{33}}{\partial x_{12}} & \frac{\partial y_{33}}{\partial x_{22}} & \frac{\partial y_{33}}{\partial x_{23}} \\
\frac{\partial y_{13}}{\partial x_{13}} & \frac{\partial y_{13}}{\partial x_{23}} & \frac{\partial y_{13}}{\partial x_{33}} & \frac{\partial y_{23}}{\partial x_{13}} & \frac{\partial y_{23}}{\partial x_{23}} & \frac{\partial y_{23}}{\partial x_{33}} & \frac{\partial y_{33}}{\partial x_{13}} & \frac{\partial y_{33}}{\partial x_{23}} & \frac{\partial y_{33}}{\partial x_{33}}
\end{bmatrix} .$$

The derivative of this matrix with respect to itself becomes

$$\frac{\partial X}{\partial X}_{Symmetric} = \begin{bmatrix}
1 & 0 & 0 & 0 & 1 & 0 & 0 & 0 & 1 \\
0 & 0 & 0 & 1 & 0 & 0 & 0 & 0 & 0 \\
0 & 0 & 0 & 0 & 0 & 0 & 1 & 0 & 0 \\
0 & 1 & 0 & 0 & 0 & 0 & 0 & 0 & 0 \\
1 & 0 & 0 & 0 & 1 & 0 & 0 & 0 & 1 \\
0 & 0 & 0 & 0 & 0 & 0 & 0 & 1 & 0 \\
0 & 0 & 1 & 0 & 0 & 0 & 0 & 0 & 0 \\
0 & 0 & 0 & 0 & 0 & 1 & 0 & 0 & 0 \\
1 & 0 & 0 & 0 & 1 & 0 & 0 & 0 & 1
\end{bmatrix} .$$

C.3 A Generalized (Weighted) Matrix Derivative

This gives us a helpful hint how to develop a generalized matrix derivative. Rather than using the standard notion of a derivative,

$$\frac{\partial Y}{\partial X} = \sum_{i=1}^{m} \sum_{j=1}^{m} \sum_{k=1}^{m} \sum_{\ell=1}^{m} \frac{\partial y_{ij}}{\partial x_{k\ell}} \left(d_i {e_j}^T\right) \otimes \left(f_k {g_\ell}^T\right) \quad ,$$

suppose we modify it slightly to write

$$\begin{aligned}
\frac{\partial Y}{\partial X} &= \sum_{i=1}^{m} \sum_{j=1}^{m} \sum_{k=1}^{m} \sum_{\ell=1}^{m} \frac{\partial y_{ij}}{\partial x_{k\ell}} \left(d_i {e_j}^T\right) \otimes \varepsilon_{k\ell} \left(f_k {g_\ell}^T\right) \\
&= \sum_{i=1}^{m} \sum_{j=1}^{m} \sum_{k=1}^{m} \sum_{\ell=1}^{m} \frac{\partial y_{ij}}{\partial x_{k\ell}} \left(E_{ij} \otimes \varepsilon_{k\ell} E_{k\ell}\right) \qquad \text{(C.1)} \\
\varepsilon_{k\ell} &= \begin{cases} 1 & k = \ell \\ \frac{1}{2} & k \neq \ell \end{cases} .
\end{aligned}$$

Under this definition, it is useful to loop through each and every index and generate the matrix for a given case. Assuming $m = 3$, the derivative matrix for a generally shaped matrix X and Y is

$$\frac{\partial Y}{\partial X} =$$

$$\begin{bmatrix}
\frac{\partial y_{11}}{\partial x_{11}} & \frac{1}{2}\frac{\partial y_{11}}{\partial x_{12}} & \frac{1}{2}\frac{\partial y_{11}}{\partial x_{13}} & \frac{\partial y_{12}}{\partial x_{11}} & \frac{1}{2}\frac{\partial y_{12}}{\partial x_{12}} & \frac{1}{2}\frac{\partial y_{12}}{\partial x_{13}} & \frac{\partial y_{13}}{\partial x_{11}} & \frac{1}{2}\frac{\partial y_{13}}{\partial x_{12}} & \frac{1}{2}\frac{\partial y_{13}}{\partial x_{13}} \\
\frac{1}{2}\frac{\partial y_{11}}{\partial x_{21}} & \frac{\partial y_{11}}{\partial x_{22}} & \frac{1}{2}\frac{\partial y_{11}}{\partial x_{23}} & \frac{1}{2}\frac{\partial y_{12}}{\partial x_{21}} & \frac{\partial y_{12}}{\partial x_{22}} & \frac{1}{2}\frac{\partial y_{12}}{\partial x_{23}} & \frac{1}{2}\frac{\partial y_{13}}{\partial x_{21}} & \frac{\partial y_{13}}{\partial x_{22}} & \frac{1}{2}\frac{\partial y_{13}}{\partial x_{23}} \\
\frac{1}{2}\frac{\partial y_{11}}{\partial x_{31}} & \frac{1}{2}\frac{\partial y_{11}}{\partial x_{32}} & \frac{\partial y_{11}}{\partial x_{33}} & \frac{1}{2}\frac{\partial y_{12}}{\partial x_{31}} & \frac{1}{2}\frac{\partial y_{12}}{\partial x_{32}} & \frac{\partial y_{12}}{\partial x_{33}} & \frac{1}{2}\frac{\partial y_{13}}{\partial x_{31}} & \frac{1}{2}\frac{\partial y_{13}}{\partial x_{32}} & \frac{\partial y_{13}}{\partial x_{33}} \\
\frac{\partial y_{21}}{\partial x_{11}} & \frac{1}{2}\frac{\partial y_{21}}{\partial x_{12}} & \frac{1}{2}\frac{\partial y_{21}}{\partial x_{13}} & \frac{\partial y_{22}}{\partial x_{11}} & \frac{1}{2}\frac{\partial y_{22}}{\partial x_{12}} & \frac{1}{2}\frac{\partial y_{22}}{\partial x_{13}} & \frac{\partial y_{23}}{\partial x_{11}} & \frac{1}{2}\frac{\partial y_{23}}{\partial x_{12}} & \frac{1}{2}\frac{\partial y_{23}}{\partial x_{13}} \\
\frac{1}{2}\frac{\partial y_{21}}{\partial x_{21}} & \frac{\partial y_{21}}{\partial x_{22}} & \frac{1}{2}\frac{\partial y_{21}}{\partial x_{23}} & \frac{1}{2}\frac{\partial y_{22}}{\partial x_{21}} & \frac{\partial y_{22}}{\partial x_{22}} & \frac{1}{2}\frac{\partial y_{22}}{\partial x_{23}} & \frac{1}{2}\frac{\partial y_{23}}{\partial x_{21}} & \frac{\partial y_{23}}{\partial x_{22}} & \frac{1}{2}\frac{\partial y_{23}}{\partial x_{23}} \\
\frac{1}{2}\frac{\partial y_{21}}{\partial x_{31}} & \frac{1}{2}\frac{\partial y_{21}}{\partial x_{32}} & \frac{\partial y_{21}}{\partial x_{33}} & \frac{1}{2}\frac{\partial y_{22}}{\partial x_{31}} & \frac{1}{2}\frac{\partial y_{22}}{\partial x_{32}} & \frac{\partial y_{22}}{\partial x_{33}} & \frac{1}{2}\frac{\partial y_{23}}{\partial x_{31}} & \frac{1}{2}\frac{\partial y_{23}}{\partial x_{32}} & \frac{\partial y_{23}}{\partial x_{33}} \\
\frac{\partial y_{31}}{\partial x_{11}} & \frac{1}{2}\frac{\partial y_{31}}{\partial x_{12}} & \frac{1}{2}\frac{\partial y_{31}}{\partial x_{13}} & \frac{\partial y_{32}}{\partial x_{11}} & \frac{1}{2}\frac{\partial y_{32}}{\partial x_{12}} & \frac{1}{2}\frac{\partial y_{32}}{\partial x_{13}} & \frac{\partial y_{33}}{\partial x_{11}} & \frac{1}{2}\frac{\partial y_{33}}{\partial x_{12}} & \frac{1}{2}\frac{\partial y_{33}}{\partial x_{13}} \\
\frac{1}{2}\frac{\partial y_{31}}{\partial x_{21}} & \frac{\partial y_{31}}{\partial x_{22}} & \frac{1}{2}\frac{\partial y_{31}}{\partial x_{23}} & \frac{1}{2}\frac{\partial y_{32}}{\partial x_{21}} & \frac{\partial y_{32}}{\partial x_{22}} & \frac{1}{2}\frac{\partial y_{32}}{\partial x_{23}} & \frac{1}{2}\frac{\partial y_{33}}{\partial x_{21}} & \frac{\partial y_{33}}{\partial x_{22}} & \frac{1}{2}\frac{\partial y_{33}}{\partial x_{23}} \\
\frac{1}{2}\frac{\partial y_{31}}{\partial x_{31}} & \frac{1}{2}\frac{\partial y_{31}}{\partial x_{32}} & \frac{\partial y_{31}}{\partial x_{33}} & \frac{1}{2}\frac{\partial y_{32}}{\partial x_{31}} & \frac{1}{2}\frac{\partial y_{32}}{\partial x_{32}} & \frac{\partial y_{32}}{\partial x_{33}} & \frac{1}{2}\frac{\partial y_{33}}{\partial x_{31}} & \frac{1}{2}\frac{\partial y_{33}}{\partial x_{32}} & \frac{\partial y_{33}}{\partial x_{33}}
\end{bmatrix}$$

In the special case of symmetric matrices X and Y, this becomes

$$\frac{\partial Y}{\partial X}_{Symmetric} =$$

$$\begin{bmatrix}
\frac{\partial y_{11}}{\partial x_{11}} & \frac{1}{2}\frac{\partial y_{11}}{\partial x_{12}} & \frac{1}{2}\frac{\partial y_{11}}{\partial x_{13}} & \frac{\partial y_{12}}{\partial x_{11}} & \frac{1}{2}\frac{\partial y_{12}}{\partial x_{12}} & \frac{1}{2}\frac{\partial y_{12}}{\partial x_{13}} & \frac{\partial y_{13}}{\partial x_{11}} & \frac{1}{2}\frac{\partial y_{13}}{\partial x_{12}} & \frac{1}{2}\frac{\partial y_{13}}{\partial x_{13}} \\
\frac{1}{2}\frac{\partial y_{11}}{\partial x_{12}} & \frac{\partial y_{11}}{\partial x_{22}} & \frac{1}{2}\frac{\partial y_{11}}{\partial x_{23}} & \frac{1}{2}\frac{\partial y_{12}}{\partial x_{12}} & \frac{\partial y_{12}}{\partial x_{22}} & \frac{1}{2}\frac{\partial y_{12}}{\partial x_{23}} & \frac{1}{2}\frac{\partial y_{13}}{\partial x_{12}} & \frac{\partial y_{13}}{\partial x_{22}} & \frac{1}{2}\frac{\partial y_{13}}{\partial x_{23}} \\
\frac{1}{2}\frac{\partial y_{11}}{\partial x_{13}} & \frac{1}{2}\frac{\partial y_{11}}{\partial x_{23}} & \frac{\partial y_{11}}{\partial x_{33}} & \frac{1}{2}\frac{\partial y_{12}}{\partial x_{13}} & \frac{1}{2}\frac{\partial y_{12}}{\partial x_{23}} & \frac{\partial y_{12}}{\partial x_{33}} & \frac{1}{2}\frac{\partial y_{13}}{\partial x_{13}} & \frac{1}{2}\frac{\partial y_{13}}{\partial x_{23}} & \frac{\partial y_{13}}{\partial x_{33}} \\
\frac{\partial y_{12}}{\partial x_{11}} & \frac{1}{2}\frac{\partial y_{12}}{\partial x_{12}} & \frac{1}{2}\frac{\partial y_{12}}{\partial x_{13}} & \frac{\partial y_{22}}{\partial x_{11}} & \frac{1}{2}\frac{\partial y_{22}}{\partial x_{12}} & \frac{1}{2}\frac{\partial y_{22}}{\partial x_{13}} & \frac{\partial y_{23}}{\partial x_{11}} & \frac{1}{2}\frac{\partial y_{23}}{\partial x_{12}} & \frac{1}{2}\frac{\partial y_{23}}{\partial x_{13}} \\
\frac{1}{2}\frac{\partial y_{12}}{\partial x_{12}} & \frac{\partial y_{12}}{\partial x_{22}} & \frac{1}{2}\frac{\partial y_{12}}{\partial x_{23}} & \frac{1}{2}\frac{\partial y_{22}}{\partial x_{12}} & \frac{\partial y_{22}}{\partial x_{22}} & \frac{1}{2}\frac{\partial y_{22}}{\partial x_{23}} & \frac{1}{2}\frac{\partial y_{23}}{\partial x_{12}} & \frac{\partial y_{23}}{\partial x_{22}} & \frac{1}{2}\frac{\partial y_{23}}{\partial x_{23}} \\
\frac{1}{2}\frac{\partial y_{12}}{\partial x_{13}} & \frac{1}{2}\frac{\partial y_{12}}{\partial x_{23}} & \frac{\partial y_{12}}{\partial x_{33}} & \frac{1}{2}\frac{\partial y_{22}}{\partial x_{13}} & \frac{1}{2}\frac{\partial y_{22}}{\partial x_{23}} & \frac{\partial y_{22}}{\partial x_{33}} & \frac{1}{2}\frac{\partial y_{23}}{\partial x_{13}} & \frac{1}{2}\frac{\partial y_{23}}{\partial x_{23}} & \frac{\partial y_{23}}{\partial x_{33}} \\
\frac{\partial y_{13}}{\partial x_{11}} & \frac{1}{2}\frac{\partial y_{13}}{\partial x_{12}} & \frac{1}{2}\frac{\partial y_{13}}{\partial x_{13}} & \frac{\partial y_{23}}{\partial x_{11}} & \frac{1}{2}\frac{\partial y_{23}}{\partial x_{12}} & \frac{1}{2}\frac{\partial y_{23}}{\partial x_{13}} & \frac{\partial y_{33}}{\partial x_{11}} & \frac{1}{2}\frac{\partial y_{33}}{\partial x_{12}} & \frac{1}{2}\frac{\partial y_{33}}{\partial x_{13}} \\
\frac{1}{2}\frac{\partial y_{13}}{\partial x_{12}} & \frac{\partial y_{13}}{\partial x_{22}} & \frac{1}{2}\frac{\partial y_{13}}{\partial x_{23}} & \frac{1}{2}\frac{\partial y_{23}}{\partial x_{12}} & \frac{\partial y_{23}}{\partial x_{22}} & \frac{1}{2}\frac{\partial y_{23}}{\partial x_{23}} & \frac{1}{2}\frac{\partial y_{33}}{\partial x_{12}} & \frac{\partial y_{33}}{\partial x_{22}} & \frac{1}{2}\frac{\partial y_{33}}{\partial x_{23}} \\
\frac{1}{2}\frac{\partial y_{13}}{\partial x_{13}} & \frac{1}{2}\frac{\partial y_{13}}{\partial x_{23}} & \frac{\partial y_{13}}{\partial x_{33}} & \frac{1}{2}\frac{\partial y_{23}}{\partial x_{13}} & \frac{1}{2}\frac{\partial y_{23}}{\partial x_{23}} & \frac{\partial y_{23}}{\partial x_{33}} & \frac{1}{2}\frac{\partial y_{33}}{\partial x_{13}} & \frac{1}{2}\frac{\partial y_{33}}{\partial x_{23}} & \frac{\partial y_{33}}{\partial x_{33}}
\end{bmatrix}$$

We want to examine what the derivative of X is with respect to X. We do so by substituting the matrix X in for the matrix Y and write

$$\frac{\partial X}{\partial X}_{Symmetric} = \begin{bmatrix}
1 & 0 & 0 & 0 & \frac{1}{2} & 0 & 0 & 0 & \frac{1}{2} \\
0 & 0 & 0 & \frac{1}{2} & 0 & 0 & 0 & 0 & 0 \\
0 & 0 & 0 & 0 & 0 & 0 & \frac{1}{2} & 0 & 0 \\
0 & \frac{1}{2} & 0 & 0 & 0 & 0 & 0 & 0 & 0 \\
\frac{1}{2} & 0 & 0 & 0 & 1 & 0 & 0 & 0 & \frac{1}{2} \\
0 & 0 & 0 & 0 & 0 & 0 & 0 & \frac{1}{2} & 0 \\
0 & 0 & \frac{1}{2} & 0 & 0 & 0 & 0 & 0 & 0 \\
0 & 0 & 0 & 0 & 0 & \frac{1}{2} & 0 & 0 & 0 \\
\frac{1}{2} & 0 & 0 & 0 & \frac{1}{2} & 0 & 0 & 0 & 1
\end{bmatrix}$$

$$= \frac{1}{2}\left\{vec(I_m)\left[vec(I_m)\right]^T + K_{mm}\right\} \quad .$$

We rewrite the final equation for future reference

$$\text{(C.2)} \qquad \frac{\partial X}{\partial X_{Symmetric}} = \frac{1}{2}\left\{vec\left(I_m\right)\left[vec\left(I_m\right)\right]^T + K_{mm}\right\} \quad .$$

This is an illustration of equation (vii) on p. 142 of Kollo[57], which we will prove shortly.

C.3.1 Kronecker Derivative of Matrix Times Matrix

This section proves the relationship

$$\frac{\partial\left(ZY\right)}{\partial X} = \frac{\partial Z}{\partial X}\left(Y \otimes I\right) + \left(Z \otimes I\right)\frac{\partial Y}{\partial X} \quad .$$

To do so, we begin by examining an outer product of two unit vectors, the first with a 1 in the ith position, for example $i = 2$, and the second with a 1 in the jth position, for example $j = 1$. That outer product is represented

$$r_2 s_3^T = \begin{bmatrix} 0 \\ 1 \\ 0 \end{bmatrix} \begin{bmatrix} 0 & 0 & 1 \end{bmatrix} = \begin{bmatrix} 0 & 0 & 0 \\ 0 & 0 & 1 \\ 0 & 0 & 0 \end{bmatrix}$$

Notice we get a matrix that is all zeros except for a 1 in the i,j position, i.e., the 2,3 position. We next do exactly the same thing for two different unit vectors, the first with a 1 in the 1 position and the second with a 1 in the 2 position as follows

$$f_1 g_2^T = \begin{bmatrix} 1 \\ 0 \\ 0 \end{bmatrix} \begin{bmatrix} 0 & 1 & 0 \end{bmatrix} = \begin{bmatrix} 0 & 1 & 0 \\ 0 & 0 & 0 \\ 0 & 0 & 0 \end{bmatrix} \quad .$$

Now consider the Kronecker product of the foregoing two matrices

$$r_2 s_3^T \otimes \varepsilon_{12} f_1 g_2^T = \begin{bmatrix} 0 & 0 & 0 \\ 0 & 0 & 1 \\ 0 & 0 & 0 \end{bmatrix} \otimes \begin{bmatrix} 0 & 1 & 0 \\ 0 & 0 & 0 \\ 0 & 0 & 0 \end{bmatrix}$$

$$= \begin{bmatrix} & & & & & & 0 & & \\ & & & & & & 0 & & \\ & & & & & & 0 & & \\ 0 & 0 & 0 & 0 & 0 & 0 & 0 & \varepsilon_{12} & 0 \\ & & & & & & 0 & 0 & 0 \\ & & & & & & 0 & 0 & 0 \\ & & & & & & & & \\ & & & & & & & & \\ & & & & & & & & \end{bmatrix} \quad .$$

The integers 2,3 are going to represent the ZY matrix element, and the integers 1,2 are going to represent the X matrix element. If we could write the equation for ZY in the foregoing matrix right next to the term ε, we would have the variable we need to differentiate there. That is easy; the product of the ZY matrix in its i,j position will be

$$[ZY]_{23} = \sum_{M=1}^{m} z_{2M} y_{M3} \quad .$$

If we make this multiplication, the matrix above will become

$$\begin{bmatrix} & & & & & & 0 & & \\ & & & & & & 0 & & \\ & & & & & & 0 & & \\ 0 & 0 & 0 & 0 & 0 & 0 & 0 & \varepsilon_{12} \sum\limits_{M=1}^{m} z_{2M} y_{M3} & 0 \\ & & & & & & 0 & 0 & 0 \\ & & & & & & 0 & 0 & 0 \\ & & & & & & & & \\ & & & & & & & & \end{bmatrix} \quad .$$

This is where we can now start our development. We write the foregoing matrix equation and differentiate it

$$\begin{aligned} \frac{\partial (ZY)}{\partial X} &= \sum_{i=1}^{m} \sum_{j=1}^{m} \sum_{k=1}^{m} \sum_{\ell=1}^{m} \frac{\partial}{\partial x_{k\ell}} \left[\sum_{M=1}^{m} z_{iM} y_{Nj} \right] \left[\left(r_i s_j^T \right) \otimes \left(\varepsilon_{k\ell} f_k g_\ell^T \right) \right] \\ &= \sum_{i=1}^{m} \sum_{j=1}^{m} \sum_{k=1}^{m} \sum_{\ell=1}^{m} \sum_{M=1}^{m} \left\{ \begin{array}{c} \left[\frac{\partial z_{iM}}{\partial x_{k\ell}} y_{Mj} + z_{iM} \frac{\partial y_{Mj}}{\partial x_{k\ell}} \right] \\ \left[\left(r_i s_j^T \right) \otimes \left(\varepsilon_{k\ell} f_k g_\ell^T \right) \right] \end{array} \right\} \quad . \end{aligned}$$

Let us define a delta function as the inner (not outer) product of two unit vectors, one with 1 in position M and one with 1 in position N. The delta function is

$$\delta_{MN} = \begin{cases} 1 \; if \; M = N \\ 0 \; otherwise \end{cases} \quad .$$

We can write this as the inner product of two unit vectors

$$\delta_{MN} = {e_M}^T e_N \quad ,$$

in which e_M is the unit vector with 1 in the Mth position and 0 otherwise. We are wanting to write the product at the end of the foregoing five summation equation as follows. First we examine what is inside the summation

$$\begin{aligned} r_i s_j^T \otimes \varepsilon_{k\ell} f_k g_\ell^T &= r_i \left(e_M^T e_N \right) s_j^T \otimes \varepsilon_{k\ell} f_k g_\ell^T \\ &= \left(r_i e_M^T \right) \left(e_N s_j^T \right) \otimes \varepsilon_{k\ell} f_k g_\ell^T \end{aligned}$$

$$= \left(r_i e_M^T\right)\left(e_N s_j^T\right) \otimes I\left(\varepsilon_{k\ell} f_k g_\ell^T\right) \quad .$$

We now apply the Kronecker product associative condition to write

$$\begin{aligned} r_i s_j^T \otimes \varepsilon_{k\ell} f_k g_\ell^T &= \left(r_i e_M^T\right)\left(e_N s_j^T\right) \otimes I\left(\varepsilon_{k\ell} f_k g_\ell^T\right) \\ &= \left(r_i e_M^T \otimes I\right)\left(e_N s_j^T \otimes \varepsilon_{k\ell} f_k g_\ell^T\right) \quad . \end{aligned}$$

$$\left(r_i {s_j}^T\right) \otimes \left(\varepsilon_{k\ell} f_k {g_\ell}^T\right) = \sum_{N=1}^{m} \left[r_i \left({e_M}^T e_N\right) {s_j}^T\right] \otimes \left(\varepsilon_{k\ell} f_k {g_\ell}^T\right) \quad .$$

This motivates us to write

$$\frac{\partial (ZY)}{\partial X} = \sum_{i=1}^{m}\sum_{j=1}^{m}\sum_{k=1}^{m}\sum_{\ell=1}^{m} \left\{ \begin{array}{l} \left[\sum\limits_{M=1}^{m} \frac{\partial z_{iM}}{\partial x_{k\ell}} y_{Nj} + z_{iM} \frac{\partial y_{Nj}}{\partial x_{k\ell}}\right] \\ \left[r_i \left(e_M^T e_N\right) s_j^T\right] \otimes \left(\varepsilon_{k\ell} f_k g_\ell^T\right) \end{array} \right\} \quad .$$

Notice the change in index we have made in the y terms, which become M terms ONLY when $M = N$. We now work on the Kronecker product

$$\begin{aligned} r_i \left({e_M}^T e_N\right) {s_j}^T \otimes \left(\varepsilon_{k\ell} f_k {g_\ell}^T\right) &= \left(r_i {e_M}^T\right)\left(e_N {s_j}^T\right) \otimes \left(\varepsilon_{k\ell} f_k {g_\ell}^T\right) \\ &= \left(r_i {e_M}^T\right)\left(e_N {s_j}^T\right) \otimes I\left(\varepsilon_{k\ell} f_k {g_\ell}^T\right) \\ &= \left(r_i {e_M}^T \otimes I\right)\left(e_N {s_j}^T \otimes \varepsilon_{k\ell} f_k {g_\ell}^T\right) \quad . \end{aligned}$$

Substituting this back into the definition of the derivative, noting the addition of the additional sum over N

$$\frac{\partial (ZY)}{\partial X} = \sum_{i=1}^{m}\sum_{j=1}^{m}\sum_{k=1}^{m}\sum_{\ell=1}^{m}\sum_{M=1}^{m} \left\{ \begin{array}{l} \left[\frac{\partial z_{iM}}{\partial x_{k\ell}} y_{Mj} + z_{iM} \frac{\partial y_{Mj}}{\partial x_{k\ell}}\right] \\ \left[\left(r_i e_M^T \otimes I\right)\left(e_N s_j^T \otimes \varepsilon_{k\ell} f_k g_\ell^T\right)\right] \end{array} \right\} \quad .$$

We first insert the foregoing result and thereafter divide the terms into two summations.

$$\begin{aligned} &\frac{\partial (ZY)}{\partial X} \\ &= \sum_{i=1}^{m}\sum_{j=1}^{m}\sum_{k=1}^{m}\sum_{\ell=1}^{m}\sum_{M=1}^{m}\sum_{N=1}^{m} \left\{ \begin{array}{l} \left[\frac{\partial z_{iM}}{\partial x_{k\ell}} y_{Mj} + z_{iM} \frac{\partial y_{Mj}}{\partial x_{k\ell}}\right] \\ \left[\left(r_i e_M^T \otimes I\right)\left(e_N s_j^T \otimes \varepsilon_{k\ell} f_k g_\ell^T\right)\right] \end{array} \right\} \\ &= \sum_{i=1}^{m}\sum_{j=1}^{m}\sum_{k=1}^{m}\sum_{\ell=1}^{m}\sum_{M=1}^{m}\sum_{N=1}^{m} \left[\frac{\partial z_{iM}}{\partial x_{k\ell}} y_{Mj} \left(r_i e_M^T \otimes I\right)\left(e_N s_j^T \otimes \varepsilon_{k\ell} f_k g_\ell^T\right)\right] \\ &+ \sum_{i=1}^{m}\sum_{j=1}^{m}\sum_{k=1}^{m}\sum_{\ell=1}^{m}\sum_{M=1}^{m}\sum_{N=1}^{m} \left[z_{iM} \frac{\partial y_{Mj}}{\partial x_{k\ell}} \left(r_i e_M^T \otimes I\right)\left(e_N s_j^T \otimes \varepsilon_{k\ell} f_k g_\ell^T\right)\right] \quad . \end{aligned}$$

Let us focus on the first of the final two terms. If you painstakingly sweep through all the six indices, you will find that the first term can be written in matrix form

$$\frac{\partial Z}{\partial X}\left(Y \otimes I\right) \quad .$$

and the second term can be written in matrix form

$$\left(Z \otimes I\right)\frac{\partial Y}{\partial X} \quad .$$

We rewrite the sum of the two terms for easy future reference

$$\frac{\partial\left(ZY\right)}{\partial X} = \frac{\partial Z}{\partial X}\left(Y \otimes I\right) + \left(Z \otimes I\right)\frac{\partial Y}{\partial X} \quad .$$

C.3.2 Kronecker Derivative of Matrix Times Scalar Function

This section proves the relationship

$$\frac{\partial\left(Zy\right)}{\partial X} = \frac{\partial Z}{\partial X}y + Z \otimes \frac{\partial y}{\partial X} \quad .$$

We can write this in the following form for the i, j^{th} term

$$\left[\frac{\partial\left(Zy\right)}{\partial X}\right]_{ij} = \sum_{i=1}^{m}\sum_{j=1}^{m}\sum_{k=1}^{m}\sum_{\ell=1}^{m}\frac{\partial\left(z_{ij}y\right)}{\partial x_{k\ell}}\left(r_i {s_j}^T\right) \otimes \left(\varepsilon_{k\ell} f_k {g_\ell}^T\right) \quad .$$

This means that the sought after matrix derivative is

$$\begin{aligned}
\frac{\partial\left(Zy\right)}{\partial X} &= \sum_{i=1}^{m}\sum_{j=1}^{m}\sum_{k=1}^{m}\sum_{\ell=1}^{m}\left(\frac{\partial z_{ij}}{\partial x_{k\ell}}y + z_{ij}\frac{\partial y}{\partial x_{k\ell}}\right)\left(r_i {s_j}^T\right) \otimes \left(\varepsilon_{k\ell} f_k {g_\ell}^T\right) \\
&= \sum_{i=1}^{m}\sum_{j=1}^{m}\sum_{k=1}^{m}\sum_{\ell=1}^{m}\frac{\partial z_{ij}}{\partial x_{k\ell}}y\left[\left(r_i {s_j}^T\right) \otimes \left(\varepsilon_{k\ell} f_k {g_\ell}^T\right)\right] \\
&+ \sum_{i=1}^{m}\sum_{j=1}^{m}\sum_{k=1}^{m}\sum_{\ell=1}^{m} z_{ij}\frac{\partial y}{\partial x_{k\ell}}\left[\left(r_i {s_j}^T\right) \otimes \left(\varepsilon_{k\ell} f_k {g_\ell}^T\right)\right] \\
&= y\sum_{i=1}^{m}\sum_{j=1}^{m}\sum_{k=1}^{m}\sum_{\ell=1}^{m}\frac{\partial z_{ij}}{\partial x_{k\ell}}\left[\left(r_i {s_j}^T\right) \otimes \left(\varepsilon_{k\ell} f_k {g_\ell}^T\right)\right] \\
&+ \sum_{i=1}^{m}\sum_{j=1}^{m}\sum_{k=1}^{m}\sum_{\ell=1}^{m} z_{ij}\frac{\partial y}{\partial x_{k\ell}}\left[\left(r_i {s_j}^T\right) \otimes \left(\varepsilon_{k\ell} f_k {g_\ell}^T\right)\right] \\
&= y\frac{\partial Z}{\partial X} + \sum_{i=1}^{m}\sum_{j=1}^{m}\sum_{k=1}^{m}\sum_{\ell=1}^{m}\frac{\partial y}{\partial x_{k\ell}}\left[\left(z_{ij} r_i {s_j}^T\right) \otimes \left(\varepsilon_{k\ell} f_k {g_\ell}^T\right)\right]
\end{aligned}$$

$$= y\frac{\partial Z}{\partial X} + \sum_{i=1}^{m}\sum_{j=1}^{m}\sum_{k=1}^{m}\sum_{\ell=1}^{m}\left[\left(z_{ij}r_i {s_j}^T\right)\otimes`\left(\frac{\partial y}{\partial x_{k\ell}}\varepsilon_{k\ell}f_k {g_\ell}^T\right)\right] \quad .$$

This puts z_{ij} into the ij^{th} spot and puts $\frac{dy}{dx_{k\ell}}$ into the $k\ell^{th}$ spot. Thus the second term is determined to be

$$\frac{\partial(Zy)}{\partial X} = y\frac{\partial Z}{\partial X} + Z\otimes\frac{\partial y}{\partial X} \quad . \tag{C.3}$$

C.3.3 Kronecker Derivative of Triple Product of Matrices

This section proves the relationship

$$\frac{\partial\left(AYB\right)}{\partial X} = \left(A\otimes I\right)\frac{\partial Y}{\partial X}\left(B\otimes I\right) \quad .$$

We appeal to the product of two matrices above

$$\frac{\partial\left(ZY\right)}{\partial X} = \frac{\partial Z}{\partial X}\left(Y\otimes I\right) + \left(Z\otimes I\right)\frac{\partial Y}{\partial X} \quad .$$

If we substitute $Z \leftarrow A$ and $Y \leftarrow YB$, we can write as a first step

$$\frac{\partial\left[A\left(YB\right)\right]}{\partial X} = \frac{\partial A}{\partial X}\left[\left(YB\right)\otimes I\right] + \left(A\otimes I\right)\frac{\partial\left(YB\right)}{\partial X} \quad .$$

Because A is a constant, the first term is zero, and this term simplifies

$$\frac{\partial\left[A\left(YB\right)\right]}{\partial X} = \left(A\otimes I\right)\frac{\partial\left(YB\right)}{\partial X} \quad .$$

We now apply the formula for the product of two matrices to YB, making the substitutions $Y \leftarrow B$ and $Z \leftarrow Y$, we obtain

$$\frac{\partial\left(YB\right)}{\partial X} = \frac{\partial Y}{\partial X}\left(B\otimes I\right) + \left(Y\otimes I\right)\frac{\partial B}{\partial X} \quad .$$

Because B is a constant, its derivative is 0, meaning we can write the previous equation

$$\frac{\partial\left(YB\right)}{\partial X} = \frac{\partial Y}{\partial X}\left(B\otimes I\right) \quad .$$

Substituting, the equation we began with is proven.

$$\frac{\partial\left(AYB\right)}{\partial X} = \left(A\otimes I\right)\frac{\partial Y}{\partial X}\left(B\otimes I\right) \quad .$$

C.3.4 Kronecker Derivative of the Inverse of a Matrix Y with Respect to X

This section begins with the product of a matrix

$$\begin{aligned}
\frac{\partial\left(Y^{-1}Y\right)}{\partial X} &= 0 = \frac{\partial Y^{-1}}{\partial X}\left(Y\otimes I\right) + \left(Y^{-1}\otimes I\right)\frac{\partial Y}{\partial X} \\
\Rightarrow \frac{\partial Y^{-1}}{\partial X}\left(Y\otimes I\right) &= -\left(Y^{-1}\otimes I\right)\frac{\partial Y}{\partial X} \\
\Rightarrow \frac{\partial Y^{-1}}{\partial X}\left(Y\otimes I\right)\left(Y\otimes I\right)^{-1} &= -\left(Y^{-1}\otimes I\right)\frac{\partial Y}{\partial X}\left(Y\otimes I\right)^{-1} \\
\Rightarrow \frac{\partial Y^{-1}}{\partial X} &= -\left(Y^{-1}\otimes I\right)\frac{\partial Y}{\partial X}\left(Y\otimes I\right)^{-1} \\
\Rightarrow \frac{\partial Y^{-1}}{\partial X} &= -\left(Y^{-1}\otimes I\right)\frac{\partial Y}{\partial X}\left(Y^{-1}\otimes I\right) \quad .
\end{aligned}$$

We rewrite this equation for easy future reference

$$\frac{\partial Y^{-1}}{\partial X} = -\left(Y^{-1}\otimes I\right)\frac{\partial Y}{\partial X}\left(Y^{-1}\otimes I\right) \quad . \tag{C.4}$$

C.3.5 Kronecker Derivative of the Trace of a Matrix $A^T X$ with Respect to X

The trace of a matrix B can be written

$$Trace\left(B\right) = \sum_{k=1}^{m} {e_k}^T B e_k \quad .$$

The matrix whose derivative we are interested in is

$$\begin{aligned}
Trace\left(A^T X\right) &= \sum_{k=1}^{m} {e_k}^T\left(A^T X\right)e_k \\
&= \sum_{k=1}^{m} {e_k}^T A^T X e_k \\
&= \sum_{k=1}^{m} \left({e_k}^T A^T\right) X e_k \quad .
\end{aligned}$$

We expand this into two key terms

(C.5)

$$\frac{\partial}{\partial X} Trace\left(A^T X\right) = \frac{\partial}{\partial X}\sum_{k=1}^{m}\left({e_k}^T A^T\right) X e_k$$
$$= \sum_{k=1}^{m}\frac{\partial}{\partial X}\left[\left({e_k}^T A^T\right) X e_k\right]$$
$$= \sum_{k=1}^{m}\left({e_k}^T A^T \otimes I\right)\frac{\partial X}{\partial X}\left(e_k \otimes I\right)$$
$$= \frac{1}{2}\sum_{k=1}^{m}\left({e_k}^T A^T \otimes I\right)\left[vec\left(I\right) vec(I)^T + K_{mm}\right]\left(e_k \otimes I\right)$$
$$= \frac{1}{2}\sum_{k=1}^{m}\left({e_k}^T A^T \otimes I\right) vec\left(I\right) vec(I)^T \left(e_k \otimes I\right)$$
$$+ \frac{1}{2}\sum_{k=1}^{m}\left({e_k}^T A^T \otimes I\right) K_{mm}\left(e_k \otimes I\right) \quad .$$

We initially focus on the first term in equation (C.5)

$$First = \frac{1}{2}\sum_{k=1}^{m}\left({e_k}^T A^T \otimes I\right) vec\left(I\right) vec(I)^T \left(e_k \otimes I\right)$$
$$= \frac{1}{2}\sum_{k=1}^{m}\left[\left({e_k}^T A^T \otimes I\right) vec\left(I\right)\right]\left[vec(I)^T \left(e_k \otimes I\right)\right] \quad .$$

The Kronecker product theorem is

$$\left({B_0}^T \otimes A_0\right) vec\left(X_0\right) = vec\left(A_0 X_0 B_0\right) \quad .$$

If we make the substitutions, we get the left hand term in the product

$$\begin{aligned} {B_0}^T &= {e_k}^T A^T \Rightarrow B_0 = A e_k \\ A_0 &= I \\ X_0 &= I \quad . \end{aligned}$$

If we make these substitutions, we have the left hand term

$$\left({e_k}^T A^T \otimes I\right) vec\left(I\right) = vec\left(I \times I \times A e_k\right) = vec\left(A e_k\right) \quad .$$

Turning to the second term in equation (C.5) in the final expression, we begin with the Kronecker product theorem,[2] rewritten here for convenience

[2] https://en.wikipedia.org/wiki/Kronecker_product.

$$\left(B_0{}^T \otimes A_0\right) vec\left(X_0\right) = vec\left(A_0 X_0 B_0\right) \quad .$$

We want to make the following substitution to apply the theorem to our context

$$\begin{aligned} B_0 &= e_k \\ A_0 &= I \\ X_0 &= I \quad . \end{aligned}$$

We can apply the Kronecker product theorem as follows

$$\begin{aligned} \left[vec(I)^T \left(e_k \otimes I\right)\right]^T &= vec\left(e_k\right) \\ \Rightarrow vec(I)^T \left(e_k \otimes I\right) &= vec(e_k)^T \quad . \end{aligned}$$

The first term in equation (C.5) therefore becomes

$$\begin{aligned} First &= \frac{1}{2} \sum_{k=1}^{m} vec(Ae_k) vec(e_k)^T \\ &= \frac{1}{2} \sum_{k=1}^{m} a^k vec(e_k)^T \\ &= \frac{1}{2} \sum_{k=1}^{m} \begin{bmatrix} 0 & . & . & . & 0 & a_{1k} & 0 & . & . & . & 0 \\ 0 & . & . & . & 0 & a_{2k} & 0 & . & . & . & 0 \\ . & & & & & . & & & & & . \\ . & & & & & . & & & & & . \\ . & & & & & . & & & & & . \\ 0 & . & . & . & 0 & a_{pk} & 0 & . & . & . & 0 \end{bmatrix} \\ &= \frac{1}{2} A \quad . \end{aligned}$$

Now we turn to the second term in equation (C.5) in the final expression

$$Second = \frac{1}{2} \sum_{k=1}^{m} \left(e_k{}^T A^T \otimes I\right) K_{mm} \left(e_k \otimes I\right) \quad .$$

Consider the right two terms

$$K_{mm} \left(e_k \otimes I\right) \quad .$$

If we consider the case of $m = 3$, we can figure out what is occurring here. The K_{mm} matrix when $m = 3$ is

$$K_{mm} = \begin{bmatrix} 1 & 0 & 0 & 0 & 0 & 0 & 0 & 0 & 0 \\ 0 & 0 & 0 & 1 & 0 & 0 & 0 & 0 & 0 \\ 0 & 0 & 0 & 0 & 0 & 0 & 1 & 0 & 0 \\ 0 & 1 & 0 & 0 & 0 & 0 & 0 & 0 & 0 \\ 0 & 0 & 0 & 0 & 1 & 0 & 0 & 0 & 0 \\ 0 & 0 & 0 & 0 & 0 & 0 & 0 & 1 & 0 \\ 0 & 0 & 1 & 0 & 0 & 0 & 0 & 0 & 0 \\ 0 & 0 & 0 & 0 & 0 & 1 & 0 & 0 & 0 \\ 0 & 0 & 0 & 0 & 0 & 0 & 0 & 0 & 1 \end{bmatrix}.$$

The matrix $e_k \otimes I$ consists of all zeroes except for in the kth partition, which is the identity matrix of size m. For the case of $k = 3$, this matrix is

$$e_k \otimes I = \begin{bmatrix} 0 & 0 & 0 \\ 0 & 0 & 0 \\ 0 & 0 & 0 \\ 0 & 0 & 0 \\ 0 & 0 & 0 \\ 0 & 0 & 0 \\ 1 & 0 & 0 \\ 0 & 1 & 0 \\ 0 & 0 & 1 \end{bmatrix}.$$

We see the identity matrix in the third partition ($k = 3$) and zero everywhere else. When we multiply the commutation matrix by this matrix for $k = 3$, we obtain a matrix that has as its third partition ($k = 3$) ONLY the commutation matrix

$$K_{33} \times (e_3 \otimes I) = \begin{bmatrix} 0 & 0 & 0 \\ 0 & 0 & 0 \\ 1 & 0 & 0 \\ 0 & 0 & 0 \\ 0 & 0 & 0 \\ 0 & 1 & 0 \\ 0 & 0 & 0 \\ 0 & 0 & 0 \\ 0 & 0 & 1 \end{bmatrix}.$$

Turning to the expression $e_3^T A^T$, we know that it contains the third ($k = 3$) column of A^T written as a row. Thus it is the 3rd row of A^T, i.e.,

$$e_3{}^T A^T = \begin{bmatrix} a_{13} & a_{23} & a_{33} \end{bmatrix} = \begin{bmatrix} a_{13} \\ a_{23} \\ a_{33} \end{bmatrix}^T.$$

Therefore the Kronecker product of this with the identity matrix I is

$$\begin{aligned} {e_3}^T A^T \otimes I &= \left[\begin{array}{ccc} a_{13}I & a_{23}I & a_{33}I \end{array} \right] \\ &= \left[\begin{array}{ccccccccc} a_{13} & 0 & 0 & a_{23} & 0 & 0 & a_{33} & 0 & 0 \\ 0 & a_{13} & 0 & 0 & a_{23} & 0 & 0 & a_{33} & 0 \\ 0 & 0 & a_{13} & 0 & 0 & a_{23} & 0 & 0 & a_{33} \end{array} \right] . \end{aligned}$$

When we multiply this on the right by the matrix $K_{33}(e_3 \otimes I)$ derived immediately above, we obtain the answer

$$\left({e_3}^T A^T \otimes I\right) K_{33} \left(e_3 \otimes I\right) = \left[\begin{array}{ccc} 0 & 0 & 0 \\ 0 & 0 & 0 \\ a_{13} & a_{23} & a_{33} \end{array} \right] .$$

Clearly when we sum over all $k = 1, ..., m$, we have the following expression for the second term in equation (C.5) in the final expression

$$Second = \frac{1}{2} A^T \quad .$$

Therefore, the final answer is

$$\frac{\partial}{\partial X} Trace \left(A^T X\right) = \frac{1}{2}\left(A + A^T\right) \quad . \tag{C.6}$$

C.3.6 The Kronecker Derivative of the Determinant of a Matrix Raised to a Power

Let us illustrate the calculation for the standard 2 matrix X. The determinant of that matrix is

$$\Delta = |X| = x_{11}x_{22} - x_{12}x_{21} \quad .$$

The determinant raised to a power is

$$\Delta^\theta = |X|^\theta = \left(x_{11}x_{22} - x_{12}x_{21}\right)^\theta \quad .$$

The four partial derivatives of the determinant with respect to each of the four elements of X are

$$\begin{aligned} \frac{\partial |X|^\theta}{\partial X} &= \left[\begin{array}{cc} \theta x_{22}\Delta^{\theta-1} & -\theta x_{21}\Delta^{\theta-1} \\ -\theta x_{12}\Delta^{\theta-1} & \theta x_{11}\Delta^{\theta-1} \end{array} \right] = \theta\Delta^{\theta-1} \left[\begin{array}{cc} x_{22} & -x_{21} \\ -x_{12} & x_{11} \end{array} \right] \\ &= \theta\Delta^{\theta-1}\Delta \left[\begin{array}{cc} \frac{1}{\Delta}x_{22} & -\frac{1}{\Delta}x_{21} \\ -\frac{1}{\Delta}x_{12} & \frac{1}{\Delta}x_{11} \end{array} \right] = \theta\Delta^\theta X^{-T} = \theta |X|^\theta X^{-T} \quad . \end{aligned} \tag{C.7}$$

We have calculated the matrix derivative of the determinant of a matrix, and perhaps amazingly this equation generalizes to arbitrary higher orders.

C.3.7 Kronecker Derivative of the Inverse of a Matrix X^{-1} with Respect to X

We substitute X for Y in equation (C.4) in a previous section immediately above to write the derivative of the inverse of a matrix with respect to itself

$$\frac{\partial X^{-1}}{\partial X} = -\left(X^{-1} \otimes I\right) \frac{\partial X}{\partial X} \left(X^{-1} \otimes I\right) \quad .$$

This derivative is the primary reason we have gone to all the trouble in this section. Let us examine and decompose this derivative, substituting in the process the derivative of the matrix X with respect to X from equation (C.2)

$$\begin{aligned} \frac{\partial X^{-1}}{\partial X} &= -\left(X^{-1} \otimes I\right) \frac{\partial X}{\partial X} \left(X^{-1} \otimes I\right) \\ &= -\left(X^{-1} \otimes I\right) \frac{1}{2} \left\{vec\left(I\right)\left[vec\left(I\right)\right]^T + K_{mm}\right\} \left(X^{-1} \otimes I\right) \\ &= -\frac{1}{2}\left(X^{-1} \otimes I\right) vec\left(I\right)\left[vec\left(I\right)\right]^T \left(X^{-1} \otimes I\right) \\ &- \frac{1}{2}\left(X^{-1} \otimes I\right) K_{mm} \left(X^{-1} \otimes I\right) \quad . \end{aligned} \tag{C.8}$$

This is the expression we want to derive and simplify in the forthcoming section, the derivative of the inverse of a matrix with respect to itself. It is central to the derivation of the variance-covariance matrix in Chapter 15 through 17.

The first part of the two part term in equation (C.8) is

$$-\frac{1}{2}\left(X^{-1} \otimes I\right) vec\left(I\right)\left[vec\left(I\right)\right]^T \left(X^{-1} \otimes I\right) \quad .$$

The Kronecker product theorem is

$$\left({B_0}^T \otimes A_0\right) vec\left(X_0\right) = vec\left(A_0 X_0 B_0\right) \quad .$$

We apply this theorem to the term

$$\left(X^{-1} \otimes I\right) vec\left(I\right) \quad .$$

We make the substitutions

$$\begin{aligned} &{B_0}^T \leftarrow X^{-1} \Rightarrow {B_0}^T \leftarrow X \Rightarrow B_0 \leftarrow X \\ &A_0 \leftarrow I \\ &X_0 \leftarrow I \quad . \end{aligned}$$

which allow us to apply the Kronecker theorem

$$\left(X^{-1} \otimes I\right) vec\left(I\right) = vec\left(I \times I \times X^{-1}\right) = vec\left(X^{-1}\right) \quad .$$

Turning to the right hand term, we write and substitute

$$\begin{aligned}
&Y = [vec\,(I)]^T \left(X^{-1} \otimes I\right) \\
&\Rightarrow Y^T = \left(X^{-1} \otimes I\right)^T [vec\,(I)] \\
&\Rightarrow Y^T = \left(X^{-1} \otimes I\right)^T [vec\,(I)] = vec\left(X^{-1}\right) \\
&\Rightarrow Y = \left[vec\left(X^{-1}\right)\right]^T \quad .
\end{aligned}$$

This means that the first term is

$$-\frac{1}{2} vec\left(X^{-1}\right) \left[vec\left(X^{-1}\right)\right]^T \quad .$$

We move to the second term of the two part term in equation (C.8), which is

$$-\frac{1}{2}\left(X^{-1} \otimes I\right) K_{mm} \left(X^{-1} \otimes I\right) \quad .$$

We need to understand precisely what is occurring with this term, so we need to postulate the case in which $m = 3$ and illustrate the requisite terms, emphasizing how those terms generalize to values of m higher than 3. We need to calculate the commutation matrix for the $m = 3$ case, which appears in Table C.1. It has a characteristic array of columns

Table C.1: The Matrix K_{mm}

1	0	0	0	0	0	0	0	0
0	0	0	1	0	0	0	0	0
0	0	0	0	0	0	1	0	0
0	1	0	0	0	0	0	0	0
0	0	0	0	1	0	0	0	0
0	0	0	0	0	0	0	1	0
0	0	1	0	0	0	0	0	0
0	0	0	0	0	1	0	0	0
0	0	0	0	0	0	0	0	1

The commutation matrix has a precisely analogous shape for orders higher than 3. We then write the matrix $(X^{-1} \otimes I)$. The matrix is comprised of a patchwork quilt of identity matrices, the diagonals of which contain elements of X^{-1}, which we have designated by the notation y. The matrix $(X^{-1} \otimes I)$ appears in Table C.2.

We note the structure of the identity matrixes embedded within the matrix. Thereafter, we calculate the product of the commutation matrix and the initial matrix $(X^{-1} \otimes I)$ in Table C.2 to obtain the result in Table C.3. We get to the final answer by multiplying the matrix $(X^{-1} \otimes I)$ on the left by the matrix

Table C.2: The Matrix $(X^{-1} \otimes I)$

y_{11}	0	0	y_{12}	0	0	y_{13}	0	0
0	y_{11}	0	0	y_{12}	0	0	y_{13}	0
0	0	y_{11}	0	0	y_{12}	0	0	y_{13}
y_{21}	0	0	y_{22}	0	0	y_{23}	0	0
0	y_{21}	0	0	y_{22}	0	0	y_{23}	0
0	0	y_{21}	0	0	y_{22}	0	0	y_{23}
y_{31}	0	0	y_{32}	0	0	y_{33}	0	0
0	y_{31}	0	0	y_{32}	0	0	y_{33}	0
0	0	y_{31}	0	0	y_{32}	0	0	y_{33}

$K_{mm}(X^{-1} \otimes I)$ in Table C.3 to obtain the final answer shown in Table C.4.

Table C.3: The Product $K_{mm}(X^{-1} \otimes I)$

y_{11}	0	0	y_{12}	0	0	y_{13}	0	0
y_{21}	0	0	y_{22}	0	0	y_{23}	0	0
y_{31}	0	0	y_{32}	0	0	y_{33}	0	0
0	y_{11}	0	0	y_{12}	0	0	y_{13}	0
0	y_{21}	0	0	y_{22}	0	0	y_{23}	0
0	y_{31}	0	0	y_{32}	0	0	y_{33}	0
0	0	y_{11}	0	0	y_{12}	0	0	y_{13}
0	0	y_{21}	0	0	y_{22}	0	0	y_{23}
0	0	y_{31}	0	0	y_{32}	0	0	y_{33}

Table C.4: The Product $(X^{-1} \otimes I)K_{mm}(X^{-1} \otimes I)$

$y_{11}y_{11}$	$y_{12}y_{11}$	$y_{13}y_{11}$	$y_{11}y_{12}$	$y_{12}y_{12}$	$y_{13}y_{12}$	$y_{11}y_{13}$	$y_{12}y_{13}$	$y_{13}y_{13}$
$y_{11}y_{21}$	$y_{12}y_{21}$	$y_{13}y_{21}$	$y_{11}y_{22}$	$y_{12}y_{22}$	$y_{13}y_{22}$	$y_{11}y_{23}$	$y_{12}y_{23}$	$y_{13}y_{23}$
$y_{11}y_{31}$	$y_{12}y_{31}$	$y_{13}y_{31}$	$y_{11}y_{32}$	$y_{12}y_{32}$	$y_{13}y_{32}$	$y_{11}y_{33}$	$y_{12}y_{33}$	$y_{13}y_{33}$
$y_{21}y_{11}$	$y_{22}y_{11}$	$y_{23}y_{11}$	$y_{21}y_{12}$	$y_{22}y_{12}$	$y_{23}y_{12}$	$y_{21}y_{13}$	$y_{22}y_{13}$	$y_{23}y_{13}$
$y_{21}y_{21}$	$y_{22}y_{21}$	$y_{23}y_{21}$	$y_{21}y_{22}$	$y_{22}y_{22}$	$y_{23}y_{22}$	$y_{21}y_{23}$	$y_{22}y_{23}$	$y_{23}y_{23}$
$y_{21}y_{31}$	$y_{22}y_{31}$	$y_{23}y_{31}$	$y_{21}y_{32}$	$y_{22}y_{32}$	$y_{23}y_{32}$	$y_{21}y_{33}$	$y_{22}y_{33}$	$y_{23}y_{33}$
$y_{31}y_{11}$	$y_{32}y_{11}$	$y_{33}y_{11}$	$y_{31}y_{12}$	$y_{32}y_{12}$	$y_{33}y_{12}$	$y_{31}y_{13}$	$y_{32}y_{13}$	$y_{33}y_{13}$
$y_{31}y_{21}$	$y_{32}y_{21}$	$y_{33}y_{21}$	$y_{31}y_{22}$	$y_{32}y_{22}$	$y_{33}y_{22}$	$y_{31}y_{23}$	$y_{32}y_{23}$	$y_{33}y_{23}$
$y_{31}y_{31}$	$y_{32}y_{31}$	$y_{33}y_{31}$	$y_{31}y_{32}$	$y_{32}y_{32}$	$y_{33}y_{32}$	$y_{31}y_{33}$	$y_{32}y_{33}$	$y_{33}y_{33}$

It will become clear why we are doing so in a moment, but let us left multiply this matrix by the commutation matrix K_{mm} to obtain the result in Table C.5.

If we examine the upper left $m \times m$ block, we see rather obviously for the first panel that

Table C.5: The Product $K_{mm}(X^{-1} \otimes I)K_{mm}(X^{-1} \otimes I)$

$y_{11}y_{11}$	$y_{11}y_{12}$	$y_{11}y_{13}$	$y_{12}y_{11}$	$y_{12}y_{12}$	$y_{12}y_{13}$	$y_{13}y_{11}$	$y_{13}y_{12}$	$y_{13}y_{13}$
$y_{11}y_{21}$	$y_{11}y_{22}$	$y_{11}y_{23}$	$y_{12}y_{21}$	$y_{12}y_{22}$	$y_{12}y_{23}$	$y_{13}y_{21}$	$y_{13}y_{22}$	$y_{13}y_{23}$
$y_{11}y_{31}$	$y_{11}y_{32}$	$y_{11}y_{33}$	$y_{12}y_{31}$	$y_{12}y_{32}$	$y_{12}y_{33}$	$y_{13}y_{31}$	$y_{13}y_{32}$	$y_{13}y_{33}$
$y_{21}y_{11}$	$y_{21}y_{12}$	$y_{21}y_{13}$	$y_{22}y_{11}$	$y_{22}y_{12}$	$y_{22}y_{13}$	$y_{23}y_{11}$	$y_{23}y_{12}$	$y_{23}y_{13}$
$y_{21}y_{21}$	$y_{21}y_{22}$	$y_{21}y_{23}$	$y_{22}y_{21}$	$y_{22}y_{22}$	$y_{22}y_{23}$	$y_{23}y_{21}$	$y_{23}y_{22}$	$y_{23}y_{23}$
$y_{21}y_{31}$	$y_{21}y_{32}$	$y_{21}y_{33}$	$y_{22}y_{31}$	$y_{22}y_{32}$	$y_{22}y_{33}$	$y_{23}y_{31}$	$y_{23}y_{32}$	$y_{23}y_{33}$
$y_{31}y_{11}$	$y_{31}y_{12}$	$y_{31}y_{13}$	$y_{32}y_{11}$	$y_{32}y_{12}$	$y_{32}y_{13}$	$y_{33}y_{11}$	$y_{33}y_{12}$	$y_{33}y_{13}$
$y_{31}y_{21}$	$y_{31}y_{22}$	$y_{31}y_{23}$	$y_{32}y_{21}$	$y_{32}y_{22}$	$y_{32}y_{23}$	$y_{33}y_{21}$	$y_{33}y_{22}$	$y_{33}y_{23}$
$y_{31}y_{31}$	$y_{31}y_{32}$	$y_{31}y_{33}$	$y_{32}y_{31}$	$y_{32}y_{32}$	$y_{32}y_{33}$	$y_{33}y_{31}$	$y_{33}y_{32}$	$y_{33}y_{33}$

$$y_{11}\begin{bmatrix} y_{11} & y_{12} & y_{13} \\ y_{21} & y_{22} & y_{23} \\ y_{31} & y_{32} & y_{33} \end{bmatrix} = y_{11}Y \quad .$$

The second panel immediately to the right is

$$y_{12}\begin{bmatrix} y_{11} & y_{12} & y_{13} \\ y_{21} & y_{22} & y_{23} \\ y_{31} & y_{32} & y_{33} \end{bmatrix} = y_{11}Y \quad ,$$

and similarly for all other panels. This means that the overall $K_{mm}(X^{-1} \otimes I)K_{mm}(X^{-1} \otimes I)$ matrix is

$$(Y \otimes Y) = (X^{-1} \otimes X^{-1}) \quad ,$$

and very importantly the following key relationship holds that

$$K_{mm}\left(X^{-1} \otimes I\right) K_{mm}\left(X^{-1} \otimes I\right) = \left(X^{-1} \otimes X^{-1}\right) \quad .$$

We left multiply this relationship by the commutation matrix K_{mm} to obtain the following equation, which is manipulated in the succeeding lines under the originally stated equation

$$\begin{aligned} &K_{mm}K_{mm}\left(X^{-1} \otimes I\right) K_{mm}\left(X^{-1} \otimes I\right) = K_{mm}\left(X^{-1} \otimes X^{-1}\right) \\ &\Rightarrow \left(X^{-1} \otimes I\right) K_{mm}\left(X^{-1} \otimes I\right) = K_{mm}\left(X^{-1} \otimes X^{-1}\right) \quad , \end{aligned}$$

because it is well known that the commutation matrix is its own inverse, i.e.,

$$K_{mm}K_{mm} = I \quad .$$

We rewrite the key result for future reference

$$\left(X^{-1} \otimes I\right) K_{mm}\left(X^{-1} \otimes I\right) = K_{mm}\left(X^{-1} \otimes X^{-1}\right) \quad .$$

Using the first and second term thus developed, we have proven the difficult and not obvious relationship that

$$\text{(C.9)} \qquad \frac{\partial X^{-1}}{\partial X} = vec\left(X^{-1}\right)\left[vec\left(X^{-1}\right)\right]^T + K_{mm}\left(X^{-1} \otimes X^{-1}\right) \quad .$$

This particular equation lies at the heart of the development of the variance-covariance matrix over the error term. This was first developed by von Rosen[101] without proof, and it was a breakthrough in thinking about variance-covariance (and higher-order moments). It was later proven in Kollo and von Rosen[57]. We have seen the need to expand and extend the proof for clarity and a direct focus on the problem at hand. This equation will be invoked at the outset of the development of the variance-covariance matrix of the error term.

C.4 Second Type of Matrix Derivative–Kronecker Type 2

We want to extend the foregoing 2×2 case to a slightly different orientation, one that is given by a different equation from that in the previous section

$$\frac{\partial Y}{\partial X} = \frac{\partial}{\partial X} \otimes Y \quad .$$

In particular, let us revise the matrix building block in the previous section that was indexed tu that held the y term constant. Let us instead conceive a block that holds the x term constant

$$\left[\frac{\partial Y}{\partial X}\right]_{ij} = \begin{bmatrix} \frac{\partial y_{11}}{\partial x_{ij}} & \cdot \;\cdot\;\cdot & \frac{\partial y_{1m}}{\partial x_{ij}} \\ \cdot & & \cdot \\ \cdot & & \cdot \\ \cdot & & \cdot \\ \frac{\partial y_{m1}}{\partial x_{ij}} & \cdot \;\cdot\;\cdot & \frac{\partial y_{mm}}{\partial x_{ij}} \end{bmatrix} \qquad i = 1, ..., m; \; j = 1, ..., m; \quad .$$

We are going to "mosaic" matrices for different ij pairs to create the full matrix derivative. When we do so, we obtain the following equation for the matrix derivative of the second type as characterized in Kollo and von Rosen[57] p. 125

$$\frac{\partial Y}{\partial X} = \frac{\partial}{\partial X} \otimes Y = \sum_{i=1}^{m}\sum_{j=1}^{m}\sum_{k=1}^{m}\sum_{\ell=1}^{m} \frac{\partial y_{ij}}{\partial x_{k\ell}} f_k {g_\ell}^T \otimes r_i {s_j}^T \quad .$$

in which r_i is a unit vector of length m with a 1 in the ith position, s_j is a unit vector of length m with a 1 in the jth position, f_k is a unit vector of length m with a 1 in the kth position, and g_ℓ is a unit vector of length m with a 1 in the ℓth position. We see that this summation enumerates each and every partial derivative once (and only once) and places it into a matrix in a uniquely defined position. Therein lies the power (and concept) of the matrix derivative. We now structure a very small example in the 2×2 case, the key terms of which are

$$X = \begin{bmatrix} x_{11} & x_{12} \\ x_{21} & x_{22} \end{bmatrix} \quad \frac{\partial}{\partial X} = \begin{bmatrix} \frac{\partial}{\partial x_{11}} & \frac{\partial}{\partial x_{12}} \\ \frac{\partial}{\partial x_{21}} & \frac{\partial}{\partial x_{22}} \end{bmatrix} \quad Y = \begin{bmatrix} y_{11} & y_{12} \\ y_{21} & y_{22} \end{bmatrix} .$$

When we substitute these terms into the foregoing definition, we obtain the equation

$$\frac{\partial}{\partial X} \otimes Y = \begin{bmatrix} \frac{\partial}{\partial x_{11}} \begin{bmatrix} y_{11} & y_{12} \\ y_{21} & y_{22} \end{bmatrix} & \frac{\partial}{\partial x_{12}} \begin{bmatrix} y_{11} & y_{12} \\ y_{21} & y_{22} \end{bmatrix} \\ \frac{\partial}{\partial x_{21}} \begin{bmatrix} y_{11} & y_{12} \\ y_{21} & y_{22} \end{bmatrix} & \frac{\partial}{\partial x_{22}} \begin{bmatrix} y_{11} & y_{12} \\ y_{21} & y_{22} \end{bmatrix} \end{bmatrix}$$

$$= \begin{bmatrix} \frac{\partial y_{11}}{\partial x_{11}} & \frac{\partial y_{12}}{\partial x_{11}} & \frac{\partial y_{11}}{\partial x_{12}} & \frac{\partial y_{12}}{\partial x_{12}} \\ \frac{\partial y_{21}}{\partial x_{11}} & \frac{\partial y_{22}}{\partial x_{11}} & \frac{\partial y_{21}}{\partial x_{12}} & \frac{\partial y_{22}}{\partial x_{12}} \\ \frac{\partial y_{11}}{\partial x_{21}} & \frac{\partial y_{12}}{\partial x_{21}} & \frac{\partial y_{11}}{\partial x_{22}} & \frac{\partial y_{12}}{\partial x_{22}} \\ \frac{\partial y_{21}}{\partial x_{21}} & \frac{\partial y_{22}}{\partial x_{21}} & \frac{\partial y_{21}}{\partial x_{22}} & \frac{\partial y_{22}}{\partial x_{22}} \end{bmatrix} .$$

This is the second method of matrix differentiation. Notice how the ij blocks are laid out as the left hand multiplier in the Kronecker product would lay them out. That is why we get this form of matrix derivative. This is another way to enumerate the partial derivative with respect to every matrix element. This generalizes to any level of m, not just $m = 2$.

C.5 Third Type of Matrix Derivative

The third type of matrix derivative is a bit different in concept from the former two. It is, in fact, a bit more direct. Suppose we vectorize the matrix Y and vectorize the matrix X and write the partial derivative of all the $Y's$ with respect to all the $X's$. If we did so, we would define the derivative.

$$\text{(C.10)} \qquad \frac{\partial}{\partial\left[vec\left(X\right)\right]}\left[vec\left(Y\right)\right]^T = \left[vec\left(Y\right)\right]^T \otimes \frac{\partial}{\partial\left[vec\left(X\right)\right]} \quad ,$$

in which

$$\frac{\partial}{\partial\left[vec\left(X\right)\right]} = \begin{bmatrix} \frac{\partial}{\partial x_{11}} \\ \cdots \\ \frac{\partial}{\partial x_{m1}} \\ \frac{\partial}{\partial x_{12}} \\ \cdots \\ \frac{\partial}{\partial x_{m2}} \\ \cdots \\ \frac{\partial}{\partial x_{1m}} \\ \cdots \\ \frac{\partial}{\partial x_{mm}} \end{bmatrix} .$$

If we apply this for the 2×2 case, we can write the key matrixes

$$\frac{\partial}{\partial\left[vec\left(X\right)\right]} = \begin{bmatrix} \frac{\partial}{\partial x_{11}} \\ \frac{\partial}{\partial x_{21}} \\ \frac{\partial}{\partial x_{12}} \\ \frac{\partial}{\partial x_{22}} \end{bmatrix} \qquad \frac{\partial}{\partial[vec\left(X\right)]^T} = \begin{bmatrix} \frac{\partial}{\partial x_{11}} & \frac{\partial}{\partial x_{21}} & \frac{\partial}{\partial x_{12}} & \frac{\partial}{\partial x_{22}} \end{bmatrix} \quad ,$$

and

$$Y = \begin{bmatrix} y_{11} & y_{12} \\ y_{21} & y_{22} \end{bmatrix} \quad .$$

Substituting into the former equation gives

$$\frac{\partial}{\partial\left[vec\left(X\right)\right]}\left[vec\left(Y\right)\right]^T = \begin{bmatrix} \frac{\partial}{\partial x_{11}} \\ \frac{\partial}{\partial x_{21}} \\ \frac{\partial}{\partial x_{12}} \\ \frac{\partial}{\partial x_{22}} \end{bmatrix} \begin{bmatrix} y_{11} & y_{21} & y_{12} & y_{22} \end{bmatrix}$$

$$= \begin{bmatrix} \frac{\partial y_{11}}{\partial x_{11}} & \frac{\partial y_{21}}{\partial x_{11}} & \frac{\partial y_{12}}{\partial x_{11}} & \frac{\partial y_{22}}{\partial x_{11}} \\ \frac{\partial y_{11}}{\partial x_{21}} & \frac{\partial y_{21}}{\partial x_{21}} & \frac{\partial y_{12}}{\partial x_{21}} & \frac{\partial y_{22}}{\partial x_{21}} \\ \frac{\partial y_{11}}{\partial x_{12}} & \frac{\partial y_{21}}{\partial x_{12}} & \frac{\partial y_{12}}{\partial x_{12}} & \frac{\partial y_{22}}{\partial x_{12}} \\ \frac{\partial y_{11}}{\partial x_{22}} & \frac{\partial y_{21}}{\partial x_{22}} & \frac{\partial y_{12}}{\partial x_{22}} & \frac{\partial y_{22}}{\partial x_{22}} \end{bmatrix} \quad .$$

We can write this in the following way (correcting a typo in Kollo and von Rosen[57])

$$\frac{\partial Y}{\partial X} = \sum_{i=1}^{m}\sum_{j=1}^{m}\sum_{k=1}^{m}\sum_{\ell=1}^{m} \frac{\partial y_{ij}}{\partial x_{k\ell}} \left(g_\ell \otimes f_k\right)\left(s_j \otimes r_i\right)^T$$

$$= \sum_{i=1}^{m}\sum_{j=1}^{m}\sum_{k=1}^{m}\sum_{\ell=1}^{m} \frac{\partial y_{ij}}{\partial x_{k\ell}} \left(g_\ell {s_j}^T\right) \otimes \left(f_k {r_i}^T\right) \quad .$$

On p. 125 of Kollo and von Rosen[57], we encounter the notion of differentiating the vectorization of a matrix with respect to the vectorization of a matrix X. This definition offers a very unequivocal and unique notion of matrix differentiation. Following their logic, we define a very specific matrix equation that characterizes this matrix derivative.

$$\frac{\partial Y}{\partial X} = \frac{\partial}{\partial vec\left(X\right)} vec(Y)^T = vec(Y)^T \otimes \frac{\partial}{\partial vec\left(X\right)^T} \quad ,$$

in which

$$\frac{\partial}{\partial vec\left(X\right)} = \begin{bmatrix} \frac{\partial}{\partial vec(x_{11})} \\ \cdots \\ \frac{\partial}{\partial vec(x_{p1})} \\ \cdots \\ \frac{\partial}{\partial vec(x_{1p})} \\ \cdots \\ \frac{\partial}{\partial vec(x_{pp})} \end{bmatrix} .$$

(The final substitution is based on the Kronecker theorem for two vectors $uv^T = u \otimes v$.) This leads us to the following definition of the "vec(.) over vec(.)" derivative.

$$\frac{\partial Y}{\partial X} = \sum_{i=1}^{m}\sum_{j=1}^{m}\sum_{k=1}^{m}\sum_{\ell=1}^{m} \frac{\partial y_{ij}}{\partial x_{k\ell}} \left(g_\ell \otimes f_k\right)\left(e_j \otimes d_i\right)^T \quad .$$

C.5.1 Proof of the Chain Rule for "Vec over Vec" Derivatives

The standard chain rule extended to matrices would have the form

$$\frac{\partial Z}{\partial X} = \frac{\partial Y}{\partial X}\frac{\partial Z}{\partial Y} \quad ,$$

in which the noted derivatives are "vec over vec" derivatives as defined previously. We shall illustrate why this is correct using a 2×2 setting, which we shall see immediately generalizes to any higher order of matrix. The ij^{th} element of this derivative is well known analytically

$$\frac{\partial z_{ij}}{\partial x_{k\ell}} = \left(\sum_{M=1}^{m}\sum_{N=1}^{m} \frac{\partial z_{ij}}{\partial y_{MN}}\frac{\partial z_{MN}}{\partial x_{k\ell}}\right) \quad .$$

We can write the number 1 in a very clever way using unit vectors in positions M and N. Consider the following Kronecker product of unit vectors

$$\left(e_N \otimes d_M\right)^T \left(e_N \otimes d_M\right) \quad .$$

Let us write out the term in the foregoing equation in the specific case of $M = 1$ and $N = 2$

$$e_N \otimes d_M = e_2 \otimes d_1 = \begin{bmatrix} 0 \\ 1 \\ 0 \end{bmatrix} \otimes \begin{bmatrix} 1 \\ 0 \\ 0 \end{bmatrix} = \begin{bmatrix} 0 \begin{bmatrix} 1 \\ 0 \\ 0 \end{bmatrix} \\ 1 \begin{bmatrix} 1 \\ 0 \\ 0 \end{bmatrix} \\ 0 \begin{bmatrix} 1 \\ 0 \\ 0 \end{bmatrix} \end{bmatrix} = \begin{bmatrix} 0 \\ 0 \\ 0 \\ 1 \\ 0 \\ 0 \\ 0 \\ 0 \\ 0 \end{bmatrix} .$$

We see that the resulting vector has M sectors of length M. The Nth sector of the overall resulting vector contains the vector d_M. All other sectors of the overall resulting vector contain zeroes. It is clear that the inner product of this vector with itself is 1.

$$(e_2 \otimes d_1)^T (e_2 \otimes d_1) = 1 \quad .$$

Returning to the partial derivative $\frac{\partial Z}{\partial X}$, we rewrite with the 1 inserted strategically into the equation and manipulate

$$\frac{\partial Z}{\partial X} = \sum_{i=1}^{m} \sum_{j=1}^{m} \sum_{k=1}^{m} \sum_{\ell=1}^{m} \left[\sum_{M=1}^{m} \sum_{N=1}^{m} \frac{\partial z_{ij}}{\partial x_{MN}} \frac{\partial y_{MN}}{\partial x_{k\ell}} (e_\ell \otimes d_k) \times 1 \times (e_j \otimes d_i)^T \right]$$

$$= \sum_{i=1}^{m} \sum_{j=1}^{m} \sum_{k=1}^{m} \sum_{\ell=1}^{m} \sum_{M=1}^{m} \sum_{N=1}^{m} \begin{bmatrix} \frac{\partial z_{ij}}{\partial x_{MN}} \frac{\partial y_{MN}}{\partial x_{k\ell}} \\ \times (e_\ell \otimes d_k) \\ (e_N \otimes d_M)^T \times (e_N \otimes d_M) \\ (e_j \otimes d_i)^T \end{bmatrix}$$

$$= \sum_{i=1}^{m} \sum_{j=1}^{m} \sum_{k=1}^{m} \sum_{\ell=1}^{m} \sum_{M=1}^{m} \sum_{N=1}^{m} \begin{bmatrix} \frac{\partial z_{ij}}{\partial x_{MN}} \frac{\partial y_{MN}}{\partial x_{k\ell}} \\ \times \left[(e_\ell \otimes d_k) (e_N \otimes d_M)^T \right] \\ \times \left[(e_N \otimes d_M) (e_j \otimes d_i)^T \right] \end{bmatrix} .$$

If we carefully examine the ranges of the summations in this equation, we can rearrange them to write

$$\frac{\partial Z}{\partial X} = \sum_{M=1}^{m} \sum_{N=1}^{m} \left\{ \begin{array}{c} \left[\sum_{k=1}^{m} \sum_{\ell=1}^{m} \frac{\partial y_{MN}}{\partial x_{k\ell}} (e_\ell \otimes d_k) (e_N \otimes d_M)^T \right] \\ \left[\sum_{i=1}^{m} \sum_{j=1}^{m} \frac{\partial z_{ij}}{\partial x_{MN}} (e_N \otimes d_M) (e_j \otimes d_i)^T \right] \end{array} \right\} .$$

The matrix $(e_\ell \otimes d_k) (e_N \otimes d_M)^T$ contains a zero in every position of the matrix except for the single cell $(\ell - 1) p + k, (N - 1) p + M$. We use the notation

$$E_{(\ell-1)p+k,(N-1)p+M} \quad ,$$

to represent a matrix that is all zeros save for the element represented by the subscript. Using this notation, we can write the equation for the sought after matrix derivative

$$\frac{\partial Z}{\partial X} = \sum_{M=1}^{m}\sum_{N=1}^{m}\left\{\begin{array}{l}\left[\sum_{k=1}^{m}\sum_{\ell=1}^{m}\frac{\partial y_{MN}}{\partial x_{k\ell}}E_{(\ell-1)p+k,(N-1)p+M}\right]\\ \left[\sum_{i=1}^{m}\sum_{j=1}^{m}\frac{\partial z_{ij}}{\partial x_{MN}}E_{(N-1)p+M,(j-1)p+i}\right]\end{array}\right\} .$$

We need to evaluate this equation for the four possible combinations of M and N and then add them up. The development, although specific to the 2×2 situation expands directly to arbitrary dimensionality.

C.5.1.1 $M = 1$ and $N = 1$

We begin by considering the term in the summation for $M = 1$ and $N = 1$, which is

$$\left[\sum_{k=1}^{m}\sum_{\ell=1}^{m}\frac{\partial y_{11}}{\partial x_{k\ell}}E_{(\ell-1)p+k,1}\right]\left[\sum_{i=1}^{m}\sum_{j=1}^{m}\frac{\partial z_{ij}}{\partial y_{11}}E_{1,(j-1)p+i}\right] .$$

The calculation of the index values in the left hand term and the right hand term of the foregoing summations in the equation are

$$\begin{bmatrix} k & \ell & \ell-1 & (\ell-1)p & left\,index & E \\ 1 & 1 & 0 & 0 & 1 & E_{1,1} \\ 1 & 2 & 1 & 2 & 3 & E_{3,1} \\ 2 & 1 & 0 & 0 & 2 & E_{2,1} \\ 2 & 2 & 1 & 2 & 4 & E_{4,1} \end{bmatrix}$$

$$\begin{bmatrix} i & j & j-1 & (j-1)p & right\,index & E \\ 1 & 1 & 0 & 0 & 1 & E_{1,1} \\ 1 & 2 & 1 & 2 & 3 & E_{1,3} \\ 2 & 1 & 0 & 0 & 2 & E_{1,2} \\ 2 & 2 & 1 & 2 & 4 & E_{1,4} \end{bmatrix} .$$

Expanding out the summations in the right and left terms, we write

$$\begin{bmatrix} \frac{\partial y_{11}}{\partial x_{11}}E_{1,1} \\ +\frac{\partial y_{11}}{\partial x_{12}}E_{3,1} \\ +\frac{\partial y_{11}}{\partial x_{21}}E_{2,1} \\ +\frac{\partial y_{11}}{\partial x_{22}}E_{4,1} \end{bmatrix}\begin{bmatrix} \frac{\partial z_{11}}{\partial y_{11}}E_{1,1} \\ +\frac{\partial z_{12}}{\partial y_{11}}E_{1,3} \\ +\frac{\partial z_{21}}{\partial y_{11}}E_{1,2} \\ +\frac{\partial z_{22}}{\partial y_{11}}E_{1,4} \end{bmatrix} .$$

We write this in matrix format

$$\begin{bmatrix} \frac{\partial y_{11}}{\partial x_{11}} & 0 & 0 & 0 \\ \frac{\partial y_{11}}{\partial x_{21}} & 0 & 0 & 0 \\ \frac{\partial y_{11}}{\partial x_{12}} & 0 & 0 & 0 \\ \frac{\partial y_{11}}{\partial x_{22}} & 0 & 0 & 0 \end{bmatrix} \begin{bmatrix} \frac{\partial z_{11}}{\partial y_{11}} & \frac{\partial z_{21}}{\partial y_{11}} & \frac{\partial z_{12}}{\partial y_{11}} & \frac{\partial z_{22}}{\partial y_{11}} \\ 0 & 0 & 0 & 0 \\ 0 & 0 & 0 & 0 \\ 0 & 0 & 0 & 0 \end{bmatrix} .$$

We perform the matrix multiplication to obtain the final result for this term

$$\begin{bmatrix} \frac{\partial y_{11}}{\partial x_{11}}\frac{\partial z_{11}}{\partial y_{11}} & \frac{\partial y_{11}}{\partial x_{11}}\frac{\partial z_{21}}{\partial y_{11}} & \frac{\partial y_{11}}{\partial x_{11}}\frac{\partial z_{12}}{\partial y_{11}} & \frac{\partial y_{11}}{\partial x_{11}}\frac{\partial z_{22}}{\partial y_{11}} \\ \frac{\partial y_{11}}{\partial x_{21}}\frac{\partial z_{11}}{\partial y_{11}} & \frac{\partial y_{11}}{\partial x_{21}}\frac{\partial z_{21}}{\partial y_{11}} & \frac{\partial y_{11}}{\partial x_{21}}\frac{\partial z_{12}}{\partial y_{11}} & \frac{\partial y_{11}}{\partial x_{21}}\frac{\partial z_{22}}{\partial y_{11}} \\ \frac{\partial y_{11}}{\partial x_{12}}\frac{\partial z_{11}}{\partial y_{11}} & \frac{\partial y_{11}}{\partial x_{12}}\frac{\partial z_{21}}{\partial y_{11}} & \frac{\partial y_{11}}{\partial x_{12}}\frac{\partial z_{12}}{\partial y_{11}} & \frac{\partial y_{11}}{\partial x_{12}}\frac{\partial z_{22}}{\partial y_{11}} \\ \frac{\partial y_{11}}{\partial x_{22}}\frac{\partial z_{11}}{\partial y_{11}} & \frac{\partial y_{11}}{\partial x_{22}}\frac{\partial z_{21}}{\partial y_{11}} & \frac{\partial y_{11}}{\partial x_{22}}\frac{\partial z_{12}}{\partial y_{11}} & \frac{\partial y_{11}}{\partial x_{22}}\frac{\partial z_{22}}{\partial y_{11}} \end{bmatrix} .$$

C.5.1.2 $M = 1$ and $N = 2$

For these values of M and N, the term in the summation is

$$\left[\sum_{k=1}^{m}\sum_{\ell=1}^{m}\frac{\partial y_{12}}{\partial x_{k\ell}}E_{(\ell-1)p+k,3}\right]\left[\sum_{i=1}^{m}\sum_{j=1}^{m}\frac{\partial z_{ij}}{\partial y_{12}}E_{3,(j-1)p+i}\right] .$$

Performing all the indexing calculations yields

$$\begin{bmatrix} \frac{\partial y_{12}}{\partial x_{11}}E_{1,3} \\ +\frac{\partial y_{12}}{\partial x_{12}}E_{3,3} \\ +\frac{\partial y_{12}}{\partial x_{21}}E_{2,3} \\ +\frac{\partial y_{12}}{\partial x_{22}}E_{4,3} \end{bmatrix} \begin{bmatrix} \frac{\partial z_{11}}{\partial y_{12}}E_{3,1} \\ +\frac{\partial z_{12}}{\partial y_{12}}E_{3,3} \\ +\frac{\partial z_{21}}{\partial y_{12}}E_{3,2} \\ +\frac{\partial z_{22}}{\partial y_{12}}E_{3,4} \end{bmatrix} .$$

If we calculate and multiply the matrices, we obtain the final answer for these values of $M = 1$ and $N = 2$

$$\begin{bmatrix} \frac{\partial y_{12}}{\partial x_{11}}\frac{\partial z_{11}}{\partial y_{12}} & \frac{\partial y_{12}}{\partial x_{11}}\frac{\partial z_{21}}{\partial y_{12}} & \frac{\partial y_{12}}{\partial x_{11}}\frac{\partial z_{12}}{\partial y_{12}} & \frac{\partial y_{12}}{\partial x_{11}}\frac{\partial z_{22}}{\partial y_{12}} \\ \frac{\partial y_{12}}{\partial x_{21}}\frac{\partial z_{11}}{\partial y_{12}} & \frac{\partial y_{12}}{\partial x_{21}}\frac{\partial z_{21}}{\partial y_{12}} & \frac{\partial y_{12}}{\partial x_{21}}\frac{\partial z_{12}}{\partial y_{12}} & \frac{\partial y_{12}}{\partial x_{21}}\frac{\partial z_{22}}{\partial y_{12}} \\ \frac{\partial y_{12}}{\partial x_{12}}\frac{\partial z_{11}}{\partial y_{12}} & \frac{\partial y_{12}}{\partial x_{12}}\frac{\partial z_{21}}{\partial y_{12}} & \frac{\partial y_{12}}{\partial x_{12}}\frac{\partial z_{12}}{\partial y_{12}} & \frac{\partial y_{12}}{\partial x_{12}}\frac{\partial z_{22}}{\partial y_{12}} \\ \frac{\partial y_{12}}{\partial x_{22}}\frac{\partial z_{11}}{\partial y_{12}} & \frac{\partial y_{12}}{\partial x_{22}}\frac{\partial z_{21}}{\partial y_{12}} & \frac{\partial y_{12}}{\partial x_{22}}\frac{\partial z_{12}}{\partial y_{12}} & \frac{\partial y_{12}}{\partial x_{22}}\frac{\partial z_{22}}{\partial y_{12}} \end{bmatrix} .$$

C.5.1.3 $M = 2$ and $N = 1$

For these values of M and N, we write the term in the summation

$$\left[\sum_{k=1}^{m}\sum_{\ell=1}^{m}\frac{\partial y_{21}}{\partial x_{k\ell}}E_{(\ell-1)p+k,2}\right]\left[\sum_{i=1}^{m}\sum_{j=1}^{m}\frac{\partial z_{ij}}{\partial y_{21}}E_{2,(j-1)p+i}\right] .$$

Proceeding through precisely the same simplifications as in the previous two sections, we obtain the final answer

$$\begin{bmatrix} \frac{\partial y_{21}}{\partial x_{11}}\frac{\partial z_{11}}{\partial y_{21}} & \frac{\partial y_{21}}{\partial x_{11}}\frac{\partial z_{21}}{\partial y_{21}} & \frac{\partial y_{21}}{\partial x_{11}}\frac{\partial z_{12}}{\partial y_{21}} & \frac{\partial y_{21}}{\partial x_{11}}\frac{\partial z_{22}}{\partial y_{21}} \\ \frac{\partial y_{21}}{\partial x_{12}}\frac{\partial z_{11}}{\partial y_{21}} & \frac{\partial y_{21}}{\partial x_{12}}\frac{\partial z_{21}}{\partial y_{21}} & \frac{\partial y_{21}}{\partial x_{12}}\frac{\partial z_{12}}{\partial y_{21}} & \frac{\partial y_{21}}{\partial x_{12}}\frac{\partial z_{22}}{\partial y_{21}} \\ \frac{\partial y_{21}}{\partial x_{21}}\frac{\partial z_{11}}{\partial y_{21}} & \frac{\partial y_{21}}{\partial x_{21}}\frac{\partial z_{21}}{\partial y_{21}} & \frac{\partial y_{21}}{\partial x_{21}}\frac{\partial z_{12}}{\partial y_{21}} & \frac{\partial y_{21}}{\partial x_{21}}\frac{\partial z_{22}}{\partial y_{21}} \\ \frac{\partial y_{21}}{\partial x_{22}}\frac{\partial z_{11}}{\partial y_{21}} & \frac{\partial y_{21}}{\partial x_{22}}\frac{\partial z_{21}}{\partial y_{21}} & \frac{\partial y_{21}}{\partial x_{22}}\frac{\partial z_{12}}{\partial y_{21}} & \frac{\partial y_{21}}{\partial x_{22}}\frac{\partial z_{22}}{\partial y_{21}} \end{bmatrix}.$$

C.5.1.4 $M = 2$ and $N = 2$

The procedure is clear to get to the final answer for $M = 2$ and $N = 2$

$$\begin{bmatrix} \frac{\partial y_{22}}{\partial x_{11}}\frac{\partial z_{11}}{\partial y_{22}} & \frac{\partial y_{22}}{\partial x_{11}}\frac{\partial z_{12}}{\partial y_{22}} & \frac{\partial y_{22}}{\partial x_{11}}\frac{\partial z_{21}}{\partial y_{22}} & \frac{\partial y_{22}}{\partial x_{11}}\frac{\partial z_{22}}{\partial y_{22}} \\ \frac{\partial y_{22}}{\partial x_{12}}\frac{\partial z_{11}}{\partial y_{22}} & \frac{\partial y_{22}}{\partial x_{12}}\frac{\partial z_{12}}{\partial y_{22}} & \frac{\partial y_{22}}{\partial x_{12}}\frac{\partial z_{21}}{\partial y_{22}} & \frac{\partial y_{22}}{\partial x_{12}}\frac{\partial z_{22}}{\partial y_{22}} \\ \frac{\partial y_{22}}{\partial x_{21}}\frac{\partial z_{11}}{\partial y_{22}} & \frac{\partial y_{22}}{\partial x_{21}}\frac{\partial z_{12}}{\partial y_{22}} & \frac{\partial y_{22}}{\partial x_{21}}\frac{\partial z_{21}}{\partial y_{22}} & \frac{\partial y_{22}}{\partial x_{21}}\frac{\partial z_{22}}{\partial y_{22}} \\ \frac{\partial y_{22}}{\partial x_{22}}\frac{\partial z_{11}}{\partial y_{22}} & \frac{\partial y_{22}}{\partial x_{22}}\frac{\partial z_{12}}{\partial y_{22}} & \frac{\partial y_{22}}{\partial x_{22}}\frac{\partial z_{21}}{\partial y_{22}} & \frac{\partial y_{22}}{\partial x_{22}}\frac{\partial z_{22}}{\partial y_{22}} \end{bmatrix}.$$

C.5.1.5 Completion of the Proof

The sum of the foregoing four individual matrices is shown in Figure C.1.

Figure C.1: The Final Summation Matrix

$$\begin{bmatrix} \frac{\partial y_{11}}{\partial x_{11}}\frac{\partial z_{11}}{\partial y_{11}}+\frac{\partial y_{12}}{\partial x_{11}}\frac{\partial z_{11}}{\partial y_{12}}+\frac{\partial y_{21}}{\partial x_{11}}\frac{\partial z_{11}}{\partial y_{21}}+\frac{\partial y_{22}}{\partial x_{11}}\frac{\partial z_{11}}{\partial y_{22}} & \frac{\partial y_{11}}{\partial x_{11}}\frac{\partial z_{21}}{\partial y_{11}}+\frac{\partial y_{12}}{\partial x_{11}}\frac{\partial z_{21}}{\partial y_{12}}+\frac{\partial y_{21}}{\partial x_{11}}\frac{\partial z_{21}}{\partial y_{21}}+\frac{\partial y_{22}}{\partial x_{11}}\frac{\partial z_{12}}{\partial y_{22}} & \frac{\partial y_{11}}{\partial x_{11}}\frac{\partial z_{12}}{\partial y_{11}}+\frac{\partial y_{12}}{\partial x_{11}}\frac{\partial z_{12}}{\partial y_{12}}+\frac{\partial y_{21}}{\partial x_{11}}\frac{\partial z_{12}}{\partial y_{21}}+\frac{\partial y_{22}}{\partial x_{11}}\frac{\partial z_{21}}{\partial y_{22}} & \frac{\partial y_{11}}{\partial x_{11}}\frac{\partial z_{22}}{\partial y_{11}}+\frac{\partial y_{12}}{\partial x_{11}}\frac{\partial z_{22}}{\partial y_{12}}+\frac{\partial y_{21}}{\partial x_{11}}\frac{\partial z_{22}}{\partial y_{21}}+\frac{\partial y_{22}}{\partial x_{11}}\frac{\partial z_{22}}{\partial y_{22}} \\ \frac{\partial y_{11}}{\partial x_{21}}\frac{\partial z_{11}}{\partial y_{11}}+\frac{\partial y_{12}}{\partial x_{21}}\frac{\partial z_{11}}{\partial y_{12}}+\frac{\partial y_{21}}{\partial x_{12}}\frac{\partial z_{11}}{\partial y_{21}}+\frac{\partial y_{22}}{\partial x_{12}}\frac{\partial z_{11}}{\partial y_{22}} & \frac{\partial y_{11}}{\partial x_{21}}\frac{\partial z_{21}}{\partial y_{11}}+\frac{\partial y_{12}}{\partial x_{21}}\frac{\partial z_{21}}{\partial y_{12}}+\frac{\partial y_{21}}{\partial x_{12}}\frac{\partial z_{21}}{\partial y_{21}}+\frac{\partial y_{22}}{\partial x_{12}}\frac{\partial z_{12}}{\partial y_{22}} & \frac{\partial y_{11}}{\partial x_{21}}\frac{\partial z_{12}}{\partial y_{11}}+\frac{\partial y_{12}}{\partial x_{21}}\frac{\partial z_{12}}{\partial y_{12}}+\frac{\partial y_{21}}{\partial x_{12}}\frac{\partial z_{12}}{\partial y_{21}}+\frac{\partial y_{22}}{\partial x_{12}}\frac{\partial z_{21}}{\partial y_{22}} & \frac{\partial y_{11}}{\partial x_{21}}\frac{\partial z_{22}}{\partial y_{11}}+\frac{\partial y_{12}}{\partial x_{21}}\frac{\partial z_{22}}{\partial y_{12}}+\frac{\partial y_{21}}{\partial x_{12}}\frac{\partial z_{22}}{\partial y_{21}}+\frac{\partial y_{22}}{\partial x_{12}}\frac{\partial z_{22}}{\partial y_{22}} \\ \frac{\partial y_{11}}{\partial x_{12}}\frac{\partial z_{11}}{\partial y_{11}}+\frac{\partial y_{12}}{\partial x_{12}}\frac{\partial z_{11}}{\partial y_{12}}+\frac{\partial y_{21}}{\partial x_{21}}\frac{\partial z_{11}}{\partial y_{21}}+\frac{\partial y_{22}}{\partial x_{21}}\frac{\partial z_{11}}{\partial y_{22}} & \frac{\partial y_{11}}{\partial x_{12}}\frac{\partial z_{21}}{\partial y_{11}}+\frac{\partial y_{12}}{\partial x_{12}}\frac{\partial z_{21}}{\partial y_{12}}+\frac{\partial y_{21}}{\partial x_{21}}\frac{\partial z_{21}}{\partial y_{21}}+\frac{\partial y_{22}}{\partial x_{21}}\frac{\partial z_{12}}{\partial y_{22}} & \frac{\partial y_{11}}{\partial x_{12}}\frac{\partial z_{12}}{\partial y_{11}}+\frac{\partial y_{12}}{\partial x_{12}}\frac{\partial z_{12}}{\partial y_{12}}+\frac{\partial y_{21}}{\partial x_{21}}\frac{\partial z_{12}}{\partial y_{21}}+\frac{\partial y_{22}}{\partial x_{21}}\frac{\partial z_{21}}{\partial y_{22}} & \frac{\partial y_{11}}{\partial x_{12}}\frac{\partial z_{22}}{\partial y_{11}}+\frac{\partial y_{12}}{\partial x_{12}}\frac{\partial z_{22}}{\partial y_{12}}+\frac{\partial y_{21}}{\partial x_{21}}\frac{\partial z_{22}}{\partial y_{21}}+\frac{\partial y_{22}}{\partial x_{21}}\frac{\partial z_{22}}{\partial y_{22}} \\ \frac{\partial y_{11}}{\partial x_{22}}\frac{\partial z_{11}}{\partial y_{11}}+\frac{\partial y_{12}}{\partial x_{22}}\frac{\partial z_{11}}{\partial y_{12}}+\frac{\partial y_{21}}{\partial x_{22}}\frac{\partial z_{11}}{\partial y_{21}}+\frac{\partial y_{22}}{\partial x_{22}}\frac{\partial z_{11}}{\partial y_{22}} & \frac{\partial y_{11}}{\partial x_{22}}\frac{\partial z_{21}}{\partial y_{11}}+\frac{\partial y_{12}}{\partial x_{22}}\frac{\partial z_{21}}{\partial y_{12}}+\frac{\partial y_{21}}{\partial x_{22}}\frac{\partial z_{21}}{\partial y_{21}}+\frac{\partial y_{22}}{\partial x_{22}}\frac{\partial z_{12}}{\partial y_{22}} & \frac{\partial y_{11}}{\partial x_{22}}\frac{\partial z_{12}}{\partial y_{11}}+\frac{\partial y_{12}}{\partial x_{22}}\frac{\partial z_{12}}{\partial y_{12}}+\frac{\partial y_{21}}{\partial x_{22}}\frac{\partial z_{12}}{\partial y_{21}}+\frac{\partial y_{22}}{\partial x_{22}}\frac{\partial z_{21}}{\partial y_{22}} & \frac{\partial y_{11}}{\partial x_{22}}\frac{\partial z_{22}}{\partial y_{11}}+\frac{\partial y_{12}}{\partial x_{22}}\frac{\partial z_{22}}{\partial y_{12}}+\frac{\partial y_{21}}{\partial x_{22}}\frac{\partial z_{22}}{\partial y_{21}}+\frac{\partial y_{22}}{\partial x_{22}}\frac{\partial z_{22}}{\partial y_{22}} \end{bmatrix}$$

Using the information therein, let us write expressions for two matrix derivatives as follows

$$\frac{\partial Y}{\partial X} = \begin{bmatrix} \frac{\partial y_{11}}{\partial x_{11}} & \frac{\partial y_{21}}{\partial x_{11}} & \frac{\partial y_{12}}{\partial x_{11}} & \frac{\partial y_{22}}{\partial x_{11}} \\ \frac{\partial y_{11}}{\partial x_{21}} & \frac{\partial y_{21}}{\partial x_{21}} & \frac{\partial y_{12}}{\partial x_{21}} & \frac{\partial y_{22}}{\partial x_{21}} \\ \frac{\partial y_{11}}{\partial x_{12}} & \frac{\partial y_{21}}{\partial x_{12}} & \frac{\partial y_{12}}{\partial x_{12}} & \frac{\partial y_{22}}{\partial x_{12}} \\ \frac{\partial y_{11}}{\partial x_{22}} & \frac{\partial y_{21}}{\partial x_{22}} & \frac{\partial y_{12}}{\partial x_{22}} & \frac{\partial y_{22}}{\partial x_{22}} \end{bmatrix}$$

$$\frac{\partial Z}{\partial Y} = \begin{bmatrix} \frac{\partial z_{11}}{\partial y_{11}} & \frac{\partial z_{21}}{\partial y_{11}} & \frac{\partial z_{12}}{\partial y_{11}} & \frac{\partial z_{22}}{\partial y_{11}} \\ \frac{\partial z_{11}}{\partial y_{21}} & \frac{\partial z_{21}}{\partial y_{21}} & \frac{\partial z_{12}}{\partial y_{21}} & \frac{\partial z_{22}}{\partial y_{21}} \\ \frac{\partial z_{11}}{\partial y_{12}} & \frac{\partial z_{21}}{\partial y_{12}} & \frac{\partial z_{12}}{\partial y_{12}} & \frac{\partial z_{22}}{\partial y_{12}} \\ \frac{\partial z_{11}}{\partial y_{22}} & \frac{\partial z_{21}}{\partial y_{22}} & \frac{\partial z_{12}}{\partial y_{22}} & \frac{\partial z_{22}}{\partial y_{22}} \end{bmatrix} \quad .$$

If we multiply these two matrices, we can see term by term that we obtain the terms in Figure C.1. This proves the chain rule, which is expressed as follows

$$\frac{\partial Z}{\partial X} = \frac{\partial Z}{\partial Y}\frac{\partial Y}{\partial X} \quad .$$

Keep in mind, these are both "vec over vec" definitions of matrix derivatives.

C.5.2 "Vec over Vec" Derivative of a Matrix X with Respect to Itself

We want to establish the relationship

$$\frac{\partial X}{\partial X} = I \quad .$$

This is easily done by starting with the definition of a "vec over vec" derivative and borrowing the proof of Kollo and von Rosen[57]

$$\begin{aligned} \frac{\partial X}{\partial X} &= \sum_{i=1}^{m}\sum_{i=j}^{m}\sum_{k=1}^{m}\sum_{\ell=1}^{m} \frac{\partial x_{ij}}{\partial x_{k\ell}} \left(e_\ell \otimes d_k\right)\left(e_j \otimes d_i\right)^T \\ &= \sum_{i=1}^{m}\sum_{i=j}^{m} \frac{\partial x_{ij}}{\partial x_{ij}} \left(e_j \otimes d_i\right)\left(e_j \otimes d_i\right)^T \\ &= \sum_{i=1}^{m}\sum_{i=j}^{m} \left(e_j e_j{}^T\right) \otimes \left(d_i d_i{}^T\right) \\ &= I \otimes I = I \quad . \end{aligned}$$

C.5.3 "Vec over Vec" Derivative of a Triple Matrix Product

This section begins by writing the Kronecker vectorization theorem[3]

$$vec(AXB) = (B^T \otimes A)vec(X) \quad .$$

We take the transpose of this relationship

$$\begin{aligned} vec\left(AXB\right) &= \left(B^T \otimes A\right) vec\left(X\right) \\ \left[vec\left(AXB\right)\right]^T &= \left[vec\left(X\right)\right]^T \left(B^T \otimes A\right)^T \end{aligned}$$

[3] See https://en.wikipedia.org/wiki/Kronecker_product.

$$= [vec(X)]^T \left(B \otimes A^T\right) \quad .$$

We are going to need the transpose because we are going to apply equation (C.10) to calculate the derivative. Substituting the foregoing relationship into the definition of the derivative

$$\begin{aligned}\frac{\partial}{\partial vec(X)}[vec(AXB)]^T &= \frac{\partial}{\partial vec(X)}[vec(X)]^T \left(B \otimes A^T\right) \\ &= B \otimes A^T \quad .\end{aligned}$$

We now turn to the triple matrix product in which we are interested. We apply the chain rule to start the evaluation

$$\frac{\partial (AYB)}{\partial X} = \frac{\partial Y}{\partial X}\frac{\partial (AYB)}{\partial Y} \quad .$$

We then apply the foregoing result to the term on the right

$$\frac{\partial (AYB)}{\partial X} = \frac{\partial Y}{\partial X}\left(B \otimes A^T\right) \quad .$$

C.5.4 "Vec over Vec" Derivative of a Matrix Product

In Proposition 1.4.6 in Kollo and von Rosen[57] on page 130, we see their statement and subsequent proof of the "vec over vec" derivative of a matrix product, well known in the literature. It is worthy of summarizing here the result they need to prove and how they have proved it:

$$\frac{\partial (ZY)}{\partial X} = \frac{\partial Z}{\partial X}(Y \otimes I) + \frac{\partial Y}{\partial X}\left(I \otimes Z^T\right) \quad .$$

We assume of course that both Z and Y are a function of X. We are first going to develop the key result in the case where $m = 2$. We can write ZY in the $m = 2$ case

$$\begin{aligned}&\begin{bmatrix} z_{11} & z_{12} \\ z_{21} & z_{22} \end{bmatrix}\begin{bmatrix} y_{11} & y_{12} \\ y_{21} & y_{22} \end{bmatrix} \\ &= \begin{bmatrix} z_{11}y_{11} + z_{12}y_{21} & z_{11}y_{12} + z_{12}y_{22} \\ z_{21}y_{11} + z_{22}y_{21} & z_{21}y_{12} + z_{22}y_{22} \end{bmatrix} \quad .\end{aligned}$$

Let us take the vectorization of ZY

$$vec(ZY) = \begin{bmatrix} z_{11}y_{11} + z_{12}y_{21} \\ z_{21}y_{11} + z_{22}y_{21} \\ z_{11}y_{12} + z_{12}y_{22} \\ z_{21}y_{12} + z_{22}y_{22} \end{bmatrix} \quad ,$$

and the transpose of this vectorization

$$vec(ZY)^T = \begin{bmatrix} z_{11}y_{11} & z_{21}y_{11} & z_{11}y_{12} & z_{21}y_{12} \\ +z_{12}y_{21} & +z_{22}y_{21} & +z_{12}y_{22} & +z_{22}y_{22} \end{bmatrix}.$$

The matrix X is

$$\begin{bmatrix} x_{11} & x_{12} \\ x_{21} & x_{22} \end{bmatrix}.$$

The vectorization of X is

$$vec\,(X) = \begin{bmatrix} x_{11} \\ x_{21} \\ x_{12} \\ x_{22} \end{bmatrix}.$$

Therefore the X derivative operator for the vec over vec derivative is

$$\begin{bmatrix} \frac{\partial}{\partial x_{11}} \\ \frac{\partial}{\partial x_{21}} \\ \frac{\partial}{\partial x_{12}} \\ \frac{\partial}{\partial x_{22}} \end{bmatrix}.$$

Applying the definition of the vec over vec derivative, we can write

$$\frac{\partial ZY}{\partial X} \triangleq \frac{\partial [vecZY]^T}{\partial vecX}$$
$$= \begin{bmatrix} \frac{\partial}{\partial x_{11}} \\ \frac{\partial}{\partial x_{21}} \\ \frac{\partial}{\partial x_{12}} \\ \frac{\partial}{\partial x_{22}} \end{bmatrix} \begin{bmatrix} z_{11}y_{11} & z_{21}y_{11} & z_{11}y_{12} & z_{11}y_{12} \\ +z_{12}y_{21} & +z_{22}y_{21} & +z_{12}y_{22} & +z_{12}y_{22} \end{bmatrix}.$$

This can be written

$$\begin{bmatrix} \frac{\partial}{\partial x_{11}} \begin{smallmatrix} z_{11}y_{11} \\ +z_{12}y_{21} \end{smallmatrix} & \frac{\partial}{\partial x_{11}} \begin{smallmatrix} z_{21}y_{11} \\ +z_{22}y_{21} \end{smallmatrix} & \frac{\partial}{\partial x_{11}} \begin{smallmatrix} z_{11}y_{12} \\ +z_{12}y_{22} \end{smallmatrix} & \frac{\partial}{\partial x_{11}} \begin{smallmatrix} z_{11}y_{12} \\ +z_{12}y_{22} \end{smallmatrix} \\ \frac{\partial}{\partial x_{21}} \begin{smallmatrix} z_{11}y_{11} \\ +z_{12}y_{21} \end{smallmatrix} & \frac{\partial}{\partial x_{21}} \begin{smallmatrix} z_{21}y_{11} \\ +z_{22}y_{21} \end{smallmatrix} & \frac{\partial}{\partial x_{21}} \begin{smallmatrix} z_{11}y_{12} \\ +z_{12}y_{22} \end{smallmatrix} & \frac{\partial}{\partial x_{21}} \begin{smallmatrix} z_{11}y_{12} \\ +z_{12}y_{22} \end{smallmatrix} \\ \frac{\partial}{\partial x_{12}} \begin{smallmatrix} z_{11}y_{11} \\ +z_{12}y_{21} \end{smallmatrix} & \frac{\partial}{\partial x_{12}} \begin{smallmatrix} z_{21}y_{11} \\ +z_{22}y_{21} \end{smallmatrix} & \frac{\partial}{\partial x_{12}} \begin{smallmatrix} z_{11}y_{12} \\ +z_{12}y_{22} \end{smallmatrix} & \frac{\partial}{\partial x_{12}} \begin{smallmatrix} z_{11}y_{12} \\ +z_{12}y_{22} \end{smallmatrix} \\ \frac{\partial}{\partial x_{22}} \begin{smallmatrix} z_{11}y_{11} \\ +z_{12}y_{21} \end{smallmatrix} & \frac{\partial}{\partial x_{22}} \begin{smallmatrix} z_{21}y_{11} \\ +z_{22}y_{21} \end{smallmatrix} & \frac{\partial}{\partial x_{22}} \begin{smallmatrix} z_{11}y_{12} \\ +z_{12}y_{22} \end{smallmatrix} & \frac{\partial}{\partial x_{22}} \begin{smallmatrix} z_{11}y_{12} \\ +z_{12}y_{22} \end{smallmatrix} \end{bmatrix}.$$

We apply the product rule in every term and reorder

$$
\begin{bmatrix}
\begin{pmatrix}
+\frac{\partial z_{11}}{\partial x_{11}}y_{11} \\ +\frac{\partial z_{12}}{\partial x_{11}}y_{21} \\ +z_{11}\frac{\partial y_{11}}{\partial x_{11}} \\ +z_{12}\frac{\partial y_{21}}{\partial x_{11}} \\
+\frac{\partial z_{11}}{\partial x_{21}}y_{11} \\ +\frac{\partial z_{12}}{\partial x_{21}}y_{21} \\ +z_{11}\frac{\partial y_{11}}{\partial x_{21}} \\ +z_{12}\frac{\partial y_{21}}{\partial x_{21}} \\
+\frac{\partial z_{11}}{\partial x_{12}}y_{11} \\ +\frac{\partial z_{12}}{\partial x_{12}}y_{21} \\ +z_{11}\frac{\partial y_{11}}{\partial x_{12}} \\ +z_{12}\frac{\partial y_{21}}{\partial x_{12}} \\
+\frac{\partial z_{11}}{\partial x_{22}}y_{11} \\ +\frac{\partial z_{12}}{\partial x_{22}}y_{21} \\ +z_{11}\frac{\partial y_{11}}{\partial x_{22}} \\ +z_{12}\frac{\partial y_{21}}{\partial x_{22}}
\end{pmatrix}
&
\begin{pmatrix}
+\frac{\partial z_{21}}{\partial x_{11}}y_{11} \\ +\frac{\partial z_{22}}{\partial x_{11}}y_{21} \\ +z_{21}\frac{\partial y_{11}}{\partial x_{11}} \\ +z_{22}\frac{\partial y_{21}}{\partial x_{11}} \\
+\frac{\partial z_{21}}{\partial x_{21}}y_{11} \\ +\frac{\partial z_{22}}{\partial x_{21}}y_{21} \\ +z_{21}\frac{\partial y_{11}}{\partial x_{21}} \\ +z_{22}\frac{\partial y_{21}}{\partial x_{21}} \\
+\frac{\partial z_{21}}{\partial x_{12}}y_{11} \\ +\frac{\partial z_{22}}{\partial x_{12}}y_{21} \\ +z_{21}\frac{\partial y_{11}}{\partial x_{12}} \\ +z_{22}\frac{\partial y_{21}}{\partial x_{12}} \\
+\frac{\partial z_{21}}{\partial x_{22}}y_{11} \\ +\frac{\partial z_{22}}{\partial x_{22}}y_{21} \\ +z_{21}\frac{\partial y_{11}}{\partial x_{22}} \\ +z_{22}\frac{\partial y_{21}}{\partial x_{22}}
\end{pmatrix}
&
\begin{pmatrix}
+\frac{\partial z_{11}}{\partial x_{11}}y_{12} \\ +\frac{\partial z_{12}}{\partial x_{11}}y_{22} \\ +z_{11}\frac{\partial y_{12}}{\partial x_{11}} \\ +z_{12}\frac{\partial y_{22}}{\partial x_{11}} \\
+\frac{\partial z_{11}}{\partial x_{21}}y_{12} \\ +\frac{\partial z_{12}}{\partial x_{21}}y_{22} \\ +z_{11}\frac{\partial y_{12}}{\partial x_{21}} \\ +z_{12}\frac{\partial y_{22}}{\partial x_{21}} \\
+\frac{\partial z_{11}}{\partial x_{12}}y_{12} \\ +\frac{\partial z_{12}}{\partial x_{12}}y_{22} \\ +z_{11}\frac{\partial y_{12}}{\partial x_{12}} \\ +z_{12}\frac{\partial y_{22}}{\partial x_{12}} \\
+\frac{\partial z_{11}}{\partial x_{22}}y_{12} \\ +\frac{\partial z_{12}}{\partial x_{22}}y_{22} \\ +z_{11}\frac{\partial y_{12}}{\partial x_{22}} \\ +z_{12}\frac{\partial y_{22}}{\partial x_{22}}
\end{pmatrix}
&
\begin{pmatrix}
+\frac{\partial z_{11}}{\partial x_{11}}y_{12} \\ +\frac{\partial z_{12}}{\partial x_{11}}y_{22} \\ +z_{11}\frac{\partial y_{12}}{\partial x_{11}} \\ +z_{12}\frac{\partial y_{22}}{\partial x_{11}} \\
+\frac{\partial z_{11}}{\partial x_{21}}y_{12} \\ +\frac{\partial z_{12}}{\partial x_{21}}y_{22} \\ +z_{11}\frac{\partial y_{12}}{\partial x_{21}} \\ +z_{12}\frac{\partial y_{22}}{\partial x_{21}} \\
+\frac{\partial z_{11}}{\partial x_{12}}y_{12} \\ +\frac{\partial z_{12}}{\partial x_{12}}y_{22} \\ +z_{11}\frac{\partial y_{12}}{\partial x_{12}} \\ +z_{12}\frac{\partial y_{22}}{\partial x_{12}} \\
+\frac{\partial z_{11}}{\partial x_{22}}y_{12} \\ +\frac{\partial z_{12}}{\partial x_{22}}y_{22} \\ +z_{11}\frac{\partial y_{12}}{\partial x_{22}} \\ +z_{12}\frac{\partial y_{22}}{\partial x_{22}}
\end{pmatrix}
\end{bmatrix}.
$$

We can write this matrix in two parts, each of which we are going to analyze

$$
\begin{bmatrix}
\begin{pmatrix}
+\frac{\partial z_{11}}{\partial x_{11}}y_{11} \\ +\frac{\partial z_{12}}{\partial x_{11}}y_{21} \\
+\frac{\partial z_{11}}{\partial x_{21}}y_{11} \\ +\frac{\partial z_{12}}{\partial x_{21}}y_{21} \\
+\frac{\partial z_{11}}{\partial x_{12}}y_{11} \\ +\frac{\partial z_{12}}{\partial x_{12}}y_{21} \\
+\frac{\partial z_{11}}{\partial x_{22}}y_{11} \\ +\frac{\partial z_{12}}{\partial x_{22}}y_{21}
\end{pmatrix}
&
\begin{pmatrix}
+\frac{\partial z_{21}}{\partial x_{11}}y_{11} \\ +\frac{\partial z_{22}}{\partial x_{11}}y_{21} \\
+\frac{\partial z_{21}}{\partial x_{21}}y_{11} \\ +\frac{\partial z_{22}}{\partial x_{21}}y_{21} \\
+\frac{\partial z_{21}}{\partial x_{12}}y_{11} \\ +\frac{\partial z_{22}}{\partial x_{12}}y_{21} \\
+\frac{\partial z_{21}}{\partial x_{22}}y_{11} \\ +\frac{\partial z_{22}}{\partial x_{22}}y_{21}
\end{pmatrix}
&
\begin{pmatrix}
+\frac{\partial z_{11}}{\partial x_{11}}y_{12} \\ +\frac{\partial z_{12}}{\partial x_{11}}y_{22} \\
+\frac{\partial z_{11}}{\partial x_{21}}y_{12} \\ +\frac{\partial z_{12}}{\partial x_{21}}y_{22} \\
+\frac{\partial z_{11}}{\partial x_{12}}y_{12} \\ +\frac{\partial z_{12}}{\partial x_{12}}y_{22} \\
+\frac{\partial z_{11}}{\partial x_{22}}y_{12} \\ +\frac{\partial z_{12}}{\partial x_{22}}y_{22}
\end{pmatrix}
&
\begin{pmatrix}
+\frac{\partial z_{11}}{\partial x_{11}}y_{12} \\ +\frac{\partial z_{12}}{\partial x_{11}}y_{22} \\
+\frac{\partial z_{11}}{\partial x_{21}}y_{12} \\ +\frac{\partial z_{12}}{\partial x_{21}}y_{22} \\
+\frac{\partial z_{11}}{\partial x_{12}}y_{12} \\ +\frac{\partial z_{12}}{\partial x_{12}}y_{22} \\
+\frac{\partial z_{11}}{\partial x_{22}}y_{12} \\ +\frac{\partial z_{12}}{\partial x_{22}}y_{22}
\end{pmatrix}
\end{bmatrix}
$$

$$+$$

$$
\begin{bmatrix}
\begin{pmatrix}
+z_{11}\frac{\partial y_{11}}{\partial x_{11}} \\ +z_{12}\frac{\partial y_{21}}{\partial x_{11}} \\
+z_{11}\frac{\partial y_{11}}{\partial x_{21}} \\ +z_{12}\frac{\partial y_{21}}{\partial x_{21}} \\
+z_{11}\frac{\partial y_{11}}{\partial x_{12}} \\ +z_{12}\frac{\partial y_{21}}{\partial x_{12}} \\
+z_{11}\frac{\partial y_{11}}{\partial x_{22}} \\ +z_{12}\frac{\partial y_{21}}{\partial x_{22}}
\end{pmatrix}
&
\begin{pmatrix}
+z_{21}\frac{\partial y_{11}}{\partial x_{11}} \\ +z_{22}\frac{\partial y_{21}}{\partial x_{11}} \\
+z_{21}\frac{\partial y_{11}}{\partial x_{21}} \\ +z_{22}\frac{\partial y_{21}}{\partial x_{21}} \\
+z_{21}\frac{\partial y_{11}}{\partial x_{12}} \\ +z_{22}\frac{\partial y_{21}}{\partial x_{12}} \\
+z_{21}\frac{\partial y_{11}}{\partial x_{22}} \\ +z_{22}\frac{\partial y_{21}}{\partial x_{22}}
\end{pmatrix}
&
\begin{pmatrix}
+z_{11}\frac{\partial y_{12}}{\partial x_{11}} \\ +z_{12}\frac{\partial y_{22}}{\partial x_{11}} \\
+z_{11}\frac{\partial y_{12}}{\partial x_{21}} \\ +z_{12}\frac{\partial y_{22}}{\partial x_{21}} \\
+z_{11}\frac{\partial y_{12}}{\partial x_{12}} \\ +z_{12}\frac{\partial y_{22}}{\partial x_{12}} \\
+z_{11}\frac{\partial y_{12}}{\partial x_{22}} \\ +z_{12}\frac{\partial y_{22}}{\partial x_{22}}
\end{pmatrix}
&
\begin{pmatrix}
+z_{11}\frac{\partial y_{12}}{\partial x_{11}} \\ +z_{12}\frac{\partial y_{22}}{\partial x_{11}} \\
+z_{11}\frac{\partial y_{12}}{\partial x_{21}} \\ +z_{12}\frac{\partial y_{22}}{\partial x_{21}} \\
+z_{11}\frac{\partial y_{12}}{\partial x_{12}} \\ +z_{12}\frac{\partial y_{22}}{\partial x_{12}} \\
+z_{11}\frac{\partial y_{12}}{\partial x_{22}} \\ +z_{12}\frac{\partial y_{22}}{\partial x_{22}}
\end{pmatrix}
\end{bmatrix}.
$$

In thinking about the first term, consider the matrix

$$Y \otimes I = \begin{bmatrix} y_{11} & y_{12} \\ y_{21} & y_{22} \end{bmatrix} \otimes \begin{bmatrix} 1 & 0 \\ 0 & 1 \end{bmatrix}$$
$$= \begin{bmatrix} y_{11}\begin{bmatrix} 1 & 0 \\ 0 & 1 \end{bmatrix} & y_{12}\begin{bmatrix} 1 & 0 \\ 0 & 1 \end{bmatrix} \\ y_{21}\begin{bmatrix} 1 & 0 \\ 0 & 1 \end{bmatrix} & y_{22}\begin{bmatrix} 1 & 0 \\ 0 & 1 \end{bmatrix} \end{bmatrix}$$
$$= \begin{bmatrix} y_{11} & 0 & y_{12} & 0 \\ 0 & y_{11} & 0 & y_{12} \\ y_{21} & 0 & y_{22} & 0 \\ 0 & y_{21} & 0 & y_{22} \end{bmatrix}.$$

Consider the following matrix in the vec over vec form that is the subject of this section

$$\frac{\partial Z}{\partial X} \triangleq \frac{\partial [vec\,(Z)]^T}{\partial vec\,(X)}$$
$$= \begin{bmatrix} \frac{\partial}{\partial x_{11}} \\ \frac{\partial}{\partial x_{21}} \\ \frac{\partial}{\partial x_{12}} \\ \frac{\partial}{\partial x_{22}} \end{bmatrix} \begin{bmatrix} z_{11} & z_{21} & z_{12} & z_{22} \end{bmatrix}$$
$$= \begin{bmatrix} \frac{\partial z_{11}}{\partial x_{11}} & \frac{\partial z_{21}}{\partial x_{11}} & \frac{\partial z_{12}}{\partial x_{11}} & \frac{\partial z_{22}}{\partial x_{11}} \\ \frac{\partial z_{11}}{\partial x_{21}} & \frac{\partial z_{21}}{\partial x_{21}} & \frac{\partial z_{12}}{\partial x_{21}} & \frac{\partial z_{22}}{\partial x_{21}} \\ \frac{\partial z_{11}}{\partial x_{12}} & \frac{\partial z_{21}}{\partial x_{12}} & \frac{\partial z_{12}}{\partial x_{12}} & \frac{\partial z_{22}}{\partial x_{12}} \\ \frac{\partial z_{11}}{\partial x_{22}} & \frac{\partial z_{21}}{\partial x_{22}} & \frac{\partial z_{12}}{\partial x_{22}} & \frac{\partial z_{22}}{\partial x_{22}} \end{bmatrix}.$$

We multiply the two matrices

$$\frac{\partial Z}{\partial X}(Y \otimes I) = \begin{bmatrix} \frac{\partial z_{11}}{\partial x_{11}} & \frac{\partial z_{21}}{\partial x_{11}} & \frac{\partial z_{12}}{\partial x_{11}} & \frac{\partial z_{22}}{\partial x_{11}} \\ \frac{\partial z_{11}}{\partial x_{21}} & \frac{\partial z_{21}}{\partial x_{21}} & \frac{\partial z_{12}}{\partial x_{21}} & \frac{\partial z_{22}}{\partial x_{21}} \\ \frac{\partial z_{11}}{\partial x_{12}} & \frac{\partial z_{21}}{\partial x_{12}} & \frac{\partial z_{12}}{\partial x_{12}} & \frac{\partial z_{22}}{\partial x_{12}} \\ \frac{\partial z_{11}}{\partial x_{22}} & \frac{\partial z_{21}}{\partial x_{22}} & \frac{\partial z_{12}}{\partial x_{22}} & \frac{\partial z_{22}}{\partial x_{22}} \end{bmatrix} \begin{bmatrix} y_{11} & 0 & y_{12} & 0 \\ 0 & y_{11} & 0 & y_{12} \\ y_{21} & 0 & y_{22} & 0 \\ 0 & y_{21} & 0 & y_{22} \end{bmatrix} =$$
$$\begin{bmatrix} \begin{pmatrix} y_{11}\frac{\partial z_{11}}{\partial x_{11}} \\ +y_{21}\frac{\partial z_{12}}{\partial x_{11}} \end{pmatrix} & \begin{pmatrix} y_{11}\frac{\partial z_{21}}{\partial x_{11}} \\ +y_{21}\frac{\partial z_{22}}{\partial x_{11}} \end{pmatrix} & \begin{pmatrix} y_{12}\frac{\partial z_{11}}{\partial x_{11}} \\ +y_{22}\frac{\partial z_{12}}{\partial x_{11}} \end{pmatrix} & \begin{pmatrix} y_{12}\frac{\partial z_{21}}{\partial x_{11}} \\ +y_{22}\frac{\partial z_{22}}{\partial x_{11}} \end{pmatrix} \\ \begin{pmatrix} y_{11}\frac{\partial z_{11}}{\partial x_{21}} \\ +y_{21}\frac{\partial z_{12}}{\partial x_{21}} \end{pmatrix} & \begin{pmatrix} y_{11}\frac{\partial z_{21}}{\partial x_{21}} \\ +y_{21}\frac{\partial z_{22}}{\partial x_{21}} \end{pmatrix} & \begin{pmatrix} y_{12}\frac{\partial z_{11}}{\partial x_{21}} \\ +y_{22}\frac{\partial z_{12}}{\partial x_{21}} \end{pmatrix} & \begin{pmatrix} y_{12}\frac{\partial z_{21}}{\partial x_{21}} \\ +y_{22}\frac{\partial z_{22}}{\partial x_{21}} \end{pmatrix} \\ \begin{pmatrix} y_{11}\frac{\partial z_{11}}{\partial x_{12}} \\ +y_{21}\frac{\partial z_{12}}{\partial x_{12}} \end{pmatrix} & \begin{pmatrix} y_{11}\frac{\partial z_{21}}{\partial x_{12}} \\ +y_{21}\frac{\partial z_{22}}{\partial x_{12}} \end{pmatrix} & \begin{pmatrix} y_{12}\frac{\partial z_{11}}{\partial x_{12}} \\ +y_{22}\frac{\partial z_{12}}{\partial x_{12}} \end{pmatrix} & \begin{pmatrix} y_{12}\frac{\partial z_{21}}{\partial x_{12}} \\ +y_{22}\frac{\partial z_{22}}{\partial x_{12}} \end{pmatrix} \\ \begin{pmatrix} y_{11}\frac{\partial z_{11}}{\partial x_{22}} \\ +y_{21}\frac{\partial z_{12}}{\partial x_{22}} \end{pmatrix} & \begin{pmatrix} y_{11}\frac{\partial z_{21}}{\partial x_{22}} \\ +y_{21}\frac{\partial z_{22}}{\partial x_{22}} \end{pmatrix} & \begin{pmatrix} y_{12}\frac{\partial z_{11}}{\partial x_{22}} \\ +y_{22}\frac{\partial z_{12}}{\partial x_{22}} \end{pmatrix} & \begin{pmatrix} y_{12}\frac{\partial z_{21}}{\partial x_{22}} \\ +y_{22}\frac{\partial z_{22}}{\partial x_{22}} \end{pmatrix} \end{bmatrix}.$$

and we discover we have created precisely the first matrix in the foregoing sum that defines the sought after matrix derivative of the product.

We next ponder the expression

$$I \otimes Z^T = \begin{bmatrix} 1 & 0 \\ 0 & 1 \end{bmatrix} \otimes \begin{bmatrix} z_{11} & z_{21} \\ z_{12} & z_{22} \end{bmatrix}$$

$$= \begin{bmatrix} 1\begin{bmatrix} z_{11} & z_{21} \\ z_{12} & z_{22} \end{bmatrix} & 0\begin{bmatrix} z_{11} & z_{21} \\ z_{12} & z_{22} \end{bmatrix} \\ 0\begin{bmatrix} z_{11} & z_{21} \\ z_{12} & z_{22} \end{bmatrix} & 1\begin{bmatrix} z_{11} & z_{21} \\ z_{12} & z_{22} \end{bmatrix} \end{bmatrix}$$

$$= \begin{bmatrix} z_{11} & z_{21} & 0 & 0 \\ z_{12} & z_{22} & 0 & 0 \\ 0 & 0 & z_{11} & z_{21} \\ 0 & 0 & z_{12} & z_{22} \end{bmatrix}.$$

We multiply the partial derivative by this matrix to write

$$\frac{\partial Y}{\partial X}\left(I \otimes Z^T\right)$$

$$= \begin{bmatrix} \frac{\partial y_{11}}{\partial x_{11}} & \frac{\partial y_{21}}{\partial x_{11}} & \frac{\partial y_{12}}{\partial x_{11}} & \frac{\partial y_{22}}{\partial x_{11}} \\ \frac{\partial y_{11}}{\partial x_{21}} & \frac{\partial y_{21}}{\partial x_{21}} & \frac{\partial y_{12}}{\partial x_{21}} & \frac{\partial y_{22}}{\partial x_{21}} \\ \frac{\partial y_{11}}{\partial x_{12}} & \frac{\partial y_{21}}{\partial x_{12}} & \frac{\partial y_{12}}{\partial x_{12}} & \frac{\partial y_{22}}{\partial x_{12}} \\ \frac{\partial y_{11}}{\partial x_{22}} & \frac{\partial y_{21}}{\partial x_{22}} & \frac{\partial y_{12}}{\partial x_{22}} & \frac{\partial y_{22}}{\partial x_{22}} \end{bmatrix} \begin{bmatrix} z_{11} & z_{21} & 0 & 0 \\ z_{12} & z_{22} & 0 & 0 \\ 0 & 0 & z_{11} & z_{21} \\ 0 & 0 & z_{12} & z_{22} \end{bmatrix}$$

$$= \begin{bmatrix} z_{11}\frac{\partial y_{11}}{\partial x_{11}} + z_{12}\frac{\partial y_{21}}{\partial x_{11}} & z_{21}\frac{\partial y_{11}}{\partial x_{11}} + z_{22}\frac{\partial y_{21}}{\partial x_{11}} & z_{11}\frac{\partial y_{12}}{\partial x_{11}} + z_{12}\frac{\partial y_{22}}{\partial x_{11}} & z_{21}\frac{\partial y_{12}}{\partial x_{11}} + z_{22}\frac{\partial y_{22}}{\partial x_{11}} \\ z_{11}\frac{\partial y_{11}}{\partial x_{21}} + z_{12}\frac{\partial y_{21}}{\partial x_{21}} & z_{21}\frac{\partial y_{11}}{\partial x_{21}} + z_{22}\frac{\partial y_{21}}{\partial x_{21}} & z_{11}\frac{\partial y_{12}}{\partial x_{21}} + z_{12}\frac{\partial y_{22}}{\partial x_{21}} & z_{21}\frac{\partial y_{12}}{\partial x_{21}} + z_{22}\frac{\partial y_{22}}{\partial x_{21}} \\ z_{11}\frac{\partial y_{11}}{\partial x_{12}} + z_{12}\frac{\partial y_{21}}{\partial x_{12}} & z_{21}\frac{\partial y_{11}}{\partial x_{12}} + z_{22}\frac{\partial y_{21}}{\partial x_{12}} & z_{11}\frac{\partial y_{12}}{\partial x_{12}} + z_{12}\frac{\partial y_{22}}{\partial x_{12}} & z_{21}\frac{\partial y_{12}}{\partial x_{12}} + z_{22}\frac{\partial y_{22}}{\partial x_{12}} \\ z_{11}\frac{\partial y_{11}}{\partial x_{22}} + z_{12}\frac{\partial y_{21}}{\partial x_{22}} & z_{21}\frac{\partial y_{11}}{\partial x_{22}} + z_{22}\frac{\partial y_{21}}{\partial x_{22}} & z_{11}\frac{\partial y_{12}}{\partial x_{22}} + z_{12}\frac{\partial y_{22}}{\partial x_{22}} & z_{21}\frac{\partial y_{12}}{\partial x_{22}} + z_{22}\frac{\partial y_{22}}{\partial x_{22}} \end{bmatrix},$$

and we discover we have created precisely the second matrix in the foregoing sum that defines the sought-after matrix derivative of the product. We have the first and the second, and we therefore have verified the matrix product relationship.

Even though this development pertains only to the case of $m = 2$, it can be easily generalized to any m. It is important to think about a more general development as well. The general development begins with the result in the previous section that

$$\frac{\partial (AZB)}{\partial X} = \frac{\partial Z}{\partial X} \left(B \otimes A^T\right) \quad ,$$

in which A and B are constant matrices with respect to X. To get the first term, we substitute $A = I$ and assume that B is constant with respect to Z (this is how first partial derivatives are defined) and write

$$\frac{\partial (ZB)}{\partial X} = \frac{\partial Z}{\partial X} (B \otimes I) \quad .$$

If we substitute Y in place of B, we obtain the first term

$$\frac{\partial (ZY)}{\partial X} = \frac{\partial Z}{\partial X} (Y \otimes I) \quad .$$

To get the second term, we substitute $B = I$ and assume that A is constant with respect to Z and write

$$\frac{\partial (AZ)}{\partial X} = \frac{\partial Z}{\partial X} \left(I \otimes A^T\right) \quad .$$

We substitute Z for A and Y for Z and write

$$\frac{\partial (ZY)}{\partial X} = \frac{\partial Y}{\partial X} \left(I \otimes Z^T\right) \quad .$$

Therefore the derivative is the sum of the two terms with the various terms held constant:

$$\frac{\partial (ZY)}{\partial X} = \frac{\partial Z}{\partial X} (Y \otimes I) + \frac{\partial Y}{\partial X} \left(I \otimes Z^T\right) \quad .$$

C.5.5 "Vec over Vec" Derivative of a Matrix Times a Scalar

We are going to need the product of a vector times a matrix for our development in Chapter 15 through 17. We develop it here in two dimensions and note that it expands to all dimensions greater than 2. We begin by repeating the derivative operator

$$\frac{\partial}{\partial vec(x)} \triangleq \frac{\partial}{\partial X} = \begin{bmatrix} \frac{\partial}{\partial x_{11}} \\ \frac{\partial}{\partial x_{p1}} \\ \frac{\partial}{\partial x_{1q}} \\ \frac{\partial}{\partial x_{pq}} \end{bmatrix} \quad .$$

In two dimensions, this is

$$\frac{\partial}{\partial X}=\frac{\partial}{\partial vec\,(x)}=\begin{bmatrix}\frac{\partial}{\partial x_{11}}\\ \frac{\partial}{\partial x_{21}}\\ \frac{\partial}{\partial x_{12}}\\ \frac{\partial}{\partial x_{22}}\end{bmatrix}.$$

We want to know the derivative of Yz

$$Zy=\begin{bmatrix} z_{11} & z_{12}\\ z_{21} & z_{22}\end{bmatrix}y=\begin{bmatrix} z_{11}y & z_{12}y\\ z_{21}y & z_{22}y\end{bmatrix}.$$

The vectorization of Yz is

$$vec\,(Zy)=\begin{bmatrix} z_{11}y\\ z_{21}y\\ z_{12}y\\ z_{22}y\end{bmatrix}\qquad [vec\,(Zy)]^T=\begin{bmatrix} z_{11}y & z_{21}y & z_{12}y & z_{22}y\end{bmatrix}.$$

We set up to calculate the matrix derivative

$$\frac{\partial[vec\,(Yz)]^T}{\partial vec\,(x)}=\begin{bmatrix}\frac{\partial}{\partial x_{11}}\\ \frac{\partial}{\partial x_{21}}\\ \frac{\partial}{\partial x_{12}}\\ \frac{\partial}{\partial x_{22}}\end{bmatrix}\begin{bmatrix} y_{11}z & y_{21}z & y_{12}z & y_{22}z\end{bmatrix}$$

$$=\begin{bmatrix}\frac{\partial}{\partial x_{11}}(y_{11}z) & \frac{\partial}{\partial x_{11}}(y_{21}z) & \frac{\partial}{\partial x_{11}}(y_{12}z) & \frac{\partial}{\partial x_{11}}(y_{22}z)\\ \frac{\partial}{\partial x_{21}}(y_{11}z) & \frac{\partial}{\partial x_{21}}(y_{21}z) & \frac{\partial}{\partial x_{21}}(y_{12}z) & \frac{\partial}{\partial x_{21}}(y_{22}z)\\ \frac{\partial}{\partial x_{12}}(y_{11}z) & \frac{\partial}{\partial x_{12}}(y_{21}z) & \frac{\partial}{\partial x_{12}}(y_{12}z) & \frac{\partial}{\partial x_{12}}(y_{22}z)\\ \frac{\partial}{\partial x_{22}}(y_{11}z) & \frac{\partial}{\partial x_{22}}(y_{21}z) & \frac{\partial}{\partial x_{22}}(y_{12}z) & \frac{\partial}{\partial x_{22}}(y_{22}z)\end{bmatrix}.$$

We expand

$$\begin{bmatrix}\frac{\partial y_{11}}{\partial x_{11}}z+y_{11}\frac{\partial z}{\partial x_{11}} & \frac{\partial y_{21}}{\partial x_{11}}z+y_{21}\frac{\partial z}{\partial x_{11}} & \frac{\partial y_{12}}{\partial x_{11}}z+y_{12}\frac{\partial z}{\partial x_{11}} & \frac{\partial y_{22}}{\partial x_{11}}z+y_{22}\frac{\partial z}{\partial x_{11}}\\ \frac{\partial y_{11}}{\partial x_{21}}z+y_{11}\frac{\partial z}{\partial x_{21}} & \frac{\partial y_{21}}{\partial x_{21}}z+y_{21}\frac{\partial z}{\partial x_{21}} & \frac{\partial y_{12}}{\partial x_{21}}z+y_{12}\frac{\partial z}{\partial x_{21}} & \frac{\partial y_{22}}{\partial x_{21}}z+y_{22}\frac{\partial z}{\partial x_{21}}\\ \frac{\partial y_{11}}{\partial x_{12}}z+y_{11}\frac{\partial z}{\partial x_{12}} & \frac{\partial y_{21}}{\partial x_{12}}z+y_{21}\frac{\partial z}{\partial x_{12}} & \frac{\partial y_{12}}{\partial x_{12}}z+y_{12}\frac{\partial z}{\partial x_{12}} & \frac{\partial y_{22}}{\partial x_{12}}z+y_{22}\frac{\partial z}{\partial x_{12}}\\ \frac{\partial y_{11}}{\partial x_{22}}z+y_{11}\frac{\partial z}{\partial x_{22}} & \frac{\partial y_{21}}{\partial x_{22}}z+y_{21}\frac{\partial z}{\partial x_{22}} & \frac{\partial y_{12}}{\partial x_{22}}z+y_{12}\frac{\partial z}{\partial x_{22}} & \frac{\partial y_{22}}{\partial x_{22}}z+y_{22}\frac{\partial z}{\partial x_{22}}\end{bmatrix}$$

We split this matrix into two parts (represented by the part before the + and the part after the +)). The first of the two matrices is

$$=\begin{bmatrix}\frac{\partial y_{11}}{\partial x_{11}}z & \frac{\partial y_{21}}{\partial x_{11}}z & \frac{\partial y_{12}}{\partial x_{11}}z & \frac{\partial y_{22}}{\partial x_{11}}z\\ \frac{\partial y_{11}}{\partial x_{21}}z & \frac{\partial y_{21}}{\partial x_{21}}z & \frac{\partial y_{12}}{\partial x_{21}}z & \frac{\partial y_{22}}{\partial x_{21}}z\\ \frac{\partial y_{11}}{\partial x_{12}}z & \frac{\partial y_{21}}{\partial x_{12}}z & \frac{\partial y_{12}}{\partial x_{12}}z & \frac{\partial y_{22}}{\partial x_{12}}z\\ \frac{\partial y_{11}}{\partial x_{22}}z & \frac{\partial y_{21}}{\partial x_{22}}z & \frac{\partial y_{12}}{\partial x_{22}}z & \frac{\partial y_{22}}{\partial x_{22}}z\end{bmatrix}=z\begin{bmatrix}\frac{\partial y_{11}}{\partial x_{11}} & \frac{\partial y_{21}}{\partial x_{11}} & \frac{\partial y_{12}}{\partial x_{11}} & \frac{\partial y_{22}}{\partial x_{11}}\\ \frac{\partial y_{11}}{\partial x_{21}} & \frac{\partial y_{21}}{\partial x_{21}} & \frac{\partial y_{12}}{\partial x_{21}} & \frac{\partial y_{22}}{\partial x_{21}}\\ \frac{\partial y_{11}}{\partial x_{12}} & \frac{\partial y_{21}}{\partial x_{12}} & \frac{\partial y_{12}}{\partial x_{12}} & \frac{\partial y_{22}}{\partial x_{12}}\\ \frac{\partial y_{11}}{\partial x_{22}} & \frac{\partial y_{21}}{\partial x_{22}} & \frac{\partial y_{12}}{\partial x_{22}} & \frac{\partial y_{22}}{\partial x_{22}}\end{bmatrix}$$

$$= z\frac{\partial[vec\,(Y)]^T}{\partial vec\,(X)} = z\frac{\partial Y}{\partial X} \quad .$$

The second matrix is

$$\begin{bmatrix} y_{11}\frac{\partial z}{\partial x_{11}} & y_{21}\frac{\partial z}{\partial x_{11}} & y_{12}\frac{\partial z}{\partial x_{11}} & y_{22}\frac{\partial z}{\partial x_{11}} \\ y_{11}\frac{\partial z}{\partial x_{21}} & y_{21}\frac{\partial z}{\partial x_{21}} & y_{12}\frac{\partial z}{\partial x_{21}} & y_{22}\frac{\partial z}{\partial x_{21}} \\ y_{11}\frac{\partial z}{\partial x_{12}} & y_{21}\frac{\partial z}{\partial x_{12}} & y_{12}\frac{\partial z}{\partial x_{12}} & y_{22}\frac{\partial z}{\partial x_{12}} \\ y_{11}\frac{\partial z}{\partial x_{22}} & y_{21}\frac{\partial z}{\partial x_{22}} & y_{12}\frac{\partial z}{\partial x_{22}} & y_{22}\frac{\partial z}{\partial x_{22}} \end{bmatrix}$$

$$= \frac{\partial z}{\partial vec\,(X)}[vec\,(Y)]^T = \frac{\partial z}{\partial X}[vec\,(Y)]^T \quad .$$

Thus we have shown the fundamental relationship here that we will need to develop thc third equation

$$\frac{\partial}{\partial X}(Yz) = \frac{\partial Y}{\partial X}z + \frac{\partial z}{\partial X}[vec\,(Y)]^T \quad . \tag{C.11}$$

Even though we eschewed this alternative approach, the derivation of this result depends on the analytical equation

$$\frac{\partial\,(Yz)}{\partial X} = \sum_{i=1}^{m}\sum_{j=1}^{m}\left(\frac{\partial y_{ij}}{\partial x_{k\ell}}z_j + y_{ij}\frac{\partial z_j}{\partial x_{k\ell}}\right) \quad ,$$

which reduces to

$$\frac{\partial\,(Yz)}{\partial X} = z\frac{\partial Y}{\partial X} + \frac{\partial z}{\partial X} \otimes Y^T \quad .$$

C.5.6 "Vec over Vec" Derivative of the Inverse of a Matrix with Respect to Itself

Starting with the relationship

$$\frac{\partial\,(YZ)}{\partial X} = \frac{\partial Y}{\partial X}(Z \otimes I) + \frac{\partial Z}{\partial X}\left(I \otimes Y^T\right) \quad .$$

If we substitute $Z = Y^{-1}$, we can write

$$\frac{\partial\left(YY^{-1}\right)}{\partial X} = \frac{\partial Y}{\partial X}\left(Y^{-1} \otimes I\right) + \frac{\partial Y^{-1}}{\partial X}\left(I \otimes Y^T\right) \quad .$$

The left hand side is the identity matrix, so the derivative is zero, meaning

$$0 = \frac{\partial Y}{\partial X}\left(Y^{-1} \otimes I\right) + \frac{\partial Y^{-1}}{\partial X}\left(I \otimes Y^T\right) \quad .$$

We can rearrange

$$\frac{\partial Y^{-1}}{\partial X}\left(I \otimes Y^T\right) = -\frac{\partial Y}{\partial X}\left(Y^{-1} \otimes I\right) \quad .$$

Right multiplying by the inverse of the matrix on the left

$$\frac{\partial Y^{-1}}{\partial X} = -\frac{\partial Y}{\partial X}\left(Y^{-1} \otimes I\right)\left(I \otimes Y^T\right)^{-1} \quad .$$

We apply the Kronecker product of inverse theorem on the right[4] to write

$$\frac{\partial Y^{-1}}{\partial X} = -\frac{\partial Y}{\partial X}\left(Y^{-1} \otimes I\right)\left(I^{-1} \otimes Y^{-T}\right) \quad .$$

We apply the Kronecker product of product of products theorem[5] to write

$$\frac{\partial Y^{-1}}{\partial X} = -\frac{\partial Y}{\partial X}\left(Y^{-1} I \otimes I Y^{-T}\right) \quad .$$

This implies for symmetric matrices that

$$\frac{\partial Y^{-1}}{\partial X} = -\frac{\partial Y}{\partial X}\left(Y^{-1} \otimes Y^{-1}\right) \quad .$$

If we now substitute X for Y, we have the final result for this "vec over vec" matrix derivative

(C.12)
$$\frac{\partial X^{-1}}{\partial X} = -\frac{\partial X}{\partial X}\left(X^{-1} \otimes X^{-1}\right) \quad .$$

C.5.7 Matrix Derivative of the Trace of a Matrix

In this section, we consider

$$\frac{\partial Trace\left(A^T X\right)}{\partial vec\left(X\right)} \quad .$$

The matrix A and its transpose are

$$A = \begin{bmatrix} a_{11} & a_{12} \\ a_{21} & a_{22} \end{bmatrix} \Rightarrow A^T = \begin{bmatrix} a_{11} & a_{21} \\ a_{12} & a_{22} \end{bmatrix} \quad .$$

The trace of $A^T X$ is

$$\begin{bmatrix} a_{11} & a_{21} \\ a_{12} & a_{22} \end{bmatrix}\begin{bmatrix} x_{11} & x_{12} \\ x_{21} & x_{22} \end{bmatrix} = \begin{bmatrix} a_{11}x_{11} + a_{21}x_{21} & * \\ * & a_{12}x_{12} + a_{22}x_{22} \end{bmatrix} \quad .$$

[4] See https://en.wikipedia.org/wiki/Kronecker_product.
[5] See https://en.wikipedia.org/wiki/Kronecker_product.

Because we are looking for the trace, we need not calculate the off diagonal elements. The trace of the matrix is thus

$$Trace\left(A^T X\right) = a_{11}x_{11} + a_{21}x_{21} + a_{12}x_{12} + a_{22}x_{22} \quad .$$

The sought after derivative is

$$\frac{\partial Trace\left(A^T X\right)}{\partial vec\left(X\right)} = \begin{bmatrix} \frac{\partial}{\partial x_{11}} \\ \frac{\partial}{\partial x_{21}} \\ \frac{\partial}{\partial x_{12}} \\ \frac{\partial}{\partial x_{22}} \end{bmatrix} \left(a_{11}x_{11} + a_{21}x_{21} + a_{12}x_{12} + a_{22}x_{22}\right)$$

$$= \begin{bmatrix} a_{11} \\ a_{21} \\ a_{12} \\ a_{22} \end{bmatrix} = vec\left(A\right)$$

.

The final result, which extends to orders higher than 2, is

$$\frac{\partial Trace\left(A^T X\right)}{\partial vec\left(X\right)} = vec\left(A\right) \quad . \tag{C.13}$$

C.5.8 Matrix Derivative of the Power of a Determinant

The determinant of a 2×2 matrix is

$$\Delta = \left(x_{11}x_{22} - x_{12}x_{21}\right) \quad .$$

The determinant raised to a power is

$$\Delta^\theta = \left(x_{11}x_{22} - x_{12}x_{21}\right)^\theta \quad .$$

The sought after matrix derivative is therefore

$$\frac{\partial \Delta^\theta}{\partial vec\left(X\right)} = \begin{bmatrix} \frac{\partial}{\partial x_{11}} \\ \frac{\partial}{\partial x_{21}} \\ \frac{\partial}{\partial x_{12}} \\ \frac{\partial}{\partial x_{22}} \end{bmatrix} \Delta^\theta \quad .$$

We can reduce this to the final form as follows

$$\frac{\partial \Delta^\theta}{\partial vec\,(X)} = \begin{bmatrix} \frac{\partial}{\partial x_{11}} \\ \frac{\partial}{\partial x_{21}} \\ \frac{\partial}{\partial x_{12}} \\ \frac{\partial}{\partial x_{22}} \end{bmatrix} \Delta^\theta = \begin{bmatrix} \frac{\partial \Delta^\theta}{\partial x_{11}} \\ \frac{\partial \Delta^\theta}{\partial x_{21}} \\ \frac{\partial \Delta^\theta}{\partial x_{12}} \\ \frac{\partial \Delta^\theta}{\partial x_{22}} \end{bmatrix} = \begin{bmatrix} \frac{\partial \Delta}{\partial x_{11}} \theta \Delta^{\theta-1} \\ \frac{\partial \Delta}{\partial x_{21}} \theta \Delta^{\theta-1} \\ \frac{\partial \Delta}{\partial x_{12}} \theta \Delta^{\theta-1} \\ \frac{\partial \Delta}{\partial x_{22}} \theta \Delta^{\theta-1} \end{bmatrix}$$

$$= \theta \Delta^{\theta-1} \begin{bmatrix} x_{22} \\ -x_{12} \\ -x_{21} \\ x_{11} \end{bmatrix} = \theta \Delta^{\theta-1} \Delta \begin{bmatrix} \frac{x_{22}}{\Delta} \\ -\frac{x_{12}}{\Delta} \\ -\frac{x_{21}}{\Delta} \\ \frac{x_{11}}{\Delta} \end{bmatrix} = \theta \Delta^\theta vec\left[\left(X^{-1}\right)^T\right] \quad .$$

When X is symmetric,

$$\frac{\partial |X|^\theta}{\partial vec\,(X)} = \theta |X|^\theta vec\left(X^{-1}\right) \quad . \tag{C.14}$$

C.6 A Weighted Vec Over Vec Derivative

To develop the results used in Chapters 15 through 17, we need a weighted definition of the matrix derivative, much akin to the weighted definition of the Kronecker derivative. To do so, we need to expand this "vec over vec" matrix derivative out in the case of $m = 3$ to note the location of the partial derivatives and the magnitudes of the coefficients throughout the matrix.

$$\frac{\partial Y}{\partial X} =$$

$$\begin{bmatrix}
\frac{\partial y_{11}}{\partial x_{11}} & \frac{\partial y_{21}}{\partial x_{11}} & \frac{\partial y_{31}}{\partial x_{11}} & \frac{\partial y_{12}}{\partial x_{11}} & \frac{\partial y_{22}}{\partial x_{11}} & \frac{\partial y_{32}}{\partial x_{11}} & \frac{\partial y_{13}}{\partial x_{11}} & \frac{\partial y_{23}}{\partial x_{11}} & \frac{\partial y_{33}}{\partial x_{11}} \\
\frac{\partial y_{11}}{\partial x_{21}} & \frac{\partial y_{21}}{\partial x_{21}} & \frac{\partial y_{31}}{\partial x_{21}} & \frac{\partial y_{12}}{\partial x_{21}} & \frac{\partial y_{22}}{\partial x_{21}} & \frac{\partial y_{32}}{\partial x_{21}} & \frac{\partial y_{13}}{\partial x_{21}} & \frac{\partial y_{23}}{\partial x_{21}} & \frac{\partial y_{33}}{\partial x_{21}} \\
\frac{\partial y_{11}}{\partial x_{31}} & \frac{\partial y_{21}}{\partial x_{31}} & \frac{\partial y_{31}}{\partial x_{31}} & \frac{\partial y_{12}}{\partial x_{31}} & \frac{\partial y_{22}}{\partial x_{31}} & \frac{\partial y_{32}}{\partial x_{31}} & \frac{\partial y_{13}}{\partial x_{31}} & \frac{\partial y_{23}}{\partial x_{31}} & \frac{\partial y_{33}}{\partial x_{31}} \\
\frac{\partial y_{11}}{\partial x_{12}} & \frac{\partial y_{21}}{\partial x_{12}} & \frac{\partial y_{31}}{\partial x_{12}} & \frac{\partial y_{12}}{\partial x_{12}} & \frac{\partial y_{22}}{\partial x_{12}} & \frac{\partial y_{32}}{\partial x_{12}} & \frac{\partial y_{13}}{\partial x_{12}} & \frac{\partial y_{23}}{\partial x_{12}} & \frac{\partial y_{33}}{\partial x_{12}} \\
\frac{\partial y_{11}}{\partial x_{22}} & \frac{\partial y_{21}}{\partial x_{22}} & \frac{\partial y_{31}}{\partial x_{22}} & \frac{\partial y_{12}}{\partial x_{22}} & \frac{\partial y_{22}}{\partial x_{22}} & \frac{\partial y_{32}}{\partial x_{22}} & \frac{\partial y_{13}}{\partial x_{22}} & \frac{\partial y_{23}}{\partial x_{22}} & \frac{\partial y_{33}}{\partial x_{22}} \\
\frac{\partial y_{11}}{\partial x_{32}} & \frac{\partial y_{21}}{\partial x_{32}} & \frac{\partial y_{31}}{\partial x_{32}} & \frac{\partial y_{12}}{\partial x_{32}} & \frac{\partial y_{22}}{\partial x_{32}} & \frac{\partial y_{32}}{\partial x_{32}} & \frac{\partial y_{13}}{\partial x_{32}} & \frac{\partial y_{23}}{\partial x_{32}} & \frac{\partial y_{33}}{\partial x_{32}} \\
\frac{\partial y_{11}}{\partial x_{13}} & \frac{\partial y_{21}}{\partial x_{13}} & \frac{\partial y_{31}}{\partial x_{13}} & \frac{\partial y_{12}}{\partial x_{13}} & \frac{\partial y_{22}}{\partial x_{13}} & \frac{\partial y_{32}}{\partial x_{13}} & \frac{\partial y_{13}}{\partial x_{13}} & \frac{\partial y_{23}}{\partial x_{13}} & \frac{\partial y_{33}}{\partial x_{13}} \\
\frac{\partial y_{11}}{\partial x_{23}} & \frac{\partial y_{21}}{\partial x_{23}} & \frac{\partial y_{31}}{\partial x_{23}} & \frac{\partial y_{12}}{\partial x_{23}} & \frac{\partial y_{22}}{\partial x_{23}} & \frac{\partial y_{32}}{\partial x_{23}} & \frac{\partial y_{13}}{\partial x_{23}} & \frac{\partial y_{23}}{\partial x_{23}} & \frac{\partial y_{33}}{\partial x_{32}} \\
\frac{\partial y_{11}}{\partial x_{33}} & \frac{\partial y_{21}}{\partial x_{33}} & \frac{\partial y_{31}}{\partial x_{33}} & \frac{\partial y_{12}}{\partial x_{33}} & \frac{\partial y_{22}}{\partial x_{33}} & \frac{\partial y_{32}}{\partial x_{33}} & \frac{\partial y_{13}}{\partial x_{33}} & \frac{\partial y_{23}}{\partial x_{33}} & \frac{\partial y_{33}}{\partial x_{33}}
\end{bmatrix} .$$

Assuming the matrix Y and the matrix X is symmetric, this matrix derivative in the case of $m = 3$ specializes to

$$\frac{\partial Y}{\partial X}_{Symmetric} =$$

$$\begin{bmatrix}
\frac{\partial y_{11}}{\partial x_{11}} & \frac{\partial y_{12}}{\partial x_{11}} & \frac{\partial y_{13}}{\partial x_{11}} & \frac{\partial y_{12}}{\partial x_{11}} & \frac{\partial y_{22}}{\partial x_{11}} & \frac{\partial y_{23}}{\partial x_{11}} & \frac{\partial y_{13}}{\partial x_{11}} & \frac{\partial y_{23}}{\partial x_{11}} & \frac{\partial y_{33}}{\partial x_{11}} \\
\frac{\partial y_{11}}{\partial x_{12}} & \frac{\partial y_{12}}{\partial x_{12}} & \frac{\partial y_{13}}{\partial x_{12}} & \frac{\partial y_{12}}{\partial x_{12}} & \frac{\partial y_{22}}{\partial x_{12}} & \frac{\partial y_{23}}{\partial x_{12}} & \frac{\partial y_{13}}{\partial x_{12}} & \frac{\partial y_{23}}{\partial x_{12}} & \frac{\partial y_{33}}{\partial x_{12}} \\
\frac{\partial y_{11}}{\partial x_{13}} & \frac{\partial y_{12}}{\partial x_{13}} & \frac{\partial y_{13}}{\partial x_{13}} & \frac{\partial y_{12}}{\partial x_{13}} & \frac{\partial y_{22}}{\partial x_{13}} & \frac{\partial y_{23}}{\partial x_{13}} & \frac{\partial y_{13}}{\partial x_{13}} & \frac{\partial y_{23}}{\partial x_{13}} & \frac{\partial y_{33}}{\partial x_{13}} \\
\frac{\partial y_{11}}{\partial x_{12}} & \frac{\partial y_{12}}{\partial x_{12}} & \frac{\partial y_{13}}{\partial x_{12}} & \frac{\partial y_{12}}{\partial x_{12}} & \frac{\partial y_{22}}{\partial x_{12}} & \frac{\partial y_{23}}{\partial x_{12}} & \frac{\partial y_{13}}{\partial x_{12}} & \frac{\partial y_{23}}{\partial x_{12}} & \frac{\partial y_{33}}{\partial x_{12}} \\
\frac{\partial y_{11}}{\partial x_{22}} & \frac{\partial y_{12}}{\partial x_{22}} & \frac{\partial y_{13}}{\partial x_{22}} & \frac{\partial y_{12}}{\partial x_{22}} & \frac{\partial y_{22}}{\partial x_{22}} & \frac{\partial y_{23}}{\partial x_{22}} & \frac{\partial y_{13}}{\partial x_{22}} & \frac{\partial y_{23}}{\partial x_{22}} & \frac{\partial y_{33}}{\partial x_{22}} \\
\frac{\partial y_{11}}{\partial x_{23}} & \frac{\partial y_{12}}{\partial x_{23}} & \frac{\partial y_{13}}{\partial x_{23}} & \frac{\partial y_{12}}{\partial x_{23}} & \frac{\partial y_{22}}{\partial x_{23}} & \frac{\partial y_{23}}{\partial x_{23}} & \frac{\partial y_{13}}{\partial x_{23}} & \frac{\partial y_{23}}{\partial x_{23}} & \frac{\partial y_{33}}{\partial x_{23}} \\
\frac{\partial y_{11}}{\partial x_{13}} & \frac{\partial y_{12}}{\partial x_{13}} & \frac{\partial y_{13}}{\partial x_{13}} & \frac{\partial y_{12}}{\partial x_{13}} & \frac{\partial y_{22}}{\partial x_{13}} & \frac{\partial y_{23}}{\partial x_{13}} & \frac{\partial y_{13}}{\partial x_{13}} & \frac{\partial y_{23}}{\partial x_{13}} & \frac{\partial y_{33}}{\partial x_{13}} \\
\frac{\partial y_{11}}{\partial x_{23}} & \frac{\partial y_{12}}{\partial x_{23}} & \frac{\partial y_{13}}{\partial x_{23}} & \frac{\partial y_{12}}{\partial x_{23}} & \frac{\partial y_{22}}{\partial x_{23}} & \frac{\partial y_{23}}{\partial x_{23}} & \frac{\partial y_{13}}{\partial x_{23}} & \frac{\partial y_{23}}{\partial x_{23}} & \frac{\partial y_{33}}{\partial x_{32}} \\
\frac{\partial y_{11}}{\partial x_{33}} & \frac{\partial y_{12}}{\partial x_{33}} & \frac{\partial y_{13}}{\partial x_{33}} & \frac{\partial y_{12}}{\partial x_{33}} & \frac{\partial y_{22}}{\partial x_{33}} & \frac{\partial y_{23}}{\partial x_{33}} & \frac{\partial y_{13}}{\partial x_{33}} & \frac{\partial y_{23}}{\partial x_{33}} & \frac{\partial y_{33}}{\partial x_{33}}
\end{bmatrix} .$$

We need to consider the symmetric matrix in the case in which $Y = X$, i.e., the matrix derivative with respect to itself. The result is

$$\frac{\partial X}{\partial X}_{Symmetric} = \begin{bmatrix}
1 & 0 & 0 & 0 & 0 & 0 & 0 & 0 & 0 \\
0 & 1 & 0 & 1 & 0 & 0 & 0 & 0 & 0 \\
0 & 0 & 1 & 0 & 0 & 0 & 1 & 0 & 0 \\
0 & 1 & 0 & 1 & 0 & 0 & 0 & 0 & 0 \\
0 & 0 & 0 & 0 & 1 & 0 & 0 & 0 & 0 \\
0 & 0 & 0 & 0 & 0 & 1 & 0 & 1 & 0 \\
0 & 0 & 1 & 0 & 0 & 0 & 1 & 0 & 0 \\
0 & 0 & 0 & 0 & 0 & 1 & 0 & 1 & 0 \\
0 & 0 & 0 & 0 & 0 & 0 & 0 & 0 & 1
\end{bmatrix} .$$

This gives us a strong hint how to redefine another matrix derivative with different coefficients in the various entries as follows:

$$\text{(C.15)} \qquad \frac{\partial Y}{\partial X} = \sum_{i=1}^{m} \sum_{j=1}^{m} \sum_{k=1}^{m} \sum_{\ell=1}^{m} \frac{\partial y_{ij}}{\partial x_{k\ell}} \varepsilon_{k\ell} \left(g_\ell \otimes f_k\right) \left(s_j \otimes r_i\right)^T \quad ,$$

the epsilon term weights different pairs of elements of Y and X at one half the level of same pairs of elements. When we use this weighted matrix derivative formula, we obtain the result

$$\frac{\partial Y}{\partial X} =$$

$$
\begin{bmatrix}
\frac{\partial y_{11}}{\partial x_{11}} & \frac{\partial y_{21}}{\partial x_{11}} & \frac{\partial y_{31}}{\partial x_{11}} & \frac{\partial y_{12}}{\partial x_{11}} & \frac{\partial y_{22}}{\partial x_{11}} & \frac{\partial y_{32}}{\partial x_{11}} & \frac{\partial y_{13}}{\partial x_{11}} & \frac{\partial y_{23}}{\partial x_{11}} & \frac{\partial y_{33}}{\partial x_{11}} \\
\frac{1}{2}\frac{\partial y_{11}}{\partial x_{21}} & \frac{1}{2}\frac{\partial y_{21}}{\partial x_{21}} & \frac{1}{2}\frac{\partial y_{31}}{\partial x_{21}} & \frac{1}{2}\frac{\partial y_{12}}{\partial x_{21}} & \frac{1}{2}\frac{\partial y_{22}}{\partial x_{21}} & \frac{1}{2}\frac{\partial y_{32}}{\partial x_{21}} & \frac{1}{2}\frac{\partial y_{13}}{\partial x_{21}} & \frac{1}{2}\frac{\partial y_{23}}{\partial x_{21}} & \frac{1}{2}\frac{\partial y_{33}}{\partial x_{21}} \\
\frac{1}{2}\frac{\partial y_{11}}{\partial x_{31}} & \frac{1}{2}\frac{\partial y_{21}}{\partial x_{31}} & \frac{1}{2}\frac{\partial y_{31}}{\partial x_{31}} & \frac{1}{2}\frac{\partial y_{12}}{\partial x_{31}} & \frac{1}{2}\frac{\partial y_{22}}{\partial x_{31}} & \frac{1}{2}\frac{\partial y_{32}}{\partial x_{31}} & \frac{1}{2}\frac{\partial y_{13}}{\partial x_{31}} & \frac{1}{2}\frac{\partial y_{23}}{\partial x_{31}} & \frac{1}{2}\frac{\partial y_{33}}{\partial x_{31}} \\
\frac{1}{2}\frac{\partial y_{11}}{\partial x_{12}} & \frac{1}{2}\frac{\partial y_{21}}{\partial x_{12}} & \frac{1}{2}\frac{\partial y_{31}}{\partial x_{12}} & \frac{1}{2}\frac{\partial y_{12}}{\partial x_{12}} & \frac{1}{2}\frac{\partial y_{22}}{\partial x_{12}} & \frac{1}{2}\frac{\partial y_{32}}{\partial x_{12}} & \frac{1}{2}\frac{\partial y_{13}}{\partial x_{12}} & \frac{1}{2}\frac{\partial y_{23}}{\partial x_{12}} & \frac{1}{2}\frac{\partial y_{33}}{\partial x_{12}} \\
\frac{\partial y_{11}}{\partial x_{22}} & \frac{\partial y_{21}}{\partial x_{22}} & \frac{\partial y_{31}}{\partial x_{22}} & \frac{\partial y_{12}}{\partial x_{22}} & \frac{\partial y_{22}}{\partial x_{22}} & \frac{\partial y_{32}}{\partial x_{22}} & \frac{\partial y_{13}}{\partial x_{22}} & \frac{\partial y_{23}}{\partial x_{22}} & \frac{\partial y_{33}}{\partial x_{22}} \\
\frac{1}{2}\frac{\partial y_{11}}{\partial x_{32}} & \frac{1}{2}\frac{\partial y_{21}}{\partial x_{32}} & \frac{1}{2}\frac{\partial y_{31}}{\partial x_{32}} & \frac{1}{2}\frac{\partial y_{12}}{\partial x_{32}} & \frac{1}{2}\frac{\partial y_{22}}{\partial x_{32}} & \frac{1}{2}\frac{\partial y_{32}}{\partial x_{32}} & \frac{1}{2}\frac{\partial y_{13}}{\partial x_{32}} & \frac{1}{2}\frac{\partial y_{23}}{\partial x_{32}} & \frac{1}{2}\frac{\partial y_{33}}{\partial x_{32}} \\
\frac{1}{2}\frac{\partial y_{11}}{\partial x_{13}} & \frac{1}{2}\frac{\partial y_{21}}{\partial x_{13}} & \frac{1}{2}\frac{\partial y_{31}}{\partial x_{13}} & \frac{1}{2}\frac{\partial y_{12}}{\partial x_{13}} & \frac{1}{2}\frac{\partial y_{22}}{\partial x_{13}} & \frac{1}{2}\frac{\partial y_{32}}{\partial x_{13}} & \frac{1}{2}\frac{\partial y_{13}}{\partial x_{13}} & \frac{1}{2}\frac{\partial y_{23}}{\partial x_{13}} & \frac{1}{2}\frac{\partial y_{33}}{\partial x_{13}} \\
\frac{1}{2}\frac{\partial y_{11}}{\partial x_{23}} & \frac{1}{2}\frac{\partial y_{21}}{\partial x_{23}} & \frac{1}{2}\frac{\partial y_{31}}{\partial x_{23}} & \frac{1}{2}\frac{\partial y_{12}}{\partial x_{23}} & \frac{1}{2}\frac{\partial y_{22}}{\partial x_{23}} & \frac{1}{2}\frac{\partial y_{32}}{\partial x_{23}} & \frac{1}{2}\frac{\partial y_{13}}{\partial x_{23}} & \frac{1}{2}\frac{\partial y_{23}}{\partial x_{23}} & \frac{1}{2}\frac{\partial y_{33}}{\partial x_{32}} \\
\frac{\partial y_{11}}{\partial x_{33}} & \frac{\partial y_{21}}{\partial x_{33}} & \frac{\partial y_{31}}{\partial x_{33}} & \frac{\partial y_{12}}{\partial x_{33}} & \frac{\partial y_{22}}{\partial x_{33}} & \frac{\partial y_{32}}{\partial x_{33}} & \frac{\partial y_{13}}{\partial x_{33}} & \frac{\partial y_{23}}{\partial x_{33}} & \frac{\partial y_{33}}{\partial x_{33}}
\end{bmatrix}
$$

If we assume that Y and X are both symmetric, this specializes to the form

$$
\frac{\partial Y}{\partial X}_{Symmetric} =
$$

$$
\begin{bmatrix}
\frac{\partial y_{11}}{\partial x_{11}} & \frac{\partial y_{12}}{\partial x_{11}} & \frac{\partial y_{13}}{\partial x_{11}} & \frac{\partial y_{12}}{\partial x_{11}} & \frac{\partial y_{22}}{\partial x_{11}} & \frac{\partial y_{23}}{\partial x_{11}} & \frac{\partial y_{13}}{\partial x_{11}} & \frac{\partial y_{23}}{\partial x_{11}} & \frac{\partial y_{33}}{\partial x_{11}} \\
\frac{1}{2}\frac{\partial y_{11}}{\partial x_{12}} & \frac{1}{2}\frac{\partial y_{12}}{\partial x_{12}} & \frac{1}{2}\frac{\partial y_{13}}{\partial x_{12}} & \frac{1}{2}\frac{\partial y_{12}}{\partial x_{12}} & \frac{1}{2}\frac{\partial y_{22}}{\partial x_{12}} & \frac{1}{2}\frac{\partial y_{23}}{\partial x_{12}} & \frac{1}{2}\frac{\partial y_{13}}{\partial x_{12}} & \frac{1}{2}\frac{\partial y_{23}}{\partial x_{12}} & \frac{1}{2}\frac{\partial y_{33}}{\partial x_{12}} \\
\frac{1}{2}\frac{\partial y_{11}}{\partial x_{13}} & \frac{1}{2}\frac{\partial y_{12}}{\partial x_{13}} & \frac{1}{2}\frac{\partial y_{13}}{\partial x_{13}} & \frac{1}{2}\frac{\partial y_{12}}{\partial x_{13}} & \frac{1}{2}\frac{\partial y_{22}}{\partial x_{13}} & \frac{1}{2}\frac{\partial y_{23}}{\partial x_{13}} & \frac{1}{2}\frac{\partial y_{13}}{\partial x_{13}} & \frac{1}{2}\frac{\partial y_{23}}{\partial x_{13}} & \frac{1}{2}\frac{\partial y_{33}}{\partial x_{13}} \\
\frac{1}{2}\frac{\partial y_{11}}{\partial x_{12}} & \frac{1}{2}\frac{\partial y_{12}}{\partial x_{12}} & \frac{1}{2}\frac{\partial y_{13}}{\partial x_{12}} & \frac{1}{2}\frac{\partial y_{12}}{\partial x_{12}} & \frac{1}{2}\frac{\partial y_{22}}{\partial x_{12}} & \frac{1}{2}\frac{\partial y_{23}}{\partial x_{12}} & \frac{1}{2}\frac{\partial y_{13}}{\partial x_{12}} & \frac{1}{2}\frac{\partial y_{23}}{\partial x_{12}} & \frac{1}{2}\frac{\partial y_{33}}{\partial x_{12}} \\
\frac{\partial y_{11}}{\partial x_{22}} & \frac{\partial y_{12}}{\partial x_{22}} & \frac{\partial y_{13}}{\partial x_{22}} & \frac{\partial y_{12}}{\partial x_{22}} & \frac{\partial y_{22}}{\partial x_{22}} & \frac{\partial y_{23}}{\partial x_{22}} & \frac{\partial y_{13}}{\partial x_{22}} & \frac{\partial y_{23}}{\partial x_{22}} & \frac{\partial y_{33}}{\partial x_{22}} \\
\frac{1}{2}\frac{\partial y_{11}}{\partial x_{23}} & \frac{1}{2}\frac{\partial y_{12}}{\partial x_{23}} & \frac{1}{2}\frac{\partial y_{13}}{\partial x_{23}} & \frac{1}{2}\frac{\partial y_{12}}{\partial x_{23}} & \frac{1}{2}\frac{\partial y_{22}}{\partial x_{23}} & \frac{1}{2}\frac{\partial y_{23}}{\partial x_{23}} & \frac{1}{2}\frac{\partial y_{13}}{\partial x_{23}} & \frac{1}{2}\frac{\partial y_{23}}{\partial x_{23}} & \frac{1}{2}\frac{\partial y_{33}}{\partial x_{23}} \\
\frac{1}{2}\frac{\partial y_{11}}{\partial x_{13}} & \frac{1}{2}\frac{\partial y_{12}}{\partial x_{13}} & \frac{1}{2}\frac{\partial y_{13}}{\partial x_{13}} & \frac{1}{2}\frac{\partial y_{12}}{\partial x_{13}} & \frac{1}{2}\frac{\partial y_{22}}{\partial x_{13}} & \frac{1}{2}\frac{\partial y_{23}}{\partial x_{13}} & \frac{1}{2}\frac{\partial y_{13}}{\partial x_{13}} & \frac{1}{2}\frac{\partial y_{23}}{\partial x_{13}} & \frac{1}{2}\frac{\partial y_{33}}{\partial x_{13}} \\
\frac{1}{2}\frac{\partial y_{11}}{\partial x_{23}} & \frac{1}{2}\frac{\partial y_{12}}{\partial x_{23}} & \frac{1}{2}\frac{\partial y_{13}}{\partial x_{23}} & \frac{1}{2}\frac{\partial y_{12}}{\partial x_{23}} & \frac{1}{2}\frac{\partial y_{22}}{\partial x_{23}} & \frac{1}{2}\frac{\partial y_{23}}{\partial x_{23}} & \frac{1}{2}\frac{\partial y_{13}}{\partial x_{23}} & \frac{1}{2}\frac{\partial y_{23}}{\partial x_{23}} & \frac{1}{2}\frac{\partial y_{33}}{\partial x_{32}} \\
\frac{\partial y_{11}}{\partial x_{33}} & \frac{\partial y_{12}}{\partial x_{33}} & \frac{\partial y_{13}}{\partial x_{33}} & \frac{\partial y_{12}}{\partial x_{33}} & \frac{\partial y_{22}}{\partial x_{33}} & \frac{\partial y_{23}}{\partial x_{33}} & \frac{\partial y_{13}}{\partial x_{33}} & \frac{\partial y_{23}}{\partial x_{33}} & \frac{\partial y_{33}}{\partial x_{33}}
\end{bmatrix}
$$

If we now consider the case $Y = X$, we see the derivatives of a matrix X with respect to itself

$$\frac{\partial X}{\partial X}_{Symmetric} = \begin{bmatrix} 1 & 0 & 0 & 0 & 0 & 0 & 0 & 0 & 0 \\ 0 & \frac{1}{2} & 0 & \frac{1}{2} & 0 & 0 & 0 & 0 & 0 \\ 0 & 0 & \frac{1}{2} & 0 & 0 & 0 & \frac{1}{2} & 0 & 0 \\ 0 & \frac{1}{2} & 0 & \frac{1}{2} & 0 & 0 & 0 & 0 & 0 \\ 0 & 0 & 0 & 0 & 1 & 0 & 0 & 0 & 0 \\ 0 & 0 & 0 & 0 & 0 & \frac{1}{2} & 0 & \frac{1}{2} & 0 \\ 0 & 0 & \frac{1}{2} & 0 & 0 & 0 & \frac{1}{2} & 0 & 0 \\ 0 & 0 & 0 & 0 & 0 & \frac{1}{2} & 0 & \frac{1}{2} & 0 \\ 0 & 0 & 0 & 0 & 0 & 0 & 0 & 0 & 1 \end{bmatrix}$$

$$\begin{bmatrix} \frac{1}{2} & 0 & 0 & 0 & 0 & 0 & 0 & 0 & 0 \\ 0 & 0 & 0 & \frac{1}{2} & 0 & 0 & 0 & 0 & 0 \\ 0 & 0 & 0 & 0 & 0 & 0 & \frac{1}{2} & 0 & 0 \\ 0 & \frac{1}{2} & 0 & 0 & 0 & 0 & 0 & 0 & 0 \\ 0 & 0 & 0 & 0 & \frac{1}{2} & 0 & 0 & 0 & 0 \\ 0 & 0 & 0 & 0 & 0 & 0 & 0 & \frac{1}{2} & 0 \\ 0 & 0 & \frac{1}{2} & 0 & 0 & 0 & 0 & 0 & 0 \\ 0 & 0 & 0 & 0 & 0 & \frac{1}{2} & 0 & 0 & 0 \\ 0 & 0 & 0 & 0 & 0 & 0 & 0 & 0 & \frac{1}{2} \end{bmatrix} \tag{C.16}$$

$$+ \begin{bmatrix} \frac{1}{2} & 0 & 0 & 0 & 0 & 0 & 0 & 0 & 0 \\ 0 & \frac{1}{2} & 0 & 0 & 0 & 0 & 0 & 0 & 0 \\ 0 & 0 & \frac{1}{2} & 0 & 0 & 0 & 0 & 0 & 0 \\ 0 & 0 & 0 & \frac{1}{2} & 0 & 0 & 0 & 0 & 0 \\ 0 & 0 & 0 & 0 & \frac{1}{2} & 0 & 0 & 0 & 0 \\ 0 & 0 & 0 & 0 & 0 & \frac{1}{2} & 0 & 0 & 0 \\ 0 & 0 & 0 & 0 & 0 & 0 & \frac{1}{2} & 0 & 0 \\ 0 & 0 & 0 & 0 & 0 & 0 & 0 & \frac{1}{2} & 0 \\ 0 & 0 & 0 & 0 & 0 & 0 & 0 & 0 & \frac{1}{2} \end{bmatrix}$$

$$= \frac{1}{2}\left(I + K_{mm}\right) \quad .$$

This is going to be an important relationship in developing the variance-covariance matrix, especially the third equation, in Chapter 15 through 17.

Appendix D

APPENDIX D–WHY WE USE THE SPECIFIC REPRESENTATION OF VARIANCE-COVARIANCE MATRIX IN CHAPTER 15 THROUGH 17 UNDER THE INVERSE WISHART

All statistics have outliers.

Nenia Campbell

This section illustrates exactly why we use the Kronecker product oriented representation of the variance-covariance matrix from Chapter 15 through 17, which is repeated here for easy reference

$$\left(\Sigma - \frac{1}{\nu_0 - m - 1} M_0\right) \otimes \left(\Sigma - \frac{1}{\nu_0 - m - 1} M_0\right) \quad .$$

If we write

- μ_{ij} as the ij^{th} element of $\frac{1}{\nu_0 - m - 1} M_0$
- σ_{ij} as the ij^{th} element of Σ .

We expand the foregoing equation using the Kronecker distributive property

$$\begin{aligned}
&(\Sigma - \eta M_0) \otimes (\Sigma - \eta M_0) \\
&= (\Sigma \otimes \Sigma) - \eta\,(M_0 \otimes \Sigma) \\
&- \eta\,(\Sigma \otimes M_0) + \eta^2\,(M_0 \otimes M_0) \quad ,
\end{aligned}$$

in which

$$\eta = \frac{1}{\nu_0 - m - 1} \quad .$$

This means that

$$\begin{aligned}
&E\left[(\Sigma - \eta M_0) \otimes (\Sigma - \eta M_0)\right] \\
&= E\left[\begin{array}{l} (\Sigma \otimes \Sigma) - \eta\,(M_0 \otimes \Sigma) \\ -\eta\,(\Sigma \otimes M_0) + \eta^2\,(M_0 \otimes M_0) \end{array}\right] \\
&= \left[\begin{array}{l} E\,(\Sigma \otimes \Sigma) - \eta\,(M_0 \otimes \eta M_0) \\ -\eta E\,(\eta M_0 \otimes M_0) + \eta^2 E\,(M_0 \otimes M_0) \end{array}\right] \\
&= E(\Sigma \otimes \Sigma) - \eta^2\,(M_0 \otimes M_0) \quad .
\end{aligned}$$

Let us take the expectation of this equation with respect to the inverse Wishart distribution to write

$$E\,(\Sigma \otimes \Sigma) - \frac{1}{(\nu_0 - m - 1)^2}\,(M_0 \otimes M_0) \quad .$$

In the specific case in which $p = 3.$ we write the $M_0 \otimes M_0$ term, the matrix in Table D.1.

Table D.1: $M_0 \otimes M_0$–General Case (Non Symmetric)

$\mu_{11}\mu_{11}$	$\mu_{11}\mu_{12}$	$\mu_{11}\mu_{13}$	$\mu_{12}\mu_{11}$	$\mu_{12}\mu_{12}$	$\mu_{12}\mu_{13}$	$\mu_{13}\mu_{11}$	$\mu_{13}\mu_{12}$	$\mu_{13}\mu_{13}$
$\mu_{11}\mu_{21}$	$\mu_{11}\mu_{22}$	$\mu_{11}\mu_{23}$	$\mu_{12}\mu_{21}$	$\mu_{12}\mu_{22}$	$\mu_{12}\mu_{23}$	$\mu_{13}\mu_{21}$	$\mu_{13}\mu_{22}$	$\mu_{13}\mu_{23}$
$\mu_{11}\mu_{31}$	$\mu_{11}\mu_{32}$	$\mu_{11}\mu_{33}$	$\mu_{12}\mu_{31}$	$\mu_{12}\mu_{32}$	$\mu_{12}\mu_{33}$	$\mu_{13}\mu_{31}$	$\mu_{13}\mu_{32}$	$\mu_{13}\mu_{33}$
$\mu_{21}\mu_{11}$	$\mu_{21}\mu_{12}$	$\mu_{21}\mu_{13}$	$\mu_{22}\mu_{11}$	$\mu_{22}\mu_{12}$	$\mu_{22}\mu_{13}$	$\mu_{23}\mu_{11}$	$\mu_{23}\mu_{12}$	$\mu_{23}\mu_{13}$
$\mu_{21}\mu_{21}$	$\mu_{21}\mu_{22}$	$\mu_{21}\mu_{23}$	$\mu_{22}\mu_{21}$	$\mu_{22}\mu_{22}$	$\mu_{22}\mu_{23}$	$\mu_{23}\mu_{21}$	$\mu_{23}\mu_{22}$	$\mu_{23}\mu_{23}$
$\mu_{21}\mu_{31}$	$\mu_{21}\mu_{32}$	$\mu_{21}\mu_{33}$	$\mu_{22}\mu_{31}$	$\mu_{22}\mu_{32}$	$\mu_{22}\mu_{33}$	$\mu_{23}\mu_{31}$	$\mu_{23}\mu_{32}$	$\mu_{23}\mu_{33}$
$\mu_{31}\mu_{11}$	$\mu_{31}\mu_{12}$	$\mu_{31}\mu_{13}$	$\mu_{32}\mu_{11}$	$\mu_{32}\mu_{12}$	$\mu_{32}\mu_{13}$	$\mu_{33}\mu_{11}$	$\mu_{33}\mu_{12}$	$\mu_{33}\mu_{13}$
$\mu_{31}\mu_{21}$	$\mu_{31}\mu_{22}$	$\mu_{31}\mu_{23}$	$\mu_{32}\mu_{21}$	$\mu_{32}\mu_{22}$	$\mu_{32}\mu_{23}$	$\mu_{33}\mu_{21}$	$\mu_{33}\mu_{22}$	$\mu_{33}\mu_{23}$
$\mu_{31}\mu_{31}$	$\mu_{31}\mu_{32}$	$\mu_{31}\mu_{33}$	$\mu_{32}\mu_{31}$	$\mu_{32}\mu_{32}$	$\mu_{32}\mu_{33}$	$\mu_{33}\mu_{31}$	$\mu_{33}\mu_{32}$	$\mu_{33}\mu_{33}$

Notice that every panel of the matrix represents a different "mosaic" area of Y, and the ij^{th} cell represents a fixed y_{ij} with respect an entry for x_{mn} for every possible value of mn. That is the whole point of a Kronecker product like this. the current context, Y and X are symmetric, so this matrix becomes that in Table D.2.

The expression for the expectation of the Kronecker product $\Sigma \otimes \Sigma$ for the symmetric case is given in Table D.3. Notice that every sector of the matrix is symmetric, and the matrix as a whole is symmetric.

Table D.2: $M_0 \otimes M_0$–Symmetric

μ_{11}^2	$\mu_{11}\mu_{12}$	$\mu_{11}\mu_{13}$	$\mu_{12}\mu_{11}$	μ_{12}^2	$\mu_{12}\mu_{13}$	$\mu_{13}\mu_{11}$	$\mu_{13}\mu_{12}$	μ_{13}^2
$\mu_{11}\mu_{12}$	$\mu_{11}\mu_{22}$	$\mu_{11}\mu_{23}$	μ_{12}^2	$\mu_{12}\mu_{22}$	$\mu_{12}\mu_{23}$	$\mu_{13}\mu_{12}$	$\mu_{13}\mu_{22}$	$\mu_{13}\mu_{23}$
$\mu_{11}\mu_{13}$	$\mu_{11}\mu_{23}$	$\mu_{11}\mu_{33}$	$\mu_{12}\mu_{13}$	$\mu_{12}\mu_{23}$	$\mu_{12}\mu_{33}$	μ_{13}^2	$\mu_{13}\mu_{23}$	$\mu_{13}\mu_{33}$
$\mu_{12}\mu_{11}$	μ_{12}^2	$\mu_{12}\mu_{13}$	$\mu_{22}\mu_{11}$	$\mu_{22}\mu_{12}$	$\mu_{22}\mu_{13}$	$\mu_{23}\mu_{11}$	$\mu_{23}\mu_{12}$	$\mu_{23}\mu_{13}$
μ_{12}^2	$\mu_{12}\mu_{22}$	$\mu_{12}\mu_{23}$	$\mu_{22}\mu_{12}$	μ_{22}^2	$\mu_{22}\mu_{23}$	$\mu_{23}\mu_{12}$	$\mu_{23}\mu_{22}$	μ_{23}^2
$\mu_{12}\mu_{13}$	$\mu_{12}\mu_{23}$	$\mu_{12}\mu_{33}$	$\mu_{22}\mu_{13}$	$\mu_{22}\mu_{23}$	$\mu_{22}\mu_{33}$	$\mu_{23}\mu_{13}$	μ_{23}^2	$\mu_{23}\mu_{33}$
$\mu_{13}\mu_{11}$	$\mu_{13}\mu_{12}$	μ_{13}^2	$\mu_{23}\mu_{11}$	$\mu_{23}\mu_{12}$	$\mu_{23}\mu_{13}$	$\mu_{33}\mu_{11}$	$\mu_{33}\mu_{12}$	$\mu_{33}\mu_{13}$
$\mu_{13}\mu_{12}$	$\mu_{13}\mu_{22}$	$\mu_{13}\mu_{23}$	$\mu_{23}\mu_{12}$	$\mu_{23}\mu_{22}$	μ_{23}^2	$\mu_{33}\mu_{12}$	$\mu_{33}\mu_{22}$	$\mu_{33}\mu_{23}$
μ_{13}^2	$\mu_{13}\mu_{23}$	$\mu_{13}\mu_{33}$	$\mu_{23}\mu_{13}$	μ_{23}^2	$\mu_{23}\mu_{33}$	$\mu_{33}\mu_{13}$	$\mu_{33}\mu_{23}$	μ_{33}^2

Table D.3: $\Sigma \otimes \Sigma$–Symmetric

$E(\sigma_{11}^2)$	$E(\sigma_{11}\sigma_{12})$	$E(\sigma_{11}\sigma_{13})$	$E(\sigma_{12}\sigma_{11})$	$E(\sigma_{12}^2)$	$E(\sigma_{12}\sigma_{13})$	$E(\sigma_{13}\sigma_{11})$	$E(\sigma_{13}\sigma_{12})$	$E(\sigma_{13}^2)$
$E(\sigma_{11}\sigma_{12})$	$E(\sigma_{11}\sigma_{22})$	$E(\sigma_{11}\sigma_{23})$	$E(\sigma_{12}^2)$	$E(\sigma_{12}\sigma_{22})$	$E(\sigma_{12}\sigma_{23})$	$E(\sigma_{13}\sigma_{12})$	$E(\sigma_{13}\sigma_{22})$	$E(\sigma_{13}\sigma_{23})$
$E(\sigma_{11}\sigma_{13})$	$E(\sigma_{11}\sigma_{23})$	$E(\sigma_{11}\sigma_{33})$	$E(\sigma_{12}\sigma_{13})$	$E(\sigma_{12}\sigma_{23})$	$E(\sigma_{12}\sigma_{33})$	$E(\sigma_{13}^2)$	$E(\sigma_{13}\sigma_{23})$	$E(\sigma_{13}\sigma_{33})$
$E(\sigma_{12}\sigma_{11})$	$E(\sigma_{12}^2)$	$E(\sigma_{12}\sigma_{13})$	$E(\sigma_{22}\sigma_{11})$	$E(\sigma_{22}\sigma_{12})$	$E(\sigma_{22}\sigma_{13})$	$E(\sigma_{23}\sigma_{11})$	$E(\sigma_{23}\sigma_{12})$	$E(\sigma_{23}\sigma_{13})$
$E(\sigma_{12}^2)$	$E(\sigma_{12}\sigma_{22})$	$E(\sigma_{12}\sigma_{23})$	$E(\sigma_{22}\sigma_{12})$	$E(\sigma_{22}^2)$	$E(\sigma_{22}\sigma_{23})$	$E(\sigma_{23}\sigma_{12})$	$E(\sigma_{23}\sigma_{22})$	$E(\sigma_{23}^2)$
$E(\sigma_{12}\sigma_{13})$	$E(\sigma_{12}\sigma_{23})$	$E(\sigma_{12}\sigma_{33})$	$E(\sigma_{22}\sigma_{13})$	$E(\sigma_{22}\sigma_{23})$	$E(\sigma_{22}\sigma_{33})$	$E(\sigma_{23}\sigma_{13})$	$E(\sigma_{23}^2)$	$E(\sigma_{23}\sigma_{33})$
$E(\sigma_{13}\sigma_{11})$	$E(\sigma_{13}\sigma_{12})$	$E(\sigma_{13}^2)$	$E(\sigma_{23}\sigma_{11})$	$E(\sigma_{23}\sigma_{12})$	$E(\sigma_{23}\sigma_{13})$	$E(\sigma_{33}\sigma_{11})$	$E(\sigma_{33}\sigma_{12})$	$E(\sigma_{33}\sigma_{13})$
$E(\sigma_{13}\sigma_{12})$	$E(\sigma_{13}\sigma_{22})$	$E(\sigma_{13}\sigma_{23})$	$E(\sigma_{23}\sigma_{12})$	$E(\sigma_{23}\sigma_{22})$	$E(\sigma_{23}^2)$	$E(\sigma_{33}\sigma_{12})$	$E(\sigma_{33}\sigma_{22})$	$E(\sigma_{33}\sigma_{23})$
$E(\sigma_{13}^2)$	$E(\sigma_{13}\sigma_{23})$	$E(\sigma_{13}\sigma_{33})$	$E(\sigma_{23}\sigma_{13})$	$E(\sigma_{23}^2)$	$E(\sigma_{23}\sigma_{33})$	$E(\sigma_{33}\sigma_{13})$	$E(\sigma_{33}\sigma_{23})$	$E(\sigma_{33}^2)$

Appendix E

APPENDIX E–THE UNIVARIATE AND MULTIVARIATE CHANGE OF VARIABLES PROBLEM

Let probability and sample size do the heavy lifting.

Donald Miller

E.1 The Univariate Change of Variables Problem

We start with a known univariate density over x with a density g(x) and transform to a univariate density over f(y) in which y is a specified function of x. We write a relationship that the density of y must satisfy:

$$F(Y) \triangleq \int_{L}^{Y} f(y)\, dy = \{L \leq y \leq Y\} \quad .$$

The transformation of variables we want to consider is the transformation from the x variable to the y variable, which is written

$$u(x) = y \quad .$$

We will be needing the inverse of this transformation

$$u^{-1}(y) = x \quad .$$

Put the foregoing transformation into the previous cumulative integral

$$\{L \le u(x) \le Y\} \quad .$$

Take the inverse transformation of this equation, recognizing that u is monotonic, to write

$$\left\{u^{-1}(L) \le u^{-1}\left[u(x)\right] \le u^{-1}(Y)\right\} \quad .$$

Simplifying this inverse transformation

$$\left\{u^{-1}(L) \le x \le u^{-1}(Y)\right\} = \int\limits_{u^{-1}(L)}^{u^{-1}(Y)} g(x)\,dx \quad .$$

This allows us to write the cumulative distribution

$$F(Y) = \int\limits_{u^{-1}(L)}^{u^{-1}(Y)} g(x)\,dx \quad .$$

Using the Fundamental Theorem of the Calculus to differentiate the integral, we have the final answer

$$\frac{d}{dY}F(Y) = \int\limits_{u^{-1}(L)}^{u^{-1}(Y)} g(x)\,dx = \frac{d\left[u^{-1}(Y)\right]}{dY} f\left[u^{-1}(Y)\right] \quad .$$

The process is simple. We begin with the distribution that we want to modify. For example

$$g(x) = \frac{1}{\sigma\sqrt{2\pi}} e^{-\frac{1}{2}\left(\frac{x-\mu}{\sigma}\right)^2} \quad .$$

We then postulate the monotonic transformation

$$y = u(x) = e^x \quad .$$

The inverse of this transformation is, trivially

$$x = ln(y)$$
$$\Rightarrow x = u^{-1}(y) = ln(y) \quad .$$

We need the derivative of this inverse transformation, which is

$$\frac{d\left[u^{-1}(y)\right]}{dy} = \frac{1}{y} \quad .$$

Therefore the transformed distribution is

$$f(y) = f\left[u^{-1}(y)\right]\frac{1}{y} = f\left[\ln(y)\right]\frac{1}{y} = \frac{1}{y\sigma\sqrt{2\pi}}e^{-\frac{1}{2}\left[\frac{\ln(y)-\mu}{\sigma}\right]^2} \quad .$$

Voila; we have the lognormal distribution.

E.2 Bivariate Change of Variables Problem

E.2.1 A Simple Illustration

To illustrate how to transform a bivariate distribution, suppose we begin with two independent, identical exponential distributions[1]

$$f(x_1, x_2) = \lambda e^{-\lambda x_1}\lambda e^{-\lambda x_2} \quad .$$

We want to transform to two variables y_1 and y_2 as follows

$$[y_1, y_2] = [g_1(x_1, x_2), g_2(x_1, x_2)] \quad ,$$

in which

$$\begin{aligned} g_1(x_1, x_2) &= x_1 + x_2 \\ g_2(x_1, x_2) &= \frac{x_1}{x_1 + x_2} \quad . \end{aligned}$$

We are going to need the partial derivatives for the Jacobian with respect to g

$$\begin{aligned} \frac{\partial g_1}{\partial x_1} &= 1\frac{\partial g_2}{\partial x_1} = \frac{x_2}{(x_1 + x_2)^2} \\ \frac{\partial g_1}{\partial x_2} &= 1\frac{\partial g_2}{\partial x_2} = \frac{x_1}{(x_1 + x_2)^2} \quad . \end{aligned}$$

The Jacobian with respect to g is therefore

$$J_g = \begin{vmatrix} 1 & \frac{x_2}{(x_1+x_2)^2} \\ 1 & \frac{-x_1}{(x_1+x_2)^2} \end{vmatrix} = \left|-\frac{1}{x_1 + x_2}\right| = \frac{1}{x_1 + x_2} \quad .$$

We are going to have to invert the foregoing variable transformations, which we accomplish as follows

$$\begin{aligned} y_1 &= x_1 + x_2 \\ y_2 &= \frac{x_1}{x_1 + x_2} \Rightarrow y_2 = \frac{x_1}{y_1} \Rightarrow x_1 = y_1 y_2 \quad . \end{aligned}$$

[1]The example is taken from http://dept.stat.lsa.umich.edu/ moulib/426-notes-3.pdf.

Substitute in for x_1 to get the inverse transformation

$$y_1 = x_1 + x_2 \Rightarrow y_1 = y_1 y_2 + x_2 \Rightarrow x_2 = y_1 - y_1 y_2 \quad .$$

Rewriting for convenience

$$\begin{aligned} x_1 &= y_1 y_2 \Rightarrow h_1(y_1, y_2) = y_1 y_2 \\ x_2 &= y_1 - y_1 y_2 \Rightarrow h_2(y_1, y_2) = y_1 - y_1 y_2 \quad . \end{aligned}$$

We need the partial derivatives for the Jacobian with respect to h

$$\begin{array}{ll} \frac{\partial h_1(y_1,y_2)}{\partial y_1} = y_2 & \frac{\partial h_2(y_1,y_2)}{\partial y_1} = 1 - y_2 \\ \frac{\partial h_1(y_1,y_2)}{\partial y_2} = y_1 & \frac{\partial h_2(y_1,y_2)}{\partial y_2} = -y_1 \end{array} \quad .$$

The Jacobian with respect to h is therefore

$$\begin{vmatrix} y_2 & 1 - y_2 \\ y_1 & y_1 \end{vmatrix} = |-y_1| = y_1 \quad .$$

We are now able to write the joint density on y_1, y_2 as follows

$$\begin{aligned} g(y_1, y_2) &= f[h_1(y_1, y_2), h_2(y_1, y_2)] \, J_h(y_1, y_2) \\ &= \lambda e^{-\lambda(y_1 y_2)} \lambda e^{-\lambda(y_1 - y_1 y_2)} y_1 \\ &= \lambda^2 y_1 e^{-\lambda y_1} \quad . \end{aligned}$$

E.2.2 A Hybrid Normal-Lognormal Bivariate Illustration

Let us apply the same principles to a bivariate normal distribution. We want to create a hybrid distribution in which one of the variables is normal and therefore whose domain is the entire real line, and the other variable is lognormal and therefore whose domain is only the positive orthant. We begin with a p-bivariate normal distribution

$$N(x; \mu, \Sigma, p) = \left(\frac{1}{2\pi}\right)^{\frac{p}{2}} |\Sigma|^{-\frac{1}{2}} e^{-\frac{1}{2}(x-\mu)^T \Sigma^{-1}(x-\mu)} \quad .$$

When $p = 2$, this specializes to

$$N(x_1, x_2) = \frac{1}{(2\pi)} |\Sigma|^{-\frac{1}{2}} e^{-\frac{1}{2}(x-\mu)^T \Sigma^{-1}(x-\mu)} \quad .$$

Let us not change the variable x_1 but change the variable x_2 to an exponential. In particular, let us make the transformation

$$\begin{aligned} y_1 &= x_1 \Rightarrow x_1 = y_1 \\ y_2 &= e^{x_2} \Rightarrow x_2 = \ln(y_2) \quad . \end{aligned}$$

The derivatives are

$$\begin{bmatrix} \frac{dx_1}{dy_1} = 1 & \frac{dx_1}{dy_2} = 0 \\ \frac{dx_2}{dy_1} = 0 & \frac{dx_2}{dy_2} = \frac{1}{y_2} \end{bmatrix} .$$

The Jacobian is therefore

$$\begin{vmatrix} 1 & 0 \\ 0 & \frac{1}{y_2} \end{vmatrix} = \frac{1}{y_2} .$$

We obtain the distribution over y_1, y_2 by making the transformation substitutions into the normal and multiplying by the Jacobian

$$\begin{aligned} N(y_1, y_2) &= \frac{1}{2\pi}|\Sigma|^{-\frac{1}{2}} e^{-\frac{1}{2}\left(\begin{bmatrix} y_1 \\ \ln(y_2) \end{bmatrix} - \mu\right)^T \Sigma^{-1}\left(\begin{bmatrix} y_1 \\ \ln(y_2) \end{bmatrix} - \mu\right)} J \\ &= \frac{1}{2\pi}|\Sigma|^{-\frac{1}{2}} e^{-\frac{1}{2}\begin{bmatrix} y_1 - \mu_1 \\ \ln(y_2) - \mu_2 \end{bmatrix}^T \Sigma^{-1}\begin{bmatrix} y_1 - \mu_1 \\ \ln(y_2) - \mu_2 \end{bmatrix}} \frac{1}{y_2} . \end{aligned}$$

It is convenient, but not fully general, to assume that the distribution over the coefficients and the error term are independent. That is, the covariances in the Σ matrix are zero. That means that we can write the variance-covariance matrix as in the first equation and calculate a number of its implications:

$$\begin{aligned} \Sigma &= \begin{bmatrix} {\sigma_{11}}^2 & 0 \\ 0 & {\sigma_{22}}^2 \end{bmatrix} \\ \Rightarrow |\Sigma| &= {\sigma_{11}}^2 {\sigma_{22}}^2 \\ \Rightarrow |\Sigma|^{-\frac{1}{2}} &= \frac{1}{\sigma_{11}\sigma_{22}} \\ \Rightarrow \Sigma^{-1} &= \begin{bmatrix} \frac{1}{{\sigma_{11}}^2} & 0 \\ 0 & \frac{1}{{\sigma_{22}}^2} \end{bmatrix} . \end{aligned}$$

If we substitute the specific structure of the Σ matrix, we can write the exponent of the former equation as follows:

$$\begin{aligned} &[y_1 - \mu_1, \ln y_2 - \mu_2] \begin{bmatrix} \frac{1}{{\sigma_{11}}^2} & 0 \\ 0 & \frac{1}{{\sigma_{22}}^2} \end{bmatrix} \begin{bmatrix} y_1 - \mu_1 \\ \ln y_2 - \mu_2 \end{bmatrix} \\ &= [y_1 - \mu_1, \ln y_2 - \mu_2] \begin{bmatrix} \frac{y_1 - \mu_1}{{\sigma_{11}}^2} \\ \frac{\ln y_2 - \mu_2}{{\sigma_{22}}^2} \end{bmatrix} \\ &= \frac{(y_1 - \mu_1)^2}{{\sigma_{11}}^2} + \frac{(\ln y_2 - \mu_2)^2}{{\sigma_{22}}^2} . \end{aligned}$$

We can regroup the terms to write

(E.1)

$$MN\left(y_1, y_2\right) = \frac{1}{2\pi}|\Sigma|^{-\frac{1}{2}} e^{-\frac{1}{2}\begin{bmatrix} y_1 - \mu_1 \\ \ln\left(y_2\right) - \mu_2 \end{bmatrix}^T \Sigma^{-1} \begin{bmatrix} y_1 - \mu_1 \\ \ln\left(y_2\right) - \mu_2 \end{bmatrix}} \frac{1}{y_2}$$

$$= \frac{1}{2\pi y_2}\left(\frac{1}{\sigma_{11}\sigma_{22}}\right) e^{-\frac{1}{2}\left(\frac{(y_1-\mu_1)^2}{\sigma_{11}{}^2} + \frac{(\ln y_2 - \mu_2)^2}{\sigma_{22}{}^2}\right)}$$

$$= \left[\frac{1}{\sigma_{11}\sqrt{2\pi}} e^{-\frac{(y_1-\mu_1)^2}{2\sigma_{11}{}^2}}\right]\left[\frac{1}{y_2\sigma_{22}\sqrt{2\pi}} e^{-\frac{(\ln y_2-\mu_2)^2}{2\sigma_{22}{}^2}}\right] .$$

in which MN stands for "mixed normal." This is the product of a univariate normal times a univariate lognormal. We could use a distribution like this as a prior for a variable that is unsigned and a variable that is strictly positive (like a variance).

In the lognormal distribution (the second bracket), the parameters of that distribution are μ_2 and σ_{22}. If the mean were designated M_{22} and the variances were designated V_{22}, we could write that distribution in terms of the mean and variance. To see how, let us write the density, the mean, and the variance equations respectively:[2]

$$f\left(x\right) = \frac{1}{x\sigma_{22}\sqrt{2\pi}} e^{-\frac{[\ln(x)-\mu_2]^2}{2\sigma_{22}{}^2}} ,$$

in which the mean and variance are respectively

$$M_{22} = e^{\mu_2 + \frac{\sigma_{22}{}^2}{2}} ,$$

and

$$V_{22} = \left(e^{\sigma_{22}{}^2} - 1\right) e^{2\mu_2 + \sigma_{22}{}^2} .$$

We begin by manipulating the equation for the mean

$$M_{22} = e^{\mu_2 + \frac{\sigma_{22}^2}{2}}$$

$$\Rightarrow M_{22} = e^{\mu_2} e^{\frac{\sigma_{22}^2}{2}}$$

$$\Rightarrow e^{\mu_2} = M_{22} e^{-\frac{\sigma_{22}^2}{2}}$$

$$\Rightarrow e^{2\mu_2} = M_{22}^2 e^{-\sigma_{22}^2}$$

We then substitute this into the variance equation above

$$V_{22} = \left(e^{\sigma_{22}^2} - 1\right) e^{\sigma_{22}^2} e^{2\mu_2}$$

[2] See https://en.wikipedia.org/wiki/Log-normal_distribution.

$$
\begin{aligned}
\Rightarrow V_{22} &= \left(e^{\sigma_{22}^2}-1\right)e^{\sigma_{22}^2}M_{22}^2e^{-\sigma_{22}^2}=\left(e^{\sigma_{22}^2}-1\right)M_{22}^2\\
\Rightarrow e^{\sigma_{22}^2} &= \frac{V_{22}+M_{22}^2}{M_{22}^2}\equiv\theta_{22}\\
\Rightarrow \sigma_{22}^2 &= \ln\left(\frac{V_{22}+M_{22}^2}{M_{22}^2}\right)=\ln\left(\theta_{22}\right)
\end{aligned}
$$

We substitute into the equation for the mean

$$
\begin{aligned}
M_{22} &= e^{\mu_2+\frac{\sigma_{22}^2}{2}}=e^{\mu_2}e^{\frac{\sigma_{22}^2}{2}}\\
M_{22} &= e^{\mu_2}e^{\ln\left(\sqrt{\theta_{22}}\right)}=e^{\mu_2}\sqrt{\theta_{22}}\\
\Rightarrow e^{\mu_2} &= \frac{M_{22}}{\sqrt{\theta_{22}}}\Rightarrow\mu_2=\ln\left(\frac{M_{22}}{\sqrt{\theta_{22}}}\right)
\end{aligned}
$$

We summarize

$$
\begin{aligned}
\mu_2 &= \ln\left(\frac{M_{22}}{\sqrt{\theta_{22}}}\right)\\
\sigma_{22}^2 &= \ln\left(\theta_{22}\right)
\end{aligned}
$$

If we substitute this into the density, we can write it as a function of its mean and variance (and by square root its standard deviation)

$$
f\left(x\right)=\frac{1}{x\sqrt{2\pi\ln\left(\theta_{22}\right)}}e^{-\frac{\left[\ln\left(x\frac{\sqrt{\theta_{22}}}{M_{22}}\right)\right]^2}{2\ln\left(\theta_{22}\right)}}\quad.
$$

Bibliography

[1] Alan Agresti. *Categorical Data Analysis*. Wiley, New York, third edition, 2013.

[2] Murray Aitkin. *Statistical Inference–An Integrated Bayesian/Likelihood Approach*. Chapman and Hall/CRC, Boca Raton, September 5, 2019.

[3] T. W. Anderson. *An Introduction to Multivariate Statistical Analysis*. John Wiley and Sons, New York, 1958.

[4] Coryn A. L. Bailer-Jones. *Practical Bayesian Inference–A Primer for Physical Scientists*. Cambridge University Press, UK, 2017.

[5] Thomas Bayes. An essay towards solving a problem in the doctrine of chances. *Philosophical Transactions of the Royal Society*, 1763.

[6] James O. Berger. *Statistical Decision Theory and Bayesian Analysis*. Springer, New York, second edition, 1980.

[7] Bernardo, J. M., Bayarri, M. J., Berger, J. O., Dawid. A. P., Heckerman. D., Smith. A. F. M., and West, M. *Bayesian Statistics 8*. Oxford University Press, Oxford, UK, first edition, 2007.

[8] Bernardo, J. M., M. J. Bayarri, and J. O. Berger. *Bayesian Statistics 9–Proceedings of the Ninth Valencia International Meeting*. Oxford University Press, Oxford, UK, first edition, 2011.

[9] Donald A. Berry. *Statistics: A Bayesian Perspective*. Duxbury Press, New York, first edition, 1995.

[10] Christopher M. Bishop. *Pattern Recognition and Machine Learning*. New York, springer edition, 2006.

[11] William M. Bolstad. *Introduction to Bayesian Statistics*. Wiley, third edition, 2004.

[12] William M. Bolstad. *Understanding Computational Bayesian Statistics*. Wiley and Sons, Hoboken, NJ, 2010.

[13] Bolstad, William M. and Curran, James M. *Introduction to Bayesian Statistics.* Wiley, New York, third edition, 25 August 2016.

[14] Bovens, Luc and Hartmann, Stephan. *Bayesian Epistemology.* Oxford University Press, 2004.

[15] Bozza, Silvia, Franco Taroni, and Alex Biederman. *Bayes Factors for Forensic Decision Analysis with R.* Springer, New York, first edition, 2022.

[16] Davidson-Pilon Cameron. *Bayesian Methods for Hackers.* Addison-Wesley, New York, 2016.

[17] Carlin, Bradley P. and Louis, Thomas A. *Bayesian Methods for Data Analysis.* Chapman and Hall/CRC Texts in Statistical Science, Boca Raton, third edition, 2009.

[18] Ron Chernow. *Washington: A Life.* Penguin Press, United States, first edition, 2010.

[19] Christensen, Ronald, Johnson, Wesley, Branscum, Adam, and Hanson, Timothy E. *Bayesian Ideas and Data Analysis: An Introduction for Scientists and Statisticians.* Chapman and Hall/CRC Texts in Statistical Science, first edition, 2011.

[20] Robert T. Clemen. *Making Hard Decisions: An Introduction to Decision Analysis.* Duxbury Press, Belmont, California, 1996.

[21] Peter Congdon. *Bayesian Statistical Modelling.* John Wiley and Sons Inc., second edition, 2007.

[22] Cook, Dennis and Forzani, Linda. On the mean and variance of the generalized inverse of a singular wishart matrix. *Electronic Journal of Statistics,* 5:146–158, 2011.

[23] Bruno de Finetti. *Theory of Probability, Volumes 1 and 2.* Wiley, New York, 1970.

[24] Eugene Demidenko. *Advanced Statistics with Applications in R.* Wiley, New York, 2020.

[25] Dey, Dipak K. and Rao, C. R. *Bayesian Thinking, Modeling and Computation.* North Holland, New York, 2020.

[26] Donovan, Therese M. and Mikey, Ruth M. *Bayesian Statistics for Beginners.* Oxford, London, first edition, 2019.

[27] Allen B. Downey. *Think Bayes.* O'Reilly, New York, second edition, May 2021.

[28] eGyanKosh. Unit 14 identification problem. Online; accessed 14 January 2021.

[29] Fairfield, Tasha and Charman, Andrew E. *Social Inquiry and Bayesian Inference*. Cambridge, UK, 2022.

[30] Florens, Jean-Pierre, Mouchart, Michel, and Rolin, Jean-Marie. *Elements of Bayesian Statistics*. M. Dekker, New York, 1990.

[31] Roman Garnett. *Bayesian Optimization*. Cambridge University Press, UK, first edition, 2021.

[32] Gatsonis, Constantine, Hodges, James S., Kass, Robert E., and Singpurwalla, Nozer D. *Case Studies in Bayesian Statistics, Volume II*. Springer Lecture Notes in Statistics (LNS, volume 105), 1995.

[33] Gelman, Andrew, Carlin, John B., Stern, Hal S., Dunson, David B., Vehtari, Aki, and Rubin. Donald B. *Bayesian Data Analysis*. CRC Press, New York, third edition, 2013.

[34] Gelman, Andrew, Goodrich, Ben, Gabry, Jonah, and Vehtari. Aki. R-squared for bayesian regression models, November 2018. Online; accessed 17 September 2022.

[35] Ghosh, Jayanta K. , Delampady, Mohan, and Samanta, Tapas. *An Introduction to Bayesian Analysis: Theory and Methods*. Springer, New York, 2010.

[36] Jeff Gill. *Bayesian Methods–A Social and Behavioral Approach*. CRC, Boca Raton, third edition, 2015.

[37] Alexander Graham. *Kronecker Products and Matrix Calculcus with Applications*. Dover, Mineola, New York, 1981, 2018.

[38] Damodar N. Gujrati. *Basic Econometrics*. McGraw Hill, Singapore, 2016.

[39] L. R. Haff. Identities for the inverse wishart distribution with computational results in linear andquadratic discrimination. *Sankhyā: The Indian Journal of Statistics, Series B (1960-2002)*, 44, No. 3:245–258, Dec. 1982.

[40] Robert C. Heinlein. *Stranger in a Strange Land*. New York, ace publishing edition, 1961.

[41] Held, Leonhard and Bové, Daniel Sabanés. *Likelihood and Bayesian Inference: With Applications in Biology and Medicine (Statistics for Biology and Health)*. Springer, New York, second edition, 2020.

[42] Hill, R. Carter, Griffiths, William E., and Lim. Guay C. *Principles of Econometrics, Truffles Data Set.* Wiley, fourth edition, 2011. Online; accessed 17 January 2021.

[43] Peter D. Hoff. *A First Course in Bayesian Statistical Methods.* Springer, New York, 2009.

[44] Horn, Roger A. and Johnson, Charles R. *Matrix Analysis (Second Edition).* Cambridge University Press, New York, 1990.

[45] Howard, Ronald A. and Abbas, Ali E. *Foundations of Decision Analysis.* Pearson, New York, 2016.

[46] Howard, Ronald A. and Matheson, James E. Influence diagrams. *Decision Analysis*, 2, No. 3:127–143, Sept. 2005.

[47] Howson, Colin and Urbach, Peter. *Scientific Reasoning: The Bayesian Approach.* Open Court, Chicago and LaSalle Illinois, third edition, 2006.

[48] Jack Johnston and John Dinardo. *Econometric Methods.* McGraw-Hill, Singapore, 1997.

[49] Edwin Jaynes. *Probability Theory: The Logic of Science.* Cambridge University Press, New York, 2001.

[50] Harold Jeffreys. Online; accessed 25 March 2021.

[51] Kahneman, Daniel, Slovik, Paul, and Tversky, Amos. *Judgment Under Uncertainty: Heuristics and Biases.* Cambridge University Press, New York, 1982.

[52] David Kaplan. *Bayesian Statistics for the Social Sciences.* Guilford Press, New York, first edition, 2013.

[53] Thomas W. Keelin. The metalog distributions. *Decision Analysis*, 5(4):243–277, 2016.

[54] Stuart A. Klugman. *Bayesian Statistics in Actuarial Science: With Emphasis on Credibility.* Kluwer Academic publishers, 1991.

[55] Jan Kmenta. *Elements of Econometrics.* University of Michigan Press, Ann Arbor, 1997.

[56] Karl-Rudolf Koch. *Introduction to Bayesian Statistics.* Springer-Verlag, Heidelburg, second edition, 2007.

[57] Kollo, Tonu and von Rosen, Dietrich. *Advanced Multivariate Statistics with Matrices (Manuscript).* Springer, Dordrecht, Netherlands, 2010.

[58] Gary Koop. *Bayesian Econometrics.* Wiley, Sussex, England, 2003.

[59] Samuel Kotz and Saralees Nadarajah. *Multivariate t Distributions and Their Applications.* Cambridge University Press, Cambridge, 2004.

[60] Will Kurt. *Bayesian Statistics the Fun Way.* No Starch Press, San Francisco, 2019.

[61] Peter M. Lee. *Bayesian Statistics, An Introduction.* Abe Books, fourth edition, 2012.

[62] Dennis V. Lindley. *Introduction to Probability and Statistics from a Bayesian Viewpoint.* Cambridge University Press, UK, first edition, 1990.

[63] Dennis V. Lindley. *Understanding Uncertainty.* Wiley, New York, 2014.

[64] David G. Luenberger. *Linear and Nonlinear Programming.* Addison Wesley, second edition edition, May 1989.

[65] Scott M. Lynch. *Introduction to Applied Bayesian Statistics and Estimation for Social Scientists.* Springer, New York, first edition, 2007.

[66] Martin, Osvaldo A., Kumar, Raum, and Loo, Junpeng. *Bayesian Modeing and Computation in Python.* CRC, Boca Raton, 2022.

[67] Richard McElreath. *Statistical Rethinking: A Bayesian Course with Examples in R and STAN.* CRC, Boca Raton, second edition, 2020.

[68] McFadden, Daniel. Quantal choice analysis: A survey. *Annals of Economic and Social Measurement*, 5(4), 1976.

[69] Kevin P. Murphy. *Probabilistic Machine Learning: An Introduction.* Cambridge, MA, mit press (kindle) edition, 2022.

[70] Radford M. Neal. *Bayesian Learning for Neural Networks.* Springer, New York, 1996.

[71] Jill E. Nesbitt. *Modeling and Estimating Market Shares in Multiattribute, Multicommodity Markets.* Ph.D. Thesis, Rice University, Houston, Texas, 2008.

[72] Nielsen, Thomas Dyhre and Jensen, Finn Verner. *Bayesian Networks and Decision Graphs.* Springer, New York, second edition, 2010.

[73] T. Papagiannakis. *Pavement Performance Models for the Trans-Canada Highway in Newfoundland–Revised Final Report, Part I: RCI Models.* Pavement Management Systems and Newfoundland Department of Works, Services, and Transportation, Cambridge, Ontario, February 1990.

[74] Pindyck, Robert S. and Rubinfield, Daniel L. *Econometric Models and Economic Forecasts.* Irwin/McGraw-Hill, Singapore, 1998.

[75] Dale J. Poirier. Frequentist and subjectivist perspectives on the problems of model building in economics. *Journal of Economic Perspectives*, 2(1):121–144, Winter, 1988.

[76] S. James Press. *Applied Multivariate Analysis–Using Bayesian and Frequentist Methods of Inference–Second Edition*. Krieger, Malabar, Florida, 1972, 1982, 2005.

[77] S. James Press. *Bayesian Statistics: Principles, Models, and Applications*. Wiley Interscience, New York, 1989.

[78] S. James Press. *Subjective and Objective Bayesian Statistics: Principles, Models, and Applications*. Wiley, New York, second edition, 2003.

[79] Press, S. James and Shigemasu, K. A note on choosing the number of factors. *Communications in Statistics–Theory and Methods*, 29(7), 1999.

[80] Howard Raiffa. *Decision Analysis: Introductory Lectures on Choices under Uncertainty*. Addison-Wesley, Reading, Massachusetts, July 1970.

[81] Howard Raiffa and Robert Schlaiffer. *Applied Statistical Decision Theory*. Harvard University Press, Cambridge, Massachusetts, 1961.

[82] Reich, Brian J. and Ghosh, Sujit K. *Bayesian Statistical Methods*. Chapman and Hall/CRC Texts in Statistical Science (Ebook), first publicaion kindle edition edition, 2019.

[83] Christian P. Robert. *The Bayesian Choice: From Decision-Theoretic Foundations to Computational Implementation*. Springer, New York, second edition, 2007.

[84] Robert, Christian and Marin, Jean-Michel. *A Practical Approach to Computational Bayesian Statistics*. Springer, New York.

[85] Gerald S. Rogers. *Matrix Derivatives*. Dekker, New York, 1980.

[86] Daniell B. Rowe. *Multivariate Bayesian Statistics: Models for Source Separation and Signal Unmixing*. CRC Press, Boca Raton, FL, USA, 2003.

[87] David Salsburg. *The Lady Tasting Tea: How Statistics Revolutionized Science in the Twentieth Century*. Henry Holt Publishing, New York, 2001.

[88] Leonard Savage. *Foundations of Statistics*. Wiley, New York, 1954.

[89] Samuel Schmitt. *Measuring Uncertainty: An Elementary Introduction to Bayesian Statistics*. Addison-Wesley, New York, 1969.

[90] Scutari, Marco and Denis, Jean-Baptiste. *Bayesian Networks: With Examples in R.* Chapman and Hall/CRC Texts in Statistical Science, Boca Raton, first edition, 2015.

[91] Shemayakin, Arkady and Kniazev, Alexander. *Introduction to Bayesian Estimation and Copula Models of Dependence.* Wiley, New York, 2017.

[92] Sivia, Devinderjit and Skilling, John. *Data Analysis: A Bayesian Tutorial.* Oxford, UK, second edition, 2009.

[93] Mel Slater. *Bayesian Methods in Statistics.* Sage, London, first edition, 2022.

[94] Spetzler, Carl S. and Stael von Holstein, Carl-Axel. Probability encoding in decision analysis. *Management Science*, 22, Issue 3:340–358, 1975.

[95] Spiegelhalter, David, Abrams, Keith R., and Myles, Johnathan P. *Bayesian Approaches to Clinical Trials and Health-Care Evaluation.* Wiley, UK, first edition, 2004.

[96] James V. Stone. *Bayes' Rule: A Tutorial Introduction to Bayesian Analysis.* Ebook Self Published, 2013.

[97] Sun, Dongchu and Berger, James O. *Objective Bayesian Analysis for the Multivariate Normal Model–Proceedings of Valencia ISBA 8th World Meeting on Bayesian Statistics.* ISBA, Benidorm (Alicante, Spain), June 1-6, 2006.

[98] Turkman, M. Antónia Amaral, Paulino, Carlos Daniel, and Müller, Peter. *Computational Bayesian Statistics: An Introduction.* Cambridge University Press, UK, first edition, 2019.

[99] von der Linden, Wolfgang, Dose, Volker, and von Toussaint, Udo. *Bayesian Probability Theory: Applications in the Physical Sciences.* Cambridge University Press, UK, first edition, 2014.

[100] Von Neumann, John and Morgenstern, Oskar. *Theory of Games and Economic Behavior.* Princeton University Press, Princeton, New Jersey, 1953.

[101] Dietrich von Rosen. Moments for the inverted wishart distribution. *Scand J Statist*, 15:97–109, 1988.

[102] Jon Wakefield. *Bayesian and Frequentist Regression Methods.* Springer, New York, first edition, 2013.

[103] Sumio Watanabe. *Mathematical Theory of Bayesian Statistics.* CRC, Boca Raton, first edition, 2018.

[104] Jeffrey M. Woodridge. *Introductory Econometrics: A Modern Approach.* Irwin/McGraw-Hill, Southwestern, 2003.

[105] Chi Yau. *Measuring Uncertainty: An Elementary Introduction to Bayesian Statistics*. Kindle, second edition, 2020.

[106] Czar Yobero. Simultaneous equation models: Estimating supply and demand functions, 2016. Online; accessed 17 January 2021.

[107] Arnold Zellner. *An Introduction to Bayesian Inference in Econometrics*. Robert E. Krieger Publishing Company, Malabar, Florida, 1971, reprinted 1987.

www.ingramcontent.com/pod-product-compliance
Ingram Content Group UK Ltd.
Pitfield, Milton Keynes, MK11 3LW, UK
UKHW020427061026
15625UKWH00005B/311

9 798349 419720